NUMERICAL
METHODS
IN ENGINEERING
AND SCIENCE

NUMERICAL METHODS IN ENGINEERING AND SCIENCE

C, C++, AND MATLAB®

B. S. Grewal

MERCURY LEARNING AND INFORMATION

Dulles, Virginia
Boston, Massachusetts
New Delhi

Publisher: David Pallai
Mercury Learning and Information
22841 Quicksilver Drive
Dulles, VA 20166
info@merclearning.com
www.merclearning.com
(800) 232-0223

B. S. Grewal. *Numerical Methods in Engineering and Science: C, C++, and MATLAB®.*
ISBN: 978-1-68392-128-8

Library of Congress Control Number: 2018935002

181920321   This book is printed on acid-free paper in the United States of America.

CONTENTS

CHAPTER 1

APPROXIMATIONS AND ERRORS IN COMPUTATION

Chapter Objectives

- Introduction
- Accuracy of numbers
- Errors
- Useful rules for estimating errors
- Error propagation
- Error in the approximation of a function
- Error in a series approximation
- Order of approximation
- Growth of error
- Objective type of questions

1.1   Introduction

The limitations of analytical methods in practical applications have led scientists and engineers to evolve numerical methods. We know that exact methods often fail in drawing plausible inferences from a given set of tabulated data or in finding roots of transcendental equations or in solving non-linear differential equations. There are many more such situations where analytical methods are unable to produce desirable results. Even if analytical solutions are available, these are not amenable to direct numerical interpretation.

The aim of numerical analysis is therefore, to provide constructive methods for obtaining answers to such problems in a numerical form.

With the advent of high speed computers and increasing demand for numerical solution to various problems, numerical techniques have become indispensible tools in the hands of engineers and scientists.

The input information is rarely exact since it comes from some measurement or the other and the method also introduces further error. As such, the error in the final result may be due to an error in the initial data or in the method or both. Our effort will be to minimize these errors, so as to get the best possible results. We therefore begin by explaining various kinds of approximations and errors which may occur in a problem and derive some results on error propagation in numerical calculations.

1.2   Accuracy of Numbers

1. *Approximate numbers.* There are two types of numbers: *exact* and *approximate. Exact numbers* are 2, 4, 9, 13, 7/2, 6.45,... etc. But there are numbers such as 4/3 (= 1.33333...), $\sqrt{2}$ (= 1.414213...) and π (= 3.141592...) which cannot be expressed by a finite number of digits. These may be approximated by numbers 1.3333, 1.4142 and 3.1416, respectively. Such numbers which represent the given numbers to a certain degree of accuracy are called *approximate numbers.*

2. *Significant figures.* The digits used to express a number are called *significant digits (figures).* Thus each of the numbers 7845, 3.589, and 0.4758 contains four significant figures while the numbers 0.00386, 0.000587, and 0.0000296 contain only three significant figures since zeros only help to fix the position of the decimal point. Similarly the numbers 45000 and 7300.00 have two significant figures only.

3. *Rounding off.* There are numbers with large number of digits, e.g., 22/7 = 3.142857143. In practice, it is desirable to limit such numbers to a manageable number of digits such as 3.14 or 3.143. This process of dropping unwanted digits is called *rounding off.*

4. *Rule* to round off a number to n significant figures:
   (*i*)   Discard all digits to the right of the nth digit.
   (*ii*) If this discarded number is
           (a) less than half a unit in the nth place, leave the nth digit unchanged;

(b) greater than half a unit in the nth place, increase the nth digit by unity;

(c) exactly half a unit in the nth place, increase the nth digit by unity if it is odd other wise leave it unchanged.

For instance, the following numbers rounded off to three significant figures are:

7.893 to 7.89	3.567 to 3.57
12.865 to 12.9	84767 to 84800
6.4356 to 6.44	5.8254 to 5.82

Also the numbers 6.284359, 9.864651, and 12.464762 are rounded off to four places of decimal at 6.2844, 9.8646, 12.4648; respectively.

———
NOTE **Obs.** *The numbers thus rounded off to n significant figures (or n decimal places) are said to be correct to n significant figures (or n decimal places).*

1.3 Errors

In any numerical computation, we come across the following types of errors:

1. *Inherent errors.* Errors which are already present in the statement of a problem before its solution, are called *inherent errors.* Such errors arise either due to the given data being approximate or due to the limitations of mathematical tables, calculators, or the digital computer. Inherent errors can be minimized by taking better data or by using high precision computing aids.

2. *Rounding errors* arise from the process of rounding off the numbers during the computation. Such errors are unavoidable in most of the calculations due to the limitations of the computing aids. Rounding errors can, however, be reduced:

(i) by changing the calculation procedure so as to avoid subtraction of nearly equal numbers or division by a small number; or

(ii) by retaining at least one more significant figure at each step than that given in the data and rounding off at the last step.

3. *Truncation errors* are caused by using approximate results or on replacing an infinite process by a finite one. If we are using a decimal computer having a fixed word length of four digits, rounding off 13.658 gives 13.66 whereas truncation gives 13.65.

For example, if $e^x = 1 + x + \dfrac{x^2}{2!} + \dfrac{x^3}{3!} + \dfrac{x^4}{4!} + \ldots \infty = X$ (say)

is replaced by $1 + x + \dfrac{x^2}{2!} + \dfrac{x^3}{3!} = X'$ (say), then the truncation error is $X - X'$.

Truncation error is a type of algorithm error.

4. *Absolute, Relative, and Percentage errors.* If X is the true value of a quantity and X' is its approximate value, then $|X - X'|$ *i.e,* |Error| is called the *absolute error* E_a..

The *relative error* is defined by $E_r = \left| \dfrac{X - X'}{X} \right|$ i.e., $\left| \dfrac{\text{Error}}{\text{True value}} \right|$

and the percentage error *is* $E_p = 100 E_r = 100 \left| \dfrac{X - X'}{X} \right|$.

If $\bar{X}$ be such a number that $|X - X'| \leq \bar{X}$, then $\bar{X}$ is an upper limit on the magnitude of absolute error and measures the *absolute accuracy.*

NOTE

Obs. 1. *The relative and percentage errors are independent of the units used while absolute error is expressed in terms of these units.*

Obs. 2. *If a number is correct to n decimal places then the error* $= \dfrac{1}{2} 10^{-n}$. *For example, if the number is 3.1416 correct to 4 decimal places, then the error* $= \dfrac{1}{2} 10^{-4} = 0.00005$.

1.4 Useful Rules for Estimating Errors

To estimate the errors which creep in when the numbers in a calculation are truncated or rounded off to a certain number of digits, the following rules are useful.

If the approximate value of a number X having n decimal digits is X', then

1. Absolute error due to truncation to k digits

$$= |X - X| < 10^{n-k}$$

2. Absolute error due to rounding off to k digits

$$= |X - X'| < \frac{1}{2}10^{n-k}$$

3. Relative error due to truncation to k digits

$$= \left|\frac{X - X'}{X}\right| < 10^{1-k}$$

4. Relative error due to rounding off to k digits

$$= \left|\frac{X - X'}{X}\right| < \frac{1}{2}10^{1-k}$$

NOTE

Obs. 1. If a number is correct to n significant digits, then the maximum relative error $\leq \frac{1}{2}10^{-n}$. If a number is correct to d decimal places, then the absolute error $\leq \frac{1}{2}10^{-d}$.

Obs. 2. If the first significant figure of a number is k and the number is correct to n significant figures, then the relative error < 1/ $(k \times 10^{n-1})$.

Let us verify this result by finding the relative error in the number 864.32 correct to five significant figures.

Here $k = 8$, $n = 5$ and

Absolute error $0.01 \times - = 0.005$.

∴ Relative error

$$\leq \frac{0.005}{864.32} = \frac{5}{864320} = \frac{1}{2 \times 86432} < \frac{1}{2 \times 80000}$$

$$= \frac{1}{2 \times 8 \times 10^4} < \frac{1}{8 \times 10^4} \ i.e., \ \frac{1}{k \times 10^{n-1}}.$$

Hence the result is verified.

EXAMPLE 1.1

Round off the numbers 865250 and 37.46235 to four significant figures and compute E_a, E_r, E_p in each case.

Solution:

(*i*) Number rounded off to four significant figures = 865200

$$E_a = |X - X_1| = |865250 - 865200| = 50$$

$$E_r = \left|\frac{X - X_1}{X}\right| = \frac{0.00235}{37.46235} = 6.27 \times 10^{-5}$$

$$E^p = E_r \times 100 = 6.71 \times 10^{-3}$$

(*ii*) Number rounded off to four significant figures = 37.46

$$\therefore E_a = |X - X_1| = |37.46235 - 37.46000| = 0.00235$$

$$E_r = \left|\frac{X - X_1}{X}\right| = \frac{0.00235}{37.46235} = 6.27 \times 10^{-5}$$

$$E_p = E_r \times 100 = 6.27 \times 10^{-3}$$

EXAMPLE 1.2

Find the absolute error if the number $X = 0.00545828$ is

(i)   truncated to three decimal digits.
(ii) rounded off to three decimal digits.

Solution: We have $X = 0.00545828 = 0.545828 \times 10^{-2}$

(i)   After truncating to three decimal places, its approximate value
      $X' = 0.545 \times 10^{-2}$
$\therefore$ Absolute error $= |X - X'| = 0.000828 \times 10^{-2}$

$$= 0.828 \text{ x } 10\text{-}5 < 10^{-2-3}$$

This proves rule (1).

(ii) After rounding off to three decimal places, its approximate value
      $X' = 0.546 \times 10^{-2}$
$\therefore$ Absolute error $= |X - X'|$

$$= |0.545828 - 0.546| \times 10^{-2}$$

$$= 0.000172 \times 10^{-2} = 0.172 \times 10^{-5}$$

which is $< 0.5 \times 10^{-2-3}$. This proves rule (2).

EXAMPLE 1.3

Find the relative error if the number $X = 0.004997$ is

(i) truncated to three decimal digits

(ii) rounded off to three decimal digits.

Solution: We have $X = 0.004997 = 0.4997 \times 10^{-2}$

(i) After truncating to three decimal places, its approximate value $X = 0.499 \times 10^{-2}$.

$$\therefore \text{Relative error} = \left| \frac{X - X'}{X} \right| = \left| \frac{0.4997 \times 10^{-2} - 0.499 \times 10^{-2}}{0.4997 \times 10^{-2}} \right|$$

$$= 0.140 \times 10^{-2} < 10^{\,1-3}$$

This proves rule (3).

(ii) After rounding off to three decimal places, the approximate value of the given number
$X' = 0.500 \times 10^{-2}$

$$\therefore \text{Relative error} = \left| \frac{X - X'}{X} \right| = \left| \frac{0.4997 \times 10^{-2} - 0.500 \times 10^{-2}}{0.4997 \times 10^{-2}} \right|$$

$$= 0.600 \times 10^{-3} = 0.06 \times 10^{-3+1}$$

which is less than $0.5 \times 10^{-3+1}$. This proves rule (4).

Exercises 1.1

1. Round off the following numbers correct to four significant figures: 3.26425, 35.46735, 4985561, 0.70035, 0.00032217, and 18.265101.

2. Round off the number 75462 to four significant digits and then calculate the absolute error and percentage error.

3. If 0.333 is the approximate value of 1/3, find the absolute and relative errors.

4. Find the percentage error if 625.483 is approximated to three significant figures.

5. Find the relative error in taking $n = 3.141593$ as 22/7.

6. The height of an observation tower was estimated to be 47 m, whereas its actual height was 45 m. Calculate the percentage relative error in the measurement.

7. Suppose that you have a task of measuring the lengths of a bridge and a rivet, and come up with 9999 and 9 cm, respectively. If the true values are 10,000 and 10 cm, respectively, compute the percentage relative error in each case.

8. Find the value of ex using series expansion $e^x = 1 + x + \dfrac{x^2}{2!} + \dfrac{x^3}{3!} + \ldots$ for $x = 0.5$ with an absolute error less than 0.005.

9. $\sqrt{29} = 5.385$ and $\sqrt{\pi} = 3.317$ correct to 4 significant figures. Find the relative errors in their sum and difference.

10. Given: $a = 9.00 \pm 0.05$, $b = 0.0356 \pm 0.0002$, $c = 15300 \pm 100$, $d = 62000 \pm 500$. Find the maximum value of absolute error in $a + b + c + d$.

11. Two numbers are 3.5 and 47.279 both of which are correct to the significant figures given. Find their product.

12. Find the absolute error and the relative error in the product of 432.8 and 0.12584 using four digit mantissa.

13. The discharge Q over a notch for head H is calculated by the formula $Q = kH^{5/2}$ where k is a given constant. If the head is 75 cm and an error of 0.15 cm is possible in its measurement, estimate the percentage error in computing the discharge.

14. If the number p is correct to 3 significant digits, what will be the maximum relative error?

1.5   Error Propagation

A number of computational steps are carried out for the solution of a problem. It is necessary to understand the way the error propagates with progressive computation.

If the approximate values of two numbers X and Y be X' and Y, respectively, then the absolute error

$$E_{ax} = X - X' \text{ and } E_{ay} = Y - Y'$$

1. Absolute error in addition operation

$$X + Y = (X' + E_{ax}) + (Y' + E_{ay})$$
$$= X' + Y' + E_{ax} + E_{ay}$$

$$\therefore \ |(X+Y)-(X'+Y')| = |E_{ax}+E_{ay}| \leq |E_{ax}|+|E_{ay}|$$

Thus the absolute error in taking $(X' + Y')$ as an approximation to $(X + Y)$ is less than or equal to the sum of the absolute errors in taking X' as an approximation to X and Y' as an approximation to Y.

2. Absolute error in subtraction operation

$$X - Y = (X' + E_{ax}) - (Y' + E_{ay})$$
$$= (X' - Y') + (E_{ax} - E_{ay})$$
$$\therefore \qquad |(X - Y) - (X' - Y')| = |E_{ax} - E_{ay}| < |E_{ax}| + |E_{ay}|$$

Thus the absolute error in taking $(X' - Y')$ as an approximation to $(X - Y)$ is less than or equal to the sum of the absolute errors in taking X' as an approximation to X and Y' as an approximation to Y.

3. Absolute error in multiplication operation

To find the absolute error E_a in the product of two numbers X and Y, we write

$$E_a = (X + E_{ax})\,(Y + E_{ay}) - XY$$

where E_{ax} and E_{ay} are the absolute errors in X and Y, respectively. Then

$$E_a = XE_{ay} + YE_{ax} + E_{ax}E_{ay}$$

Assuming E_{ax} and E_{ay} are reasonably small so that $E_{ax}E_{xy}$ can be ignored. Thus $E_a = XE_{ay} + YE_{ax}$ approximately.

4. Absolute error in division operation

Similarly the absolute error E_a in the quotient of two numbers X and Y is given by

$$E_a = \frac{X + E_{ax}}{Y + E_{ay}} - \frac{X}{Y} = \frac{YE_{ax} - XE_{ay}}{Y(Y + E_{ay})}$$

$$= \frac{YE_{ax} - XE_{ay}}{Y^2\left(1 + E_{ay}/Y\right)}$$

$$= \frac{YE_{ax} - XE_{ay}}{Y^2}, \ \text{assuming } E_{ay}/Y \text{ to be small.}$$

$$= \frac{X}{Y}\left(\frac{E_{ax}}{X} - \frac{E_{ay}}{Y}\right)$$

EXAMPLE 1.4

Find the absolute error and relative error in $\sqrt{6} + \sqrt{7} + \sqrt{8}$ correct to 4 significant digits.

Solution:

We have $\sqrt{6} = 2.449, \sqrt{7} = 2.646, \sqrt{8} = 2.828$

$\therefore \qquad\qquad S = \sqrt{6} + \sqrt{7} + \sqrt{8} = 7.923.$

Then the absolute error E_a in S, is

$$E_a = 0.0005 + 0.0007 + 0.0004 = 0.0016$$

This shows that S is correct to 3 significant digits only. Therefore, we take $S = 7.92$ Then the relative error E_r is

$$E_r = \frac{0.0016}{7.92} = 0.0002.$$

EXAMPLE 1.5

The area of cross-section of a rod is desired up to 0.2% error. How accurately should the diameter be measured?

Solution:

If A is the area and D is the diameter of the rod, then $A = \pi\left(\frac{D}{2}\right)^2 = \frac{\pi}{4}D \times D.$

Now error in area A is 0.2%, i.e., 0.002 which is due to the error in the product $D \times D$.

We know that if E_a is the absolute error in the product of two numbers X and Y, then

$$E_a = X_{aY}E + YE_{aX}$$

Here, $X = Y = D$ and $E_{aX} = E_{aY} = E_D$, therefore

$$E_a = DE_D + DE_D \quad \text{or} \quad 0.002 = 2DE_D$$

Thus, $E_a = 0.001/D$, i.e., the error in the diameter should not exceed $0.001\ D^{-1}$.

EXAMPLE 1.6

Find the product of the numbers 3.7 and 52.378 both of which are correct to given significant digits.

Solution:

Since the absolute error is greatest in 3.7, therefore we round off the other number to 3 significant figures, i.e., 52.4.

$\therefore$ Their product $P = 3.7 \times 52.4 = 193.88 = 1.9388 \times 10^2$.

Since the first number contains only two significant figures, therefore retaining only two significant figures in the product, we get

$$P = 1.9 \times 10^2.$$

1.6 Error in the Approximation of a Function

Let $y = f(x_1, x_2)$ be a function of two variables x_1, x_2. If δx_1, δx_2 be the errors in x_1, x_2, then the error δy in y is given by

$$y + \delta y = f(x_1 + \delta x_1, x_2 + \delta x_2)$$

Expanding the right hand side by Taylor's series, we get

$$y + \delta y = f(x_1, x_2) + \left(\frac{\partial f}{\partial x_1} \delta x_1 + \frac{\partial f}{\partial x_2} \delta x_2 \right)$$

+ terms involving higher powers of δx_1 and δx_2 (i)

If the errors δx_1, δx_2 are so small that their squares and higher powers can be neglected, then (i) gives

$$\delta y = \frac{\partial f}{\partial x_1} \delta x_1 + \frac{\partial f}{\partial x_2} \delta x_2 \text{ approximately.}$$

Hence, $\delta y \approx \dfrac{\partial y}{\partial x_1} \delta x_1 + \dfrac{\partial y}{\partial x_2} \delta x_2$

In general, the error δy in the function $y = f(x_1, x_2, \ldots x_n)$ corresponding to the errors δx_i in $_{xi}$ $(i = 1, 2, \ldots n)$ is given by

$$\delta y \approx \frac{\partial y}{\partial x_1} \delta x_1 + \frac{\partial y}{\partial x_2} \delta x_2 + \ldots + \frac{\partial y}{\partial x_n} \delta x_n$$

and the relative error in y is $E_r = \dfrac{\delta y}{y} \approx \dfrac{\partial y}{\partial x_1} \dfrac{\delta x_1}{y} + \dfrac{\partial y}{\partial x_2} \dfrac{\delta x_2}{y} + \ldots + \dfrac{\partial y}{\partial x_n} \dfrac{\delta x_n}{y}$.

EXAMPLE 1.7

If $u = 4x^2 y^3 / z^4$ and errors in x, y, z are 0.001, compute the relative maximum error in u when $x = y = z = 1$.

Solution:

Since $\dfrac{\partial u}{\partial x} = \dfrac{8xy^3}{z^4}$, $\dfrac{\partial u}{\partial y} = \dfrac{12x^2 y^2}{z^4}$, $\dfrac{\partial u}{\partial z} = -\dfrac{16x^2 y^3}{z^5}$

$$\therefore \ \delta u = \frac{\partial u}{\partial x}\delta x + \frac{\partial u}{\partial y}\delta y + \frac{\partial u}{\partial z}\delta z = \frac{8xy^3}{z^4}\delta x + \frac{12x^2y^2}{z^4}\delta y - \frac{16x^2y^3}{z^5}\delta z.$$

Since the errors δx, δy, δz may be positive or negative, we take the absolute values of the terms on the right side, giving

$$(\delta u)_{\text{max}} \approx \left|\frac{8xy^3}{z^4}\delta x\right| + \left|\frac{12x^2y^2}{z^4}\delta y\right| - \left|\frac{16x^2y^3}{z^5}\delta z\right|$$

$$= 8(0.001) + 12(0.001) + 16(0.001) = 0.036$$

Hence the maximum relative error $= (\delta u)_{\text{max}}/u = 0.036/4 = 0.009$.

EXAMPLE 1.8

Find the relative error in the function $y = ax_1{}^{m_1} x_2{}^{m_2} \cdots x_n{}^{m_n}$.

Solution:

We have $\log y = \log a + m_1 \log x_1 + m_2 \log x_2 + \ldots + m_n \log x_n$

$$\therefore \ \frac{1}{y}\frac{\partial y}{\partial x_1} = \frac{m_1}{x_1}, \frac{1}{y}\frac{\partial y}{\partial x_2} = \frac{m_2}{x_2} \ \text{etc.}$$

Hence $\ E_r \approx \dfrac{\partial y}{\partial x_1}\dfrac{\delta x_1}{y} + \dfrac{\partial y}{\partial x_2}\dfrac{\delta x_2}{y} + \ldots + \dfrac{\partial y}{\partial x_n}\dfrac{\delta x_n}{y}$

$$= m_1\frac{\delta x_1}{x_1} + m_2\frac{\delta x_2}{x_2} + \ldots + m_n\frac{\delta x_n}{x_n}$$

Since the errors δx_1, δx_2, ..., δx_n may be positive or negative, we take the absolute values of the terms on the right side. This gives:

$$(E_r)_{\text{max}} \le m_1\left|\frac{\delta x_1}{x_1}\right| + m_2\left|\frac{\delta x_2}{x_2}\right| + \ldots + m_n\left|\frac{\delta x_n}{x_n}\right|$$

Cor. Taking $a = 1$, $m_1 = m_2 = \ldots = m_n = 1$, we have

$$y = x_1 x_2 \ldots x_n.$$

then $\ E_r \approx \dfrac{\delta x_1}{x_1} + \dfrac{\delta x_2}{x_2} + \ldots + \dfrac{\delta x_n}{x_n}$

Thus the relative error of a product of n numbers is approximately equal to the algebraic sum of their relative errors.

1.7 Error in a Series Approximation

We know that the Taylor's series for $f(x)$ at $x = a$ with a remainder after n terms is

$$f(x) = f(a + \overline{x - a}) = f(a) + (x - a)f'(a) + \frac{(x - a)^2}{2!}f''(a)$$

$$+ \ldots + \frac{(x - a)^{n-1}}{(n - 1)!}f^{n-1}(a) + R_n(x)$$

where $R_n(x) = \dfrac{(x - a)^n}{n!}f^n(\theta), a < \theta < x.$

If the series is convergent, $R_n(x) \to 0$ as $n \to \infty$ and hence if $f(x)$ is approximated by the first n terms of this series, then the maximum error will be given by the remainder term $R_n(x)$. On the other hand, if the accuracy required in a series approximation is preassigned, then we can find n, the number of terms which would yield the desired accuracy.

EXAMPLE 1.9

Find the number of terms of the exponential series such that their sum gives the value of e^x correct to six decimal places at $x = 1$.

Solution: We have $e^x = 1 + x + \dfrac{x^2}{2!} + \dfrac{x^3}{3!} + \ldots + \dfrac{x^{n-1}}{(n-1)!} + R_n(x)$ \hfill (i)

where $R_n(x) = \dfrac{x^n}{n!}e^\theta, 0 < \theta < x.$

∴ Maximum absolute error (at $\theta = x$) $= \dfrac{x^n}{n!}e^x$ and the maximum relative error $= \dfrac{x^n}{n!}$

Hence $(E_r)_{max}$ at $x = 1$ is $\dfrac{1}{n!}$.

For a six decimal accuracy at $x = 1$, we have

$\dfrac{1}{n!} < \dfrac{1}{2}10^{-6}$, i.e., $n! > 2 \times 10^6$ which gives $n = 10$.

Thus we need 10 terms of the series (i) in order that its sum is correct to six decimal places.

EXAMPLE 1.10

The function $f(x) = \tan^{-1} x$ can be expanded as

$$\tan^{-1} x = x - \frac{x^3}{3} + \frac{x^5}{5} - \cdots + (-1)^{n-1} \frac{x^{2n-1}}{2n-1} + \cdots,$$

Find n such that the series determine $\tan^{-1} x$ correct to eight significant digits at $x = 1$.

Solution:

If we retain n terms in the expansion of $\tan^{-1} x$ then $(n + 1)$th term

$$= (-1)^n \frac{x^{2n+1}}{2n+1}$$

$$= \frac{(-1)^n}{2n+1} \text{ for } x = 1.$$

To determine $\tan^{-1}(1)$ correct to eight significant digits accuracy

$$\left| \frac{(-1)^n}{2n+1} \right| < \frac{1}{2} \times 10^{-8} \text{ i.e., } 2n + 1 > 2 \times 108 \text{ or } n > 10^{-8} - \frac{1}{2}$$

Hence $n = 10^8 + 1$.

1.8 Order of Approximation

We often replace a function $f(h)$ with its approximation $\phi(h)$ and the error bound is known to be $\mu(h^n)$, n being a positive integer so that

$$|f(h) - \phi(h)| \leq \mu |h^n| \text{ for sufficiently small } h.$$

Then we say that $\phi(h)$ *approximates* $f(h)$ *with order of approximation* $O(h^n)$ and write $f(h) = \phi(h) + O(h^n)$.

For instance, $\dfrac{1}{1-h} = 1 + h + h^2 + h^3 + h^4 + h^5 + \cdots$

is written as $\dfrac{1}{1-h} = 1 + h + h^2 + h^3 + O(h^4)$ (*i*)

to the 4th order of approximation.

Similarly $\cos(h) = 1 - \dfrac{h^2}{2!} + \dfrac{h^4}{4!} - \dfrac{h^6}{6!} + \dfrac{h^8}{8!} - \cdots$

to the 6th order of approximation becomes

$$\cos(h) = 1 - \frac{h^2}{2!} + \frac{h^4}{4!} + O(h^6) \qquad\qquad (ii)$$

The sum of (i) and (ii) gives

$$(1-h)^{-1} + \cos(h) = 2 + h + \frac{h^2}{2!} + h^3 + O(h^4) + \frac{h^4}{4!} O(h^6) \qquad (iii)$$

Since $O(h^4) + \dfrac{h^4}{4!} = O(h^4)$ and $O(h^4) + O(h^6) = O(h^4)$

$\therefore$ (iii) takes the form $(1-h)^{-1} + \cos(h) = 2 + h + \dfrac{h^2}{2} + h^3 + O(h^4)$, which is of the 4th order of approximation.

Similarly the product of (i) and (ii) yields

$$(1-h)^{-1}\cos(h) = \left(1 + h + h^2 + h^3\right)\left(1 - \frac{h^2}{2!} + \frac{h^4}{4!}\right) + \left(1 + h + h^2 + h^3\right)O(h^6)$$

$$+ \left(1 - \frac{h^2}{2!} + \frac{h^4}{4!}\right)O(h^4) + O(h^4)O(h^6)$$

$$= 1 + h + \frac{h^2}{2} + \frac{h^3}{2} - \frac{11h^4}{24} + \frac{11}{24}h^5 + \frac{h^6}{24} + \frac{h^7}{24} + O(h^4) + O(h^6) + O(h^4)O(h^6) \qquad (iv)$$

Since $O(h^4)\, O(h^6) = O(h^{10})$

and $-\dfrac{11h^4}{24} + \dfrac{11}{24}h^5 + \dfrac{h^6}{24} + \dfrac{h^7}{24} + O(h^4) + O(h^6) + O(h^{10}) = O(h^4)$

$\therefore$ (iv) is reduced to $(1-h)^{-1}\cos(h) = 1 + h + \dfrac{h^2}{2} + \dfrac{h^3}{2} + O(h^4)$, which is of the 4th order of approximation.

1.9 Growth of Error

Let $e(n)$ represent the growth of error after n steps of a computation process.

If $|e(n)| \sim n\,\varepsilon$, we say that the growth of error is **linear.**

If $|e(n)| \sim \delta^n\,\varepsilon'$ we say that the growth of error is exponential.

If $\delta > 1$, the exponential error grows indefinitely as $n \to \infty$, and

if $0 < \delta < 1$, the exponential error decreases to zero as $n \to \infty$.

Exercises 1.2

1. Find the smaller root of the equation $x^2 - 400x + 1 = 0$, correct to four decimal places.

2. If $r = h(4h^5 - 5)$, find the percentage error in r at $h = 1$, if the error in h is 0.04.

3. If $R = 10 x^3 y^2 z^2$ and errors in x, y, z are 0.03, 0.01 and 0.02, respectively at $x = 3, y = 1, z = 2$. Calculate the absolute error and % relative error in evaluating R.

4. If $R = 4xy^2/z^3$ and errors in x, y, z are 0.001, show that the maximum relative error at $x = y = z = 1$ is 0.006.

5. If $V = \dfrac{1}{2}\left(\dfrac{r^2}{h} + h\right)$ and the error in V is at the most 0.4%, find the percentage error allowable in r and h when $r = 5.1$ cm and $h = 5.8$ cm.

6. Find the value of $I = \int_0^{0.8} \dfrac{\sin x}{x} dx$ correct to four decimal places.

7. Using the series $\sin x = x - \dfrac{x^3}{3!} + \dfrac{x^5}{5!} - \cdots$, evaluate $\sin 25°$ with an accuracy of 0.001.

8. Determine the number of terms required in the series for $\log (1 + x)$ to evaluate $\log 1.2$ correct to six decimal places.

9. Use the series $\log_e\left(\dfrac{1+x}{1-x}\right) = 2\left(x + \dfrac{x^3}{3} + \dfrac{x^5}{5} + \cdots\right)$ to compute the value of $\log (1.2)$ correct to seven decimal places and find the number of terms retained.

10. Find the order of approximation for the sum and product of the following expansions:

$$e^h = 1 + h + \dfrac{h^2}{2} + \dfrac{h^3}{3!} + O(h^4) \quad \text{and} \quad \cos(h) = 1 - \dfrac{h^2}{2!} + \dfrac{h^4}{4!} + O(h^6).$$

11. Given the expansions:

$$\sin(t) = t - \dfrac{t^3}{3!} + \dfrac{t^5}{5!} + O(t^7) \text{ and } \cos(t) = 1 - \dfrac{t^2}{2!} + \dfrac{t^3}{4!} + O(t^6)$$

Determine the order of approximation for their sum and product.

1.10   Objective Type of Questions

Exercises 1.3

Select the correct answer or fill up the blanks in the following questions:

1. If x is the true value of a quantity and $x1$ is its approximate value, then the relative error is

(a) $|x_1 - x|/x_1$ (b) $|x - x_1|/x$ (c) $|x_1/x|$ (d) $x/|x_1 - x|$.

2. The relative error in the number 834.12 correct to five significant figures is ...

3. If a number is rounded to k decimal places, then the absolute error is

(a) $\frac{1}{2}10^{k-1}$ (b) $\frac{1}{2}10^{-k}$ (c) -10 (d) $\frac{1}{4}10^{-k}$.

4. If π is taken = 3.14 in place of 3.14156, then the relative error is

5. Given $x = 1.2$, $y = 25.6$, and $z = 4.5$, then the relative error in evaluating $w = x^2 + y/z$ is...

6. Round off values of 43.38256, 0.0326457, and 0.2537623 to four significant digits: ...

7. Round relative maximum error in $3x2y/z$ when $dx = dy = dz = 0.001$ at $x = y = z = 1$: ...

8. If both the digits of the number 8.6 are correct, then the relative error is...

9. If a number is correct to n significant digits, then the relative error is

(a) -10 (b) $\frac{1}{2}10^{n-1}$ (c) $\leq \frac{1}{2}10^{-n}$ (d) $< \frac{1}{2}10^{n-1}$.

10. If $\left(\sqrt{3} + \sqrt{5} + \sqrt{7}\right)$ is rounded to four significant digits, then the absolute error is

11. $\left(\sqrt{102} - \sqrt{101}\right)$ correct to three significant figures is...

12. Approximate values of 1/3 are given as 0.3, 0.3, and 0.34. Out of these the best approximation is ...

13. The relative error if 2/3 is approximated to 0.667, is…

14. If the first significant digit of a number is p and the number is correct to n significant digits, then the relative error is …

SOLUTION OF ALGEBRAIC AND TRANSCENDENTAL EQUATIONS

Chapter Objectives

- Introduction
- Basic properties of equations
- Transformation of equations
- Synthetic division; to diminish the roots of an equation by h
- Iterative methods
- Graphical solution of equations
- Convergence
- Bisection method
- Method of false position
- Secant method
- Iteration method; Aitken's $\Delta 2$ method.
- Newton-Raphson method
- Some deductions from Newton-Raphson formula
- Muller's method
- Roots of polynomial equations; approximate solution of polynomial equations-Horner's method
- Multiple roots
- Complex roots
- Lin-Bairstow's method
- Graeffe's root squaring method

- Comparison of Iterative methods
- Objective type of questions

2.1   Introduction

An expression of the form $f(x) = a_0 x^n + a_1 x^{n-1} + \cdots + a_{n-1} x + a_n$

where a's are constants $(a_0 \neq 0)$ and n is a positive integer, is called a *polynomial* in x of degree n. The polynomial $f(x) = 0$ is called an algebraic *equation* of degree *n*. If $f(x)$ contains some other functions such as trigonometric, logarithmic, exponential etc., then $f(x) = 0$ is called a *transcendental equation.*

   Def. *The value α of x which satisfies $f(x) = 0$*   (1)

is called a **root** *of* $f(x) = 0$. Geometrically, a root of (1) is that value of x where the graph of $y = f(x)$ crosses the x-axis.

   The process of finding the roots of an equation is known as the *solution of that equation.* This is a problem of basic importance in applied mathematics.

   If $f(x)$ is a quadratic, cubic, or a biquadratic expression, algebraic solutions of equations are available. But the need often arises to solve higher degree or transcendental equations for which no direct methods exist. Such equations can best be solved by approximate methods. In this chapter, we shall discuss some numerical methods for the solution of algebraic and transcendental equations.

2.2   Basic Properties of Equations

   I.   *If $f(x)$ is exactly divisible by x − α, then α is a root of $f(x) = 0$.*

   II. *Every equation of the nth degree has only n roots* (real or imaginary).

   Conversely if α_1, α_2, ..., α_n are the roots of the *n*th degree equation $f(x) = 0$, then

$$f(x) = A(x - \alpha_1)(x - \alpha_2) \ldots | (x - \alpha_n)$$

where *A* is a constant.

NOTE   **Obs.** *If a polynomial of degree n vanishes for more than n values of x, it must be identically zero.*

EXAMPLE 2.1

Solve the equation $2x^3 + x^2 - 13x + 6 = 0$.

Solution: By inspection, we find $x = 2$ satisfies the given equation.

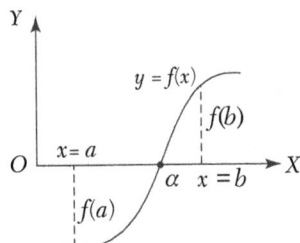

FIGURE 2.1

∴ 2 is its root, i.e., $x - 2$ is a factor of $2x^3 + x^2 - 13x + 6$.

Dividing this polynomial by $x - 2$, we get the quotient $2x^2 + 5x - 3$ and remainder 0.

Equating this quotient to zero, we get $2x^2 + 5x - 3 = 0$.

Solving this quadratic, we get

$$x = \frac{-5 \pm \sqrt{[5^2 - 4.2.(-3)]}}{2.2} = -3, 1/2.$$

Hence the roots of the given equation are 2, − 3, 1/2.

III. **Intermediate value property.** If $f(x)$ *is continuous in the interval* $[a, b]$ *and* $f(a), f(b)$ *have different signs, then the equation* $f(x) = 0$ *has at least one root between* $x = a$ *and* $x = b$.

Since $f(x)$ is continuous between a and b, so while x changes from a to b, $f(x)$ must pass through all the values from $f(a)$ to $f(b)$ [Figure 2.1]. But one of these quantities $f(a)$ or $f(b)$ is positive and the other negative, it follows that at least for one value of x (say α) lying between a and b, $f(x)$ must be zero. Then α is the required root.

IV. *In an equation with real coefficients, imaginary roots occur in conjugate pairs, i.e., if $\alpha + i\beta$ is a root of the equation $f(x) = 0$, then $\alpha - i\beta$ must also be its root.*

Similarly if $a + \sqrt{b}$ is an irrational root of an equation, then $a - \sqrt{b}$ must also be its root.

NOTE

Obs. *Every equation of the odd degree has at least one real root.*

This follows from the fact that imaginary roots occur in conjugate pairs.

EXAMPLE 2.2

Solve the equation $3x^3 - 4x^2 + x + 88 = 0$, one root being $2 + \sqrt{7}i$.

Sol. Since one root is $2 + \sqrt{7}i$, the other root must be $2 - \sqrt{7}i$.

∴ The factors corresponding to these roots are $(x - 2 - \sqrt{7}i)$ and $(x - 2 + \sqrt{7}i)$

or $\left(x - 2 - \sqrt{7}i\right)\left(x - 2 + \sqrt{7}i\right) = (x-2)^2 + 7 = x^2 - 4x + 11$

is a divisor of $\qquad 3x^3 - 4x^2 + x + 88 \qquad$ (i)

∴ Division of (i) by $x^2 - 4x + 11$ gives $3x + 8$ as the quotient.

Thus the depressed equation is $3x + 8 = 0$. Its root is $-8/3$.

Hence the roots of the given equation are $2 \pm \sqrt{7}i - 8/3$.

V. **Descarte's rule of signs.** *The equation $f(x) = 0$ cannot have more positive roots than the changes of signs in $f(x)$; and more negative roots than the changes of signs in $f(-x)$.*

For instance, consider the equation $f(x) = 2x^7 - x^5 + 4x^3 - 5 = 0$ (i)

Signs of $f(x)$ are $+ \quad - \quad + \quad -$

Clearly $f(x)$ has 3 changes of signs (from + to – or – to +).

Thus (i) cannot have more than 3 positive roots.

Also $\qquad f(-x) = 2(-f)^7 \ (-x)^5 + 4(-x)^3 - 5$
$$= -2x^7 + x^5 - 4x^3 - 5$$

This shows that f(x) has 2 changes of signs.

Thus (i) cannot have more than 2 negative roots.

NOTE **Obs.** *Existence of imaginary roots. If an equation of the nth degree has at the most p positive roots and at the most q negative roots, then it follows that the equation has at least n − (p + q) imaginary roots.*

Evidently (i) above is an equation of the 7th degree and has at the most 3 positive roots and 2 negative roots. Thus (i) has at least 2 imaginary roots.

VI. Relations between roots and coefficients. *If $\alpha_1, \alpha_2 \alpha_3 \cdots, \alpha_n$ are the roots of the equation*

$$a_0 x^n + a_1 x^{n-1} + a_2 x^{n-2} + \cdots + a_{n-1} x + a_n = 0 \qquad (1)$$

then $\displaystyle\sum \alpha_1 = -\frac{a_1}{a_0}$

$$\sum \alpha_1 \alpha_2 = \frac{a_2}{a_0}$$

$$\sum \alpha_1 \alpha_2 \alpha_3 = \frac{a_3}{a_0}, \cdots$$

$$\alpha_1 \alpha_2 \alpha_3 \cdots \alpha_n = (-1)^n \frac{a_n}{a_0}$$

EXAMPLE 2.3

Solve the equation $x^3 - 7x^2 + 36 = 0$, given that one root is double of another.

Solution:

Let the roots be α, β, γ such that $\beta = 2\alpha$.

Also $\quad \alpha + \beta + \gamma = 7, \ \alpha\beta + \beta\gamma + \gamma\alpha = 0, \ \alpha\beta\gamma = -36$

$\therefore \qquad\qquad 3\alpha + \gamma = 7 \qquad\qquad\qquad\qquad (i)$

$\qquad\qquad\qquad 2\alpha^2 + 3\alpha\gamma = 0 \qquad\qquad\qquad (ii)$

$\qquad\qquad\qquad\quad 2\alpha^2\gamma = -36 \qquad\qquad\qquad (iii)$

Solving (i) and (ii), we get $\alpha = 2, \gamma = -2$.

[The values $\alpha = 0, \gamma = 7$ are inadmissible, as they do not satisfy (iii)].

Hence the roots are 3, 6 and -2.

EXAMPLE 2.4

Solve the equation $x^4 - 2x^3 + 4x^2 + 6x - 21 = 0$, given that the sum of two its roots is zero.

Solution:

Let the roots be $\alpha, \beta, \gamma, \delta$ such that $\alpha + \beta = 0$.

Also $\quad \alpha + \beta + \gamma + \delta = 2$

$\therefore \qquad\qquad \gamma + \delta = 2$

Thus the quadratic factor corresponding to α, β is of the form $x^2 - 0x + p$ and that corresponding to γ, δ is of the form of $x^2 - 2x + q$.

$\therefore \qquad x^4 - 2x^3 + 4x^2 + 6x - 21 = (x^2 + p)(x^2 - 2x + q) \qquad (i)$

Equating coefficients of x^2 and x from both sides of (i), we get

$$4 = p + q \qquad 6 = -2p$$

$\therefore \qquad p = -3, \qquad q = 7.$

Hence the given equation is equivalent to

$$(x^2 - 3)(x^2 - 2x + 7) = 0$$

$\therefore$ The roots are $x = \pm \sqrt{3}, 1 \pm i\sqrt{6}$.

EXAMPLE 2.5

Find the condition that the cubic $x^3 - lx^2 + mx - n = 0$ should have its roots in

(a) Arithmetical progression   (b) Geometrical progression.

Solution:

(a)   Let the roots be $a - d, a, a + d$ so that the sum of the roots $= 3a = l$ i.e., $a = l/3$.

Since a is the root of the given equation $a^3 - la^2 + ma - n = 0$

Substituting $a = l/3$, we get $2l^3 - 9lm + 27n = 0$ which is the required condition.

(b)   Let the roots be $a/r, a, ar$, then the product of the roots $= a^3 = n$.

Since a is a root of the given equation

$\therefore \quad a^3 - la^2 + ma - n = 0$

Putting $a = (n)^{1/3}$, we get $n - ln^{2/3} + mn^{1/3} - n = 0$ or $m = ln^{1/3}$

Cubing both sides, we get $m^3 = l^3n$

which is the required condition.

EXAMPLE 2.6

If α, β, γ are the roots of the equation $x^3 + px + q = 0$, find the value of

(a) $\Sigma\alpha^2\beta$,   (b) $\Sigma\alpha^4$.

Solution:

We have
$$\alpha + \beta + \gamma = 0 \tag{i}$$
$$\alpha\beta + \beta\gamma + \gamma\alpha = p \tag{ii}$$
$$\alpha\beta\gamma = -q \tag{iii}$$

(a) Multiplying (i) and (ii), we get
$$\alpha^2\beta + \alpha^2\gamma + \beta^2\gamma + \beta^2\alpha + \gamma^2\alpha + \gamma^2\beta + 3\alpha\beta\gamma = 0$$

or $\qquad\qquad \Sigma\alpha^2\beta = -3\alpha\beta\gamma = 3q \qquad$ [by (iii)]

(b) Multiplying the given equation by x, we get
$$x^4 + px^2 + qx = 0$$

Putting $x = \alpha, \beta, \gamma$ successively and adding, we get
$$\Sigma\alpha^4 + p\Sigma\alpha^2 + q\Sigma\alpha = 0 \quad \text{or} \quad \Sigma\alpha^4 = -p\Sigma\alpha^2 - q(0) \tag{iv}$$

Now squaring (i), we get
$$\alpha^2 + \beta^2 + \gamma^2 + 2(\alpha\beta + \beta\gamma + \gamma\alpha) = 0 \quad \text{or} \quad \Sigma\alpha^2 = -2p \quad \text{[by (ii)]}$$

Hence, substituting the value of $\Sigma\alpha^2$ in (iv), we obtain
$$\Sigma\alpha^4 = -p(-2p) = 2p^2$$

Exercises 2.1

1. Form the equation of the fourth degree whose roots are $3 + i$ and $\sqrt{7}$.

2. Solve the equation:

(i) $x^3 + 6x + 20 = 0$, one root being $1 + 3i$.

(ii) $x^4 - 2x^3 - 22x^2 + 62x - 15 = 0$ given that $2 + \sqrt{3}$ is a root.

3. Show that $x^7 - 3x^4 + 2x^3 - 1 = 0$ has at least four imaginary roots.

4. The equation $x^4 - 4x^3 + ax^2 + 4x + b = 0$ has two pairs of equal roots. Find the values of a and b.

Solve the equations (5–7):

5. $2x^4 - 3x^3 - 9x^2 + 15x - 5 = 0$, given that the sum of two of its roots is zero.

6. $x^3 - 4x^2 - 20x + 48 = 0$ given that the roots α and β are connected by the relation $\alpha + 2\beta = 0$.

7. $x^3 - 12x^2 + 39x - 28 = 0$, roots being in arithmetical progression.

8. O, A, B, C are the four points on a straight line such that the distances of A, B, C from O are the roots of equation $ax^3 + 3bx^2 + 3cx + d = 0$. If B is the middle point of AC, show that $a^2d - 3abc + 2b^3 = 0$.

9. If α, β, γ are the roots of the equation $x^3 + 4x - 3 = 0$, find the value of $\alpha^{-1} + \beta^{-1} + \gamma^{-1}$.

2.3   Transformation of Equations

1. To find an equation whose roots are m times the roots of the given equation, *multiply the second term by m, third term by m^2 and so on* (all missing terms supplied with zero coefficients).

For instance, let the given equation be

$$3x^4 + 6x^3 + 4x^2 - 8x + 11 = 0 \qquad\qquad (i)$$

To multiply its roots by m, put $y = mx$ (or x = y/m) in (i). Then

$$3(y/m)^4 + 6(y/m)^3 + 4(y/m)^2 - 8(y/m) + 11 = 0$$

or multiplying by m^4, we get

$$3y^4 + m(6y^3) + m^2(4y^2) - m^3(y) + m^4(11) = 0$$

This is same as multiplying the second term by m, third term by m^2, and so on in (i).

Cor. To find an equation whose roots are with opposite signs to those of the given equation, change the signs of every alternative term of the given equation beginning with the second.

Changing the signs of roots of (i) is same as multiplying its roots by $- 1$.

$\therefore$ The required equation will be

$$3x^4 + (-1)6x^3 + (-1)24x^2 - (-1)3\,8x + (-1)^4\,11 = 0$$

or $\qquad\qquad\qquad\qquad 3x^4 - 6x^3 + 4x^3 + 8x + 11 = 0$

which is (i) with signs of every alternate term changed beginning with the second.

---

2. To find an equation whose roots are reciprocal of the roots of the given equation, *change x to 1/x*.

EXAMPLE 2.7

Solve $6x^3 - 11x^2 - 3x + 2 = 0$, given that its roots are in harmonic progression.

Solution:

Since the roots of the given equation are in H.P., the roots of the equation having reciprocal roots will be in A.P.

∴ The equation with reciprocal roots is

$$6(1/x)^3 - 11(1/x)^2 - 3(1/x) + 2 = 0$$

or $\qquad\qquad\qquad 2x^3 - 3x^2 - 11x + 6 = 0 \qquad\qquad\qquad\qquad$ (i)

Since the roots of the given equation are in H.P., therefore, the roots of (i) are in A.P.

Let the roots be $a - d, a, a + d$. Then $3a = 3/2$ and $a(a^2 - d^2) = -3$.

Solving these equations, we get $a = 1/2, d = 5/2$. Thus the roots of (i) are $-2, 1/2, 3$.

Hence the roots of the given equation are $-1/2, 2, 1/3$.

EXAMPLE 2.8

If α, β, γ be the roots of the cubic $x^3 - px^2 + qx - r = 0$, form the equation whose roots are $\beta\gamma + 1/\alpha, \gamma\alpha + 1/\beta, \alpha\beta + 1/\gamma$.

Solution:

If x is a root of the given equation and y, a root of the required equation, then

$$y = \beta\gamma + \frac{1}{\alpha} = \frac{\alpha\beta\gamma + 1}{\alpha} = \frac{r+1}{\alpha} = \frac{r+1}{x} \qquad (\because \ \alpha\beta\gamma = r)$$

Thus $x = (r+1)/y$.

Substituting this value of x in the given equation, we get

$$\left(\frac{r+1}{y}\right)^3 - p\left(\frac{r+1}{y}\right)^2 + q\left(\frac{r+1}{y}\right) - r = 0$$

or $\qquad ry^3 - q(r+1)y^2 + p(r+1)^2 y - (r+1)^3 = 0$

which is the required equation.

3. Reciprocal equations. *If an equation remains unaltered on changing x to be 1/x, it is called a* reciprocal equation.

Such equations are of the following types:

(*i*) *A reciprocal equation of an odd degree having coefficients of terms equidistant from the beginning and end equal.* It has a root = – 1.

(*ii*) *A reciprocal equation of an odd degree having coefficients of terms equidistant from the beginning and end equal but opposite in sign.* It has a root = 1.

(*iii*) *A reciprocal equation of an even degree having coefficients of terms equidistant from the beginning and end equal but opposite in sign.* Such an equation has two roots = 1 and –1.

The substitution $x + 1/x = y$ reduces the degree of the equation to half its former degree.

EXAMPLE 2.9

Solve: (*i*) $6x^5 - 41x^4 + 97x^3 - 97x^2 + 41x - 6 = 0$

(*ii*) $6x^6 - 25x^5 + 31x^4 - 31x^2 + 25x - 6 = 0$.

Solution:

(*i*) This is a reciprocal equation of odd degree with opposite signs.

∴ $x = 1$ is a root.

Dividing L.H.S. by $x - 1$, the equation reduces to

$$6x^4 - 35x^3 + 62x^2 - 35x + 6 = 0$$

Dividing by x^2, we have $6\left(x^2 + \dfrac{1}{x^2}\right) - 35\left(x + \dfrac{1}{x}\right) + 62 = 0$

Putting $x + \dfrac{1}{x} = y$ and $x^2 + \dfrac{1}{x^2} = y^2 - 2$, we get

$6(y^2 - 2) - 35y + 62 = 0$ or $6y2 - 35y + 50 = 0$

or $(3y - 10)(2y - 5) = 0$ ∴ $x + \dfrac{1}{x} = \dfrac{10}{3}$ or $\dfrac{5}{2}$

i.e., $3x^2 - 10x + 3 = 0$ or $2x^2 - 5x + 2 = 0$

i.e., $(3x - 1)(x - 3) = 0$ or $(2x - 1)(x - 2) = 0$

∴ $x = \dfrac{1}{3}, 3$ or $x = \dfrac{1}{2}, 2$

Hence the roots are $1, \frac{1}{3}, 3\frac{1}{2}, 2$.

(*ii*) This is a reciprocal equation of even degree with opposite signs.

$\therefore x = 1, -1$ are its roots.

Dividing L.H.S. by $x - 1$ and $x + 1$, the given equation reduces to

$6x^4 - 25x^3 + 37x^2 - 25x + 6 = 0$.

Dividing by x^2, we get $6\left(x^2 + \frac{1}{x^2}\right) - 25\left(x + \frac{1}{x}\right) + 37 = 0$

Putting $x + \frac{1}{x} = y$ and $x^2 + \frac{1}{x^2} = y^2 - 2$, it becomes

$6y^2 - 25y + 25 = 0 \qquad \text{or} \qquad (2y - 5)(3y - 5) = 0$

$\therefore \qquad x + \frac{1}{x} = y = \frac{5}{2} \quad \text{or} \qquad \frac{5}{3}$

$i.e., \qquad 2x^2 - 5x + 2 = 0 \qquad \text{or} \qquad 3x^2 - 5x + 3 = 0$

$\therefore \qquad x = 2, \frac{1}{2} \text{ or } x = \dfrac{5 \pm \sqrt{(-11)}}{6}$

Hence the roots are $1, -1, 2, \dfrac{1}{2}, \dfrac{5 \pm i\sqrt{11}}{6}$

2.4 Synthetic Division of a Polynomial by A Linear Expression

The division of the polynomial $f(x) = a_0 x^n + a_1 x^{n-1} + a_2 x^{n-2} + \cdots + a_{n-1} x + a_n$ by a binomial $x - \alpha$ is affected compactly by synthetic division as follows:

a_0	a_1	$a_2 \cdots a_{n-1}$	a_n	
	αb_0	$\alpha b_1 \cdots \alpha b_{n-2}$	αb_{n-1}	α
a_0	$a_1 + \alpha b_0$	$a_2 + \alpha b_1 \cdots a_{n-1} + \alpha b_{n-2}$	$a_n + \alpha b_{n-1}$	
$(= b_0)$	$(= b_1)$	$(= b_{n-1})$	$(= R)$	

Hence quotient $= b_0 x^{n-1} + b_1 x^{n-2} + \cdots + b_{n-1}$ and remainder $= R$

Explanation:

(*i*) Write down the coefficients of the powers of x (supplying missing powers of x by zero coefficients) and write α on extreme right.

(*ii*) Put $a_0 \ (= b_0)$ as the first term of 3rd row and multiply it by a and write the product under a_1 and add, giving $a_1 + ab_0 \ (= b_1)$.

(*iii*) Multiply b_1 by a and write the product under a_2 and add, giving $a_2 + ab_1 \ (= b_2)$ and so on.

(*iv*) Continue this process until we get R.

1. To diminish the roots of an equation $f(x) = 0$ by h, *divide $f(x)$ by $x - h$ successively.* Then with the *successive remainders, determine the coefficients of the required equation.*

Let the given equation be

$$a_0 x^n + a_1 x^{n-1} + \cdots + a_{n-1} x + a_n = 0 \qquad (i)$$

To diminish its roots h, put $y = x - h$ (or $x = y + h$) in (*i*) so that

$$a_0 (y + h)^n + a_1 (y + h)^{n-1} + \cdots + a_n = 0 \qquad (ii)$$

On simplification, it takes the form

$$A_0 y^n + A_1 y^{n-1} + \cdots + A_n = 0 \qquad (iii)$$

Its coefficients $A_0, A_1, \cdots, A_n$ can easily be found with help of *synthetic division.* For this, we put $y = x - h$ in (*iii*) so that

$$A_0 (x - h)^n + A_1 (x - h)^{n-1} + \cdots + A_n = 0$$

Clearly (*i*) and (*iv*) are identical. If we divide L.H.S. of (*iv*) by $x - h$, the remainder is A_n and the quotient $Q = A_0 (x - h)^{n-1} + A_1 (x - h)^{n-1} + \cdots + A_{n-1}$. Similarly if we divide Q by $x - h$, the remainder is A_{n-1} and the quotient is Q_1 (say). Again dividing Q_1 by $x - h$, A_{n-2} will be obtained and so on.

NOTE   **Obs.** *To increase the roots by h, we take h negative.*

EXAMPLE 2.10

Transform the equation $x^3 - 6x^2 + 5x + 8 = 0$ into another in which the second term is missing.

Solution:

Sum of the roots of the given equation = 6.

Due to the fact that the second term in the transformed equation is missing, the sum of the roots will be zero.

Since the equation has 3 roots, if we decrease each root by 2, the sum of the roots of the equation will become zero. To diminish the roots by 2, we divide $x^3 - 6x^2 + 5x + 8$ by $x - 2$ successively.

1	−6	5	8	2
	2	−8	−6	
	−4	−3	**2**	
	2	−4		
	−2	**−7**		
	2			
1	**0**			

Thus the transformed equation is $x^3 - 7x + 2 = 0$.

2. Synthetic division of a polynomial by a quadratic expression. The division of the polynomial $f(x)$ by the quadratic $x^2 - \alpha x - \beta$ is carried out by the following synthetic scheme:

a_0	a_1	a_2	$a_3 \cdots a_{n-1}$	a_n	α
	αb_0	αb_1	$\alpha b_2 \cdots \alpha b_{n-2}$		β
		βb_0	$\beta b_1 \cdots \beta b_{n-3}$	βb_{n-2}	
a_0	$a_1 + \alpha b_0$	$a_2 + \alpha b_1 + \beta b_0$,	$a_3 + \alpha b_2 + \beta b_1, \cdots$	$a_n + \beta b_{n-2}$	
$(= b_0)$	$(= b_1)$	$(= b_2)$	$(= b_3) \cdots (= b_{n-1})(= b_n)$		

Hence the quotient $= b_0 x^{n-2} + b_1 x^{n-3} + \cdots + b_{n-2}$ and the remainder $= b_{n-1} x + b_n$.

EXAMPLE 2.11

Divide $2x^5 - 3x^4 + 4x^3 - 5x^2 + 6x - 9$ by $x^2 - x + 2$ synthetically.

Solution:

2	−3	4	−5	6	−9	1
	2	−1	−1	−4		
		−4	2	2	8	−2
2	−1	−1	−4	4	−1	

Hence the quotient $= 2x^3 - x^2 - x - 4$ and the remainder $= 4x - 1$.

Exercises 2.2

1. Find the equation whose roots are 3 times the roots of $x^3 + 2x^2 - 4x + 1 = 0$.

2. Change the sign of the roots of the equation $x^7 + 3x^5 + x^3 - x^2 + 7x + 1 = 0$.

3. Find the equation whose roots are the negative reciprocals of the roots of $x^4 + 7x^3 + 8x^2 - 9x + 10 = 0$.

4. Solve the equation $81x^3 - 18x^2 - 36x + 8 = 0$, given that its roots are in H.P.

5. Solve: (*i*) $6x^5 + x^4 - 43x^3 - 43x^2 + x + 6 = 0$.
(*ii*) $4x^4 - 20x^3 + 33x^2 - 20x + 4 = 0$.

6. Find the equation whose roots are the roots of: $x^4 + x^3 - 3x^2 - x + 2 = 0$ each diminished by 3.

7. Show that the equation $x^4 - 10x^3 + 23x^2 - 6x - 15 = 0$ can be transformed into a reciprocal equation by diminishing the roots by 2. Hence solve the equation.

8. Find the equation of squared differences of the roots of the cubic $x^3 + 6x^2 + 7x + 2 = 0$.

9. If α, β, γ are the roots of the equation $2x^3 + 3x^2 - x - 1 = 0$, form the equation whose roots are $(1 - \alpha)^{-1}$, $(1 - \beta)^{-1}$, and $(1 - \gamma)^{-1}$.

10. Divide $15x^7 - 16x^6 + 30x^5 - 3x^4 - 5x^3 - 2x^2 + 5x + 8$ by $x^2 - x + 1$ synthetically.

2.5 Iterative Methods

The limitations of analytical methods for the solution of equations have necessitated the use of *iterative methods*. An iterative method begins with an approximate value of the root which is generally obtained with the help of *Intermediate value property* of the equation (Section 2.2). This initial approximation is then successively improved iteration by iteration and this process stops when the desired level of accuracy is achieved. The various iterative methods begin their process with one or more initial approximations. Based on the number of initial approximations used, these iterative methods are divided into two categories: *Bracketing Methods* and *Open-end Methods*.

Bracketing methods begin with two initial approximations which bracket the root. Then the width of this bracket is systematically reduced until the root is reached to desired accuracy. The commonly used methods in this category are:

1. Graphical method

2. Bisection method

3. Method of False position.

Open-end methods are used on formulae which require a single starting value or two starting values which do not necessarily bracket the root. The following methods fall under this category:

1. Secant method

2. Iteration method

3. Newton-Raphson method

4. Muller's method

5. Horner's method

6. Lin-Bairstow method.

2.6 Graphical Solution of Equations

Let the equation be $f(x) = 0$.

(i) Find the interval (a, b) in which a root of $f(x) = 0$ lies.

(ii) Write the equation $f(x) = 0$ as $\phi(x) = \psi(x)$

where $\psi(x)$ contains only terms in x and the constants.

(iii) Draw the graphs of $y = \phi(x)$ and $y = \phi(x)$ on the same scale and with respect to the same axes.

(iv) Read the abscissae of the points of intersection of the curves $y = \phi$ (x) *and* $y = \psi$

These are the initial approximations to the roots of $f(x) = 0$.

Sometimes it may not be convenient to write the given equation $f(x) = 0$ in the form $\phi(x) = \psi(x)$. In such cases, we proceed as follows:

(i) Form a table for the value of x and $y = f(x)$ directly.

(*ii*) Plot these points and pass a smooth curve through them.

(*iii*) Read the abscissae of the points where this curve cuts the *x*-axis. These are rough approximations to the roots of $f(x) = 0$.

EXAMPLE 2.12

Find graphically an approximate value of the root of the equation $3 - x = e^{x-1}$.

Solution:

Let $f(x) = e^{x-1} + x - 3 = 0$ (*i*)

$$f(1) = 1 + 1 - 3 = -\text{ ve and}$$

$$f(2) = e + 2 - 3 = 2.718 - 1 = +\text{ ve}$$

∴ A root of (*i*) lies between $x = 1$ and $x = 2$.

Let us write (i) as $e^{x-1} = 3 - x$.

The abscissa of the point of intersection of the curves

$$y = e^{x-1} \tag{ii}$$

and $\qquad\qquad\qquad\qquad y = 3 - x \qquad\qquad\qquad\qquad$ (*iii*)

will give the required root.

To plot (*ii*), we form the following table of values:

x	1.1	1.2	1.3	1.4	1.5	1.6	1.7	1.8	1.9	2.0
$y = e^{x-1}$	1.11	1.22	1.35	1.49	1.65	1.82	2.01	2.23	2.46	2.72

Taking the origin at (1, 1) and 1 small unit along either axis = 0.02, we plot these points and pass a smooth curve through them as shown in Figure 2.2.

To draw the line (*iii*), we join the points (1, 2) and (2, 1) on the same scale and with the same axes.

From the figure, we get the required root to be $x = 1.44$ nearly.

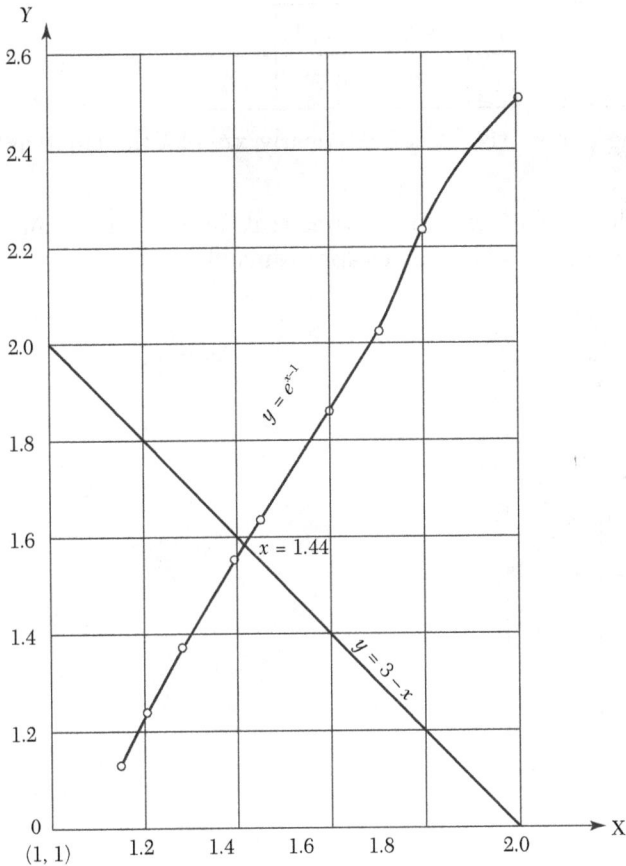

FIGURE 2.2

EXAMPLE 2.13

Obtain graphically an approximate value of the root of $x = \sin x + \pi/2$.

Solution:

Let us write the given equation as $\sin x = x - \pi/2$.

The abscissa of the point of intersection of the curve $y = \sin x$ and the line $y = x - \pi/2$ will give a rough estimate of the root.

To draw the curve $y = \sin x$, we form the following table:

x	0	$\pi/4$	$\pi/2$	$3\pi/4$	π
y	0	0.71	1	0.71	0

Taking 1 unit along either axis $= \pi/4 = 0.8$ nearly, we plot the curve as shown in Figure 2.3.

Also we draw the line $y = x - \pi/2$ to the same scale and with the same axes. From the graph, we get $x = 2.3$ radians approximately.

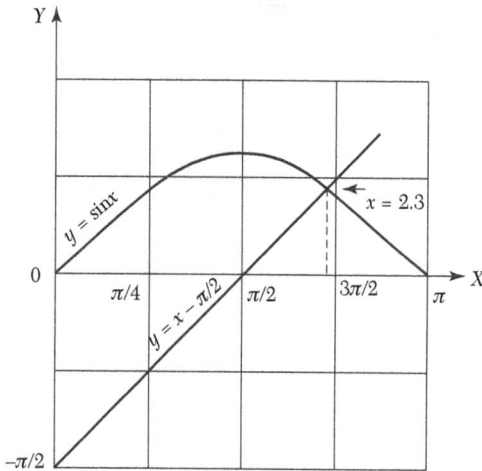

FIGURE 2.3

EXAMPLE 2.14

Obtain graphically an approximate value of the lowest root of $\cos x \cosh x = -1$.

Solution:

Let $f(x) = \cos x \cosh x + 1 = 0$ (i)

∴ $f(0) = +\text{ve}, f(\pi/2) = +\text{ve}$ and $\pi = -\text{ve}$.

∴ The lowest root of (i) lies between $x = \pi/2$ and $x = \pi$.

Let us write (i) as $\cos x = -\text{sech } x$.

The abscissa of the point of intersection of the curves

$$y = \cos x \qquad\qquad\qquad (ii)$$

and $$y = -\text{sech } x \qquad\qquad\qquad (iii)$$

will give the required root.

To draw (*ii*), we form the following table:

$x =$	$\pi/2 = 1.57$	$3\pi/4 = 2.36$	$\pi = 3.14$
$y = \cos x$	0	−0.71	−1

Taking the origin at (1.57, 0) and 1 unit along either axis = $\pi/8$, = 0.4 nearly, we plot the cosine curve as shown in Figure 2.4.

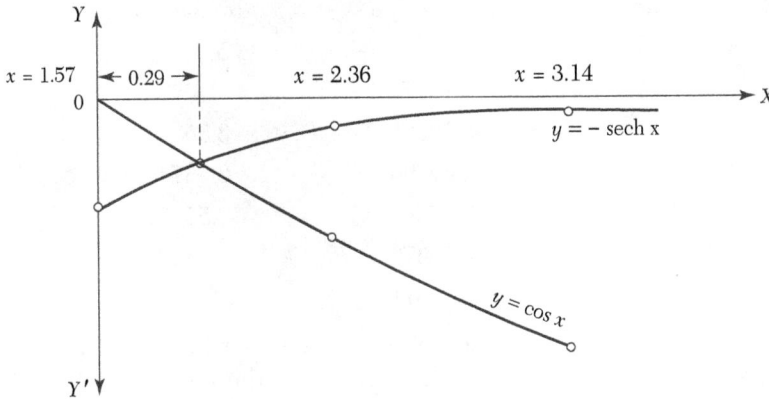

FIGURE 2.4

To draw (*iii*), we form the following table:

x	1.57	2.36	3.14
cosh x	2.51	5.34	11.57
$y = -$ sech x	−0.4	−0.19	−0.09

Then we plot the curve (*iii*) to the same scale with the same axes.

From the above figure, we get the lowest root to be approximately $x = 1.57 + 0.29 = 1.86$.

Exercises 2.3

Find the approximate value of the root of the following equations graphically (1–4):

1. $x^3 - x - 1 = 0$

2. $x^3 - 6x^2 + 9x - 3 = 0$

3. $\tan x = 1.2\,x$

4. $x = 3 \cos (x - \pi/4)$.

2.7 Rate of Convergence

Let $x_0, x_1, x_2, \ldots\ldots$ be the values of a root (α) of an equation at the 0th, 1st, 2nd $\ldots\ldots$ iterations while its actual value is 3.5567. The values of this root calculated by three different methods, are as given below:

Root	1st method	2nd method	3rd method
x_0	5	5	5
x_1	5.6	3.8527	3.8327
x_2	6.4	3.5693	3.56834
x_3	8.3	3.55798	3.55743
x_4	9.7	3.55687	3.55672
x_5	10.6	3.55676	
x_6	11.9	3.55671	

The values in the 1st method do not converge toward the root 3.5567. In the 2nd and 3rd methods, the values converge to the root after 6th and 4th iterations, respectively. Clearly 3rd method converges faster than the 2nd method. This *fastness of convergence in any method is represented by its* rate of convergence.

If e be the error then $e_i = \alpha - x_i = x_{i+1} - x_i$.

If e_{i+1}/e_i is almost constant, convergence is said to be linear, *i.e., slow.*

If e_{i+1}/e_i is *nearly constant, convergence is said to be of order p, i.e., faster.*

2.8 Bisection Method

This method is based on the repeated application of the *intermediate value property*. Let the function $f(x)$ be continuous between a and b. For definiteness, let $f(a)$ be negative and $f(b)$ be positive. Then the first approximation to the root is $x_1 = \dfrac{1}{2}(a + b)$.

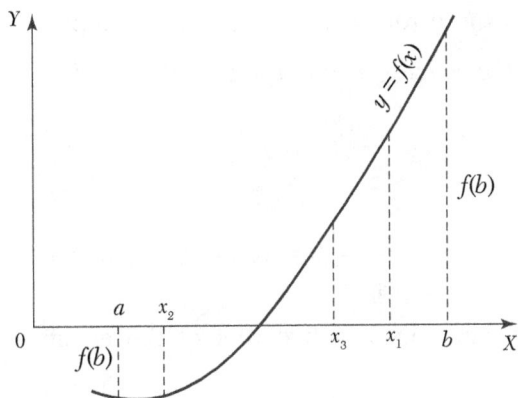

FIGURE 2.5

If $f(x_1) = 0$, then x_1 is a root of $f(x) = 0$. Otherwise, the root lies between a and x_1 or x_1 and b according as $f(x_1)$ is positive or negative. Then we bisect the interval as before and continue the process until the root is found to desired accuracy.

In the Figure 2.4, $f(x_1)$ is + ve, so that the root lies between a and x_1. Then the second approximation to the root is $x_2 = \dfrac{1}{2}(a + x_1)$. If $f(x_2)$ is – ve, the root lies between x_1 and x_2. Then the third approximation to the root is $x_3 = \dfrac{1}{2}(x_1 + x_2)$ and so on.

NOTE **Obs.** *1. Since the new interval containing the root, is exactly half the length of the previous one, the interval width is reduced by a factor of $\dfrac{1}{2}$ at each step. At the end of the nth step, the new interval will therefore, be of length $(b - a)/2^n$. If on repeating this process n times, the latest interval is as small as given ε then*

$(b - a)/2^n \leq \varepsilon$

or $\qquad n \geq [\log(b - a) - \log\varepsilon]/\log 2.$

This gives the number of iterations required for achieving an accuracy ε.

In particular, the minimum number of iterations required for converging to a root in the interval $(0, 1)$ for a given ε are as under:

ε:	10^{-2}	10^{-3}	10^{-4}
n:	7	10	14

Rate of Convergence. As the error decreases with each step by a factor of $\frac{1}{2}$, (*i.e.,* $e_{n+1}/e_n = \frac{1}{2}$), *the convergence in the bisection method is* **linear**.

EXAMPLE 2.15

(*a*) Find a root of the equation $x^3 - 4x - 9 = 0$, using the bisection method correct to three decimal places.

(*b*) Using bisection method, find the negative root of the equation $x^3 - 4x + 9 = 0$.

Solution:

(*a*) Let $f(x) = x^3 - 4x - 9$

Since $f(2)$ is – ve and $f(3)$ is + ve, a root lies between 2 and 3.

∴ First approximation to the root is

$$x_1 = \frac{1}{2}\ (2+3) = 2.5.$$

Thus $\qquad f(x_1) = (2.5)^3 - 4(2.5) - 9 = -3.375\ i.e.,\ \text{–ve}.$

∴ The root lies between x_1 and 3. Thus the second approximation to the root is

$$x_2 = \frac{1}{2}\ (x_1 + 3) = 2.75.$$

Then $\qquad f(x_2) = (2.75)^3 - 4(2.75) - 9 = 0.7969\ i.e.,\ \text{+ve}.$

∴ The root lies between x_1 and x_2. Thus the third approximation to the root is

$$x_3 = \frac{1}{2}\ (x_1 + x_2) = 2.625.$$

Then $\qquad f(x_3) = (2.625)^3 - 4(2.625) - 9 = -1.4121\ i.e.,\ \text{–ve}.$

The root lies between x_2 and x_3. Thus the fourth approximation to the root is

$$x_4 = \frac{1}{2}(x_2 + x_3) = 2.6875.$$

Repeating this process, the successive approximations are

$x_5 = 2.71875,\qquad x_6 = 2.70313,\qquad x_7 = 2.71094$
$x_8 = 2.70703,\qquad x_9 = 2.70508,\qquad x_{10} = 2.70605$
$x_{11} = 2.70654,\qquad x_{12} = 2.70642$

Hence the root is 2.7064.

(*b*)If α, β, γ are the roots of the given equation, then $-\alpha, -\beta, -\gamma$ are the roots of $(-x)^3 - 4(-x) + 9 = 0$

$\therefore$ The negative root of the given equation is the positive root of $x^3 - 4x - 9 = 0$ which we have found above to be 2.7064.

EXAMPLE 2.16

Using the bisection method, find an approximate root of the equation $\sin x = 1/x$, that lies between $x = 1$ and $x = 1.5$ (measured in radians). Carry out computations up to the 7th stage

Solution:

Let $f(x) = x \sin x - 1$. We know that $1^r = 57.3°$.

Since $f(1) = 1 \times \sin(1) - 1 = \sin(57.3°) - 1 = -0.15849$

and $f(1.5) = 1.5 \times \sin(1.5)r - 1 = 1.5 \times \sin(85.95)° - 1 = 0.49625$;

a root lies between 1 and 1.5.

$\therefore$ First approximation to the root is $x_1 = \dfrac{1}{2}(1 + 1.5) = 1.25$

Then $f(x_1) = (1.25) \sin(1.25) - 1 = 1.25 \sin(71.625°) - 1 = 0.18627$ and $f(1) < 0$.

$\therefore$ A root lies between 1 and $x_1 = 1.25$.

Thus the second approximation to the root is $x2 = \dfrac{1}{2}(1 + 1.25) = 1.125$.

Then $f(x_2) = 1.125 \sin(1.125) - 1 = 1.125 \sin(64.46)° - 1 = 0.01509$ and $f(1) < 0$.

$\therefore$ A root lies between 1 and $x_2 = 1.125$.

Thus the third approximation to the root is $x_3 = \dfrac{1}{2}(1 + 1.125) = 1.0625$.

Then $f(x_3) = 1.0625 \sin(1.0625) - 1 = 1.0625 \sin(60.88) - 1 = -0.0718$ < 0 and $f(x_2) > 0$, *i.e.*, now the root lies between $x_3 = 1.0625$ and $x_2 = 1.125$.

$\therefore$ Fourth approximation to the root is $x_4 = \dfrac{1}{2}(1.0625 + 1.125) = 1.09375$

Then $f(x_4) = -0.02836 < 0$ and $f(x_2) > 0$,

i.e., The root lies between $x_4 = 1.09375$ and $x_2 = 1.125$.

$\therefore$ Fifth approximation to the root is $x_5 = \dfrac{1}{2}(1.09375 + 1.125) = 1.10937$

Then $f(x_5) = -0.00664 < 0$ and $f(x_2) > 0$.

∴   The root lies between $x_5 = 1.10937$ and $x_2 = 1.125$.

Thus the sixth approximation to the root is

$$x_6 = \frac{1}{2}(1.10937 + 1.125) = 1.11719$$

Then $f(x_6) = 0.00421 > 0$.

But $f(x_5) < 0$.

∴   The root lies between $x_5 = 1.10937$ and $x_6 = 1.11719$.

Thus the seventh approximation to the root is

$$x_7 = \frac{1}{2}(1.10937 + 1.11719) = 1.11328$$

Hence the desired approximation to the root is 1.11328.

EXAMPLE 2.17

Find the root of the equation $\cos x = xe^x$ using the bisection method correct to four decimal places.

Solution:

Let $f(x) = \cos x - xe^x$.

Since $f(0) = 1$ and $f(1) = -2.18$, so a root lies between 0 and 1.

∴ First approximation to the root is $x_1 = \frac{1}{2}(0 + 1) = 0.5$

Now $f(x_1) = 0.05$ and $f(1) = -2.18$, therefore the root lies between 1 and $x_1 = 0.5$.

∴ Second approximation to the root is $x_2 = \frac{1}{2}(0.5 + 1) = 0.75$

Now $f(x_2) = -0.86$ and $f(0.5) = 0.05$, therefore the root lies between 0.5 and 0.75.

∴ Third approximation to the root is $x_3 = \frac{1}{2}(0.5 + 0.75) = 0.625$

Now $f(x_3) = -0.36$ and $f(0.5) = 0.05$, therefore the root lies between 0.5 and 0.625.

∴ Fourth approximation to the root is $x_4 = \frac{1}{2}(0.5 + 0.625) = 0.5625$

Now $f(x_4) = -0.14$ and $0.5 = 0.05$, therefore the root lies between 0.5 and 0.5625.

∴ Fifth approximation is $x_5 = 1 \dfrac{1}{2} (0.5 + 0.5625) = 0.5312$

Now $f(x_5) = -0.04$ and $f(0.5) = 0.05$, therefore the root lies between 0.5 and 0.5312.

∴ Sixth approximation is $x_6 = \dfrac{1}{2} (0.5 + 0.5312) = 0.5156$

Hence the desired approximation to the root is 0.5156.

EXAMPLE 2.18

Find a positive real root of $x \log_{10}{}^x = 1.2$ using the bisection method.

Solution:

Let $f(x) = x \log_{10}{}^x - 1.2$.

Since $f(2) = -0.598$ and $f(3) = 0.231$, so a root lies between 2 and 3.

∴ First approximation to the root is $x_1 = \dfrac{1}{2} (2 + 3) = 2.5$.

Now $f(2.5) = -0.205$ and $f(3) = 0.231$, therefore a root lies between 2.5 and 3.

∴ Second approximation to the root is $x_2 = 1 \dfrac{1}{2} (2.5 + 3) = 2.75$.

Now $f(2.75) = 0.008$ and $f(2.5) = -0.205$, therefore, a root lies between 2.5 and 2.75.

∴ Third approximation to the root is $x_3 = \dfrac{1}{2} (2.5 + 2.75) = 2.625$

Now $f(2.625) = -0.1$ and $f(2.75) = 0.008$, therefore a root lies between 2.625 and 2.75.

∴ Fourth approximation to the root is $x4 = \dfrac{1}{2} (2.625 + 2.75) = 2.687$

Hence the desired root is 2.687.

2.9   Method of False Position *or* Regula-Falsi Method *or* Interpolation Method

This is the oldest method of finding the real root of an equation $f(x) = 0$ and closely resembles the bisection method.

Here we choose two points x_0 and x_1 such that $f(x_0)$ and $f(x_1)$ are of opposite signs *i.e.*, the graph of $y = f(x)$ crosses the x-axis between these points (Figure 2.6). This indicates that a root lies between x_0 and x_1 and consequently $f(x_0)f(x1) < 0$.

Equation of the chord joining the points $A[x_0, f(x_0)]$ and $B[x1, f(x_1)]$ is

$$y - f(x_0) = \frac{f(x_1) - f(x_0)}{x_1 - x_0}(x - x_0)$$

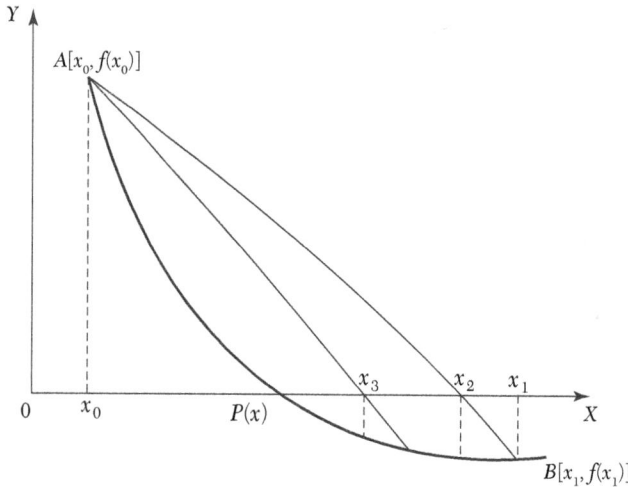

FIGURE 2.6

The method consists in replacing the curve AB by means of the chord AB and taking the point of intersection of the chord with the x-axis as an approximation to the root. So the abscissa of the point where the chord cuts the x-axis ($y = 0$) is given by

$$x_2 - x_0 = \frac{x_1 - x_0}{f(x_1) - f(x_0)}f(x_0) \tag{1}$$

which is an approximation to the root.

If now $f(x_0)$ and $f(x_2)$ are of opposite signs, then the root lies between x_0 and x_2. So replacing x_1 by x_2 in (1), we obtain the next approximation x_3. (The root could as well lie between x_1 and x_2 and we would obtain x_3 accordingly). This procedure is repeated until the root is found to the desired accuracy. The iteration process based on (1) is known as the *method of false position*.

Rate of Convergence. This method has linear rate of convergence which is faster than that of the bisection method.

EXAMPLE 2.19

Find a real root of the equation $x^3 - 2x - 5 = 0$ by the method of false position correct to three decimal places.

Solution:

Let $$f(x) = x^3 - 2x - 5$$

so that $$f(2) = -1 \text{ and } f(3) = 16,$$

i.e., A root lies between 2 and 3.

$\therefore$ Taking $x_0 = 2$, $x_1 = 3$, $f(x_0) = -1$, $f(x_1) = 16$, in the method of false position, we get

$$x_2 = x_0 - \frac{x_1 - x_0}{f(x_1) - f(x_0)} f(x_0) = 2 + \frac{1}{17} = 2.0588 \qquad (i)$$

Now $f(x^2) = f(2.0588) = -0.3908$

i.e., The root lies between 2.0588 and 3.

$\therefore$ Taking $x_0 = 2.0588$, $x_1 = 3$, $f(x_0) = -0.3908$, $f(x_1) = 16$, in (i), we get

$$x_3 = 2.0588 - \frac{0.9412}{19.3908}(-0.3908) = 2.0813$$

Repeating this process, the successive approximations are

$x_4 = 2.0862$, $\qquad x_5 = 2.0915$, $\qquad x_6 = 2.0934$,

$x_7 = 2.0941$, $\qquad x_8 = 2.0943$ etc.

Hence the root is 2.094 correct to three decimal places.

EXAMPLE 2.20

Find the root of the equation $\cos x = xe^x$ using the regula-falsi method correct to four decimal places.

Solution:

Let $\quad f(x) = \cos x - xe^x = 0$

so that $f(0) = 1, f(1) = \cos 1 - e = -2.17798$

i.e., the root lies between 0 and 1.

∴ Taking $x_0 = 0$, $x_1 = 1$, $f(x_0) = 1$ and $f(x_1) = -2.17798$ in the regula-falsi method, we get

$$x_2 = x_0 - \frac{x_1 - x_0}{f(x_1) - f(x_0)} f(x_0) = 0 + \frac{1}{3.17798} \times 1 = 0.31467 \tag{i}$$

Now $f(0.31467) = 0.51987$ *i.e.*, the root lies between 0.31467 and 1.

∴ Taking $x_0 = 0.31467$, $x_1 = 1$, $f(x_0) = 0.51987$, $f(x_1) = -2.17798$ in (*i*), we get

$$x_3 = 0.31467 + \frac{0.68533}{2.69785} \times 0.51987 = 0.44673$$

Now $f(0.44673) = 0.20356$ *i.e.*, the root lies between 0.44673 and 1.

∴ Taking $x_0 = 0.44673$, $x_1 = 1$, $f(x_0) = 0.20356$, $f(x_1) = -2.17798$ in (*i*), we get

$$x_4 = 0.44673 + \frac{0.55327}{2.38154} \times 0.20356 = 0.49402$$

Repeating this process, the successive approximations are

$x_5 = 0.50995$, $x_6 = 0.51520$, $x_7 = 0.51692$

$x_8 = 0.51748$, $x_9 = 0.51767$, $x_{10} = 0.51775$ etc.

Hence the root is 0.5177 correct to four decimal places.

EXAMPLE 2.21

Find a real root of the equation $x \log_{10} x = 1.2$ by regula-falsi method correct to four decimal places.

Solution:

Let $f(x) = x \log 10\, x - 1.2$

so that $f(1) = -\text{ve}$, $f(2) = -\text{ve}$ and $f(3) = +\text{ve}$.

∴ A root lies between 2 and 3.

Taking $x_0 = 2$ and $x_1 = 3$, $f(x_0) = -0.59794$ and $f(x_1) = 0.23136$, in the method of false position, we get

$$x_2 = x_0 - \frac{x_1 - x_0}{f(x_1) - f(x_0)} f(x_0) = 2.72102 \tag{i}$$

Repeating this process, the successive approximations are

$x_4 = 2.74024$, $x_5 = 2.74063$ etc.

Hence the root is 2.7406 correct to 4 decimal places.

EXAMPLE 2.22

Use the method of false position, to find the fourth root of 32 correct to three decimal places.

Solution:

Let $x = (32)^{1/4}$ so that $x^4 - 32 = 0$

Take $f(x) = x^4 - 32$. Then $f(2) = -16$ and $f(3) = 49$, *i.e.*, a root lies between 2 and 3.

∴ *Taking* $x_0 = 2$, $x_1 = 3$, $f(x_0) = -16$, $f(x_1) = 49$ in the method of false position, we get

$$x_2 = x_0 - \frac{x_1 - x_0}{f(x_1) - f(x_0)} f(x_0) = 2 + \frac{16}{65} = 2.2462 \qquad (i)$$

Now $f(x_2) = f(2.2462) = -6.5438$ *i.e.*, the root lies between 2.2462 and 3.

∴ Taking $x_0 = 2.2462$, $x_1 = 3$, $f(x_0) = -6.5438$, $f(x_1) = 49$ in (i), we get

$$x_3 = 2.2462 - \frac{3 - 2.2462}{49 + 6.5438}(-6.5438) = 2.335$$

Now $f(x_3) = f(2.335) = -2.2732$ *i.e.*, the root lies between 2.335 and 3.

∴ Taking $x_0 = 2.335$ and $x_1 = 3$, $f(x_0) = -2.2732$ and $f(x_1) = 49$ in (i), we obtain

$$x_4 = 2.335 - \frac{3 - 2.335}{49 + 2.2732}(-2.2732) = 2.3645$$

Repeating this process, the successive approximations are $x_5 = 2.3770$, $x_6 = 2.3779$ etc. Since $x_5 = x_6$ upto three decimal places, we take $(32)^{1/4} = 2.378$.

2.10   Secant Method

This method is an improvement over the method of false position as it does not require the condition $f(x_0) f(x_1) < 0$ of that method (Figure 2.5).

Here also the graph of the function $y = f(x)$ is approximated by a secant line but at each iteration, two most recent approximations to the root are used to find the next approximation. Also it is not necessary that the interval must contain the root.

Taking x_0, x_1 as the initial limits of the interval, we write the equation of the chord joining these as

$$y - f(x_1) = \frac{f(x_1) - f(x_0)}{x_1 - x_0}(x - x_1)$$

Then the abscissa of the point where it crosses the x-axis $(y = 0)$ is given by

$$x_2 = x_1 - \frac{x_1 - x_o}{f(x_1) - f(x_0)} f(x_1)$$

which is an approximation to the root. The general formula for successive approximations is, therefore, given by

$$x_{n+1} = x_n - \frac{x_n - x_{n-1}}{f(x_n) - f(x_{n-1})} f(x_n), n \geq 1.$$

Rate of Convergence. If at any interation $f(x_n) = f(x_{n-1})$, this method fails and shows that it does not converge necessarily. This is a drawback of secant method over the method of false position which always converges. But if the secant method once converges, its rate of convergence is 1.6 which is faster than that of the method of false position.

EXAMPLE 2.23

Find a root of the equation $x^3 - 2x - 5 = 0$ using the secant method correct to three decimal places.

Solution:

Let $f(x) = x^3 - 2x - 5$ so that $f(2) = -1$ and $f(3) = 16$.

∴ Taking initial approximations $x_0 = 2$ and $x_1 = 3$, by the secant method, we have

$$x_2 = x_1 - \frac{x_1 - x_0}{f(x_1) - f(x_0)} f(x_1) = 3 - \frac{3-2}{16+1} 16 = 2.058823$$

Now $f(x_2) = -0.390799$

$$\therefore \qquad x_3 = x_2 - \frac{x_2 - x_1}{f(x_2) - f(x_1)} f(x_2) = 2.081263$$

and $\qquad f(x_3) = -0.147204$

$$\therefore \qquad x_4 = x_3 - \frac{x_3 - x_2}{f(x_3) - f(x_2)} f(x_3) = 2.094824$$

and $\qquad f(x_4) = 0.003042$

$$\therefore \qquad x_5 = x_4 - \frac{x_4 - x_3}{f(x_4) - f(x_3)} f(x_4) = 2.094549$$

Hence the root is 2.094 correct to three decimal places

EXAMPLE 2.24

Find the root of the equation $xe^x = \cos x$ using the secant method correct to four decimal places.

Solution:

Let $\qquad f(x) = \cos x - xe^x = 0.$

Taking the initial approximations $x_0 = 0, x_1 = 1$

so that $f(x_0) = 1, f(x_1) = \cos 1 - e = -2.17798$

Then by the secant method, we have

$$x_2 = x_1 - \frac{x_1 - x_0}{f(x_1) - f(x_0)} f(x_1) = 1 + \frac{1}{3.17798}(-2.17798) = 0.31467$$

Now $\quad f(x_2) = 0.51987$

$$\therefore x_3 = x_2 - \frac{x_2 - x_1}{f(x_2) - f(x_1)} f(x_2) = 0.44673 \; and \; f(x_3) = 0.20354$$

$$\therefore \qquad x_4 = x_3 - \frac{x_3 - x_2}{f(x_3) - f(x_2)} f(x_3) = 0.53171$$

Repeating this process, the successive approximations are $x_5 = 0.51690$, $x_6 = 0.51775, x_7 = 0.51776$ etc.

Hence the root is 0.5177 correct to four decimal places.

NOTE

Obs. *Comparing Examples 2.18 and 2.21, we notice that the rate of convergence in the secant method is definitely faster than that of the method of false position*

2.11   Iteration Method

To find the roots of the equation $f(x) = 0$ (i)

by successive approximations, we rewrite (i) in the form $x = \phi(x)$ (ii)

The roots of (i) are the same as the points of intersection of the straight line $y = x$ and the curve representing $y = \phi(x)$. Figure 2.7 illustrates the working of the iteration method which provides a spiral solution.

Let $x = x_0$ be an initial approximation of the desired root α. Then the first approximation x_1 is given by $x_1 = \phi(x_0)$

Now treating x_1 as the initial value, the second approximation is $x_2 = \phi(x_1)$

Proceeding in this way, the nth approximation is given by $x_n = \phi(x_{n-1})$

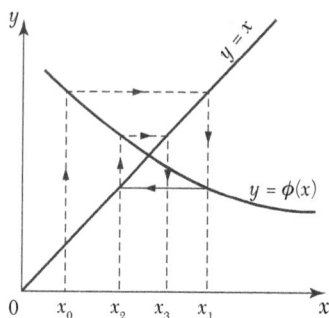

FIGURE 2.7

Sufficient condition for convergence of iterations. It is not certain whether the sequence of approximations $x_1, x_2,..., x_n$ always converges to the same number which is a root of (1) or not. As such, we have to choose the initial approximation x_0 suitably so that the successive approximations $x_1, x_2,..., x_n$ converge to the root α. The following theorem helps in making the right choice of x_0:

Theorem:

If (i) α be a root of $f(x) = 0$ which is equivalent to $x = \phi(x)$,

(ii) I, be any interval containing the point $x = \alpha$,

(iii) $|\phi'(x)| < 1$ for all x in I,

then the sequence of approximations $x_0, x_1, x_2,..., x_n$ will converge to the root α provided the initial approximation x_0 is chosen in I.

Proof. Since α is a root of $x = \phi(x)$, we have $\alpha = \phi(\alpha)$

If x_{n-1} and xn be 2 successive approximations to α, we have $xn = \phi(x_{n-1})$

$$\therefore \qquad x_n - \alpha = \phi(x_{n-1}) - \phi(\alpha) \qquad\qquad (i)$$

By mean value theorem, $\dfrac{\phi(x_{n-1}) - \phi(\alpha)}{x_{n-1} - \alpha} = \phi'(\xi)$ where $x_{n-1} < \xi < \alpha$

Hence (1) becomes $x_n - \alpha = (x_{n-1} - \alpha)\,\phi'(\alpha)$

If $|\phi'(x_i)| \le k < 1$ for all i, then

$$|x_n - \alpha| \le k\,|x_{n-1} - \alpha| \qquad\qquad (2)$$

Similarly $\qquad |x_{n-1} - \alpha| \le k\,|x_{n-2} - \alpha|$

i.e., $\qquad |x_n - \alpha| \le k^2\,|x_{n-2} - \alpha|$

Proceeding in this way, $|x_n - \alpha| \le k^n\,|x_0 - \alpha|$

As $n \to \infty$, the R.H.S. tends to zero, therefore, the sequence of approximations converges to the root α.

NOTE

Obs. 1. *The smaller the value of $\phi'(x)$, the more rapid will be the convergence.*
2. This method of iteration is particularly useful for finding the real roots of an equation given in the form of an infinite series.

Acceleration of convergence. From (2), we have

$$|x_n - \alpha| \le k\,|x_{n-1} - \alpha|, k < 1.$$

It is clear from this relation that the iteration method is linearly convergent. *This slow rate of convergence can be improved by using the following method:*

Aitken's Δ^2 method. Let x_{i-1}, x_i, x_{i+1} be three successive approximations to the desired root α of the equation $x = \phi(x)$. Then we know that

$$\alpha - x_i = k(\alpha - x_{i-1}),\ \alpha - x_{i+1} = k(\alpha - x_i)$$

Dividing, we get $\qquad \dfrac{\alpha - x_i}{\alpha - x_{i+1}} = \dfrac{\alpha - x_{i-1}}{\alpha - x_1}$

Whence $\alpha = x_{i+1} - \dfrac{(x_{i+1} - x_i)^2}{x_{i+1} - 2x_i = x_{i-1}}$ $\qquad\qquad (3)$

But in the sequence of successive approximations, we have

$$\Delta x_i = x_{i+1} - x_i$$

$$\Delta^2 x_i = \Delta(\Delta x_i) = \Delta(x_{i+1} - x_i) = \Delta x_{i+1} - \Delta x_i$$

$$= x_{i+2} - x_{i+1} - (x_{i+1} - x_i) = x_{i+2} - 2x_{i+1} + x_i$$

$$\therefore \qquad \Delta^2 x_{i-1} = x_{i+1} - 2x_i + x_{i-1}$$

Hence (3) can be written as $\alpha = x_{i+1} - \dfrac{(\Delta x_i)^2}{\Delta^2 x_{i-1}}$ \hfill (4)

which yields successive approximations to the root α.

EXAMPLE 2.25

Find a real root of the equation $\cos x = 3x - 1$ correct to three decimal places using

(i) Iteration method

(ii) Aitken's Δ^2 method.

Solution:

(i) We have $f(x) = \cos x - 3x + 1 = 0$

$$f(0) = 2 = + \text{ve and } f(\pi/2) = -3\pi/2 + 1 = -\text{ve}$$

$\therefore$ A root lies between 0 and $\pi/2$.

Rewriting the given equation as $x = \dfrac{1}{3}(\cos x + 1) = \phi(x)$, we have

$$\phi'(x) = \frac{\sin x}{3} \text{ and } |\phi'(x)| = \frac{1}{3}|\sin x| < 1 \text{ in } (0, \pi/2).$$

Hence the iteration method can be applied and we start with $x0 = 0$. Then the successive approximations are,

$$x_1 = \phi(x0) = \frac{1}{3}(\cos 0 + 1) = 0.6667$$

$$x_2 = \phi(x1) = \frac{1}{3}(\cos 0.6667 + 1) = 0.5953$$

$$x_3 = \phi(x2) = \frac{1}{3}(\cos 0.5953 + 1) = 0.6093$$

$$x_4 = \phi(x3) = \frac{1}{3}(\cos 0.6093 + 1) = 0.6067$$

$$x_5 = \phi(x4) = \frac{1}{3}(\cos 0.6067 + 1) = 0.6072$$

$$x_6 = \phi(x5) = \frac{1}{3}(\cos 0.6072 + 1) = 0.6071$$

Hence x_5 and x_6 being almost the same, the root is 0.607 correct to three decimal places. (*ii*) We calculate x_1, x_2, x_3 as above. To use Aitken's method, we have

x	Δx	$\Delta^2 x$
$x_1 = 0.667$		
	-0.0714	
$x_2 = 0.5953$		0.0854
	0.014	
$x_3 = 0.6093$		

Hence

$$x_4 = x_3 - \frac{(\Delta x_2)^2}{\Delta^2 x_1} = 0.6093 - \frac{(0.014)^2}{0.0854} = 0.607$$

which corresponds to six iterations in normal form.

Thus the required root is 0.607.

EXAMPLE 2.26

Using iteration method, find a root of the equation $x^3 + x^2 - 1 = 0$ correct to four decimal places.

Solution:

We have $f(x) = x^3 + x^2 - 1 = 0$

Since $f(0) = -1$ and $f(1) = 1$, a root lies between 0 and 1.

Rewriting the given equation as $x = (x + 1)^{-1/2} = \phi(x)$, we have $\phi'(x) = -\frac{1}{2}(x + 1)^{-3/2}$ and $|\phi'(x)| < 1$ for $x < 1$. Hence the iteration method can be applied. Starting with $x_0 = 0.75$, the successive approximations are

$$x_1 = \phi(x_0) = \frac{1}{\sqrt{(x_0 + 1)}} = 0.7559$$

$$x_2 = \phi(x_1) = \frac{1}{\sqrt{(0.7559 + 1)}} = 0.75466$$

$$x_3 = 0.75492, \; x_4 = 0.75487, \; x_5 = 0.75488$$

Hence x_4 and x_5 being almost the same, the root is 0.7548 correct to four decimal places.

EXAMPLE 2.27

Apply iteration method to find the negative root of the equation $x^3 - 2x + 5 = 0$ correct to four decimal places.

Solution:

If α, β, γ are the roots of the given equation, then $-\alpha, -\beta, -\gamma$ are the roots of

$$(-x)^3 - 2(-x) + 5 = 0$$

∴ The negative root of the given equation is the positive root of

$$f(x) = x^3 - 2x - 5 = 0. \tag{i}$$

Since $f(2) = -1$ and $f(3) = 16$, a root lies between 2 and 3.

Rewriting (i) as $x = (2x + 5)^{1/3} = \phi(x)$,

we have $\phi'(x) = \dfrac{1}{3}(2x + 5)^{-2/3} \cdot 2$ and $|\phi'(x)| < 1$ for $x < 3$.

∴ The iteration method can be applied:

Starting with $x_0 = 2$. The successive approximations are

$$x_1 = \phi x_0) = (2x_0 + 5)^{1/3} = 2.08008$$

$$x_2 = \phi(x_1) = 2.09235, \qquad x_3 = 2.09422$$

$$x_4 = 2.09450, \qquad x_5 = 2.09454$$

Since x_4 and x_5 being almost the same, the root of (i) is 2.0945 correct to four decimal places.

Hence the negative root of the given equation is -2.0945.

EXAMPLE 2.28

Find a real root of $2x - \log_{10} x = 7$ correct to four decimal places using the iteration method.

Solution:

We have $f(x) = 2x - \log_{10} x - 7$

$$f(3) = 6 - \log_{10} 3 - 7 = 6 - 0.4771 - 7 = -1.4471$$

$$f(4) = 8 - \log_{10} 4 - 7 = 8 - 0.602 - 7 = 0.398$$

∴ A root lies between 3 and 4.

Rewriting the given equation as $x = \dfrac{1}{2}(\log_{10}x + 7) = \phi(x)$, we have

$$\phi'(x) = \frac{1}{2}\left(\frac{1}{x}\log_{10}e\right)$$

$\therefore |\phi'(x)| < 1$ when $3 < x < 4$ $\hspace{2cm}$ $[\because \log_{10}e = 0.4343]$

Since $|f(4)| < |f(3)|$, the root is near to 4.

Hence the iteration method can be applied. Taking $x_0 = 3.6$, the successive approximations are

$$x_1 = \phi\,(x_0) = \frac{1}{2}\,(\log10\ 3.6 + 7) = 3.77815$$

$$x_2 = \phi\,(x_1) = \frac{1}{2}\,(\log10\ 3.77815 + 7) = 3.78863$$

$$x_3 = \phi\,(x_2) = \frac{1}{2}\,(\log 3.78863 + 7) = 3.78924$$

$$x_4 = \phi\,(x_3) = \frac{1}{2}\,(\log 3.78924 + 7) = 3.78927$$

Hence x_3 and x_4 being almost equal, the root is 3.7892 correct to four decimal places.

EXAMPLE 2.29

Find the smallest root of the equation

$$1 - x + \frac{x^2}{(2!)^2} - \frac{x^3}{(3!)^2} + \frac{x^4}{(4!)^2} - \frac{x^5}{(5!)^2} + \cdots = 0$$

Solution:

Writing the given equation as

$$x = 1 + \frac{x^2}{(2!)^2} - \frac{x^3}{(3!)^2} + \frac{x^4}{(4!)^2} - \frac{x^5}{(5!)^2} + \cdots = \phi(x)$$

Omitting x^2 and higher powers of x, we get $x = 1$ approximately.

Taking $x_0 = 1$, we obtain

$$x_1 = \phi(x_0) = 1 + \frac{1}{(2!)^2} - \frac{1}{(3!)^2} + \frac{1}{(4!)^2} - \frac{1}{(5!)^2} + \cdots = 1.2239$$

$$x_2 = \phi(x_1) = 1 + \frac{(1.2239)^2}{(2!)^2} - \frac{(1.2239)^3}{(3!)^2} + \frac{(1.2239)^4}{(4!)^2} - \frac{(1.2239)^5}{(5!)^2} + \cdots = 1.3263$$

Similarly $x_3 = 1.38,$ $x_4 = 1.409,$ $x_5 = 1.425$

$x_6 = 1.434,$ $x_7 = 1.439,$ $x_8 = 1.442.$

The values of x_7 and x_8 indicate that the root is 1.44 correct to two decimal places.

Exercises 2.4

1. Find a root of the following equations, using the bisection method correct to three decimal places:

$(i)\ x^3 - x - 1 = 0$ $(ii)\ x^3 - x^2 - 1 = 0$

$(iii)\ 2x^3 + x^2 - 20x + 12 = 0$ $(iv)\ x^4 - x - 10 = 0.$

2. Evaluate a real root of the following equations by bisection method:

$(i)\ x - \cos x = 0$ $(ii)\ e^{-x} - x = 0$

$(iii)\ e^x = 4 \sin x.$

3. Find a real root of the following equations correct to three decimal places, by the method of false position:

$(i)\ x^3 - 5x + 1 =$ $(ii)\ x^3 - 4x - 9 = 0$

$(iii)\ x^6 - x^4 - x^3 - 1 = 0.$

4. Using the regula falsi method, compute the real root of the following equations correct to three decimal places:

$(i)\ xe^x = 2$ $(ii)\ \cos x = 3x - 1$

$(iii)\ xe^x = \sin x$ $(iv)\ x \tan x = -1$

$(v)\ 2x - \log x = 7$

$(vi)\ 3x + \sin x = e^x.$

5. Find the fourth root of 12 correct to three decimal places by the interpolation method.

6. Locate the root of $f(x) = x^{10} - 1 = 0$, between 0 and 1.3 using the bisection method and method of false position. Comment on which method is preferable.

7. Find a root of the following equations correct to three decimal places by the secant method:

$(i)\ x^3 + x^2 + x + 7 = 0$ $(ii)\ x - e^{-x} = 0$

$(iii)\ x\ \log_{10} x = 1.9.$

8. Use the iteration method to find a root of the equations to four decimal places:

$(i)\ x^3 + x^2 - 100 = 0$ $(ii)\ x^3 - 9x + 1 = 0$

$(iii)\ x = \dfrac{1}{2} + \sin x$ $(iv)\ \tan x = x$

$(v)\ e^x = 5x$ $(vi)\ 2^x - x - 3 = 0$ which lies between $(-3, -2)$.

9. Evaluate $\sqrt{30}$ by (i) secant method (ii) iteration method correct to four decimal places.

10. Find the root of the equation $2x = \cos x + 3$ correct to three decimal places using (i) iteration method, (ii) Aitken's Δ^2 method.

11. Find the real root of the equation

$$x - \frac{x^3}{3} + \frac{x^5}{10} - \frac{x^7}{42} + \frac{x^9}{216} - \frac{x^{11}}{1320} + \ldots = 0.443$$

correct to three decimal places using iteration method

2.12   Newton-Raphson Method

Let x_0 be an approximate root of the equation $f(x) = 0$. If $x_1 = x_0 + h$ be the exact root, then $f(x_1) = 0$.

$\therefore$ Expanding $f(x_0 + h)$ by Taylor's series $f(x_0) + hf'(x_0) + \dfrac{h^2}{2!} f''(x_0) + \cdots = 0$

Since h is small, neglecting h^2 and higher powers of h, we get $f(x_0) + h f'(x_0) = 0$

or $$h = -\frac{f(x_0)}{f'(x_0)} \tag{1}$$

$\therefore$ A closer approximation to the root is given by

$$x_1 = x_0 - \frac{f(x_0)}{f'(x_0)}.$$

Similarly starting with x_1, a still better approximation x_2 is given by

$$x_2 = x_1 - \frac{f(x_0)}{f'(x_1)}.$$

In general, $X_{n+1} = X_n - \dfrac{f(X_n)}{f'(X_n)}$ $(n = 0, 1, 2....)$ \hfill (2)

which is known as the *Newton-Raphson formula* or *Newton's iteration formula*.

NOTE

Obs. 1. *Newton's method is useful in cases of large values of $f'(x)$ i.e., when the graph of $f(x)$ while crossing the x-axis is nearly vertical.*

For if $f'(x)$ is small in the vicinity of the root, then by (1), h will be large and the computation of the root is slow or may not be possible. Thus this method is not suitable in those cases where the graph of $f(x)$ is nearly horizontal while crossing the x-axis.

Obs. 2. *Geometrical interpretation. Let x_0 be a point near the root α of the equation $f(x) = 0$ (Figure 2.8). Then the equation of the tangent at $A_0[x_0, f(x_0)]$ is*

$$y - f(x_0) = f'(x_0) (x - x_0).$$

It cuts the x-axis at $x_1 = x_0 - \dfrac{f(x_0)}{f'(x_0)}$

which is a first approximation to the root α. If A_1 is the point corresponding to x_1 on the curve, then the tangent at A_1 will cut the x-axis at x_2 which is nearer to α and is, therefore, a second approximation to the root. Repeating this process, we approach the root α quite rapidly. Hence the method consists in replacing

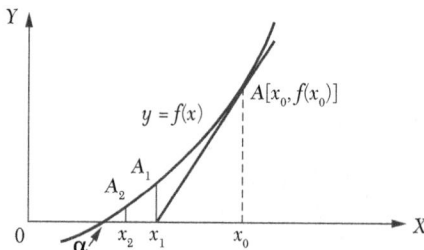

FIGURE 2.8

the part of the curve between the point A_0 and the x-axis by means of the tangent to the curve at A_0.

Obs. 3. *Newton's method is generally used to improve the result obtained by other methods. It is applicable to the solution of both algebraic and transcendental equations.*

Convergence of Newton-Raphson Method. Newton's formula converges provided the initial approximation x_0 is chosen sufficiently close to the root.

If it is not near the root, the procedure may lead to an endless cycle. A bad initial choice will lead one astray. *Thus a proper choice of the initial guess is very important for the success of Newton's method.*

Comparing (2) with the relation $xn_{+1} = \phi(xn)$ of the iteration method, we get

$$\phi(x_n) = x_{n+1} = x_n - \frac{f(x_n)}{f'(x_n)}$$

In general, $\phi(x) = x - \dfrac{f(x)}{f'(x)}$ which gives $\phi'(x) = \dfrac{f(x)f''(x)}{\left[f'(x)\right]^2}$

Since the iteration method (Section 2.10) converges if $|\phi'(x)| < 1$

$\therefore$ Newton's formula will converge if $|f(x)f''(x)| < |f'(x)|^2$ in the interval considered. Assuming $f(x), f'(x)$ and $f''(x)$ to be continuous, we can select a small interval in the vicinity of the root α, in which the above condition is satisfied. Hence the result.

Newtons method converges conditionally while the regula-falsi method always converges. However when the Newton-Raphson method converges, it converges faster and is preferred.

Newton's method has a quadratic convergence.

Suppose x_n differs from the root α by a small quantity ε_n so that

$x_0 = \alpha + \varepsilon_n$ and $x_{n+1} = \alpha + \varepsilon_{n+1}$.

Then (2) becomes

$$\alpha + \varepsilon_{n+1} = \alpha + \varepsilon_n - \frac{f(\alpha + \varepsilon_n)}{f'(\alpha + \varepsilon_n)}$$

i.e., $\varepsilon_{n+1} = \varepsilon_n - \dfrac{f(\alpha + \varepsilon_n)}{f'(\alpha + \varepsilon_n)}$

$$= \varepsilon_n - \frac{f(\alpha) + \varepsilon_n f'(\alpha) + \frac{1}{2!}\varepsilon_n 2 f''(\alpha) + \cdots}{f'(\alpha) + \varepsilon_n f'(\alpha) + \cdots} \quad \text{by Taylor's expansion}$$

$$= \varepsilon_n - \frac{\varepsilon_n f'(\alpha) + \frac{1}{2}\varepsilon_n 2 f''(\alpha) + \cdots}{f'(\alpha) + \varepsilon_n f''(\alpha) + \cdots} = \frac{\varepsilon_n 2}{2}\frac{f''(\alpha)}{f'(\alpha)}. \quad [\because f(\alpha) = 0]$$

This shows that the subsequent error at each step is proportional to the square of the previous error and as such the convergence is quadratic. Thus the Newton-Raphson method has second order convergence.

EXAMPLE 2.30

Find the positive root of $x^4 - x = 10$ correct to three decimal places, using the Newton-Raphson method.

Solution:

Let $f(x) = x4 - x - 10$

so that $f(1) = -10 = -ve$, $f(2) = 16 - 2 - 10 = 4 = +ve$.

$\therefore$ A root of $f(x) = 0$ lies between 1 and 2.

Let us take $x_0 = 2$

Also $f'(x) = 4x^3 - 1$

Newton-Raphson's formula is

$$x_{n+1} = x_n - \frac{f(x_n)}{f'(x_n)}$$

Putting $n = 0$, the first approximation x_1 is given by

$$x_1 = x_0 - \frac{f(x_0)}{f'(x_0)} = 2 - \frac{f(2)}{f'(2)}$$

$$= 2 - \frac{4}{4 \times 2^3 - 1} = 2 - \frac{4}{31} = 1.871$$

Putting $n = 1$ in (i), the second approximation is

$$x_2 = x_1 - \frac{f(x_1)}{f'(x_1)} = 1.871 - \frac{f(1.871)}{f'(1.871)}$$

$$= 1.871 - \frac{(1.871)^4 - (1.871) - 10}{4(1.871)^3 - 1}$$

$$= 1.871 - \frac{0.3835}{25.199} = 1.856$$

Putting $n = 2$ in (ii), the third approximation is

$$x_3 = x_2 - \frac{f(x_2)}{f'(x_2)} = 1.856 - \frac{(1.856)^4 - (1.856) - 10}{4(1.856)^3 - 1}$$

$$= 1.856 - \frac{0.010}{24.574} = 1.856$$

Here $x_2 = x_3$. Hence the desired root is 1.856 correct to three decimal places.

EXAMPLE 2.31

Find by Newton's method, the real root of the equation $3x = \cos x + 1$, correct to four decimal places.

Solution:

Let $f(x) = 3x - \cos x - 1$

$$f(0) = -2 = -\text{ve}, f(1) = 3 - 0.5403 - 1 = 1.4597 = +\text{ve}.$$

So a root of $f(x) = 0$ lies between 0 and 1. It is nearer to 1. Let us take $x_0 = 0.6$.

Also $\quad f'(x) = 3 + \sin x$

∴ Newton's iteration formula gives

$$x_{n+1} = x_n - \frac{f(x_n)}{f'(x_n)} = x_n - \frac{3x_n - \cos x_n - 1}{3 + \sin x_n}$$

$$= \frac{x_n \sin x_n + \cos x_n + 1}{3 + \sin x_n} \qquad (i)$$

Putting $n = 0$, the first approximation x_1 is given by

$$x_1 = \frac{x_0 \sin x_0 + \cos x_0 + 1}{3 + \sin x_0} = \frac{(0.6)\sin(0.6) + \cos(0.6) + 1}{3 + \sin(0.6)}$$

$$= \frac{0.6 \times 0.5729 + 0.82533 + 1}{3 + 0.5729} = 0.6071$$

Putting $n = 1$ in (i), the second approximation is

$$x_2 = \frac{x_1 \sin x_1 + \cos x_1 + 1}{3 + \sin x_1} = \frac{0.6071\sin(0.6071) + \cos(0.6071) + 1}{3 + \sin(0.6071)}$$

$$= \frac{0.6071 \times 0.57049 + 0.8213 + 1}{3 + 0.57049} = 0.6071$$

Here $x_1 = x_2$. Hence the desired root is 0.6071 correct to four decimal places.

EXAMPLE 2.32

Using Newton's iterative method, find the real root of $x \log_{10} x = 1.2$ correct to five decimal places.

Solution:

Let $f(x) = x \log_{10} x - 1.2$

$f(1) = -1.2 = -$ ve, $f(2) = 2 \log_{10} 2 - 1.2 = 0.59794 = -$ ve

and $f(3) = 3_{\log} 10\ 3 - 1.2 = 1.4314 - 1.2 = 0.23136 = +$ ve.

So a root of $f(x) = 0$ lies between 2 and 3. Let us take $x_0 = 2$.

Also $f'(x) = \log_{10} x + x.\dfrac{1}{x}\log_{10} e = \log_{10} x + 0.43429$

$\therefore$ Newton's iteration formula gives

$$x_{n+1} = x_n - \frac{f(x_n)}{f'(x_n)} = \frac{0.43429 x_n + 12}{\log 10 x_n + 0.43429} \qquad (i)$$

Putting $n = 0$, the first approximation is

$$x_1 = \frac{0.43429 \times x_0 + 12}{\log_{10} x_0 + 0.43429} = \frac{0.43429 \times 2 + 12}{\log_{10} 2 + 0.43429}$$

$$= \frac{0.86858 \times 12}{0.30103 + 0.43429} = 2.81$$

Similarly putting $n = 1, 2, 3, 4$ in (i), we get

$$x_2 = \frac{0.43429 \times 2.81 + 1.2}{\log_{10} 2.81 + 0.43429} = 2.741$$

$$x3 = \frac{0.43429 \times 2.741 + 1.2}{\log_{10} 2.741 + 0.43429} = 2.74064$$

$$x4 = \frac{0.43429 \times 2.741 + 1.2}{\log_{10} 2.74064 + 0.43429} = 2.74065$$

$$x5 = \frac{0.43429 \times 2.74065 + 1.2}{\log_{10} 2.74065 + 0.43429} = 2.74065$$

Here $x_4 = x_5$. Hence the required root is 2.74065 correct to five decimal places.

2.13   Some Deductions From Newton-Raphson Formula

We can derive the following useful results from the Newton's iteration formula:

(1) Iterative formula to find $1/N$ is $x_{n+1} = x_n(2 - Nx_n)$

(2) Iterative formula to find $\sqrt{N}$ is $x_{n+1} = \dfrac{1}{2}(x_n + N/x_n)$

(3) Iterative formula to find $1/\sqrt{N}$ is $x_{n+1} = \dfrac{1}{2}(x_n + 1/Nx_n)$

(4) Iterative formula to find $\sqrt[k]{N}$ is $x_{n+1} = \dfrac{1}{k}[(k-1)x_n + N/x_n^{k-1})]$

Proofs. (1) Let $x = 1/N$ or $1/x - N = 0$

Taking $f(x) = 1/x - N$, we have $f'(x) = -x-2$.

Then Newton's formula gives

$$x_{n+1} = x_n - \frac{f(x_n)}{f'(x_n)} = x_n - \frac{(1/x_n - N)}{-x_n - 2} = x_n + \left(\frac{1}{x_n} - N\right)x_n 2.$$

$$= x_n + x_n - Nx_n^2 = x_n(2 - Nx_n)$$

(2) Let $x = \sqrt{N}$ or $x^2 - N = 0$.

Taking $f(x) = x^2 - N$, we have $f'(x) = 2x$.

Then Newton's formula gives

$$x_{n+1} = x_n - \frac{f(x_n)}{f'(x_n)} = x_n - \frac{x_n^2 - N}{2x_n} = \frac{1}{2}(x_n N/x_n)$$

(3) Let $x = \dfrac{1}{\sqrt{N}}$ or $x^2 - \dfrac{1}{N} = 0$

Taking $f(x) = x^2 - 1/N$, we have $f'(x) = 2x$.

Then Newton's formula gives

$$x_{n+1} = x_n - \frac{f(x_n)}{f'(x_n)} = x_n - \frac{x_n^2 - 1/N}{2x_n} = \frac{1}{2}\left(x_n + \frac{1}{Nx_n}\right)$$

(4) Let $x = \sqrt[k]{N}$ or $x^k - N = 0$

Taking $f(x) = x^k - N$, we have $f'(x) = kx^{k-1}$

Then Newton's formula gives

$$x_{n+1} = x_n - \frac{f(x_n)}{f'(x_n)} = x_n \frac{x_n^h - N}{k x_n^{k-1}} = \frac{1}{k}\left[(k-1)x_n + \frac{N}{x_n^{k-1}}\right].$$

EXAMPLE 2.33

Evaluate the following (correct to four decimal places) by Newton's iteration method:

(*i*) 1/31   (*ii*) $\sqrt{5}$ (*iii*) $1/\sqrt{14}$ (*iv*) 24 3

(*v*) $(30)^{-1/5}$.

Solution:

(*i*) Taking $N = 31$, the above formula (1) becomes

$$x_{n+1} = x_n(2 - 31x_n)$$

Since an approximate value of 1/31 = 0.03, we take $x_0 = 0.03$.

Then $\quad x_1 = x_0(2 - 31x_0) = 0.03(2 - 31 \times 0.03) = 0.0321$

$\qquad x_2 = x_1(2 - 31x_1) = 0.0321(2 - 31 \times 0.0321) = 0.032257$

$\qquad x_3 = x_2(2 - 31x_2) = 0.032257(2 - 31 \times 0.032257) = 0.03226$

Since $x_2 = x_3$ upto four decimal places, we have 1/31 = 0.0323.

(*ii*) Taking $N = 5$, the above formula (2), becomes $x_{n+1} = \frac{1}{2}\left(x_n + 5/x_n\right)$.

Since an approximate value of $\sqrt{5} = 2$, = 2, we take $x_0 = 2$.

Then $\quad x_1 = \frac{1}{2}\left(x_0 + 5/x_0\right) = \frac{1}{2}\left(2 + 5/2\right) = 2.25$

$\qquad x_2 = \frac{1}{2}\left(x_1 + 5/x_1\right) = 2.2361$

$\qquad x_3 = \frac{1}{2}\left(x_2 + 5/x_2\right) = 2.2361$

Since $x_2 = x_3$ upto four decimal places, we have $\sqrt{5} = 2.2361$.

(*iii*) Taking $N = 14$, the above formula (3), becomes $x_{n+1} = \frac{1}{2}[x_n + 1/(14x_n)]$

Since an approximate value of $1/\sqrt{14} = 1/\sqrt{16} = \frac{1}{4} = 0.25$, we take $x_0 = 0.25$,

Then $x_1 = \frac{1}{2}[x_0 + (14x_0)^{-1}] = \frac{1}{2}\left[0.25 + (14 \times 0.25)^{-1}\right] = 0.26785$

$x_2 = \frac{1}{2}\left[x_1 + (14x_1)^{-1}\right] = \frac{1}{2}\left[0.26785 + (14 \times 0.26785)^{-1}\right] = 0.2672618$

$x_3 = \frac{1}{2}\left[x_2 + (14x_2)^{-1}\right] = \frac{1}{2}\left[0.2672618 + (14 \times 0.2672618)^{-1}\right] = 0.2672612$

Since $x_2 = x_3$ upto four decimal places, we take $1/\sqrt{14} = 0.2673$.

(iv) Taking $N = 24$ and $k = 3$, the above formula (4) becomes

$$X_{n+1} = \frac{1}{3}\left[2X_n + 24/X_n^{2}\right]$$

Since an approximate value of $(24)^{1/3} = (27)^{1/3} = 3$, we take $x_0 = 3$.

Then $X_1 = \frac{1}{3}\left(2X_0 + 24X_0^{2}\right) = \frac{1}{3}\left(6 + 24/9\right) = 2.88889$

$X_2 = \frac{1}{3}\left(2X_1 + 24/X_1^{2}\right) = \frac{1}{3}\left[(2 \times 2.88889) + 24/(2.88889)^{2}\right] = 2.88451$

$X_3 = \frac{1}{3}\left(2X_2 + 24/X_2^{2}\right) = \frac{1}{3}\left[2 \times 2.88451 + 24/(2.88451)^{2}\right] = 2.8845$

Since $X_2 = X_3$ up to four decimal places, we take $(24)^{1/3} = 2.8845$.

(v) Taking $N = 30$ and $k = -5$, the above formula (4) becomes

$$X_{n+1} = \frac{1}{-5}\left(6X_n + 30/X_n^{-6}\right) = \frac{X_n}{5}\left(6 - 30X_n^{5}\right)$$

Since an approximate value of $(30)^{-1/5} = (32)^{-1/5} = 1/2$, we take $x_0 = 1/2$

Then $X_1 = \frac{X_0}{5}\left(6 - 30X_0^{5}\right) = \frac{1}{10}\left(6 - 30/2^{5}\right) = 0.506495$

$X_2 = \frac{X_1}{5}(6 - 30x_1^{5}) = \frac{0.50625}{5}[6 - 30(0.50625)^{5}] = 0.506495$

$X_3 = \frac{X_2}{5}(6 - 30X_2^{5}) = \frac{0.506495}{5}[6 - 30(0.506495)^{5}] = 0.506496$

Since $x_2 = x_3$ up to four decimal places, we take $(30)^{-1/5} = 0.5065$.

Exercises 2.5

1. Find by Newton-Raphson method, a root of the following equations correct to three decimal places:

   (i) $x^3 - 3x + 1 = 0$ (ii) $x^3 - 2x - 5 = 0$

   (iii) $x^3 - 5x + 3 = 0$ (iv) $3x^3 - 9x^2 + 8 = 0$.

2. Using Newton's iterative method, find a root of the following equations correct to four decimal places:

   (i) $x^4 + x^3 - 7x^2 - x + 5 = 0$ which lies between 2 and 3.

   (ii) $x^5 - 5x^2 + 3 = 0$.

3. Find the negative root of the equation $x^3 - 21x + 3500 = 0$ correct to 2 decimal places by Newton's method.

4. Using Newton-Raphson method, find a root of the following equations correct to three decimal places:

   (i) $x^2 + 4 \sin x = 0$

   (ii) $x \sin x + \cos x = 0$ or $x \tan x + 1 = 0$

   (iii) $e^x = x^3 + \cos 25x$ which is near 4.5

   (iv) $x \log_{10} x = 12.34$, start with $x_0 = 10$.

   (v) $\cos x = xe^x$ (vi) $10^x + x - 4 = 0$.

5. The equation $2e^{-x} = \dfrac{1}{x+2} + \dfrac{1}{x+1}$ has two roots greater than -1. Calculate these roots correct to five decimal places.

6. The bacteria concentration in a reservoir varies as $C = 4e^{-2t} + e^{-0.1t}$. Using the Newton Raphson (N.R.) method, calculate the time required for the bacteria concentration to be 0.5.

7. Use Newton's method to find the smallest root of the equation $e^x \sin x = 1$ to four decimal places.

8. The current i in an electric circuit is given by $i = 10e^{-t} \sin 2\pi t$ where t is in seconds. Using Newton's method, find the value of t correct to three decimal places for $i = 2$ amp.

9. Find the iterative formulae for finding $\sqrt{N}, \sqrt[3]{N}$ where N is a real number, using the Newton-Raphson formula.

Hence evaluate:

(*a*) $\sqrt{10}$

(*b*) $\sqrt{21}$

(*c*) the cube-root of 17 to three decimal places.

10. Develop an algorithm using the N.R. method, to find the fourth root of a positive number N and hence find $\sqrt[4]{32}$

11. Evaluate the following (correct to three decimal places) by using the Newton-Raphson method.

(*i*) 1/18   (*ii*) $1/\sqrt{15}$ (*iii*) $(28)^{-1/4}$.

12. Obtain *Newton-Raphson extended formula*

$$x_1 = x_0 - \frac{f(x_0)}{f'(x_0)} - \frac{1}{2}\frac{|f(x_0)|^2\, f''(x_0)}{\{f'(x_0)\}^2}$$

for the root of the equation $f(x) = 0$.

Hence find the root of the equation $\cos x = xe^x$ correct to five decimal places.

Solution:

Expanding $f(x)$ in the neighborhood of x_0 by Taylor's series; we have

$$0 = f(fx) = f(x_0 + \overline{x - x_0}) = f(x_0) + (x - x_0)f'(x_0) \quad \text{to first approxi-}$$
mately.

Hence the first approximation to the root is given by

$$x_1 - x_0 = -f(x_0)/f'(x) \tag{i}$$

Again by Taylor's series to the second approximation, we get

$$f(x_1) = f(x_0) + (x_1 - x_0)f'(x_0) + \frac{1}{2!}\ (x_1 - x_0)2f''(x_0)$$

Since x_1 is an approximation to the root, $f(x_1) = 0$

$$\therefore \qquad f(x_0) + (x_1 - x_0)f'(x_0) + \frac{1}{2}(x_1 - x_0)^2 f''(x_0) = 0$$

or $\qquad\qquad x_1 - x_0 = -\dfrac{f(x_0)}{f'(x_0)} - \dfrac{1}{2}\left\{\dfrac{-f(x_0)}{f'(x_0)}\right\}f''(x_0)$ [by (*i*)]

whence follows the desired formula. [This is known as the **Chebyshev formula** of third order.]

2.14   Muller's Method

1. This method is a generalization of the secant method as it doesn't require the derivative of the function. It is an iterative method that requires three starting points. Here, $y = f(x)$ is approximated by a second degree parabola passing through these three points (x_{i-2}, y_{i-2}), (x_{i-1}, y_{i-1}) and (x_i, y_i) in the vicinity of the root. Then a root of this quadratic is taken as the next approximation x_{i+1} to the root of $f(x) = 0$.

2. Let x_{i-2}, x_{i-1}, x_i be three approximations to the root α of the equation $f(x) = 0$ and y_{i-2}, y_{i-1}, y_i be the corresponding values of $f(x)$.

   Assuming the equation of the parabola through the points (x_{i-2}, y_{i-2}), (x_{i-1}, y_{i-1}) and (x_i, y_i) to be

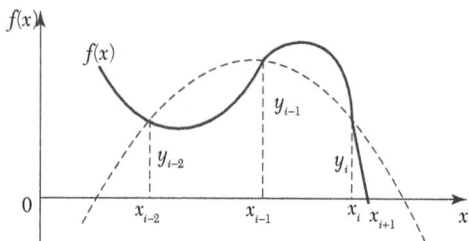

FIGURE 2.9

$$y = ax^{2} + bx + c, \tag{1}$$

we get

$$\left.\begin{array}{l} y_{i-2} = ax_{i-2}^2 + bx_{i-2} + c \\ y_{i-1} = ax_{i-1}^2 + bx_{i-1} + c \\ y_i = ax_i^2 + bx_i + c \end{array}\right\} \tag{2}$$

and

Eliminating a, b, c from (1) and (2), we obtain

$$\begin{vmatrix} y & x^2 & x & 1 \\ y_{i-2} & x_i - 2^2 & x_{i-2} & 1 \\ y_{i-1} & x_i - 1^2 & x_{i-1} & 1 \\ y_i & x_i^2 & x_i & 1 \end{vmatrix} = 0$$

which can be written as

$$y = \frac{(x - x_{i-1})(x - xi)}{(x_{i-2} - x_{i-1})(x_{i-2} - x_i)} y_{i-2} + \frac{(x - x_{i-2})(x - x_i)}{(x_{i-1} - x_{i-2})(x_{i-1} - x_i)} y_{i-1} \tag{3}$$

$$+ \frac{(x - x_{i-2})(x - x_{i-1})}{(x_i - x_{i-2})(x_i - x_{i-1})} y_i$$

We now define $\lambda = \dfrac{x - x_i}{x_i - x_{i-1}}, \lambda_i = \dfrac{x_i - x_{i-1}}{x_{i-1} - x_{i-2}}$ and $\delta_i = \dfrac{x_i - x_{i-2}}{x_{i-1} - x_{i-2}}$ \hfill (4)

Then (3) simplifies to

$$y = \frac{\left(y_{i-2}\lambda_i + y_{i-1}\delta_i + y_i\right)\lambda_i\lambda^2}{\delta_i} + \frac{y_i - 2^{\lambda}i^2 - y_{i-1}\delta i^2 + y_i\left(\lambda i + \delta\right)}{\delta_i}\lambda + y_i \quad (5)$$

From (4), we get $x = x_i + \lambda(xi - x_{i-1})$ \hfill (6)

Now to find a better approximation to the root, we need the unknown quantity λ. To determine λ, we put $y = 0$ in (5) giving

$$(y_{i-2}\lambda_i - y_{i-1}\delta_i + y_i)\,\lambda_i\lambda^2 + \mu_i\lambda + \delta_i\,y_i = 0 \quad (7)$$

Where $\quad \mu_i = y_{i-2}\,\lambda_i^2 - y_{i-1}\,\delta_i^2 + y_i(\lambda_i + \delta_i)$

Dividing throughout by $\lambda i\lambda^2$ and solving for $1/\lambda^*$, we get

$$\frac{1}{\lambda} = \frac{-\mu_i \pm \sqrt{\left[\mu_i^2 - 4y_i\delta_i\lambda_i\left(y_i - 2\lambda_i - y_i - 1\delta_i + y_i\right)\right]}}{2y_i\delta_i}$$

Since x is close to xi, λ should be small in magnitude. Therefore the sign should be so chosen to make the numerator largest in magnitude. Then (6) gives a better approximation to the root.

NOTE

Obs. *This method is iterative and converges for almost all initial approximations quadratically. In case no better approximations are known, we take,* $x_{i-2} = -1, x_{i-1} = 0,$ *and* $X_i = 1.$

EXAMPLE 2.34

Apply Muller's method to find the root of the equation $cos\ x = xe^x$ which lies between 0 and 1.

Solution:

Let $\quad y = \cos x - xe^x$

Taking the initial approximations as

$x_{i-2} = -1, x_{i-1} = 0, x_i = 1$

We obtain $\quad y_{i-2} = \cos 1 + e^{-1}, y_{i-1} = 1, y_i = \cos 1 - e$

---

*As a direct solution of (7) usually leads to inaccurate results, we solve it for $1/\lambda$.

$$\lambda = x - 1, \lambda_i = 1, \delta_i = 2$$

and $\quad \mu_i = (\cos 1 + e^{-1}) - 4 + 3(\cos 1 - e)$.

∴ From (7), we get two values of λ^{-1}. $\hspace{3cm}$ (i)

We choose the –ve sign so that the numerator in (i) is largest in magnitude and obtain $\lambda = -0.5585$.

∴ The next approximation to the root is given by (6) as

$$x_{i+1} = x_i + \lambda(x_i - x_{i-1}) = 1 - 0.5585 = 0.4415.$$

Repeating the above process, we get

$$x_{i+2} = 0.5125, x_{i+3} = 0.5177, x_{i+4} = 0.5177$$

Hence the root is 0.518 correct to three decimal places.

Exercises 2.6

Using Muller's method, find a root of the following equations, correct to three decimal places:

1. $x^3 - 2x - 1 = 0$ $\hspace{2cm}$ 2. $x^3 - x^2 - x - 1 = 0$.

3. $x^3 + 2x^2 + 10x - 20 = 0$ taking $x_0 = 0$, $x_1 = 1$ and $x_2 = 2$.

4. $\log x = x - 3$ taking $x_0 = 0.25$, $x_1 = 0.5$ and $x_2 = 1$.

2.15 Roots of Polynomials Equations

The methods so far discussed for finding the roots of equations can also be applied to polynomials. These methods, however, do not work well when the polynomial equations contain multiple or complex roots. We now discuss methods for finding all the real and complex roots of polynomials. These methods are especially designed for polynomials and cannot be applied to transcendental equations. We begin with Horner's method which is the best for finding the approximate values of real roots of a numerical polynomial equation.

Approximate Solution of Polynomial Equations—Horner's Method

This method consists in diminution of the roots of an equation by successive digits occurring in the roots.

If the root of an equation lies between a and $a + 1$, then the value of this root will be $a.bcd$......, where b, c, d...... are digits in its decimal part. To obtain these, we proceed as follows:

(i) Diminish the roots of the given equation by a so that the root of the new equation is $o.\,bcd$......

(ii) Then multiply the roots of the transformed equation by 10 so that the root of the new equation is $b.\,cd$......

(iii) Now diminish the root by b and multiply the roots of the resulting equation by 10 so that the root is $c.d$......

(iv) Next diminish the root by c and so on. By continuing this process, the root may be evaluated to any desired degree of accuracy digit by digit. The method will be clear from the following example:

EXAMPLE 2.35

Find by Horner's method, the positive root of the equation $x^3 + x^2 + x - 100 = 0$ correct to three decimal places.

Solution:

Step I. Let $\quad f(x) = x^3 + x^2 + x - 100$

By Descartes' rule of signs, there is only one positive root. Also $f(4) = -$ ve and $f(5) = +$ve, therefore, the root lies between 4 and 5.

Step II. Diminish the roots of given equation by 4 so that the transformed equation is

$$x^3 + 13x^2 + 57x - 16 = 0 \qquad\qquad (i)$$

Its root lies between 0 and 1. (We draw a zig-zag line above the set of figures 13, 57,– 16 which are the coefficients of the terms in (i) as shown below.) Now multiply the roots of (i) by 10 for which attach one zero to the second term, two zeros to the third term, and three zeros to the fourth term. Then we get the equation

$$f_1(x) = x^3 + 130x^2 + 5700x - 16000 = 0 \qquad\qquad (ii)$$

1	1	1	−100	(4.264)

$$
\begin{array}{llll}
1 & 1 & 1 & -100 \qquad (4.264)\\
 & 4 & \underline{20} & 84\\
 & 5 & 21 & -16000\\
 & 4 & 36 & 11928\\
 & 9 & 5700 & -4072000\\
 & 4 & \underline{264} & 3788376\\
 & 130 & 5964 & -28362400\\
 & \underline{2} & 268 &\\
 & 132 & 623200 &\\
 & \underline{2} & \underline{8196} &\\
 & 134 & 631396 &\\
 & 2 & 8232 &\\
 & 1360 & 63962800 &\\
 & \underline{6} & &\\
 & 1366 & &\\
 & \underline{6} & &\\
 & 1372 & &\\
 & 6 & &\\
 & 13780 & &
\end{array}
$$

Its root lies between 0 and 10.

Clearly $f_1(2) = -\text{ve}, f_1(3) = +\text{ve}$.

∴ The root of (*ii*) lies between 2 and 3, *i.e.,*. first figure after the decimal is 2.

Step III. Diminish the roots of $f_1(x) = 0$ by 2 so that the next transformed equation is

$$x^3 + 136x^2 + 6232x - 4072 = 0. \qquad (iii)$$

Its root lies between 0 and 1. (We draw the second zig-zag line above the set of figures 136, 6232, − 4072). Multiply the roots of (*iii*) by 10, *i.e.,* attach one zero to second term, two zeros to the third term, and three zeros to the fourth term. Then the new equation is

$$f_2(x) = x^3 + 1360x^2 + 623200x - 4072000 = 0$$

Its root lies between 0 and 10, which is nearly $= \dfrac{4072000}{623200} = 6$.

Hence the second figure after the decimal place is 6.

Step IV. Diminish the roots of $f_2(x) = 0$ by 6, so that the transformed equation is

$$x^3 + 1378x^2 + 639628x - 283624 = 0.$$

Its root lies between 0 and 1. (We draw the third zig-zag line above the set of figures 1378, 639628, − 283624.) As before multiply its roots by 10, *i.e.*, attach one zero to the second term, two zeros to the third term, and three zeros to the fourth term. Then the equation becomes

$$f_3(x) = x^3 + 13780x^2 + 63962800x - 283624000 = 0$$

Its root lies between 0 and 10, which is nearly $= \dfrac{283624000}{63962800} = 4$. Thus

the roots of $f_3(x) = 0$ are to be diminished by 4, *i.e.*, the third figure after the decimal place is 4. But there is no need to proceed further as the root is required correct to three decimal places only.

Hence the root is 4.264.

Obs. 1. *After two steps of diminishing, we apply the* principle of trial divisor *in which we divide the last coefficient by the last but one coefficient to get the next integer by which the roots are to be diminished.* These last two coefficients should have opposite signs.

Obs. 2. *At any stage if the trial divisor suggests the next integer to be zero, then we should again multiply the roots by 10 and write zero in the decimal place of the root.*

EXAMPLE 2.36

Find the cube root of 30 correct to three decimal places, using Horner's method.

Solution:

Step I. Let $x = \sqrt[3]{30}$ *i.e.* $f(x) = x^3 - 30 = 0$

Now $f(3) = - 3$ (–ve), $f(4) = 34$ (+ve)

∴ The root lies between 3 and 4.

Step II. Diminish the roots of the given equation by 3 so that the transformed equation is

$$x^3 + 9x^2 + 27x - 3 = 0 \qquad\qquad (i)$$

Its roots lies between 0 and 1. (We draw a zig-zag line above the set of numbers 9, 27, − 3 which are the coefficients of the terms in (i)). Now multiply the roots of (i) by 10 for which attach one zero to the second term, two zeros to the third term, and three zeros to the fourth term. Then we get the equation

$$f_1(x) = x^3 + 90x^2 + 2700x - 3000 = 0 \qquad (ii)$$

Its roots lies between 0 and 10.

Clearly $f_1(1) = -ve, \qquad f_2(2) = +ve$

∴ The root of (ii) lies between 1 and 2, i.e., first figure after the decimal place is 1.

Step III. Diminish the roots of $f_1(x) = 0$ by 1, so that the next transformed equation is

$$x^3 + 93x^2 + 2883x - 209 = 0. \qquad (iii)$$

Its root lies between 0 and 1. (We draw a second zig-zag line above the set of figures 93, 2883, − 209). Multiply the roots of (iii) by 10, i.e., attach one zero to second term, two zeros to the third term, and three zeros to the fourth term. Then the new equation is

$$f_2(x) = x^3 + 930x^2 + 288300x - 209000 = 0.$$

Its root lies between 0 and 10, which is nearly = 209000/288300 = 0.724 > 0 and < 1.

Hence second figure after the decimal place is 0.

1	0	0	−30	(3.107
	3	9	27	
	3	9	−30000	
	3	18	2791	
	6	2700	−209000000	
	3	91		
	90	2791		
	1	92		
	91	28830000		
	1			
	92			
	1			
	9300			

Step IV. Diminish the root of $f_2(x) = 0$ by 0 and then multiply its roots by 10 so that

$$f_3(x) = x^3 + 9300x^2 + 28830000x - 209000000 = 0$$

Its root lies between 0 and 10, which is nearly

$$= 209000000/28830000 = 7.2 > 7 \text{ and } < 8.$$

Thus the roots of $f_3(x) = 0$ are to be diminished by 7, *i.e.*, the third figure after the decimal is 7.

Hence the required root is 3.107.

Exercises 2.7

1. Find by Horner's method, the root (correct to three decimal places) of the equations (*i*) $x^3 - 3x + 1 = 0$ which lies between 1 and 2. (*ii*) $x^3 + x - 1 = 0$. (*iii*) $x^3 - 3x^2 + 2.5 = 0$ which lies between 1 and 2.

2. Using Horner's method, find the largest real root of $x^3 - 4x + 2 = 0$ correct to three decimal places.

3. Show that a root of the equation $x^4 + x^3 - 4x^2 - 16 = 0$ lies between 2 and 3. Find its value correct up to two decimal places by Horner's method.

4. Find the negative root of the equation $x^3 - 9x^2 + 18 = 0$ correct to two decimal places by Horner's method.

5. Find the cube root of 25, correct to four decimal places, using Horner's method

2.16 Multiple Roots

If α is a root of $f(x) = 0$ of order m, then $f(\alpha) = 0$, $f'(\alpha) = 0,\cdots,$ $f^{m-1}(\alpha) = 0$ and $f^m(\alpha) \neq 0$. Such an equation can be written as $f(x) = (x - \alpha)^m$ $\phi(x) = 0$. In other words, if α is a root of $f(x) = 0$ repeated m times, then it is also a root of $f'(x) = 0$ repeated $(m - 1)$ times, of $f''(x) = 0$ repeated $(m - 2)$ times and so on.

Multiple roots by Newton's method. Let α be a root of the polynomial equation $f(x) = 0$ which is repeated m times, If $x_0, x_1, x_2,\cdots, x_{n+1}$, be its successive approximations then on the lines of Newton's iterative method,

we have $$x_{n+1} = x_n - m\frac{f(x_n)}{f'(x_n)}$$

which is called the *generalized Newton's formula*. It reduces to Newton-Raphson formula for $m = 1$.

NOTE

Obs. 1. *If initial approximation x_0 is sufficiently close to the root α, then the expressions*

$$x_0 = m\frac{f(x_0)}{f'(x_0)}, x_0 - (m-1)\frac{f'(x_0)}{f''(x_0)}, x_0 - (m-2)\frac{f''(x_0)}{f'''(x_0)}, \cdots$$

will have the same value.

Obs. 2. *Generalized Newton's formula has a second order convergence for determining a multiple root. (see Example 2.38).*

EXAMPLE 2.37

Find the double root of the equation $x^3 - x^2 - x + 1 = 0$.

Solution:

Let $\qquad\qquad f(x) = x^3 - x^2 - x + 1$

So that $\qquad f'(x) = 3x^2 - 2x - 1, f''(x) = 6x - 2$

Starting with $x_0 = 0.9$, we have

$$x_0 - 2\frac{f(x_0)}{f'(x_0)} = 0.9 - \frac{2 \times 0.019}{-0.37} = 1.003$$

and $\qquad x_0 - (2-1)\frac{f'(x_0)}{f''(x_0)} = 0.9 - \frac{(-0.37)}{3.4} = 1.009$

The closeness of these values implies that there is a double root near $x = 1$.

$\therefore$ Choosing $x_1 = 1.01$ for the next approximation, we get

$$x_1 - 2\frac{f(x_1)}{f'(x_1)} = 1.01 - \frac{2 \times 0.002}{0.0403} = 1.0001$$

$$x_1 - (2-1)\frac{f'(x_1)}{f''(x_1)} = 1.01 - \frac{0.0403}{4.06} = 1.0001$$

This shows that there is a double root at $x = 1.0001$ which is quite near the actual root $x = 1$.

EXAMPLE 2.38

Show that the generalized Newton's formula $x_{n+1} = x_n - 2f(x_n)/f'(x_n)$ gives a quadratic convergence when the equation $f(x) = 0$ *has a pair of double roots in the neighborhood of $x = x_n$.*

Solution:

Suppose $x = \alpha$ is a double root near $x = x_n$.

Then $f(\alpha) = 0, f'(\alpha) = 0$ $\hspace{3cm}$ (i)

We have $\hspace{2cm} \varepsilon_{n+1} = \varepsilon_n - \dfrac{2f(\alpha + \varepsilon_n)}{f'(\alpha + \varepsilon_n)}$

Expanding $f(\alpha + \varepsilon)$ and $f'(\alpha + \varepsilon)$ in powers of ε_n and using (i), we get

$$\varepsilon \quad = \varepsilon - \frac{\left[\dfrac{\varepsilon_n{}^2}{2!} f''(\alpha) + \dfrac{\varepsilon_n{}^2}{3!} f'''(\alpha) + \cdots\cdots\right]}{\left[\varepsilon\ f''(\alpha) + \dfrac{1}{2!} f'''(\alpha) + \cdots\cdots\right]}$$

$$= \varepsilon \ - \frac{\varepsilon_n\left[f''(\alpha) + -\varepsilon_n f''(\alpha)\right]}{f''(\alpha) + - f'''(\alpha)}\text{approx.}$$

$$= \frac{1}{6}\varepsilon_n{}^2 \frac{f'''(\alpha)}{f''(\alpha) + - f'''(\alpha)} = \frac{1}{6}\varepsilon_n{}^2 \frac{f'''(\alpha)}{''(\ )}$$

which shows that $\varepsilon_{n+1} \propto \varepsilon_n{}^2$ and so the convergence is of second order.

2.17 Complex Roots

We know that the complex roots of an equation occur in conjugate pairs, *i.e.*, if $\alpha + i\beta$ is a root of $f(x) = 0$, $\alpha - i\beta$ is also its root. In other words, $[x - (\alpha + i\beta)]$ and $[x - (\alpha - i\beta)]$ are factors of $f(x)$ or $(x - \alpha - i\beta)(x - \alpha + i\beta) = x^2 - 2x\alpha + \alpha^2 + \beta^2$ is a factor of $f(x)$. This implies that we should try to isolate complex roots by finding the appropriate quadratic factors of the original polynomial. A method which is often used for finding such quadratic factors of polynomials is the Lin-Bairstow's method. However Newton's method can also be used to find the complex roots of a polynomial equation which we illustrate below:

EXAMPLE 2.39

Solve $x^4 - 5x^3 + 20x^2 - 40x + 60 = 0$, by Newton's method given that all the roots of the given equation are complex.

Solution:

Let $\quad f(x) = x^4 - 5x^3 + 20x^2 - 40x + 60 = 0 \qquad\qquad (i)$

so that $f'(x) = 4x^3 - 15x^2 + 40x - 40$

∴ Newton-Raphson method gives

$$x_{n+1} = x_n - \frac{f(x_n)}{f'(x_n)} = x_n - \frac{x_n^4 - 5x_n^3 + 20x_n^2 - 40x_n + 60}{4x_n^3 - 15x_n^2 + 40x_n - 40}$$

$$= \frac{3x_n^4 - 10x_n^3 + 20x_n^2 - 60}{4x_n^3 - 15x_n^2 + 40x_n - 40}$$

Putting $n = 0$ and taking $x_0 = 2(1 + i)$ by trial, we get

$$x_1 = \frac{3(2+2i)^4 - 10(2+2i)^3 + 20(2+2i)^2 - 60}{4(2+2i)^3 - 15(2+2i)^2 + 40(2+2i) - 40} = 1.92(1+i)$$

Similarly

$$x_2 = \frac{3(1.92+1.92i)^4 - 10(1.92+1.92i)^3 + 20(1.92+1.92i) - 60}{4(1.92+1.92i)^3 - 15(1.92+1.92i)^2 + 40(1.92+1.92i) - 40}$$

$$= 1.915 + 1.908i$$

Since complex roots occur in conjugate pairs so the roots of (i) are 1.915 ±1.908i up to three places of decimals. Assuming that the other pair of roots of (i) is $\alpha \pm i\beta$, we have

Sum of the roots $= (\alpha + i\beta) + (\alpha - i\beta) + (1.915 + 1.908i) + (1.915 - 1.908i) = 5$

i.e., $\quad 2\alpha + 3.83 = 5$ or $\alpha = 0.585$.

Also the product of roots $= (\alpha^2 + \beta^2) \{(1.915)^2 + (1.908)^2\} = 60$

which gives $\beta = 2.805$. Hence the other two roots are 0.585±2.805i.

2.18   Lin-Bairstow's Method

This method is often used for finding the complex roots of a polynomial equation with real coefficients, such as

$$f(x) = x^n + a_1 x^{n-1} + a_2 x^{n-2} + \cdots + a_{n-1} x + a_n = 0. \qquad (1)$$

Since complex roots occur in pairs as $\alpha \pm i\beta$, each pair corresponds to a quadratic factor

$$\{x - (\alpha + i\beta)\}\{x - (\alpha - i\beta)\} = x^2 - 2\alpha x + \alpha^2 + \beta^2$$

which is of the form $x^2 + px + q$.

If we divide $f(x)$ by $x^2 + px + q$, we obtain the quotient $Q_{n-2} = x^{n-2} + b_1 x^{n-3} + \cdots + b_{n-2}$ and the remainder $Rn = Rx + S$.

Thus $f(x) = (x^2 + px + q)(x^{n-2} + b_1 x^{n-3} + \cdots + b_{n-2}) + Rx + S.$ (2)

If $x^2 + px + q$ divides $f(x)$ completely, the remainder $Rx + S = 0$, *i.e.*, $R = 0, S = 0$. Obviously R and S both depend upon p and q. So our problem is to find p and q such that

$$R(p, q) = 0, S(p, q) = 0.$$ (3)

Let $p + \Delta p, q + \Delta q$ be the actual values of p and q which satisfy (3). Then

$$R(p + \Delta p, q + \Delta q) = 0, S(p + \Delta p, q + \Delta q) = 0.$$ (4)

To find the corrections $\Delta p, \Delta q$, we expand these by Taylor's series and neglect second and higher order terms.

$$\therefore \quad \left. \begin{array}{l} R(p,q) + \dfrac{\partial R}{\partial p}\Delta p + \dfrac{\partial R}{\partial q}\Delta q = 0 \\[2mm] S(p,q) + \dfrac{\partial S}{\partial p}\Delta p + \dfrac{\partial S}{\partial q}\Delta q = 0 \end{array} \right\}$$ (5)

We solve these simultaneous equations for Δp and Δq and then the procedure is repeated with the corrected values for p and q. Now to compute the coefficients bi, R, and S, we compare the coefficients of like powers of x in (2) giving

$$b_1 = a_1 - p$$
$$b_2 = a_2 - pb_1 - q$$

$$\cdots\cdots\cdots\cdots\cdots\cdots\cdots\cdots$$

$$b_i = a_i - pb_{i-1} - qb_{i-2}$$ (6)

$$\cdots\cdots\cdots\cdots\cdots\cdots\cdots\cdots$$

$$R = a_{n-1} - pb_{n-2} - qb_{n-3}, \; S = a_n - qb_{n-2}$$

We now introduce b_{n-1} and bn and define

$$b_i = a_i - p\, b_{i-1} - q\, b_{i-2}, \; i = 1, 2, \cdots n$$ (7)

where $\quad b_0 = 1, b_{-1} = 0 = b_{-2}$

Comparing the last two equations with those of (6), we get

$$b_{n-1} = a_{n-1} - p\, b_{n-2} - q\, b_{n-3} = R$$

$$b_n = a_n - p\,b_{n-1} - q\,b_{n-2} = S - p\,b_{n-1}$$

giving $\qquad R = b_{n-1}$ and $S = b_n + p\,b_{n-1}$ $\hspace{3cm}$ (8)

Substituting these values in (5), we get

$$b_{n-1} + \frac{\partial b_{n-1}}{\partial p}\Delta p + \frac{\partial b_{n-1}}{\partial p}\Delta q = 0$$

$$b_n + pb_{n-1} + \left(\frac{\partial b_n}{\partial p} + p\frac{\partial b_{n-1}}{\partial p} + b_{n-1}\right)\Delta p + \left(\frac{\partial b_n}{\partial q} + p\frac{\partial b_{n-1}}{\partial q}\right)\Delta q = 0$$

Multiplying the first of these equations by p and subtracting from the second, we get

$$\left.\begin{array}{l} \dfrac{\partial b_{n-1}}{\partial p}\Delta p \, \dfrac{\partial b_{n-1}}{\partial q}\Delta q + b_{n-1} = 0 \\[3mm] \left(\dfrac{\partial b_n}{\partial p} + b_{n-1}\right)\Delta p + \dfrac{\partial b_n}{\partial q} + b_n = 0 \end{array}\right\} \hspace{2cm} (9)$$

Now differentiating (7) w.r.t. p and q partially and noting that all ai's are constants and all bi's are functions of p and q, we have

$$\left.\begin{array}{l} \dfrac{\partial bi}{\partial p} = -b_{i-1} - p\dfrac{\partial b_{i-1}}{\partial p} - q\dfrac{\partial b_{i-2}}{\partial p};\ \dfrac{\partial b_{-1}}{\partial p} = 0 = \dfrac{\partial b_{-1}}{\partial p} \\[3mm] \dfrac{\partial bi}{\partial q} = -b_{i-2} - p\dfrac{\partial b_{i-1}}{\partial q} - q\dfrac{\partial b_{i-2}}{\partial q};\ \dfrac{\partial b_{-1}}{\partial q} = 0 = \dfrac{\partial b_{-1}}{\partial q} \end{array}\right\} \hspace{1.5cm} (10)$$

Also from (6), we get

$$\frac{\partial b_0}{\partial p} = 0 = \frac{\partial b_1}{\partial q};\ \frac{\partial b_1}{\partial p} = b_0, \frac{\partial b_2}{\partial q} = -b_0 - p;\ \frac{\partial b_1}{\partial q} = -b_0$$

$$\frac{\partial b_2}{\partial p} = -b_1 - p\frac{\partial b_1}{\partial p} = -b_1 + pb_o$$

$$\frac{\partial b_2}{\partial q} = -b_1 - p\frac{\partial b_2}{\partial q} - p\frac{\partial b_1}{\partial q} = -b_1 + pb_o$$

Thus we have $\dfrac{\partial b_2}{\partial q} = \dfrac{\partial b_1}{\partial p}$ and $\dfrac{\partial b_3}{\partial q} = \dfrac{\partial b_2}{\partial p}$

By mathematical induction, we shall prove that $\dfrac{\partial b_{i+1}}{\partial q} = \dfrac{\partial b_i}{\partial p}$, for all i

Let the result be true for $i = r$, then $\dfrac{\partial b_r}{\partial q} \quad \dfrac{\partial b_r}{\partial q}$ (11)

But using (10)

$$\frac{\partial b_{r+2}}{\partial q} = -b_r - p\frac{\partial b_{r+1}}{\partial q} - q\frac{\partial b_r}{\partial q}$$

and $\quad \dfrac{\partial b_{r+1}}{\partial p} = b_r, -p\dfrac{\partial b_r}{\partial p} - q\dfrac{\partial b_{r-1}}{\partial p} = -br - p\dfrac{\partial b_{r+1}}{\partial q} - q\dfrac{\partial b_r}{\partial q}$ [by (11)]

This shows that $\dfrac{\partial b_{r+2}}{\partial q} = \dfrac{\partial b_{r+1}}{\partial q}$ *i.e.*, the result is true for $i = r + 1$. But it is for $i = 1$ and should this be $i = 2$. Hence by induction, it is true for all values of i.

Now writing $\dfrac{\partial b_{i+1}}{\partial q} = \dfrac{\partial b_i}{\partial p} - c_{i-1}, i = 0,1,2,\cdots,n-1$ (12)

the equations in (10) can be expressed as

$$c_{i-1} = b_{i-1} - p\,c_{i-2} - q\,c_{i-3}, \; c_{i-2} = b_{i-2} - p\,c_{i-3} - q\,c_{i-4}$$

These can be compressed into a single equation

$$c_i = b_i - p\,c_{i-1} - q\,c_{i-2}$$

with $\quad c_0 = 0, \, c_{-1} = 0, \, i = 1, \, 2,..., \, (n-1)$ (13)

Thus c_i is computed from b_i in exactly the same way as b_i from ai in (7).

Differentiating the relations in (8) and using (12), we get

$$\frac{\partial R}{\partial p} = \frac{\partial b_{n-1}}{\partial p} = c_{n-2}, \frac{\partial R}{\partial q} = \frac{\partial b_{n-1}}{\partial q} = -c_{n-3}$$

and

$$\frac{\partial S}{\partial p} = \frac{\partial b_n}{\partial p} + b_{n-1} + p\frac{\partial bn-1}{\partial p} = -c_{n-1} - pc_{n-2} + b_{n-1}$$

$$\frac{\partial S}{\partial p} = \frac{\partial bn}{\partial q} + p\frac{\partial b_{n-1}}{\partial q} = c_{n-2} - P^c_{n-3}$$

Substituting these in (5), we get

$$b_{n-1} - c_{n-2}\,\Delta p - c_{n-3}\,\Delta q = 0$$

and $\quad b_n + pb_{n-1} + (-c_{n-1} - pc_{n-2} + b_{n-1})\,\Delta p + (-c_{n-2} - pc_{n-3})\,\Delta q = 0$

or

$$\left.\begin{array}{l} c_{n-2}\Delta p + c_{n-3}\Delta q = b_{n-1} \\ (c_{n-1} - b_{n-1})\,\Delta p + c_{n-2}\Delta q = b_n \end{array}\right\} \qquad (14)$$

After finding the values of b_i's and c_i's from (7) and (13) and putting in (14), we obtain the approximate values of Δp and Δq, say Δp_0 and Δq_0. If p_0, q_0 are the initial approximations then their improved values are $p_1 = p_0 + \Delta p_0$, $q_1 = q_0 + \Delta q_0$. Now taking p_1 and q_1 as the initial values and repeating the process, we can get better values of p and q.

NOTE

Obs. *The values of bi's and ci's are found by the following (synthetic division) scheme:*

$a_0(=1)$	a_1	a_2	$a_3 \ldots a_{n-2}$	a_{n-1}	a_n	
	$-pb_0$	$-pb_1$	$-pb_2 \ldots -pb_{n-3}$	$-pb_{n-2}$	$-pb_{n-1}$	$-p$
		$-qb_0$	$-qb_1 \ldots -qb_{n-1}$	$-qb_{n-3}$	$-qb_{n-2}$	$-q$
$b_0(=1)$	b_1	b_2	$b_3 \ldots b_{n-2}$	b_{n-1}	b_n	
	$-pc_0$	$-pc_1$	$-pc_2 \ldots -pc_{n-3}$	$-pc_{n-2}$		$-p$
		$-qc_0$	$-qc_1 \ldots -qc_{n-4}$	$-qc_{n-3}$		$-q$
$c_0(=1)$	c_1	c_2	$c_3 \ldots$	c_{n-2}	c_{n-1}	

EXAMPLE 2.40

Solve $x^4 - 5x^3 + 20x^2 - 40x + 60 = 0$, given that all the roots of $f(x) = 0$ are complex, by using the Lin-Bairstow method

Solution:

Starting with the values $p_0 = -4$, $q_0 = 8$, we have

1	-5	20	-40	60	
–	4	-4	32	0	4
–	–	-8	8	-64	--8
1	-1	8	$0(= b_{n-1})$	$-4(= b_n)$	
	4	12	48		4
		-8	-24		-8
1	$3(= c_{n-3})$	$12(= c_{n-2})$	$24(= c_{n-1})$		

$$\therefore \quad c_{n-1} - b_{n-1} = 24 - 0 = 24$$

Corrections Δp_0 and Δq_0 are given by

$$c_{n-2}\,\Delta p_0 + c_{n-3}\,\Delta q_0 = b_{n-1}\ i.e.,\ 12\,\Delta p_0 + 3\,\Delta q_0 = 0$$
$$(c_{n-1} - b_{n-1})\,\Delta p_0 + c_{n-2}\,\Delta q_0 = b_n\ i.e.,\ 24\,\Delta p_0 + 12\,\Delta q_0 = -4$$

Solving, we get $\Delta p_0 = 0.1667$, $\Delta q_0 = -0.6667$

$$\therefore \qquad\qquad p_1 = p_0 + \Delta p_0 = -3.8333$$
$$q_1 = q_0 + \Delta q_0 = 7.333$$

Now repeating the same process, *i.e.*, dividing $f(x)$ by $x^2 - 3.8333x + 7.3333$, we get

1	−5	20	−40	60	
	3.8333	−4.4723	31.4116	−0.125	3.8333
		−7.3333	8.5558	−60.092	−7.3333
1	−1.1667	8.1944	−0.0326	−0.217	
			$(= b_{n+1})$	$(= b_n)$	
	3.8333	10.2219	42.4845		3.8333
		−7.3333	−19.555		−7.3333

1    $2.6666(= c_{n-3})$ $11.083(= c_{n-2})$ $22.8969(= c_{n-1})$

$$\therefore \qquad c_{n-1} - b_{n-1} = 22.8969 + 0.0326 = 22.9295$$

Corrections Δp_1 and Δq_1 are given by

$$11.083\,\Delta p_1 + 2.6666\,\Delta q_1 = -0.0326$$
$$22.9295\,\Delta p_1 + 11.083\,\Delta q_1 = -0.217$$

Solving, we get $\Delta p_1 = 0.0033$ and $\Delta q_1 = -0.0269$

$$p_2 = p_1 + \Delta p_1 = -3.83,\ q_2 = q_1 + \Delta q_1 = 7.3064.$$

So one of the quadratic factors of $f(x)$ is

$$x^2 - 3.83\,x + 7.3064. \qquad\qquad (i)$$

If $\alpha \pm i\beta$ be its roots, then $2\alpha = 3.83$, $\alpha^2 + \beta^2 = 7.3064$ giving $\alpha = 1.9149$ and $\beta = 1.9077$.

Hence a pair of roots is $1.9149 \pm 1.9077\,i$

To find the remaining two roots of $f(x) = 0$, we divide $f(x)$ by (i) as follows [by Section 2.5 (3)]:

1	−5	20	−40	60		
	3.83	−4.4811	31.4539		3.83	
		−7.3064	8.5485	−60.0038	−7.3064	
1	−1.17	8.2125	0.0024	−0.0038		
			≈ 0	≈ 0		

∴ The other quadratic factor is $x^2 - 1.17x + 8.2125$.

If $\gamma \pm i\,\delta$ be its roots, then $2\delta = 1.17$, $\gamma^2 + \delta^2 = 8.2125$ giving $\gamma = 0.585$ and $\delta = 2.8054$.

Hence the other pair of roots is $0.585 \pm 2.8054\,i$.

2.19   Graeffe's Root Squaring Method

This method has an advantage over the other methods in that it does not require any prior information about the roots. But it is applicable to polynomial equations only and is capable of giving all the roots. Consider the polynomial equation

$$x^n + a_1 x^{n-1} + a_2 x^{n-2} + \ldots + a^{n-1} x + a_n = 0 \tag{1}$$

Separating the even and odd powers of x and squaring, we get

$$(x^n + a_2 x^{n-2} + a_4 x^{n-4} + \cdots)^2 = (a_1 x^{n-1} + a_3 x^{n-3} + \cdots)^2$$

Putting $x^2 = y$ and simplifying, the new equation becomes

$$y^n + b_1 y^{n-1} + \ldots + b_{n-1} y + b_n = 0 \tag{2}$$

where
$$\left.\begin{aligned}
b_1 &= -a_1^2 + 2a_2 \\
b_2 &= a_2^2 - 2a_1 a_3 + 2a_4 \\
&\cdots\cdots\cdots\cdots\cdots\cdots\cdots\cdots \\
b_n &= (-1)^n a_n^2
\end{aligned}\right\} \tag{3}$$

If $\alpha_1, \alpha_2, \ldots \alpha_n$ be the roots of (1) then the roots of (2) are $\alpha_1^2, \alpha_2^2, \ldots \alpha_n^2$. After m squarings, let the new transformed equation be

$$z^n + c_1 z^{n-1} + \ldots + c_{n-1} z + c_n = 0 \tag{4}$$

whose roots $\gamma_1, \gamma_2, \ldots, \gamma n$ are such that $\gamma i = \alpha i^{2m}$, $i = 1, 2, \ldots n$.

Assuming that $|\alpha_1| > |\alpha_2| > \ldots > |\alpha n|$, then $|\gamma_1| >> |\gamma_2| >> \ldots >> |\gamma_n|$ where >> stands for "much greater than."

Thus $\dfrac{|\gamma_2|}{|\gamma_1|} = \dfrac{\gamma_2}{\gamma_1}, \cdots, \dfrac{|\gamma_n|}{|\gamma_{n-1}|} = \dfrac{\gamma_n}{\gamma_{n-1}}$ are negligible as compared to unity. (5)

Also γ_i being an even power of α_i is always positive.

$\therefore$ From (4), we have

$$\sum \gamma_1 = -c_1 \ i.e. \ c_1 = \gamma_1 \left(1 + \dfrac{\gamma_2}{\gamma_1} + \dfrac{\gamma_3}{\gamma_1} + \cdots \right)$$

$$\sum \gamma_1 \gamma_2 = -c_2 \ i.e. \ c_2 = \gamma_1 \gamma_2 \left(1 + \dfrac{\gamma_3}{\gamma_1} + \cdots \right)$$

$$\sum \gamma_1 \gamma_2 \gamma_3 = -c_3 \quad i.e., \quad c_3 = -\gamma_1 \gamma_2 \gamma_3 \left(1 + \dfrac{\gamma_4}{\gamma_1} + \cdots \right)$$

....

$$\gamma_1 \gamma_2 \cdots \gamma_n = (-1)^n c_n i.e. \ c_n = (-1)^n \gamma_1 \gamma_2 \cdots \gamma_n$$

Hence by (5), we get $c_1 \approx -\gamma_1, c_2 \approx \gamma_1 \gamma_2, c_3 \approx \gamma_1 \gamma_2 \gamma_3, \cdots$

i.e., $\quad \gamma_1 \approx -c_1, \gamma_2 \approx -c_2/c_1, \gamma_3 \approx -c_3/c_2, \cdots, \gamma_n \approx -c_n/c_{n-1}$

Now since $\gamma_1 = \alpha_i^{2m}, \qquad \therefore \alpha_i = (\gamma_i)^{1/2m} = |c_i/c_{i-1}|$ (6)

Thus we can determine $\alpha_1, \alpha_2, \ldots \alpha n$, the roots of (1).

NOTE

Obs. 1. *Double root. If the magnitude of ci is half the square of the magnitude of the corresponding coefficient in the previous equation after a few squarings, then it shows that αi is a double root of (1). We find this double root as follows:*

$$\gamma_k \approx \dfrac{c_k}{c_{k-1}} \qquad \text{and} \qquad \gamma_{k+1} \approx -\dfrac{c_{k+1}}{ck}$$

$$\therefore \ \gamma_k \gamma_{k+1} \approx \gamma_k^2 \approx \left| \dfrac{c_{k+1}}{c_{k-1}} \right| \qquad i.e., \qquad \alpha_k^{2m} = \gamma_k^2 = \left| \dfrac{c_{k+1}}{c_{k-1}} \right| \quad (7)$$

This gives the magnitude of the double root and substituting in (1), we can find its sign.

Obs. 2. *Complex roots. If α_r and α_{r+1} form a complex pair $\rho_r e^{\pm i\phi_r}$, then the coefficients of x^{n-r} in successive squarings would fluctuate both in magnitude and sign by an amount $2\rho_r^m \cos m\phi_r$.*

For m sufficiently large ρ_r and ϕ_r can be determined by

$$p r^{2^{(2m)}} \approx \left|\frac{c_{r+1}}{c_{r-1}}\right|, 2p_r^m \cos m\phi r \approx \frac{c_{r+1}}{c_{r-1}} \tag{8}$$

If (1) has only one pair of complex roots say: $\rho_r e^{\pm i\phi r} = \xi + i\,\eta$, then we can find all the real roots. Thereafter ξ is given by

$$\alpha_1 + \alpha_2 + \cdots + \alpha_{r-1} + 2\xi + \alpha_{r+2} + \cdots + \alpha_n = -a_1 \tag{9}$$

and η is given by $\rho_r^2 = \xi^2 + n^2$ or $\eta = \sqrt{\left(\rho r^2 - \xi^2\right)}$

EXAMPLE 2.41

Find all roots of the equation $x^3 - 2x^2 - 5x + 6 = 0$ by Graeffe's method, squaring three times

Solution:

Let
$$f(x) = x^3 - 2x^2 - 5x + 6 = 0 \tag{i}$$
$$+ \quad - \quad - \quad +$$

By Descartes rule of signs, there being two changes of sign, (i) has two positive roots.

Also
$$f(-x) = -x^3 - 2x^2 + 5x + 6$$
$$- \quad - \quad + \quad +$$

i.e., one change in sign, there is one negative root.

Rewriting (i) as $x^3 - 5x = 2x^2 - 6$ and squaring,

we get $y(y-5)^2 = (2y-6)^2$ where $y = x^2$

or
$$y(y^2 + 49) = 14y^2 + 36 \qquad ...(ii)$$

Squaring again and putting $y^2 = z$, we obtain $z(z+49)^2 = (14z+36)^2$

or
$$z(z^2 + 1393) = 98z^2 + 1296 \tag{iii}$$

Squaring once again and putting $z^2 = u$,

we get $u(u+1393)^2 = (98u+1296)^2$

or $u^3 - 6818u^2 + 1686433u - 1679616 = 0 \tag{iv}$

If the roots of (iv) are $\gamma_1, \gamma_2, \gamma_3$, then $\gamma_1 = -c_1 = 6818$,

$$\gamma_2 = -\frac{c_2}{c_1} = \frac{1686433}{6818} = 247 - 3501.$$

$$\gamma_3 = -\frac{c3}{c2} = \frac{1679616}{1686433} = 0.996.$$

If $\alpha_1, \alpha_2, \alpha_3$ be the roots of (i), then

$$|\alpha_1| = (\gamma_1)^{1/8} = 3.014443 \approx 3$$

$$|\alpha_2| = (\gamma_2)^{1/8} = 1.991425 \approx 2$$

$$|\alpha_3| = (\gamma_3)^{1/8} = 0.999499 \approx 1$$

The sign of a root is found by substituting the root in $f(x) = 0$. We find $f(3) = 0, f(-2) = 0, f(1) = 0$.

Hence the roots are $3, -2, 1$.

EXAMPLE 2.42

Apply Graeffe's method to find all the roots of the equation $x^4 - 3x + 1 = 0$.

Solution:

We have
$$f(x) = x^4 - 3x + 1 = 0 \qquad\qquad (i)$$
$$+ - +$$

$\therefore$ There being two changes in sign, (i) has two positive real roots and no negative real root.

Thus the remaining two roots are complex.

Rewriting (i) as $x^4 + 1 = 3x$, and squaring, we get $(y^2 + 1)^2 = 9y$ where $y = x^2$.

Squaring again and putting $y^2 = z$, we obtain
$$(z + 1)^4 = 81z \quad \text{or,} \ z^4 + 4z^3 + 6z^2 - 77z + 1 = 0 \qquad (ii)$$

or $\qquad z^4 + 6z^2 + 1 = -z(4z^2 - 77)$

Squaring once again and putting $z^2 = u$, we get $(u^2 + 6u + 1)^2 = u(4u - 77)^2$

or $\qquad\qquad u^4 - 4u^3 + 654u^2 - 5917u + 1 = 0 \qquad\qquad (iii)$

If $\alpha_1, \alpha_2, \alpha_3, \alpha_4$ be the roots of (i), then the roots of (iii) are $\alpha_1^8, \alpha_2^8, \alpha_3^8, \alpha_4^8$. Thus (iii) gives

$$\alpha_1^8 = 4 \qquad\qquad\qquad i.e., \alpha_1 = 1.1892$$

$$\alpha_2^8 = \frac{654}{4} = 163.5 \qquad i.e., \alpha_2 = 1.891$$

$$\alpha_3^8 = \frac{5917}{654} = 9.0474 \qquad i.e., \alpha_3 = 1.3169$$

$$\alpha_4^8 = \frac{1}{5917} = 0.00017 \qquad i.e., \alpha_4 = 0.3379$$

From (ii) and (iii), we observe that the magnitudes of the coefficients c_1 and c_4 have become constant. This indicates that α_1 and α_4 are the real roots whereas α_2 and α_3 are a pair of complex roots.

∴ The real roots $\alpha_1 = 1.1892$ and $\alpha_4 = 0.3379$.

Now let us find the complex roots $\rho_2^{e \pm i\phi 2} = \xi + i\eta$.

From (iii), its magnitude is given by

$$\rho_2^{2(2^3)} = \frac{c_{2+1}}{c_{2-1}} \quad \text{or} \quad \rho_2^{16} = \frac{5917}{4} = 1479.25$$

Where $\rho_2 = 1.5781$.

Also from (i), $\alpha_1 + 2\zeta + \alpha_4 = 0$

This gives $\qquad \xi = -\dfrac{1}{2}(\alpha_1 + \alpha_4) = -0.7636$ and $\eta = \sqrt{(p_2^2 - \xi^2)} = 1.381$

Hence the complex roots are $-0.7636 \pm 1.381\,i$.

Exercises 2.8

1. Find a double root of the equation $x^3 - 5x^2 + 8x - 4 = 0$ which is near 1.8.

2. Find the multiplicity and the multiple root of the equation $x^4 - 11x^3 + 36x^2 - 16x - 64 = 0$ which is near 3.9.

3. Apply the Newton's method to find a pair of complex roots of the equation $x^4 + x^3 + 5x^2 + 4x + 4 = 0$ starting with $x_0 = i$.

4. Apply Lin-Bairstow method to find a quadratic factor of the equation $x^4 + 5x^3 + 3x^2 - 5x - 9$ close to $x^2 + 3x - 5$.

5. Find the roots of the equation $x4 + 9x^3 + 36x^2 + 51x + 27 = 0$ to three decimal places using the Bairstow iterative method.

6. Find the quadratic factors of the equation $x^4 - 8x^3 + 39x^2 - 62x + 50 = 0$ by using the Lin-Bairstow method (up to the third iteration) starting with $p_0 = 0$, $q_0 = 0$.

7. Solve $x^3 - 8x^2 + 17x - 10 = 0$ by Graeffe's method.

8. Apply Graeffe's method to find all the roots of the equation $x^3 - 6x^2 + 11x - 6 = 0$.

9. Solve the equation $x^3 - 5x^2 - 17x + 20 = 0$ by Graeffe's method, squaring three times.

10. Find all the roots of the equation $x^3 - 4x^2 + 5x - 2 = 0$ by Graeffe's method, squaring thrice.

11. Determine all roots of the equation $x^3 - 9x^2 + 18x - 6 = 0$ by Graeffe's method.

2.20 Comparison of Iterative Methods

1. Convergence in the case of *the bisection method* is slow but steady. It is, however, the simplest method and it never fails.

2. *The method of false position* is slow and it is first order convergent. Convergence however, is guaranteed. Most often, it is found superior to the bisection method.

3. *The secant method* is not guaranteed to converge. But its order of convergence being 1.62, it converges faster than the method of false position. This method is considered most economical giving reasonably rapid convergence at a low cost.

4. Of all the above methods, *Newton-Raphson method* has the fastest rate of convergence. The method is quite sensitive to the starting value. Also it may diverge if $f'(x)$ is near zero during the iterative cycle.

5. For locating the complex roots, *Newton's method* can be used. *Muller's method* is also effective for finding complex roots.

6. If all the roots of the given equation are required then the *Lin-Bairstow method* is recommended. After a quadratic factor has been found, then the Lin-Bairstow method must be applied on the reduced polynomial. If the location of some roots is known, first find these roots to a desired accuracy and then apply the Lin-Bairstow method on the reduced polynomial.

7. If the roots of the given polynomial are real and distinct then *Graeffe's root squaring method* is quite useful.

2.21   Objective Type of Questions

Exercises 2.9

Select the correct answer or fill up the blanks in the following questions:

1. The order of convergence in the Newton-Raphson method is
   (*a*)   (*b*) 3   (*c*) 0   (*d*) none.

2. The Newton-Raphson algorithm for finding the cube root of N is...........

3. The bisection method for finding the roots of an equation $f(x) = 0$ is.........

4. In theRegula-falsi method, the first approximation is given by............

5. If $f(x) = 0$ is an algebraic equation, the Newton-Raphson method is given by $xn_{+1} = xn - f(xn)/?$
   $(a) f(x_{n-1})$ $(b) f'(x_{n-1})$ $(c) f'(x_n)$ $(d) f''(x_n)$.

6. In the Regula-falsi method of finding the real root of an equation, the curve AB is replaced by......

7. Newton's iterative formula to find the value of $\sqrt{N}$ is..............

8. A root of $x^3 - x + 4 = 0$ obtained using the bisection method correct to two places, is......... .

9. Newton-Raphson formula converges when............ .

10. In the case of bisection method, the convergence is
   (*a*) linear   (*b*) quadratic   (*c*) very slow.

11. Out of the method of false position and the Newton-Raphson method, the rate of convergence is faster for............ .

12. Using Newton's method, the root of $x^3 = 5x - 3$ between 0 and 1 correct to two decimal places, is......... .

13. The Newton-Raphson method fails when
   $(a) f'(x)$ is negative $(b) f'(x)$ is too large
   $(c) f'(x)$ is zero (d) Never fails.

14. The condition for the convergence of the iteration method for solving $x = \phi(x)$ is......

15. While finding a root of an equation by the Regula-falsi method, the number of iterations can be reduced......... .

16. Newton's method is useful when the graph of the function while crossing the x-axis is nearly vertical. (True or False)

17. The difference between a Transcendental equation and polynomial equation is......... .

18. The interval in which a real root of the equation $x^3 - 2x - 5 = 0$ lies is....... .

19. The iterative formula for finding the reciprocal of N is $x_{n+1} =$......... .

20. While finding the root of an equation by the method of false position, the number of iterations can be reduced...... .

3

SOLUTION OF SIMULTANEOUS ALGEBRAIC EQUATIONS

Chapter Objectives

- Introduction to determinants
- Introduction to matrices
- Solution of linear simultaneous equations
- Direct methods of solution: Cramer's rule, Matrix inversion method, Gauss elimination method, Gauss-Jordan method, Factorization method
- Iterative methods of solution: Jacobi's method, Gauss-Seidal method, Relaxation method
- Ill-conditioned equations
- Comparison of various methods
- Solution of non-linear simultaneous equations—Newton-Raphson method
- Objective type of questions

3.1   Introduction t to Determinants

1. Definition. The expression $\begin{vmatrix} a_1 & b_1 \\ a_2 & b_2 \end{vmatrix}$ is called a *determinant of the second order* and stands for '$a_1b_2 - a_2b_1$'. It contains four numbers a_1, b_1, a_2, b_2 (called *elements*) which are arranged along two horizontal lines (called *rows*) and two vertical lines (called *columns*).

Similarly
$$\begin{vmatrix} a_1 & b_1 & c_1 \\ a_2 & b_2 & c_2 \\ a_3 & b_3 & c_3 \end{vmatrix} \qquad (i)$$

is called a *determinant of the third order.* It consists of nine elements which are arranged in three rows and three columns.

In general, a *determinant of the nth order is* of the form

$$\begin{vmatrix} a_{11} & a_{12} & a_{13}........a_{1n} \\ a_{21} & a_{22} & a_{23}........a_{2n} \\ .... \\ .... \\ a_{n1} & a_{n2} & a_{n3}........a_{nn} \end{vmatrix}$$

which is a block of n^2 *elements* in the form of a square along n *rows* and n *columns.* The diagonal through the left-hand top corner which contains the elements $a_{11}, a_{22}, a_{33}, ..., a_{nn}$ is called the *leading diagonal.*

Expansion of a determinant. *The* **cofactor** *of an element in a determinant is the determinant obtained by deleting the row and the column which intersect at that element, with the proper sign.* The sign of an element in the ith row and jth column is $(-1)^{i+j}$. The cofactor of an element is usually denoted by the corresponding capital letter.

For instance, the cofactor of b_3 in (i) is $B_3 = (-1)^{3+2} \begin{vmatrix} a_1 & c_1 \\ a_2 & c_2 \end{vmatrix}$

A determinant can be expanded in terms of any row or column as follows:

Multiply each element of the row (or column) in terms of which we intend expanding the determinant, by its cofactor and then add up all these products.

∴ Expanding (i) by $R_1(i.e.$ 1st row),

$$\Delta = a_1 A_1 + b_1 B_1 + c_1 C_1$$
$$= a_1 \begin{vmatrix} b_2 & c_2 \\ b_3 & c_3 \end{vmatrix} - b_1 \begin{vmatrix} a_2 & c_2 \\ a_3 & c_3 \end{vmatrix} + c1 \begin{vmatrix} a_2 & b_2 \\ b_3 & b_3 \end{vmatrix}$$
$$= a_1 (b_2 c_3 - b_3 c_2) - b_1 (a_2 c_3 - a_3 c_2) + c_1 (a_2 b_3 - a_3 b_2)$$

Similarly expanding by C_2 (*i.e.* 2nd column),

$$\Delta = b_1 B_1 + b_2 B_2 + b_3 B_3$$

$$= -b_1 \begin{vmatrix} a_2 & c_2 \\ a_3 & c_3 \end{vmatrix} - b_2 \begin{vmatrix} a_1 & c_1 \\ a_3 & c_3 \end{vmatrix} + b3 \begin{vmatrix} a_1 & c_1 \\ a_2 & c_2 \end{vmatrix}$$

$$= b_1(a_2 c_3 - a_3 c_2) - b_2(a_1 c_3 - a_3 c_1) - b_3(a_1 c_2 - a_2 c_1)$$

EXAMPLE 3.1

Find the value of $\Delta = \begin{vmatrix} 0 & 1 & 2 & 3 \\ 1 & 0 & 3 & 0 \\ 2 & 3 & 0 & 1 \\ 3 & 0 & 1 & 2 \end{vmatrix}$

Solution:

Since there are two zeros in the second row, therefore, expanding by R_2, we get

$$\Delta = - \begin{vmatrix} 1 & 2 & 3 \\ 3 & 0 & 1 \\ 0 & 1 & 2 \end{vmatrix} + 0 - 3 \begin{vmatrix} 0 & 1 & 3 \\ 2 & 3 & 1 \\ 3 & 0 & 2 \end{vmatrix} + 0$$

$$\text{(Expand By } C_1) \qquad \text{(Expand by } R_1)$$

$$= -[1(0 \times 2 - 1 \times 1) - 3(2 \times 2 - 1 \times 3) + 0]$$

$$- 3[0 - (2 \times 2 - 3 \times 1) + 3(2 \times 0 - 3 \times 3)]$$

$$= -(-1 - 3) - 3(-1 - 27) = 4 + 84 = 88.$$

Basic properties. The following properties enable us to simplify and evaluate a given determinant without expanding it:

    I. A determinant remains unaltered by changing its rows into columns and columns into rows.

   II. If two parallel lines of a determinant are interchanged, the determinant retains its numerical value but changes in sign.

  III. A determinant vanishes if two of its parallel lines are identical.

  IV. If each element of a line is multiplied by the same factor, the whole determinant is multiplied by that factor.

   V. If each element of a line consists of m terms, the determinant can be expressed as the sum of m determinants.

  VI. If to each element of a line there can be added equi-multiples of the corresponding elements of one or more parallel lines, the determinant remains unaltered.

For instance $\begin{vmatrix} a_1 + pb_1 - qc_1 & b_1 & c_1 \\ a_2 + pb_2 - qc_2 & b_2 & c_2 \\ a3 + pb_3 - qc_3 & b3 & c_3 \end{vmatrix}$

$$= \begin{vmatrix} a_1 & b_1 & c_1 \\ a_2 & b_2 & c_2 \\ a_3 & b_3 & c_3 \end{vmatrix} + p \begin{vmatrix} b_1 & b_1 & c_1 \\ b_2 & b_2 & c_2 \\ b_3 & b_3 & c_3 \end{vmatrix} - q \begin{vmatrix} c_1 & b_1 & c_1 \\ c_2 & b_2 & c_2 \\ c_3 & b_3 & c_3 \end{vmatrix}$$

$$= \Delta + 0 + 0 = \Delta \qquad\qquad\qquad\qquad [\text{From } (iv)]$$

Rule for multiplication of determinants:

$$\begin{vmatrix} a_1 & b_1 & c_1 \\ a_2 & b_2 & c_2 \\ a_3 & b_3 & c_3 \end{vmatrix} \times \begin{vmatrix} l_1 & m_1 & n_1 \\ l_2 & m_2 & n_2 \\ l_3 & m_3 & n_3 \end{vmatrix}$$

$$= \begin{vmatrix} a_1l_1 + b_1m_1 + c_1n_1 & a_2l_1 + b_2m_1 + c_2n_1 & a_3l_1 + b_3m_1 + c_3n_1 \\ a_1l_2 + b_1m_2 + c_1n_2 & a_2l_2 + b_2m_2 + c_2n_2 & a_3l_2 + b_3m_2 + c_3n_2 \\ a_1l_3 + b_1m_3 + c_1n_3 & a_2l_3 + b_2m_3 + c_2n_3 & a_3l_3 + b_3m_3 + c_3n_3 \end{vmatrix}$$

i.e., the product of two determinants of the same order is itself a determinant of that order.

EXAMPLE 3.2.

If $\begin{vmatrix} a & a^2 & a^3 - 1 \\ b & b^2 & b^3 - 1 \\ c & c^2 & c^3 - 1 \end{vmatrix} = 0$ in which a, b, c are different, show that $abc = 1$.

Solution:

As each term of C_3 in the given determinant consists of two terms, we express it as a sum of two determinants.

$$\begin{vmatrix} a & a^2 & a^3 - 1 \\ b & b^2 & b^3 - 1 \\ c & c^2 & c^3 - 1 \end{vmatrix} = \begin{vmatrix} a & a^2 & a^3 \\ b & b^2 & b^3 \\ c & c^2 & c^3 \end{vmatrix} + \begin{vmatrix} a & a^2 & -1 \\ b & b^2 & -1 \\ c & c^2 & -1 \end{vmatrix}$$

$$= abc \begin{vmatrix} a & a^2 & a2 \\ b & b^2 & b^2 \\ c & c^2 & c^2 \end{vmatrix} - \begin{vmatrix} a & a^2 & 1 \\ b & b^2 & 1 \\ c & c^2 & 1 \end{vmatrix}$$

[Taking common a, b, c from R_1, R_2, R_3 respectively of the first determinant and -1 from C_3 of the second determinant]

$$= abc \begin{vmatrix} 1 & a & a^2 \\ 1 & b & b^2 \\ 1 & c & c^2 \end{vmatrix} - \begin{vmatrix} 1 & a & a^2 \\ 1 & b & b^2 \\ 1 & c & c^2 \end{vmatrix}$$

[Passing C_3 over C_2 and C_1 in the second determinant]

$$\therefore \qquad \begin{vmatrix} 1 & a & a^2 \\ 1 & b & b^2 \\ 1 & c & c^2 \end{vmatrix} (abc - 1) = 0$$

Hence $abc = 1$, since $\begin{vmatrix} 1 & a & a^2 \\ 1 & b & b^2 \\ 1 & c & c^2 \end{vmatrix} \neq 0$ as a, b, c are all different.

EXAMPLE 3.3

Solve the equation $\begin{vmatrix} x+2 & 2x+3 & 3x+4 \\ 2x+3 & 3x+4 & 4x+5 \\ 3x+5 & 5x+8 & 10x+17 \end{vmatrix} = 0$

Solution:

Operating $R_3 - (R_1 + R_2)$, we get

$$\begin{vmatrix} x+2 & 2x+3 & 3x+4 \\ 2x+3 & 3x+4 & 4x+5 \\ 0 & 1 & 3x+8 \end{vmatrix} = 0 \quad \text{(Operate } R_2 - R_1 \text{ and } R_1 + R_3)$$

or $\begin{vmatrix} x+2 & 2x+4 & 6x+12 \\ x+1 & x+1 & x+1 \\ 0 & 1 & 3x+8 \end{vmatrix} = 0$

or $(x+1)(x+2) \begin{vmatrix} 1 & 2 & 6 \\ 1 & 1 & 1 \\ 0 & 1 & 3x+8 \end{vmatrix} = 0$

To bring one more zero in C_1, operate $R_1 - R_2$.

$$\therefore \qquad (x+1)(x+2)\begin{vmatrix} 0 & 1 & 5 \\ 1 & 1 & 1 \\ 0 & 1 & 3x+8 \end{vmatrix} = 0$$

Now expand by C_1.

$$\therefore \qquad -(x+1)(x+2)(3x+8-5) = 0 \text{ or } -3(x+1)(x+2)(x+1) = 0.$$

Thus $x = -1, -1, -2$.

EXAMPLE 3.4

Prove that $\begin{vmatrix} 1+a & 1 & 1 & 1 \\ 1 & 1+b & 1 & 1 \\ 1 & 1 & 1+c & 1 \\ 1 & 1 & 1 & 1+d \end{vmatrix} = abc\left(1 + \dfrac{1}{a} + \dfrac{1}{b} + \dfrac{1}{c} + \dfrac{1}{d}\right)$

Solution:

Let Δ be the given determinant.

Taking a, b, c, d common from R_1, R_2, R_3, R_4 respectively, we get

$$\Delta = abcd \begin{vmatrix} a^{-1}+1 & a^{-1} & a^{-1} & a^{-1} \\ b^{-1} & b^{-1}+1 & b^{-1} & b^{-1} \\ c^{-1} & c^{-1} & c^{-1}+1 & c^{-1} \\ d^{-1} & d^{-1} & d^{-1} & d^{-1}+1 \end{vmatrix}$$

[Operate $R_1 + (R_2 + R_3 + R_4)$ and take out the common factor from R_1]

$$= abcd(1 + a^{-1} + b^{-1} + c^{-1} + d^{-1}) \begin{vmatrix} 1 & 1 & 1 & 1 \\ b^{-1} & b^{-1}+1 & b^{-1} & b^{-1} \\ c^{-1} & c^{-1} & c^{-1}+1 & c^{-1} \\ d^{-1} & d^{-1} & d^{-1} & d^{-1}+1 \end{vmatrix}$$

[Operate $C_2 - C_1, C_3 - C_1, C_4 - C_1$]

$$= abcd\left(1 + \frac{1}{a} + \frac{1}{b} + \frac{1}{c} + \frac{1}{d}\right) \begin{vmatrix} 1 & 0 & 0 & 0 \\ b^{-1} & 1 & 0 & 0 \\ c^{-1} & 0 & 1 & 0 \\ d^{-1} & 0 & 0 & 1 \end{vmatrix}$$

$$= abcd\left(1 + \frac{1}{a} + \frac{1}{b} + \frac{1}{c} + \frac{1}{d}\right)$$

EXAMPLE 3.5

Evaluate $\begin{vmatrix} a^2 + \lambda^2 & ab + c\lambda & ca - b\lambda \\ ab - c\lambda & b^2 + \lambda^2 & bc + a\lambda \\ ca + b\lambda & bc - a\lambda & c^2 + \lambda^2 \end{vmatrix} \times \begin{vmatrix} \lambda & c & -b \\ -c & \lambda & a \\ b & -a & \lambda \end{vmatrix}$

Solution:

By the rule of multiplication of determinants, the resulting determinant

$$\Delta = \begin{vmatrix} d_{11} & d_{12} & d_{13} \\ d_{21} & d_{22} & d_{23} \\ d_{31} & d_{32} & d_{33} \end{vmatrix}$$

where $d_{11} = (a^2 + \lambda^2)\lambda + (ab + c\lambda)c + (ca - b\lambda)(-b) = \lambda(a^2 + b^2 + c^2 + \lambda^2)$

$d_{12} = (a^2 + \lambda^2)(-c) + (ab + c\lambda)\lambda + (ca - b\lambda)a = 0, d_{13} = 0,$

$d_{21} = 0, d_{22} = \lambda(a^2 + b^2 + c^2 + \lambda^2), d_{23} = 0.$

$d_{31} = 0, d_{32} = 0, d_{33} = \lambda(a^2 + b^2 + c^2 + \lambda^2).$

Hence,

$$\Delta = \begin{vmatrix} \lambda(a^2 + b^2 + c^2 + \lambda^2) & 0 & 0 \\ 0 & \lambda(a^2 + b^2 + c^2 + \lambda^2) & 0 \\ 0 & 0 & \lambda(a^2 + b^2 + c^2 + \lambda^2) \end{vmatrix}$$

$$= \lambda^3 \left(a^2 + b^2 + c^2 + \lambda^2\right)^3.$$

Exercises 3.1

1. If $\begin{vmatrix} x & x^2 & 1+x^3 \\ y & y^2 & 1+y^3 \\ z & z^2 & 1+z^3 \end{vmatrix} = 0$, then prove, without expansion, that $xyz = -1$ where

x, y, z are unequal.

Prove the following results: (2 and 3)

2. $\begin{vmatrix} a+b & a & b \\ a & a+c & c \\ b & c & b+c \end{vmatrix} = 4abc$

3. $\begin{vmatrix} 4 & 5 & 6 & x \\ 5 & 6 & 7 & y \\ 6 & 7 & 8 & z \\ x & y & z & 0 \end{vmatrix}$ .is a perfect square.

4. $\begin{vmatrix} 1 & a & a^2 & a^3+bcd \\ 1 & b & b^2 & b^3+cda \\ 1 & c & c^2 & c^3+dab \\ 1 & c & d^3 & d^3+abc \end{vmatrix}$ vanishes.

5. Solve the equation $\begin{vmatrix} x+1 & 2x+1 & 3x+1 \\ 2x & 4x+3 & 6x+3 \\ 4x+1 & 6x+4 & 8x+4 \end{vmatrix} = 0$.

6. Find the value of the determinant (M) if $M = 3A^2 + AB + B^2$

where $A = \begin{vmatrix} 2 & 1 & 1 \\ 1 & 2 & 1 \\ 0 & -1 & 0 \end{vmatrix}, B = \begin{vmatrix} 5 & 0 & -1 \\ -1 & 0 & 1 \\ 0 & 2 & 3 \end{vmatrix}$

without evaluating A and B independently.

3.2 Introduction to Matrices

Definition. *A system of mn numbers arranged in a rectangular array of m rows and n columns is called an m × n* **matrix.** Such a matrix is denoted by

$$A = \begin{bmatrix} a_{11} & a_{12} \cdots a_{1n} \\ a_{21} & a_{22} \cdots a_{2n} \\ \cdots & \cdots \\ a_{m1} & a_{m2} \cdots a_{mn} \end{bmatrix} = \begin{bmatrix} a_{ij} \end{bmatrix}$$

Special matrices

1. *Row and column matrices.* A matrix having a single row is called a *row matrix* while a matrix having a single column is called a *column matrix.*

2. *Square matrix.* A matrix having n rows and n columns is called a *square matrix.* A square matrix is said to be *singular* if its determinant is zero otherwise it is called *non-singular.*

The elements a_{ii} in a square matrix form the *leading diagonal* and their sum Σa_{ii} is called the *trace* of the matrix.

3. *Unit matrix.* A diagonal matrix of order n which has unity for all its diagonal elements is called a *unit matrix of order n* and is denoted by I_n.

4. *Null matrix.* If all the elements of a matrix are zero, it is called a *null matrix.*

5. *Symmetric and skew-symmetric matrices.* A square matrix $[a_{ij}]$ is said to be *symmetric* when $a_{ij} = a_{ji}$ for all i and j.

If $a_{ij} = -a_{ji}$ for all i and j so that all the leading diagonal elements are zero, then the matrix is called *skew-symmetric.* Examples of symmetric and skew-symmetric matrices are respectively

$$\begin{bmatrix} a & h & g \\ h & b & f \\ g & f & c \end{bmatrix} \text{and} \begin{bmatrix} 0 & h & -g \\ -h & 0 & f \\ g & -f & 0 \end{bmatrix}$$

6. *Triangular matrix.* A square matrix all of whose elements below the leading diagonal are zero is called an *upper triangular matrix.* A square matrix all of whose elements above the leading diagonal are zero is called a *lower triangular matrix.*

Operations on matrices

1. *Equality of matrices.* Two matrices A and B are said to be equal if and only if (i) they are of the same order,

and (ii) each element of A is equal to the corresponding element of B.

2. *Addition and subtraction of matrices.* If A and B are two matrices of the same order, then their sum $A + B$ is defined as the matrix each element of which is the sum of the corresponding elements of A and B.

Similarly $A - B$ is defined as the matrix whose elements are obtained by subtracting the elements of B from the corresponding elements of A.

3. *Multiplication of a matrix by a scalar.* The product of a matrix A by a scalar k is a matrix whose each element is k times the corresponding elements of A.

4. *Multiplication of matrices.* Two matrices can be multiplied only when the number of columns in the first is equal to the number of rows in the second. Such matrices are said to be *conformable.* Thus if A and B be $(m \times n)$ and $(n \times p)$ matrices, then their product $C = AB$ is defined and will be a $(m \times p)$ matrix. *The elements of C are obtained by the following rule: Element c_{ij} of C = sum of the products of corresponding elements of the ith row of A with those of the jth column of B.*

For example, if $A = \begin{bmatrix} a_{11} & a_{12} & a_{13} \\ a_{21} & a_{22} & a_{23} \\ a_{31} & a_{32} & a_{33} \\ a_{41} & a_{42} & a_{43} \end{bmatrix}$ and $B = \begin{bmatrix} b_{11} & b_{12} \\ b_{21} & b_{22} \\ b_{31} & b_{32} \end{bmatrix}$

then $AB = \begin{bmatrix} a_{11}b_{11} + a_{12}b_{21} + a_{13}b_{31} & a_{11}b_{12} + a_{12}b_{22} + a_{13}b_{32} \\ a_{21}b_{11} + a_{22}b_{21} + a_{23}b_{31} & a_{21}b_{12} + a_{22}b_{22} + a_{23}b_{32} \\ a_{31}b_{11} + a_{32}b_{21} + a_{33}b_{31} & a_{31}b_{12} + a_{32}b_{22} + a_{33}b_{32} \\ a_{41}b_{11} + a_{42}b_{21} + a_{43}b_{32} & a_{41}b_{12} + a_{42}b_{22} + a_{43}b_{32} \end{bmatrix}$

NOTE

Obs. 1. *In general AB ≠ BA even if both exist.*
2. *If A be a square matrix, then the product AA is defined as A^2. Similarly $A.A^2 = A^3$ etc.*

EXAMPLE 3.6

Evaluate $3A - 4B$, where $A = \begin{bmatrix} 3 & -4 & 6 \\ 5 & 1 & 7 \end{bmatrix}$ and $4B = \begin{bmatrix} 4 & 0 & 4 \\ 8 & 0 & 12 \end{bmatrix}$

Solution:

We have $3A = \begin{bmatrix} 9 & -12 & 18 \\ 15 & 3 & 21 \end{bmatrix}$ and $4B = \begin{bmatrix} 4 & 0 & 4 \\ 8 & 0 & 12 \end{bmatrix}$

EXAMPLE 3.7

If $A = \begin{bmatrix} 0 & 1 & 2 \\ 1 & 2 & 3 \\ 2 & 3 & 4 \end{bmatrix}$ and $B = \begin{bmatrix} 1 & -2 \\ -1 & 0 \\ 2 & -1 \end{bmatrix}$, form the product AB. Is BA Defined?

Solution:

Since the number of columns of A = the number of rows of B (each being = 3). The product AB is defined and

$$2 = \begin{bmatrix} 0.1 + 1.-1 + 2.2 & 0.-2 + 1.0 + 2.-1 \\ 1.1 + 2 - 1 + 3.2 & 1.-2 - 2.0 + 3 - 1 \\ 2.1 + 3.-1 + 4.2 & 2.-2 + 3.0 + 4.-1 \end{bmatrix} = \begin{bmatrix} 3 & -2 \\ 5 & -5 \\ 7 & -8 \end{bmatrix}$$

Again since the number of columns of $B \neq$ the number of rows of A.

$\therefore$ The product BA is not defined.

EXAMPLE 3.8.

If $A = \begin{bmatrix} 3 & 2 & 2 \\ 1 & 3 & 1 \\ 5 & 4 & 4 \end{bmatrix}$, find the matrix B, such that $AB = \begin{bmatrix} 3 & 4 & 2 \\ 1 & 6 & 1 \\ 5 & 6 & 4 \end{bmatrix}$

Solution:

Let $AB = \begin{bmatrix} 3 & 2 & 2 \\ 1 & 3 & 1 \\ 5 & 3 & 4 \end{bmatrix} \begin{bmatrix} l & m & n \\ p & q & r \\ u & v & w \end{bmatrix}$

$$= \begin{bmatrix} 3l + 2p + 2u & 3m + 2q + 2v & 3n + 2r + 2w \\ l + 3p + u & m + 3q + v & n + 3r + w \\ 5l + 33p + 4u & 5m + 3q + 4v & 5n + 3r + 4w \end{bmatrix}$$

$$= \begin{bmatrix} 3 & 4 & 2 \\ 1 & 6 & 1 \\ 5 & 6 & 4 \end{bmatrix} \qquad \text{(given)}$$

Equating corresponding elements, we get

$3l + 2p + 2u = 3,$	$l + 3p + u = 1,$	$5l + 3p + 4u = 5$ (*i*)
$3m + 2q + 2v = 4,$	$m + 3q + v = 6,$	$5m + 3q + 4v = 6$ (*ii*)
$3n + 2r + 2w = 2,$	$n + 3r + w = 1,$	$5n + 3r + 4w = 4$ (*iii*)

Solving the equations (*i*), we get $l = 1, p = 0, u = 0$

Similarly equations (*ii*) give $m = 0, q = 2, v = 0$

and equations (*iii*) give $n = 0, r = 0, w = 1$

$$\text{Thus } B = \begin{bmatrix} 1 & 0 & 0 \\ 0 & 2 & 0 \\ 0 & 0 & 1 \end{bmatrix}$$

Related matrices

I. **Transpose of a matrix.** *The matrix obtained from a given matrix A, by interchanging rows and columns, is called the* **transpose** *of A and is denoted by A¢.*

NOTE

Obs. 1. *For a symmetric matrix, A¢ = A and for a skew-symmetric matrix, A¢ = – A.*

2. *The transpose of the product of two matrices is the product of their transposes taken in the reverse order*

i.e., $\qquad (AB)' = B'A'.$

3. *Any square matrix A can be written as*

$$A = \frac{1}{2}(A + A') + \frac{1}{2}(A - A') = B + c \text{ (say)}$$

such that $\qquad B' = \frac{1}{2}(A + A')' = \frac{1}{2}(A' + A) = B$

i.e., B is a symmetric matrix

and $\qquad c' = \frac{1}{2}(A - A')' = \frac{1}{2}(A' - A) = -c$

i.e.,. C is a skew-symmetric matrix.

Thus every square matrix can be expressed as the sum of a symmetric and a skew-symmetric matrix.

II. **Adjoint of a square matrix A** *is the transposed matrix of cofactors of A and is written as adj A.* Thus the adjoint of the matrix

$$\begin{bmatrix} a_1 & b_1 & c_1 \\ a_2 & b_2 & c_2 \\ a_3 & b_3 & c_3 \end{bmatrix} \text{ is } \begin{bmatrix} A_1 & A_2 & A_3 \\ B_1 & B_2 & B_3 \\ C_1 & C_2 & C_3 \end{bmatrix}$$

III. **Inverse of a matrix.** *If A is a non-singular square matrix of order n, then a square matrix B of the same order such that AB = BA = I, is then called the* **inverse** *of A, I being a unit matrix.*

The inverse of A is written as A^{-1} so that $A A^{-1} = A^{-1} A = I$

Also $A^{-1} = \dfrac{AdjA}{|A|}$

NOTE **Obs. 1.** *Inverse of a matrix, when it exists, is unique.*

2. $(A^{-1})^{-1} = A.$

3. $(AB)^{-1} = B^{-1} A^{-1}.$

EXAMPLE 3.9

Find the inverse of $A = \begin{bmatrix} 1 & 1 & 3 \\ 1 & 3 & -3 \\ -2 & -4 & -4 \end{bmatrix}$

Solution:

Here $|A| = \begin{bmatrix} 1 & 1 & 3 \\ 1 & 3 & -3 \\ -2 & -4 & -4 \end{bmatrix} = \begin{bmatrix} a_1 & b_1 & c_1 \\ a_2 & b_2 & c_2 \\ a_3 & b_3 & c_3 \end{bmatrix}$ (say)

and $\quad$ adj $A = \begin{bmatrix} A_1 & A_2 & A_3 \\ B_1 & B_2 & B_3 \\ C_1 & C_2 & C_3 \end{bmatrix} = \begin{bmatrix} -24 & -8 & -12 \\ 10 & 2 & 6 \\ 2 & 2 & 2 \end{bmatrix}$

Hence $\quad A^{-1} = \dfrac{adj\ A}{|A|} = \dfrac{1}{8} \begin{bmatrix} -24 & -8 & -12 \\ 10 & 2 & 6 \\ 2 & 2 & 2 \end{bmatrix}$

Note: For other methods of finding the inverse of a matrix refer to chapter 4.

Rank of a matrix. If we select any r rows and r columns from any matrix A, deleting all other rows and columns, then the determinant formed by these $r \times r$ elements is called the *minor of A of order r*. Clearly there will be a number of different minors of the same order, got by deleting different rows and columns from the same matrix.

Def. *A matrix is said to be of rank r when*

I. *it has at least one non-zero minor of order r, and*

II. *every minor of order higher than r vanishes.*

Elementary transformations of a matrix. The following operations, three of which refer to rows and three to columns are known as *elementary transformations*:

    I. *The interchange of any two rows (columns).*

    II. *The multiplication of any row (column) by a non-zero number.*

    III. *The addition of a constant multiple of the elements of any row (column) to the corresponding elements of any other row (column).*

Notation. The *elementary row transformations* will be denoted by the following symbols:

    (*i*) R_{ij} for the interchange of the *i*th and *j*th rows.

    (*ii*) kR_i for multiplication of the *i*th row by k.

    (*iii*) $R_i + pR_j$ for addition to the *i*th row, p times the *i*th row.

The corresponding column transformation will be denoted by writing C in place of R. These transformations, being precisely those performed on the rows (columns) of a determinant, need no explanation.

NOTE

Obs. 1. *Elementary transformations do not change either the order or rank of a matrix. While the value of the minors may get changed by the transformations I and II, their zero or non-zero character remains unaffected.*

Equivalent matrix. *Two matrices A and B are said to be equivalent if one can be obtained from the other by a sequence of elementary transformations.* Two equivalent matrices have the same order and the same rank. The symbol ~ is used for equivalence.

Elementary matrix. *An elementary matrix is that, which is obtained from a unit matrix, by subjecting it to any of the elementary transformations.*

Normal form of a matrix. *Every non-zero matrix A of rank r, can be reduced by a sequence of elementary transformations, to the form $\begin{bmatrix} I_r & 0 \\ 0 & 0 \end{bmatrix}$ which is called the* **normal form** *of A.*

EXAMPLE 3.10

Determine the rank of the following matrices:

(i) $\begin{bmatrix} 1 & 2 & 3 \\ 1 & 4 & 2 \\ 2 & 6 & 5 \end{bmatrix}$ (ii) $\begin{bmatrix} 0 & 1 & -3 & -1 \\ 0 & 0 & 1 & 1 \\ 3 & 1 & 0 & 2 \\ 1 & 1 & 2 & 0 \end{bmatrix}$

Solution:

(i) Operate $R_2 - R_1$ and $R_3 - 2R_1$ so that the given matrix

$$\sim \begin{bmatrix} 1 & 2 & 3 \\ 0 & 2 & -1 \\ 0 & 2 & -1 \end{bmatrix} = A \text{ (say)}$$

Obviously, the third order minor of A vanishes. Also its second order minors formed by its second and third rows are all zero. But another second order minor is

$$\begin{bmatrix} 1 & 3 \\ 0 & -1 \end{bmatrix} = -1 \neq 0.$$

Hence $R(A)$, the rank of the given matrix, is 2.

(ii) Given matrix

$$\sim \begin{bmatrix} 0 & 1 & -3 & -1 \\ 1 & 0 & 0 & 0 \\ 3 & 1 & -3 & -1 \\ 1 & 1 & -3 & -1 \end{bmatrix} \qquad \sim \begin{bmatrix} 0 & 1 & -3 & -1 \\ 1 & 0 & 0 & 0 \\ 3 & 0 & 0 & 0 \\ 1 & 0 & 0 & 0 \end{bmatrix}$$

[Operating $C_3 - C_1, C_4 - C_1$]   [Operating $R_3 - R_1, R_4 - R_1$]

$$\sim \begin{bmatrix} 0 & 1 & -3 & -1 \\ 1 & 0 & 0 & 0 \\ 0 & 0 & 0 & -1 \\ 0 & 0 & 0 & -1 \end{bmatrix} \qquad \sim \begin{bmatrix} 0 & 1 & 0 & 0 \\ 1 & 0 & 0 & 0 \\ 0 & 0 & 0 & 0 \\ 0 & 0 & 0 & 0 \end{bmatrix} = A \text{ (say)}$$

[Operating $R_3 - 3R_2, R_4 - R_2$]   [Operating $C_3 + 3C_2, C_4 + C_2$]

Obviously, the fourth order minor of A is zero. Also every third order minor of A is zero.

But, of all the second order minors, only $\begin{bmatrix} 0 & 1 \\ 1 & 0 \end{bmatrix} = -1 \neq 0$.

Hence $R(A)$, the rank of the given matrix, is 2.

Consistency of a system of linear equations. Consider the system of m linear equations in n unknowns

$$\left. \begin{array}{l} a_{11}x_1 + a_{12}x_2 + \ldots + a_{1n}x_n = k_1 \\ a_{21}x_1 + a_{22}x_2 + \ldots + a_{2n}x_n = k_2 \\ \ldots\ldots\ldots\ldots\ldots\ldots\ldots\ldots\ldots\ldots\ldots\ldots\ldots\ldots\ldots \\ am_1x_1 + a_{m2}x_2 + \ldots + a_{mn}x_n = km \end{array} \right\}$$

To determine whether these equations are consistent or not, we find the ranks of the matrices

$$A = \begin{bmatrix} a_{11} & a_{12}\ldots a_{1n} \\ a_{21} & a_{22}\ldots a_{2n} \\ \ldots & \ldots \\ a_{m1} & a_{m2}\ldots a_{m3} \end{bmatrix} \text{ and } K = \begin{bmatrix} a_{11} & a_{12}\ldots a_{1n} & k_1 \\ a_{21} & a_{22}\ldots a_{2n} & k_2 \\ \ldots & \ldots & \ldots \\ a_{m1} & a_{m2}\ldots a_{mn} & k_m \end{bmatrix}$$

A is the *coefficient matrix* and K is called the *augmented matrix*.

If $R(A) \neq R(K)$, the equations (i) are inconsistent, i.e., have no solution.

If $R(A) = R(K) = n$, the equations (i) are consistent and have a unique solution.

If $R(A) = R(K) < n$, the equations are consistent but have an infinite number of solutions.

System of linear homogeneous equations. Consider the homogeneous linear equations

$$\left. \begin{array}{l} a_{11}x_1 + a_{12}x_2 + \ldots + a_{1n}x_n = 0 \\ a_{21}x_1 + a_{22}x_2 + \ldots + a_{2n}x_n = 0 \\ \ldots\ldots\ldots\ldots\ldots\ldots\ldots\ldots\ldots\ldots\ldots\ldots\ldots\ldots\ldots \\ am_1x_1 + a_{m2}x_2 + \ldots + a_{mn}x_n = 0 \end{array} \right\}$$

Find the rank r of the coefficient matrix A by reducing it to the triangular form by elementary row operations.

I. If $r = n$, the equations (i) have only a trivial solution $x_1 = x_2 = \ldots = x_n = 0$.

If $r < n$, the equations have $(n - r)$ independent solutions. (r cannot be $> n$) The number of linearly independent solutions of (i) is $(n - r)$ means, if arbitrary values are assigned to $(n - r)$ of the variables, the values of the remaining variables can be uniquely found.

II. When $m < n$ (i.e., the number of equations is less than the number of variables) the solution is always other than $x_1 = x_2 = \ldots = xn = 0$.

III. When $m = n$ (i.e., the number of equations = the number of variables) the necessary and sufficient condition for solutions other than $x_1 = x_2 = \ldots = x_n = 0$ is that $|A| = 0$ (i.e., the determinant of the coefficient matrix is zero).

EXAMPLE 3.11

Test for consistency and solve

$$5x + 3y + 7z = 4, \quad 3x + 26y + 2z = 9, \quad 7x + 2y + 10z = 5.$$

Solution:

We have
$$\begin{bmatrix} 5 & 3 & 7 \\ 3 & 26 & 2 \\ 7 & 2 & 20 \end{bmatrix} \begin{bmatrix} x \\ y \\ z \end{bmatrix} = \begin{bmatrix} 4 \\ 9 \\ 5 \end{bmatrix}$$

Operate $3R_1, 5R_2,$
$$\begin{bmatrix} 15 & 9 & 21 \\ 15 & 130 & 10 \\ 7 & 2 & 10 \end{bmatrix} \begin{bmatrix} x \\ y \\ z \end{bmatrix} = \begin{bmatrix} 12 \\ 45 \\ 5 \end{bmatrix}$$

Operate $R_2 - R_1,$
$$\begin{bmatrix} 15 & 9 & 21 \\ 0 & 121 & -11 \\ 7 & 2 & 10 \end{bmatrix} \begin{bmatrix} x \\ y \\ z \end{bmatrix} = \begin{bmatrix} 12 \\ 33 \\ 5 \end{bmatrix}$$

Operate $\dfrac{7}{8}R_1, 5R_3, \dfrac{1}{11}R_2,$
$$\begin{bmatrix} 35 & 21 & 49 \\ 0 & 11 & -1 \\ 35 & 10 & 50 \end{bmatrix} \begin{bmatrix} x \\ y \\ z \end{bmatrix} = \begin{bmatrix} 28 \\ 3 \\ 25 \end{bmatrix}$$

Operate $R_3 - R_1 + R_2, \dfrac{1}{7}R_1,$
$$\begin{bmatrix} 5 & 3 & 7 \\ 0 & 11 & -1 \\ 0 & 0 & 0 \end{bmatrix} \begin{bmatrix} x \\ y \\ z \end{bmatrix} = \begin{bmatrix} 4 \\ 3 \\ 0 \end{bmatrix}$$

In the last set of equations, the number of non-zero rows in the coefficient matrix is two, and its rank is two. Also the number of non-zero rows in the augmented matrix being, its rank of two.

Now, the ranks of coefficient matrix and augmented matrix being equal, the equations are consistent. Also the given system is equivalent to

$$5x + 3y + 7z = 4, \quad 11y - z = 3.$$

$$\therefore \quad y = \frac{33}{11} + \frac{z}{11} \text{ and } x = \frac{7}{11} - \frac{16}{11}z$$

where z is a parameter.

Hence $x = \dfrac{7}{11}, y = \dfrac{3}{11}$ and $z = 0$ is a particular solution.

EXAMPLE 3.12

Examine the system of equations $3x + 3y + 2z = 1$, $x + 2y = 4$, $10y + 3z = -2$, $2x - 3y - z = 5$ for consistency and then solve it.

Solution:

We have

$$\begin{bmatrix} 3 & 3 & 2 \\ 1 & 2 & 0 \\ 0 & 10 & 3 \\ 2 & -3 & -1 \end{bmatrix} \begin{bmatrix} x \\ y \\ z \end{bmatrix} = \begin{bmatrix} 1 \\ 4 \\ -2 \\ 5 \end{bmatrix} \text{ or } \begin{bmatrix} 1 & 2 & 0 \\ 3 & 3 & 2 \\ 0 & 10 & 3 \\ 2 & -3 & -1 \end{bmatrix} \begin{bmatrix} x \\ y \\ z \end{bmatrix} = \begin{bmatrix} 4 \\ 1 \\ -2 \\ 5 \end{bmatrix}$$

$$[\text{Interchanging } R_1 \text{ and } R_2]$$

$$\text{or } \begin{bmatrix} 1 & 2 & 0 \\ 0 & -3 & 2 \\ 0 & 10 & 3 \\ 0 & -7 & -1 \end{bmatrix} \begin{bmatrix} x \\ y \\ z \end{bmatrix} = \begin{bmatrix} 4 \\ -11 \\ -2 \\ -3 \end{bmatrix} \qquad [\text{Operating } R_2 - 3R_1, R_4 - 2R_1]$$

$$\text{or } \begin{bmatrix} 1 & 2 & 0 \\ 0 & 1 & -2/3 \\ 0 & 0 & 29/3 \\ 0 & 0 & -17/3 \end{bmatrix} \begin{bmatrix} x \\ y \\ z \end{bmatrix} = \begin{bmatrix} 4 \\ 11/3 \\ -116/3 \\ 68/3 \end{bmatrix} \qquad \left[\text{Operating } R_3 + \frac{10}{4}R_2, R_4 - \frac{7}{3}R_2\right]$$

$$\text{or } \begin{bmatrix} 1 & 2 & 0 \\ 0 & 1 & -2/3 \\ 0 & 0 & 29/3 \\ 0 & 0 & 0 \end{bmatrix} \begin{bmatrix} x \\ y \\ z \end{bmatrix} = \begin{bmatrix} 4 \\ 11/3 \\ -116/3 \\ 0 \end{bmatrix} \qquad \left[\text{Operating } R_4 + \frac{17}{29}R_3\right]$$

Now in the last set of equations, the number of non-zero rows in the coefficient matrix is three, and its rank is three.

Also the number of non-zero rows in the augmented matrix is three, and its rank is three.

Since the ranks of the coefficient and the augmented matrices are equal, the given equations are consistent.

Also number of unknowns = rank of the coefficient matrix.

Hence the given equations have a unique solution given by

$$x + 2y = 4, y - \frac{2}{3}z = \frac{11}{3}, \frac{29}{3}z = -\frac{116}{3},$$

These equations show $z = -4$, $y = 1$, $x = 2$.

EXAMPLE 3.13

Investigate the values of λ and μ so that the equations

$$2x + 3y + 5z = 9, \quad 7x + 3y - 2z = 8, \quad 2x + 3y + \lambda z = \mu,$$

have (i) no solution, (ii) a unique solution, and (iii) *an infinite number of solutions*.

Solution:

We have $\begin{bmatrix} 2 & 3 & -5 \\ 7 & 3 & -2 \\ 2 & 3 & \lambda \end{bmatrix} \begin{bmatrix} x \\ y \\ z \end{bmatrix} = \begin{bmatrix} 9 \\ 8 \\ \mu \end{bmatrix}$

The system admits a unique solution if and only if, the coefficient matrix has the rank of 3. This requires that

$$\begin{bmatrix} 2 & 3 & 5 \\ 7 & 3 & -2 \\ 2 & 3 & \lambda \end{bmatrix} = 15(5 - \lambda) \neq 0$$

Thus for a unique solution $\lambda \neq 5$ and μ may have any value. If $\lambda = 5$, the system will have no solution for those values of μ for which the matrices

$$A = \begin{bmatrix} 2 & 3 & 5 \\ 7 & 3 & -1 \\ 2 & 3 & \lambda \end{bmatrix} \text{ and } K = \begin{bmatrix} 2 & 3 & 5 & 9 \\ 7 & 3 & -2 & 8 \\ 2 & 3 & 5 & \mu \end{bmatrix}$$

are not of the same rank. But A has the rank 2 and K does not have the rank of 2 unless $\mu = 9$. Thus if $\lambda = 5$ and $\mu \neq 9$, the system will have no solution.

If $\lambda = 5$ and $\mu = 9$, the system will have an infinite number of solutions.

EXAMPLE 3.14

Solve the equations

$$4x + 2y + z + 3w = 0,\ 6x + 3y + 4z + 7w = 0,\ 2x + y + w = 0.$$

Solution:

Rank of the coefficient matrix

$$\begin{bmatrix} 4 & 2 & 1 & 3 \\ 6 & 3 & 4 & 7 \\ 2 & 1 & 0 & 1 \end{bmatrix} \sim \begin{bmatrix} 4 & 2 & 1 & 3 \\ 0 & 0 & 5/2 & 5/2 \\ 0 & 0 & -1/2 & -1/2 \end{bmatrix} \quad \left[\text{Operating } R_2 - \frac{3}{3}R_1, R_3 - \frac{1}{2}R1\right]$$

$$\sim \begin{bmatrix} 4 & 2 & 1 & 3 \\ 0 & 0 & 5/2 & 5/2 \\ 0 & 0 & 2 & 0 \end{bmatrix} \quad \left[\text{Operating } R_3 + \frac{1}{5}R_2\right]$$

is 2 which is less than the number of variables.

$\therefore$ The number of independent solutions $= 4 - 2 = 2$.

Also the given system is equivalent to

$$4x + 2y + z + 3w = 0$$

$$z + w = 0$$

$$\therefore \qquad\qquad z = -w, x = -\frac{1}{2}(y + w).$$

Choosing $w = k_1$ and $x = k_2$, we have $y = -2k_2 - k_1$ and $z = -k_1$.

EXAMPLE 3.15

Find the values of k for which the system of equations $(3k - 8)\,x + 3y + 3z = 0,\ 3x + (3k - 8)y + 3z = 0,\ 3x + 3y + (3k - 8)\,z = 0$ has a non-trivial solution.

Solution

For the given system of equations to have a non-trivial solution, the determinant of coefficient matrix should be zero.

$$i.e., \quad \begin{vmatrix} 3k-8 & 3 & 3 \\ 3 & 3k-8 & 3 \\ 3 & 3 & 3k-8 \end{vmatrix} = 0 \text{ or } \begin{vmatrix} 3k-2 & 3 & 3 \\ 3k-2 & 3k-8 & 3 \\ 3k-2 & 3 & 3k-8 \end{vmatrix} = 0$$

$$\text{or} \quad (3k-2)\begin{vmatrix} 1 & 3 & 3 \\ 1 & 3k-8 & 3 \\ 1 & 3 & 3k-8 \end{vmatrix} = 0 \qquad [\text{Operating } C_1 + (C_2 + C_3)]$$

$$\text{or} \quad (3k-2)\begin{vmatrix} 1 & 3 & 3 \\ 0 & 3k-11 & 0 \\ 0 & 0 & 3k-11 \end{vmatrix} = 0 \qquad [\text{Operating } R_2 - R_1, R_3 - R_1]$$

or $(3k - 2)(3k - 11)^2 = 0$ where $k = 2/3$ or $11/3$.

Exercises 3.2

1. Find x, y, z and w given that $3\begin{bmatrix} x & y \\ z & w \end{bmatrix} = \begin{bmatrix} x & 6 \\ -1 & 2w \end{bmatrix} + \begin{bmatrix} 4 & x+y \\ z+w & 3 \end{bmatrix}$

2. If $A = \begin{bmatrix} 1 & 3 & 0 \\ -1 & 2 & 1 \\ 0 & 0 & 2 \end{bmatrix}$, $B = \begin{bmatrix} 2 & 3 & 4 \\ 1 & 2 & 3 \\ -1 & 1 & 2 \end{bmatrix}$, compute AB, BA and show that $AB \neq BA$.

3. Express the matrix $\begin{bmatrix} 0 & 5 & -3 \\ 1 & 1 & 1 \\ 4 & 5 & 9 \end{bmatrix}$ as the sum of symmetric and skew-symmetric matrices.

4. If $A = \begin{bmatrix} 2 & 5 & 3 \\ 3 & 1 & 2 \\ 1 & 2 & 1 \end{bmatrix}$, find adj A and $A-1$.

5. If $A = \dfrac{1}{3}\begin{bmatrix} 1 & 2 & 2 \\ 2 & 1 & -2 \\ -2 & 2 & -1 \end{bmatrix}$, prove that $A^{-1} = A'$.

6. Factorize the matrix $\begin{bmatrix} 5 & -2 & 1 \\ 7 & 1 & -5 \\ 3 & 7 & 4 \end{bmatrix}$ into the form LU, where L is the

lower triangular and U is the upper triangular matrix.

7. Determine the ranks of the following matrices:

(i) $\begin{bmatrix} 1 & 3 & 4 & 3 \\ 3 & 9 & 12 & 3 \\ 1 & 3 & 4 & 1 \end{bmatrix}$ (ii) $\begin{bmatrix} 1 & 2 & 3 & 0 \\ 2 & 4 & 3 & 2 \\ 3 & 2 & 1 & 3 \\ 6 & 8 & 7 & 5 \end{bmatrix}$

8. Examine for consistency the following equations and then solve these:
   $(i)\ x + 2y = 1,\ 7x + 14y = 12.$
   $(ii)\ 2x - 3y + 7z = 5,\ 3x + y - 3z = 13,\ 2x + 19y - 47z = 32.$
   $(iii)\ x + 2y + z = 3,\ 2x + 3y + 2z = 5,\ 3x - 5y + 5z = 2,\ 3x + 9y - z = 4.$

9. Investigate for what values of λ and μ the simultaneous equations
   $x + y + z = 6,\ x + 2y + 3z = 10,\ x + 2y + \lambda z = \mu$, have (i) no solution, (ii) a
   unique solution, (iii) an infinite number of solutions.

10. Determine the values of λ for which the following set of equations may
   possess non-trivial solutions
   $$3x_1 + x_2 - \lambda x_3 = 0,\ 4x_1 - 2x_2 - 3x_3 = 0,\ 2\lambda x_1 + 4x_2 + \lambda x_3 = 0.$$
   For each permissible value of ë, determine the general solution.

3.3 Solution of Linear Simultaneous Equations

   Simultaneous linear equations occur quite often in engineering and science. The analysis of electronic circuits consisting of invariant elements, analysis of a network under sinusoidal steady-state conditions, determination of the output of a chemical plant, and finding the cost of chemical reactions are some of the Exercises which depend on the solution of simultaneous linear algebraic equations. The solution of such equations can be obtained by *Direct* or *Iterative methods*. We describe below some such methods of solution.

3.4   Direct Methods of Solution

(1) Method of determinants—Cramer's rule. Consider the equations

$$\left.\begin{array}{c} a_1 x + b_1 y + c_1 z = d_1 \\ a_2 x + b_2 y + c_2 z = d_2 \\ a_3 x + b_3 y + c_3 z = d_3 \end{array}\right\}$$

If the determinant of coefficients is

$$\Delta = \begin{vmatrix} a_1 & b_1 & c_1 \\ a_2 & b_2 & c_2 \\ a_3 & b_3 & c_3 \end{vmatrix}$$

then
$$x\Delta = \begin{vmatrix} xa_1 & b_1 & c_1 \\ xa_2 & b_2 & c_2 \\ xa_3 & b_3 & c_3 \end{vmatrix} \qquad [\text{Operate } C_1 + yC_2 + zC_3]$$

$$= \begin{vmatrix} a_1 x + b_1 y + c_1 z & b_1 & c_1 \\ a_2 x + b_2 y + c_2 z & b_2 & c_2 \\ a_3 x + b_3 y + c_3 z & b_3 & c_3 \end{vmatrix} = \begin{vmatrix} d_1 & b_1 & c_1 \\ d_2 & b_2 & c_2 \\ d_3 & b_3 & c_3 \end{vmatrix}$$

Thus
$$x = \begin{vmatrix} d_1 & b_1 & c_1 \\ d_2 & b_2 & c_2 \\ d_3 & b_3 & c_3 \end{vmatrix} = \begin{vmatrix} a_1 & b_1 & c_1 \\ a_2 & b_2 & c_2 \\ a_3 & b_3 & c_3 \end{vmatrix} \text{ provided } \Delta \neq 0 \qquad (2)$$

Similarly
$$y = \begin{vmatrix} a_1 & d_1 & c_1 \\ a_2 & d_2 & c_2 \\ a_3 & d_3 & c_3 \end{vmatrix} \div \begin{vmatrix} a_1 & b_1 & c_1 \\ a_2 & b_2 & c_2 \\ a_3 & b_3 & c_3 \end{vmatrix} \qquad (3)$$

and
$$z = \begin{vmatrix} a_1 & b_1 & d_1 \\ a_2 & b_2 & d_2 \\ a_3 & b_3 & d_3 \end{vmatrix} \div \begin{vmatrix} a_1 & b_1 & c_1 \\ a_2 & b_2 & c_2 \\ a_3 & b_3 & c_3 \end{vmatrix} \qquad (4)$$

The equations (2), (3), and (4) giving the values of x, y, z constitute the *Cramer's rule*[1] which reduces the solution of the linear system (1) to a problem in evaluation of determinants.

[1.] Gabriel Cramer (1704—1752), was a Swiss mathematician.

NOTE

Obs. 1. *Cramer's rule fails for* $\Delta = 0$.

2. This method is quite general but involves a lot of labor when the number of equations exceeds four. For a 10×10 system, Cramer's rule requires about $70,000,000$ multiplications. We shall explain another method which requires only 333 multiplications, for the same 10×10 system. As such, Cramer's rule is not at all suitable for large systems.

EXAMPLE 3.16

Apply Cramer's rule to solve the questions

$$3x + y + 2z = 3, \ 2x - 3y - z = -3, \ x + 2y + z = 4.$$

Solution:

Here
$$\Delta = \begin{vmatrix} 3 & 1 & 2 \\ 2 & -3 & -1 \\ 1 & 2 & 1 \end{vmatrix} = 8$$

$\therefore$
$$x = \frac{1}{\Delta} \begin{vmatrix} 3 & 1 & 2 \\ -3 & -3 & -1 \\ 4 & 2 & 1 \end{vmatrix} = \frac{1}{8}(8) = 1,$$

$$y = \frac{1}{\Delta} \begin{vmatrix} 3 & 3 & 2 \\ 2 & -3 & -1 \\ 1 & 4 & 1 \end{vmatrix} = \frac{1}{8}(16) = 2$$

and
$$z = \frac{1}{\Delta} \begin{vmatrix} 3 & 1 & 3 \\ 2 & -3 & -3 \\ 1 & 2 & 4 \end{vmatrix} = \frac{1}{8}(-8) = -1$$

Hence $x = 1$, $y = 2$ and $z = -1$.

(2) Matrix inversion method. Consider the equations

$$\left. \begin{array}{l} a_1 x + b_1 y + c_1 z = d_1 \\ a_2 x + b_2 y + c_2 z = d_2 \\ a_3 x + b_3 y + c_3 z = d_3 \end{array} \right\} \tag{1}$$

If $A = \begin{bmatrix} a_1 & b_1 & c_1 \\ a_2 & b_2 & c_2 \\ a_3 & b_3 & c_3 \end{bmatrix}$, $X = \begin{bmatrix} x \\ y \\ z \end{bmatrix}$ and $D = \begin{bmatrix} d_1 \\ d_2 \\ d_3 \end{bmatrix}$,

then the equations (1) are equivalent to the matrix equation

$$AX = D. \tag{2}$$

Multiplying both sides of (2) by the inverse matrix A^{-1}, we get

$$A^{-1} AX = A^{-1} D \text{ or } IX = A^{-1} D \qquad [\because A^{-1}A = I]$$

or $\qquad\qquad X = A^{-1} D$

i.e., $\qquad \begin{bmatrix} x \\ y \\ z \end{bmatrix} = \frac{1}{|A|} \begin{bmatrix} A_1 & A_2 & A_3 \\ B_1 & B_2 & B_3 \\ C_1 & C_2 & C_3 \end{bmatrix} \times \begin{bmatrix} d_1 \\ d_2 \\ d_3 \end{bmatrix}$

where A_1, B_1, etc. are the cofactors of a_1, b_1, etc. in the determinant $|A|$.

Hence equating the values of x, y, z to the corresponding elements in the product on the right side of (3) we get the desired solution.

NOTE

Obs. *This method fails when A is a singular matrix, i.e., $|A| = 0$. Although this method is quite general, it is not suitable for large systems since the evaluation of A^{-1} by cofactors becomes very cumbersome. We shall now explain some methods which can be applied to any number of equations.*

EXAMPLE 3.17

Solve the equations $3x + y + 2z = 3$; $2x - 3y - z = - 3$; $x + 2y + z = 4$ by matrix inversion method. (cf. Example 3.16)

Solution:

Here $A = \begin{bmatrix} 3 & 1 & 2 \\ 2 & -3 & -1 \\ 1 & 2 & 1 \end{bmatrix} = \begin{bmatrix} a_1 & b_1 & c_1 \\ a_2 & b_2 & c_2 \\ a_3 & b_3 & c_3 \end{bmatrix}$ (say)

$\therefore \quad \begin{bmatrix} x \\ y \\ z \end{bmatrix} = \frac{1}{|A|} \begin{bmatrix} A_1 & A_2 & A_3 \\ B_1 & B_2 & B_3 \\ C_1 & C_2 & C_3 \end{bmatrix} \times \begin{bmatrix} d_1 \\ d_2 \\ d_3 \end{bmatrix} = \frac{1}{8} \begin{bmatrix} -1 & 3 & 5 \\ -3 & 1 & 7 \\ 7 & -5 & -11 \end{bmatrix} \times \begin{bmatrix} 3 \\ -3 \\ 4 \end{bmatrix} = \begin{bmatrix} 1 \\ 2 \\ -1 \end{bmatrix}$

Hence $x = 1$, $y = 2$, $z = -1$.

Gauss elimination method. In this method, the unknowns are eliminated successively and the system is reduced to an upper triangular system from which the unknowns are found by back substitution. The method is quite general and is well-adapted for computer operations. Here we shall explain it by considering a system of three equations for the sake of clarity.

Consider the equations

$$\left. \begin{array}{l} a_1 x + b_1 y + c_1 z = d_1 \\ a_2 x + b_2 y + c_2 z = d_2 \\ a_3 x + b_3 y + c_3 z = d_3 \end{array} \right\} \tag{1}$$

Step I. To eliminate x from the second and third equations.

Assuming $a_1 \neq 0$, we eliminate x from the second equation by subtracting (a_2/a_1) times the first equation from the second equation. Similarly we eliminate x from the third equation by eliminating (a_3/a_1) times the first equation from the third equation. We thus, get the new system

Assuming $a_1 \neq 0$, we eliminate x from the second equation by subtracting (a_2/a_1) times the first equation from the second equation. Similarly we eliminate x from the third equation by eliminating (a_3/a_1) times the first equation from the third equation. We thus, get the new system

$$\left. \begin{array}{l} a_1 x + b_1 y + c_1 z = d_1 \\ b_2' y + c_2' z = d_2' \\ b_3' y + c_3' z = d_3' \end{array} \right\} \tag{2}$$

Here the first equation is called the *pivotal equation* and a_1 is called the *first pivot*.

Step II. To eliminate y from third equation in (2).

Assuming $' \neq 0$, we eliminate y from the third equation of (2), by subtracting $(b_3'b_2')$ multiplied by times the second equation from the third equation. We thus, get the new system

$$\left. \begin{array}{l} a_1 x + b_1 y + c_1 z = d_1 \\ b_2' y + c_2' z = d_2' \\ c_3'' z = d_3'' \end{array} \right\} \tag{3}$$

Here the second equation is the *pivotal equation* and b_2' is the *new pivot*.

Step III. To evaluate the unknowns.

The values of x, y, z are found from the reduced system (3) by back substitution.

NOTE

Obs. 1. *On writing the given equations as*

$$\begin{bmatrix} a_1 & b_1 & c_1 \\ a_2 & b_2 & c_2 \\ a_3 & b_3 & c_3 \end{bmatrix} \begin{bmatrix} x \\ y \\ z \end{bmatrix} = \begin{bmatrix} d_1 \\ d_2 \\ d_3 \end{bmatrix} \ i.e., AX = D,$$

this *method consists in* transforming the coefficient matrix A to the upper triangular matrix *by elementary row transformations only.*

2. *Clearly the method will fail if any one of the pivots a_1, b_2', or c_3'' becomes zero. In such cases, we rewrite the equations in a different order so that the pivots are non-zero.*

3. *Partial and complete pivoting. In the first step, the numerically largest coefficient of x is chosen from all the equations and brought as the first pivot by interchanging the first equation with the equation having the largest coefficient of x. In the second step, the numerically largest coefficient of y is chosen from the remaining equations (leaving the first equation) and brought as the second pivot by interchanging the second equation with the equation having the largest coefficient of y. This process is continued until we arrive at the equation with the single variable. This modified procedure is called partial pivoting.*

If we are not keen about the elimination of x, y, z in a specified order, then we can choose at each stage the numerically largest coefficient of the entire matrix of coefficients. This requires not only an interchange of equations but also an interchange of the position of the variables. This method of elimination is called *complete pivoting*. It is more complicated and does not appreciably improve the accuracy.

EXAMPLE 3.18

Apply Gauss elimination method to solve the equations $x + 4y - z = -5$; $x + y - 6z = -12$; $3x - y - z = 4$.

Solution:

We have

	Check sum	
$x + 4y - z = -5$	-1	(i)
$x + y - 6z = -12$	-16	(ii)
$3x - y - z = 4$	5	(iii)

Step I. To eliminate x, operate $(ii) - (i)$ and $(iii) - 3(i)$:

	Check sum	
$-3y - 5z = -7$	-15	(iv)
$-13y + 2z = 19$	8	(v)

Step II. To eliminate y, operate $(v) - \dfrac{13}{3} (iv)$:

	Check sum	
$\dfrac{71}{3} z = \dfrac{148}{3}$	73	(vi)

Step III. By back-substitution, we get

From (vi): $z = \dfrac{148}{71} = 2.0845$

From (iv): $y = \dfrac{7}{3} - \dfrac{5}{3}\left(\dfrac{148}{71}\right) = -\dfrac{81}{71} = -1.1408$

From (i): $x = -5 - 4\left(-\dfrac{81}{71}\right) + \left(\dfrac{148}{71}\right) = \dfrac{117}{71} = 1.6479$

Hence, $x = 1.6479$, $y = -1.1408$, $z = 2.0845$.

Note. A useful check is provided by noting the sum of the coefficients and terms on the right, operating on those numbers as on the equations and checking that the derived equations have the correct sum.

Otherwise: We have
$$\begin{bmatrix} 1 & 4 & -1 \\ 1 & 1 & -6 \\ 3 & -1 & -1 \end{bmatrix} \begin{bmatrix} x \\ y \\ z \end{bmatrix} = \begin{bmatrix} -5 \\ -12 \\ 4 \end{bmatrix}$$

Operate $R_2 - R_1$ and $R_3 - 3R_1$,
$$\begin{bmatrix} 1 & 4 & -1 \\ 0 & -3 & -5 \\ 0 & -13 & 2 \end{bmatrix} \begin{bmatrix} x \\ y \\ z \end{bmatrix} = \begin{bmatrix} -5 \\ -7 \\ 19 \end{bmatrix}$$

Operate $R_3 - \dfrac{13}{3}R_2,$ $\begin{bmatrix} 1 & 4 & -1 \\ 0 & -3 & -5 \\ 0 & 0 & 71/3 \end{bmatrix} \begin{bmatrix} x \\ y \\ z \end{bmatrix} = \begin{bmatrix} -5 \\ -7 \\ 148/3 \end{bmatrix}$

Thus, we have $z = 148/71 = 2.0845,$

$3y = 7 - 5z = 7 - 10.4225 = -3.4225,$ *i.e.,* $y = -1.1408$

and $x = -5 - 4y + z = -5 + 4(1.1408) + 2.0845 = 1.6479$

Hence $x = 1.6479,\ y = -1.1408,\ z = 2.0845.$

EXAMPLE 3.19

Solve $10x - 7y + 3z + 5u = 6,\ -6x + 8y - z - 4u = 5,\ 3x + y + 4y + 11u = 2,\ 5x - 9y - 2z + 4u = 7$ by the Gauss elimination method.

Solution:

	Check sum	
We have $10x - 7y + 3z + 5u = 6$	17	(*i*)
$-6x + 8y - z - 4u = 5$	2	(*ii*)
$3x + y + 4z + 11u = 2$	21	(*iii*)
$5x - 9y - 2z + 4u = 7$	5	(*iv*)

Step I. To eliminate x, operate

$$\left[(ii) - \left(\frac{-6}{10} \right)(i), \right] \left[(iii) - \frac{3}{10}(i) \right] \left[(iv) - \frac{5}{10}(i) \right]:$$

	Check sum	
$3.8y + 0.8z - u = 8.6$	12.2	(*v*)
$3.1y + 3.1z + 9.5u = 0.2$	15.9	(*vi*)
$-5.5y - 3.5z + 1.5u = 4$	-3.5	(*vii*)

Step II. To eliminate y, operate $\left[(vi) - \dfrac{3.1}{3.8}(v) \right] \left[(vii) - \left(-\dfrac{5.5}{3.8} \right)(v) \right]:$

$2.4473684z + 10.315789u = -6.8157895 \qquad\qquad (viii)$

$-2.3421053z + 0.0526315u = 16.447368 \qquad\qquad (ix)$

Step III. To eliminate z, operate $\left[(ix) - \left(\dfrac{-2.3421053}{2.4473684} \right)(viii) \right]:$

$9.9249319u = 9.9245977$

Step IV. By back-substitution, we get

$$u = 1, z = -7, y = 4 \text{ and } x = 5.$$

EXAMPLE 3.20

Using the Gauss elimination method, solve the equations: $x + 2y + 3z - u = 10$, $2x + 3y - 3z - u = 1$, $2x - y + 2z + 3u = 7$, $3x + 2y - 4z + 3u = 2$.

Solution:

We have
$$\begin{bmatrix} 1 & 2 & 3 & -1 \\ 2 & 3 & -3 & -1 \\ 2 & -1 & 2 & 3 \\ 3 & 2 & -4 & 3 \end{bmatrix} \begin{bmatrix} x \\ y \\ z \\ u \end{bmatrix} = \begin{bmatrix} 10 \\ 1 \\ 7 \\ 2 \end{bmatrix}$$

Operate $R_2 - 2R_1$, $R3 - 2R_1$, $R_4 - 3R_1$

$$\begin{bmatrix} 1 & 2 & 3 & -1 \\ 0 & -1 & -9 & 1 \\ 0 & -5 & -4 & 5 \\ 0 & -4 & -13 & 6 \end{bmatrix} \begin{bmatrix} x \\ y \\ z \\ u \end{bmatrix} = \begin{bmatrix} -10 \\ -19 \\ -13 \\ -28 \end{bmatrix}$$

Operate $R_3 - 5R_2$, $R_4 - 4R_2$
$$\begin{bmatrix} 1 & 2 & 3 & -1 \\ 0 & -1 & -9 & 1 \\ 0 & 0 & 41 & 0 \\ 0 & 0 & 23 & 2 \end{bmatrix} \begin{bmatrix} x \\ y \\ z \\ u \end{bmatrix} = \begin{bmatrix} 10 \\ -19 \\ 82 \\ 48 \end{bmatrix}$$

Thus, we have $41z = 82$, *i.e.*, $z = 2$.

$$23z + 2u = 48, \text{ } i.e., \text{ } 46 + 2u = 48, \qquad \therefore u = 1$$

$$-y - 9z + u = -19, \text{ } i.e., \text{ } -y - 18 + 1 = -19, \qquad \therefore y = 2$$

$$x + 2y + 3z - u = 10, \text{ } i.e., \text{ } x + 4 + 6 - 1 = 10, \qquad \therefore x = 1$$

Hence $x = 1, y = 2, z = 2, u = 1$.

Gauss-Jordan method. This is a modification of the Gauss elimination method. In this method, elimination of unknowns is performed not in the equations below but in the equations above also, ultimately reducing the system to a diagonal matrix form, *i.e., each equation involving only one unknown. From these equations, the unknowns x, y, z can be obtained readily.*

Thus in this method, the labor of back-substitution for finding the unknowns is saved at the cost of additional calculations.

—————
NOTE

Obs. *For a system of 10 equations, the number of multiplications required for the Gauss-Jordan method is about 500 whereas for the Gauss elimination method we need only 333 multiplications. This shows that though the Gauss-Jordan method appears to be easier but requires 50 percent more operations than the Gauss elimination method. As such, the Gauss elimination method is preferred for large systems.*

EXAMPLE 3.21

Apply the Gauss-Jordan method to solve the equations

$x + y + z = 9$; $2x - 3y + 4z = 13$; $3x + 4y + 5z = 40$.

Solution:

We have

$$x + y + z = 9 \qquad (i)$$

$$2x - 3y + 4z = 13 \qquad (ii)$$

$$3x + 4y + 5z = 40 \qquad (iii)$$

Step I. To eliminate x from (ii) and (iii), operate $(ii) - 2(i)$ and $(iii) - 3(i)$:

$$x + y + z = 9 \qquad (iv)$$

$$-5y + 2z = -5 \qquad (v)$$

$$y + 2z = 13 \qquad (vi)$$

Step II. To eliminate y from (iv) and (vi), operate $(iv) + \dfrac{1}{5}\ (v)$ and $(vi) + \dfrac{1}{5}\ (v)$:

$$x + \frac{7}{5} z = 8 \qquad (vii)$$

$$-5y + 2z = -5 \qquad (viii)$$

$$\frac{12}{5} z = 12 \qquad (ix)$$

Step III. To eliminate z from (vii) and $(viii)$, operate $(vii) - \dfrac{7}{12}(ix)$ and $(viii) - \dfrac{5}{6}(ix)$:

$$x = 1$$

$$-5y = -15$$

$$\frac{12}{5} z = 12$$

Hence the solution is $x = 1$, $y = 3$, $z = 5$.

Otherwise: Rewriting the equations as $\begin{bmatrix} 1 & 1 & 1 \\ 2 & -3 & 4 \\ 3 & 4 & 5 \end{bmatrix} \begin{bmatrix} x \\ y \\ z \end{bmatrix} = \begin{bmatrix} 9 \\ 13 \\ 40 \end{bmatrix}$

Operate $R_2 - 2R_1$, $R_3 - 3R_1$, $\begin{bmatrix} 1 & 1 & 1 \\ 0 & -5 & 2 \\ 0 & 1 & 2 \end{bmatrix} \begin{bmatrix} \\ \\ \end{bmatrix} = \begin{bmatrix} 9 \\ -5 \\ 13 \end{bmatrix}$

Operate $R_3 + \dfrac{1}{5}R_2$, $\begin{bmatrix} 1 & 1 & 1 \\ 0 & -5 & 2 \\ 0 & 0 & 12/5 \end{bmatrix} \begin{bmatrix} x \\ y \\ z \end{bmatrix} = \begin{bmatrix} 9 \\ -5 \\ 12 \end{bmatrix}$

Operate $-R_2$, $5R_3$ $\begin{bmatrix} 1 & 1 & 1 \\ 0 & 5 & -2 \\ 0 & 0 & 12 \end{bmatrix} \begin{bmatrix} x \\ y \\ z \end{bmatrix} = \begin{bmatrix} 9 \\ 5 \\ 60 \end{bmatrix}$

Operate $R_2 + \dfrac{1}{6}R_3, \dfrac{1}{12}R_3$ $\begin{bmatrix} 1 & 1 & 1 \\ 0 & 5 & 0 \\ 0 & 0 & 1 \end{bmatrix} \begin{bmatrix} x \\ y \\ z \end{bmatrix} = \begin{bmatrix} 9 \\ 15 \\ 5 \end{bmatrix}$

Operate $\dfrac{1}{5}R_2$ $\begin{bmatrix} 1 & 1 & 1 \\ 0 & 1 & 0 \\ 0 & 0 & 1 \end{bmatrix} \begin{bmatrix} x \\ y \\ z \end{bmatrix} = \begin{bmatrix} 9 \\ 3 \\ 5 \end{bmatrix}$

Operate $R_1 - R_2 - R_3$ $\begin{bmatrix} 1 & 0 & 0 \\ 0 & 1 & 0 \\ 0 & 0 & 1 \end{bmatrix} \begin{bmatrix} x \\ y \\ z \end{bmatrix} = \begin{bmatrix} 1 \\ 3 \\ 5 \end{bmatrix}$

Hence $x = 1$, $y = 3$, $z = 5$.

NOTE

Obs. *Here the process of elimination of variables amounts to reducing the given coefficient metric to a* **diagonal matrix** *by elementary row transformations only.*

EXAMPLE 3.22

Solve the equations $10x - 7y + 3z + 5u = 6$; $-6x + 8y - z - 4u = 5$; $3x + y + 4z + 11u = 2$; and $5x - 9y - 2z + 4u = 7$ by the Gauss-Jordan method. (cf. Example 3.19)

Solution:

We have
$$10x - 7y + 3z + 5u = 6 \qquad\qquad (i)$$
$$-6x + 8y - z - 4u = 5 \qquad\qquad (ii)$$
$$3x + y + 4z + 11u = 2 \qquad\qquad (iii)$$
$$5x - 9y - 2z + 4u = 7 \qquad\qquad (iv)$$

Step I. To eliminate x, operate

$$\left[(ii) - \left(\frac{-6}{10}\right)(i)\right], \left[(iii) - \left(\frac{3}{10}\right)(i)\right], \left[(iv) - \left(\frac{5}{10}\right)(i)\right]:$$

$$10x - 7y + 3z + 5u = 6 \qquad\qquad (v)$$
$$3.8y + 0.8z - u = 8.6 \qquad\qquad (vi)$$
$$3.1y + 3.1z + 9.5u = 0.2 \qquad\qquad (vii)$$
$$-5.5y - 3.5z + 1.5u = 4 \qquad\qquad (viii)$$

Step II. To eliminate y, operate

$$\left[(v) - \left(\frac{-7}{3.8}\right)(vi)\right], \left[(vii) - \left(\frac{3.1}{3.8}\right)(vi)\right], \left[(viii) - \left(\frac{5.5}{3.8}\right)(vi)\right]:$$

$$10x + 4.4736842z + 3.1578947u = 21.842105 \qquad\qquad (ix)$$
$$3.8y + 0.8z - u = 8.6 \qquad\qquad (x)$$
$$2.4473684z + 10.315789u = -6.8157895 \qquad\qquad (xi)$$
$$-2.3421053z + 0.0526315u = 16.447368 \qquad\qquad (xii)$$

Step III. To eliminate z, operate

$$\left[(ix) - \left(\frac{4.473684}{2.4473684}\right)(xi)\right], \left[(x) - \left(\frac{0.8}{2.4473684}\right)(xi)\right], \left[(xii) - \left(\frac{-2.3421053}{2.4473684}\right)(xi)\right]:$$

$$10x - 15.698923u = 34.301075$$
$$3.8y - 4.3720429u = 10.827957$$
$$2.4473684z + 10.315789u = -6.8157895$$
$$9.9247309u = 9.9245975$$

Step IV. From the last equation $u = 1$ nearly.

Substitution of $u = 1$ in the above three equations gives $x = 5$, $y = 4$, $z = -7$.

Factorization method[2]. This method is based on the fact that every square matrix A can be expressed as the product of a lower triangular matrix and an upper triangular matrix, provided all the principal minors of A are non-singular, *i.e.*, if $A = [a_{ij}]$, then

$$a_{11} \neq 0, \begin{vmatrix} a_{11} & a_{12} \\ a_{21} & a_{22} \end{vmatrix} \neq 0, \begin{vmatrix} a_{11} & a_{12} & a_{13} \\ a_{21} & a_{22} & a_{23} \\ a_{31} & a_{32} & a_{33} \end{vmatrix} \neq 0, \text{ etc.}$$

Also such a factorization if it exists, is unique.

Now consider the equations

$$a_{11}x_1 + a_{12}x_2 + a_{13}x_3 = b_1$$
$$a_{21}x_1 + a_{22}x_2 + a_{23}x_3 = b_2$$
$$a_{31}x_1 + a_{32}x_2 + a_{33}x_3 = b_3$$

which can be written as $AX = B$ $\qquad$ (1)

where $\quad A = \begin{bmatrix} a_{11} & a_{12} & a_{13} \\ a_{21} & a_{22} & a_{23} \\ a_{31} & a_{32} & a_{33} \end{bmatrix}, X = \begin{bmatrix} x_1 \\ x_2 \\ x_3 \end{bmatrix}$ and $B = \begin{bmatrix} b_1 \\ b_2 \\ b_3 \end{bmatrix}$

Let $A = LU$,

where $L = \begin{bmatrix} 1 & 0 & 0 \\ l_{21} & 1 & 0 \\ l_{31} & l_{32} & 1 \end{bmatrix}$ and $U = \begin{bmatrix} u_{11} & u_{12} & u_{13} \\ 0 & u_{22} & u_{23} \\ 0 & 0 & u_{33} \end{bmatrix}$

Then (1) becomes $\qquad LUX = B$ $\qquad$ (3)

Writing $\qquad\qquad\qquad UX = V,$ $\qquad$ (4), (3) becomes $LV = B$

which is equivalent to the equations $v1 = b_1, l_{21}v_1 + v_2 = b_2, l_{31}v_1 + l_{32}v_2 + v_3 = b_3$

Solving these for v_1, v_2, v_3, we know V. Then, (4) becomes

$$u_{11}x_1 + u_{12}x_2 + u_{13}x_3 = v_1, u_{22}x_2 + u_{23}x_3 = v_2, u_{33}x_3 = v_3,$$

from which $x_3, x_2,$ and x_1 can be found by *back-substitution*.

[2.] Another name given to this decomposition is **Triangulization method.**

To compute the matrices L and U, we write (2) *as*

$$\begin{bmatrix} 1 & 0 & 0 \\ l_{21} & 1 & 0 \\ l_{31} & l_{32} & 1 \end{bmatrix} \begin{bmatrix} u_{11} & u_{12} & u_{13} \\ 0 & u_{22} & u_{23} \\ 0 & 0 & u_{33} \end{bmatrix} = \begin{bmatrix} a_{11} & a_{12} & a_{13} \\ a_{21} & a_{22} & a_{23} \\ a_{31} & a_{32} & a_{33} \end{bmatrix}$$

Multiplying the matrices on the left and equating corresponding elements from both sides, we obtain

(*i*) $u_{11} = a_{11}, u_{12} = a_{12}, u_{13} = a_{13}$

(*ii*) $l_{21}u_{11} = a_{21}$ or $l_{21} = a_{21}/a_{11}; l_{31}u_{11} = a_{31}$ or $l_{31} = a_{31}/a_{11}$

(*iii*) $l_{21}u_{12} + u_{22} = a_{22}$ or $u_{22} = a_{23} - \dfrac{a_{21}}{a_{11}}a_{13}$

(*iv*) $l_{31}u_{12} + l_{32}u_{22} = a_{32}$ or $l_{32} = \dfrac{1}{u_{22}}\left[a_{32} - \dfrac{a_{31}}{a_{11}}a_{12}\right]$

(*v*) $l_{31}u_{13} + l_{32}u_{23} + u_{33} = a_{33}$ which gives u_{33}.

Thus we compute the elements of L and U in the following set order:

(*i*) First row of U,   (*ii*) First column of L,

(*iii*) Second row of U,   (*iv*) Second column of L,

(*v*) Third row of U.

This procedure can easily be generalized.

NOTE   **Obs.** *This method is superior to the Gauss elimination method and is often used for the solution of linear systems and for finding the inverse of a matrix. The number of operations involved in terms of multiplications for a system of 10 equations by this method is about 110 as compared with 333 operations of the Gauss method. Among the direct methods, the factorization method is also preferred as the software for computers.*

EXAMPLE 3.23

Apply the factorization method to solve the equations:

$3x + 2y + 7z = 4; 2x + 3y + z = 5; 3x + 4y + z = 7.$

Solution:

Let $\begin{bmatrix} 1 & 0 & 0 \\ l_{21} & 1 & 0 \\ l_{31} & l_{32} & 1 \end{bmatrix} \begin{bmatrix} u_{11} & u_{12} & u_{13} \\ 0 & u_{22} & u_{23} \\ 0 & 0 & u_{33} \end{bmatrix} = \begin{bmatrix} 3 & 2 & 7 \\ 2 & 3 & 1 \\ 3 & 4 & 1 \end{bmatrix}$ (*i.e.*, A), so that

(i) R_1 of U : $u_{11} = 3$, $u_{12} = 2$, $u_{13} = 7$.

(ii) C_1 of L : $l_{21}u_{11} = 2$, $\therefore\ l_{21} = 2/3$,

                 $l_{31}u_{11} = 3$, $\therefore\ l_{31} = 1$.

(iii) R_2 of U : $l_{21}u_{12} + u_{22} = 3$, $\therefore\ u_{22} = 5/3$,

                 $l_{21}u_{13} + u_{23} = 1$, $\therefore\ u_{23} = -11/3$.

(iv) C_2 of L : $l_{31}u_{12} + l_{32}u_{22} = 4$ $\therefore\ l_{32} = 6/5$.

(v) R_3 of U : $l_{31}u_{13} + l_{32}u_{23} + u_{33} = 1$ $\therefore\ u_{33} = -8/5$.

Thus $$A = \begin{bmatrix} 1 & 0 & 0 \\ 2/3 & 1 & 0 \\ 1 & 6/5 & 1 \end{bmatrix} = \begin{bmatrix} 3 & 2 & 7 \\ 0 & 5/3 & -11/3 \\ 0 & 0 & -8/5 \end{bmatrix}$$

Writing $UX = V$, the given system becomes

$$\begin{bmatrix} 1 & 0 & 0 \\ 2/3 & 1 & 0 \\ 1 & 6/5 & 1 \end{bmatrix} \begin{bmatrix} v_1 \\ v_2 \\ v_3 \end{bmatrix} = \begin{bmatrix} 4 \\ 5 \\ 7 \end{bmatrix}$$

Solving this system, we have $v_1 = 4$,

$$\frac{2}{3}v_1 + v_2 = 5 \qquad \text{or} \qquad v_2 = \frac{7}{3}$$

$$v_1 + \frac{6}{5}v_2 + v_3 = 7 \qquad \text{or} \qquad -$$

Hence the original system becomes

$$\begin{bmatrix} 3 & 2 & 7 \\ 0 & 5/3 & -11/3 \\ 0 & 0 & -8/5 \end{bmatrix} \begin{bmatrix} x \\ y \\ z \end{bmatrix} = \begin{bmatrix} 4 \\ 7/3 \\ 1/5 \end{bmatrix}$$

i.e., $3x + 2y + 7z = 4, \dfrac{5}{3}y - \dfrac{11}{3}z = \dfrac{7}{3}, \quad -\dfrac{8}{5}z = \dfrac{1}{5}$

By back-substitution, we have

$$z = -1/8, y = 9/8 \quad \text{and} \quad x = 7/8.$$

EXAMPLE 3.24

Solve the equations $10x - 7y + 3z + 5u = 6$; $-6x + 8y - z - 4u = 5$; $3x + y + 4z + 11u = 2$; $5x - 9y - 2z + 4u = 7$ by factorization method.

(cf. Example 3.19)

Solution:

$$\text{Let} \begin{bmatrix} 1 & 0 & 0 & 0 \\ l_{21} & 1 & 0 & 0 \\ l_{31} & l_{32} & 1 & 0 \\ l_{41} & l_{42} & l_{43} & 1 \end{bmatrix} \begin{bmatrix} u_{11} & u_{12} & u_{13} & u_{14} \\ 0 & u_{22} & u_{22} & u_{24} \\ 0 & 0 & 0 & u_{34} \\ 0 & 0 & 0 & u_{44} \end{bmatrix} = \begin{bmatrix} 10 & -7 & 3 & 5 \\ -6 & 8 & -1 & -4 \\ 3 & 1 & 4 & 11 \\ 5 & -9 & 2 & 4 \end{bmatrix} \quad (i.e., A)$$

so that

(i) R_1 of U: $u_{11} = 10$, $u_{12} = -7$, $u_{13} = 3$, $u_{14} = 5$

(ii) C_1 of L: $l_{21} = -0.6$, $l_{31} = 0.3$, $l_{41} = 0.5$

(iii) R_2 of U: $u_{22} = 3.8$, $u_{23} = 0.8$, $u_{24} = -1$

(iv) C_2 of L: $l_{32} = 0.81579$, $l_{42} = -1.44737$

(v) R_3 of U: $u_{33} = 2.44737$, $u_{34} = 10.31579$

(vi) C_3 of L: $l_{43} = -0.95699$

(vii) R_4 of U: $u_{44} = 9.92474$

$$\text{Thus } A = \begin{bmatrix} 1 & 0 & 0 & 0 \\ -0.6 & 1 & 0 & 0 \\ 0.3 & 0.81579 & 1 & 0 \\ 0.5 & -1.44737 & -0.95699 & 1 \end{bmatrix} \begin{bmatrix} 10 & -7 & 3 & 5 \\ 0 & 3.8 & 0.8 & -1 \\ 0 & 0 & 2.44737 & 10.31579 \\ 0 & 0 & 0 & 9.92474 \end{bmatrix}$$

Writing $UX = V$, the given system becomes

$$\begin{bmatrix} 1 & 0 & 0 & 0 \\ -0.6 & 1 & 0 & 0 \\ 0.3 & 0.81577 & 1 & 0 \\ 0.5 & -1.44737 & -0.95699 & 1 \end{bmatrix} \begin{bmatrix} v1 \\ v2 \\ v3 \\ v4 \end{bmatrix} = \begin{bmatrix} 6 \\ 5 \\ 2 \\ 7 \end{bmatrix}$$

Solving this system, we get

$$v_1 = 6, \ v_2 = 8.6, \ v_3 = -6.81579, \ v_4 = 9.92474.$$

Hence the original system becomes

$$
\begin{bmatrix}
10 & -7 & 3 & 5 \\
0 & 3.8 & 0.8 & -1 \\
0 & 0 & 2.44737 & 10.31579 \\
0 & 0 & 0 & 9.92474
\end{bmatrix}
\begin{bmatrix}
x \\ y \\ z \\ u
\end{bmatrix}
=
\begin{bmatrix}
6 \\ 8.6 \\ -6.81579 \\ 9.92474
\end{bmatrix}
$$

i.e., $10x - 7y + 3z + 5u = 6,\ 3.8y + 0.8z - u = 8.6,$

$$2.44737z + 10.31579u = -6.81579,\ u = 1.$$

By back-substitution, we get

$$u = 1, z = -7, y = 4, x = 5.$$

Exercises 3.3

Solve the following equations by Cramer's rule:

1. $x + 3y + 6z = 2;\ 3x - y + 4z = 9;\ x - 4y + 2z = 7.$

2. $x + y + z = 6.6;\ x - y + z = 2.2;\ x + 2y + 3z = 15.2.$

3. $x^2 z^3/y = e^8;\ y^2 z/x = e^4;\ x^3 y/z^4 = 1.$

4. $2vw - wu + uv = 3uvw;\ 3vw + 2wu + 4uv = 19uv;\ 6vw + 7wu - uv = 17uvw.$

5. $3x + 2y - z + t = 1;\ x - y - 2z + 4t = 3;\ 2x - 3y + z - 2t = -2;\ 5x - 2y + 3z + 2t = 0.$

Solve the following equations by the matrix inversion method:

6. $x + y + z = 3;\ x + 2y + 3z = 4;\ x + 4y + 9z = 6.$

7. $x + y + z = 1;\ x + 2y + 3z = 6;\ x + 3y + 4z = 6.$

8. $2x - y + 3z = 8;\ x - 2y - z = -4;\ 3x + y - 4z = 0.$

9. $2x_1 + x_2 + 2x_3 + x_4 = 6;\ 4x_1 + 3x_2 + 3x_3 - 3x_4 = -1;\ 6x_1 - 6x_2 + 6x_3 + 12x_4 = 36,\ 2x_1 + 2x_2 - x_3 + x_4 = 10.$

10. In a given electrical network, the equations for the currents $i_1, i_2,$ and i_3 are

$$3i_1 + i_2 + i_3 = 8;\ 2i_1 - 3i_2 - 2i_3 = -5;\ 7i_1 + 2i_2 - 5i_3 = 0.$$

Calculate i_1 and i_3 by (a) Cramer's rule, (b) matrix inversion.

Solve the following equations by the Gauss elimination method:

11. $x + y + z = 9; 2x - 3y + 4z = 13; 3x + 4y + 5z = 40$

12. $2x + 2y + z = 12; 3x + 2y + 2z = 8; 5x + 10y - 8z = 10.$

13. $2x - y + 3z = 9; x + y + z = 6; x - y + z = 2.$

14. $2x_1 + 4x_2 + x_3 = 3; 3x_1 + 2x_2 - 2x_3 = -2; x_1 - x_2 + x_3 = 6.$

15. $5x_1 + x_2 + x_3 + x_4 = 4; x_1 + 7x_2 + x_3 + x_4 = 12; x_1 + x_2 + 6x_3 + x_4 = -5; x_1 + x_2 + x_3 + 4x_4 = -6.$

Solve the following equations by the Gauss-Jordan method:

16. $2x + 5y + 7z = 52; 2x + y - z = 0; x + y + z = 9.$

17. $2x - 3y + z = -1; x + 4y + 5z = 25; 3x - 4y + z = 2.$

18. $x + y + z = 9; 2x + y - z = 0; 2x + 5y + 7z = 52.$

19. $x + 3y + 3z = 16; x + 4y + 3z = 18, x + 3y + 4z = 19$

20. $2x_1 + x_2 + 5x_3 + x_4 = 5; x_1 + x_2 - 3x_3 + 4x_4 = -1;$

21. $3x_1 + 6x_2 - 2x_3 + x_4 = 8; 2x_1 + 2x_2 + 2x_3 - 3x_4 = 2.$

Solve the following equations by the factorization method:

22. $2x + 3y + z = 9; x + 2y + 3z = 6; 3x + y + 2z = 8.$

23. $10x + y + z = 12; 2x + 10y + z = 13; 2x + 2y + 10z = 14.$

24. $10x + y + 2z = 13; 3x + 10y + z = 14; 2x + 3y + 10z = 15.$

25. $2x_1 - x_2 + x_3 = -1; 2x_2 - x_3 + x_4 = 1; x1 + 2x_3 - x_4 = -1; x_1 + x_2 + 2x_4 = 3.$

3.5   Iterative Methods of Solution

The preceding methods of solving simultaneous linear equations are known as *direct methods*, as these methods yield the solution after a certain amount of fixed computations. On the other hand, an iterative method is that in which we start from an approximation to the true solution and obtain better and better approximations from a computation cycle repeated as often as may be necessary for achieving a desired accuracy. Thus in an iterative method, the amount of computation depends on the degree of accuracy required.

For large systems, iterative methods may be faster than the direct methods. Even the round-off errors in iterative methods are smaller. In fact, iteration is a self correcting process and any error made at any stage of computation gets automatically corrected in the subsequent steps.

Simple iterative methods can be devised for systems in which the coefficients of the leading diagonal are large as compared to others. We now describe three such methods:

(1) Jacobi's iteration method. Consider the equations

$$\left.\begin{array}{l} a_1x + b_1y + c_1z = d_1 \\ a_2x + b_2y + c_2z = d_2 \\ a_3x + b_3y + c_3z = d_3 \end{array}\right\} \tag{1}$$

If a_1, b_2, c_3 are large as compared to other coefficients, solve for x, y, z, respectively.

Then the system can be written as

$$\left.\begin{array}{l} x = \dfrac{1}{a_1}(d_1 - b_1y - c_1z) \\[2mm] y = \dfrac{1}{b_2}(d_2 - a_2x - c_2z) \\[2mm] z = \dfrac{1}{c_3}(d_3 - a_3x - b_3y) \end{array}\right\} \tag{2}$$

Let us start with the initial approximations x_0, y_0, z_0 for the values of x, y, z, respectively. Substituting these on the right sides of (2), the first approximations are given by

$$x_1 = \frac{1}{a_1}(d_1 - b_1y_0 - c_1z_0)$$

$$y_1 = \frac{1}{b_2}(d_2 - a_2x_0 - c_2z_0)$$

$$z_1 = \frac{1}{c_3}(d_3 - a_3x_0 - b_3y_o)$$

Substituting the values x_1, y_1, z_1 on the right sides of (2), the second approximations are given by

$$x_2 = \frac{1}{a_1}(d_1 - b_1y_1 - c_1z_1)$$

$$y_2 = \frac{1}{b_2}(d_2 - a_2x_1 - c_2z_1)$$

$$z_2 = \frac{1}{c_3}(d_3 - a_3x_1 - b_3y_1)$$

This process is repeated until the difference between two consecutive approximations is negligible.

Obs. *In the absence of any better estimates for x_0, y_0, z_0, these may each be taken as zero.*

EXAMPLE 3.25

Solve, by Jacobi's iteration method, the equations

$20x + y - 2z = 17; 3x + 20y - z = -18; 2x - 3y + 20z = 25.$

Solution:

We write the given equations in the form

$$x = \frac{1}{20}(17 - y + 2z)$$
$$y = \frac{1}{20}(-18 - 3x + z) \Big\} \qquad (i)$$
$$z = \frac{1}{20}(25 - 2x + 3y)$$

We start from an approximation $x_0 = y_0 = z_0 = 0.$

Substituting these on the right sides of the equations (i), we get

$$x_1 = \frac{17}{20} = 0.85, \quad y_1 = \frac{18}{20} = -0.9, \quad z_1 = \frac{25}{20} = 1.25$$

Putting these values on the right sides of the equations (i), we obtain

$$x_2 = \frac{1}{20}(17 - y1 + 2z_1) = 1.02$$
$$y_2 = \frac{1}{20}(-18 - 3x + z1) = -0.965$$
$$z_2 = \frac{1}{20}(25 - 2x1 + 3y_1) = 1.03$$

Substituting these values on the right sides of the equations (i), we have

$$x_3 = \frac{1}{20}(17 - y_2 + 2z_2) = 1.00125$$
$$y_3 = \frac{1}{20}(-18 - 3x_2 + z_2) = 1.0015$$
$$z_3 = \frac{1}{20}(25 - 2x_2 + 3y_2) = 1.00325$$

Substituting these values, we get

$$x_4 = \frac{1}{20}(17 - y_3 + 2z_3) = 1.0004$$

$$y_4 = \frac{1}{20}(-18 - 3x_3 + z_3) = -1.000025$$

$$z_4 = \frac{1}{20}(25 - 2x_3 + 3y_3) = 0.9965$$

Putting these values, we have

$$x_5 = \frac{1}{20}(-17 - y_4 + 2z_4) = 0.999966$$

$$y_5 = \frac{1}{20}(-18 - 3x_4 + z_4) = -1.000078$$

$$z_5 = \frac{1}{20}(25 - 2x_4 + 3y_4) = 0.999956$$

Again substituting these values, we get

$$x_6 = \frac{1}{20}(-17 - y_5 + 2z_5) = 1.0000$$

$$y_6 = \frac{1}{20}(-18 - 3x_5 + z_5) = 0.999997$$

$$z_6 = \frac{1}{20}(25 - 2x_5 + 3y_5) = 0.999992$$

The values in the fifth and sixth iterations being practically the same, we can stop. Hence the solution is

$$x = 1, y = -1, z = 1.$$

EXAMPLE 3.26

Solve by Jacobi's iteration method, the equations $10x + y - z = 11.19$, $x + 10y + z = 28.08, -x + y + 10z = 35.61$, correct to two decimal places.

Solution:

Rewriting the given equations as

$$x = \frac{1}{10}(11.19 - y + z), y = \frac{1}{10}(28.08 - x - z), z = \frac{1}{10}(35.61 + x - y)$$

We start from an approximation, $x_0 = y_0 = z_0 = 0$.

First iteration

$$x_1 = \frac{11.19}{10} = 1.119, y_1 = \frac{28.08}{10} = 2.808, z_1 = \frac{35.61}{10} = 3.561$$

Second iteration

$$x_2 = \frac{1}{10}(11.19 - y_1 + z_1) = 1.19$$

$$y_2 = \frac{1}{10}(28.08 - x_1 - z_1) = 2.34$$

$$z_2 = \frac{1}{10}(35.61 + x_1 - y_1) = 3.39$$

Third iteration

$$x_3 = \frac{1}{10}(11.19 - y_2 + z_2) = 1.22$$

$$y_3 = \frac{1}{10}(28.08 - x_2 - z_2) = 2.35$$

$$z_3 = \frac{1}{10}(35.61 + x_2 - y_2) = 3.45$$

Fourth iteration

$$x_4 = \frac{1}{10}(11.19 - y_3 + z_3) = 1.23$$

$$y_4 = \frac{1}{10}(28.08 - x_3 - z_3) = 2.34$$

$$z_4 = \frac{1}{10}(35.61 + x_3 - y_3) = 3.45$$

Fifth iteration

$$x_5 = \frac{1}{10}(11.19 - y_4 + z_4) = 1.23$$

$$y_5 = \frac{1}{10}(28.08 - x_4 - z_4) = 2.34$$

$$z_5 = \frac{1}{10}(35.61 + x_4 - y_4) = 3.45$$

Hence $x = 1.23, y = 2.34, z = 3.45$

EXAMPLE 3.27

Solve the equations

$$10x - 2x_2 - x_3 - x_4 = 3$$
$$-2x_1 + 10x_2 - x_3 - x_4 = 15$$
$$-x_1 - x_2 + 10x_3 - 2x_4 = 27$$
$$-x_1 - x_2 - 2x_3 + 10x_4 = -9, \text{ by the Gauss-Jacobi iteration method.}$$

Solution:

Rewriting the given equation as

$$x_1 = \frac{1}{10}(3 + 2x_2 + x_3 + x_4)$$

$$x_2 = \frac{1}{10}(15 + 2x_1 + x_3 + x_4)$$

$$x_3 = \frac{1}{10}(27 + x_1 + x_2 + 2x_4)$$

$$x_4 = \frac{1}{10}(-9 + x_1 + x_2 + 2x_3)$$

We start from an approximation $x_1 = x_2 = x_3 = x_4 = 0$.

First iteration

$$x_1 = 0.3, x_2 = 1.5, x_3 = 2.7, x_4 = -0.9.$$

Second iteration

$$x_1 = \frac{1}{10}[3 + 2(1.5) + 2.7 + (-0.9)] = 0.78$$

$$x_2 = \frac{1}{10}[15 + 2(0.3) + 2.7 + (-0.9)] = 1.74$$

$$x_3 = \frac{1}{10}[27 + 0.3 + 1.5 + 2(-0.9)] = 2.7$$

$$x_4 = \frac{1}{10}[-9 + 0.3 + 1.5 + 2(-0.9)] = 0.18$$

Proceeding in this way, we get

Third iteration	$x_1 = 0.9, x_2 = 1.908, x_3 = 2.916, x_4 = -0.108$
Fourth iteration	$x_1 = 0.9624, x_2 = 1.9608, x_3 = 2.9592, x_4 = -0.036$
Fifth iteration	$x_1 = 0.9845, x_2 = 1.9848, x_3 = 2.9851, x_4 = -0.0158$
Sixth iteration	$x_1 = 0.9939, x_2 = 1.9938, x_3 = 2.9938, x_4 = -0.006$

Seventh iteration $x_1 = 0.9939, x_2 = 1.9975, x_3 = 2.9976, x_4 = -0.0025$

Eighth iteration $x_1 = 0.999, x_2 = 1.999, x_3 = 2.999, x_4 = -0.001$

Ninth iteration $x_1 = 0.9996, x_2 = 1.9996, x_3 = 2.9996, x_4 = -0.004$

Tenth iteration $x_1 = 0.9998, x_2 = 1.9998, x_3 = 2.9998, x_4 = -0.0001$

Hence $x_1 = 1, x_2 = 2, x_3 = 3, x_4 = 0$.

Gauss-Seidal iteration method. This is a modification of Jacobi's method. As before the system of equations:

$$\left.\begin{array}{l} a_1 x + b_1 y + c_1 z = d_1 \\ a_2 x + b_2 y + c_2 z = d_2 \\ a_3 x + b_3 y + c_3 z = d_3 \end{array}\right\} \qquad (1)$$

is written as

$$\left.\begin{array}{l} x = \dfrac{1}{a_1}(d_1 - b_1 y - c_1 z) \\[2ex] y = \dfrac{1}{b_2}(d_2 - a_2 x - c_2 z) \\[2ex] z = \dfrac{1}{c_3}(d_3 - a_3 x - b_3 y) \end{array}\right\} \qquad (2)$$

Here also we start with the initial approximations x_0, y_0, z_0 for x, y, z, respectively which may each be taken as zero. Substituting $y = y_0, z = z_0$ in the first of the equations (2), we get

$$x_1 = \frac{1}{a_1}(d_1 - b_1 y_0 - c_1 z_0)$$

Then putting $x = x_1, z = z_0$ in the second of the equations (2), we have

$$y_1 = \frac{1}{b_2}(d_2 - a_2 x_1 - c_2 z_0)$$

Next substituting $x = x_1, y = y_1$ in the third of the equations (2), we obtain

$$z_1 = \frac{1}{c_3}(d_3 - a_3 x_1 - b_3 y_1)$$

and so on, *i.e.*, as soon as a new approximation for an unknown is found, it is immediately used in the next step.

This process of iteration is repeated until the values of x, y, z are obtained to a desired degree of accuracy.

NOTE

Obs. 1. *Since the most recent approximations of the unknowns are used while proceeding to the next step, the convergence in the Gauss-Seidal method is twice as fast as in Jacobi's method.*

2. *Jacobi and Gauss-Seidal methods converge for any choice of the initial approximations if in each equation of the system, the absolute value of the largest co-efficient is almost equal to or is at least one equation greater than the sum of the absolute values of all the remaining coefficients.*

EXAMPLE 3.28

Apply the Gauss-Seidal iteration method to solve the equations $20x + y - 2z = 17$; $3x + 20y - z = -18$; $2x - 3y + 20z = 25$. (cf. Example 3.25)

Solution:

We write the given equations in the form

$$x = \frac{1}{20}(17 - y + 2z) \qquad (i)$$

$$y = \frac{1}{20}(-18 - 3x + z) \qquad (ii)$$

$$z = \frac{1}{20}(25 - 2x + 3y) \qquad (iii)$$

First iteration

Putting $y = y_0$, $z = z_0$ in (i), we get $\qquad x_1 = \frac{1}{2}(17 - y_0 + 2z_0) = 0.8500$

Putting $x = x_1$, $z = z_0$ in (ii), we have $\qquad y_1 = \frac{1}{20}(-18 - 3x_1 + z_0) = -1.0275$

Putting $x = x_1$, $y = y_1$ in (iii), we obtain $z_1 = \frac{1}{20}(25 - 2x_1 + 3y_1) = 1.0109$

Second iteration

Putting $y = y_1$, $z = z_1$ in (i), we get $\qquad x_2 = \frac{1}{20}(17 - y_1 + 2z_1) = 1.0025$

Putting $x = x_2$, $z = z_1$ in (ii), we obtain $y_2 = \frac{1}{20}(-18 - 3x_2 + z_1) = -0.9998$

Putting $x = x_2$, $y = y_2$ in (iii), we get $\qquad z_2 = \frac{1}{20}(25 - 2x_2 + 3y_2) = 0.9998$

Third iteration, we get

$$x_3 = \frac{1}{20}(17 - y_2 + 2z_2) = 1.0000$$

$$y_3 = \frac{1}{20}(-18 - 3x_3 + z_2) = -1.0000$$

$$z_3 = \frac{1}{20}(25 - 2x_3 + 3y_3) = 1.0000$$

The values in the second and third iterations being practically the same, we can stop.

Hence the solution is $x = 1$, $y = -1$, $z = 1$.

EXAMPLE 3.29

Solve the equations $27x + 6y - z = 85$; $x + y + 54z = 110$; $6x + 15y + 2z = 72$ by the Gauss-Jacobi method and the Gauss-Seidel method.

Solution:

Rewriting the given equations as

$$x = \frac{1}{27}(85 - 6y + z) \qquad (i)$$

$$y = \frac{1}{15}(72 - 6x - 2z) \qquad (ii)$$

$$z = \frac{1}{54}(110 - x - y) \qquad (iii)$$

(a) Gauss-Jacobi's method

We start from an approximation $x_0 = y_0 = z_0 = 0$

First iteration

$$x_1 = \frac{85}{27} = 3.148, y_1 = \frac{72}{15} = 4.8, z_1 = \frac{110}{54} = 2.037$$

Second iteration

$$x_2 = \frac{1}{27}(85 - 6y_1 + z_1) = 2.157$$

$$y_2 = \frac{1}{15}(72 - 6x_1 - 2z_1) = 3.269$$

$$z_2 = \frac{1}{54}(110 - x_1 - y_1) = 1.890$$

Third iteration

$$x_3 = \frac{1}{27}(85 - 6y_2 + 7z_2) = 2.492$$

$$y_3 = \frac{1}{15}(72 - 6x_2 - 2z_2) = 3.685$$

$$z_3 = \frac{1}{54}(110 - x_2 - y_2) = 1.937$$

Fourth iteration

$$x_4 = \frac{1}{27}(85 - 6y_3 + z_3) = 2.401$$

$$y_4 = \frac{1}{15}(72 - 6x_3 - 2z_3) = 3.545$$

$$z_4 = \frac{1}{54}(110 - x_3 - y_3) = 1.923$$

Fifth iteration

$$x_5 = \frac{1}{27}(85 - 6y_4 + z_4) = 2.432$$

$$y_5 = \frac{1}{15}(72 - 6x_4 - 2z_4) = 3.583$$

$$z_4 = \frac{1}{54}(110 - x_3 - y_3) = 1.927$$

Repeating this process, the successive iterations are:

$$x_6 = 2.423, \ y_6 = 3.570, \ z_6 = 1.926$$
$$x_7 = 2.426, \ y_7 = 3.574, \ z_7 = 1.926$$
$$x_8 = 2.425, \ y_8 = 3.573, \ z_8 = 1.926$$
$$x_9 = 2.426, \ y_9 = 3.573, \ z_9 = 1.926$$

Hence $x = 2.426, \ y = 3.573, \ z = 1.926$

(b) Gauss-Seidal method
First iteration

Putting $y = y_0 = 0, \ z = z_0 = 0$ in (*i*), $\quad x_1 = \frac{1}{27}(85 - 6y_0 + z_0) = 3.14$

Putting $x = x_1, \ z = z_0$ in (*ii*), $\qquad y1 = \frac{1}{15}(72 - 6x_1 - 2z_0) = 3.541$

Putting $x = x_1, \ y = y_1$ in (*iii*), $\qquad z_1 = \frac{1}{54}(110 - x_1 - y_1) = 1.913$

Second iteration

$$x_2 = \frac{1}{27}(85 - 6y_1 + z_1) = 2.432$$

$$y2 = \frac{1}{15}(72 - 6x_2 - 2z_1) = 3.572$$

$$z_2 = \frac{1}{54}(110 - x_2 - y_2) = 1.926$$

Third iteration

$$x_3 = \frac{1}{27}(85 - 6y_2 + z_2) = 2.426$$

$$y_3 = \frac{1}{15}(72 - 6x_3 - 2z_2) = 3.573$$

$$z_3 = \frac{1}{54}(110 - x_3 - y_3) = 1.926$$

Fourth iteration

$$x_4 = \frac{1}{27}(85 - 6y_3 + z_3) = 2.426$$

$$y_4 = \frac{1}{15}(72 - 6x_4 - 2z_3) = 3.573$$

$$z_4 = \frac{1}{54}(110 - x_4 - y_4) = 1.926.$$

Hence $x = 2.426$, $y = 3.573$, $z = 1.926$.

NOTE **Obs.** *We have seen that the convergence is quite fast in the Gauss-Seidal method as compared to the Gauss-Jacobi method.*

EXAMPLE 3.30

Apply the Gauss-Seidal iteration method to solve the equations:
$10x_1 - 2x_2 - x_3 - x_4 = 3; -2x_1 + 10x_2 - x_3 - x_4 = 15; -x_1 - x_2 + 10x_3 + 2x_4 = 27;$
$-x_1 - x_2 - 2x_3 + 10x_4 = -9.$ (cf. Example 3.27)

Solution:

Rewriting the given equations as

$$x_1 = 0.3 + 0.2x_2 + 0.1x_3 + 0.1x_4 \tag{i}$$

$$x_2 = 1.5 + 0.2x_1 + 0.1x_3 + 0.1x_4 \qquad (ii)$$

$$x_3 = 2.7 + 0.1x_1 + 0.1x_2 + 0.2x_4 \qquad (iii)$$

$$x_4 = -0.9 + 0.1x_1 + 0.1x_2 + 0.2x_3 \qquad (iv)$$

First iteration

Putting $x_2 = 0$, $x_3 = 0$, $x_4 = 0$ in (i), we get $x_1 = 0.3$

Putting $x_1 = 0.3$, $x_3 = 0$, $x_4 = 0$ in (ii), we obtain $x_2 = 1.56$

Putting $x_1 = 0.3$, $x_2 = 1.56$, $x_4 = 0$ in (iii), we obtain $x_3 = 2.886$

Putting $x_1 = 0.3$, $x_2 = 1.56$, $x_3 = 2.886$ in (iv), we get $x_4 = -0.1368$.

Second iteration

Putting $x_2 = 1.56$, $x_3 = 2.886$, $x_4 = -0.1368$ in (i), we obtain $x_1 = 0.8869$

Putting $x_1 = 0.8869$, $x_3 = 2.886$, $x_4 = -0.1368$ in (ii), we obtain $x_2 = 1.9523$

Putting $x_1 = 0.8869$, $x_2 = 1.9523$, $x_4 = -0.1368$ in (iii), we have $x_3 = 2.9566$

Putting $x_1 = 0.8869$, $x_2 = 1.9523$, $x_3 = 2.9566$ in (iv), we get $x_4 = -0.0248$.

Third iteration

Putting $x_2 = 1.9523$, $x_3 = 2.9566$, $x_4 = -0.0248$ in (i), we obtain $x_1 = 0.9836$

Putting $x_1 = 0.9836$, $x_3 = 2.9566$, $x_4 = -0.0248$ in (ii), we obtain $x_2 = 1.9899$

Putting $x_1 = 0.9836$, $x_2 = 1.9899$, $x_4 = -0.0248$ in (iii), we get $x_3 = 2.9924$

Putting $x_1 = 0.9836$, $x_2 = 1.9899$, $x_3 = 2.9924$ in (iv), we get $x_4 = -0.0042$.

Fourth iteration. Proceeding as above

$x_1 = 0.9968$, $x_2 = 1.9982$, $x_3 = 2.9987$, $x_4 = -0.0008$.

Fifth iteration is $x_1 = 0.9994$, $x_2 = 1.9997$, $x_3 = 2.9997$, $x_4 = -0.0001$.

Sixth iteration is $x_1 = 0.9999$, $x_2 = 1.9999$, $x_3 = 2.9999$, $x_4 = -0.0001$

Hence the solution is $x_1 = 1$, $x_2 = 2$, $x_3 = 3$, $x_4 = 0$.

(3) Relaxation method[3]. Consider the equations

$$a_1 x + b_1 y + c_1 z = d_1$$
$$a_2 x + b_2 y + c_2 z = d_2$$
$$a_3 x + b_3 y + c_3 z = d_3$$

[3.] This method was originally developed by R.V. Southwell in 1935, for application to structural engineering Exercises

We define the residuals R_x, R_y, R_z by the relations

$$\left.\begin{array}{l} R_x = d_1 - a_1 x - b_1 y - c_1 z \\ R_y = d_2 - a_2 x - b_2 y - c_2 z \\ R_z = d_3 - a_3 x - b_3 y - c_3 z \end{array}\right\} \qquad (1)$$

To start with we assume $x = y = z = 0$ and calculate the initial residuals. Then the residuals are reduced step by step, by giving increments to the variables. For this purpose, we construct the following *operation table*:

	δR_x	δR_y	δR_z
$\delta x = 1$	$-a_1$	$-a_2$	$-a_3$
$\delta y = 1$	$-b_1$	$-b_2$	$-b_3$
$\delta z = 1$	$-c_1$	$-c_2$	$-c_3$

We note from the equations (1) that if x is increased by 1 (keeping y and z constant), R_x, R_y, and R_z decrease by a_1, a_2, $a_{3,}$ respectively. This is shown in the above table along with the effects on the residuals when y and z are given unit increments. (The table is the transpose of the coefficient matrix).

At each step, the numerically largest residual is reduced to almost zero. To reduce a particular residual, the value of the corresponding variable is changed; *e.g.*, to reduce R_x by p, x should be increased by $p/a1$.

When all the residuals have been reduced to almost zero, the increments in x, y, z are added separately to give the desired solution.

——————
NOTE
——————

Obs. 1. *As a check, the computed values of x, y, z are substituted in (1) and the residuals are calculated. If these residuals are not all negligible, then there is some mistake and the entire process should be rechecked.*

2. *Relaxation method can be applied successfully only if the diagonal elements of the coefficient matrix dominate the other coefficients in the corresponding row, i.e., if in the equations (1)*

$$|a_1| \geq |b_1| + |c_1|$$
$$|b_2| \geq |a2| + |c_2|$$
$$|c_3| \geq |a_3| + |b_3|$$

where > sign should be valid for at least one row.

EXAMPLE 3.31

Solve, by the Relaxation method, the equations:

$$9x - 2y + z = 50; x + 5y - 3z = 18; -2x + 2y + 7z = 19.$$

Solution:

The residuals are given by

$$R_x = 50 - 9x + 2y - z;$$
$$R_y = 18 - x - 5y + 3z;$$
$$R_z = 19 + 2x - 2y - 7z$$

The operations table is

	δR_x	δR_y	δR_z
$\delta x = 1$	−9	−1	2
$\delta y = 1$	2	−5	−2
$\delta z = 1$	−1	−3	−7

The relaxation table is

	R_x	R_y	R_z	
$x=y=z=0$	50	18	19	(i)
$\delta x = 5$	5	13	29	(ii)
$\delta z = 14$	1	25	1	(iii)
$\delta y = 5$	11	0	−9	(iv)
$\delta x = 1$	2	−1	−7	(v)
$\delta z = -1$	3	−4	0	(vi)
$\delta y = -0.8$	1.4	0	1.6	(vii)
$\delta y = 0.23$	1.17	0.69	−0.69	(viii)
$\delta y = 0.13$	0	0.56	0.17	(ix)
$\delta y = 0.112$	0.224	0	−0.054	(x)

$$\Sigma \delta x = 6.13, \Sigma \delta y = 4.31, \Sigma \delta z = 3.23.$$

Thus $\qquad x = 6.13, y = 4.31, z = 3.23$

[**Explanation.** In (i), the largest residual is 50. To reduce it, we give an increment $\delta x = 5$ and the resulting residuals are shown in (ii). Of these $R_x = 29$ is the largest and we give an increment $\delta z = 4$ to get the results in (iii). In (vi) $R_y = -4$ is the (numerically) largest and we give an increment $\delta y = -4/5 = -0.8$ to obtain the results in (vii). Similarly the other steps have been carried out.]

EXAMPLE 3.32

Solve the equations:

$10x - 2y - 3z = 205$; $- 2x + 10y - 2z = 154$; $- 2x - y + 10z = 120$ by Relaxation method.

Solution:

The residuals are given by

$$R_x = 205 - 10x + 2y + 3z;$$
$$R_y = 154 + 2x - 10y + 2z;$$
$$R_z = 120 + 2x + y - 10z.$$

The operations table is

	δR_x	δR_y	δR_z
$\delta x = 1$	-10	2	2
$\delta y = 1$	2	-10	1
$\delta z = 1$	3	2	-10

The relaxation table is

	R_x	R_y	R_z
$x = y = z = 0$	205	154	120
$\delta x = 20$	5	194	160
$\delta y = 19$	43	4	179
$\delta z = 18$	97	40	-1
$\delta x = 10$	-3	60	19
$\delta y = 6$	9	0	25
$\delta z = 2$	15	4	5
$\delta x = 2$	-5	8	9
$\delta z = 1$	-2	10	-1
$\delta y = 1$	0	0	0

$$\Sigma \delta x = 32, \; \Sigma \delta y = 26, \; \Sigma \delta z = 21.$$

Hence $x = 32, y = 26, z = 21$.

Exercises 3.4

1. Solve by Jacobi's method, the equations: $5x - y + z = 10$; $2x + 4y = 12$; $x + y + 5z = -1$; starting with the solution $(2, 3, 0)$.

2. Solve by Jacobi's method the equations:

$13x + 5y - 3z + u = 18; 2x + 12y + z - 4u = 13; x - 4y + 10z + u = 29;$
$2x + y - 3z + 9u = 31.$

3. Solve the equations $27x + 6y - z = 85; x + y + 54z = 40; 6x + 15y + 2z = 72$ by

(a) Jacobi's method (b) Gauss-Seidal method.

Solve the following equations by Gauss-Seidal method:

4. $2x + y + 6z = 9; 8x + 3y + 2z = 13; x + 5y + z = 7.$

5. $28x + 4y - z = 32; x + 3y + 10z = 24; 2x + 17y + 4z = 35$

6. $10x + y + z = 12; 2x + 10y + z = 13; 2x + 2y + 10z = 14.$

7. $7x_1 + 52x_2 + 13x_3 = 104; 83x_1 + 11x_2 - 4x_3 = 95; 3x_1 + 8x_2 + 29x_3 = 71.$

8. $3x_1 - 0.1x_2 - 0.2x_3 = 7.85; 0.1x_1 + 7x_2 - 0.3x_3 = -19.3; 0.3x_1 - 0.2x_2 + 10x_3 = 71.4.$

Solve, by the Relaxation method, the following equations:

9. $3x + 9y - 2z = 11; 4x + 2y + 13z = 24; 4x - 4y + 3z = -8.$

10. $10x - 2y - 2z = 6; -x + 10y - 2z = 7; -x - y + 10z = 8.$

11. $-9x + 3y + 4z + 100 = 0; x - 7y + 3z + 80 = 0; 2x + 3y - 5z + 60 = 0.$

12. $54x + y + z = 110; 2x + 15y + 6z = 72; -x + 6y + 27z = 85$

3.6 Ill-Conditioned Equations

A linear system is said to be *ill-conditioned* if small changes in the co-efficients of the equations result in large changes in the values of the un-knowns. On the contrary, a system is *well-conditioned* if small changes in the coefficients of the system also produce small changes in the solution. We often come across ill-conditioned systems in practical applications. Ill-conditioning of a system is usually expected when the determinant of the coefficient matrix is small. The coefficient matrix of an ill-conditioned system is called an *ill-conditioned matrix*.

While solving simultaneous equations, we also come across two forms of *instabilities: Inherent* and *Induced*. Inherent instability of a system is a property of the given problem and occurs due to the problem being ill-conditioned. It can be avoided by reformulation of the problem suitably.

Induced instability occurs because of the incorrect choice of method.

(2) Iterative method to improve accuracy of an ill-conditioned system. Consider the system of equations

$$
\left.\begin{array}{l}
a_1x + b_1y + c_1z = d_1 \\
a_2x + b_2y + c_2z = d_2 \\
a_3x + b_3y + c_3z = d_3
\end{array}\right\} \tag{1}
$$

Let x', y', z' be an approximate solution. Substituting these values on the left-hand sides, we get new values of d_1, d_2, d_3 as $d1', d2', d3'$ so that the new system is

$$
\left.\begin{array}{l}
a_1x' + b_1y' + c_1z' = d_1' \\
a_2x' + b_2y' + c_2z' = d_2' \\
a_3x' + b_3y' + c_{33}z' = d_3'
\end{array}\right\} \tag{2}
$$

Subtracting each equation in (2) from the corresponding equations in (1), we obtain

$$
\left.\begin{array}{l}
a_1x_e + b_1y_e + c_1z_e = k_1 \\
a_2x_e + b_2y_e + c_2z_e = k_2 \\
a_3x_e + b_3y_e + c_3z_e = k_3
\end{array}\right\} \tag{3}
$$

where $x_e = x - x'$, $y_e = y - y'$, $ze = z - z'$ and $ki = di - di'$

We now solve the system (3) for x_e, y_e, z_e giving $x = x' + x_e$, $y = y' + y_e$ and $z = z' + z_e$, which will be better approximations for x, y, z. We can repeat the procedure for improving the accuracy.

EXAMPLE 3.33

Establish whether the system $1.01x + 2y = 2.01$; $x + 2y = 2$ is well conditioned or not?

Solution:

Its solution is $x = 1$ and $y = 0.5$.

Now consider the system $x + 2.01y = 2.04$ and $x + 2y = 2$

which has the solution $x = -6$ and $y = 4$.

Hence the system is ill-conditioned.

EXAMPLE 3.34

An approximate solution of the system $2x + 2y - z = 6; x + y + 2z = 8; -x + 3y + 2z = 4$ is given by $x = 2.8$, $y = 1$, and $z = 1.8$. Using the above iterative method, improve this solution.

Solution:

Substituting the approximate values $x' = 2.8$, $y' = 1$, $z' = 1.8$ in the given equations, we get

$$\left.\begin{array}{l} 2(2.8) + 2(1) - 1.8 = 5.8 \\ 2.8 + 2 + 2(1.8) = 7.4 \\ -2.8 + 3(1) + 2(1.8) = 3.8 \end{array}\right\} \qquad (i)$$

Subtracting each equation in (i) from the corresponding given equations, we obtain

$$\left.\begin{array}{l} 2x_e + 2y_e - z_e = 0.2 \\ x_e + y_e + 2z_e = 0.6 \\ -x_e + 3y_e + 2z_e = 0.2 \end{array}\right\} \qquad (ii)$$

where $x_e = x - 2.8$, $y_e = y - 1$, $z_e = z - 1.8$.

Solving the equations (ii), we get $x_e = 0.2$, $y_e = 0$, $z_e = 0.2$.

This gives the better solution $x = 3$, $y = 1$, $z = 2$, which incidently is the exact solution.

Exercises 3.5

1. Establish whether the system of equations
$$10x + 8y + 9z + 6w = 33,$$
$$6x + 7y + 5z + 5w = 23,$$
$$8x + 10y + 7z + 7w = 32,$$
$$9x + 7y + 10z + 5w = 31$$
is well-conditioned or not?

2. An approximate solution of the equations $x + 4y + 7z = 5$; $2x + 5y + 8z = 7$; $3x + 6y + 9.1z = 9.1$ is given by $x = 1.8$, $y = -1.2$, $z = 1$. Improve this solution by using the iterative method.

3.7   Comparison of Various Methods

Direct and iterative methods have their advantages and disadvantages and a choice of method depends on a particular system of equations. The direct methods yield a solution in a finite number of steps for any non-singular set of equations, while in an iterative method the amount of computation depends on the accuracy desired. In general, it is preferable to use a direct method for the solution of a linear system. However for large systems, an iterative method yields the solution faster and should therefore be preferred.

Gauss elimination method requires more of recording and is quite time consuming for operations. As such it is more expensive from the programming point of view. Among the direct methods, Crout's triangularization method is used more often for the solution of a linear system and as software for computers.

The rounding off errors also get propagated in the elimination method whereas in the iteration techniques only the rounding off errors committed in the final iteration have any effect. In general, the iteration methods have smaller round-off errors for iteration since it is a self- correcting technique. Thus the use of an iterative method for ill-conditioned system is preferable.

On the other hand, an iterative method may not always converge. When it converges, the iterative method is definitely better than the direct methods.

We come across two types of instabilities while solving a linear system of equations, *i.e.,*

Inherent instability and Induced instability.

Inherent instability occurs due to the set of equations being ill-conditioned and as such is a property of the problem itself. It can, however, be avoided by a suitable reformulation of the problem.

On the other hand, induced instability occurs due to an incorrect choice of the method of solution.

3.8   Solution of Non-Linear Simultaneous Equations

Newton-Raphson method. Consider the equations

$$f(x, y) = 0, g(x, y) = 0 \tag{1}$$

If an initial approximation (x_0, y_0) to a solution has been found by a graphical method or otherwise, then a better approximation (x_1, y_1) can be obtained as follows:

Let $x_1 = x_0 + h, y_1 = y_0 + k$, so that

$$f(x_0 + h, y_0 + k) = 0, g(x_0 + h, y_0 + k) = 0 \qquad (2)$$

Expanding each of the functions in (2) by Taylor's series to first degree terms, we get approximately

$$\left. \begin{aligned} f_0 + h\frac{\partial f}{\partial x_0} + k\frac{\partial f}{\partial y_0} = 0 \\ g_0 + h\frac{\partial g}{\partial x_0} + k\frac{\partial g}{\partial y_0} = 0 \end{aligned} \right\} \qquad (3)$$

where $f_0 = f(x_0, y_0), \dfrac{\partial f}{\partial x_0} = \left(\dfrac{\partial f}{\partial x}\right)_{x_0, y_0}$ etc.

Solving the equations (3) for h and k, we get a new approximation to the root as

$$x_1 = x_0 + h, y_1 = y_0 + k.$$

This process is repeated until we get the values to the desired accuracy.

NOTE   **Obs. 1.** *This method will not converge unless the starting values of the roots chosen are close to the actual roots.*

2. *The method can be extended to three equations in three variables. But it is very cumbersome to obtain a meaningful solution unless the entire information about the equations and their physical context is available.*

Otherwise. Whenever it is possible, one of the variables may be eliminated from the given equations giving a single polynomial equation in the other variable. Then find this variable to a desired degree of accuracy by the Newton-Raphson method. Sometimes the above polynomial equation is seen to have a root by trial. If so, reduce this equation to the next lower degree equation and find its other root by the Newton-Raphson method. Having found this variable to a required degree of accuracy, the other variable can at once, be found from one of the given equations.

EXAMPLE 3.35

Solve the system of non-linear equations:

$$x^2 + y = 11, \, y^2 + x = 7.$$

Solution:

An initial approximation to the solution is obtained from a rough graph of (1), as $x_0 = 3.5$ and $y_0 = -1.8$.

We have $f = x^2 + y - 11$ and $g = y^2 + x - 7$ so that

$$\frac{\partial}{\partial} \quad 2 \, , \qquad \frac{\partial g}{\partial x} = 1$$

$$\frac{\partial g}{\partial x} = 1, \qquad \frac{\partial g}{\partial y} = 2y.$$

Then Newton-Raphson's equations (3) above will be

$$7h + k = 0.55, \, h - 3.6k = 0.26.$$

Solving these, we get $h = 0.0855, \, k = -0.0485$

∴ The better approximation to the root is

$$x_1 = x_0 + h = 3.5855, \, y_1 = y_0 + k = -1.8485.$$

Repeating the above process, replacing (x_0, y_0) by (x_1, y_1), we obtain $x_2 = 3.5844, \, y_2 = -1.8482$.

Otherwise. Eliminating y from the given equations, we get

$$x_4 - 22x^2 + x + 114 = 0$$

By trial, $x = 3$ is its root.

∴ The reduced equation is $x^3 + 3x^2 - 13x - 38 = 0$

To find the other root, we apply the Newton-Raphson method to

$$f(x) = x^3 + 3x^2 - 13x - 38.$$

Taking $x_0 = 3.5$, we get $x_2 = 3.5844$.

Thus $y = 11 - x^2$ gives $y = -1.848$ for $x = 3.5844$

Also $y = -2$ for $x = 3$.

EXAMPLE 3.36

Solve the equations $2x^2 + 3xy + y^2 = 3$, $4x^2 + 2xy + y^2 = 30$ correct to three decimal places, using Newton-Raphson method, given that $x_0 = -3$ and $y_0 = 2$.

Solution:

We have $f = 2x^2 + 3xy + y^2 - 3$ and $g = 4x^2 + 2xy + y^2 - 30$,

So that $\dfrac{\partial f}{\partial x} = 4x + 3y, \dfrac{\partial f}{\partial y} = 3x + 2y$

$\dfrac{\partial g}{\partial x} = 8x + 2y. \dfrac{\partial g}{\partial y} = 2x + 2y$

Now
$$f_0 = 2x_0^2 + 3x_0y_0 + y_0^2 - 3 = 1$$
$$g_0 = 4x_0^2 + 2x_0y_0 + y_0^2 - 30 = -2$$
$$\frac{\partial f}{\partial x_0} = -6, \frac{\partial f}{\partial y_0} = -5; \frac{\partial g}{\partial x_0} = -20, \frac{\partial g}{\partial y_0} = -2$$

Then Newton-Raphson equations (3) above will be
$$20h + 2k = -2; 6h + 5k = 1$$

Solving these equations, we get $h = -\dfrac{3}{22} - -0.1364, k = \dfrac{4}{11} = 0.3636$

∴ The better approximation is

$x_1 = x_0 + h = -3 - 0.1364 = -3.1364$

$y_1 = y_0 + k = 2 + 0.3636 = 2.3636$

Repeating the above process and replacing (x_0, y_0) by (x_1, y_1), we obtain $x_2 = -3.131, y_2 = 2.362$

Again proceeding as above and replacing (x_1, y_1) by (x_2, y_2), we obtain $x_3 = -3.1309, y_3 = 2.3617$

Since the values x_2, y_2 and x_3, y_3 are approximately equal, the solution correct to three decimal places is $x = -3.131, y = 2.362$.

Exercises 3.6

1. Solve the equations $x^2 + y = 5$, $y^2 + x = 3$.

2. Solve the non-linear equations $x = 2(y + 1)$, $y^2 = 3xy - 7$ correct to three decimals.

3. Find a root of the equations $xy = x + 9$, $y^2 = x^2 + 7$.

4. Use the Newton-Raphson method to solve the equations $x = x^2 + y^2$, $y = x^2 - y^2$ correct to two decimals, starting with the approximation $(0.8, 0.4)$.

5. Solve the non-linear equations $x^2 - y^2 = 4$, $x^2 + y^2 = 16$ numerically with $x_0 = y_0 = 2.828$ using the Newton-Raphson method. Carry out two iterations.

3.9 Objective Type of Questions

Exercises 3.7

Select the correct answer or fill up the blanks in the following questions:

1. As soon as a new value of a variable is found by iteration, it is used immediately in the following equations, this method is called
   (*a*) Gauss-Jordan method   (*b*) Gauss-Seidal method
   (*c*) Jacobi's method   (*d*) Relaxation method.

2. The difference between direct and iterative method of solving simultaneous linear equations is ....

3. In solving simultaneous equations by the Gauss-Jordan method, the coefficient matrix is reduced to .... matrix.

4. The condition for the convergence of the Gauss-Seidel matrix is that in each equation of the system ....

5. A matrix in which $a_{ij} = 0$ for $i \neq j$ is called ....

6. Solutions of simultaneous non-linear equations can be obtained using
   (*a*) Method of iteration   (*b*) Newton-Raphson method
   (*c*) None of the above.

7. To which form is the coefficient matrix is transformed when $AX = B$ is solved by Gauss elimination method?

8. Guass-Seidal iteration converges only if the coefficient matrix is diagonally dominant.   (True or False)

9. What is "partial pivoting" and "complete pivoting" in the solution of linear simultaneous

equations.

10. The convergence in the Gauss-Seidal method is ...... than that in Jacobi's method:

    (a) more fast   (b) more slow

    (c) slow   (d) equal.

11. By the Gauss elimination method, solve $x + y = 2$ and $2x + 3y = 5$.

CHAPTER 4

MATRIX INVERSION AND EIGENVALUE PROBLEM

Chapter Objectives

- Introduction
- Matrix inversion
- Gauss elimination method
- Gauss-Jordan method
- Factorization method
- Partition method
- Iterative method
- Eigenvalues and eigenvectors
- Properties of eigenvalues
- Bounds for eigenvalues
- Power method
- Jacobi's method
- Given's method
- House-holder's method
- Objective type of questions

4.1 Introduction

There are two main numerical exercises which arise in connection with the matrices. One of these is the problem of finding the inverse of a matrix. The other problem is that of finding the eigenvalues and the corresponding eigenvectors of a matrix. When a student first encounters an eigenvalue problem, it appears to him somewhat artificial and theoretical only. In fact the computation of eigenvalues is required in many engineering and scientific problems. For instance, the frequencies of the vibrations of beams are the eigenvalues of a matrix. Eigenvalues are also required while finding the frequencies associated with

(i) the vibrations of a system of masses and springs,

(ii) the symmetric vibrations of an annular membrane,

(iii) the oscillations of a triple pendulum,

(iv) the torsional oscillations of a uniform cantilever,

(v) the torsional oscillations of a multi-cylinder engine etc.

Once the physical formulation in any of the above situations is completed, all these Exercises have the same mathematical approach: that of finding an eigenvalue for a numerical matrix.

4.2   Matrix Inversion

In Section 3.2(4), we have already defined the inverse of a non-singular square matrix A, to be another matrix B of the same order such that $AB = BA = I$, I being a unit matrix of the same order.

The inverse of a matrix A is written as A^{-1} so that $AA^{-1} = A^{-1}A = I$.

Thus the inverse of a matrix exists if and only if it is a non-singular square matrix. Also inverse of a matrix, when it exists is unique.

There are several methods of finding the inverse of a matrix. Of these, the method of obtaining the inverse with the help of an adjoint has already been illustrated by Example 3.9. But it requires a lot of calculations. As such, we shall now, describe some other methods which require less of computational labor and can be easily extended to matrices of higher order.

4.3   Gauss Elimination Method

The method involves the same procedure as explained in Section3.4(3). Here we take a unit matrix of the same order as the given matrix A and write it as AI.

Now making simultaneous row operations on AI, we try to convert A into an upper triangular matrix and then to a unit matrix. Ultimately when A is transformed into a unit matrix, the adjacent matrix (emerged out from the transformation of I) gives the inverse of A. To increase the accuracy, the largest element in A is taken as the pivot element for performing the row operations.

4.4   Gauss-Jordan Method

This is similar to the Gauss elimination method except that instead of first converting A into upper triangular form, it is directly converted into the unit matrix.

In practice, the two matrices A and I are written side by side and the same row transformations are performed on both. As soon as A is reduced to I, the other matrix represents A^{-1}.

EXAMPLE 4.1

Using Gauss-Jordan method, find the inverse of the matrix

$$\begin{bmatrix} 1 & 1 & 2 \\ 1 & 3 & -3 \\ -2 & -4 & -4 \end{bmatrix}$$

Solution:

Writing the given matrix side by side with the unit matrix of order 3, we have

$$\begin{bmatrix} 1 & 1 & 3 & : & 1 & 0 & 0 \\ 1 & 3 & -3 & : & 0 & 1 & 0 \\ -2 & -4 & -4 & : & 0 & 0 & 1 \end{bmatrix} \qquad \text{(Operate } R_2 - R_1 \text{ and } R_3 + 2R_1)$$

$$\sim \begin{bmatrix} 1 & 1 & 3 & : & 1 & 0 & 0 \\ 0 & 2 & -6 & : & -1 & 1 & 0 \\ 0 & -2 & 2 & : & 2 & 0 & 1 \end{bmatrix} \qquad \text{(Operate } \frac{1}{2}R_2 \text{ and } \frac{1}{2}R_3)$$

$$\sim \begin{bmatrix} 1 & 1 & 3 & : & 1 & 0 & 0 \\ 0 & 1 & -3 & : & -1/2 & 1/2 & 0 \\ 0 & -1 & 1 & : & 1 & 0 & 1/2 \end{bmatrix} \qquad \text{(Operate } R_1 - R_2 \text{ and } R_3 + R_2)$$

$$\sim \begin{bmatrix} 1 & 0 & 6 & : & 3/2 & 1/2 & 0 \\ 0 & 1 & -3 & : & -1/2 & 1/2 & 0 \\ 0 & 0 & -2 & : & 1/2 & 1/2 & 1/2 \end{bmatrix}$$

$$\left[\text{Operate } R_1 + 3R_3,\ R_2 - \frac{3}{2}R_3 \text{ and } \frac{1}{2}R_2\right]$$

$$\sim \begin{bmatrix} 1 & 0 & 0 & : & 3 & 1 & 3/2 \\ 0 & 1 & 0 & : & 5/4 & 1/4 & 3/4 \\ 0 & 0 & 1 & : & 1/4 & 1/4 & 1/4 \end{bmatrix}$$

Hence the inverse of the given matrix is $\begin{bmatrix} 3 & 1 & 3/2 \\ -5/4 & -1/4 & -3/4 \\ -1/4 & -1/4 & -1/4 \end{bmatrix}$

EXAMPLE 4.2

Using Gauss-Jordan method, find the inverse of the matrix $\begin{bmatrix} 2 & 2 & 3 \\ 2 & 1 & 1 \\ 1 & 3 & 5 \end{bmatrix}$

Solution:

Writing the given matrix side by side with the unit matrix of order 3, we have

$$\begin{bmatrix} 2 & 2 & 3 & : & 1 & 0 & 0 \\ 2 & 1 & 1 & : & 0 & 1 & 0 \\ 1 & 3 & 5 & : & 0 & 0 & 1 \end{bmatrix} \qquad \left(\text{Operate } \frac{1}{2}R_1\right)$$

$$\sim \begin{bmatrix} 1 & 1 & 3/2 & : & 1/2 & 0 & 0 \\ 2 & 1 & 1 & : & 0 & 1 & 0 \\ 1 & 3 & 5 & : & 0 & 0 & 1 \end{bmatrix} \qquad (\text{Operate } R_2 - 2R_1,\ R_3 - R_1)$$

$$\sim \begin{bmatrix} 1 & 1 & 3/2 & : & 1/2 & 0 & 0 \\ 0 & -1 & -2 & : & -1 & 1 & 0 \\ 0 & 2 & 7/2 & : & -1/2 & 0 & 1 \end{bmatrix} \qquad (\text{Operate } R_1 + R_2,\ R_3 + 2R_2)$$

$$\sim \begin{bmatrix} 1 & 0 & -1/2 & : & -1/2 & 1 & 0 \\ 0 & -1 & -2 & : & -1 & 1 & 0 \\ 0 & 0 & -1/2 & : & -5/2 & 2 & 1 \end{bmatrix}$$

$$\sim \begin{bmatrix} 1 & 0 & -1/2 & : & -1/2 & 1 & 0 \\ 0 & 1 & 2 & : & 1 & -1 & 0 \\ 0 & 0 & -1/2 & : & -5/2 & 2 & 1 \end{bmatrix} \qquad \text{(Operate } (-2)\, R_3)$$

$$\sim \begin{bmatrix} 1 & 0 & -1/2 & : & -1/2 & 1 & 0 \\ 0 & 1 & 2 & : & 1 & -1 & 0 \\ 0 & 0 & 1 & : & 5 & -4 & -2 \end{bmatrix} \qquad \text{(Operate } R_1 + \frac{1}{2}\, R_3,\, R_2 - 2R_3)$$

$$\sim \begin{bmatrix} 1 & 0 & 0 & : & 2 & -1 & -1 \\ 0 & 1 & 0 & : & -9 & 7 & 4 \\ 0 & 0 & 1 & : & 5 & -4 & -2 \end{bmatrix}$$

Hence the inverse of the given matrix is $\begin{bmatrix} 2 & -1 & -1 \\ -9 & 7 & 4 \\ 5 & -4 & -2 \end{bmatrix}$

4.5 Factorization Method

In this method, we factorize the given matrix as $A = LU$ (1)

where L is a lower triangular matrix with unit diagonal elements and U is an upper triangular matrix

i.e., $L = \begin{bmatrix} 1 & 0 & 0 \\ l_{21} & 1 & 0 \\ l_{31} & l_{32} & 1 \end{bmatrix}$ and $U = \begin{bmatrix} u_{11} & u_{12} & u_{13} \\ 0 & u_{22} & u_{23} \\ 0 & l_0 & u_{33} \end{bmatrix}$ [Section3.4(5)1]

Now (1) gives $A^{-1} = (LU)^{-1} = U^{-1}L^{-1}$ (2)

To find L^{-1}, let $L^{-1} = X$, where X is a lower triangular matrix.

Then $LX = I$

i.e., $\begin{bmatrix} 1 & 0 & 0 \\ l_{21} & 1 & 0 \\ l_{31} & l_{32} & 1 \end{bmatrix} \begin{bmatrix} x_{11} & 0 & 0 \\ x_{21} & x_{22} & 0 \\ x22 & x_{32} & x_{33} \end{bmatrix} = \begin{bmatrix} 1 & 0 & 0 \\ 0 & 1 & 2 & 0 \\ 0 & 0 & 1 \end{bmatrix}$

Multiplying the matrices on the L.H.S. and equating the corresponding elements, we have

$$x_{11} = 1,\, x_{22} = 1,\, x_{33} = 1 \qquad (3)$$

$$l_{21}x_{11} + x_{21} = 0,\, l_{31}x_{11} + l_{32}x_{21} + x_{31} = 0$$

and $l_{32}x_{22} + x_{32} = 0$ (4)

(3) gives $\qquad x_{11} = x_{22} = x_{33} = 1$

(4) $\qquad x_{21} = -l_{21}x_{11}, x_{31} = -(l_{31} + l_{32}x_{21})$ and $x_{32} = -l_{32}$

Thus $L^{-1} = X$ is completely determined.

To find U^{-1}, let $U^{-1} = Y$, where Y is an upper triangular matrix.

Then $YU = I$

i.e.,
$$\begin{bmatrix} y_{11} & y_{12} & y_{13} \\ 0 & y_{22} & y_{23} \\ 0 & 0 & y_{33} \end{bmatrix}\begin{bmatrix} u_{11} & u_{12} & u_{13} \\ 0 & u_{22} & u_{23} \\ 0 & 0 & u_{33} \end{bmatrix} = \begin{bmatrix} 1 & 0 & 0 \\ 0 & 1 & 0 \\ 0 & 0 & 1 \end{bmatrix}$$

Multiplying the matrices on the L.H.S. and then equating the corresponding elements, we have

$$y_{11}u_{11} = 1, y_{22}u_{22} = 1, y_{33}u_{33} = 1 \tag{5}$$

and
$$\left.\begin{array}{r} y_{11}u_{12} + y_{12}u_{22} = 0, y_{11}u_{13} + y_{12}u_{23} + y_{13}u_{33} = 0 \\ y_{22}u_{23} + y_{23}u_{33} = 0 \end{array}\right\} \tag{6}$$

From (5), $y_{11} = 1/u_{11}, y_{22} = 1/u_{22}, y_{33} = 1/u_{33}$

From (6), $y_{12} = -y_{11}u_{12}/u_{22}, y13 = -(y_{11}u_{13} + y_{12}u_{23})/u_{33}; y_{23} = -y_{22}u_{23}/u_{33}.$

∴ We get $U^{-1} = Y$, completely.

Hence, by (2), we obtain A^{-1}.

EXAMPLE 4.3

Using the factorization method, find the inverse of the matrix
$$A = \begin{bmatrix} 50 & 107 & 36 \\ 25 & 54 & 20 \\ 31 & 66 & 21 \end{bmatrix}$$

Solution:

Taking $\quad L = \begin{bmatrix} 1 & 0 & 0 \\ l_{21} & 1 & 0 \\ l_{31} & l_{32} & 1 \end{bmatrix}$ and $U = \begin{bmatrix} u_{11} & u_{12} & u_{13} \\ 0 & u_{22} & u_{23} \\ 0 & 0 & u_{33} \end{bmatrix}$

$$A = LU \begin{bmatrix} 50 & 107 & 36 \\ 25 & 54 & 20 \\ 31 & 66 & 21 \end{bmatrix} = \begin{bmatrix} 1 & 0 & 0 \\ l_{21} & 1 & 0 \\ l_{31} & l_{32} & 1 \end{bmatrix}\begin{bmatrix} u_{11} & u_{12} & u_{13} \\ 0 & u_{22} & u_{23} \\ 0 & 0 & u_{33} \end{bmatrix}$$

$\therefore$ $50 = u_{11},\ 107 = u_{12},\ 36 = u_{13};$

$25 = l_{21}u_{11},\ 54 = l_{21}u_{12} + u_{22},\ 20 = l_{21}u_{13} + u_{23};$

$31 = l_{31}u_{11},\ 66 = l_{21}u_{12} + l_{32}u_{22},\ 21 = l_{31}u_{13} + l_{32}u_{23} + u_{33}$

or $u_{11} = 50,\ u_{12} = 107,\ u_{13} = 36,\ l_{21} = 1/2,\ u_{22} = 1/2,\ u_{23} = 2,$

$l_{31} = 31/50,\ l_{32} = -17/25,\ u_{23} = 1/25.$

Thus $L = \begin{bmatrix} 1 & 0 & 0 \\ 1/2 & 1 & 0 \\ 31/50 & -17/25 & 1 \end{bmatrix}$ and $U = \begin{bmatrix} 50 & 107 & 36 \\ 0 & 1/2 & 2 \\ 0 & 0 & 1/25 \end{bmatrix}$

To find L^{-1}, let $L^{-1} = X$. Then $LX = I$

i.e., $\begin{bmatrix} 1 & 0 & 0 \\ 1/2 & 1 & 0 \\ 31/50 & -17/25 & 1 \end{bmatrix}\begin{bmatrix} x_{11} & 0 & 0 \\ x_{21} & x_{22} & 0 \\ x_{31} & x_{32} & 0 \end{bmatrix} = \begin{bmatrix} 1 & 0 & 0 \\ 0 & 1 & 0 \\ 0 & 0 & 1 \end{bmatrix}$

$\therefore$ $x_{11} = 1, \dfrac{1}{2}x_{11} + x_{21} = 0,$

$x_{22} = 1, \dfrac{31}{50}x_{11} - \dfrac{17}{25}x_{21} + x_{31} = 0,$

$-\dfrac{17}{25}x_{22} + x_{32} = 0, x_{33} = 1.$

or $x_{11} = x_{22} = x_{33} = 1, x_{21} = -\dfrac{1}{2}, x_{31} = -\dfrac{24}{25}, x_{32} = \dfrac{17}{25}$

Thus $L^{-1} = X = \begin{bmatrix} 1 & 0 & 0 \\ -1/2 & 1 & 0 \\ -24/25 & 17/25 & 1 \end{bmatrix}$

To find U^{-1}, let $U^{-1} = Y$. Then $YU = I$

i.e., $\begin{bmatrix} y_{11} & y_{12} & y_{13} \\ 0 & y_{22} & y_{23} \\ 0 & 0 & y_{33} \end{bmatrix}\begin{bmatrix} 50 & 107 & 36 \\ 0 & 1/2 & 2 \\ 0 & 0 & 1/25 \end{bmatrix} = \begin{bmatrix} 1 & 0 & 0 \\ 0 & 1 & 0 \\ 0 & 0 & 1 \end{bmatrix}$

$\therefore 50y_{11} = 1, 50y_{12} + 107y_{22} = 0, 50y_{13} + 107y_{23} + 36y_{33} = 0$

$\dfrac{1}{2}y_{22} = 1, \dfrac{1}{2}y_{33} + 2y_{33} = 0, \dfrac{1}{25}y_{33} = 1.$

or $y_{11} = 1/50, y_{22} = 2, y_{33} = 25, y_{12} = -107/25, y_{23} = -100, y_{13} = 196.$

So that $\quad U^{-1} = \begin{bmatrix} 1/50 & -107/25 & 196 \\ 0 & 2 & -100 \\ 0 & 0 & 25 \end{bmatrix}$

Hence, $A^{-1} = U^{-1}L^{-1} = \begin{bmatrix} 1/50 & -107/25 & 196 \\ 0 & 2 & -100 \\ 0 & 0 & 25 \end{bmatrix} \begin{bmatrix} 1 & 0 & 0 \\ -1/2 & 1 & 0 \\ -24/25 & 17/25 & 1 \end{bmatrix}$

$$= \begin{bmatrix} -186 & 129 & 196 \\ 95 & -66 & -100 \\ -24 & 17 & 25 \end{bmatrix}$$

4.6 Partition Method

According to this method, if the inverse of a matrix A_n of order n is known, then the inverse of a matrix A_{n+1} of order $(n + 1)$ can be determined by adding $(n + 1)$th row and $(n + 1)$th column to A_n.

$$\text{Suppose } A = \begin{bmatrix} A_1 : A_2 \\ \cdots\cdots\cdots \\ A_3' : \alpha \end{bmatrix} \text{ and } A^{-1} = \begin{bmatrix} X_1 : X_2 \\ \cdots\cdots\cdots \\ X_3' : x \end{bmatrix}$$

where A_2, X_2 are column vectors and A_3', X_3' are row vectors (*i.e.*, transposes of column vectors A_3, X_3) and α, x are ordinary numbers.

Also we assume that $A1^{-1}$ is known. Actually A_3 and X_3 are column vectors since their transposes are row vectors.

Now $AA^{-1} = I_{n+1}$ gives

$$A_1X_1 + A_1X_3' = I_n \tag{1}$$

$$A_1X_2 + A_2x = 0 \tag{2}$$

$$A_3'X_1 + \alpha X_3' = 0 \tag{3}$$

$$A_3'X_2' + \alpha x = 1 \tag{4}$$

From (2), $X_2 = -A1^{-1}A_2x$ $\tag{5}$

and using this, (4) gives $(\alpha - A_3'A_1^{-1}A_2)x = 1$. $\tag{6}$

Hence x and then X_2 can be found.

Also from (1), $X_1 = A1^{-1}(I_n - A_2X_3')$ $\tag{7}$

and using this, (3) gives $\left(\alpha - A_3'A_1^{-1}A_2\right)X_3' = -A_3'A_1^{-1}$ (8)

whence X_3' and then X_1 are determined.

Thus, having found X_1, X_2, X_3' and x, A^{-1} is completely known.

NOTE   **Obs.** *The partition method is also known as the "Escalator method."*

EXAMPLE 4.4

Using the partition method, find the inverse of $A = \begin{bmatrix} 13 & 14 & 6 & 4 \\ 8 & -1 & 13 & 9 \\ 6 & 7 & 3 & 2 \\ 9 & 5 & 16 & 11 \end{bmatrix}$

Solution:

We have $A = \begin{bmatrix} 13 & 14 & 6 & : & 4 \\ 8 & -1 & 13 & : & 9 \\ 6 & 7 & 3 & : & 2 \\ \cdots & \cdots & \cdots & \cdots & \cdots \\ 9 & 5 & 16 & : & 11 \end{bmatrix} = \begin{bmatrix} A_1 : A_2 \\ \cdots\cdots\cdots \\ A_3 : \alpha \end{bmatrix}$

so that $A_1 = \begin{bmatrix} 13 & 14 & 6 \\ 8 & -1 & 13 \\ 6 & 7 & 3 \end{bmatrix}, A_2 = \begin{bmatrix} 4 \\ 9 \\ 2 \end{bmatrix}$

$A_3' = \begin{bmatrix} 9 & 5 & 16 \end{bmatrix}$ and $\alpha = 11$.

We find $A_1^{-1} = \dfrac{1}{94}\begin{bmatrix} 94 & 0 & -188 \\ -54 & -3 & 121 \\ -62 & 7 & 125 \end{bmatrix}$

Let $A^{-1} = \begin{bmatrix} X_1 & : & X_2 \\ \cdots\cdots & \cdots & \cdots \\ X_3' & : & \alpha \end{bmatrix}$. Then $AA^{-1} = I$

Hence $A_3'A_1^{-1}A_2 = \begin{bmatrix} 9 & 5 & 16 \end{bmatrix}\dfrac{1}{94}\begin{bmatrix} 94 & 0 & -188 \\ -54 & -3 & 121 \\ -62 & 7 & 125 \end{bmatrix}\begin{bmatrix} 4 \\ 9 \\ 2 \end{bmatrix}$

$= \dfrac{1}{94}\begin{bmatrix} 9 & 5 & 16 \end{bmatrix}\begin{bmatrix} 0 \\ -1 \\ 65 \end{bmatrix} = \dfrac{1035}{94}$

$$\therefore \quad (\alpha - A_3'A_1^{-1}A_2)\, x = 1 \qquad\qquad [(6) \text{ of Section } 4.6]]$$

becomes, $\left(11 - \dfrac{1035}{95}\right)x = 1 \ \ i.e., x = -94$

$$\text{Also } X_2 = -A_1^{-1}A_2 x = -\frac{1}{94}\begin{bmatrix} 94 & 0 & -188 \\ -54 & -3 & 121 \\ -62 & 7 & 125 \end{bmatrix}\begin{bmatrix} 4 \\ 9 \\ 2 \end{bmatrix}(-94) = \begin{bmatrix} 0 \\ -1 \\ 65 \end{bmatrix}$$

$$[(5) \text{ of Section } 4.7]$$

Then $\alpha - A_3'A_1^{-1}A_2)\, X_3' = -A_3'A_1^{-1}$

becomes $\left(11 - \dfrac{1035}{94}\right)X_3' = -\dfrac{1}{94}[-416, 97, -913]$ whench $X_3' = [-416, 97, 913]$

Finally $X_1 = A_1^{-1}(I - A_2 X_3')$ $\qquad\qquad [(7) \text{ of Section } 4.6]$

$$\text{where } A_2 X_3' = \begin{bmatrix} 4 \\ 9 \\ 2 \end{bmatrix}[-416, 97, 913] = \begin{bmatrix} -1664 & 388 & 3652 \\ -3744 & 873 & 8217 \\ -832 & 194 & 1826 \end{bmatrix}$$

$$\therefore \ X_1 = \frac{1}{94}\begin{bmatrix} 94 & 0 & -188 \\ -54 & -3 & 121 \\ -62 & 7 & 125 \end{bmatrix}\begin{bmatrix} 1665 & -388 & -3652 \\ 3744 & -872 & -8217 \\ 832 & -194 & -1825 \end{bmatrix} = \begin{bmatrix} 1 & 0 & -2 \\ -5 & 1 & 11 \\ 287 & -67 & -630 \end{bmatrix}$$

$$\text{Hence} \qquad A^{-1} = \begin{bmatrix} X_1 & X_2 \\ X_3' & x \end{bmatrix} = \begin{bmatrix} 1 & 0 & -2 & 0 \\ -5 & 1 & 11 & -1 \\ 287 & -67 & -630 & 65 \\ -416 & 97 & 913 & -94 \end{bmatrix}$$

EXAMPLE 4.5

If A and C are non-singular matrices, then show that

$$\begin{bmatrix} A & 0 \\ B & C \end{bmatrix}^{-1} = \begin{bmatrix} A^{-1} & 0 \\ -C^{-1}BA^{-1} & C^{-1} \end{bmatrix}$$

Hence find inverse of $\begin{bmatrix} 1 & 0 & 0 & 0 \\ 0 & 2 & 0 & 0 \\ 3 & 0 & 4 & 0 \\ 0 & 1 & 0 & 3 \end{bmatrix}$

Solution:

Let the given matrix be $M = \begin{bmatrix} A & O \\ B & C \end{bmatrix}$ and its inverse be $M^{-1} = \begin{bmatrix} P & Q \\ R & S \end{bmatrix}$ both in the portioned form where A, B, C, P, Q, R, S are all matrices.

$$\therefore \quad MM^{-1} = \begin{bmatrix} A & 0 \\ B & C \end{bmatrix}\begin{bmatrix} P & Q \\ R & S \end{bmatrix} = I$$

or

$$\begin{bmatrix} AP + 0R & AQ + 0S \\ BP + CR & BQ + CS \end{bmatrix} = \begin{bmatrix} I & 0 \\ 0 & I \end{bmatrix}$$

$\therefore$ Equating corresponding elements, we have

$$AP + 0R = I, AQ + 0S = 0, BP + CR = 0, BQ + CS = I.$$

Second relation gives $AQ = 0$ *i.e.* $Q = 0$ as A is non-singular.

First relation gives $AP = I$, *i.e.* $P = A^{-1}$.

First third equation, $BP + CR = 0$, *i.e.*, $CR = -BP = -BA^{-1}$

$\therefore \quad C^{-1} CR = -C^{-1}BA^{-1}$ or $IR = -C^{-1} BA^{-1}$ or $R = -C^{-1} BA^{-1}$

From fourth equation, $BQ + CS = I$, or $CS = I$ or $S = C^{-1}$

Hence $M^{-1} = \begin{bmatrix} A^{-1} & 0 \\ -C^{-1}BA^{-1} & C^{-1} \end{bmatrix}$

4.7 Iterative Method

Suppose we wish to compute A^{-1} and we know that B is an approximate inverse of A. Then the error matrix is given by $E = AB - I$

or $\qquad\qquad AB = I + E$

$\therefore \qquad\qquad (AB)^{-1} = (I + E)^{-1}$ *i.e.* $B^{-1} A_{-1} = (I + E)^{-1}$

or $\qquad\qquad A^{-1} = B(I + E)^{-1} = B(I - E + E^2 - \ldots\ldots),$

provided the series converges.

Thus we can find further approximations of A^{-1}, by using $A^{-1} = B(1 - E + E^2 - \ldots)$

EXAMPLE 4.6

Using the iterative method, find the inverse of

$$A = \begin{bmatrix} 1 & 10 & 1 \\ 2 & 0 & 1 \\ 3 & 3 & 2 \end{bmatrix} \text{taking } B = \begin{bmatrix} 0.4 & 2.4 & 1.4 \\ 0.14 & 0.14 & -0.14 \\ -0.85 & -3.8 & 2.8 \end{bmatrix}$$

Solution:

Here $E = AB - I = \begin{bmatrix} 1 & 10 & 1 \\ 2 & 0 & 1 \\ 3 & 3 & 2 \end{bmatrix} \begin{bmatrix} 0.4 & 2.4 & -1.4 \\ 0.14 & 0.14 & -0.14 \\ -0.85 & -3.8 & 2.8 \end{bmatrix} \begin{bmatrix} 1 & 0 & 0 \\ 0 & 1 & 0 \\ 0 & 0 & 1 \end{bmatrix}$

$$= \begin{bmatrix} -0.05 & 0 & 0 \\ -0.05 & 0 & 0 \\ -0.08 & 0.02 & -0.02 \end{bmatrix}$$

$$\therefore \quad E^2 = \begin{bmatrix} 0.0025 & 0 & 0 \\ 0.0025 & 0 & 0 \\ 0.0064 & -0.0004 & -0.0004 \end{bmatrix}$$

To the second approximation, we have

$$A^{-1} = B(1 - E + E^2) = B - BE + BE^2$$

$$= \begin{bmatrix} 0.4 & 2.4 & -14 \\ 0.0025 & 0.14 & -0.14 \\ -0.85 & -3.8 & -2.8 \end{bmatrix} - \begin{bmatrix} -0.02 & -0.12 & 0.07 \\ -0.02 & -0.12 & 0.07 \\ -0.0122 & -0.1132 & 0.0532 \end{bmatrix}$$

$$+ \begin{bmatrix} 0.001 & 0.006 & -0.0035 \\ 0.001 & 0.006 & -0.0035 \\ 0.0014 & 0.0095 & -0.0053 \end{bmatrix} = \begin{bmatrix} 0.421 & 2526 & -1474 \\ 0..161 & 0.266 & -0.214 \\ -0.836 & -3.677 & 2.742 \end{bmatrix}$$

Exercises 4.1

Use Gauss-Jordan method to find the inverse of the following matrices:

1. $\begin{bmatrix} 2 & 3 & 4 \\ 4 & 3 & 1 \\ 1 & 2 & 4 \end{bmatrix}$

2. $\begin{bmatrix} 0 & 1 & 2 \\ 1 & 2 & 3 \\ 3 & 1 & 1 \end{bmatrix}$

3. $\begin{bmatrix} 8 & 4 & 3 \\ 2 & 1 & 1 \\ 1 & 2 & 1 \end{bmatrix}$

Use factorization method, to find the inverse of the following matrices:

4. $\begin{bmatrix} 3 & 2 & 1 \\ 2 & 3 & 2 \\ 1 & 2 & 2 \end{bmatrix}$

5. $\begin{bmatrix} 2 & -2 & 4 \\ 2 & 3 & 2 \\ -1 & 1 & -1 \end{bmatrix}$

6. $\begin{bmatrix} 5 & -2 & 1 \\ 7 & 1 & -5 \\ 3 & 7 & 4 \end{bmatrix}$

7. $\begin{bmatrix} 10 & 2 & 1 \\ 2 & 20 & -2 \\ -2 & 3 & 10 \end{bmatrix}$

Apply the partition method to obtain the inverse of the following matrices:

8. $\begin{bmatrix} 1 & 1 & 1 \\ 4 & 3 & -1 \\ 3 & 5 & 3 \end{bmatrix}$

9. $\begin{bmatrix} 1 & 3 & 3 & 2 \\ 1 & 4 & 3 & 4 \\ 1 & 3 & 4 & 5 \\ 2 & 5 & 3 & 2 \end{bmatrix}$

10. Using iterative method, find the inverse of the matrix $A = \begin{bmatrix} 5 & 2 \\ 3 & -1 \end{bmatrix}$

taking $B = \begin{bmatrix} 0.1 & 0.2 \\ 0.3 & -0.4 \end{bmatrix}$

11. Apply iterative method to find more accurate inverse of $A = \begin{bmatrix} 1 & 10 & 1 \\ 2 & 0 & 1 \\ 3 & 3 & 2 \end{bmatrix}$,

assuming the initial inverse matrix to be $\begin{bmatrix} 0.43 & 2.43 & -1.43 \\ 0.14 & 0.15 & -0.14 \\ -0.85 & -3.85 & 0.85 \end{bmatrix}$

4.8   Eigenvalues and Eigenvectors

If A is any square matrix of order n with elements a_{ij}, we can find a column matrix X and a constant λ such that $AX = \lambda X$ or $AX - \lambda IX = 0$ or $[A - \lambda I] X = 0$.

This matrix equation represents n homogeneous linear equations

$$(a_{11} - \lambda)x_1 + a_{12} x_2 + \ldots\ldots + a_{1n} x_n = 0$$
$$a_{21} x_1 + (a_{22} - \lambda) x_2 + \ldots\ldots + a_{2n}x_n = 0 \tag{1}$$
$$\ldots\ldots\ldots\ldots\ldots\ldots\ldots\ldots\ldots\ldots\ldots\ldots$$
$$a_{n1}x_1 + a_{n2}x_2 + \ldots.. + (a_{nn} - \lambda) x_n = 0$$

which will have a non-trivial solution only if the coefficient determinant vanishes, *i.e.*,

$$\begin{vmatrix} a_{11} - \lambda & a_{12}\ldots\ldots\ldots a_{1n} \\ a_{21} & a_{22} - \lambda\ldots\ldots a_{2n} \\ a_{n1} & a_{n2}\ldots\ldots\ldots a_{nn} - \lambda \end{vmatrix} = 0 \tag{2}$$

On expansion, it gives an nth degree equation in λ, called the *characteristic equation* of the matrix A. Its roots λ_i ($i = 1, 2,\ldots.. n$) are called the *eigenvalues* or *latent roots* and corresponding to each eigenvalue, the equation (2) will have a non-zero solution

$$X = [x_1, x_2,\ldots\ldots, x_n]'$$

which is known as the *eigenvector*. Such an equation can ordinarily be solved easily. However for larger systems better methods are to be applied.

Cayley-Hamilton theorem. *Every square matrix satisfies its own characteristic equations i.e.*, if the characteristic equation for the nth order square matrix A is

$$|A - \lambda I| = (-1)^n \lambda^n + k_1 \lambda^{n-1} + \ldots\ldots + k_n = 0$$

then $\qquad\qquad (-1)n A^n + k_1 A^{n-1} + k_n = 0.$

EXAMPLE 4.7

Find the eigenvalues and eigenvectors of the matrix $\begin{bmatrix} 5 & 4 \\ 1 & 2 \end{bmatrix}$

Solution:

The characteristic equation is $[A - \lambda I] = 0$

i.e.,. $\begin{bmatrix} 5-\lambda & 4 \\ 1 & 2-\lambda \end{bmatrix} = $ or $\lambda^2 - 7\lambda + 6 = 0$

or $\quad (1-6)(1-1) = 0 \;\; \therefore 1 = 6, 1.$

Thus the eigenvalues are 6 and 1.

If x, y be the components of an eigenvector corresponding to the eigenvalue λ, then

$$[A - \lambda I]X = \begin{bmatrix} 5-\lambda & 4 \\ 1 & 2-\lambda \end{bmatrix}\begin{bmatrix} x \\ y \end{bmatrix} = 0$$

Corresponding to $1 = 6$, we have $\begin{bmatrix} -1 & 4 \\ 1 & -4 \end{bmatrix}\begin{bmatrix} x \\ y \end{bmatrix} = 0$

which gives only one independent equation $-x + 4y = 0$

$\therefore \dfrac{x}{y} = \dfrac{y}{1}$ giving the eigenvector (4, 1).

Corresponding to $1 = 1$, we have $\begin{bmatrix} 4 & 4 \\ 1 & 1 \end{bmatrix}\begin{bmatrix} x \\ y \end{bmatrix} = 0$ which gives only one independent equation $x + y = 0$.

$\therefore \dfrac{x}{1} = \dfrac{y}{-1}$ giving the eigenvector (1, – 1).

EXAMPLE 4.8

Find the eigenvalues and eigenvectors of the matrix

$$A = \begin{bmatrix} 8 & -6 & 3 \\ -6 & 7 & -4 \\ 2 & -4 & 3 \end{bmatrix}$$

Solution:

The characteristic equation is

$$|A - \lambda I| = \begin{bmatrix} 8-\lambda & -6 & 2 \\ -6 & 7-\lambda & -4 \\ 2 & -4 & 3-\lambda \end{bmatrix} = \lambda^3 + 18\lambda^2 - 45\lambda = 0$$

or $\quad \lambda(\lambda-3)(\lambda-15)=0 \therefore \lambda=0,3,15.$

Thus the eigenvalues of A are 0, 3, 15.

If x, y, z be the components of an eigenvector corresponding to the eigenvalue λ, we have

$$(A-\lambda)IX = \begin{bmatrix} 8-\lambda & -6 & 2 \\ -6 & 7-\lambda & -4 \\ 2 & -4 & 3-\lambda \end{bmatrix}\begin{bmatrix} x \\ y \\ z \end{bmatrix} = 0 \qquad (i)$$

Putting $l = 0$, we have $8x - 6y + 2z = 0$, $-6x + 7y - 4z = 0$, $2x - 4y + 3z = 0$.

These equations determine a single linearly independent solution which may be taken as $(1, 2, 2)$ so that every non-zero multiple of this vector is an eigenvector corresponding to $\lambda = 0$. $\qquad (ii)$

Similarly, the eigenvectors corresponding to $\lambda = 3$ and $\lambda = 15$ are the arbitrary nonzero multiples of the vectors $(2, 1, -2)$ and $(2, -2, 1)$ which are obtained from (i).

Hence the three eigenvectors may be taken as $(1, 2, 2)$, $(2, 1, -2)$, $(2, -2, 1)$.

NOTE   **Obs.** *The eigenvector $[x, y, z]'$ such that $x^2 + y^2 + z^2 = 1$ is said to be **normalized**. In particular, if we choose $x = 1/3$, $y = 2/3$, $z = 2/3$ in (ii), the corresponding normalized eigenvector will be $(1/3, 2/3, 2/3)$.*

EXAMPLE 4.9

Using the Cayley-Hamilton theorem, find the inverse of the matrix

$(i)\; A = \begin{bmatrix} 1 & 4 \\ \lambda & 3 \end{bmatrix}$ $(ii)\; A = \begin{bmatrix} 2 & 1 & 1 \\ 0 & 1 & 0 \\ 1 & 1 & 2 \end{bmatrix}$

Solution:

(i) The characteristic equation of the matrix is

$$\begin{vmatrix} 1-\lambda & 4 \\ 2 & 3-\lambda \end{vmatrix} = 0 \text{ or } \lambda^2 - 4\lambda - 5 = 0$$

By Cayley-Hamilton theorem, we have $A^2 - 4A - 5 = 0$

Multiplying by A^{-1}, we get $A - 4I - 5A^{-1} = 0$

or $\quad A - 1 = \dfrac{1}{5}(Z - 4I) = \dfrac{1}{5}\left\{ \begin{bmatrix} 1 & 4 \\ 2 & 3 \end{bmatrix} - 4 \begin{bmatrix} 1 & 0 \\ 0 & 1 \end{bmatrix} \right\} = \dfrac{1}{5} \begin{bmatrix} 3 & 4 \\ 2 & -1 \end{bmatrix}$

(*ii*) The characteristic equation of the matrix is

$$\begin{vmatrix} 2 - \lambda & 1 & 1 \\ 0 & 1 - \lambda & 0 \\ 1 & 2 & 1 - \lambda \end{vmatrix} = 0 \text{ or } \lambda^3 - 5\lambda^2 + 7\lambda - 3 = 0$$

By the Cayley-Hamilton theorem, we have $A^3 - 5A^2 + 7A - 3I = 0 \quad (i)$

Multiplying (*i*) by A^{-1}, we get

$$A^2 - 5A + 7I - 3A^{-1} = 0 \text{ or } A^{-1} = \frac{1}{3}\left(A^2 - 5A + 7 \right) \qquad (ii)$$

But $\quad A^2 = \begin{bmatrix} 2 & 1 & 1 \\ 0 & 1 & 0 \\ 1 & 1 & 2 \end{bmatrix} \begin{bmatrix} 2 & 1 & 1 \\ 0 & 1 & 0 \\ 1 & 1 & 2 \end{bmatrix} = \begin{bmatrix} 5 & 4 & 4 \\ 0 & 1 & 0 \\ 4 & 4 & 5 \end{bmatrix}$

$\therefore \ A^2 - 5A + 7I = \begin{bmatrix} 5 & 4 & 4 \\ 0 & 1 & 0 \\ 4 & 4 & 5 \end{bmatrix} - 5 \begin{bmatrix} 2 & 1 & 1 \\ 0 & 1 & 0 \\ 1 & 1 & 2 \end{bmatrix} + 7 \begin{bmatrix} 1 & 0 & 0 \\ 0 & 1 & 0 \\ 0 & 0 & 1 \end{bmatrix} = \begin{bmatrix} 2 & -1 & -1 \\ 0 & 3 & 0 \\ -1 & -1 & 2 \end{bmatrix}$

Hence from (ii), $A^{-1} = \dfrac{1}{3} \begin{bmatrix} 2 & -1 & -1 \\ 0 & 3 & 0 \\ -1 & -1 & 2 \end{bmatrix}$

4.9   Properties of Eigenvalues

We now state, some of the important properties of eigenvalues for ready reference:

I. *The sum of the eigenvalues of matrix A is the sum of the elements of its principal diagonal.*

II. *If λ is an eigenvalue of matrix A, then $1/\lambda$ is the eigenvalue of A^{-1}.*

III. *If λ is an eigenvalue of an orthogonal matrix, then $1/\lambda$ is also its eigenvalue.*

IV. *If $\lambda_1, \lambda_2, \ldots, \lambda_n$ are the eigenvalues of matrix A, then A^m has the eigenvalues $\lambda 1^m, \lambda_2^m, \ldots, \lambda_n m$ (m being a positive integer).*

V. *If a square matrix A has n linearly independent eigenvectors, then a matrix P can be found such that $P^{-1} AP$ is a diagonal matrix whose diagonal elements are the eigenvalues of A.*

The transformation of A by a non-singular matrix P to $P^{-1} AP$ is called a *similarity transformation*.

VI. *Any similarity transformation applied to a matrix leaves its eigenvalues unchanged.*

4.10   Bounds for Eigenvalues

If λ is an eigenvalue of matrix A, then for some k $(1 \le k \le n)$,

$$|\lambda - a_{kk}| \le |a_{k1}| + |a_{k2}| + \cdots + |a_{kn}| = p_k \text{ (say)},$$

i.e., all the eigenvalues of A lie in the union of the n circles with centers a_{kk} and radii ρ_k.

Proof. Let λ be an eigenvalue of an arbitrary square matrix A and X be the corresponding eigenvector. Then $AX = \lambda X$

or
$$a_{11} x_1 + a_{12} x_2 + \ldots + a_{1n} x_n = \lambda x_1$$
$$\cdots\cdots\cdots\cdots\cdots\cdots$$
$$a_{k1} x_1 + a_{k2} x_2 + \ldots + a_{kn} x_n = \lambda x_k$$
$$\cdots\cdots\cdots\cdots\cdots\cdots$$
$$a_{n1} x_1 + a_{n2} x_2 + \ldots + a_{nn} x_n = \lambda x_n$$

If x_k be the largest component of X, then $|x_m/x_k| \le 1$ $(m = 1, 2, \cdots, n)$ ...(1)

Dividing the kth equation by x_k, we obtain

$$a_{k1}(x_1/x_k) + \ldots + a_{k, k-1}(x_{k-1}/x_k) + a_{kk} + \cdots + a_{kn}(x_n/x_k) = \lambda$$

or $$\lambda - a_{kk} = a_{k1}(x_1/x_k) + \cdots + a_{k, k-1}(x_{k-1}/x_k) + \cdots + a_{kn}(x_n/x_k)$$

Taking absolute values on both sides and using the theorem $|a + b| \le |a| + |b|$, we obtain

$$|\lambda - a_{kk}| \le |a_{k1}| + \cdots + |a_{k, k-1}| + \cdots + |a_{kn}| = \rho_k \text{ (say) [by (1)]}$$

This shows that all the eigenvalues of A lie within or on the union of the circles with centers a_{kk} and radii ρ_k.

As A and A' have the same eigenvalues, the above theorem is also true for columns. *These circles are called the* **Gerschgorin circles**

The bounds thus obtained being all independent all the eigenvalues of A must lie in the intersection of these bounds. *These bounds are called the* **Gerschgorin bounds**.

The above theorem gives us the possible location of the eigenvalues and also helps us to estimate their bounds. *If any of the Gerschgorin circles is isolated, then it contains exactly one eigenvalue.*

EXAMPLE 4.10

Using Gerschgorin circles, determine the limits of the eigenvalues of

the matrix $A = \begin{bmatrix} 1 & 3 & 2 \\ 3 & 4 & 6 \\ 2 & 6 & 1 \end{bmatrix}$

Solution:

The three Gerschgorin circles are

$(a) \, |z - 1| = |3| + |2| = 5$

$(b) \, |z - 4| = |3| + |6| = 9$

$(c) \, |z - 1| = |2| + |6| = 8$

One eigenvalue lies within the circle having the center at $(1, 0)$ and radius 5.

Second eigenvalue lies within the circle having the center at $(4, 0)$ and radius 9.

Third eigenvalue lies within the circle having the center at $(1, 0)$ and radius 8.

Since the circle (a) lies within the circles (b) and (c), therefore all the eigenvalues of A lie within the region defined by (b) and (c). thus $-5 \le \lambda \le 13$ and $-7 \le \lambda \le 9$.

Hence the limits to the eigenvalues are given by $-7 \le \lambda \le 13$.

4.11 Power Method

In many engineering problems, it is required to compute the numerically largest eigenvalue and the corresponding eigenvector. In such cases, the following *iterative method* is quite convenient which is also well-suited for machine computations.

If $X_1, X_2 \cdots X_n$ are the eigenvectors corresponding to the eigenvalues λ_1, $\lambda_2, \cdots \lambda_n$, then an arbitrary column vector can be written as

$$X = k_1 X_1 + k_2 X_2 + \cdots + k_n X_n$$

Then
$$AX = k_1 AX_1 + k_2 AX_2 + \cdots + k_n AX_n$$

$$= k_1 \lambda_1 X_1 + k_2 \lambda_2 X_2 + \cdots + k_n \lambda_n X_n$$

Similarly
$$A^2 X = k_1 \lambda_1^2 X_1 + k_2 \lambda_2^2 X_2 + \cdots + k_n \lambda_n^2 X_n$$

and
$$A^r X = k_1 \lambda_1 \, ^r X_1 + k_2 \lambda_2 \, ^r X_2 + \cdots + k_n \lambda_n \, ^r X_n$$

If $|\lambda_1| > |\lambda_2| > \cdots > |\lambda_n|$, then λ_1 is the largest root and the contribution of the term $k_1\lambda_1{}^r X_1$ to the sum on the right increases with r and, therefore, every time we multiply a column vector by A, it becomes nearer to the eigenvector X_1. Then we make the largest component of the resulting column vector unity to avoid the factor k_1.

Thus we start with a column vector X which is as near the solution as possible and evaluate AX which is written as $\lambda^{(1)} X^{(1)}$ after normalization. This gives the first approximation $\lambda^{(1)}$ to the eigenvalue and $X^{(1)}$ to the eigenvector. Similarly we evaluate $AX^{(1)} = \lambda^{(2)} X^{(2)}$ which gives the second approximation. We repeat this process until $[X^{(r)} - X^{(r-1)}]$ becomes negligible. Then $\lambda^{(r)}$ will be the largest eigenvalue and $X^{(r)}$, the corresponding eigenvector.

This iterative procedure for finding the dominant eigenvalue of a matrix is known as

Rayleigh's power method.

NOTE **Obs.** *Rewriting $AX = \lambda X$ as $A^{-1} AX = \lambda A^{-1} X$ or $X = \lambda A^{-1} X$.*

We have $A^{-1} X = \dfrac{1}{\lambda} X$

If we use this equation, then the above method yields the smallest eigenvalue.

EXAMPLE 4.10

Determine the largest eigenvalue and the corresponding eigenvector of the matrix $\begin{bmatrix} 5 & 4 \\ 1 & 2 \end{bmatrix}$

Solution:

Let the initial approximation to the eigenvector corresponding to the largest eigenvalue of A be $X = \begin{bmatrix} 1 \\ 0 \end{bmatrix}$

Then $AX = \begin{bmatrix} 5 & 4 \\ 1 & 2 \end{bmatrix}\begin{bmatrix} 1 \\ 0 \end{bmatrix} = \begin{bmatrix} 5 \\ 1 \end{bmatrix} = 5\begin{bmatrix} 1 \\ 0.2 \end{bmatrix} = \lambda^{(1)} X^{(1)}$

So the first aporoximation to the eigenvalue is $\lambda^{(1)} = 5$ and the corresponding eigenvector is $X^{(1)} = \begin{bmatrix} 1 \\ 0.2 \end{bmatrix}$

Now $AX^{(1)} = \begin{bmatrix} 5 & 4 \\ 1 & 2 \end{bmatrix}\begin{bmatrix} 1 \\ 0.2 \end{bmatrix} = \begin{bmatrix} 5.8 \\ 14 \end{bmatrix} = 5.8\begin{bmatrix} 1 \\ 0.241 \end{bmatrix} = \lambda^{(2)} X^{(2)}$

Thus the second aporoximation to the eigenvalue is $\lambda^{(2)} = 5.8$ and the corresponding eigenvector is $X^{(2)} = \begin{bmatrix} 1 \\ 0.241 \end{bmatrix}$, repeating the above process, we get

$$AX^{(2)} = \begin{bmatrix} 5 & 4 \\ 1 & 2 \end{bmatrix}\begin{bmatrix} 1 \\ 0.241 \end{bmatrix} = 5.966\begin{bmatrix} 1 \\ 0.248 \end{bmatrix} = \lambda^{(3)} X^{(3)}$$

$$AX^{(3)} = \begin{bmatrix} 5 & 4 \\ 1 & 2 \end{bmatrix}\begin{bmatrix} 1 \\ 0.249 \end{bmatrix} = 5.966\begin{bmatrix} 1 \\ 0.250 \end{bmatrix} = \lambda^{(4)} X^{(4)}$$

$$AX^{(4)} = \begin{bmatrix} 5 & 4 \\ 1 & 2 \end{bmatrix}\begin{bmatrix} 1 \\ 0.250 \end{bmatrix} = 5.999\begin{bmatrix} 1 \\ 0.25 \end{bmatrix} = \lambda^{(5)} X^{(5)}$$

$$AX^{(5)} = \begin{bmatrix} 5 & 4 \\ 1 & 2 \end{bmatrix}\begin{bmatrix} 1 \\ 0.25 \end{bmatrix} = 6\begin{bmatrix} 1 \\ 0.25 \end{bmatrix} = \lambda^{(6)} X^{(6)}$$

Clearly $\lambda^{(5)} = \lambda^{(6)}$ and $X^{(5)} = X^{(6)}$ upto 3 decimal places. Hence the largest eigenvalue is 6 and the corresponding eigenvector is $\begin{bmatrix} 1 \\ 0.25 \end{bmatrix}$

EXAMPLE 4.11

Find the largest eigenvalue and the corresponding eigenvector of the

Matrix $\begin{bmatrix} 2 & -1 & 0 \\ -1 & 2 & -1 \\ 0 & -1 & 2 \end{bmatrix}$ using the power method. Take $[1, 0, 0]^T$ as the initial eigenvector.

Solution:

Let the initial approximation to the required eigenvector be $X = [1, 0, 0]'$.

Then $AX = \begin{bmatrix} 2 & -1 & 0 \\ -1 & 2 & -1 \\ 0 & -1 & 2 \end{bmatrix}\begin{bmatrix} 1 \\ 0 \\ 0 \end{bmatrix} = \begin{bmatrix} 2 \\ -1 \\ 0 \end{bmatrix} = 2\begin{bmatrix} 1 \\ -0.5 \\ 0 \end{bmatrix} = \lambda^{(1)}X^{(1)}$

So the first approximation to the eigenvalue is 2 and the corresponding eigenvector

$X(1) = [1, -0.5, 0]'$.

Hence $AX^{(1)} = \begin{bmatrix} 2 & -1 & 0 \\ -1 & 2 & -1 \\ 0 & -1 & 2 \end{bmatrix}\begin{bmatrix} 1 \\ -0.5 \\ 0 \end{bmatrix} = \begin{bmatrix} 2.5 \\ -2 \\ 0.5 \end{bmatrix} = \begin{bmatrix} 1 \\ -0.8 \\ 0.2 \end{bmatrix} = \lambda^{(2)}X^{(2)}$

Repeating the above process, we get

$AX^{(2)} = 2.8\begin{bmatrix} 1 \\ -1 \\ 0.43 \end{bmatrix} = \lambda^{(3)}X^{(3)}; \qquad AX^{(3)} = 3.43\begin{bmatrix} 0.87 \\ -1 \\ 0.54 \end{bmatrix} = \lambda^{(4)}X^{(4)}$

$AX^{(4)} = 3.41\begin{bmatrix} 0.80 \\ -1 \\ 0.61 \end{bmatrix} = \lambda^{(5)}X^{(5)}; \qquad AX^{(5)} = 3.41\begin{bmatrix} 0.76 \\ -1 \\ 0.65 \end{bmatrix} = \lambda^{(6)}X^{(6)}$

$AX^{(6)} = 3.41\begin{bmatrix} 0.74 \\ -1 \\ 0.67 \end{bmatrix} = \lambda^{(7)}X^{(7)}$

Clearly $\lambda^{(6)} = \lambda^{(7)}$ and $X^{(6)} = X^{(7)}$ approximately.

Hence the largest eigenvalue is 3.41 and the corresponding eigenvector is $[0.74, -1, 0.67]'$

EXAMPLE 4.12

Obtain by the power method, the numerically dominant eigenvalue and eigenvector of the matrix

$$A = \begin{bmatrix} 15 & -4 & -3 \\ -10 & 12 & -6 \\ -20 & 4 & -2 \end{bmatrix}$$

Solution:

Let the initial approximation to the eigenvector be $X = [1, 1, 1]'$. Then

$$AX = \begin{bmatrix} 15 & -4 & -3 \\ -10 & 12 & -6 \\ -20 & 4 & -2 \end{bmatrix} \begin{bmatrix} 1 \\ 1 \\ 1 \end{bmatrix} = \begin{bmatrix} 8 \\ -4 \\ -18 \end{bmatrix} = -18 \begin{bmatrix} -0.444 \\ 0.222 \\ 1 \end{bmatrix} = \lambda^{(1)} X^{(1)}$$

So the first approximation to eigenvalue is -18 and the corresponding eigenvector is $[-0.444, 0.222, 1]'$.

$$\text{Now} \quad AX^{(1)} = \begin{bmatrix} 15 & -4 & -3 \\ -10 & 12 & -6 \\ -20 & 4 & -2 \end{bmatrix} \begin{bmatrix} -0.444 \\ 0.222 \\ 1 \end{bmatrix} = -10.548 \begin{bmatrix} 1 \\ -0.105 \\ -0.736 \end{bmatrix} = \lambda^{(1)} X^{(2)}$$

∴ The second approximation to the eigenvalue is -10.548 and the eigenvector is $[1, -0.105, -0.736]'$.

Repeating the above process

$$AX^{(2)} = - \begin{bmatrix} -0.930 \\ - \\ 1 \end{bmatrix} = \lambda^3 X^{(3)} \quad AX^{(3)} = - \begin{bmatrix} 1 \\ - \\ -0.981 \end{bmatrix} = \lambda^4 X^{(4)}$$

$$AX^{(4)} = -19.698 \begin{bmatrix} -0.995 \\ 0.462 \\ 1 \end{bmatrix} = \lambda^{(5)} X^{(5)}; AX^{(5)} = -19.773 \begin{bmatrix} 1 \\ -480 \\ -0.999 \end{bmatrix} = \lambda^{(6)} X^{(6)}$$

$$AX^{(6)} = -19.922 \begin{bmatrix} -0.997 \\ 0.490 \\ 1 \end{bmatrix} = \lambda^{(7)} X^{(7)}; AX^{(7)} = -19.956 \begin{bmatrix} 1 \\ -495 \\ -0.999 \end{bmatrix} = \lambda^{(8)} X^{(8)}$$

Since $\lambda^{(7)} = \lambda^{(8)}$ and $X^{(7)} = X^{(8)}$ approximately, therefore the dominant eigenvalue and the corresponding eigenvector are given by

$$\lambda^{(8)} X^{(8)} = 19.956 \begin{bmatrix} -1 \\ 0.495 \\ 0.999 \end{bmatrix} i.e., \ 20 \begin{bmatrix} -1 \\ 0.5 \\ 1 \end{bmatrix}$$

Hence the dominant eigenvalue is 20 and eigenvector is $[-1, 0.5, 1]'$.

Exercises 4.2

1. Find the eigenvalues and eigenvectors of the matrices.

$(a) \begin{bmatrix} 1 & 4 \\ 3 & 2 \end{bmatrix}$
$(b) \begin{bmatrix} 1 & -2 \\ -5 & 4 \end{bmatrix}$

2. Find the latent root and the latent vectors of the matrices

$(a) \begin{bmatrix} 2 & 0 & 1 \\ 0 & 2 & 0 \\ 1 & 0 & 2 \end{bmatrix}$
$(b) \begin{bmatrix} -2 & 2 & -3 \\ 2 & 1 & -6 \\ -1 & -2 & 0 \end{bmatrix}$
$(c) \begin{bmatrix} -6 & -2 & 2 \\ -2 & 3 & -1 \\ 2 & -1 & 3 \end{bmatrix}$

3. Using the Cayley-Hamilton theorem, find the inverse of

$(a) \begin{bmatrix} 1 & 1 & 2 \\ 0 & -2 & 0 \\ 0 & 0 & 3 \end{bmatrix}$
$(b) \begin{bmatrix} 2 & 1 & 1 \\ 0 & 1 & 0 \\ 1 & 1 & 2 \end{bmatrix}$
$(c) \begin{bmatrix} 2 & -1 & 1 \\ -1 & 2 & -1 \\ 1 & -1 & 2 \end{bmatrix}$

4. Using Gerschgorim circles, find the limits of the eigenvalues of the

$$\text{matrix } A = \begin{bmatrix} 2 & 2 & 0 \\ 2 & 5 & 0 \\ 0 & 0 & 3 \end{bmatrix}$$

5. Find, by power method, the larger eigenvalue of the following matrices:

$(a) \begin{bmatrix} 1 & 2 \\ 3 & 4 \end{bmatrix}$
$(b) \begin{bmatrix} 4 & 1 \\ 1 & 3 \end{bmatrix}$

6. Find the largest eigenvalue and the corresponding eigenvector of the matrices:

$(a) \begin{bmatrix} 1 & 3 & -1 \\ 3 & 2 & 4 \\ -1 & 4 & 10 \end{bmatrix}$
$(b) \begin{bmatrix} 1 & -3 & 2 \\ 4 & 4 & -1 \\ 6 & 3 & 5 \end{bmatrix}$

(c) $\begin{bmatrix} 25 & 1 & 2 \\ 1 & 3 & 0 \\ 2 & 0 & -4 \end{bmatrix}$ taking $[1, 0, 0]^T$ as initial eigenvector.

NOTE

Obs. *The iteration method is a special method as it gives the largest or the smallest eigenvalue only. Now we shall describe three modern methods for finding all the eigenvalues of a real symmetric matrix A.*

The eigenvalues of A are given by the diagonal elements when A is reduced to either the diagonal matrix D or the lower triangular matrix L or the upper triangular matrix U. Thus the methods of finding eigenvalues of A are based on reducing A to D or L or U.

4.12   Jacobi's Method

Let A be a given real symmetric matrix. Its eigenvalues are real and there exists a real orthogonal matrix B such that $B^{-1} AB$ is a diagonal matrix D. Jacobi's method consists of diagonalizing A by applying a series of orthogonal transformations B_1, B_2,$\cdots$, B_r such that their product B satisfies the equation $B^{-1} AB = D$.

For this purpose, we choose the numerically largest non-diagonal element a_{ij} and form a 2×2 submatrix $A_1 = \begin{bmatrix} a_{ij} & a_{ij} \\ a_{ji} & a_{jj} \end{bmatrix}$.

Where $a_{ij} = a_{ji}$, which can easily be diagonalized.

Consider an orthogonal matrix $B_1 = \begin{bmatrix} \cos\theta & -\sin\theta \\ \sin\theta & \cos\theta \end{bmatrix}$ so that $B_1^{-1} = B_1'$

Then $B^{-1} A_1 B_1 = \begin{bmatrix} \cos\theta & \sin\theta \\ -\sin\theta & \cos\theta \end{bmatrix} \begin{bmatrix} aii & aij \\ aji & ajj \end{bmatrix} \begin{bmatrix} \cos\theta & -\sin\theta \\ \sin\theta & \cos\theta \end{bmatrix}$

$$= \begin{bmatrix} a_{ii}\cos^2\theta + a_{jj}\sin^2\theta + a_{ij}\sin 2\theta, & a_{ij}\cos 2\theta + \frac{1}{2}\left(a_{jj} - a_{ii}\right)\sin 2\theta \\ a_{ij}\cos 2\theta + \frac{1}{2}\left(a_{jj} - a_{ii}\right)\sin 2\theta, & a_{ii}\sin^2\theta + a_{jj}\cos^2\theta - a_{ij}\sin 2\theta \end{bmatrix} \quad (1)$$

Now this matrix will reduce to the diagonal form, if $a_{ij}\cos 2\theta + \frac{1}{2}(a_{jj} - a_{ii})\sin 2\theta = 0$

i.e., if

$$\tan 2\theta = \frac{2aij}{a_{ii} - a_{jj}} \tag{2}$$

This equation gives four values of θ, but to get the least possible rotation, we choose $-\pi/4 \le \theta \le \pi/4$.

Thus (1) reduces to a diagonal matrix.

As a next step, the largest non-diagonal element (in magnitude) in the new rotated matrix is found and the above procedure is repeated using the orthogonal matrix B_2.

In this way, a series of such transformations are performed so as to annihiliate the non-diagonal elements. After making r transformations, we obtain

$$B_r^{-1} B_{r-1}^{-1} \cdots B_1^{-1} AB_1 \cdots B_{r-1} B_r = B^{-1} AB$$

As $r \to \infty$, $B^{-1} AB$ approaches a diagonal matrix whose diagonal elements are the eigenvalues of A.

Also the corresponding columns of $B = B_1 B_2 \cdots B_r$, are the eigenvectors of A.

EXAMPLE 4.13

Using Jacobi's method, find all the eigenvalues and the eigenvectors of the matrix

$$A = \begin{bmatrix} 1 & \sqrt{2} & 2 \\ \sqrt{2} & 3 & \sqrt{2} \\ 2 & \sqrt{2} & 1 \end{bmatrix}$$

Solution:

Here the largest non-diagonal element is $a_{13} = a_{31} = 2$. Also $a_{11} = 1$ and $a_{33} = 1$

$$\therefore \qquad \tan 2\theta = \frac{2a_{13}}{a_{11} - a_{33}} = \frac{2 \times 2}{1 - 1} \to \infty$$

i.e., $\quad 2\theta = \pi/2$ or $\theta = \pi/4$

Then $\quad B_1 = \begin{bmatrix} \cos\theta & 0 & -\sin\theta \\ 0 & 1 & 0 \\ \sin\theta & 0 & \cos\theta \end{bmatrix} = \begin{bmatrix} 1/\sqrt{2} & 0 & -1/\sqrt{2} \\ 0 & 1 & 0 \\ 1/\sqrt{2} & 0 & 1/\sqrt{2} \end{bmatrix}$ and $B_1^{-1} = B'$

∴ The first transformation gives

$$D_1 = B_1^{-1}AB_1 = \begin{bmatrix} 1/\sqrt{2} & 0 & 1/\sqrt{2} \\ 0 & 1 & 0 \\ -1/\sqrt{2} & 0 & 1/\sqrt{2} \end{bmatrix} \begin{bmatrix} 1 & 1/\sqrt{2} & 2 \\ 1/\sqrt{2} & 3 & 1/\sqrt{2} \\ 2 & 1/\sqrt{2} & 1 \end{bmatrix} \times \begin{bmatrix} 1/\sqrt{2} & 0 & -1/\sqrt{2} \\ 0 & 1 & 0 \\ 1/\sqrt{2} & 0 & 1/\sqrt{2} \end{bmatrix}$$

$$= \begin{bmatrix} 3 & 2 & 0 \\ 2 & 3 & 0 \\ 0 & 0 & -1 \end{bmatrix}$$

Now the largest non-diagonal element is $a_{12} = a_{21} = 2$. Also $a_{11} = 3$ and $a_{22} = 3$.

∴ $$\tan 2\theta = \frac{2a_{12}}{a_{11} - a_{22}} = \frac{2 \times 2}{0} \to \infty$$

i.e., $2\theta = \pi/2$ or $\theta = \pi/4$.

Then $$B_2 = \begin{bmatrix} \cos\theta & -\sin\theta & 0 \\ \sin\theta & \cos\theta & 0 \\ 0 & 0 & 1 \end{bmatrix} = \begin{bmatrix} 1/\sqrt{2} & -1/\sqrt{2} & 0 \\ 1/\sqrt{2} & 1/\sqrt{2} & 0 \\ 0 & 0 & 1 \end{bmatrix}$$

∴ The second transformation gives

$$B_2^{-1}D_1B_2 = \begin{bmatrix} 1/\sqrt{2} & 1/\sqrt{2} & 0 \\ -1/\sqrt{2} & 1/\sqrt{2} & 0 \\ 0 & 0 & 1 \end{bmatrix} \begin{bmatrix} 3 & 2 & 0 \\ 2 & 3 & 0 \\ 0 & 0 & -1 \end{bmatrix} \times \begin{bmatrix} 1/\sqrt{2} & -1/\sqrt{2} & 0 \\ 1/\sqrt{2} & 1/\sqrt{2} & 0 \\ 0 & 0 & 1 \end{bmatrix} = \begin{bmatrix} 5 & 0 & 0 \\ 0 & 1 & 0 \\ 0 & 0 & -1 \end{bmatrix}$$

Hence the eigenvalues of the given matrix are 5, 1, – 1 and the corresponding eigenvectors are the columns of

$$B = B_1B_2 = \begin{bmatrix} 1/\sqrt{2} & 0 & -1/\sqrt{2} \\ 0 & 1 & 0 \\ 1/\sqrt{2} & 0 & 1/\sqrt{2} \end{bmatrix} \begin{bmatrix} 1/\sqrt{2} & -1/\sqrt{2} & 0 \\ 1/\sqrt{2} & 1/\sqrt{2} & 0 \\ 0 & 0 & 1 \end{bmatrix} = \begin{bmatrix} 1/\sqrt{2} & -1/\sqrt{2} & -1/\sqrt{2} \\ 1/\sqrt{2} & 1/\sqrt{2} & 0 \\ 1/2 & -1/\sqrt{2} & 1/\sqrt{2} \end{bmatrix}$$

NOTE *A disadvantage of Jacobi's method is that the element annihiliated by a transformation, may not remain zero during the subsequent transformations. Given's suggested a reduction which does not disturb zeros already formed. But instead of leading to a diagonal matrix as in Jacobi's method, the Given's method leads to a tri-diagonal matrix. The eigenvalues and eigenvectors of the original matrix have to be found from those of the tri-diagonal matrix.*

EXAMPLE 4.14

Obtain using Jacobi's method, all the eigenvalues and eigenvectors of the matrix

$$A = \begin{bmatrix} 1 & 1 & 0.5 \\ 1 & 1 & 0.25 \\ 0.5 & 0.25 & 2 \end{bmatrix}$$

Solution:

Here the largest non-diagonal element is $a_{12} = 1$.

Also $a_{11} = 1$, $a_{22} = 1$.

$\therefore$
$$\tan 2\theta = \frac{2a_{12}}{a_{11} = a_{12}} = \frac{2 \times 1}{0} \to \infty.$$

i.e.,
$$2\theta = \frac{\pi}{2} \text{ or } \theta = \frac{\pi}{4}.$$

Then
$$B_1 = \begin{bmatrix} \cos\theta & -\sin\theta & 0 \\ \sin\theta & \cos\theta & 0 \\ 0 & 0 & 1 \end{bmatrix} = \begin{bmatrix} 1/\!\!\sqrt{2} & -1/\!\!\sqrt{2} & 0 \\ 1/\!\!\sqrt{2} & 1/\!\!\sqrt{2} & 0 \\ 1 & 0 & 1 \end{bmatrix} \text{ and } B_1 - 1 = B_1'$$

$\therefore$ The first transformation is

$$D_1 = B_1^{-1}AB_1 = \begin{bmatrix} 1/\!\!\sqrt{2} & 1/\!\!\sqrt{2} & 0 \\ -1/\!\!\sqrt{2} & 1/\!\!\sqrt{2} & 0 \\ 0 & 0 & 1 \end{bmatrix} \begin{bmatrix} 1 & 1 & 1/\!\!\sqrt{2} \\ 1 & 1 & 1/4 \\ 1/\!\!\sqrt{2} & 1/4 & 1 \end{bmatrix} \begin{bmatrix} 1/\!\!\sqrt{2} & -1/\!\!\sqrt{2} & 0 \\ 1/\!\!\sqrt{2} & 1/\!\!\sqrt{2} & 0 \\ 0 & 0 & 1 \end{bmatrix}$$

$$= \begin{bmatrix} 2 & 0 & 3\sqrt{2/8} \\ 0 & 0 & -\sqrt{2/8} \\ 3\sqrt{2/8} & \sqrt{2/8} & 2 \end{bmatrix}$$

Now the largest non-diagonal element of $+ D_1$ is $a_{13} = 3\sqrt{2}/8$. Also, $\alpha_{11} = 2, \alpha_{33} = 2$.

$\therefore$
$$\tan 2\theta = \frac{2\alpha_{13}}{\alpha_{11} - \alpha_{33}} \to \infty, i.e., 2\theta = \frac{\pi}{2} \text{ or } \theta \frac{\pi}{4}$$

Then
$$B_2 = \begin{bmatrix} \cos\theta & 0 & -\sin\theta \\ 0 & 1 & 0 \\ \sin\theta & 0 & \cos\theta \end{bmatrix} = \begin{bmatrix} 1/\!\!\sqrt{2} & 0 & -1/\!\!\sqrt{2} \\ 0 & 1 & 0 \\ 1/\!\!\sqrt{2} & 0 & 1/\!\!\sqrt{2} \end{bmatrix}$$

$\therefore$ The second transformation gives

$$D_2 = B_2^{-1}D_1B_2 = \begin{bmatrix} 1/\sqrt{2} & 0 & 1/\sqrt{2} \\ 0 & 1 & 0 \\ -1/\sqrt{2} & 0 & 1/\sqrt{2} \end{bmatrix} \begin{bmatrix} 2 & 0 & 3\sqrt{2}/8 \\ 0 & 0 & -\sqrt{2}/8 \\ 3\sqrt{2}/8 & -\sqrt{2}/8 & 2 \end{bmatrix} \begin{bmatrix} 1/\sqrt{2} & 0 & -1/\sqrt{2} \\ 0 & 1 & 0 \\ 1/\sqrt{2} & 0 & 1/\sqrt{2} \end{bmatrix}$$

$$= \begin{bmatrix} 2.530 & -0.125 & 0 \\ -0.125 & 0 & -0.125 \\ 0 & -0.125 & 147 \end{bmatrix}$$

Repeating the above steps, we obtain

$$B_3 = \begin{bmatrix} 0.998 & 0.049 & 0 \\ -0.049 & 0.998 & 0 \\ 0 & 0 & 1 \end{bmatrix}$$

and

$$D_3 = B_3^{-1}D_2B_3 = \begin{bmatrix} 2.536 & -0.000 & 0.006 \\ -0.000 & -0.006 & -0.125 \\ 0.006 & -0.125 & 1469 \end{bmatrix}$$

Hence the eigenvalues of A are 2.536, – 0.006, 1.469 approximately and the corresponding eigenvectors are the columns of

$$B = B_1B_2B_3 = \begin{bmatrix} 0.531 & -0.721 & -0.444 \\ 0.461 & 0.686 & -0.562 \\ 0.710 & 0.094 & 0.698 \end{bmatrix}$$

4.13   Given's Method

If A is a real symmetric matrix, then Given's method consists of the following steps:

Step I. To reduce A to a tri-diagonal symmetric matrix:

To begin with, consider the matrix $A_1 = \begin{bmatrix} a_{11} & a_{12} & a_{13} \\ a_{12} & a_{22} & a_{23} \\ a_{13} & a_{23} & a_{33} \end{bmatrix}$ (1)

and the orthogonal rotation matrix B_1 in the plane (2, 3) as

$$B_1 = \begin{bmatrix} 1 & 0 & 0 \\ 0 & \cos\theta & -\sin\theta \\ 1 & \sin\theta & \cos\theta \end{bmatrix}$$

$$\text{Then } B_1^{-1}A_1B_1 = \begin{bmatrix} 1 & 0 & 1 \\ 0 & \cos\theta & -\sin\theta \\ 0 & -\sin\theta & \cos\theta \end{bmatrix} \begin{bmatrix} a_{11} & a_{12} & a_{13} \\ a_{12} & a_{22} & a_{23} \\ a_{13} & a_{23} & a_{33} \end{bmatrix} \times \begin{bmatrix} 1 & 0 & 1 \\ 0 & \cos\theta & -\sin\theta \\ 0 & \sin\theta & \cos\theta \end{bmatrix}$$

In the resulting matrix, $(1, 3)$ element $= -a_{12}\sin\theta + a_{13}\cos\theta$. It will be zero, if $-a_{12}\sin\theta + a_{13}\cos\theta = 0$, *i.e.*, if $\tan\theta = a_{13}/a_{12}$. $\qquad$ (2)

Thus with this value of θ, the above transformation gives zeros in $(1, 3)$ and $(3, 1)$ positions.

Now we perform rotation in the plane $(2, 4)$ and put the resulting element $(1, 4) = 0$. This would not affect the zeros obtained earlier. Proceeding in this way, the transformations are applied to the matrix so as to annihilate the elements $(1, 3), (1, 4), (1, 5), \cdots, (1, n), (2, 4), (2, 5), \cdots, (2, n)$ in this order. Finally we arrive at the tri-diagonal matrix

$$p = \begin{bmatrix} p_1 & q_1 & 0 & & 0 \cdots\cdots 0 \\ q_1 & p_2 & q_2 & & 0 \cdots\cdots 0 \\ 0 & q_2 & p3 & & q_3 \cdots\cdots 0 \\ \cdots & \cdots & \cdots & & p_{n-1}\, q_{n-1} \\ 0 & 0 & 0 & 0\cdots q_{n-1} \cdots P_n \end{bmatrix}$$

Step II. To find the eigenvalues of a tri-diagonal matrix.

Let the resulting tri-diagonal matrix after first transformation be

$$\begin{bmatrix} \alpha_{11} & \alpha_{12} & 0 \\ \alpha_{12} & \alpha_{22} & \alpha_{23} \\ 0 & \alpha_{23} & \alpha_{33} \end{bmatrix} \qquad (3)$$

Then the eigenvalues of (1) and (3) are the same. To obtain the eigenvalues of (3), we have

$$\begin{bmatrix} \alpha_{11}{}^{-\lambda} & \alpha_{12} & 0 \\ \alpha_{12} & \alpha_{22}{}^{-\lambda} & \alpha_{23} \\ 0 & \alpha_{23} & \alpha_{33}{}^{-\lambda} \end{bmatrix} = 0 = f_3(\lambda) \quad [\text{say}]$$

$$\therefore \qquad f_0(\lambda) = 1, f_1(\lambda) = \alpha_{11} - \lambda = \alpha_{11} - \lambda f_0(\lambda)$$

$$f_2(\lambda) \begin{vmatrix} \alpha_{11}{}^{-\lambda} & \alpha_{12} \\ \alpha_{12} & \alpha_{22}{}^{-\lambda} \end{vmatrix} = (\alpha_{22} - \lambda) f_1(\lambda) - \alpha_{12}{}^2 f_0(\lambda)$$

Expanding $f_3(\lambda)$ in terms of the third row, we get

$$f_3^{(\lambda)} = (\alpha_{33} - \lambda) \begin{vmatrix} \alpha_{11} - \lambda & \alpha_{12} \\ \alpha_{12} & \alpha_{22} - \lambda \end{vmatrix} - \alpha_{23} \begin{vmatrix} \alpha_{11} - \lambda & 0 \\ \alpha_{12} & \alpha_{23} \end{vmatrix}$$

i.e., $f_3(\lambda) = (\alpha_{33} - \lambda) f_2(\lambda) - (\alpha_{23})^2 f_1(\lambda)$ (4)

In general, the recurrence formula is

$$f_k(\lambda) = (\alpha_{kk} - \lambda) f_{k-1}(\lambda) - (\alpha_{k-1}, k)^2 f_{k-2}(\lambda), \, 2 \le k \le n$$ (5)

The equation $f_k(\lambda) = 0$ is the characteristic equation which can be solved by any standard method. Thus the roots of (5) will be the eigenvalues of the given symmetric matrix.

Step III. To find the eigenvectors of the tri-diagonal matrix.

If Y is an eigenvector of the tri-diagonal matrix P and if $B_1, B_2, \cdots B_j$ are the orthogonal matrices employed in reducing the matrix A to the form P, then the corresponding eigenvector of A is given by $X = B_1 B_2 \cdots B_j Y$.

NOTE

Obs. 1. *The number of rotations required for the Given's method are equivalent to the number of non-tri-diagonal elements of the matrix. In the case of a 3 × 3 matrix, only one rotation is required; whereas for a 4 × 4 matrix, three rotations are needed and so on.*

The amount of computation goes on decreasing from one rotation to the next, as the order of the matrix for computation also starts reducing.

Obs. 2. *The sequence of functions $f_0(\lambda), f_1(\lambda), f_2(\lambda), \cdots f_k(\lambda)$ is called the* **Strum sequence.** *A table of this sequence for various values of λ is prepaired and the number of changes in sign of the Strum sequence is calculated. Then the difference between these number of changes of sign for consecutive values of λ gives an approximate location of the eigenvalues. Once the location of the eigenvalues is known, their exact values can be found by any iterative method, e.g., Newton-Raphson method.*

EXAMPLE 4.15

Using Given's method, reduce the following matrix to the tri-diagonal form:

$$A = \begin{bmatrix} 2 & 1 & 3 \\ 1 & 4 & 2 \\ 3 & 2 & 3 \end{bmatrix}$$

Solution:

There being only one non-tri-diagonal element $a_{13}(= 3)$ which has to be reduced to zero, only one rotation is required.

To annihilate a_{13}, we define the orthogonal matrix in the plane $(2, 3)$ as:

$$B = \begin{bmatrix} 1 & 0 & 0 \\ 0 & \cos\theta & -\sin\theta \\ 0 & \sin\theta & \cos\theta \end{bmatrix}$$

where θ is found from the formula

$$\tan\theta = \frac{a_{13}}{a_{12}} = \frac{3}{1} = 3 \text{ and hence } \sin\theta = 3/\sqrt{10} \text{ and } \cos\theta = 1/\sqrt{10}.$$

$$\therefore \qquad A_1 = B^{-1}AB = \begin{bmatrix} \alpha_{11} & \alpha_{12} & 0 \\ \alpha_{12} & \alpha_{22} & \alpha_{23} \\ 0 & \alpha_{23} & \alpha_{33} \end{bmatrix}$$

where $\quad \alpha_{11} = 2, \alpha_{12} = a_{12}\cos\theta + a_{13}\sin\theta = \sqrt{10} = 3.16,$

$\alpha_{22} = a_{22}\cos^2\theta + 2a_{23}\sin\theta\cos\theta + a_{33}\sin^2\theta = 4.3$

$\alpha_{23} = (a_{33} - a_{22})\sin\theta\cos\theta + a_{23}(\cos^2\theta - \sin^2\theta) = -1.9$

$\alpha_{33} = a_{22}\sin^2\theta + a_{33}\cos^2\theta = 3.9.$

Hence A is reduced to the tri-diagonal matrix $\begin{bmatrix} 2 & 3.16 & 0 \\ 3.16 & 4.3 & -19 \\ 0 & -19 & 3.9 \end{bmatrix}$

Note. An alternative procedure for reduction of a symmetric matrix to the tri-diagonal form has been suggested by Householder. This method, though more complicated, requires half as much computation, as the Given's method. In any case, it is a substantial improvement on the Given's procedure since it reduces an entire row and column by a single transformation. In this method, the matrix is reduced to tri-diagonal form using elementary orthogonal transformations

4.14 House-Holder's Method

Consider an nth order real symmetric matrix $A = [a_{ij}]$. This method consists in pre and post-multiplying A by a real symmetric orthogonal matrix

P such that PAP reduces to the tridiagonal form.

Let the matrix P be of the form $P = I - 2ww'$ $\quad\quad$ (1)

$ww' = w1^2 + w_2^2 + \cdots + w_n^2 = 1$ $\quad\quad$ (2)

Then $\quad\quad P' = (I - 2ww')' = I - 2ww' = P$

And $\quad\quad P'P = (I - 2ww')'(I - 2ww')$

$\quad\quad\quad\quad\quad\quad = I - 4ww' + 4ww'.\,ww' = I$ $\quad\quad$ [by (2)

Thus P is a symmetric orthogonal matrix.

Now take w with first $(k-1)$ zero components, so that

$$w_k' = [0, 0, \cdots, 0, x_k, \cdots, x+n]$$ $\quad\quad$ (3)

Since $w_k'w_k = 1$, we have $x_k^2 + x_{k+1}^2 + \cdots x_n^2 = 1$

Then $P_k^{-1}AP_k = P_k'AP_k = P_kAP_k$

We now form successively $A_k = P_kA_{k-1}P_k;\ k = 2, 3, \cdots, n-1$.

As a first transformation, we determine x's so that zeros are created in the positions

$(1, 3), (1, 4), \cdots, (1, n)$ and $(3, 1), (4, 1), \cdots, (n, 1)$

As a second transformation, we find x's so that zeros are created in the positions $(2, 4), (2, 5), \cdots, (2, n)$ and $(4, 2), (5, 2), \cdots, (n, 2)$.

After $(n-2)$ such transformations, we arrive at a tri-diagonal matrix.

EXAMPLE 4.16

Using House-holder's method, reduce the following matrix to the tridiagonal form:

$$A = \begin{bmatrix} 1 & 4 & 3 \\ 4 & 1 & 2 \\ 3 & 2 & 1 \end{bmatrix}$$

Solution:

Let $\quad\quad A = \begin{bmatrix} a_{11} & a_{12} & a_{13} \\ a_{12} & a_{22} & a_{23} \\ a_{13} & a_{23} & a_{33} \end{bmatrix}$

Here we take $\quad\quad w_2' = [0, x_2, x_3]$

so that
$$P = 1 - 2w_2w_2' = \begin{bmatrix} 1 & 0 & 0 \\ 0 & 1 - 2x_2^2 & -1 - 2x_2x_3 \\ 0 & -1 - 2x_2x_3 & 1 - 2x_3 \end{bmatrix}$$

Now the element $(1, 3)$ of PAP can become zero only if the corresponding element in AP is zero. The first row elements of AP are a_{11}, $a_{12} - 2p_1x_2$, $a_{13} - 2p_1x_3$ where $p_1 = a_{12}x_2 + a_{13}x_3$.

∴ We require that $a_{13} - 2p_1x_3 = 0$ \hfill (i)

Since the sum of the squares of the elements in any row is invariant under an orthogonal transformation, we have

$$a_{11}^2 + a_{12}^2 + a_{13}^2 = a_{11}^2 + (a_{12} - 2p_1x_2)^2 + 0$$

or
$$a_{12} - 2p_1x_2 = \pm\sqrt{\left(a\,|\,12^2 + a_{13}^2\right)}$$ \hfill (ii)

For the given matrix, (i) and (ii) become

$$3 - 2p_1x_3 = 0$$ \hfill (iii)

$$4 - 2p_1x_2 = \pm\sqrt{\left(4^2 + 3^2\right)} = \pm 5$$ \hfill (iv)

where $p_1 = 4x_2 + 3x_3$

Multiplying (iii) by x_3 and (iv) by x_2 and adding, we get

$$3x_3 + 4x_2 - 2p_1((x_3^2 + x_2^2) = \pm 5x_2 \qquad [\because x_2^2 + x_3^2\ ]$$

i.e.,
$$p_1 - 2p_1 = \pm 5x_2 \text{ or } p_1 = \mp 5x_2$$ \hfill (v)

Substituting in (iv), we obtain $4 - 2\,(\mp 5x_2)x_2 = \pm 5$

which gives $x_2 = 1/\sqrt{10}$ or $x2 = 3/\sqrt{10}$

From (iii),
$$x3 = \frac{3}{2p1} = \mp\frac{3}{10x_2}$$ \hfill [by(v)

Since x_3 contains x_2 in the denominator, we obtain best accuracy if x_2 is large

∴ Choosing
$$x_2 = \frac{3}{\sqrt{10}}, x_3 = \mp\frac{3}{10} \times \frac{\sqrt{10}}{3} = \pm\frac{1}{\sqrt{10}}$$

Taking + ve sign, we get $x_2 = \dfrac{3}{\sqrt{10}}$ and $x_3 = \dfrac{1}{\sqrt{10}}$

∴
$$P = I - 2w_2w_2'$$

$$= \begin{bmatrix} 1 & 0 & 0 \\ 0 & 1 & 0 \\ 0 & 0 & 1 \end{bmatrix} - 2 \begin{bmatrix} 0 \\ 3/\sqrt{10} \\ 1/\sqrt{10} \end{bmatrix} \left(0, 3/\sqrt{10}, 1/\sqrt{10} \right)$$

$$= \begin{bmatrix} 1 & 0 & 0 \\ 0 & -4/5 & -3/5 \\ 0 & -3/5 & -4/5 \end{bmatrix}$$

Hence $PAP = \begin{bmatrix} 1 & 0 & 0 \\ 0 & -4/5 & -3/5 \\ 0 & -3/5 & -4/5 \end{bmatrix} \begin{bmatrix} 1 & 4 & 3 \\ 4 & 1 & 2 \\ 3 & 2 & 1 \end{bmatrix} \begin{bmatrix} 1 & 0 & 0 \\ 0 & -4/5 & -3/5 \\ 0 & -3/5 & -4/5 \end{bmatrix}$

$$= \begin{bmatrix} 1 & 0 & 0 \\ 0 & -4/5 & -3/5 \\ 0 & -3/5 & -4/5 \end{bmatrix} \begin{bmatrix} 1 & -5 & 0 \\ 4 & -2 & 1 \\ 3 & -11/5 & -2/5 \end{bmatrix} = \begin{bmatrix} 1 & -5 & 0 \\ -5 & 73/25 & -14/25 \\ 0 & -3/5 & -11/25 \end{bmatrix}$$

which is the required tri-diagonal matrix.

Exercises 4.3

1. Using Jacobi's method, find all the eigenvalues and the eigenvectors of the matrices:

$(a) \begin{bmatrix} 5 & 0 & 1 \\ 0 & -2 & 0 \\ 1 & 0 & 5 \end{bmatrix} \qquad (b) \begin{bmatrix} 2 & 3 & 1 \\ 3 & 2 & 2 \\ 1 & 2 & 1 \end{bmatrix}$

2. Reduce the matrix $\begin{bmatrix} 1 & 1 & 1/2 \\ 1 & 1 & 1/4 \\ 1/2 & 1/4 & 2 \end{bmatrix}$ to the tri-diagonal form, using the Householder's method

3. Apply Householder's method, to find the eigenvalues of the matrix

$$\begin{bmatrix} 2 & -1 & -1 \\ -1 & 2 & -1 \\ -1 & -1 & 2 \end{bmatrix}$$

4. Transform the matrix $\begin{bmatrix} 1 & 2 & 2 \\ 2 & 1 & 2 \\ 2 & 2 & 1 \end{bmatrix}$ to the tri-diagonal form using Given's method. Hence find the largest eigenvalue and the corresponding eigenvector of the tri-diagonal matrix.

5. Find the eigenvalues of the matrix $\begin{bmatrix} 2 & -i & 0 \\ i & 2 & 0 \\ 0 & 0 & 3 \end{bmatrix}$

4.15 Objective Type of Questions

Select the correct answer or fill up the blanks in the following questions:

1. The eigenvalues of a triangular matrix are..........

2. Inverse of $\begin{bmatrix} 5 & 3 \\ 3 & 2 \end{bmatrix}$ is..........

3. The most suitable initial eigenvector out of $\begin{bmatrix} 1 \\ 1 \end{bmatrix}, \begin{bmatrix} 1 \\ 0 \end{bmatrix}$, and $\begin{bmatrix} 0 \\ 0 \end{bmatrix}$ to find the larger eigenvalue of the matrix $\begin{bmatrix} 1 & 4 \\ 3 & 2 \end{bmatrix}$ in one iteration, is..........

4. Two eigenvalues of the matrix $A = \begin{bmatrix} 2 & 2 & 1 \\ 1 & 3 & 1 \\ 1 & 2 & 2 \end{bmatrix}$ re equal to 1 each, then the eigenvalues of A^{-1} are..........

5. Eigenvalues of $\begin{bmatrix} 1 & 4 \\ 3 & 2 \end{bmatrix}$ are..........

6. If λ is an eigenvalue of a matrix A, then $1/\lambda$ is the eigenvalue of..........

7. The product of two eigenvalues of the matrix $\begin{bmatrix} 6 & -2 & 2 \\ -2 & 3 & -1 \\ 2 & -1 & 3 \end{bmatrix}$ is 16, then the third eigenvalue is..........

8. The Power method works satisfactorily only if the matrix A has a...... eigenvalue.

9. Eigenvalues of the matrix $\begin{bmatrix} 1 & 1-i \\ 1+i & 1 \end{bmatrix}$ are..........

10. If λ is an eigenvalue of an orthogonal matrix, then $1/\lambda$ is also its..........

11. Dominant eigenvalues of $\begin{bmatrix} 1 & 2 \\ 3 & 4 \end{bmatrix}$ by the Power method are..........

12. The eigenvalues of an idempotent matrix are..........

13. If the eigenvalues of a matrix A are $-4, 3, 1$, then the dominant eigenvalue of A is.......

14. If $A = \begin{bmatrix} 2 & 2 & 1 \\ 1 & 3 & 1 \\ 1 & 2 & 2 \end{bmatrix}$, then $A^{-1} =$..........

15. The eigenvalue that can the obtained by using the Power method is.......

16. If λ is the largest eigenvalue of the matrix A, then the relation giving the smallest eigenvalue is.......

CHAPTER 5

EMPIRICAL LAWS AND CURVE-FITTING

Chapter Objectives

- Introduction
- Graphical method
- Laws reducible to the linear law
- Principle of least squares
- Method of least squares
- Fitting a curve of the type $y = a + bx^2$, etc.
- Fitting of other curves
- Most Plausible values
- Method of group averages
- Laws containing three constants
- Method of moments
- Objective type of questions.

5.1 Introduction

In many branches of Applied Mathematics, it is required to express a given data, obtained from observations, in the form of a law connecting the two variables involved. Such a law inferred by some scheme, is known as the *empirical law*. For example, it may be desirable to obtain the law connecting the length and the temperature of a metal bar. At various temperatures, the length of the bar is measured. Then, by one of the methods explained below, a law is obtained that represents the relationship existing between temperature and length for the observed values. This relation ship can then be used to predict the length at an arbitrary temperature.

Scatter diagram. To find a relationship between the set of paired observations x and y, we plot their corresponding values on the graph, taking one of the variables along the x-axis and other along the y-axis, *i.e.*, (x_1, y_1), (x_2, y_2)..., (x_n, y_n). The resulting diagram showing a collection of dots is called a *scatter diagram*. A smooth curve that approximates the above set of points is known as the *approximating curve.*

Curve fitting. Several equations of different types can be obtained to express the given data approximately. But the problem is to find the equation of the curve of "best fit" which may be most suitable for predicting the unknown values. The process of finding such an equation of 'best fit' is known as *curve-fitting.*

If there are n pairs of observed values then it is possible to fit the given data to an equation that contains n arbitrary constants, and we can solve n simultaneous equations for n unknowns. If we desired to obtain an equation representing these data but have less then n arbitrary constants, then we can have recourse to any of these four methods: *Graphical method, method of least-squares, method of group averages, and method of moments.* The graphical method and the method of averages fail to give the values of the unknown constants uniquely and accurately, while the other methods do. The method of least squares is probably the best to fit a unique curve to a given data. It is widely used in applications and can be easily implemented on a computer.

5.2 Graphical Method

When the curve representing the given data is a **linear law** $y = mx + c$; we proceed as follows:

(*i*) *Plot the given points on the graph paper* taking a suitable *scale.*

(*ii*) *Draw the straight line of best fit* such that the points are evenly distributed about the line.

(*iii*) *Taking two suitable points* (x_1, y_1) *and* (x_2, y_2) *on the line, calculate m,* the slope of the line and *c,* its intercept on the y-axis.

When the points representing the observed values do not approximate to a straight line, a smooth curve is drawn through them. From the shape of the graph, we try to infer the law of the curve and then reduce it to the form $y = mx + c$.

5.3   Laws Reducible to the Linear Law

We give below some of the laws in common use, indicating the way these can be reduced to the linear form by suitable substitutions:

1. When the law is $y = mx^n + c$

Taking $x^n = X$ and $y = Y$, the above law becomes $Y = mX + c$

2.   When the law is $y = ax^n$.

Taking logarithms of both sides, it becomes $\log_{10} y = \log_{10} a + n \log_{10} x$

Putting $\log_{10} x = X$ and $\log_{10} y = Y$, it reduces to the form
$Y = nX + c$, where $c = \log_{10} a$.

3. When the law is $y = ax^n + b \log x$.

Writing it as $\dfrac{y}{\log x} = a \dfrac{x^n}{\log x} + b$ and taking $x^n/\log x = X$ and $y/\log x = Y$, the

given law becomes, $Y = aX + b$.

4. When the law is $y = ae^{bx}$.

Taking logarithms, it becomes $\log_{10} y = (b \log_{10} e)x + \log_{10} a$.
Putting $x = X$ and $\log_{10} y = Y$, it takes the form $Y = mX + c$
where $m = b \log_{10} e$ and $c = \log_{10} a$.

5. When the law is $xy = ax + by$.

Dividing by x, we have $y = b\dfrac{y}{x} + a$.

Putting $y/x = X$ and $y = Y$, it reduces to the form $Y = bX + a$.

EXAMPLE 5.1

R is the resistance to motion of a train at speed V; find a law of the type $R = a + bV^2$ connecting R and V, using the following data:

V (km/hr):	10	20	30	40	50
R (kg/ton):	8	10	15	21	30

Solution:

Given law is $R = a + bV^2$ (i)

Taking $V^2 = x$ and $R = y$,

(i) becomes, $y = a + bx$ (ii)

which is a *linear law*.

Table for the values of x and y is as follows:

x:	100	400	900	1600	2500
y:	8	10	15	21	30

Plot these points. Draw the straight line of best fit through these points (Figure 5.1).

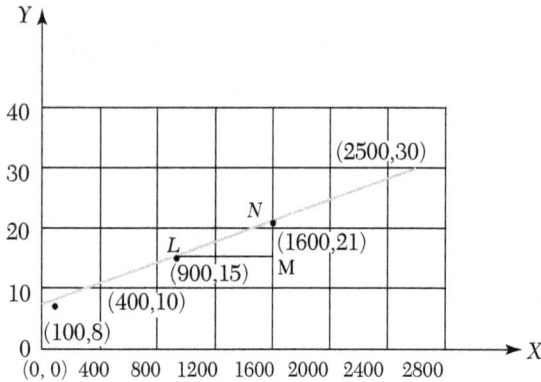

FIGURE 5.1

Slope of this line $(= b)$

$$= \frac{MN}{LM} = \frac{21-15}{1600-900} = \frac{6}{700}$$
$$= 0.0085 \text{ nearly.}$$

Since $L(900, 15)$ lies on (ii),

$\therefore$ $15 = a + 0.0085 \times 900$,

where $a = 15 - 7.65 = 7.35 \text{ nearly.}$

EXAMPLE 5.2

The following values of x and y are supposed to follow the law $y = ax^2 + b \log_{10} x$. Find graphically the most probable values of the constants a and b.

x	2.85	3.88	4.66	5.69	6.65	7.77	8.67
y	16.7	26.4	35.1	47.5	60.6	77.5	93.4

Solution:

Given law is $y = ax^2 + b \log_{10} x$

i.e., $\quad \dfrac{y}{\log_{10} x} = a\dfrac{x^2}{\log 10^x} + b$ $\hspace{2cm}$ (i)

Putting $\quad x^2/\log_{10} x = X$ and $y/\log_{10} x = Y$,

(i) becomes $\hspace{2cm} Y = aX + b$ $\hspace{2cm}$ (ii)

This is a *linear law*.

Table for the values of X and Y is as follows:

$X = x^2/\log_{10} x$	17.93	25.56	32.49	42.87	53.75	67.80	80.83
$Y = y/\log_{10} x$	35.59	44.83	52.50	69.90	73.65	87.04	99.56
Points	P_1	P_2	P_3	P_4	P_5	P_6	P_7

Plot these points and draw the straight line of best fit through these points (Figure 5.2).

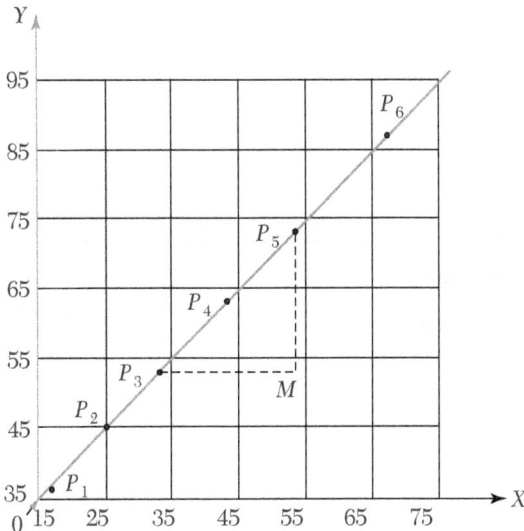

FIGURE 5.2

Slope of this line $(= a) = \dfrac{MP_5}{P_3 M} = \dfrac{73.65 - 52.50}{53.75 - 32.49} = \dfrac{21.15}{21.26} = 0.99$

Since P_3 lies on (ii), therefore,

$$52.50 = 0.99 \times 32.49 + b$$

where $\qquad b = 20.2$

Hence (i) becomes $\qquad y = (0.99) x^2 + (20.2) \log_{10}{}^x.$

EXAMPLE 5.3

The values of x and y obtained in an experiment are as follows:

x	2.30	3.10	4.00	4.92	5.91	7.20
y	33.0	39.1	50.3	67.2	85.6	125.0

The probable law is $y = ae^{bx}$. Test graphically the accuracy of this law and if the law holds good, find the best values of the constants.

Solution:

Given law is $y = ae^{bx}$ $\hspace{3cm}$ (i)

Taking logarithms to base 10, we have
$$\log_{10}{}^y = \log_{10}a + (b \log_{10}e)\, x$$

Putting $x = X$ and $\log_{10}{}^y = Y$, it becomes

$y = (b \log_{10}e)\, X + \log_{10}a$ $\hspace{3cm}$ (ii)

Table for the values of X and Y is as under:

$X = x$	2.30	3.10	4.00	4.92	5.91	7.20
$Y = \log_{10} y$	1.52	1.59	1.70	1.83	1.93	2.10
Points	P_1	P_2	P_3	P_4	P_5	P_6

Scale: 1 small division along x-axis $= 0.1$

$\qquad\qquad$ 10 small divisions along y-axis $= 0.1$.

Plot these points and draw the line of best fit. As these points are lying almost along a straight line, the given law is nearly accurate (Figure 5.3).

Now the slope of this line $(= b \log 10^e) = \dfrac{MN}{NM} = 0.12$

where $\quad b = \dfrac{0.12}{\log_{10} e} = 0.12 \times 2.303 = 0.276$

Since the point L (4, 1.71) lies on (*ii*), therefore,

$$1.71 = 0.12 \times 4 + \log_{10} a \text{ where } a = 17 \text{ nearly.}$$

Hence the curve of best fit is $y = 17\, e^{0.276\,x}$.

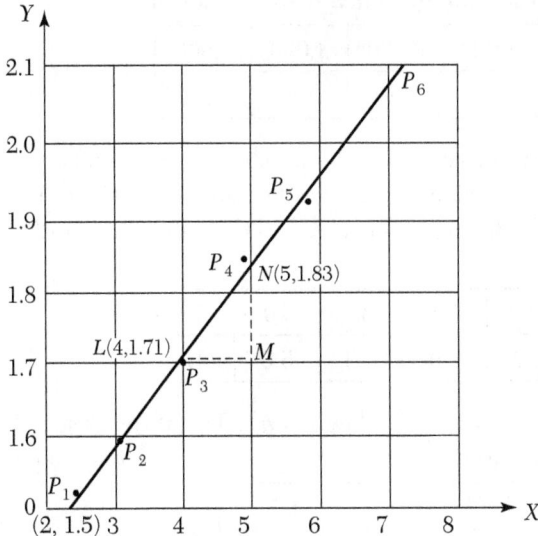

FIGURE 5.3

Exercises 5.1

1. If p is the pull required to lift the weight by means of a pulley block, find a linear law of the form $p = a + bw$, connecting p *and* w, using the following data:

w(lb):	50	70	100	120
p(lb):	12	15	21	25

Compute p, when $w = 150$ lb.

2. Convert the following equations to their linear forms:

  (*i*) $y = ax + bx^2$ (*ii*) $y = b/[x(x - a)]$.

3. The resistance R of a carbon filament lamp was measured at various values of the voltage V and the following observations were made:

Voltage V...	62	70	78	84	92
Resistance R...	73	70.7	69.2	67.8	66.3

Assuming a law of the form $R = \dfrac{a}{V} + b,$ find by graphical method the best values of a and b.

4. Verify if the values of x and y, related as shown in the following table, obey the law $y = a + b\sqrt{x}$. If so, find graphically the values of a and b.

x:	500	1,000	2,000	4,000	6,000
y:	0.20	0.33	0.38	0.45	0.51

5. The following table gives the pressure p and the volume v at various instants during the expansion of steam in a cylinder. Show that the equation of the expansion is of the form $pv^n = c$ and find the values of n and c approximately.

p:	200	100	50	30	20	10
v:	1.0	1.7	2.9	4.8	5.9	10

6. The following values of T and l follow the law $T = al^n$. Test if this is so and find the best values of a and n.

$T = 1.0$	1.5	2.0	2.5
$l = 25$	56.2	100	1.56

Fit the curve $y = ae^{bx}$ to the following data:

x:	0	2	4
y:	5.1	10	31.1

The following are the results of an experiment on friction of bearings. The speed being kept constant, corresponding values of the coefficient of friction and the temperature are shown in the table:

t:	120	110	100	90	80	70	60
μ:	0.0051	0.0059	0.0071	0.0085	0.00102	0.00124	0.00148

If μ and t are given by the law $\mu = ae^{bt}$, find the values of a and b by plotting the graph for μ and t.

5.4 Principle of Least Squares

The graphical method has the obvious drawback of being unable to give a unique curve of fit. *The principle of least squares, however, provides an elegant procedure of fitting a unique curve to a given data.*

Let the curve $y = a + bx + cx^2 + \cdots + kx^m$

be fitted to the set of data points $(x_1, y_1), (x_2, y_2),\dots, (x_n, y_n)$.

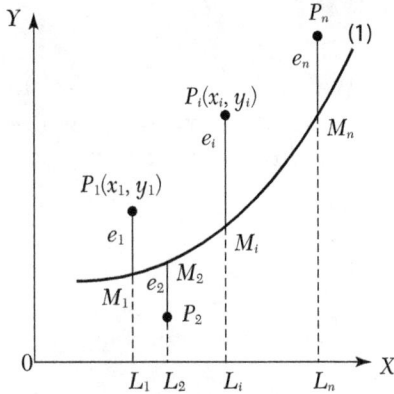

FIGURE 5.4

Now we have to determine the constants a, b, c,... k such that they represents the curve of best fit. In the case of $n = m$, when substituting the values (x_i, y_i) in (1), we get n equations from which a unique set of n constants can be found. But when $n > m$, we obtain n equations which are more than the m constants and hence cannot be solved for these constants. So we try to determine the values of a, b, c, $\cdots$ k which satisfy all the equations as nearly as possible and thus may give the best fit. In such cases, we apply the *principle of least squares*.

At $x = x_i$, the *observed (experimental)* value of the ordinate is yi and the corresponding value on the fitting curve (1) is $a + bx_i + cx_i^2 + \cdots + kx_i^m$ $(= \eta_i$, say) which is the *expected* (or *calculated*) *value* (Figure 5.4). The difference of the observed and the expected values, *i.e.*, $y_i - \eta_i (= e_i)$ is called the *error* (or *residual*) at $x = x_i$. Clearly some of the errors $e_1, e_2,\cdots, e_n$ will be positive and others negative. Thus to give equal weightage to each error, we square each of these and form their sum, *i.e.*, $E = e_1^2 + e_2^2 + \cdots + e_n^2$.

The curve of best fit is that for which e's are as small as possible, i.e., the sum of the squares of the errors is a minimum. This is known as the *principle of least squares* and was suggested by a French mathematician *Adrien Marie Legendre* in 1806.

NOTE **Obs.** *The principle of least squares does not help us to determine the form of the appropriate curve which can fit a* *given data. It only determines the best possible values of the constants in the equation when the form of the curve is known before hand. The selection of the curve is a matter of experience and practical considerations.*

5.5 Method of Least Squares

For clarity, suppose it is required to fit the curve $y = a + bx + cx^2$ to a given set of observations $(x_1, y_1), (x_2, y_2), \cdots, (x_5, y_5)$. For any x_i, the observed value is y_i and the expected value is $\eta_i = a + bx_i + cx_i^2$ so that the error $e_i = y_i - \eta_i$.

∴ The sum of the squares of these errors is

$$E = e_1^2 + e_2^2 + \ldots e_5^2$$

$$= [y_1 - (a + bx_1 + cx_1^2)]^2 + [y_2 - bx_2 + cx_2^2)]^2 + \ldots + [y_5 - (a + bx_5 + cx_5^2)]^2$$

For E to be minimum, we have

$$\frac{\partial E}{\partial a} = 0 = -2[y_1 - (a + bx_1 + cx_1^2)]^2 - 2[y_2 - (a + bx_2 + cx_2^2)]$$
$$- \cdots - 2[y_5 - (a + bx_5 + cx_5^2)]^2 \qquad (1)$$

$$\frac{\partial E}{\partial b} = 0 = -2x_1[y_1 - (a + bx_1 + cx_1^2)] - 2x_2[y_2 - (a + bx_2 + cx_2^2)]$$
$$- \cdots - 2x_2[y_5 - (a + bx_5 + cx_5^2)] \qquad (2)$$

$$\frac{\partial E}{\partial b} = 0 = -2x_1^2[y_1 - (a + bx_1 + cx_1^2)] - 2x_2^2[y_2 - (a + bx_2 + cx_2^2)]$$
$$- \cdots - 2x_5^2[y_5 - (a + bx_5 + cx_5^2)] \qquad (3)$$

Equation (1) simplifies to

$$y_1 + y_2 + \cdots + y_5 = 5a + b(x_1 + x_2 + \cdots + x_5) + c(x_1^2 + x_2^2 + \cdots + x_5^2)$$

i.e,
$$\Sigma yi = 5a + b\Sigma xi + cxi^2 \qquad (4)$$

Equation (2) becomes

$$x_1y_1 + x_2y_2 + \cdots + x_5y_5 = a(x_1 + x_2 + \cdots + x_5) + b(x_1^2 + x_2^2 + \cdots + x_5^2)$$
$$+ c(x_1^3 + x_2^3 + \cdots + x_5^3)$$

i.e., $\sum x_i y_i = a \sum x_i + b \sum x_i^2 + x \sum x_i^3$

Similarly (3) simplifies to

$$\sum x_i^2 y_i = a \sum x_i^2 + b \sum x_i^3 + c \sum x_i^4$$

The equations (4), (5) and (6) are known as *normal equations* and can be solved as simultaneous equations in a, b, c. The values of these constants when substituted in (1) give the desired curve of best fit.

_____

NOTE **Obs.** *On calculating* $\dfrac{\partial^2 E}{\partial a^2}, \dfrac{\partial^2 E}{\partial b^2}, \dfrac{\partial^2 E}{\partial c^2}$ *and substituting the values*

of a, b, c just obtained, we will observe that each is positive, i.e,. E is a minimum.

Working procedure

(a) *To fit the straight line $y = a + bx$*

   (i) Substitute the observed set of n values in this equation.

   (ii) Form normal equations for each constant, i.e., $\Sigma y = na + b\Sigma x$, $\Sigma xy = a\Sigma x + b\Sigma x^2$.

   [The normal equation for the unknown a is obtained by multiplying the equations by the coefficient of a and adding. The normal equation for b is obtained by multiplying the equations by the coefficient of b (*i.e.* x) and adding.]

   (iii) Solve these normal equations as simultaneous equations for a and b.

   (iv) Substitute the values of a and b in $y = a + bx$, which is the required line of best fit.

(b) *To fit the parabola: $y = a + bx + cx^2$*

   (i) Form the normal equations $\Sigma y = na + b\Sigma x + c\Sigma x^2$

   $\Sigma xy = a\Sigma x + b\Sigma x^2 + c\Sigma x^3$ and $\Sigma x^2 y = a\Sigma x^2 + b\Sigma x^3 + c\Sigma x^4$

   [The normal equation for c has been obtained by multiplying the equations by the coefficient of c (*i.e.*, x^2) and adding.]

   (ii) Solve these as simultaneous equations for a, b, c.

   (iii) Substitute the values of a, b, c in $y = a + bx + cx^2$, which is the required parabola of best fit.

(c) In general, the curve $y = a + bx + cx^2 + \cdots + kx^{m-1}$ can be fitted to a given data by writing m normal equations.

EXAMPLE 5.4

If P is the pull required to lift a load W by means of a pulley block, find a linear law of the form $P = mW + c$ connecting P and W, using the following data:

$P = 12$ 15   21   25

$W = 50$ 70   100   120

where P and W are taken in kg-wt. Compute P when $W = 150$ kg.

Solution:

The corresponding normal equations are

$$\left.\begin{array}{l} \sum P = 4c + m \sum W \\ \sum WP = c \sum W + m \sum W^2 \end{array}\right\} \tag{i}$$

The values of ΣW etc. are calculated by means of the following table:

W	P	W^2	WP
50	12	2500	600
70	15	4900	1050
100	21	10000	2100
120	25	14400	3000
Total = 340	73	31800	6750

∴ The equations (i) become $73 = 4c + 340m$ and $6750 = 340c + 31800m$

i.e., $2c + 170m = 365$ (ii)

and $34c + 3180m = 675$ (iii)

Multiplying (ii) by 17 and subtracting from (iii), we get $m = 0.1879$

∴ from (ii), $c = 2.2785$

Hence the line of best fit is $P = 2.2759 + 0.1879\ W$

When $W = 150$ kg, $P = 2.2785 + 0.1879 \times 150 = 30.4635$ kg.

EXAMPLE 5.5

Fit a straight line to the following data:

x	6	7	7	8	8	8	9	9	10
y	5	5	4	5	4	3	4	3	3

Solution:

Let the straight line be $y = ax + b$.

Then the normal equations are $\Sigma y = a\Sigma x + 9b$

$$\Sigma xy = a\Sigma x^2 + b\Sigma x \tag{i}$$

The values of Σx, Σy etc. are calculated below:

x	y	xy	x^2
6	5	30	36
7	5	35	49
7	4	28	16
8	5	40	64
8	4	32	64
8	3	24	64
9	4	36	81
9	3	27	81
10	3	30	100
$\Sigma x = 72$	$\Sigma y = 36$	$\Sigma xy = 282$	$\Sigma x2 = 588$

$\therefore$ The equations (i) become $36 = 72a + 9b$ and $282 = 588a + 72b$

i.e.,
$$8a + b = 4 \tag{ii}$$
$$98a + 12b = 47 \tag{iii}$$

Multiplying (ii) by 12 and subtracting from (iii), we get $a = -0.5$.

From (ii), $b = 8$.

Hence the required line of best fit is $y = -0.5x + 8$.

EXAMPLE 5.6

Fit a second degree parabola to the following data:

x	0	1	2	3	4
y	1	1.8	1.3	2.	6.3

Solution:

Let $u = x - 2$ and $v = y$ so that the parabola of fit $y = a + bx + cx^2$ becomes

$$v = A + Bu + Cu^2$$

The normal equations are

$$\Sigma v = 5A + B\Sigma u + C\Sigma u^2 \quad \text{or} \quad 12.9 = 5A + 10C$$

$$\Sigma uv = A\Sigma u + B\Sigma u^2 + C\Sigma u^2 \quad \text{or} \quad 11.3 = 10B$$

$$\Sigma u^2 v = A\Sigma u^2 + B\Sigma u^3 + C\Sigma u^4 \quad \text{or} \quad 33.5 = 10A + 34C$$

Solving these as simultaneous equations, we get

$$A = 1.48, B = 1.13, C = 0.55$$

∴ (i) becomes; $v = 1.48 + 1.13u + 0.55u^2$

$$y = 1.48 + 1.13(x - 2) + 0.55(x - 2)^2$$

Hence $y = 1.42 - 1.07x + 0.55x^2$

NOTE

Obs. *For the sake of convenience and ease in calculations, it is sometimes advisable to change the origin and scale with the substitutions X = (x − A)/h and Y = (y − B)/h, where A and B are the assumed means (or middle values) of x and y series, respectively and h is the width of the interval.*

EXAMPLE 5.7

Fit a second degree parabola to the following data:

x = 1.0 1.5   2.0   2.5   3.0   3.5   4.0

y = 1.1 1.3   1.6   2.0   2.7   3.4   4.1

Solution:

We shift the origin to $(2.5, 0)$ and take 0.5 as the new unit. This amounts to changing the variable x to X, by the relation $X = 2x - 5$.

Let the parabola of fit be $y = a + bX + cX^2$.

The values of ΣX etc. are calculated as below:

x	X	y	Xy	X^2	X^2y	X^3	X^4
1.0	−3	1.1	−3.3	9	9.9	−27	81
1.5	−2	1.3	−2.6	4	5.2	−8	16
2.0	−1	1.6	−1.6	1	1.6	−1	1
2.5	0	2.0	0.0	0	0.0	0	0
3.0	1	2.7	2.7	1	2.7	1	1
3.5	2	3.4	6.8	4	13.6	8	16
4.0	3	4.1	12.3	9	36.9	27	81
Total	0	16.2	14.3	28	69.9	0	196

The normal equations are

$$7a + 28c = 16.2, \ 28b = 14.3, \ 28a + 196c = 69.9$$

Solving these as simultaneous equations, we get

$$a = 2.07, b = 0.511, c = 0.061.$$

∴ $y = 2.07 + 0.511X + 0.061 \ X^2$

Replacing X by $2x - 5$ in the above equation, we get

$$y = 2.07 + 0.511 (2x - 5) + 0.061 (2x - 5)^2$$

which simplifies to $y = 1.04 - 0.198x + 0.244x^2$

This is the required parabola of best fit.

EXAMPLE 5.8

Fit a second degree parabola to the following data:

x:	1989	1990	1991	1992	1993	1994	1995	1996	1997
y:	352	356	357	358	360	361	361	360	359

Solution:

Taking $u = x - 1993$ and $v = y - 357$, the equation $y = a + bx + cx^2$ becomes

$$v = A + Bu + Cu^2 \qquad (i)$$

x	$u = $ $x - 1993$	y	$v = $ $y - 357$	uv	u^2	u^2v	u^2	u^4
1989	−4	352	−5	20	16	−80	−64	256
1990	−3	356	−1	3	9	−9	−27	81
1991	−2	357	0	0	4	0	−8	16
1992	−1	358	1	−1	1	1	−1	1
1993	0	360	3	0	0	0	0	0
1994	1	361	4	4	1	4	1	1
1995	2	361	4	8	4	16	8	16
1996	3	360	3	9	9	27	27	81
1997	4	359	2	8	16	32	64	256
Total	$\Sigma u = 0$		$\Sigma v = 11$	$\Sigma uv = 51$	$\Sigma u^2 = 60$	$\Sigma u^2v = -9$	$\Sigma u^3 = 0$	$\Sigma u^4 = 708$

The normal equations are

$$\Sigma v = 9A + B\Sigma u + C\Sigma u^2 \text{ or } 11 = 9A + 60C$$

$$\Sigma uv = A\Sigma u + B\Sigma u^2 + C\Sigma u^3 \text{ or } 51 = 60B \text{ or } B = \frac{17}{20}$$

$$\Sigma u^2 v = A\Sigma u^2 + B\Sigma u^3 + C\Sigma u^4 \text{ or } -9 = 60A + 708C$$

On solving these equations, we get $A = \frac{694}{231}, B = \frac{17}{20}, C = \frac{247}{924}$

$\therefore$ (i) becomes $v = \frac{694}{231} + \frac{17}{20}u - \frac{247}{924}u^2$

or $y - 357 = \dfrac{694}{231} + \dfrac{17}{20}(x - 1993) - \dfrac{247}{924}(x - 1993)^2$

or $y = \dfrac{694}{231} - \dfrac{33881}{20} - \dfrac{247}{924}(1993)^2 + \dfrac{17}{20}x + \dfrac{247 \times 3986}{924}x - \dfrac{247}{924}x^2$

or $y = 3 - 1694.05 - 1061792.32 + 357 + 0.85x + 1065.52x - 0.267x^2$

Hence $y = -1062526.37 + 1066.37x - 0.267x^2$

Exercises 5.2

1. By the method of least squares, find the straight line that best fits the following data:

x:	1	2	3	4	5
y:	14	27	40	55	68

2. In some determinations of the value v of carbon dioxide dissolved in a given volume of water at different temperatures θ, the following pairs of values were obtained:

$\theta = 0$	5	10	15
$v = 1.80$	1.45	1.18	1.00

   Obtain by the method of least squares, a relation of the form $v = a + b\theta$ which best fits to these observations.

3. A simply supported beam carries a concentrated load P(lb) at its midpoint. Corresponding to various values of P, the maximum deflection Y (in) is measured. The data are given below:

P:	100	120	140	160	180	200
Y:	0.45	0.55	0.60	0.70	0.80	0.85

   Find a law of the form $Y = a + bP$.

4. The result of measurement of electric resistance R of a copper bar at various temperatures $t°$ C are listed below:

t:	19	25	30	36	40	45	50
R:	76	77	79	80	82	83	85

   Find a relation $R = a + bt$ when a and b are constants to be determined by you.

5. A chemical company, wishing to study the effect of extraction time (t) on the efficiency of an extraction operation (e) obtained the data shown in the following table:

t:	27	45	41	19	3	39	19	49	15	31
e:	57	64	80	46	62	72	52	77	57	68

Fit a straight line to the given data by the method of least squares.

6. Find the parabola of the form $y = a + bx + cx2$ which fits most closely with the observations:

x:	-3	-2	-1	0	1	2	3
y:	4.63	2.11	0.67	0.09	0.63	2.15	4.58

7. By the method of least squares, fit a parabola of the form $y = a + bx + cx^2$, to the following data:

x:	2	4	6	8	10
y:	6.07	12.85	31.47	57.38	91.29

Fit a parabola $y = a + bx + cx^2$ to the following data:

x:	1	2	3	4	5	6	7	8	9
y:	2	6	7	8	10	11	11	10	9

8. The velocity V of a liquid is known to vary with temperature T according to a quadratic law $V = a + b\,T + CT^2$. Find the best values of a, b, c for the following table:

T:	1	2	3	4	5	6	7
V:	2.31	2.01	3.80	1.66	1.55	1.46	1.41

9. The following table gives the results of the measurements of train resistance, V is the velocity in miles per hour, R is the resistance in pounds per ton:

V:	20	40	60	80	100	120
R:	5.5	9.1	14.9	22.8	33.3	46.0

If R is related to V by the relation $R = a + bV + cV^2$, find a, b and c.

5.6   Fitting A Curve of the Type

(1) $y = a + bx^2$ (2) $y = ax + bx^2$

(3) $y = ax + b/x$ (4) $ax^2 + b/x$.

(1) $y = a + bx^2$

Putting $x^2 = X$, we have $y = a + bX$ *(i)*

which is a linear equation. Its normal equations are

$$\Sigma y = na + b\Sigma X; \ \Sigma yX = a\Sigma X + b\Sigma X2$$

Solving these, we get a and b. Substituting these values of a, b and replacing X by x^2 in *(i)*, we obtain the desired equation of best fit.

(2) $y = ax + bx^2$

Rewriting this equation as $y/x = a + bx$ and putting $y/x = Y$, we have

$$Y = a + bx \quad\quad\quad (i)$$

Its normal equations are

$$\Sigma Y = na + b\Sigma x; \ \Sigma Yx = a\Sigma x + b\Sigma x^2$$

Solving these we get a and b. Replacing Y by y/x in *(i)*, we obtain the desired equation of best fit.

(3) $y = ax + b/x$

Rewriting this equation as $xy = ax^2 + b$

and putting $x^2 = X$ and $xy = Y$, we have $Y = b + aX$ *(i)*

Its normal equations are

$$\Sigma Y = nb + \alpha\Sigma X; \ \Sigma XY = b\Sigma X + a\Sigma X^2$$

Solving these equations, we get a and b. Replacing X by x^2 and Y by xy in *(i)*, we obtain the desired equation of best fit.

(4) $y = ax^2 + b/x$

Rewriting this equation as $xy = ax3 + b$ and putting $x3 = X$ and $xy = Y$, we have

$$Y = b + aX \quad\quad\quad (i)$$

Its normal equations are

$$\Sigma Y = bn + a\Sigma X; \ \Sigma XY = b\Sigma X + X^2$$

Solving these equations, we get a and b. Replacing X by x^3 and Y by xy, we obtain the desired equation of best fit.

EXAMPLE 5.9

Find the least squares fit of the form $y = a_0 + a_1 x^2$ to the following data

x:	-1	0	1	2
y:	2	5	3	0

Solution:

Putting $x^2 = X$, we have $y = a_0 + a_1 X$ (i)

∴ The normal equations are

$$\Sigma y = 4a + a_1 \Sigma X; \ \Sigma Xy = a_0 \Sigma X + a_1 \Sigma X^2.$$

The values of ΣX, ΣX^2 etc. are calculated below:

X	y	X	X^2	XY
-1	2	1	1	2
0	5	0	0	0
1	3	1	1	3
2	0	4	16	0
	$\Sigma y = 10$	$\Sigma X = 6$	$\Sigma X^2 = 18$	$\Sigma XY = 5$

∴ The normal equations become $10 = 400 + 6a_1$; $5 = 600 + 18a_1$

Solving these equations we get, $a_0 = 4.167$, $a_1 = -1.111$.

Hence the curve of best fit is

$$y = 4.167 - 1.111X \ i.e., \ y = 4.167 - 1.111x^2.$$

EXAMPLE 5.10

Using the method of least squares, fit the curve $y = ax^2 + b/x$ to the following data:

x:	1	2	3	4
y:	-1.51	0.99	8.88	7.66

Solution:

Rewriting the given equation as $xy = ax^3 + b$ and putting $x^3 = X$ and $xy = Y$, we get

$$Y = aX + b \qquad\qquad (i)$$

∴ The normal equations are

$$\Sigma Y = a\Sigma X + 4b; \ \Sigma XY = a\Sigma X^2 + b\Sigma X$$

The values of ΣX, ΣY etc. are calculated below:

X	y	$X = x^3$	$Y = xy$	XY	X^2
1	−1.51	1	−1.51	−1.51	1
2	0.99	8	1.98	15.84	64
3	3.88	27	11.64	314.28	729
4	7.66	64	30.64	1960.96	4096
		$\Sigma X = 100$	$\Sigma Y = 42.\,75$	$\Sigma XY = 2289.57$	$\Sigma X^2 = 4890$

$\therefore$ The normal equations become

$$42.75 = 100a + 4b$$
$$2289.57 = 4890a + 100b$$

Solving these equations, we get $a = 0.51$, $b = -2.06$

Hence the curve of best-fit is $Y = 0.51X \div 2.06$

i.e., $\qquad xy = 0.51x^3 - 2.06$ or $y = 0.51x^2 - \dfrac{2.06}{x}$

5.7   Fitting of Other Curves

(1) $y = ax^b$

Taking logarithms, $\log 10\, y = \log_{10} a + b\log_{10} x$

i.e., $\qquad\qquad\qquad Y = A + bX$ $\qquad\qquad\qquad\qquad$ (i)

where $X = \log_{10} x$, $Y = \log_{10} y$ and $A = \log_{10} a$.

$\therefore$ The normal equations for (i) are

$$\Sigma Y = nA + b\Sigma X, \ \Sigma XY = A\Sigma X + b\Sigma X^2$$

from which A and b can be determined. Then a can be calculated from $A = \log_{10} a$.

(2) $y = ae^{bx}$ (*Exponential curve*)

Taking logarithms, $\log_{10} y = \log_{10} a + bx \log_{10} e$

i.e., $Y = A + Bx$ where $Y = \log_{10} y$, $A = \log_{10} a$ and $B = b \log_{10} e$

Here the normal equations are $\Sigma Y = nA + B\Sigma x$, $\Sigma xY = A\Sigma x + B\Sigma x^2$ from which A, B can be found and consequently a, b can be calculated.

(3) $xy^a = b$ (*or $pv^\gamma = k$*) $\qquad\qquad\qquad\qquad$ (*Gas equation*)

Taking logarithms, $\log_{10} x + a \log_{10} y = \log_{10} b$

or
$$\log_{10} y = \frac{1}{a} \log_{10} b - \frac{1}{a} \log 10^x$$

This is of the form $Y = A + BX$

where $\qquad X = \log_{10} x, Y = \log_{10} y, A = \frac{1}{a} \log_{10} b, B = -\frac{1}{a}$

Here also the problem reduces to finding a straight line of best fit through the given data.

EXAMPLE 5.11

An experiment gave the following values:

v (ft/min):	350	400	50	600
t (min):	61	26	7	2.6

It is known that v and t are connected by the relation $v = at^b$. Find the best possible values of a and b.

Solution:

We have $\log 10\ v = \log_{10} a + b \log 10^t$

or $Y = A + bX$ where $X = \log_{10} t,\ Y = \log_{10} v,\ A = \log_{10} a$.

∴ The normal equations are

$\Sigma Y = 4A + b\Sigma X$ $\qquad\qquad\qquad\qquad\qquad\qquad\qquad (i)$

$\Sigma XY = A\Sigma X + b\Sigma X^2$ $\qquad\qquad\qquad\qquad\qquad\qquad (ii)$

Now ΣX etc. are calculated as in the following table:

v	t	$X = \log_{10} t$	$Y = \log_{10} v$	XY	X^2
350	61	1.7853	2.5441	4.542	3.187
400	26	1.4150	2.6021	3.682	2.002
500	7	0.8451	2.6990	2.281	0.714
600	2.6	0.4150	2.7782	1.153	0.172
Total		4.4604	10.6234	11.658	6.075

∴ Equations (i) and (ii) become

$$4A + 4.46b = 10.623;\ 4.46A + 6.075b = 11.658$$

Solving these, $A = 2.845,\ b = -0.1697$

∴ $a = $ antilog $A = $ antilog $2.845 = 699.8$.

EXAMPLE 5.12

Predict the mean radiation dose at an altitude of 3000 feet by fitting an exponential curve to the given data:

Altitude (x):	50	450	780	1200	4400	4800	5300
Dose of radiation (y):	28	30	32	36	51	58	69

Solution:

Let $y = ab^x$ be the exponential curve.

Then $\log 10^y = \log_{10} a + x \log_{10} b$

or $\quad Y = A + Bx$ where $Y = \log_{10} y$, $A = \log_{10} a$, $B = \log_{10} b$

∴ The normal equations are

$$\Sigma Y = 7A + B \Sigma x \qquad (i)$$

$$\Sigma x\, Y = A\Sigma x + B \Sigma x^2 \qquad (ii)$$

Now Σx etc., are calculated as follows:

x	Y	$Y = \log_{10} y$	xY	x^2
50	28	1.447158	72.3579	2500
450	30	1.477121	664.7044	202500
780	32	1.505150	1174.0170	608400
1200	36	1.556303	1867.5636	1440000
4400	51	1.707570	7513.3080	19360000
4800	58	1.763428	8464.4544	23040000
5300	69	1.838849	9745.8997	28090000
$\Sigma = 16980$		11.295579	29502.305	72743400

∴ Equations (i) and (ii) become

$$11.295579 = 7A + 16980B$$

$$29502.305 = 16980A + 72743400B$$

Solving these equations, we get $A = 1.4521015$, $B = 0.0000666289$

∴ $\quad \log_{10} y = Y = 1.4521015 + 0.0000666289\, x$

Hence y(at $x = 3000$) = 44.874 $i.e.$, 44.9 approx.

EXAMPLE 5.13

Fit a curve of the form $y = ae^{bx}$ to the following data:

x:	0	1	2	3
y:	1.05	2.10	3.85	8.30

Solution:

Taking logarithms of both sides, the given equation becomes

$$\log_{10} y = \log_{10} a + bx \log_{10} e$$

i.e., $Y = A + bx$ where $Y = \log_{10} y$, $A = \log_{10} a$, $B = b \log_{10} e$

∴   The normal equations are

$$\Sigma Y = 4A + B\Sigma x; \ \Sigma xY = A\Sigma x + B\Sigma x^2.$$

Now Σx, ΣY etc. are calculated as in the table below:

x	Y	Y	x^2	xY
0	1.05	0.0212	0	0
1	2.10	0.3222	1	0.3222
2	3.85	0.5855	4	1.1710
3	8.30	0.9191	9	2.7573
$\Sigma x = 6$		$\Sigma Y = 1.8480$	$\Sigma x^2 = 14$	$\Sigma xY = 4.2505$

Substituting these values in the normal equations, we get

$$4A + 6B = 1.848; \ 6A + 14B = 4.2505.$$

Solving these equations, $A = 0.0185$, $B = 0.2956$

∴ $a = $ antilog $A = 1.0186$, $b = B/\log 10 \ e = 0.6806$

Hence the required curve of best fit is $y = 1.0186 \ e^{0.6806x}$.

EXAMPLE 5.14

The pressure and volume of a gas are related by the equation $pV^\gamma = k$, γ and k being constants. Fit this equation to the following set of observations:

p (kg/cm²):	0.5	1.0	1.5	2.0	2.5	3.0
V (litres):	1.62	1.00	0.75	0.62	0.52	0.46

Solution:

We have $\log_{10} p + \gamma \log_{10} V = \log_{10} k$

or $$\log_{10} V = \frac{1}{\gamma} \log_{10} k - \frac{1}{\gamma} \log_{10} P$$

or $Y = A + BX$ where $X = \log_{10} P, Y = \log_{10} V, A = \frac{1}{\gamma} \log_{10} k, B = -\frac{1}{\gamma}$

$\therefore$ The normal equations are

$$\Sigma Y = 6A + B\Sigma X \qquad\qquad (i)$$

$$\Sigma XY = A\Sigma X + B\Sigma X^2 \qquad\qquad (ii)$$

Now ΣX etc. are calculated as follows:

p	V	$X = \log_{10} p$	$Y = \log_{10} V$	XY	X^2
0.5	1.62	−0.3010	0.2095	−0.0630	0.0906
1.0	1.00	0.0000	0.0000	−0.0000	0.0000
1.5	0.75	0.1761	−0.1249	−0.0220	0.0310
2.0	0.62	0.3010	−0.2076	−0.0625	0.0906
2.5	0.52	0.3979	−0.2840	−0.1130	0.1583
3.0	0.46	0.4771	−0.3372	−0.1609	0.2276
Total		1.0511	− 0.7442	− 0.4214	0.5981

$\therefore$ Equations (i) and (ii) become

$$6A + 1.0511B = - 0.7442; \; 1.0511A + 0.5981B = - 0.4214$$

Solving these, we get $A = 0.0132, B = - 0.7836$.

$\therefore \qquad \gamma = - 1/B = 1.276$

and $k = $ antilog $(A\gamma) = $ antilog $(0.0168) = 1.039$.

Hence the equation of best fit is $pV1.276 = 1.039$.

5.8   Most Plausible Values of Unknowns

Consider m linear equations in n unknowns:

$$\left.\begin{array}{l} a_{11}x_1 + a_{12}x_2 + \ldots a_{1n}x_n = k_1 \\ a_{12}x_1 + a_{22}x_2 + \ldots x_{2n}x_n = k_2 \\ \ldots\ldots\ldots\ldots\ldots\ldots\ldots\ldots \\ am_1x_1 + am_2x_2 + \ldots a_{mn}x_n = k_m \end{array}\right\} \qquad (1)$$

When $m = n$, we can find a set of values of the unknowns uniquely.

When $m > n$, *i.e.*, the number of equations is greater than the number of unknowns, it may not be possible to find these values uniquely. Then we find those values of $x_1, x_2, \ldots xn$ which satisfy (1) as nearly as possible. Applying the principle of least squares, these values

can be obtained by minimizing $E = \sum_{x=1}^{m} (a_{i1}x_1 + a_{i2}x_2 + \ldots + a_{in}x_n - k_i)^2$

using the conditions of minima, *i.e.*,

$$\frac{\partial E}{\partial x_1} = 0, \frac{\partial E}{\partial x_2} = 0, \ldots, \frac{\partial E}{\partial x_n} = 0,$$

we get n equations. Solving these equations, we get most plausible values of $x_1, x_2, \ldots x_n$.

EXAMPLE 5.15

Find the most plausible values of x, y, and z from the equations $x - 3y - 3z = -14$, $4x + y + 4z = 21$, $3x + 2y - 5z = 5$ and $x - y + 2z = 3$, by forming the normal equations.

Solution:

Let $E = (x - 3y - 3z + 14)^2 + (4x + y + 4z - 21)^2$
$$+ (3x + 2y - 5z - 5)^2 + (x - y + 2z - 3)^2$$

The most plausible values of x, y, z will be those which make E minimum. These will be given by

$$\frac{\partial E}{\partial x} = 0, \frac{\partial E}{\partial y} = 0, \frac{\partial E}{\partial z} = 0$$

$\therefore \dfrac{\partial E}{\partial x} = 2(x - 3y - 3z + 14) + 2(4x + y + 4z - 21)4 + 2(3x + 2y - 5z - 5)3 \quad (i)$

$+ 2\left(x - y + 2z - 3\right) = 0$ *i.e.*, $27x + 6y = 88$

Similarly $\qquad \dfrac{\partial E}{\partial y} = 0$ gives $6x + 15y + z = 70$

and $\qquad \dfrac{\partial E}{\partial z} = 0$ gives $y + 54z = 107$

Solving (*i*), (*ii*), and (*iii*) we get the desired values $x = 2.47$, $y = 3.55$, $z = 1.92$.

Exercises 5.3

1. If V (km/hr) and R(kg/ton) are related by a relation of the type $R = a + bV^2$, find by the method of least squares a and b with the help of the following table:

$V =$	10	20	30	40	50
$R =$	8	10	15	21	30

2. Using the method of least squares fit the curve $y = ax + bx^2$ to following observations:

x:	1	2	3	4	5
y:	1.8	5.1	8.9	14.1	19.8

3. Fit the curve $y = ax + b/x$ to the following data:

x:	1	2	3	4	5	6	7	8
y:	5.4	6.3	8.2	10.3	12.6	14.9	17.3	19.5

4. Estimate y at $x = 2.25$ by fitting the *indifference curve* of the form $xy = Ax + B$ to the following data:

x:	1	2	3	4
y:	3	1.5	6	7.5

5. Fit a least square *geometric curve* $y = ax^b$ to the following data:

x:	1	2	3	4	5
y:	0.5	2	4.5	8	12.5

6. Predict y at $x = 3.75$, by fitting a *power curve* $y = ax^b$ to the given data:

x:	1	2	3	4	5	6
y:	2.98	4.26	5.21	6.10	6.80	7.50

7. Obtain a relation of the form $y = kx^m$ for the following data by the method of least squares:

x:	1	2	3	4	5
y:	7.1	27.8	62.1	110	161

8. Fit the *exponential curve* $y = ae^{bx}$ to the following data:

x:	2	4	6	8
y:	25	38	56	84

9. Fit the curve of the form $y = ae^{bx}$ to the following data:

x:	77	100	185	239	285
y:	2.4	3.4	7.0	11.1	19.6

10. Growth of bacteria (N) in a culture after t hrs. is given in the following table:

t:	0	1	2	3	4	5	6
N:	32	47	65	92	132	190	275

Fit a curve of the form $N = ab^t$ and estimate N when $t = 7$.

11. The voltage v across a capacitor at time t seconds is given by the following table:

t:	0	2	4	6	8
v:	15	0 63	28	12	5.6

Use the method of least squares to fit a curve of the form $v = aekt$ to this data.

12. Obtain the least square fit of the form $f(t) = ae^{-3t} + be^{-2t}$ for the data:

x:	0.1	0.2	0.3	0.4
$f(t)$:	0.76	0.58	0.44	0.35

13. Find the most plausible values of x and y from the equations $x + 3y = 7.03$, $x + y = 3.01$, $2x - y = 0.03$, $3x + y = 4.97$, by forming the normal equations.

14. Obtain the most plausible values of x, y and z from the equations:

$$x + 2y + z = 1, \ -x + y + 2z = 3, \ 2x + y + z = 4, \ 4x + 2y - 5z = -7$$

5.9 Method of Group Averages

Let the straight line $y = a + bx$ (1)

fit the set of n observations (x_1, y_1), (x_2, y_2), ..., (x_n, y_n) quite closely. (Figure 5.5).

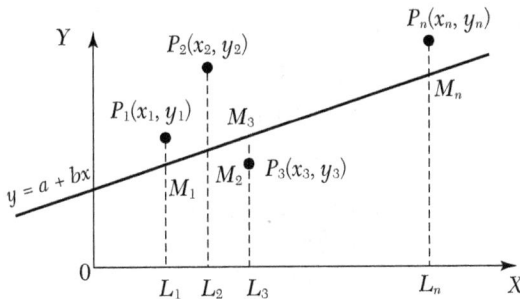

FIGURE 5.5

When $x = x_1$, the *observed (or experimental) value of* $y = y_1 = L_1P_1$

and from (1), $y = a + bx_1 = L_1M_1$, which is known as the *expected* (or *calculated*) *value* of y at L_1.

Then $\quad e_1 =$ observed value at L_1 –expected value at L_1

$$= y_1 - (a + bx_1) = M_1P_1,$$

which is called the *error* (or *residual*) at x_1. Similarly the errors for the other observations are

$$e_2 = y_2 - (a + bx_2) = M_2P_2$$

$$\dots\dots\dots\dots\dots$$

$$e_n = y_n - (a + bx_n) = M_nP_n$$

Some of these errors may be positive and others negative.

The method of group averages is based on the assumption that the sum of the residuals is zero. To find the constants a and b in (1), we require two equations. As such we divide the data into two groups: the first containing k observations $(x_1, y_1), (x_2, y_2)\dots (x_k, y_k)$; and the second group having the remaining $n - k$ observations $(x_{k+1}, y_{k+1}), (x_{k+2}, y_{k+2}),\dots,(x_n, y_n)$.

Assuming that the sum of the errors in each group is zero, we get

$$\{y_1 - (a + bx_1)\} + \{y_2 - (a + bx_2)\} + \dots + \{yk - (a + bx_k)\} = 0$$
$$\{y_{k+1} - (a + bx_{k+1})\} + \{y_{k+2} - (a + bx_{k+2})\} + \dots + \{yn - (a + bx_n)\} = 0$$

On simplification, we obtain

$$\frac{y_1 + y_2 + \dots y_k}{k} = a + b\frac{x_1 + x_2 + \dots x_k}{k}$$

$$\frac{y_{k+1} + y_{K+2} + \dots yn}{n - k} = a + b\frac{x_{k+1} + x_{K+2} + \dots xn}{n - k}$$

In (2), $\frac{1}{k}(x_1 + x_2 + \dots + x_k)$ and $\frac{1}{k}(y_1 + y_2 + \dots + y_k)$ are simply the average values of x's and y's of the first group. Hence the equations (2) and (3) are obtained from (1) by replacing x and y by their respective averages of the two groups. Solving (2) and (3), we get a and b.

NOTE **Obs.** *The main drawback of this method is that a different grouping of the observations will give different values of a and b. In practice, we divide the data in such a way that each group contains almost an equal number of observations.*

EXAMPLE 5.16

The latent heat of vaporization of steam r, is given in the following table at different temperatures t:

t:	40	50	60	70	80	90	100	110
r:	1069.1	1063.6	1058.2	1052.7	1049.3	1041.8	1036.	1030.8

For this range of temperature, a relation of the form $r = a + bt$ is known to fit the data. Find the values of a and b by the method of group averages.

Solution:

Let us divide the data into two groups each containing four readings. Then we have

t	r	t	r
40	1069.1	80	1049.3
50	1063.6	90	1041.8
60	1058.2	100	1036.3
70	1052.7	110	1030.8
$\Sigma t = 220$	$\Sigma r = 4243.6$	$\Sigma t = 380$	$\Sigma r = 4158.2$

Substituting the averages of t's and r's of the two groups in the given relation, we get

$$\frac{4243.6}{4} = a + b\frac{220}{4} \quad i.e.\ 1060.9 = a + 55b \tag{i}$$

$$\frac{4158.2}{4} = a + b\frac{380}{4} \quad i.e.\ 1039.55 = a + 95b \tag{ii}$$

Solving (i) and (ii), we obtain

$$a = 1090.26, b = -0.534.$$

EXAMPLE 5.17

The observations in the following table fit a law of the form $y = ax^n$. Estimate a and n by the method of group averages.

x:	10	20	30	40	50	60	70	80
y:	1.06	1.33	1.52	1.68	1.81	1.91	2.01	2.11

Solution:

We have $\qquad\qquad y = a^{xn}$

Taking logarithms, $\qquad \log_{10} y = \log_{10} a + n \log_{10} x$

i.e. $\qquad\qquad\qquad Y = A + nX$ $\qquad\qquad\qquad$ (i)

where $\qquad\qquad X = \log_{10} x,\ Y = \log_{10} y,\ A = \log_{10} a$

Divide the data into two groups each containing four pairs of values, so that

x	y	$X = \log_{10} x$	$Y = \log_{10} y$
10	1.06	1.0253	0.0253
20	1.33	1.3010	0.1238
30	1.52	1.4771	0.1818
40	1.68	1.6021	0.2253
		$\Sigma X = 5.4055$	$\Sigma Y = 0.5562$
50	1.81	1.6990	0.2577
60	1.91	1.7782	0.2810
70	2.01	1.8451	0.3032
80	2.11	1.9031	0.3243
		$\Sigma X = 7.2254$	$\Sigma Y = 1.1662$

Substituting the averages of X's and Y's of the two groups in (i), we get

$$\frac{0.5562}{4} = A + n\frac{5.4055}{4} \quad i.e.,\ 0.1390 = A + 1.3514\,n$$
$\qquad\qquad\qquad\qquad\qquad\qquad\qquad\qquad\qquad\qquad\qquad$ (ii)

$$\frac{11662}{4} = A + n\frac{7.2254}{4} \quad i.e.,\ 0.2916 = A + 1.8064\,n$$
$\qquad\qquad\qquad\qquad\qquad\qquad\qquad\qquad\qquad\qquad\qquad$ (iii)

(iii) – (ii) gives $0.1526 = 0.455\,n$ i.e., $n = 0.3354$

From (ii), $A = -0.3142$ i.e., $\log 10\ a = -0.3142$

where $a = $ antilog $(-0.3142) = 0.4851$

5.10 Laws Containing Three Constants

We have so far applied the above method to fit the data to laws involving two constants only. But at times we come across laws of the form

$$y = a + bx + cx^2,\ y = a + bx^c \quad \text{and} \quad y = a + be^{cx}$$

each of which contains three constants. To fit such laws to a set of observations, we devise the following procedures to reduce these to laws previously discussed.

(1) Equation $y = a + bx + cx^2$

Let (x_1, y_1) be a point on the curve satisfying the given data so that

$$y_1 = a + bx_1 + cx_1^2$$

Then
$$y - y_1 = b(x - x_1) + c(x^2 - x_1^2)$$

or
$$\frac{y - y1}{x - x1} = b + c(x + x_1)$$

Putting $x + x_1 = X$ and $(y - y_1)/(x - x_1) = Y$, it takes the linear form $Y = b + cX$.

Now b and c can be found by the graphical method or the method of averages.

(2) Equation $y = a + bx^c$

It can be rewritten as $y - a = bx^c$ $\qquad$ (1)

To find a, let (x_1, y_1), (x_2, y_2), (x_3, y_3) be three particular points on the curve (1) such that x_1, x_2, x_3 are in geometric progression

i.e.,
$$x_1 x_3 = x_2^2 \qquad (2)$$

Then
$$y_1 - a = bx_1^c$$
$$y_2 - a = bx_2^c$$

and
$$y_3 - a = bx_3^c$$

$\therefore$
$$(y_1 - a)(y_3 - a) = b^2(x_1 x_3)^c = b^2(x_2^2)^c \qquad \text{[by (2)]}$$
$$= (bx_2^c)^2 = (y_2 - a)^2$$

or
$$a(y_1 + y_3 - 2y_2) = y_1 y_3 - y_2^2$$

which gives a. Now (1) reduces to a law containing two constants b and c only.

Taking logarithms, (1) becomes

$$\log_{10}(y - a) = \log_{10} b + c \log_{10} x$$

or
$$Y = B + cX \qquad (3)$$

where $X = \log_{10} x$, $Y = \log_{10}(y - a)$, $B = \log_{10} b$.

Hence we can find b and c as before from (3).

(3) Equation $y = a + be^{cx}$

It can be written as $y - a = be^{cx}$ (1)

To find a, let (x_1, y_1), (x_2, y_2), (x_3, y_3) be three particular points on the curve (1) such that x_1, x_2, x_3 are in arithmetic progression

i.e., $$x_1 + x_3 = 2x_2 \qquad (2)$$

Then $$y_1 - a = be^{cx_1}, \quad y_2 - a = be^{cx_2} \quad \text{and} \quad y_3 - a = be^{cx_3}$$

$$\therefore \quad (y_1 - a)(y_3 - a) = b^2 e^{c(x_1 + x_3)} = (be^{cx_2})^2 = (y_2 - a)^2$$

or $$a(y_1 + y_3 - 2y_2) = y_1 y_3 - y_2^2$$

which gives a. Now (1) reduces to a law containing two constants b and c only.

Taking logarithms, (1) becomes

$$\log_{10}(y - a) = \log_{10} b + cx \log_{10} e$$

or $$Y = B + Cx \qquad (3)$$

where $Y = \log_{10}(y - a)$, $B = \log_{10} b$, $C = c \log_{10} e$.

Hence we can find b and c as before from (3).

EXAMPLE 5.18

The corresponding values of x and y are given by the following table:

x:	87.5	84.0	77.8	63.7	46.7	36.9
y:	292	283	270	235	197	181

Fit a parabola of the form $y = a + bx + cx^2$, by the method of group averages.

Solution:

Taking $x = 84$, $y = 283$ as a particular point on $y = a + bx + cx^2$, we get

$$283 = a + b(84) + c(84)^2 \qquad (i)$$

$$y - 283 = b(x - 84) + c[x^2 - (84)^2]$$

or $$\frac{y - 283}{x - 84} = b + c(x + 84) \quad i.e,. \quad Y = b + cX \qquad (ii)$$

where $X = x + 84$, $Y = (y - 283)/(x - 84)$.

Now we have the following table of values:

x	y	$X = x + 84$	$Y = (y - 283)/(x - 84)$
87.5	292	171.5	2.571
84.0	283	—	—
77.8	270	161.8	2.097
		$\Sigma X = 333.3$	$\Sigma Y = 4.668$
63.7	235	147.7	2.364
46.7	197	130.7	2.306
36.9	181	120.9	2.166
		$\Sigma X = 399.3$	$\Sigma Y = 6.836$

Substituting the averages of X and Y in (ii), we get

$$\frac{4.668}{2} = b + c\frac{333.3}{2} \text{ i.e. } 2.33 = b + 166.65 \text{ c} \qquad (iii)$$

$$\frac{6.836}{3} = b + c\frac{399.3}{3} \text{ i.e. } 2.28 = b + 131.1 \text{ c} \qquad (iv)$$

$(iv) - (iii)$ gives $c = 0.0014$

and (iii) gives $b = 2.0967$ i.e., 2.1 nearly

From (i), we get $a = 96.9988$ i.e., 97 nearly.

Hence the parabola of fit is $y = 97 + 2.1x + 0.0014x^2$

EXAMPLE 5.19

The train resistance R (lbs/ton) is measured for the following values of its velocity V (km/hr):

V:	20	40	60	80	100
R:	5	9	14	25	36

If R is related to V by the formula $R = a + bV^n$, find a, b, and n.

Solution:

To find a, we take the following three values of v which are in G.P.:

$$v_1 = 20, \qquad v_2 = 40, \qquad v_3 = 80$$

Then $\qquad R_1 = 5, \qquad R_2 = 9, \qquad R_3 = 25$

$$\therefore \quad (R_1 - a)(R_3 - a) = (R_2 - a)^2$$

where $a = \dfrac{R_1 R_3 - R_2^2}{R_1 + R_3 - 2R_2} = 3.67$

Thus $R - 3.67 = bV^n$ or $\log_{10}(R - 3.67) = \log_{10} b + n \log_{10} V$

$$Y = k + nX \qquad\qquad\qquad (i)$$

where $X = \log_{10} V$, $Y = \log_{10}(R - 3.67)$, $k = \log_{10} b$.

Now we have the following table of values:

V	R	$X = \log_{10} V$	$Y = \log_{10}(R - 3.67)$
20	5	1.3010	0.1238
40	9	1.602	1 0.7267
60	14	1.7782	1.0141
		$\Sigma X = 4.6813$	$\Sigma Y = 1.8646$
80	25	1.9031	1.3290
100	36	2.0000	1.5096
		$\Sigma X = 3.9031$	$\Sigma Y = 2.8386$

Substituting the averages of X's and Y's in (1), we obtain

$$\frac{1.8646}{2} = k + n\frac{4.6813}{2} \quad i.e.\ 0.6215 = k + 1.5604\,n \qquad (ii)$$

$$\frac{2.8386}{2} = k + n\frac{3.9031}{2} \quad i.e.\ 1.4193 = k + 1.9516\,n \qquad (iii)$$

Solving (ii) and (iii), we get $n = 2.04$, $k = -2.56$ approx.

$$\therefore \quad b = \text{antilog } k = \text{antilog}(-2.56) = 0.0028.$$

Exercises 5.4

1. Fit a straight line of the form $y = a + bx$ to the following data by the method of group averages:

x:	0	5	10	15	20	25
y:	12	15	17	22	24	30

2. Apply the method of group averages to work out Example 4.13.

3. The weights of a calf taken at weekly intervals are given below:

Age:	1	2	3	4	5	6	7	8	9	10
Weight:	52.5	58.7	65.0	70.2	75.4	81.1	87.2	95.5	102.2	108.4

Find a straight line of best fit.

4. Work out Example 5.1, by the method of group averages.

5. The head of water H (ft) and the quantity of water Q(ft^3) flowing per second are related by the law $Q = CH^n$. Find the best values of C and n by the method of group averages for the following data:

H:	1.2	1.4	1.6	1.8	2.0	2.4
Q:	4.2	6.1	8.5	11.5	14.9	23.5.

6. Using the method of averages, fit a parabola $y = ax2 + bx + c$ to the following data:

x:	20	40	60	80	100	120
y:	5.5	9.1	14.9	22.8	33.3	46.0

7. While testing a centrifugal pump, the following data is obtained. It is assumed to fit the equation $y = a + bx + cx^2$, where x is the discharge in liter/sec and y, head in meter of water. Find the values of the constants a, b, c by the method of group averages.

x:	2	2.5	3	3.5	4	4.5	5	5.5	6
y:	18	17.8	17.5	17	15.8	14.8	13.3	11.7	9

8. By the method of averages, fit a curve of the form $y = ae^{bx}$ to the following data:

x:	5	15	20	30	35	40
y:	10	14	25	40	50	62

9. In an experiment, the voltage v is observed for the following values of the current i:

i:	0.5	1	2	4	8	12
v:	160	120	94	75	62	56

If v and i are connected by the relation $v = a + bi^k$, find $a, b,$ and k.

10. The variables s and t are connected by the relation $s = a + be^{nt}$ and their corresponding values are given in the following table:

t:	1	2	6	8	11
s:	12.7	12.5	11.6	11.3	11

Find the best possible values of $a, b,$ and n.

5.11   Method of Moments

Let (x_1, y_1), (x_2, y_2), $\cdots$, (x_n, y_n) be the set of n observations such that

$$x_2 - x_1 = x_3 - x_2 = \cdots = x_n - x_{n-1} = h \text{ (say)}$$

We define the *moments of the observed values* of y as follows:

m_1, the 1st moment $= h\Sigma y$

m_2, the 2nd moment $= h\Sigma xy$

m_3, the third moment $= h\Sigma x^2 y$ and so on.

Let the curve fitting the given data be $y = f(x)$. Then the *moments of the calculated values* of y are

μ_1, the 1st moment $= \int y dx$

μ_2, the 2nd moment $= \int xy \, dx$

μ_3, the 3rd moment $= \int x^2 y \, dx$ and so on.

This method is based on the assumption that the moments of the observed values of v are respectively equal to the moments of the calculated values of y, i.e., $m_1 = \mu_1$, $m_2 = \mu_2$, $m_3 = \mu_3$ etc. These equations (known as *observation equations*) are used to determine the constants in $f(x)$.

m's are calculated from the tabulated values of x and y while μ's are computed as follows:

In Figure 5.6, y_1 the ordinate of $P_1(x = x_1)$, can be taken as the value of y at the mid-point of the interval $(x_1 - h/2, x_1 + h/2)$. Similarly y_n, the ordinate of $P_n(x = x_n)$, can be taken as the value of y at the mid-point of the interval $(x_n - h/2, x_n + h/2)$. If A and B be the points such that

$$OA = x_1 - h/2 \text{ and } OB = x_n + h/2,$$

then $\quad \mu_1 = \int y \, dx = \int_{x_1 - h/2}^{x_n h/2} f(x) dxa$

$$\mu_2 = \int_{x_1 = h/2}^{x_n + h/2} xf(x) dx \qquad \mu_3 = \int_{x_1 - h/2}^{x_n + h/2} x^2 f(x) dx \text{ and so on.}$$

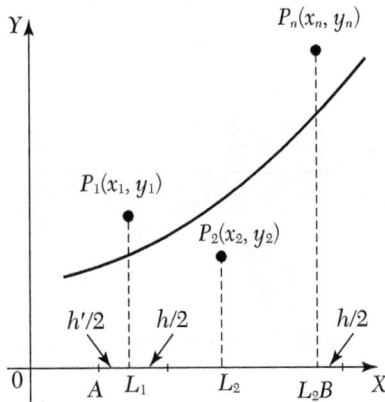

FIGURE 5.6

EXAMPLE 5.20

Fit a straight line $y = a + bx$ to the following data by the method of moments:

x:	1	2	3	4
y:	16	19	23	26

Solution:

Since only two constants a and b are to be found, it is sufficient to calculate the first two moments in each case. Here $h = 1$.

$$m_1 = h\Sigma y = 1(16 + 19 + 23 + 26) = 84$$

$$m_2 = h\Sigma xy = 1(1 \times 16 + 2 \times 19 + 3 \times 22 + 4 \times 26) = 227.$$

To compute the moments of calculated values of $y = a + bx$, the limits of integration will be $1 - h/2$ and $4 + h/2$, i.e., 0.5 and 4.5.

$$\therefore \quad \mu_1 = \int_{0.5}^{4.5} (a + bx)dx = \left| ax + b\frac{x^2}{2} \right|_{0.5}^{4.5} = 4a + 10b$$

$$\mu_2 = \int_{0.5}^{4.5} x(a + bx)dx = 10a + \frac{91}{3}b.$$

Thus, the observation equations $mr = \eta r$ $(r = 1, 2)$ are

$$4a + 10b = 84; \quad 10a + \frac{91}{3}b = 227$$

Solving these, $a = 13.02$ and $b = 3.19$.

Hence the required equation is $y = 13.02 + 3.19x$.

EXAMPLE 5.21

Given the following data:

x :	0	1	2	3	4
y :	1	5	10	22	38

find the parabola of best fit by the method of moments.

Solution:

Let the parabola of best fit be $y = a + bx + cx^2$ (i)

Since three constants are to be found, we calculate the first three moments in each case. Here $h = 1$.

$$m_1 = h\Sigma y = 1(1 + 5 + 10 + 22 + 38) = 76$$
$$m_2 = h\Sigma xy = 1(0 + 5 + 20 + 66 + 152) = 243$$
$$m_3 = h\Sigma x2y = 1(0 + 5 + 40 + 198 + 608) = 851$$

For computing the moments of calculated values of (i), the limits of integration will be $0 - h/2$ and $4 + h/2$, i.e., -0.5 and 4.5.

$$\mu_1 = \int_{-0.5}^{4.5} (a + bx + cx^2)dx = 5a + 10b + 30.4c$$
$$\mu_2 = \int_{-0.5}^{4.5} x(a + bx + cx^2)dx = 10a + 30.4b + 102.5c$$
$$\mu_3 = \int_{-0.5}^{4.5} x^2(a + bx + cx^2)dx = 30.4a + 102.5b + 369.1c$$

Thus the observation equations $mr = \mu_r$ $(r = 1, 2, 3)$ are

$$5a + 10b + 30.4c = 76$$
$$10a + 30.4b + 102.5c = 243$$
$$30.4a + 102.5b + 369.1c = 851$$

Solving these equations, we get $a = 0.4$, $b = 3.15$, $c = 1.4$.

Hence the parabola of best fit is $y = 0.4 + 3.15x + 1.4x^2$.

Exercises 5.5

1. Use the method of moments to fit the straight line $y = a + bx$ to the data:

x:	1	2	3	4
y:	0.17	0.18	0.23	0.32

Empirical Laws and Curve-Fitting • 231

2. Fit a straight line to the following data, using the method of moments:

x:	1	3	5	7	9
y:	1.5	2.8	4.0	4.7	6.0

3. Fit a parabola of the form $y = a + bx + cx^2$ to the data:

x:	1	2	3	4
y:	1.7	1.8	2.3	3.2

by the method of moments.

4. By using the method of moments, fit a parabola to the following data:

x:	1	2	3	4
y:	0.30	0.64	1.32	5.40

5.12 Objective Type of Questions

Exercises 5.6

Select the correct answer or fill up the blanks in the following questions:

1. The *method of group averages* is based on the assumption that the sum of the residuals is…

2. $y = ax^b + c$ in linear form is…..

3. To fit the straight line $y = mx + c$ to n observations, the normal equations are
   (i) $\Sigma y = n\Sigma x + \Sigma cm$, $\Sigma xy = c\,\Sigma x^2 + c\Sigma n$.
   (ii) $\Sigma y = m\Sigma x + nc$, $\Sigma xy = m\Sigma x^2 + c\Sigma x$.
   (iii) $\Sigma y = c\Sigma x + m\Sigma n$, $\Sigma xy = c\Sigma x^2 + m\Sigma x$.

4. To fit $y = ab^x$ by least square method, normal equations are…

5. The observation equations for fitting a straight line by *method of moments* are…

6. The principle of '*least squares*' states that…

7. $y = ax^2 + b\log_{10} x$ reduced to linear law takes the form….

8. Given $\begin{bmatrix} x: & 0 & 1 & 2 \\ y: & 0 & 1.1 & 2.1 \end{bmatrix}$, then the straight line of best fit is….

9. The *method of moments* is based on the assumption that....

10. In $y = a + bx$, $\Sigma x = 50$, $\Sigma y = 80$, $\Sigma xy = 1030$, $\Sigma x^2 = 750$, and $n = 10$, then $a = ...$, $b = ...$

11. The gas equation $pvr = k$ can be reduced to $y = a + bx$ where $a = ...$ $b = ...$

12. $y = \dfrac{x}{ax + b}$ in linear form is...

13. If $y = ke^{mx}$, then the first normal equation is $\Sigma\log_{10} y = ...$

    (a) $kn + m\Sigma x$ (b) $k\Sigma x + m\Sigma x2$

    (c) $n \log_{10} k + m \log_{10}e\Sigma x$ (d) $k\Sigma\log_{10}y + m\Sigma x$.

14. If $y = a + bx + cx^2$ and

x:	0	1	2	3	4
y:	1	1.8	1.3	2.5	7.3

then the first normal equation is

    (a) $15 = 5a + 10b + 29c$ (b) $15 = 5a + 10b + 31c$

    (c) $12.9 = 5a + 10b + 30c$ (d) $34 = 5a + 10b + 27c$.

15. If $y = 2x + 5$ is the best fit for 8 pairs of values of (x, y) by the method of least squares, and $\Sigma y = 120$, then $\Sigma x = ...$

    (a) 35   (b) 40

    (c) 45   (d) 30.

16. If $y = a + bx^2$ and n is the number of observations, then the first normal equations is $\Sigma y = ...$

    (a) $na + b\Sigma x^2$ (b) $na\ \Sigma x + b\Sigma x^2$

    (c) $na + b\Sigma x^3$ (d) $na\ \Sigma y + b\Sigma yx^2$.

CHAPTER 6

FINITE DIFFERENCES

Chapter Objectives

- Introduction
- Finite differences
- Differences of a polynomial
- Factorial notation
- Reciprocal factorial function
- Inverse operator of Δ
- Effect of an error on a difference table
- Other difference operators
- Relations between the operators
- To find one or more missing terms
- Application to summation of series
- Objective type of questions

6.1   Introduction

The calculus of finite differences deals with the changes that take place in the value of the function (dependent variable), *due to finite changes in the independent variable.* Through this, we also study the relations that exist between the values assumed by the function, whenever the independent variable changes by finite jumps whether equal or unequal. On the other hand, in infinitesimal calculus, we study those changes of the function which occur when the independent variable changes continuously in a given interval. In this chapter, we shall study the variations in the function when the independent variable changes by equal intervals.

6.2 Finite Differences

Suppose that the function $y = f(x)$ is tabulated for the equally spaced values $x = x_0, x_0 + h, x_0 + 2h, \ldots, x_0 + nh$ giving $y = y_0, y_1, y_2, \cdots, y_n$. To determine the values of $f(x)$ or $f'(x)$ for some intermediate values of x, the following three types of differences are found useful:

Forward differences. The differences $y_1 - y_0, y_2 - y_1, \cdots, y_n - y_{n-1}$ when denoted by $\Delta y_0, \Delta y_1, \cdots, \Delta y_{n-1}$ respectively are called the *first forward differences* where Δ is the *forward difference operator*. Thus the first forward differences are $\Delta y_r = y_{r+1} - y_r$.

Similarly these second forward differences are defined by $\Delta^2 y_r = \Delta y_{r+1} - \Delta y_r$.

In general, $\Delta^p y_r = \Delta^{p-1} y_{r+1} - \Delta^{p-1} y_r$ defines the pth forward differences. These differences are systematically set out in Table 6.1.

In a difference table, x is called the *argument* and y the *function* or the *entry*. y_0, the first entry, is called the *leading term* and $\Delta y_0, \Delta^2 y_0, \Delta^3 y_0$ etc. are called the *leading differences*.

TABLE 6.1 Forward Difference Table

Value of x	Value of y	1st diff.	2nd diff.	3rd diff.	4th diff.	5th diff.
x_0	y_0					
		Δy_0				
$x_0 + h$	y_1		$\Delta^2 y_0$			
		Δy_1		$\Delta^3 y_0$		
$x_0 + 2h$	y_2		$\Delta^2 y_1$		$\Delta^4 y_0$	
		Δy_2		$\Delta^3 y_1$		$\Delta^5 y_0$
$x_0 + 3h$	y_3		$\Delta^2 y_2$		$\Delta^4 y_1$	
		Δy_3		$\Delta^3 y_2$		
$x_0 + 4h$	y_4		$\Delta^2 y_3$			
		Δy_4				
$x_0 + 5h$	y_5					

NOTE

Obs. 1. *Any higher order forward difference can be expressed in terms of the entries.*

We have $D^2 y_0 = Dy_1 - Dy_0 = (y_2 - y_1) - (y_1 - y_0) = y_2 - 2y_1 + y_0$

$$\Delta^3 y_0 = \Delta^2 y_1 - \Delta^2 y_0 = (y_3 - 2y_2 + y_1) - (y_2 - 2y_1 + y_0)$$

$$= y_3 - 3y_2 + 3y_1 - y_0$$

$$\Delta^4 y0 = \Delta^3 y_1 - \Delta^3 y_0$$

$$= (y_4 - 3y_3 + 3y_2 - y_1) - (y_3 - 3y_2 + 3y_1 - y_0)$$

$$= y_4 - 4y_3 + 6y_2 - 4y_1 + y_0$$

The coefficients occurring on the right-hand side being the binomial coefficients, we have in general,

$$\Delta^n y_0 = y_n - {}^n c_1 y_{n-1} + {}^n c_2 y_{n-2} - \cdots + (-1)^n y_0.$$

Obs.2. *The operator* Δ *obeys the distributive, commutative, and index laws*

i.e., (*i*) $\Delta[f(x) \pm \phi(x)] = \Delta f(x) \pm \Delta \phi(x)$.

(*ii*) $\Delta[cf(x)] = c\Delta f(x)$, c being a constant.

(*iii*) $\Delta m \, \Delta n \, f(x) = \Delta m + n \, f(x)$, m and n being positive integers. In view of (*i*) and (*ii*), Δ is a *linear operator*.

But $\Delta[f(x).\phi(x)] \neq f(x).\Delta \phi(x)$.

Backward differences. The differences $y_1 - y_0, \, y_2 - y_1, \cdots, y_n - y_{n-1}$ when denoted by $\nabla y_1, \nabla y_2, \cdots, \nabla y_n$ respectively, are called the *first backward differences* where Δ

TABLE 6.2 Backward Difference Table

Value of x	Value of y	1st diff.	2nd diff.	3rd diff.	4th diff.	5th diff.
x_0	y_0					
		∇y_1				
$x_0 + h$	y_1		$\nabla^2 y_2$			
		∇y_2		$\nabla^3 y_3$		
$x_0 + 2h$	y_2		$\nabla^2 y_3$		$\nabla^4 y_4$	
		∇y_3		$\nabla^3 y_4$		$\nabla^5 y_5$
$x_0 + 3h$	y_3		$\nabla^2 y_4$		$\nabla^4 y_5$	
		Δy_4		$\nabla^3 y_5$		
$x_0 + 4h$	y_4		$\nabla^2 y_5$			
		∇y_5				
$x_0 + 5h$	y_5					

is the *backward difference operator*. Similarly we define higher order backward differences. Thus we have $\nabla y_r = y_r - y_{r-1}$, $\nabla^2 y_r = \nabla y_r - \nabla y_{r-1}$, $\nabla^3 y_r = \nabla^2 y_r - \nabla^2_{r-1}$, etc.

These differences are exhibited in the Table 6.2.

Central differences. Sometimes it is convenient to employ another system of differences known as *central differences*. In this system, the *central difference operator* δ is defined by the relations:

$$y_1 - y_0 = \delta y_{1/2}, \, y_2 - y_1 = \delta y_{3/2}, \, \cdots, \, y_n - y_{n-1} = \delta y_{n-1/2}$$

Similarly, higher order central differences are defined as

$$\delta y_{3/2} - \delta y_{1/2} = \delta^2 y_1, \, \delta y_{5/2} - \delta y_{3/2} = \delta^2 y_2, \, \cdots, \delta^2 y_2 - \delta^2 y_1 = \delta^3 y_{3/2} \text{ and so on.}$$

These differences are shown in Table 6.3.

TABLE 6.3 Central Difference Table

Value of x	Value of y	1st diff.	2nd diff.	3rd diff.	4th diff.	5th diff.
x_0	y_0					
		$\delta y_{1/2}$				
$x_0 + h$	y_1		$\delta^2 y_1$			
		$\delta y_{3/2}$		$\delta^3 y_{3/2}$		
$x_0 + 2h$	y_2		$\delta^2 y_2$		$\delta^4 y_2$	
		$\delta y_{5/2}$		$\delta^3 y_{5/2}$		$\delta^5 y_{5/2}$
$x_0 + 3h$	y_3		$\delta^2 y_3$		$\delta^4 y_3$	
		$\delta y_{7/2}$				
$x_0 + 4h$	y_4		$\delta^2 y_4$	$\delta^3 y_{7/2}$		
		$\delta y_{9/2}$				
$x_0 + 5h$	y_5					

We see from this table that the central differences on the same horizontal line have the same suffix. Also the differences of odd order are known only for half values of the suffix and those of even order for only integral values of the suffix.

It is often required to find the mean of adjacent values in the same column of differences. We denote this mean by μ.

Thus $\mu \delta y_1 = \dfrac{1}{2}(\delta y_{1/2} + \delta y_{3/2})$, $\mu \delta^2 y_{3/2} = \dfrac{1}{2}(\delta^2 y_1 + \delta^2 y_2)$, etc.

NOTE

Obs. *The reader should note that it is only the notation which changes and not the differences. e.g.*

$$y_1 - y_0 = \nabla y_0 = \Delta y_1 = \delta y_{1/2}.$$

Of all the formulae, those involving central differences are most useful in practice as the coefficients in such formulae decrease much more rapidly.

EXAMPLE 6.1

Evaluate (i) $\Delta \tan^{-1} x$ (ii) $\Delta(e^x \log 2x)$ (iii) $\Delta(x^2/\cos 2x)$ $(iv)\Delta(^nC_{r+1})$.

Solution:

(i) $\Delta\tan^{-1}x = \tan^{-1}(x+h) - \tan^{-1}x$

$$= \tan^{-1}\left\{\frac{x+h-x}{1+(x+h)x}\right\} = \tan^{-1}\left\{\frac{h}{1+hx+x^2}\right\}$$

(ii) $\Delta(e^x \log 2x) = e^{x+h} \log 2 \ (x+h) - e^x \log 2x$

$$= e^{x+h} \log 2 \ (x+h) - e^{x+h} \log 2x + e^{x+h} \log 2x - e^x \log 2x$$

$$= e^{x+h} \log\frac{x+h}{x} (e^{x+h} - e^x)\log 2x$$

$$= e^x\left[e^h \log\left(1+\frac{h}{x}\right) + \left(e^h - 1\right)\log 2x\right]$$

(iii) $\Delta\left(\dfrac{x^2}{\cos 2x}\right) = \dfrac{(x+h)^2}{\cos 2(x+h)} - \dfrac{x^2}{\cos 2x}$

$$= \frac{(x+h)^2 \cos 2x - x^2 \cos 2(x+h)}{\cos 2(x+h)\cos 2x}$$

$$= \frac{\left[(x+h)^2 - x^2\right]\cos 2x + x^2\left[\cos 2x - \cos 2(x+h)\right]}{\cos 2(x+h)\cos 2x}$$

$$= \frac{(2hx+h^2)\cos 2x + 2x^2 \ \sin(h)\sin(2x+h)}{\cos 2(x+h)\cos 2x}$$

(iv) $\Delta(^nC_{r+1}) = {}^{n+1}C_{r+1} - {}^nC_{r+1}$

$$= \frac{(n+1)!}{(r+1)!(n-r)!} - \frac{(n)!}{(r+1)!(n-r-1)!}$$

$$= \frac{(n)!}{(r+1)!(n-r-1)!} \left(\frac{n+1}{n-r} - 1\right)$$

$$= \frac{(n)!}{(r+1)!(n-r-1)!}\frac{(r+1)}{(n-r)} = \frac{n!}{r!(n-r)!} = {}^nc_r$$

EXAMPLE 6.2

Evaluate (*i*) $\Delta^2\left(\dfrac{5x+12}{x^2+5x+16}\right)$ (*ii*) $\Delta^2\cos 2x$

(*iii*) $\Delta^2(ab^{cx})$ (*iv*) $\Delta^n(e^x)$

Interval of differencing being unity

Solution:

(*i*): $\Delta^2\left(\dfrac{5x+12}{x2+5x+16}\right) = \Delta^2\left\{\dfrac{5x+12}{(x+2)(x+3)}\right\} = \Delta^2\left\{\dfrac{2}{x+2}+\dfrac{3}{x+3}\right\}$

$$= \Delta\left\{\Delta\left(\frac{2}{x+2}\right)+\Delta\left(\frac{3}{x+3}\right)\right\}$$

$$= \Delta\left\{2\left(\frac{1}{x+3}-\frac{1}{x+2}\right)+3\left(\frac{1}{x+4}-\frac{1}{x+3}\right)\right\}$$

$$= -2\Delta\left\{\frac{1}{(x+2)(x+3)}\right\}-3\Delta\left\{\frac{1}{(x+3)(x+4)}\right\}$$

$$= -2\Delta\left\{\frac{1}{(x+3)(x+4)}-\frac{1}{(x+2)(x+3)}\right\}$$

$$\qquad -3\left\{\frac{1}{(x+4)(x+5)}-\frac{1}{(x+3)(x+4)}\right\}$$

$$= \frac{4}{(x+2)(x+3)(x+4)}+\frac{6}{(x+3)(x+4)(x+5)}$$

$$= \frac{2(5x+16)}{(x+2)(x+3)(x+4)(x+5)}$$

(*ii*) $\Delta^2\cos 2x = \Delta\{\cos 2(x+h)-\cos 2x\}$

$$= \Delta\cos 2(x+h)-\Delta\cos 2x$$

$$= [\cos 2(x+2h)-\cos 2(x+h)]-[\cos 2(x+h)-\cos 2x]$$

$$= -2\sin(2x+3h)\sin h+2\sin(2x+h)\sin h$$

$$= -2\sin h[\sin(2x+3h)-\sin(2x+h)]$$

$$= -2\sinh[2\cos(2x + 2h)\sinh] = -4\sin^2 h\cos(2x + 2h).$$

(iii) $\Delta(ab^{cx}) = a\Delta(b^{cx}) = a[b^{c(x+1)} - b^{cx}] = ab^{cx}(b^c - 1)$

$\Delta^2(ab^{cx}) = \Delta[\Delta(ab^{cx})] = a(b^c - 1)\Delta(b^{cx})$

$\qquad = a(b^c - 1)(b^{c(x+1)} - b^{cx}) = a[b^c - 1]^2 b^{cx}$

(iv) $\Delta e^x = e^{x+1} - e^x = (e - 1)e^x$

$\qquad \Delta^2 e^x = \Delta(\Delta e^x) = \Delta[(e - 1)e^x]$

$\qquad\qquad = (e - 1)\Delta e^x = (e - 1)(e - 1)e^x = (e - 1)^2 e^x$

Similarly $\Delta^3 e^x = (e - 1)^3 e^x, \Delta^4 e^x = (e - 1)^4 e^x, \quad \cdots$

and $\Delta^n e^x = (e - 1)^n e^x.$

EXAMPLE 6.3

If $y = a(3)^x + b(-2)^x$ and $h = 1$, prove that $(\Delta^2 + \Delta - 6)y = 0$.

Solution:

We have $y = a(3)^x + b(-2)^x$

$\therefore \qquad \Delta y = [a(3)^{x+1} + b(-2)^{x+1}] - [a(3)^x + b(-2)^x]$

$\qquad\qquad = 2a(3)^x - 3b(-2)^x$

and $\Delta^2 y = [2a(3)^{x+1} - 3b(-2)^{x+1}] - [2a(3)^x - 3b(-2)^x]$

$\qquad\qquad = 4a(3)^x + 9b(-2)^x$

Hence $(\Delta^2 + \Delta - 6)y = [4a(3)^x + 9b(-2)^x] + (2a(3)^x - 3b(-2)^x)$

$$- 6[a(3)^x + b(-2)^x] = 0$$

EXAMPLE 6.4

Find the missing y_x values from the first differences provided:

y_x	0	—	—	—	—	—
Δy_x	0	1	2	4	7	11

Solution:

Let the missing values be y_1, y_2, y_3, y_4, y_5. Then we have

y_x	0	y_1	y_2	y_3	y_4	y_5
Δy_x	0	1	2	4	7	11

$\therefore \qquad y_1 - 0 = 1, y_2 - y_1 = 2, y_3 - y_2 = 4, y_4 - y_3 = 7, y_5 - y_4 = 11$

$i.e., \qquad y_1 = 1, y_2 = 2 + y_1 = 3, y_3 = 4 + y_2 = 7, y_4 = 7 + y_3 = 14,$

$\qquad\qquad y_5 = 11 + y_4 = 25.$

6.3 Differences of A Polynomial

The nth differences of a polynomial of the nth degree are constant and all higher order differences are zero.

Let the polynomial of the *n*th degree in *x*, be

$$f(x) = ax^n + bx^{n-1} + cx^{n-2} + \cdots + kx + l$$

$\therefore \qquad \Delta f(x) = f(x+h) - f(x)$

$$= a[(x+h)^n - x^n] + b[(x+h)^{n-1} - x^{n-1}] + \cdots + kh$$

$$= anhx^{n-1} + b'x^{n-2} + c'x^{n-3} + \cdots + k'x + l' \qquad (1)$$

where $b', c', \cdots, l'$ are the new constant co–efficients.

Thus the first differences of a polynomial of the *n*th degree is a polynomial of degree $(n-1)$.

Similarly

$$\Delta^2 f(x) = \Delta[f(x+h) - f(x)] = \Delta f(x+h) - \Delta f(x)$$

$$= anh[(x+h)^{n-1} - x^{n-1}] + b'[(x+h)^{n-2} - x^{n-2}] + \cdots + k'h$$

$$= an(n-1)h^2 x^{n-2} + b''x^{n-3} + c''x^{n-4} + \cdots + k'', \qquad \text{by}(1)$$

$\therefore$ The second differences represent a polynomial of degree $(n-2)$

Continuing this process, for the *n*th differences we get a polynomial of degree zero *i.e.*

$$\Delta^n f(x) = an(n-1)(n-2)\ldots 1 . h^n = an\,!h^n \qquad (2)$$

which is a constant. Hence the $(n+1)$th and higher differences of a polynomial of *n*th degree

will be zero.

NOTE

Obs. *The converse of this theorem is also true, i.e., if the nth differences of a function tabulated at equally spaced intervals are constant, the function is a polynomial of degree n. This* *fact is important in numeric alanalysis as it enables us to approximate a function by a polynomial of nth degree, if it s nth order differences become nearly constant.*

EXAMPLE 6.5

Evaluate $\Delta^{10}[(1-ax)(1-bx^2)(1-cx^3)(1-cx^3)(1-dx^4)]$.

Solution:

$$\Delta^{10}[(1-ax)(1-bx^2)(1-cx^3)(1-dx^4)] = \Delta^{10}[abcdx^{10} + (\ )x^9 + (\ )x^8 + \ldots + 1]$$

$$= abcd\ \Delta^{10}(x^{10}) \qquad\qquad [\therefore \Delta^{10}(x^n) = 0 \text{ for } n<10]$$

$$= abcd(10!) \qquad\qquad\qquad [\text{by}(2)\text{above}]$$

Exercises 6.1

1. Write forward difference table if

x:	10	20	30	40
y:	1.1	2.0	4.4	7.9

2. Construct the table of differences for the data below:

x:	0	1	2	3	4
$f(x)$:	1.0	1.5	2.2	3.1	4.6

Evaluate $\Delta^3 f(2)$.

3. If $u_0 = 3$, $u_1 = 12, u_2 = 81, u_3 = 2000$, $u_4 = 100$, calculate $\Delta^4 u_0$.

4. Show that $\Delta^3 y_i = y_{i+3} - 3y_{i+2} + 3y_{i+1} - yi$.

5. If $y = x^3 + x^2 - 2x + 1$, evaluate the values of y for $x = 0,1,\ 2,\ 3,\ 4,\ 5$ from the difference table. Find the value of y at $x = 6$ by extending the table and verify that same value is obtained by substitution.

6. Form a table of differences for the function $f(x) = x^3 + 5x - 7$ for $x = -1, 0, 1, 2, 3, 4, 5$.

Continue the table to obtain $f(6)$.

7. Extend the following table to two more terms on either side by constructing the difference table:

x:	−0.2	0.0	0.2	0.4	0.6	0.8	1.0
y:	2.6	3.0	3.4	4.28	7.08	14.2	29.0.

8. Show that

(i) $\Delta\left[\dfrac{1}{f(x)}\right] = \dfrac{-\Delta f(x)}{f(x)f(x+1)}$ (ii) $\Delta \log f(x) = \log\left\{1 + \dfrac{\Delta f(x)}{f(x)}\right\}$

9. Evaluate (taking interval of differencing as unity)

(*i*) $\Delta(x + \cos x)$ (*ii*) $\Delta \tan^{-1}\left(\dfrac{n-1}{n}\right)$

(*iii*) $\Delta\,(e^{3x}\log 2x)$ (*iv*)$\Delta(2^x/x!)$

10. Evaluate:

(i) $\Delta^2 \cos 3x$ (ii) $\Delta^2\left(\dfrac{1}{x^2+5x+6}\right)$

(*iii*) $\Delta n(e^{2x+3})$ (*iv*) $\Delta^n\left(\dfrac{1}{x}\right)$

(*v*) $\Delta^n \sin\,(ax + b)$

11. If $f(x) = e^{ax+b}$, show that its leading differences form a geometric progression

12. Prove that

(*i*) $y_3 = y_2 + \Delta y_1 + \Delta^2 y_0 + \Delta^3 y_0$ (*ii*)$\Delta^2 y_8 = y_8 - 2y_7 + y_6$
(*iii*) $\delta^2 y_5 = y_6 - 2y_5 + y_4$.

13. Evaluate:

(*i*)$\Delta^4[(1-x)(1-2x)(1-3x)(1-4x)],(h=1)$.
(*ii*)$\Delta^{10}[(1-x)(1-2x^2)(1-3x^3)(1-4x^4)]$, if the interval of differencing is 2.

6.3 Factorial Notation

A product of the form $x(x-1)(x-2)\cdots(x-r+1)$ is denoted by $[x]^r$ and is called a **factorial.**

In particular $[x] = x, [x]^2 = x(x-1), [x]^3 = x(x-1)(x-2),$ etc.

In general $[x]^n = x(x-1)(x-2)\,.....(x-n+1)$

If the interval of differencing is h, then $[x]^n = x(x-h)(x-2h)\cdots(x-\overline{n-1}h)$ which is called a **factorial polynomial** or **function** of degree n.

The factorial notation is of special utility in the theory of finite differences. It helps in finding the successive differences of a polynomial directly by simple rule of differentiation.

To show that $\Delta n[x]n = n!$ and $\Delta n + 1[x]n = 0$

We have

$$\Delta[x]^n = \left[x+h\right]^n - \left[x\right]^n$$

$$= (x+h)(x+h-h)(x+h-2h)\cdots(x+h-\overline{n-1}h)$$

$$\qquad\qquad -x(x-h)(x-2h)\cdots(x-\overline{n-1}h) \qquad\qquad (i)$$

$$= x(x-h)\cdots(x-\overline{n-2}h)[x+h-(x-nh+h)]$$

$$= nh[x]^{n-1}$$

Similarly $\Delta^2[x]^n = \Delta\{nh[x]^{n-1}\} = nh\Delta[x]^{n-1}$

Replacing n by $n-1$ in (i),

we get $\Delta^2[x]^n = nh.(n-1)h[x]^{n-2} = n(n-1)h^2[x]^{n-2}$

Proceeding in this way, we obtain $\Delta^{n-1}[x]^n = n(n-1)\cdots 2h^{n-1}x$

$\therefore \qquad \Delta^n[x]^n = n(n-1)\cdots 2.h^{n-1}\Delta x$

$$= n(n-1)\cdots 2.1.h^{n-1}(x+h-x)$$

$$= n!h^n \qquad\qquad (ii)$$

Also $\Delta^{n+1}[x]^n = n!h^n - n!h^n = 0$

In particular, when $h = 1$, we have

$$\Delta[x]^n = n[x]^{n-1} \quad \text{and} \quad \Delta^n[x]^n = n! \qquad\qquad (iii)$$

Similarly $\Delta^r[ax+b]^n = n(n-1)\cdots(n-r+1)a^r h^r[ax+b]^{n-r}$

Thus we have an important result:

$$\Delta[x]^n = n[x]^{n-1}; \Delta[ax+b]^n = an[ax+b]^{n-1} \qquad\qquad (iv)$$

i.e., the result of differencing [x]n is analogous to that of differentiating x^n.

NOTE

Obs.1. *As it is easier to find* $\Delta x[x]n$ *than* Δr xn, xn *must always be expressed as a factorial polynomial before finding* Δx.

Obs.2. *Every polynomial of degree n can be expressed as a factorial polynomial of the same degree and vice versa.*

EXAMPLE 6.6

Express $y = 2x^3 - 3x^2 + 3x - 10$ in factorial notation and hence show that $\Delta^3 y = 12$.

Solution:

First method: Let $y = A[x]^3 + B[x]^2 + C[x] + D$.

Using the method of synthetic division (p.29), we divide by x, $x - 1, x - 2$, etc. successively. Then

		x^3	x^2	x	
1		2	–3	3	– 10 = D
		–	2	–1	
2		2	–1		2 = C
		–	4		
3		2			3 = B
		–			
	2 = A				

Hence $\qquad y = 2[x]^3 + 3[x]^2 + 2[x] - 10$

$\therefore \qquad \Delta_y = 2 \times 3[x]^2 + 3 \times 2[x] + 2$

$\qquad\qquad \Delta^2 y = 6 \times 2[x] + 6$

$\Delta^3 y = 12$, which shows that the third differences of y are constant, as they should be.

NOTE **Obs.** *The coefficient of the highest power of x remains unchanged while transforming a polynomial to factorial notation.*

Second method (*Direct method*):

Let $\qquad y = 2x^3 - 3x^2 + 3x - 10$

$\qquad\qquad = 2x(x - 1)(x - 2) + Bx\,(x - 1) + Cx + D$

Putting $\qquad x = 0, - 10 = D$.

Putting $\qquad x = 1, 2 - 3 + 3 - 10 = C + D$

$\therefore \qquad C = - 8 - D = - 8 + 10 = 2$

Putting $\qquad x = 2, 16 - 12 + 6 - 10 = 2B + 2C + D$

$\therefore \qquad B = \dfrac{1}{2}(-2C - D) = \dfrac{1}{2}(-4 + 10) = 3$

Hence $\qquad y = 2x(x - 1)(x - 2) + 3x(x - 1) + 2x - 10$

$\qquad\qquad = 2[x]^3 + 3[x]^2 + 2[x] - 10$

$\therefore \qquad \Delta_y = 2 \times 3[x]^2 + 3 \times 2[x] + 2, \Delta^2 y = 6 \times 2[x] + 6, \Delta^3 y = 12.$

EXAMPLE 6.7

Express $u = x^4 - 12x^3 + 24x^2 - 30x + 9$ and its successive differences in factorial notation. Hence show that $\Delta^5 u = 0$.

Solution:

Let $u = A[x]^4 + B[x]^3 + C[x]^2 + D[x] + E$.

Using the method of synthetic division, we divide by $x, x - 1,$ $x - 2, x - 3$ successively.

Then

	x^4	x^3	x^2	x	
1	1	-12	24	-30	$9 (= E)$
	0	1	-11	13	
2	1	-11	13	$-17 (= D)$	
	0	2	-18		
3	1	-9	$-5 (= C)$		
	0	3			
	$1 (= A)$	$-6 (= B)$			

Hence $\qquad u = [x]^4 - 6[x]^3 - 5[x]^2 - 17[x] + 9$

$\therefore \qquad \Delta u = 4[x]^3 - 18[x]^2 - 10[x] - 17$

$\qquad\qquad \Delta^2 u = 12[x]^2 - 36[x] - 10$

$\qquad\qquad \Delta^3 u = 24[x] - 36$

$\qquad\qquad \Delta^4 u = 24 \quad \text{and} \quad \Delta^5 u = 0.$

EXAMPLE 6.8

If $f(x) = (2x + 1)(2x + 3)(2x + 5)\cdots(2x + 15)$, find the value of $\Delta^4 f(x)$

Solution:

We have $f(x) = 2^8 \left(x + \dfrac{15}{2} \right)\left(x + \dfrac{13}{2} \right)\cdots\left(x + \dfrac{3}{2} \right)\left(x + \dfrac{1}{2} \right)$ [There are 8 factors]

$\qquad\qquad = 2^8 \left[x + \dfrac{15}{2} \right]^8$

$\therefore \qquad \Delta f(x) = 2^8 \times 8 \left[x + \dfrac{15}{2} \right]^7 ; \Delta^2 f(x) = 2^8 \times 8 \times 7 \left[x + \dfrac{15}{2} \right]^6$

$$\Delta^3 f(x) = 2^8 \times 56 \times 6 \left[x + \frac{15}{2}\right]^5$$

$$\Delta^3 f(x) = 2^8 \times 336 \times 5 \left[x + \frac{15}{2}\right]^4$$

$$= 2^8 \times 1680 \left(x + \frac{15}{2}\right)\left(x + \frac{13}{2}\right)\left(x + \frac{11}{2}\right)\left(x + \frac{9}{2}\right)$$

$$= 26880 \, (2x + 9) \, (2x + 11) \, (2x + 13) \, (2x + 15)$$

6.5 Reciprocal Factorial Function

The function $\dfrac{1}{(x+1)(x+2)\cdots(x+n)}$ is denote d by $[x]^{-n}$ and is called a

reciprocal factorial function.

If the interval of differencing is h, then

$$[x]^{-n} = \frac{1}{(x+h)(x+2h)\cdots(x+nh)}$$

Which is called a reciprocal factorial function of order n

Differences of

$$[x]^{-n} = [x+h]^{-n} - [x]^{-n}$$

$$= \frac{1}{(x+2h)(x+3h)...(x+\overline{n+1}h)} - \frac{1}{(x+h)(x+2h)...(x+nh)}$$

$$= \frac{1}{(x+2h)(x+3h)...(x+\overline{n+1}h)}(x+h-[x+\overline{n+1}h])$$

$$= -nh\,[x]^{-(n+1)} \tag{i}$$

Similarly $\Delta^2[x]^{-n} = (-1)^2 n(n+1)h^2[x]^{-(n+2)}$

In general, $\Delta^x[x]^{-h} = (-1)^r n(n+1)...(n+r+1)h^r[x]^{-(n+r)} \tag{ii}$

In particular when $h = 1, \Delta^r[x]^{-n} = (-1)^r n(n+1)...(n+r+1)(x)^{-(n+r)}$

Similarly $\Delta^r[ax+b]^{-n} = (-1)^r n(n+1)...(n+r-1)a^r h^r[ax+b]^{-(n+r)} \, (iii)$

Thus we have an important result:

$$\Delta[x]^{-n} = -n[x]^{-(n+1)} \, ; \Delta[ax+b]^{-n} = -na[ax+b]^{-(n+1)} \tag{iv}$$

6.6   Inverse Operator of Δ

The process of finding y_x when Δy_x is given is known as inverse finite difference operation.

i.e., If $\Delta y_x = u_x$ then $y_x = \Delta^{-1} u_x$

The symbol Δ^{-1} or $1/\Delta$ is called the inverse of the operator Δ.

Thus we have two important results

$$\Delta^{-1}[x]^n = \frac{[x]^{n+1}}{n+1}; \Delta^{-1}[ax + b]^n = \frac{[ax+b]^{n+1}}{a(n+1)}$$

$$\Delta^{-1}[x]^{-n} = \frac{[x]^{-(n+1)}}{-n+1}; \Delta^{-1}[ax + b]^{-n} = \frac{[ax+b]^{-(n+1)}}{a(-n+1)}$$

i.e., $\Delta - 1$ is analogous to $D - 1$ or integration in calculus.

EXAMPLE 6.9

Obtain the function whose first difference is $2x^3 - 3x^2 + 3x - 10$.

Solution:

Let $f(x)$ be the function whose first difference is given.

We first express $\Delta f(x)$ as a factorial polynomial. Referring to Example 6.6, we have

$$\Delta f(x) = 2[x]^3 + 3[x]^2 + 2[x] - 10$$

$$\therefore \qquad f(x) = \Delta^{-1}\{2[x]^3 + 3[x]^2 + 2[x] - 10\}$$

$$= 2\frac{[x]^4}{4} + 3\frac{[x]^3}{3} + 2\frac{[x]^2}{2} - 10[x]^1$$

$$= \frac{1}{2}(x-1)(x-2)(x-3) + x(x-1)(x-2) + x(x-1) - 10x$$

$$= \frac{1}{2}x^4 - 2x^3 + \frac{7}{2}x^2 - 12x$$

EXAMPLE 6.10

If $y = \dfrac{1}{(3x+1)(3x+4)(3x+7)}$ evaluate $\Delta^2 y$. Also find $\Delta^{-1}y$.

Solution:

(i) We have $y = \dfrac{1}{3^3}\dfrac{1}{(x+1/3)(x+4/3)(x+7/3)} = \dfrac{1}{27}\left[x - \dfrac{2}{3}\right]^{-3}$

$$\Delta y = \frac{1}{27}(-3)\left[x - 2/3\right]^{-4}$$

$$\Delta^2 y = \frac{1}{27}(-3)(-4)\left[x - 2/3\right]^{-5}$$

$$= \frac{4}{9} - \frac{1}{\left(x+1/3\right)\left(x+4/3\right)\left(x+7/3\right)\left(x+10/3\right)\left(x+13/3\right)}$$

$$= \frac{108}{(3x+1)(3x+4)(3x+7)(3x+10)(3x+13)}$$

(ii) $\qquad y = \dfrac{1}{27}\left[x - 2/3\right]^{-3}$

$$\Delta^{-1} y = \frac{1}{27}\frac{\left[x - 2/3\right]^{-3}}{-2} = -\frac{1}{54}\frac{1}{\left(x+1/3\right)\left(x+4/3\right)}$$

$$= -\frac{1}{6}\frac{1}{(3x+1)(3x+4)}$$

6.7   Effect of an Error on a Difference Table

Suppose there is an error ε in the entry y_5 of a table. As higher differences are formed, this error spreads out and is considerably magnified. Let us see, how it effects the difference table.

The below table shows that:

(i) The error increases with the order of differences.

(ii) The coefficients of ε's in any column are the binomial coefficients of $(1 - \varepsilon)^n$. Thus the errors in the fourth difference column are ε, $-4\varepsilon, 6\varepsilon, -4\varepsilon, \varepsilon$.

(iii) The algebraic sum of the errors in any difference column is zero.

(iv) The maximum error in each column, occurs opposite to the entry containing the error, $i.e., .y_5$.

The above facts enable us to detect errors in a difference table.

x	y	Δy	$\Delta^2 y$	$\Delta^3 y$	$\Delta^4 y$
x_0	y_0				
		Δy_0			
x_1	y_1		$\Delta^2 y_0$		
		Δy_1		$\Delta^3 y_0$	
x_2	y_2		$\Delta^2 y_1$		$\Delta^4 y_0$
		Δy_2		$\Delta^3 y_1$	
x_3	y_3		$\Delta^2 y_2$		$\Delta^4 y_1 + \varepsilon$
		Δy_3		$\Delta^3 y_2 + \varepsilon$	
x_4	y_4		$\Delta_2 y_3 + \varepsilon$		$\Delta^4 y_2 - 4\varepsilon$
		$\Delta y_4 + \varepsilon$		$\Delta^3 y_3 - 3\varepsilon$	
x_5	$y_5 + \varepsilon$		$\Delta^2 y_4 - 2\varepsilon$		$\Delta^4 y_3 + 6\varepsilon$
		$\Delta y_5 - \varepsilon$		$\Delta^3 y_4 + 3\varepsilon$	
x_6	y_6		$\Delta^2 y_5 + \varepsilon$		$\Delta^4 y_4 - 4\varepsilon$
		Δy_6		$\Delta^3 y_5 - \varepsilon$	
x_7	y_7		$\Delta^2 y_6$		$\Delta^4 y_5 + \varepsilon$
		Δy_7		$\Delta^3 y_6$	
x_8	y_8		$\Delta^2 y_7$		
		Δy_8			
x_9	y_9				

EXAMPLE 6.11

One entry in the following table is incorrect and y is a cubic polynomial in x. Use the difference table to locate and correct the error.

x:	0	1	2	3	4	5	6	7
y:	1	−1	1	−1	1	—	—	—

Solution:

The difference table is as under:

x	y	Δy	$\Delta^2 y$	$\Delta^3 y$
0	25			
		− 4		
1	21		1	
		− 3		2
2	18		3	
		0		6
3	18		9	
		9		0
4	27		9	
		18		4
5	45		13	
		31		3
6	76		16	
		47		
7	123			

y being a polynomial of the third degree, $\Delta^3 y$ must be constant, *i.e.*, .the same. The sum of the third differences being 15, each entry under $\Delta^3 y$ must be 15/5, *i.e.*, 3. Thus the four entries under $\Delta^3 y$ are in error which can be written as

$$3 + (-1), 3 - 3(-1), 3 + 3(-1), 3 - (-1)$$

Taking $\varepsilon = -1$, we find that the entry corresponding to $x = 3$ is in error.

$$\therefore \qquad y + \varepsilon = 18$$

Thus the true value of $y = 18 - \varepsilon = 18 - (-1) = 19$.

EXAMPLE 6.12

Assuming that the following values of y belong to a polynomial of degree 4, compute the next three values:

x:	0	1	2	3	4	5	6	7
y:	1	−1	1	−1	1	—	—	—

Solution:

We construct the difference table from the given data.

x	y	Δy	$\Delta^2 y$	$\Delta^3 y$	$\Delta^4 y$
0	$y_0 = 1$				
		−2			
1	$y_1 = -1$		4		
		2		−8	
2	$y_2 = 1$		−4		16
		−2		8	
3	$y_3 = -1$		4		16
		2		$\Delta^3 y_2$	
4	$y_4 = 1$		$\Delta^2 y_3$		16
		Δy_4		$\Delta^3 y_3$	
5	y_5		$\Delta^2 y_4$		16
		Δy_5		$\Delta^3 y_4$	
6	Y_6		$\Delta^2 y_5$		
		Δy_6			
7	y_7				

Since the values of y belong to a polynomial of degree 4, the fourth differences must be constant. But $\Delta^4 y = 16$.

$\therefore$ The other fourth order differences must also be 16. Thus,

$$\Delta^4 y_1 = 16 = \Delta^3 y_2 - \Delta^3 y_1$$

i.e.,
$$\Delta^3 y_2 = \Delta^3 y_1 + \Delta^4 y_1 = 8 + 16 = 24$$
$$\Delta^2 y_3 = \Delta^2 y_2 + \Delta^3 y_2 = 4 + 24 = 28$$
$$\Delta y_4 = \Delta y_3 + \Delta^2 y_3 = 2 + 28 = 30$$

and
$$y_5 = y_4 + \Delta y_4 = 1 + 30 = 31$$

Similarly starting with $\Delta^4 y_2 = 16$,

we get $\qquad \Delta^3 y_3 = 40, \Delta^2 y_4 = 68, \Delta y_5 = 98, y_6 = 129.$

Starting with $\Delta^4 y_3 = 16$,

we obtain $\Delta^3 y_4 = 56, \Delta^2 y_5 = 124, \Delta y_6 = 222, y_7 = 351.$

Exercises 6.2

1. Express $x^3 - 2x^2 + x - 1$ into factorial polynomial. Hence show that $\Delta^4 f(x) = 0$.

2. Express $3x^4 - 4x^3 + 6x^2 + 2x + 1$ as a factorial polynomial and find differences of all orders.

3. Find the first and second differences of $x^4 - 6x^3 + 11x^2 - 5x + 8$ with $h = 1$. Show that the fourth difference is constant.

4. Obtain the function whose first difference is (*i*) $2x^3 + 3x^2 - 5x + 4$. (*ii*) $x^4 - 5x^3 + 3x + 4$.

5. Show that $\Delta[x(x+1)(x+2)(x+3)] = 4(x+1)(x+2)(x+3)$.

6. Find $\Delta^4 f(x)$ when $f(x) = (2x+1)(2x+3)(2x+5)...(2x+19)$.

7. If $y = \dfrac{1}{(4x+1)(4x+5)(4x+9)}$, find $\Delta^2 y$ and $\Delta^{-1} y$.

8. Given $\log 100 = 2, \log 101 = 2.0043, \log 103 = 2.0128, \log 104 = 2.0170$, find $\log 102$.

9. Find the first term of the series whose second and subsequent terms are $8, 3, 0, -1, 0$.

10. Write down the polynomial of lowest degree which satisfies the following set of numbers: $0, 7, 26, 63, 124, 215, 342, 511$

6.8 Other Difference Operators

We have already introduced the operators $\Delta, \Delta,$ and δ. Besides these, there are the operators E and μ, which we define below:

Shift operator E *is the operation of increasing the argument x by h so that* $E f(x) = f(x + h)$, $E^2 f(x) = f(x + 2h)$, $E^3 f(x) = f(x + 3h)$ etc.

The inverse operator E^{-1} is defined by $E^{-1} f(x) = f(x - h)$

If y_x is the function $f(x)$, then $Ey_x = y_{x+h}, E^{-1} y_x = y_{x-h}, E^n y_x = y_{x+nh}$, where n may be any real number.

Averaging operator μ is defined by the equation $\mu y_x = \dfrac{1}{2}\left(y_{x+\frac{1}{2}h} + y_{x-\frac{1}{2}h}\right)$.

NOTE **Obs.** *In the difference calculus E is regarded as the fundamental operator and $\Delta, \nabla, \delta, \mu$ can be expressed in terms of E.*

6.9 Relations Between the Operators

We shall now establish the following identities:

(i) $\Delta = E - 1$ (ii) $\Delta = 1 - E^{-1}$

(iii) $\delta = E^{1/2} - E^{-1/2}$ (iv) $\mu = \dfrac{1}{2}(E^{1/2} + E^{-1/2})$

(v) $\Delta = E\Delta = \Delta E = \delta E 1/2$ (vi) $E = e^{hD}$.

Proofs. (i) $\Delta y_x = y_{x+h} - y_x = Ey_x - y_x = (E - 1)y_x$

This shows that the operators Δ and E are connected by the symbolic relation

$\Delta = E - 1$ or $E = 1 + \Delta$.

NOTE **Obs.** *These relations imply that the effect of operator E on yx is the same as that of the operators $(1 + \Delta)$ on yx. The operator's E and Δ do not have any existence as separate entities.*

(ii) $\Delta y_x = y_x - y_{x-h} = y_x - E^{-1} y_x = (1 - E^{-1})y_x$

$\therefore$ $\Delta = 1 - E^{-1}$

(iii) $\delta y_x = y_{x+\frac{1}{2}h} + y_{x-\frac{1}{2}h} = E^{1/2} y_x - E^{-1/2} y_x = \left(E^{1/2} - E^{-1/2} \right) y_x$

$\delta = E^{1/2} - E^{-1/2}$

(iv) $\mu y_x = \frac{1}{2} \left(y_{x+\frac{1}{2}h} + y_{x-\frac{1}{2}h} \right) = \frac{1}{2} \left(E^{\frac{1}{2}} y_x - E^{-\frac{1}{2}} y_x \right) = \frac{1}{2} \left(E^{\frac{1}{2}} - E^{-\frac{1}{2}} \right) y_x$

$\therefore \quad \mu = \frac{1}{2} \left(E^{1/2} - E^{-1/2} \right)$

(v) $E\Delta y_x = E(y_x - y_{x-h}) = E y_x - E y_{x-h} = y_{x+h} - y_x = \Delta y_x$

$\therefore \quad E\Delta = \Delta$

$\Delta E y_x = \Delta y_{x+h} = y_{x+h} - y_x = \Delta y_x$

$\therefore \quad \Delta E = \Delta$

$\delta E^{1/2} y x = \delta y_{x+\frac{1}{2}h} = y_{x+\frac{1}{2}h+\frac{1}{2}h} - y_{x+\frac{1}{2}h-\frac{1}{2}h} = y_{x+h} - y_x = \Delta y_x$

$\therefore \quad \delta E^{1/2} = \Delta$

Hence $\Delta = E\Delta = \Delta E = \delta E^{1/2}$

(vi) $\quad E f(x) = f(x+h) = f(x) + h f'(x) + \dfrac{h^2}{2!} f''(x) + \cdots$

[by Taylor's series]

$= f(x) + h D f(x) + \dfrac{h^2}{2!} D^2 f(x) + \dots$

$= \left(1 + hD + \dfrac{h^2 D^2}{2!} + \dfrac{h^3 D^3}{3!} + \cdots \right) f(x) = e^{hD} f(x)$

$\therefore \quad E = e^{hD}$

Cor. = $E = 1 + \Delta = e^{hD}$

NOTE

A table showing the symbolic relations between the various operators is given below for ready reference To prove such relations between the operators, always express each operator in terms of the fundamental operator E.

Relations between the various operators

In terms of	E	Δ	∇	δ	hD
E	—	Δ + 1	$(1 - \nabla)^{-1}$	$1 + \frac{1}{2}\delta^2$ $+ \Delta\sqrt{(1 + \delta^2/4)}$	e^{hD}
Δ	E − 1	—	$(1 - \nabla)^{-1} - 1$	$\frac{1}{2}\delta^2 +$ $\Delta\sqrt{(1 + \delta^2/4)}$	$e^{hD} - 1$
∇	$1 - E - 1$	$1 - (1 + \Delta)^{-1}$	—	$-\frac{1}{2}\delta^2$ $+ \Delta\sqrt{(1 + \Delta^2/4)}$	$1 - e^{-hD}$
σ	$E^{1/2} - E^{-1/2}$	$\Delta(1 + \Delta)^{-1/2}$	$\Delta(1 - \Delta)^{-1/2}$	—	$2\sinh(hD/2)$
μ	$\frac{1}{2}(E^{1/2} + E^{-1/2})$	$\frac{(1 + \Delta/2)}{(1 + \Delta)^{-1/2}}$	$\frac{(1 + \Delta/2)}{(1 + \Delta)^{-1/2}}$	$\sqrt{(1 + \Delta^2/4)}$	$\cosh(hD/2)$
hD	$\log E$	$\log(1 + \Delta)$	$\log(1 - \nabla)^{-1}$	$2\sinh^{-1}(\delta/2)$	—

EXAMPLE 6.13

Prove that $e^x = \left(\dfrac{\Delta^2}{E}\right)e^x \cdot \dfrac{Ee^x}{\Delta^2 e^x}$,, the interval of differencing being h.

EXAMPLE 6.14

Prove with the usual notations, that

(i) $hD = \log(1 + \Delta) = -\log(1 - \nabla) = \sinh^{-1}(\mu\delta)$

(ii) $(E^{1/2} + E^{-1/2})(1 + \Delta)^{1/2} = 2 + \Delta$

(iii) $\Delta - \nabla = \Delta\nabla = \delta^2$

(iv) $\Delta^3 y^2 = \Delta^3 y_5$.

Solution:

(i) We know that $e^{hD} = E = 1 + \Delta$

∴ $hD = \log(1 + \Delta)$

Also $hD = \log E = -\log(E^{-1}) = -\log(1 - \nabla)$ $[\because E^{-1} = 1 - \nabla]$

We have prove that $\mu = \dfrac{1}{2}(E^{1/2} + E^{1/2})$

and $\delta = E^{1/2} - E^{1/2}$

$\mu\delta = \dfrac{1}{2}(E^{1/2} + E^{1/2})(E^{1/2} - E^{1/2})$

$$= \frac{1}{2}(E - E^{-1}) = \frac{1}{2}(e^{hD} - e^{-hD}) = \sinh(hD)$$

i.e.,
$$hD = \sinh^{-1}(\mu\delta)$$

Hence $hD = \log(1 + \Delta) = -\log(1 - \Delta) = \sinh - 1(\mu\delta).$

(*ii*) $(E^{1/2} + E^{-1/2})(1 + \Delta)^{1/2}$

$$= (E^{1/2} + E^{-1/2})E^{1/2} = E + 1 = 1 + \Delta + 1 = 2 + \Delta.$$

We know that $\qquad \Delta = E - 1, \nabla = 1 - E^{-1} \;$ and $\;\; \Delta = E^{1/2} - E^{-1/2}$

$\therefore \qquad\qquad \Delta - \nabla = E - 2 + E^{-1} = (E^{1/2} - E^{-1/2})^2 = \delta^2$

Also $\qquad\qquad \Delta\nabla = (E - 1)(1 - E^{-1}) = E + E^{-1} - 2$

$$= (E^{1/2} - E^{-1/2})^2 = \delta^2.$$

Hence $\qquad\qquad \Delta - \nabla = \Delta\nabla = \delta^2.$

(*iv*) $\qquad\qquad \Delta^3 y_2 = (E - 1)^3 y_2 \qquad\qquad\qquad [\because \Delta = E - 1]$

$$= (E^3 - 3E^2 + 3E - 1)y_2$$

$$= y_5 - 3y_4 + 3y_3 - y_2 \qquad\qquad\qquad (1)$$

$$\Delta^3 y_5 = (1 - E^{-1})^3 y_5 \qquad\qquad\qquad [\because = 1 - E^{-1}]$$

$$= (1 - 3E^{-1} + 3E^{-2} - E^{-3})y_5$$

$$= y_5 - 3y_4 + 3y_3 - y_2 \qquad\qquad\qquad (2)$$

From (1) and (2), $\qquad\qquad \Delta^3 y_2 = \Delta^3 y_5$

EXAMPLE 6.15

Prove that

(*i*) $\qquad \Delta = \frac{1}{2}\delta^2 + \delta\sqrt{\left(1 + \frac{\delta^2}{4}\right)}$

(*ii*) $\qquad 1 + \delta^2 \mu^2 = \left(1 + \frac{1}{2}\delta^2\right)$

(*iii*) $\qquad \mu = \frac{2 + \Delta}{2\sqrt{(1 + \Delta)}} + \sqrt{\left(1 + \frac{1}{4}\delta^2\right)}$

Solution:

(i) $\dfrac{1}{2}\delta^2 + \delta\sqrt{\left(1 + \dfrac{\delta^2}{4}\right)}$

$$= \dfrac{1}{2}(E^{1/2} - E^{-1/2})^2 + (E^{1/2} - E^{-1/2})\sqrt{[1 + (E^{1/2} - E^{-1/2})^{2/4}]}$$

$$= \dfrac{1}{2}(E + E^{-1} - 2) + (E^{1/2} - E^{-1/2})\sqrt{[(E + E^{-1} + 2)/4]}$$

$$= (E + E^{-1} - 2) + (E^{1/2} - E^{-1/2})(E^{1/2} + E^{-1/2})/2$$

$$= \dfrac{1}{2}[(E + E^{-1} - 2) + (E - E^{-1})] = E - 1 = \Delta.$$

(*ii*) We know that $\Delta = E^{1/2} - E^{-1/2}$ and $\mu = (E^{1/2} + E^{-1/2})/2.$

$\therefore$ L.H.S. $= 1 + \delta^2\mu^2 = 1 + (E^{1/2} - E^{-1/2})^2(E^{1/2} + E^{-1/2})^2/4$

$$= \dfrac{1}{4}[4 + (E - E^{-1})^2] = \dfrac{1}{4}(E^2 + E^{-2} + 2) = \dfrac{1}{4}(E + E^{-1})2$$

R.H.S. $= \left(1 + \dfrac{1}{2}\delta^2\right)^2 = \left[1 + \dfrac{1}{2}(E^{1/2} - E^{-1/2})^2\right]^2 = \left[1 + \dfrac{1}{2}(E - E^{-1} - 2)\right]^2$

$$= \dfrac{1}{4}(E - E^{-1})^2$$

Hence $1 + \delta^2\mu^2 = \left(1 + \dfrac{1}{2}\delta^2\right)^2$

(*iii*) Since $\Delta = E - 1, \delta = E^{1/2} - E^{-1/2}$ and $\mu = \dfrac{1}{2}\left(E^{1/2} - E^{-1/2}\right)$

$\therefore$ $\dfrac{2 + \Delta}{2\sqrt{1 + \Delta}} = \dfrac{2 + E + 1}{2\sqrt{1 + E - 1}} = \dfrac{E + 1}{2\sqrt{E}}$

$$= \dfrac{1}{2}\left(E^{1/2} + E^{-1/2}\right) = \mu \tag{1}$$

Also $\sqrt{\left(1 + \dfrac{1}{4}\delta^2\right)} = \sqrt{\left(1 + \dfrac{1}{4}\left(E^{1/2} - E^{-1/2}\right)^2\right)} = \sqrt{\left(1 + \dfrac{1}{4}\left(E + E^{-1} - 2\right)\right)}$

$$= \dfrac{1}{2}\sqrt{(E + E^{-1} + 2)} = \dfrac{1}{2}(E^{1/2} + E^{-1/2}) = \mu \tag{2}$$

Hence from (1) and (2), we get

$$\mu = \frac{2+\Delta}{2\sqrt{1+\Delta}} = \sqrt{\left(1+\frac{1}{4}\delta^2\right)}$$

EXAMPLE 6.16

Prove that $\nabla y_{n+1} = h\left(1 + \frac{1}{2}\nabla + \frac{5}{12}\nabla^2 + \cdots\right)y'_n$

Solution:

We have $\nabla_{y+1} = y_{n+1} - y_n = (E-1)y_n$

$$= (e^{hD} - 1)y_n = \left(1 + hD + \frac{h^2 D^2}{2!} + \frac{h^3 D^3}{3!} + \cdots - 1\right)y_n$$

$$= hD\left(1 + \frac{hD}{2!} + \frac{h^2 D^2}{3!} + \cdots\right)y_n$$

$$= h\left(1 + \frac{hD}{2!} + \frac{h^2 D^2}{3!} + \cdots\right)Dy_n$$

Since $\qquad E^{-1} = 1 - \nabla = e^{-hD},$

$\therefore \qquad hD = -\log(1-\nabla) = \nabla + \frac{1}{2}\nabla^2 + \frac{1}{3}\nabla^3 + \cdots$

$\therefore \qquad \nabla y_{n+1} = h\left\{1 + \frac{1}{2}\left(\nabla + \frac{1}{2}\nabla^2 + \frac{1}{3}\nabla^3 + \cdots\right)\right.$

$$\left. + \frac{1}{6}\left(\nabla + \frac{1}{2}\nabla^2 + \frac{1}{3}\nabla^3 + \cdots\right) + \cdots\right\}y'_n$$

Hence $\nabla y_n = h\left(1 + \frac{1}{2}\nabla + \frac{5}{12}\nabla + \quad\right)y'_n$

6.10　To Find One or More Missing Terms

When one or more values of $y = f(x)$ corresponding to the equidistant values of x are missing, we can find these using any of the following two methods:

First method: We assume the missing term or terms as a, b etc. and form the difference table. Assuming the last difference as zero, we solve these equations for a, b. These give the missing term/terms.

Second method: If n entries of y are given, $f(x)$ can be represented by $a(n-1)^{\text{th}}$ degree polynomial, *i.e.*, $\Delta n\ y = 0$. Since $\Delta = E - 1$, therefore $(E-1)n\ y = 0$. Now expanding $(E-1)n$ and substituting the given values, we obtain the missing term/terms.

EXAMPLE 6.17

Find the missing term is the table:

x:	2	3	4	5	6
y:	45.0	49.2	54.1	...	67.4

Solution:

Let the missing value be a. Then the difference table is as follows:

X	y	Δy	$\Delta^2 y$	$\Delta^3 y$	$\Delta^4 y$
2	$45.0(=y_0)$				
		4.2			
3	$49.2(=y_1)$		0.7		
		4.9		$a-59.7$	
4	$54.1(=y_2)$		$a-59.0$		$240.2-4a$
		$a-54.1$		$180.5-3a$	
5	$a(=y_3)$		$121.5-2a$		
		$67.4-\alpha$			
6	$67.4(=y_4)$				

We know that $\Delta^4 y = 0$, *i.e.*, $240.2 - 4a = 0$.

Hence $\qquad a = 60.05$.

Otherwise. As only four entries y_0, y_1, y_2, y_3 are given, therefore $y = f(x)$ can be represented by a third degree polynomial.

$\therefore \quad \Delta^3 y = $ constant $\qquad$ or $\quad \Delta^4 y = 0$, *i.e.*, $(E-1)^4 y = 0$

i.e., $(E^4 - 4E^3 + 6E^2 - 4E + 1)\ y = 0 \quad$ or $\quad y_4 - 4y_3 + 6y_2 - 4y_1 + y_0 = 0$

Let the missing entry y_3 be a so that

$67.4 - 4a + 6(54.1) - 4(49.2) + 45 = 0 \quad$ or $\quad -4a = -240.2$

Hence $a = 60.05$.

EXAMPLE 6.18

Find the missing values in the following data:

x:	45	50	55	60	65
y:	3.0	...	2.0	...	− 2.4

Solution:

Let the missing values be a, b. Then the difference table is as follows:

x	y	Δy	$\Delta^2 y$	$\Delta^3 y$
45	$3(= y_0)$			
		$a - 3$		
50	$a(= y_1)$		$5 - 2a$	
		$2 - a$		$3a + b - 9$
55	$2(= y_2)$		$b + a - 4$	
		$b - 2$		$3.6 - a - 36$
60	$b(= y_3)$		$- 0.4 - 2b$	
		$- 2.4 - b$		
65	$- 2.4 \ (= y_4)$			

As only three entries y_0, y_2, y_4 are given, y can be represented by a second degree polynomial having third differences as zero.

∴ $\qquad\qquad\qquad \Delta^3 y_0 = 0$ and $\Delta^3 y_1 = 0$

i.e., $\qquad\qquad 3a + b = 9,\ a + 3b = 3.6$

Solving these, we get $a = 2.925$, $b = 0.225$.

Otherwise. As only three entries $y_0 = 3$, $y_2 = 2$, $y_4 = - 2.4$ are given, y can be represented by a second degree polynomial having third differences as zero.

∴ $\qquad\qquad \Delta^3 y_0 = 0$ and $\Delta^3 y_1 = 0$

i.e., $\qquad\quad (E - 1)^3 y_0 = 0$ and $(E - 1)^3 y_1 = 0$

i.e., $\quad (E^3 - 3E^2 + 3E - 1) y_0 = 0;\ (E^3 - 3E^2 + 3E - 1).y_1 = 0$

or $\qquad y_3 - 3y_2 + 3y_1 - y_0 = 0;\ y_4 - 3y_3 + 3y_2 - y_1 = 0$

or $\qquad y_3 + 3y_1 = 9;\ 3y_3 + y_1 = 3.6$

Solving three, we get $y_1 = 2.925$, $y_2 = 0.225$.

EXAMPLE 6.19

The following table gives the values of y which is a polynomial of degree five. It is known that $f(3)$ is in error. Correct the error.

x:	0	1	2	3	4	5	6
y:	1	2	33	254	1025	3126	7777

Solution:

Let the correct value of y when $x = 3$ be a. Then the difference table is as follows:

x:	y:	Δy	$\Delta^2 y$	$\Delta^3 y$	$\Delta^4 y$	$\Delta^5 y$	$\Delta^6 y$
0	1						
		1					
1	2		30				
		31		$a - 94$			
2	3		$a - 64$		$1216 - 4a$		
	$a - 33$			$1122 - 3a$		$-2320 - 10a$	
3	A		$1058 - 2a$		$-1104 + 6a$		$4880 - 20a$
		$1025 - a$		$18 + 3a$		$2560 - 10a$	
4	1025		$1076 + a$		$1456 - 4a$		
		2101		$1474 - a$			
5	3126		2550				
		4651					
6	7777						

Since y is a polynomial of fifth degree, the sixth difference $\Delta^6 y = 0$

i.e., $\qquad 4880 - 20a = 0$

Hence $\qquad a = 244.$

Otherwise. As y is a polynomial of fifth degree, the sixth difference $\Delta^6 y = 0$

i.e., $\qquad (E - 1)^6 y = 0$

or $\qquad (E^6 - 6E^5 + 15E^4 - 20E^3 + 15E^2 - 6E + 1)y_0 = 0$

or $\qquad y_6 - 6y_5 + 15y_4 - 20y_3 + 15y_2 - 6y_1 + y_0 = 0$

i.e., $\quad 7777 - 6(3126) + 15(1025) + 20y_3 + 15(33) - 6(2) + 1 = 0$

$\qquad \therefore \quad 4880 = 20y_3 \qquad \therefore \quad y_3 = 244$

Hence the error $= 254 - 244 = 10.$

EXAMPLE 6.20

If $y_{10} = 3$, $y_{11} = 6$, $y_{12} = 11$, $y_{13} = 18$, $y_{14} = 27$, find y_4.

Solution:

Taking y_{14} as u_0, we are required to find y_4, *i.e.*, $.u_{-10}$. Then the difference table is

x	u	Δu	$\Delta^2 u$	$\Delta^3 u$
x_{-4}	$y_{10} = u_{-4} = 3$			
		3		
x_{-3}	$y_{11} = u_{-3} = 6$		2	
		5		0
x_{-2}	$y_{12} = u_{-2} = 11$		2	
		7		0
x_{-1}	$y_{13} = u_{-1} = 18$		2	
		9		
x_0	$y_{14} = u_0 = 27$			

Then $\quad y_4 = u_{-10} = (E^{-1})^{10} u_0 = (1 - \Delta)^{10} u_8$

$$= \left(1 - 10\nabla + \frac{10.9}{2}\nabla^2 - \frac{10.9.8}{1.2.3}\nabla^3 + \cdots \right) u_0$$

$$= u_0 - 10\Delta u_0 + 45\Delta^2 u_0 - 120\Delta_3 u_0$$

$$= 27 - 10 \times 9 + 45 \times 2 - 120 \times 0 = 27.$$

EXAMPLE 6.21

If y_x is a polynomial for which fifth difference is constant and $y_1 + y_7 = -784$, $y_2 + y_6 = 686$, $y_3 + y_5 = 1088$, find y_4.

Solution:

Starting with y_1 instead of y_0, we note that $\Delta^6 y_1 = 0$ $[\because \Delta^5 y_1$ is constant$]$

i.e., $\quad (E - 1)^6 y_1 = (E^6 - 6E^5 + 15E^4 - 20E^3 + 15E^2 - 6E + 1) y_1 = 0$

$\therefore \qquad\qquad y_7 - 6y_6 + 15y_5 - 20y_4 + 15y_3 - 6y_2 + y_1 = 0$

or $\qquad\qquad (y_7 + y_1) - 6(y_6 + y_2) + 15(y_5 + y_3) - 20y_4 = 0$

i.e., $\qquad y_4 = \frac{1}{20} \Big[(y_1 + y_7) - 6(y_2 + y_6) + 15(y_3 + y_5) \Big]$

$$= \frac{1}{20} \Big[-784 - 6(686) + 15(1088) \Big] = 571$$

EXAMPLE 6.22

Using the method of separation of symbols, prove that

(i) $u_1 x + u_2 x^2 + u_3 x^3 + \cdots = \dfrac{x}{1-x} u_1 + \left(\dfrac{x}{1-x}\right)^2 \Delta u_1 + \left(\dfrac{x}{1-x}\right)^3 \Delta u_1 + \cdots$

(ii) $u_0 + \dfrac{u_1 x}{1!} + \dfrac{u_2 x^2}{2!} + \dfrac{u_3 x^3}{3!} + \cdots = e^x \left(u_0 + x \Delta u_0 + \dfrac{x^2}{2!} \Delta^2 u_0 + \dfrac{x^3}{3!} \Delta^3 u_0 + \cdots \right)$

Solution:

(i) L.H.S $= xu_1 + x^2 E u_1 + x^3 E u_1 + \cdots$

$$= x\left(1 + xE + x^2 E^2 + \cdots\right) u^1 = x \cdot \dfrac{1}{1-xE} u_1, \text{ taking sum of infinite G.P.}$$

$$= x\left[\dfrac{1}{1-x(1+\Delta)}\right] u_1 \qquad \left[\because E = 1 + \Delta\right]$$

$$= x\left(\dfrac{1}{1-x-x\Delta}\right) u_1 = \dfrac{x}{1-x}\left(1 - \dfrac{x\Delta}{1-x}\right)^{-1} u_1$$

$$= \dfrac{x}{1-x}\left(+ \dfrac{x\Delta}{1-x} + \dfrac{x^2 \Delta^2}{(\quad)} \right)$$

$$= \dfrac{x}{1-x} u_1 + \dfrac{x^2}{(1-x)^2} \Delta u_1 + \dfrac{x^3}{(1-x)^3} \Delta^2 u_1 + \cdots = \text{R.H.S}$$

(ii) L.H.S. $= u_0 + \dfrac{x}{1!} E u_0 + \dfrac{x^2}{2!} E^2 u_0 + \dfrac{x^3}{3!} E^3 u_0 + \cdots$

$$= \left(1 + \dfrac{xE}{1!} + \dfrac{x^2 E^2}{2!} + \dfrac{x^3 E^3}{3!} + \cdots\right) u_0 = e^{xE} u_0 = e^{x(1+\Delta)} u_0 = e^x . e^{x\Delta} u_0$$

$$= e^x \left(1 + \dfrac{x\Delta}{1!} + \dfrac{x^2 \Delta^2}{2!} + \dfrac{x^3 \Delta^3}{3!} + \cdots\right) u_0$$

$$= e^x \left(u_0 + \dfrac{x}{1!} \Delta u_0 + \dfrac{x^2}{2!} \Delta^2 u_0 + \dfrac{x^3}{3!} \Delta^3 u_0 + \cdots\right) = \text{R.H.S}$$

Exercises 6.3

1. Explain the difference between $\left(\dfrac{\Delta^2}{E}\right)u_x$ and $\left(\dfrac{\Delta^2 u_x}{Eu_x}\right)$.

2. Evaluate taking h as the interval of differencing:

(i) $\dfrac{\Delta^2}{E}\sin x$

(ii). $(\Delta+\nabla)^2\left(x^2+x\right),(h=1)$

(iii) $\dfrac{\Delta^2 x^3}{Ex^3}$

(iv) $\dfrac{\Delta^2}{E}\sin(x+h)+\dfrac{\Delta^2\sin(x+h)}{E\sin(x+h)}$

3. With the usual notations, show that

(i) $\Delta=1-e^{-hD}$

(ii) $D=\dfrac{2}{h}\sinh^{-1}\left(\dfrac{\delta}{2}\right)$

(iii) $(1+\Delta)(1-\nabla)=1$

(iv) $\Delta\nabla=\nabla\Delta=\delta^2$

4. Prove that

(i) $\delta=\Delta(1+\Delta)^{-1/2}=\nabla(1-\nabla)^{-1/2}$

(ii) $\mu^2=1+\dfrac{\delta^2}{4}$

(iii) $\delta\left(E^{1/2}+E^{-1/2}\right)=\Delta E^{-1}+\Delta$

5. Show that
(i) $\Delta=\Delta E^{-1/2}=\Delta E^{1/2}$

(ii) $\mu\Delta=\dfrac{1}{2}(\Delta+\nabla)$

(iii) $1+\delta^2/2=\sqrt{(1+\delta^2\mu^2)}$

6. Show that

(i) $\Delta=\mu\delta+\dfrac{\delta^2}{2}$

(ii) $E^{1/2}=\left(1+\dfrac{\delta^2}{4}\right)^{1/2}+\dfrac{\delta^2}{2}$

(iii) $E^r=\left(\mu+\dfrac{1}{2}\delta\right)^{2r}$

(iv) $\mu=\dfrac{2+\Delta}{2\lambda(1+\Delta)}=\dfrac{2+\nabla}{2\lambda(1+\nabla)}$

7. Prove that

(i) $\Delta+\nabla=\dfrac{\Delta}{\nabla}-\dfrac{\nabla}{\Delta}$

(ii) $\nabla=\Delta E^{-1}=E^{-1}\Delta=1-E^{-1}$

(iii) $E=\displaystyle\sum_{i=0}^{\infty}\nabla_i$

(iv) $\nabla^2=h^2D^2-h^3D^3+\dfrac{7}{12}h^4D^4-\cdots$

8. Prove that $\delta^2 y_5 = y_6 - 2y_5 + y_4$.

9. Prove with usual notations, that

(i) $\nabla^r f_k = \Delta^r f_{k-r}$

(ii) $\Delta\left(f^2_k\right) = \left(f_k + f_k + 1\right)\Delta f_k$ (iii) $\sum_{k=0}^{n-1} \Delta^2 f_k = \Delta f_n - \Delta f_0$

10. Estimate the missing term in the following table:

x:	0	1	2	3	4
f(x)	1	3	9	–	81

11. Find the missing terms in the following table:

x:	1	1.5	2	2.5	3	3.5
y:	6	?	10	20	?	1.5

12. Find the missing values in the following table:

0	1	2	3	4	5	6
5	11	22	40	...	140	...

13. Estimate the production for 2004 and 2006 from the following data:

Year:	2001	2002	2003	2004	2005	2006	2007
Production:	200	200	260	...	350	...	430

14. If $U_{13} = 1, U_{14} = -3, U_{15} = -1, U_{16} = 13$, find U_8

15. Evaluate y_4 from the following data (stating the assumptions you make)

$y_0 + y_8 = 1.9243,\ y_1 + y_7 = 1.9590$

$y_2 + y_6 = 1.9823,\ y_3 + y_5 = 1.9956$

Using the method of separation of symbols, prove that

16. $u_0 + u_1 + u_2 + \cdots + u_n = {}^{n+1}C_1 u_0 + {}^{n+1}C_2 \Delta u_0 + {}^{n+1}C_3 \Delta^2 u_0 + \cdots$
$$+ {}^{n+1}C_{n+1} \Delta^n u_0$$

17. $\Delta^n u_x = u_{x+n} - {}^n C_1 u_{x+n-1} + {}^n C_2 u_{x+n-2} + \cdots + (-1)^n u_x$

18. $y_x = y_n - {}^{n-x}C_1 \Delta y_{n-1} + {}^{n-x}C_2 \Delta^2 y_{n-2} - \cdots + (-1)^{n-x}\Delta^{n-x} y_{n-(n-x)}$.

6.11 Application to Summation of Series

The calculus of finite differences is very useful for finding the sum of a given series. The inverse operator Δ^{-1} (Section 6.6) is especially useful to find the sum of a series. This is explained below:

If $u_r = \Delta y_r = y_{r+1} - y_r$

then $u_1 = y_2 - y_1, u_2 = y_3 - y_2 \cdots u_{n-1} = y_n - y_{n-1}, u_n = y_{n+1} - y_1$

$\therefore u_1 + u_2 + \ldots + u_n = y_{n+1} - y = \Delta^{-1} y_r \Big|_{r=1}^{r=n+1}$

Thus $\sum_{r=1}^{n} u_r = \Delta^{-1} u_r \Big|_{r=1}^{r=n+1}$ $\qquad [\because yr = \Delta^{-1} u_r]$

The method is best illustrated by the following examples

EXAMPLE 6.23

Find the sum to n terms of the series

(i) $2.3.4 + 3.4.5 + 4.5.6 + \cdots$

(ii) $\dfrac{1}{3.4.5} + \dfrac{1}{4.5.6} + \dfrac{1}{5.6.7} + \cdots$

Solution:

(i) Let $\displaystyle\sum_{r=1}^{n} u_r = 2.3.4 + 3.4.5 + 4.5.6 + \cdots (n+1)(n+2)(n+3)$

$\therefore \qquad u_r = (r+1)(r+2)(r+3) = [r+3]^{3}$

$\therefore \qquad \displaystyle\sum_{r=1}^{n} u_r = \left[\Delta^{-1} u_r\right]_{r=1}^{r=n+1}$

$= \dfrac{1}{4}\left\{[n+4]^4 - [4]^4\right\}$

$= \dfrac{1}{4}\left\{(n+4)(n+3)(n+2)(n+1) - 4.3.2.1\right\}$

$= \dfrac{1}{4}\left\{(n+4)(n+3)(n+2)(n+1) - 24\right\}$

(ii) Let $\displaystyle\sum_{r=1}^{n} u_r = \dfrac{1}{3.4.5} + \dfrac{1}{4.5.6} + \dfrac{1}{5.6.7} + \cdots + \dfrac{1}{(n+2)(n+3)(n+4)}$

$\therefore \qquad u_r = \dfrac{1}{(r+2)(r+3)(r+4)} = [r+1]^{-3}$

$$\sum_{r=1}^{n} u_r = \left\{ \Delta^{-1} u_r \right\}_{r=1}^{r=n+1} = \left\{ \Delta^{-1} [r+1]^{-3} \right\}_{r=1}^{r=n+1}$$

$$= \left\{ \frac{[r+1]^{-2}}{-2} \right\}_{r=1}^{r=n+1} = -\frac{1}{2} \left\{ [n+2]^{-2} - [2]^{-2} \right\}$$

$$= -\frac{1}{2} \left\{ \frac{1}{(n+3)(n+4)} - \frac{1}{3.4} \right\} = \frac{1}{2} \left\{ \frac{1}{12} - \frac{1}{(n+3)(n+4)} \right\}$$

EXAMPLE 6.24

Sum the following series $1^3 + 2^3 + 3^3 + \cdots + n^3$

Solution:

Denoting $1^3, 2^3, 3^3, \cdots, n^3$ by $u_0, u_1, u_2, \cdots$ respectively, the required sum

$$S = u_0 + u_1 + u_2 + \cdots + u_{n-1}$$

$$= \left(1 + E + E^2 + \cdots + E^{n-1} \right) \qquad \left[\because u_1 = E u_0, u_2 = E^2 u_0 \right]$$

$$= \frac{E^n - 1}{E - 1} u_0 = \frac{(1+\Delta) - 1}{\Delta} u_0$$

$$= \frac{1}{\Delta} \left[1 + n\Delta + \frac{n(n-1)}{2!} \Delta^2 + \frac{n(n-1)(n-2)}{3!} \Delta^3 + \cdots \Delta^n - 1 \right] u_0$$

$$= n + \frac{n(n-1)}{2!} \Delta u_0 + \frac{n(n-1)(n-2)}{3!} \Delta^2 u_0 + \cdots$$

Now $\Delta u_0 = u_1 - u_0 = 2^3 - 1^3 = 7$, $\Delta^2 u_0 = u^2 - 2u_1 + u_0 = 3^3 - 2.2^3 + 1^3 = 12$,

$$\Delta^3 u_0 = u_3 - 3u_2 + 3u_1 - u_0 = 4^3 - 3.3^3 + 3.2^3 - 1^3 = 6$$

and $\Delta^4 u_0, \Delta^5 u_0, \cdots$ are all zero as $u_r = r^3$ is a polynomial of third degree

Hence $S = n + \dfrac{n(n-1)}{2} .7 + \dfrac{n(n-1)(n-2)}{6} 12 + \dfrac{n(n-1)(n-2)(n-3)}{24} .6$

$$= \frac{n^2}{4} \left(n^2 + 2n + 1 \right) + \left[\frac{n(n-1)}{2} \right]^2$$

EXAMPLE 6.25

Prove Montmort's theorem that

$$u_0 + u_1 x + u_2 x^2 + \cdots + \infty = \frac{u_0}{1-x} + \frac{x\Delta u_0}{(1-x)^2} + \frac{x^2 \Delta^2 u_0}{(1-x)^3} + \cdots + \infty$$

Hence find the sum of the series $1.2 + 2.3x + 3.4x^2 + \cdots + \infty$

Solution:

$$u_0 + u_1 x + u_2 x^2 + \cdots + \infty = \left(1 + xE + x^2 E^2 + x^3 E^3 + \cdots + \infty\right) u_0$$

$$= \frac{1}{1-xE} u_0 = \frac{1}{1-x(1+\Delta)} u_0$$

$$= \frac{1}{(1-x)-x\Delta} u_0 = \frac{1}{(1-x)} \left(1 - \frac{x}{1-x}\Delta\right)^{-1} u_0$$

$$= = \frac{1}{1-x} \left\{ 1 - \frac{x\Delta}{(1-x)} + \frac{x^2 \Delta^2}{(1-x^2)} + \ldots \right\} u_0$$

$$= \frac{u_0}{(1-x)} + \frac{x}{(1-x)^2} \Delta u_0 + \frac{x^2}{(1-x)^3} \Delta^2 u_0 + \cdots \infty$$

Now let us construct the difference table for the coefficients of the given series:

u	Δu	$\Delta^2 u$	$\Delta 3u$
$u_0 = 2$			
	4		
$u_1 = 6$		2	
	6		0
$u_2 = 12$	2		
	8		0
$u_3 = 20$		2	
	10		
$u_4 = 30$			

This shows that $u_0 = 2, \Delta u_0 = 4, \Delta^2 u_0 = 2, \Delta^3 u_0 = \Delta^4 u_0$ etc. all $= 0$.

Thus $\quad 1.2 + 2.3x + 3.4x^2 + \cdots + \infty$

$$= u_0 + u_1 x + u_2 x^2 + \cdots + \infty$$

$$= \frac{u_0}{(1-x)} + \frac{x}{(1-x)^2} \Delta u_0 + \frac{x^2}{(1-x)^3} \Delta^2 u_0 + \cdots \infty$$

$$= \frac{2}{1-x} + \frac{4x}{(1-x)^2} + \frac{2x^2}{(1-x)^3} = \frac{2}{(1-x)^3}$$

Exercises 6.4

Using the method of finite differences, sum the following series:

1. $2.5 + 5.8 + 8.11 + 11.14 + \cdots n$ terms.

2. $1.2.3 + 2.3.4 + 3.4.5 + \cdots$.to n terms

3. $\dfrac{1}{1.2.3} + \dfrac{1}{2.3.4} + \dfrac{1}{3.4.5} + \ldots$ to n terms

4. $\dfrac{1}{4.5.6} + \dfrac{1}{5.6.7} + \dfrac{1}{6.7.8} + \cdots$ to n terms

5. $1^2 + 2^2 + 3^2 + \cdots + n^2$

6. Show that $u_0 + \dfrac{u_1 x}{1!} + \dfrac{u_2 x^2}{2!} + \cdots + = e^x \left(= u_0 + \dfrac{x}{1!} \Delta u_0 + \dfrac{x^2}{2!} \Delta^2 u_0 + \cdots \right)$

Hence sum the series

$(i)\ 1^3 + \dfrac{2^3}{1!} x + \dfrac{3^3}{2!} x^2 + \dfrac{4^3}{3!} + x^3 + \cdots + \infty$

$(ii)\ 1 + \dfrac{4x}{1!} + \dfrac{10x^2}{2!} + \dfrac{20x^3}{3!} + \dfrac{35x^4}{4!} + \dfrac{56x^5}{5!} + \cdots + \infty$

7. Using Montmort's theorem find the sum of the series

$(i)\ 1.3 + 3.5x + 5.7x^2 + 7.9x^3 + \cdots + \infty$

$(ii)\ 1^2 + 2^2 x + 3^2 x^2 + 4^2 x^3 + \cdots + n^2$

8. show that $\displaystyle\sum_{r=1}^{n} u_r = nC_1 u_1 +^n C_2 \Delta u_1 +^n C_3 \Delta^2 u_1 + \cdots + \Delta^{n-1} u_1 r = 1$

Hence evaluate $1^4 + 2^4 + 3^4 + \cdots + n^4$.

9. Sum the series $1.2\Delta xn - 2.3\Delta^2 x^n + 3.4\Delta^3 x^n - 4.5\Delta^4 x^n + \cdots$ to n terms

10. Show that $\Delta x^n - \dfrac{1}{2}\Delta^2 x^n + \dfrac{1.3}{2.4}\Delta^3 x^n + \dfrac{1.3.5}{2.4.6}\Delta^4 x^n + \cdots$ to n terms $=$

$$\left(x+1/2\right)^n - \left(x-1/2\right)^n$$

6.12   Objective Type of Questions

Exercises 6.5

Select the correct answer or fill up the blanks in the following questions:

1. $\Delta\nabla =$

    (a) $\nabla\Delta$ (b) $\nabla + \Delta$ (c) $\nabla - \Delta$.

2. Which one of the following results is correct:

    (a) $\Delta x^n = nx^{n-1}$ (b) $\Delta x^{(n)} = nx^{(n-1)}$

    (c) $\Delta^n e^x = e^x$ (d) $\Delta\cos x = -\sin x$.

3. If $f(x) = 3x^3 - 2x^2 + 1$, then $\Delta^3 f(x) = \cdots$

4. The relationship between the operators E and D is $\cdots$.

5. The $(n+1)$th order difference of the nth degree polynomial is $\cdots$

6. If $y(x) = x(x-1)(x-2)$, then $\Delta y(x) = \cdots$.

7. $x^3 - 2x^2 + x - 1$ in factorial form $= \cdots$.

8. Taking h as the interval of differencing, $\Delta^2 x^3 = \cdots$

9. In terms of E, $\Delta = \cdots$.

10. The form of the function tabulated at equally spaced intervals with sixth differences constant, is $\cdots$

11. If the interval of differencing is unity, then $\Delta^4[(1-x)(1-2x)(1-3x)] = \cdots$

12. Taking the interval of differencing as unity, the first difference of $x^4 - 3x^3 + 2x - 1$ is $\cdots$.

13. The missing values of y in the following data:

yx:	0	...	...	...	25
Δyx:	1	2	4	7	11,

are $\cdots$.

14. $\Delta^3[(1-x)(1-3x)(1-5x)] = \cdots$ (interval of differencing being 1)

15. $\Delta \tan^{-1} x = \cdots$.

16. If $y = x^2 - 2x + 2$, taking interval of differencing as unity, $\Delta^2 y = \cdots$.

17. Relation between Δ and E is given by $\cdots$..

18. The kth difference of a polynomial of degree k is $\cdots$

19. $\Delta^r y_k$ in terms of backward differences $= \cdots$.

20. The value of $(\Delta^2/E)e^x = \cdots$.

21. The relation between the shift operator E and second order backward difference operator Δ^2 is $\cdots$

22. The value of $\Delta^n(e^x) = \cdots$ (intervalofdifferencingbeing1).

23. Relationship between E, Δ and Δ is $\cdots$

24. If the fifth and higher order differences of a function vanish, then the function represents a polynomial of degree $\cdots$.

25. The value of $E^{-1}\Delta = \cdots$.

26. If $E^2 u_x = x^2$ and $h = 1$, then $u_x = \cdots$.

27. Given $y_0 = 2$, $y_1 = 4$, $y_2 = 8$, $y_4 = 32$, then $y_3 = \cdots$.

28. $y_0 = 1$, $y_1 = 5$, $y_2 = 8$, $y_3 = 3$, $y_4 = 7$, $y_5 = 0$, then $\Delta^5 y_6 =$
   (*a*) 61   (*b*) -62
   (*c*) 62   (*d*) -61.

29. Given $x = 1\ 2\ 3$
   $f(x) = 3\ 8\ 15$, then $\Delta^2 f(1) =$
   (*a*) 3   (*b*) 4
   (*c*) 2   (*d*) 1

30. $(E^{1/2} + E^{-1/2})(1+\Delta)^{1/2} =$
   (*a*) $\Delta + 1$ (*b*) $\Delta - 1$
   (*c*) $\Delta + 2$ (*d*) $\Delta - 2$.

31. Which one is incorrect?
   (*a*) $E = 1 + \Delta$ (*b*) $\Delta(5) = 0$
   (*c*) $\Delta(f_1 + f_2) = \Delta f_1 + \Delta f_2$ (*d*) $\Delta(f_1 \cdot f_2) = \Delta f_1 + \Delta f_2$.

32. $\Delta - \nabla = \delta^2$. (True or False)

33. $\Delta + \nabla = E + E^{-1}$. (True or False)

34. $E = e^{-hD}$. (True or False)

35. If $f(x) = e^x$, then $\Delta^6 e^x = (e^h - 1)^6 e^x$. (True or False)

36. $\Delta^n = \delta^n E^{n/2}$ (True or False)

37. $(1 + \Delta)(1 - \nabla) = 1$. (True or False)

38. With the usual notations, match the items on right hand side with those in left hand side:

(i) $E\nabla$ (a) $\dfrac{1}{2}(\Delta + \nabla)$

(ii) hD (b) $\Delta - \nabla$

(iii) $\nabla\Delta$ (c) Δ

(iv) $\mu\delta$ (d) $-\log(1 - \nabla)$

CHAPTER 7

INTERPOLATION

Chapter Objectives

- Introduction
- Newton's forward interpolation formula
- Newton's backward interpolation formula
- Central difference interpolation formulae
- Gauss's forward interpolation formula
- Gauss's backward interpolation formula
- Stirling's formula
- Bessel's formula
- Everett's formula
- Choice of an interpolation formula
- Interpolation with unequal intervals
- Lagrange's interpolation formula
- Divided differences
- Newton's divided difference formula
- Relation between divided and forward differences
- Hermite's interpolation formula
- Spline interpolation—Cubic spline
- Double interpolation
- Inverse interpolation
- Lagrange's method

- Iterative method
- Objective type of questions

7.1 Introduction

Suppose we are given the following values of $y = f(x)$ for a set of values of x:

x:	x_0	x_1	$x_2 \cdots x_n$
y:	Y_0	y_1	$y_2 \cdots y_n$.

Then the process of finding the value of y corresponding to any value of $x = x_i$ between x_0 and x_n is called *interpolation*. Thus *interpolation is the technique of estimating the value of a function for any intermediate value of the independent variable* while the process of computing the value of the function outside the given range is called *extrapolation*. The term interpolation however, is taken to include extrapolation.

If the function $f(x)$ is known explicitly, then the value of y corresponding to any value of x can easily be found. Conversely, if the form of $f(x)$ is not known (as is the case in most of the applications), it is very difficult to determine the exact form of $f(x)$ with the help of tabulated set of values (x_i, y_i). In such cases, $f(x)$ is replaced by a simpler function $\phi(x)$ which assumes the same values as those of $f(x)$ at the tabulated set of points. Any other value may be calculated from $\phi(x)$ which is known as the *interpolating function* or *smoothing function*. If $\phi(x)$ is a polynomial, then it called the *interpolating polynomial* and the process is called the *polynomial interpolation*. Similarly when $\phi(x)$ is a finite trigonometric series, we have trigonometric interpolation. But we shall confine ourselves to polynomial interpolation only.

The study of interpolation is based on the calculus of finite differences. We begin by deriving two important *interpolation formulae* by means of forward and backward differences of a function. These formulae are often employed in engineering and scientific investigations.

7.2 Newton's Forward Interpolation Formula

Let the function $y = f(x)$ take the values $y_0, y_1, \cdots, y_n$ corresponding to the values $x0, x1, \cdots, x_n$ of x. Let these values of x be equispaced such that $x_i = x_0 + ih$ $(i = 0, 1, \cdots)$. Assuming $y(x)$ to be a polynomial of the nth degree in x such that $y(x_0) = y_0, y(x_1) = y_1, \cdots, y(x_n) = y_n$. We can write

$$y(x) = a_0 + a_1(x - x_0) + a_2(x - x_0)(x - x_1) + a_3(x - x_0)(x - x_1)(x - x_2)$$

$$+ \cdots + a_n(x - x_0)(x - x_1) \cdots (x - x_{n-1}) \quad (1)$$

Putting $x = x_0, x_1, \cdots, x_n$ successively in (1), we get

$$y_0 = a_0, y_1 = a_0 + a_1(x_1 - x_0), y_2 = a_0 + a_1(x_2 - x_0) + a_2(x_2 - x_0)(x_2 - x_1)$$
and so on.

From these, we find that $a_0 = y_0, \Delta y_0 = y_1 - y_0 = a_1(x_1 - x_0) = a_1 h$

$$\therefore \quad a_1 = \frac{1}{h} \Delta y_0$$

Also $\quad \Delta y_1 = y_2 - y_1 = a_1(x_2 - x_1) + a_2(x_2 - x_0)(x_2 - x_1)$

$$= a_1 h + a_2 hh = \Delta y_0 + 2h^2 a_2$$

$$\therefore \quad a_2 = \frac{1}{2h^2}\left(\Delta y_1 - \Delta y_0\right) = \frac{1}{2!h^2} \Delta^2 y_0$$

Similarly $a_3 = \dfrac{1}{3!h^3} \Delta^3 y_0$ and so on.

Substituting these values in (1), we obtain

$$y(x) = y_0 + \frac{\Delta y_0}{h}(x - x_0) + \frac{\Delta^2 y_0}{2!h^2}(x - x_0)(x - x_1) + \frac{\Delta^3 y_0}{3!h^3}(x - x_0)(x - x_1)(x - x_2) + \cdots$$

$$(2)$$

Now if it is required to evaluate y for $x = x_0 + ph$, then

$$(x - x_0) = ph, x - x_1 = x - x_0 - (x - x_0) = ph - h = (p - 1)h,$$

$$(x - x_0) = x - x_0 - (x - x_0) = (p - 1)h - h = (p - 2)h \quad \text{etc.}$$

Hence, writting $y(x) = y(x_0 + ph) = y_p$, (2) becomes

$$y_p = y_0 + p\Delta y_0 + \frac{p(p-1)}{2!} \Delta^2 y_0 + \frac{p(p-1)(p-2)}{3!} \Delta^3 y_0$$

$$+ \cdots + \frac{p(p-1)\cdots\left(p - \overline{n-1}\right)}{3!} \Delta^n y_0 \quad (3)$$

It is called Newton's forward interpolation formula as (3) contains y_0 and the forward differences of y_0

Otherwise: Let the function $y = f(x)$ take the values $y_0, y_1, y_2, \cdots$ corresponding to the values $x_0, x_0 + h, x_0 + 2h, \cdots$ of x. Suppose it is required to evaluate $f(x)$ for $x = x_0 + ph$, where p is any real number.

For any real number p, we have defined E such that
$$E^p f(x) = f(x + ph)$$

$$y_p = f(x_0 + ph) = E^p f(x_0) = (1 + \Delta)^p y_0 \qquad [\because E = 1 + \Delta]$$

$$= \left\{ 1 + p\Delta + \frac{p(p-1)}{2!} \Delta^2 + \frac{p(p-1)(p-2)}{3!} \Delta^3 y_0 + \cdots \right\} y_0 \qquad (4)$$

[Using binomial theorem]

i.e., $\quad y_p = y_0 + p\Delta y_0 + \frac{p(p-1)}{2!} \Delta^2 y_0 + \frac{p(p-1)(p-2)}{3!} \Delta^3 y_0 + \cdots$

If $y = f(x)$ is a polynomial of the nth degree, then $\Delta^{n+1} y_0$ and higher differences will be zero.

Hence (4) will become

$$y_p = y_0 + p\Delta y_0 + \frac{p(p-1)}{2!} \Delta^2 y_0 + \frac{p(p-1)(p-2)}{3!} \Delta^3 y_0 + \cdots$$

$$+ \frac{p(p-1) \cdots (p - \overline{n-1})}{3!} \Delta^n y_0$$

Which is same as (3)

NOTE

Obs. 1. *This formula is used for interpolating the values of y near the beginning of a set of tabulated values and extrapolating values of y a little backward (i.e., to the left) of y_0.*

Obs. 2. *The first two terms of this formula give the linear interpolation while the first three terms give a parabolic interpolation and so on.*

7.3   Newton's Backward Interpolation Formula

Let the function $y = f(x)$ take the values $y_0, y_1, y_2, \ldots$ corresponding to the values $x_0, x_0 + h, x_0 + 2h, \cdots$ of x. Suppose it is required to evaluate $f(x)$ for $x = x_n + ph$, where p is any real number. Then we have

$$y_p = f(x_n + ph) = Ep\, f(x_n) = (1 - \nabla)^{-p} y_n \qquad [\because E^{-1} = 1 - \nabla]$$

$$= \left[1 + p\nabla + \frac{p(p+1)}{2!} \nabla^2 + \frac{p(p+1)(p+2)}{3!} \nabla^3 y_0 + \cdots \right] y_n$$

[using binomial theorem]

i.e., $\quad y_p = y_n + p\nabla y_n + \dfrac{p(p+1)}{2!}\nabla^2 y_n + \dfrac{p(p+1)(p+2)}{3!}\nabla^3 y_n + \cdots$ (1)

It is called *Newton's backward interpolation formula* as (1) contains y_n and backward differences of y_n

NOTE **Obs.** *This formula is used for interpolating the values of y near the end of a set of tabulated values and also for extrapolating values of y a little ahead (to the right) of y_n*

EXAMPLE 7.1

The table gives the distance in nautical miles of the visible horizon for the given heights in feet above the earth's surface:

x = height:	100	150	200	250	300	350	400
y = distance:	10.63	13.03	15.04	16.81	18.42	19.90	21.27

Find the values of y when

(i) $x = 160\,ft.$ (ii) $x = 410.$

Solution:

The difference table is as under:

x	y	Δ	Δ^2	Δ^3	Δ^4
100	10.63				
		2.40			
150	**13.03**		– 0.39		
		2.01		0.15	
200	15.04		**– 0.24**		– 0.07
		1.77		**0.08**	
250	16.81		– 0.16		**– 0.05**
		1.61		0.03	
300	18.42		– 0.13		– 0.01
		1.48		**0.02**	
350	19.90		**– 0.11**		
		1.37			
400	**21.27**				

(i) If we take $x_0 = 160$, then $y_0 = 13.03$, $\Delta y_0 = 2.01$, $\Delta^2 y_0 = -0.24$, $\Delta^3 = 0.08$, $\Delta^4 y_0 = -0.05$

Since $x = 160$ and $h = 50$, $\qquad \therefore \ p = \dfrac{x - x_0}{h} = \dfrac{10}{50} = 0.2$

$\therefore$ $\qquad$ Using Newton's forward interpolation formula, we get

$$y_{218} = y_p = y_0 + p\Delta y_0 + \frac{p(p-1)}{2!}\Delta^2 y_0 + \frac{p(p-1)(p-2)}{3!}\Delta^3 y_0$$

$$+ \frac{p(p-1)(p-2)(p-3)}{4!}\Delta^4 y_0 + \cdots$$

$y_{160} = 13.03 + 0.402 + 0.192 + 0.0384 + 0.00168 = 13.46$ nautical miles

(*ii*) Since $x = 410$ is near the end of the table, we use Newton's backward interpolation formula.

$\therefore$ $\qquad$ Taking $x_n = 400$, $p = \dfrac{x - x_n}{h} = \dfrac{10}{50} = 0.2$

Using the line of backward difference

$\qquad y_n = 21.27$, $\nabla y_n = 1.37$, $\nabla^2 y_n = -0.11$, $\nabla^3 y_n = 0.02$ etc.

$\therefore$ Newton's backward formula gives

$$y_{410} = y_{400} + p\nabla y_{400} + \frac{p(p+1)}{2!}\nabla^2 y_{400}$$

$$+ \frac{p(p+1)(p+2)}{3!}\Delta^3 y_{400} + \frac{p(p+1)(p+2)(p+3)}{4!}\nabla^4 y_{400} + \cdots$$

$$= 21.27 + 0.2(1.37) + \frac{0.2(1.2)}{2!}(-0.11)$$

$$+ \frac{0.2(1.2)(2.2)}{3!}(0.02) + \frac{0.2(1.2)(2.2)(3.2)}{4!}(-0.01)$$

$$= 21.27 + 0.274 - 0.0132 + 0.0018 - 0.0007$$

$$= 21.53 \text{ nautical miles}$$

EXAMPLE 7.2

From the following table, estimate the number of students who obtained marks between 40 and 45:

Marks:	30—40	40—50	50—60	60—70	70—80
No. of students:	31	42	51	35	31

Solution:

First we prepare the cumulative frequency table, as follows:

Marks less than (x):	40	50	60	70	80
No. of students (y_x):	31	73	124	159	190

Now the difference table is

x	y_x	Δy_x	$\Delta^2 y_x$	$\Delta^3 y_x$	$\Delta 4yx$
40	**31**				
		42			
50	**73**		**9**		
		51		**−25**	
60	**124**		**−16**		**37**
		35		**12**	
70	**159**		**−4**		
		31			
80	**190**				

We shall find y_{45}, *i.e.,* *the* number of students with marks less than 45.

Taking $x_0 = 40$, $x = 45$, we have

$$p = \frac{x - x_0}{h} = \frac{5}{10} = 0.5 \qquad\qquad [\because h = 10]$$

∴   Using Newton's forward interpolation formula, we get

$$y_{45} = y_{40} + p\Delta y_{40} + \frac{p(p-1)}{2!}\Delta^2 y_{40} + \frac{p(p-1)(p-2)}{3!}\Delta^3 y_{40}$$

$$+ \frac{p(p-1)(p-2)(p-3)}{4!}\Delta^4 y_{40}$$

$$= 31 + 0.5 \times 42 + \frac{(0.5)(-0.5)}{2} \times 9 + \frac{(0.5)(-0.5)(-15)}{6} \times (-25)$$

$$+ \frac{(0.5)(-0.5)(-15)(-2.5)}{24} \times 37$$

$$= 31 + 21 - 1.125 - 1.5625 - 1.4453$$

$$= 47.87, \text{ on simplification.}$$

The number of students with marks less than 45 is 47.87, i.e., 48. But the number of students with marks less than 40 is 31.

Hence the number of students getting marks between 40 and 45 = 48 − 31 = 17.

EXAMPLE 7.3.

Find the cubic polynomial which takes the following values:

x:	0	1	2	3
f(x):	1	2	1	10

Hence or otherwise evaluate $f(4)$.

Solution:

The difference table is

x	f(x)	Δf(x)	Δ²f(x)	Δ³f(x)
0	**1**			
		1		
1	2		**−2**	
		−1		**12**
2	1		**10**	
		9		
3	**10**			

We take $x_0 = 0$ and $p = \dfrac{x-0}{h} = x$ $\qquad [\because h = 1]$

∴ Using Newton's forward interpolation formula, we get

$$f(x) = f(0) + \frac{x}{1}\Delta f(0) + \frac{x(x-1)}{1.2}\Delta^2 f(0) + \frac{x(x-1)(x-2)}{1.2.3}\Delta^3 f(0)$$

$$= 1 + x(1) + \frac{x(x-1)}{2}(-2) + \frac{x(x-1)(x-2)}{6}(12)$$

$$= 2x^3 - 7x^2 + 6x + 1$$

which is the required polynomial.

To compute $f(4)$, we take $x_n = 3$, $x = 4$ so that $p = \dfrac{x - x_n}{h} = 1$ $\quad [\because h = 1]$

NOTE **Obs.** *Using Newton's backward interpolation formula, we get*

$$f(4) = f(3) + p\nabla f(3) + \frac{p(p+1)}{1.2}\nabla^2 f(3) + \frac{p(p+1)(p+2)}{1.2.3}\nabla^3 f(3)$$

$$= 10 + 9 + 10 + 12 = 41$$

which is the same value as that obtained by substituting x = 4 in the cubic polynomial above.

The above example shows that if a tabulated function is a polynomial, then interpolation and extrapolation give the same values.

EXAMPLE 7.4

Using Newton's backward difference formula, construct an interpolating polynomial of degree 3 for the data: $f(-0.75) = -0.0718125$, $f(-0.5) = -0.02475$, $f(-0.25) = 0.3349375$, $f(0) = 1.10100$. Hence find $f(-1/3)$.

Solution:

The difference table is

x	y	Δy	Δ²y	Δ³y
– 0.75	– 0.0718125			
		0.0470625		
– 0.50	– 0.02475		0.312625	
		0.3596875		**0.09375**
– 0.25	0.3349375		**0.400375**	
		0.7660625		
0	**1.10100**			

We use Newton's backward difference formula

$$y(x) = y_3 + \frac{p}{1!}\nabla y_3 + \frac{p(p+1)}{2!}\nabla^2 y_3 + \frac{p(p+1)(p+2)}{3!}\nabla^3 y_3$$

taking $x = 0, p = \dfrac{x}{0.25} = \dfrac{x}{0.25} = 4x$ $\qquad [\because h = 0.25]$

$$y(x) = 1.10100 + 4x(0.7660625) + \frac{4x(4x+1)}{2}(0.400375)$$

$$+ \frac{4x(4x+1)(4x+2)}{6}(0.09375)$$

$$= 1.101 + 3.06425x + 3.2511x^2 + 0.81275x + x^3 + 0.75x^2 + 0.125x$$

$$= x^3 + 4.001x^2 + 4.002x + 1.101$$

Put $x = -\dfrac{1}{3}$, so that

$$y\left(-\frac{1}{3}\right) = \left(-\frac{1}{3}\right)^3 + 4.001\left(-\frac{1}{3}\right)^2 + 4.002\left(-\frac{1}{3}\right) + 1.101$$

$$= 0.1745$$

EXAMPLE 7.5

In the table below, the values of y are consecutive terms of a series of which 23.6 is the 6th term. Find the first and tenth terms of the series:

x:	3	4	5	6	7	8	9
y:	4.8	8.4	14.5	23.6	36.2	52.8	73.9

Solution:

The difference table is

x	y	Δy	$\Delta^2 y$	$\Delta^3 y$	$\Delta^4 y$
3	4.8				
		3.6			
4	8.4		2.5		
		6.1		0.5	
5	14.5		3.0		0
		9.1		0.5	
6	23.6		3.5		0
		12.6		0.5	
7	36.2		4.0		0
		16.6		0.5	
8	52.8		4.5		
		21.1			
9	73.9				

To find the first term, use Newton's forward interpolation formula with $x_0 = 3$, $x = 1$, $h = 1$, and $p = -2$. We have

$$y(1) = 4.8 + \frac{(-2)}{1} \times 3.6 + \frac{(-2)(-3)}{1.2} \times 2.5 + \frac{(-2)(-3)(-4)}{1.2.3} \times 0.5 = 3.1$$

To obtain the tenth term, u se Newton's backward interpolation formula with $x_n = 9$, $x = 10$, $h = 1$, and $p = 1$. This gives

$$y(10) = 73.9 + \frac{1}{1} \times 21.1 + \frac{1(2)}{1.2} \times 4.5 + \frac{1(2)(3)}{1.2.3} \times 0.5 = 100$$

EXAMPLE 7.6

Using Newton's forward interpolation formula show

$$\sum n^3 = \left\{ \frac{n(n+1)}{2} \right\}^2$$

Solution:

If $s_n = sn^3$, then $s_{n+1} = \Sigma(n+1)^3$

$$\therefore \qquad \Delta s_n = s_{n+1} - s_n = \sum (n+1)^3 - \sum n^3 = (n+1)^3$$

Then $\quad \Delta^2 s_n = \Delta s_{n+1} - \Delta s_n = (n+2)^3 - (n+1)^3 = 3n^2 + 9n + 7$

$$\Delta^3 s_n = \Delta^2 s_{n+1} - \Delta^2 s_n$$

$$= \left[3(n+1)^2 + 9(n+1) + 7 \right] - \left(3n^2 + 9n + 7 \right) = 6n + 12$$

$$\Delta^4 s_n = \Delta^3 s_{n+1} - \Delta^3 s_n = \left[6(n+1) + 12 \right] - \left[6n + 12 \right] = 6$$

and $\qquad \Delta^5 s_n = \Delta^5 s_n = \ldots = 0$

Since the first term of the given series is 1, therefore taking $n = 1$, $s_1 = 1$, $\Delta s_1 = 8$, $\Delta^2 s_1 = 19$, $\Delta^3 s_1 = 18$, $\Delta^4 s_1 = 6$.

Substituting these in the Newton's for war d interpolation formula, *i.e.,*

$$s = s + (n-1)\Delta s_1 + \frac{(n-1)(n-2)}{2!} \Delta^2 s_1 + \frac{(n-1)(n-2)(n-3)}{3!} \Delta^3 s_1$$

$$+ \frac{(n-1)(n-2)(n-3)(n-4)}{4!} \Delta^4 s_1$$

$$sn = 1 + 8(n-1) + \frac{19}{2}(n-1)(n-2) + 3(n-1)(n-2)(n-3)$$

$$+ \frac{1}{4}(n-1)(n-2)(n-3)(n-4) = \frac{1}{4}(n^4 + 2n^3 + n^2) = \left\{ \frac{n(n+1)}{2} \right\}^2$$

Exercises 7.1

1. Using Newton 's forward formula, fin d the value of $f(1.6)$, if

x:	1	1.4	1.8	2.2
f(x):	3.49	4.82	5.96	6.5

2. From the following table find y when $x = 1.85$ an d 2.4 by Newton's interpolation formula:

x:	1.7	1.8	1.9	2.0	2.1	2.2	2.3
$y = e^x$:	5.474	6.050	6.686	7.389	8.166	9.025	9.974

3. Express the value of θ in terms of x using the following data:

x:	40	50	60	70	80	90
θ:	184	204	226	250	276	304

Also find θ at $x = 43$.

4. Given $\sin 45° = 0.7071$, $\sin 50° = 0.7660$, $\sin 55° = 0.8192$, $\sin 60° = 0.8660$, find $\sin 52°$ using Newton's forward formula.

5. From the following table:

x:	0.1	0.2	0.3	0.4	0.5	0.6
$f(x)$:	2.68	3.04	3.38	3.68	3.96	4.21

find $f(0.7)$ approximately.

6. The area A of a circle of diameter d is given for the following values:

d:	80	85	90	95	100
A:	5026	5674	6362	7088	7854

Calculate the area of a circle of diameter 105

7. From the following table:

x°:	10	20	30	40	50	60	70	80
cos x:	0.9848	0.9397	0.8660	0.7660	0.6428	0.5000	0.3420	0.1737

Calculate cos 25° and cos 73° using the Gregory-1 Newton formula.

8. A test performed on a NPN transistor gives the following result:

Base current f (mA)	0	0.01	0.02	0.03	0.04	0.05
Collector current I_C (mA)	0	1.2	2.5	3.6	4.3	5.34

Calculate (i) the value of the collector current for the base current of 0.005 mA.

(ii) the value of base current required for a collector correct of 4.0 mA.

9. Find $f(22)$ from the following data using Newton's backward formulae.

x:	20	25	30	35	40	45
$f(x)$:	354	332	291	260	231	204

10. Find the number of men getting wages between Rs. 10 and 15 from the following data:

Wages in Rs:	0—10	10—20	20—30	30—40
Frequency:	9	30	35	42

11. From the following data, estimate the number of persons having incomes between 2000 and 2500:

Income	Below 500	500–1000	1000–2000	2000–3000	3000–4000
No. of persons	6000	4250	3600	1500	650

12. Construct Newton's forward interpolation polynomial for the following data:

x:	4	6	8	10
y:	1	3	8	16

Hence evaluate y for $x = 5$.

13. Find the cubic polynomial which takes the following values:

$y(0) = 1$, $y(1) = 0$, $y(2) = 1$ and $y(3) = 10$.

Hence or otherwise, obtain $y(4)$.

14. Construct the difference table for the following data:

x:	0.1	0.3	0.5	0.7	0.9	1.1	1.3
$f(x)$:	0.003	0.067	0.148	0.248	0.370	0.518	0.697

Evaluate $f(0.6)$

15. Apply Newton's backward difference formula to the data below, to obtain a polynomial of degree 4 in x:

x:	1	2	3	4	5
y:	1	-1	1	-1	1

16. The following table gives the population of a town during the last six censuses. Estimate the increase in the population during the period from 1976 to 1978:

Year:	1941	1951	1961	1971	1981	1991
Population: (*in thousands*)	12	15	20	27	39	52

17. In the following table, the values of y are consecutive terms of a series of which 12.5 is the fifth term. Find the first and tenth terms of the series.

x:	3	4	5	6	7	8	9
y:	2.7	6.4	12.5	21.6	34.3	51.2	72.9

18. Using a polynomial of the third degree, complete the record given below of the export of a certain commodity during five years:

Year:	1989	1990	1991	1992	1993
Export: (in tons)	443	384	—	397	467

19. Given $u_1 = 40, u_3 = 45, u_5 = 54$, find u_2 and u_4.

20. If $u_{-1} = 10, u_1 = 8, u_2 = 10, u_4 = 50$, find u_0 and u_3.

21. Given $y_0 = 3, y_1 = 12, y_2 = 81, y_3 = 200, y_4 = 100, y_5 = 8$, without forming the difference table, find $\Delta^5 y0$.

7.4 Central Difference Interpolation Formulae

In the preceding sections, we derived Newton's forward and backward interpolation formulae which are applicable for interpolation near the beginning and end of tabulated values. Now we shall develop central difference formulae which are best suited for interpolation near the middle of the table.

If x takes the values $x_0 - 2h, x_0 - h, x_0, x_0 + h, x_0 + 2h$ and the corresponding values of $y = f(x)$ are $y_{-2}, y_{-1}, y_0, y_1, y_2$, then we can write the difference table in the two notations as follows:

x	y	*1st diff.*	*2nd diff.*	*3rd diff.*	*4th diff.*
$x_0 - 2h$	y_{-2}				
		$\Delta y_{-2} (= \Delta y_{-3/2})$			
$x_0 - h$	y_{-1}		$\Delta^2 y_{-2} (= \Delta^2 y_{-1})$		
		$\Delta y_{-1} (= \Delta y_{-1/2})$		$\Delta^3 y_{-2} (= \Delta^3 y_{-1/2})$	
x_0	y_0		$\Delta^2 y_{-1} (= \Delta^2 y_0)$		$\Delta^3 y_{-2} (= \Delta^4 y_0)$
		$\Delta y_0 (= \Delta y_{1/2})$		$\Delta^3 y_{-1} (= \Delta^3 y_{1/2})$	
$x_0 + h$	y_1		$\Delta^2 y_0 (= \Delta^2 y_1)$		
		$\Delta y_1 (= \Delta y_{3/2})$			
$x_0 + 2h$	y_2				

7.5   Gauss's Forward Interpolation Formula

The Newton's forward interpolation formula is

$$y_0 = y_0 + p\Delta y_0 + \frac{p(p-1)}{1.2}\Delta^2 y_0 + \frac{p(p-1)(p-2)}{1.2.3}\Delta^3 y_0 + \cdots \qquad (1)$$

We have $\Delta^2 y_0 - \Delta^2 y_{-1} = \Delta^3 y_{-1}$

i.e., $\qquad\qquad \Delta^2 y_0 = \Delta^2 y_{-1} + \Delta^3 y_{-1} \qquad\qquad\qquad (2)$

Similarly $\qquad\quad \Delta^3 y_0 = \Delta^3 y_{-1} + \Delta^4 y_{-1} \qquad\qquad\qquad (3)$

$$\Delta^4 y_0 = \Delta^4 y_{-1} + \Delta^5 y_{-1} \text{ etc.} \qquad\qquad (4)$$

Also $\qquad \Delta^3 y_{-1} - \Delta^3 y_{-2} = \Delta^4 y_{-2}$

i.e., $\qquad\qquad \Delta^3 y_{-1} = \Delta^3 y_{-2} + \Delta^4 y_{-2}$

Similarly $\qquad \Delta^4 y_{-1} = \Delta^4 y_{-2} + \Delta^5 y_{-2} \text{etc.} \qquad\qquad (5)$

Substituting for $\Delta^2 y_0$, $\Delta^3 y_0$, $\Delta^4 y_0$ from (2), (3), (4)..in (1), we get

$$y_p = y_0 + p\Delta y_0 + \frac{p(p-1)}{1.2}\left(\Delta^2 y_{-1} + \Delta^3 y_{-1}\right) + \frac{p(p-1)(p-2)}{1.2.3}\left(\Delta^3 y_{-1} + \Delta^4 y_{-1}\right)$$

$$+ \frac{p(p-1)(p-2)(p-3)}{1.2.3.4}\left(\Delta^4 y_{-1} + \Delta^5 y_{-1}\right)$$

Hence $\quad y_p = y_0 + p\Delta y_0 + \dfrac{p(p-1)}{2!}\Delta^2 y_{-1} + \dfrac{p(p-1)(p-2)}{3!}\Delta^3 y_{-1}$

$$+ \frac{(p+1)(p-2)(p-3)}{4!}\Delta^4 y_{-2} + \cdots \text{ [using (5)]}$$

which is called Gauss's forward interpolation formula.

Cor. In the central differences notation, this formula will be

$$y_p = y_0 + p\delta y_{1/2} + \frac{p(p-1)}{2!}\delta^2 y_{1/2} + \frac{p(p-1)(p-2)}{3!}\delta^3 y_{1/2}$$

$$+ \frac{p(p-1)(p-2)(p-3)}{4!}\delta^4 y_{1/2}$$

NOTE

Obs. 1. *It employs odd differences just below the central line and even difference on the central line as shown below:*

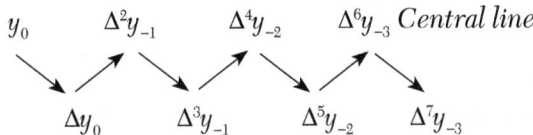

$$y_0 \qquad \Delta^2 y_{-1} \qquad \Delta^4 y_{-2} \qquad \Delta^6 y_{-3} \text{ Central line}$$

$$\Delta y_0 \qquad \Delta^3 y_{-1} \qquad \Delta^5 y_{-2} \qquad \Delta^7 y_{-3}$$

Obs. 2. *This formula is used to interpolate the values of y for p (0 < p < 1) measured forwardly from the origin.*

7.6   Gauss's Backward Interpolation Formula

The Newton's forward interpolation formula is

$$y_p = y_0 + p\Delta y_0 + \frac{p(p-1)}{1.2}\Delta^2 y_0 + \frac{p(p-1)(p-2)}{1.2.3}\Delta^3 y_0 + \cdots \qquad (1)$$

We have $\Delta y_0 - \Delta y_{-1} = \Delta^2 y_{-1}$

i.e., $\qquad\qquad \Delta y_0 = \Delta y_{-1} + \Delta^2 y_{-1} \qquad\qquad\qquad (2)$

Similarly $\qquad \Delta^2 y_0 = \Delta^2 y_{-1} + \Delta^3 y_{-1} \qquad\qquad\qquad (3)$

$\qquad\qquad\quad \Delta^3 y_0 = \Delta^3 y_{-1} + \Delta^4 y_{-1}$ etc. $\qquad\qquad (4)$

Also $\quad \Delta^3 y_{-1} - \Delta^3 y_{-2} = \Delta^4 y_{-2}$

i.e., $\qquad\qquad \Delta^3 y_{-1} = \Delta^3 y_{-2} + \Delta^4 y_{-2} \qquad\qquad\qquad (5)$

Similarly $\qquad \Delta^4 y_{-1} = \Delta^4 y_{-2} + \Delta^5 y_{-2}$ etc. $\qquad\qquad (6)$

Substituting for $\Delta y_0, \Delta^2 y_0, \Delta^3 y_0, \cdots$ from (2), (3), (4) in (1), we get

$$y_p = y_0 + p(\Delta y_{-1} + \Delta^2 y_{-1}) + \frac{p(p-1)}{1.2}\left(\Delta^2 y_{-1} + \Delta^3 y_{-1}\right)$$

$$+\frac{p(p-1)(p-2)}{1.2.3}\left(\Delta^3 y_{-1} + \Delta^4 y_{-1}\right) + \frac{p(p-1)(p-2)(p-3)}{1.2.3.4}\times\left(\Delta^4 y_{-1} + \Delta^5 y_{-1}\right) + \cdots$$

$$= y_0 + p\Delta y_{-1} + \frac{p(p+1)}{1.2}\Delta^2 y_{-1} + \frac{(p+1)p(p-1)}{1.2.3}\Delta^3 y_{-1}$$

$$+\frac{(p+1)p(p-1)(p-2)}{1.2.3.4}\Delta^4 y_{-1} + \frac{p(p-1)(p-2)(p-3)}{1.2.3.4}\Delta^5 y_{-1} + \cdots$$

$$= y_0 + p\Delta y_{-1} + \frac{(p+1)p}{1.2}\Delta^2 y_{-1} + \frac{(p+1)p(p-1)}{1.2.3}\left(\Delta^3 y_{-2} + \Delta^4 y_{-2}\right)$$

$$+\frac{(p+1)p(p-1)(p-2)}{1.2.3.4}\left(\Delta^4 y_{-2} + \Delta^5 y_{-2}\right) + \cdots$$

[using (5) and (6)

Hence $\quad yp = y_0 + p\Delta_{y-1} + \frac{p(p+1)}{2!}\Delta^2 y_{-1} + \frac{(p+1)p(p-1)}{3!}\Delta^3 y_{-2}$

$$+\frac{(p+1)p(p+1)(p-1)}{4!}\Delta^4 y_{-2} + \cdots$$

which is called Gauss's backward interpolation formula.

Cor. In the central differences notation, this formula will be

$$y_p = y_0 + p\delta y_{-1/2} + \frac{(p+1)p}{2!}\delta^2 y_0 + \frac{(p+1)p(p-1)}{3!}\delta^3 y_{-1/2}$$
$$+ \frac{(p+2)(p+1)p(p-1)}{4!}\delta^4 y_0 + \cdots$$

NOTE *Obs. 1. This formula contains odd differences above the central line and even differences on the central line as shown below:*

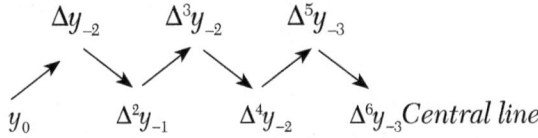

Obs. 2. It is used to interpolate the values of y for a negative value of p lying between – 1 and 0.

Obs. 3. Gauss's forward and backward formulae are not of much practical use. However, these serve as intermediate steps for obtaining the important formulae of the following sections.

7.7 Stirling's Formula

Gauss's forward interpolation formula is

$$y_p = y_0 + p\Delta y_0 + \frac{(p+1)}{2!}\Delta^2 y_{-1} + \frac{(p+1)p(p-1)}{3!}\Delta^3 y_{-1}$$
$$+ \frac{(p+1)p(p-1)(p-2)}{4!}\Delta^4 y_{-2} + \cdots \tag{1}$$

Gauss's backward interpolation formula is

$$y_p = y_0 + p\Delta y_{-1} + \frac{(p+1)p}{2!}\Delta^2 y_{-1} + \frac{(p+1)p(p-1)}{3!}\Delta^3 y_{-2}$$
$$+ \frac{(p+2)(p+1)p(p-1)}{4!}\Delta^4 y_{-2} + \cdots \tag{2}$$

Taking the mean of (1) and (2), we obtained

$$y_p = y_0 + p\left(\frac{\Delta y_0 + \Delta y_{-1}}{2}\right) + \frac{p^2}{2!}\Delta^2 y_{-1} + \frac{p(p^2-1)}{3!}$$
$$\times \left(\frac{\Delta^3 y_{-1} + \Delta^3 y_{-2}}{2}\right) + \frac{p^2(p^2-1)}{4!}\Delta^4 y_{-2} + \cdots \tag{3}$$

Which is called Stirling's formula.

Cor. *In the central difference notation, (3) takes the form*

$$y_p = y_0 + p\mu\delta y_0 + \frac{p^2}{2!}\delta^2 y_0 + \frac{p(p^2-1^2)}{3!}\mu\delta^3 y_0 + \frac{p^2(p^2-1^2)}{4!}\delta^4 y_0 + \cdots$$

For

$$\frac{1}{2}\left(\Delta y_0 + \Delta y_{-1}\right) = \frac{1}{2}\left(\delta y_{1/2} + \delta y_{-1/2}\right) = \mu\delta y_0$$

$$\frac{1}{2}\left(\Delta^3 y_{-1} + \Delta^3 y_{-2}\right) = \frac{1}{2}\left(\delta^3 y_{1/2} + \delta^3 y_{-1/2}\right) = \mu\delta^3 y_0 \text{ etc.}$$

NOTE

Obs. *This formula involves means of the odd differences just above and below the central line and even differences on this line as shown below:*

$$\cdots y_0 \cdots \begin{pmatrix} \Delta y_{-1} \\ \Delta y_0 \end{pmatrix} \cdots \Delta^2 y_{-1} \cdots \begin{pmatrix} \Delta^3 y_{-2} \\ \Delta^3 y_{-1} \end{pmatrix} \cdots \Delta^4 y_{-2} \cdots \begin{pmatrix} \Delta^5 y_{-1} \\ \Delta^5 y_0 \end{pmatrix} \cdots \Delta^6 y_{-3} \cdots$$

Central line.

7.8   Bessel's Formula

Gauss's forward interpolation formula is

$$y_p = y_0 + p\Delta y_0 + \frac{p(p-1)}{2!}\Delta^2 y_{-1} + \frac{(p+1)p(p-1)}{3!}\Delta^3 y_{-1}$$

$$+ \frac{(p+1)p(p-1)(p+1)}{4!}\Delta^4 y_{-2} + \cdots$$

We have $\Delta^2 y_0 - \Delta^2 y_{-1} = \Delta^3 y_{-1}$ (1)

i.e., $\Delta^2 y_{-1} = \Delta^2 y_0 - \Delta^3 y_{-1}$ (2)

Similarly $\Delta^4 y_{-2} = \Delta^4 y_{-1} - \Delta^4 y_{-2}$ etc.

Now (1) can be written as

$$y_p = y_0 + p\Delta y_0 + \frac{p(p-1)}{2!}\left(\frac{1}{2}\Delta^2 y_{-1} + \frac{1}{2}\Delta^2 y_{-1}\right) + \frac{p(p^2-1)}{3!}\Delta^3 y_{-1}$$

$$+ \frac{p(p^2-1)(p-2)}{4!}\left(\frac{1}{2}\Delta^4 y_{-2} + \frac{1}{2}\Delta^4 y_{-2}\right) + \cdots$$

$$= y_0 + p\Delta y_0 + \frac{1}{2}\frac{p(p+1)}{2!}\Delta^2 y_{-1} + \frac{1}{2}\frac{p(p-1)}{2!}\left(\Delta^2 y_0 + \Delta^3 y_{-1}\right)$$

$$+ \frac{p(p^2-1)}{3!}\Delta^3 y_{-1} + \frac{1}{2}\frac{p(p^2-1)(p-2)}{4!}\Delta^4 y_{-2} + \frac{1}{2}\frac{p(p^2-1)(p-2)}{4!}$$

$$\times\left(\Delta^4 y_{-1} - \Delta^5 y_{-1}\right) + \cdots$$

[Using (2), (3) etc.]

Hence $y_p = y_0 + p\Delta y_0 + \dfrac{p(p-1)}{2!}\dfrac{\Delta^2 y_{-1} + \Delta^2 y_0}{2} + \dfrac{\left(p-\dfrac{1}{2}\right)p(p-1)}{3!}\Delta^3 y_{-1}$ (4)

Which is known as Bessel's formula.

Cor. *In the central difference notation, (4) becomes*

$$y_p = y_0 + p\delta y_{1/2} + \dfrac{p(p-1)}{2!}\mu\delta^2 y_{1/2} + \dfrac{\left(p-\dfrac{1}{2}\right)p(p-1)}{2!}\delta^3 y_{1/2}$$

$$+ \dfrac{(p+1)p(p-1)(p-2)}{4!}\mu\delta^4 y_{1/2} + \cdots$$

for $\dfrac{1}{2}\left(\Delta^2 y_{-1} + \Delta^2 y_0\right) = \mu\delta^2 y_{1/2}, \dfrac{1}{2}\left(\Delta^4 y_{-2} + \Delta^4 y_{-1}\right) = \mu\delta^4 y_{1/2}$ etc.,

NOTE   **Obs.** *This is a very useful formula for practical purposes. It involves odd differences below the central line and means of even differences of and below this line as shown below*

$$y_0 \quad \begin{Bmatrix} \Delta^2 y_{-1} \\ \Delta^2 y_0 \end{Bmatrix} \quad \Delta^3 y_{-1} \begin{Bmatrix} \Delta^4 y_{-2} \\ \Delta^4 y_{-1} \end{Bmatrix} \quad \Delta^5 y_{-2} \begin{Bmatrix} \Delta^6 y_{-1} \\ \Delta^6 y_0 \end{Bmatrix} \quad \Delta^7 y_{-3} \qquad \begin{matrix} Central\ line \end{matrix}$$

7.9 Laplace-Everett's Formula

Gauss's forward interpolation formula is

$$y_p = y_0 + p\Delta y_0 + \dfrac{(p-1)p}{2!}\Delta^2 y_{-1} + \dfrac{(p+1)p(p-1)}{3!}\Delta^3 y_{-1}$$

$$+ \dfrac{(p+1)p(p-1)(p-2)}{4!}\Delta^4 y_{-2} + \dfrac{(p+2)(p+1)p(p-1)(p-2)}{5!}\times\Delta^5 y_{-2} + \cdots$$

(1)

We eliminate the odd differences in (1) by using the relations

$$\Delta y_0 = y_1 - y_0, \Delta^3 y_{-1} = \Delta^2 y_0 - \Delta^2 y_{-1}, \Delta^5 y_{-2} = \Delta^4 y_{-1} - \Delta^4 y_{-2}$$ etc.

Then (1) becomes

$$y_p = y_0 + p(y_1 - y_0) + \dfrac{p(p-1)}{2!}\Delta^2 y_{-1} + \dfrac{(p+1)p(p-1)}{3!}\left(\Delta^2 y_0 - \Delta^2 y_{-1}\right)$$

$$+ \dfrac{(p+1)p(p-1)(p-2)}{4!}\Delta^4 y_{-2} + \dfrac{(p+2)(p+1)p(p-1)(p-2)}{5!}$$

$$\times\left(\Delta^4 y_{-1} - \Delta^4 y_{-2}\right) + \cdots$$

$$= (1-p)y_0 + py_1 - \frac{p(p-1)(p-2)}{3!}\Delta^2 y_{-1} + \frac{(p+1)p(p-1)}{3!}\Delta^2 y_0$$

$$- \frac{(p+1)p(p-1)(p-2)(p-3)}{5!}\Delta^4 y_{-2}$$

$$+ \frac{(p+2)(p+1)p(p-1)(p-2)}{5!} \times \Delta^4 y_{-1} - \cdots$$

To change the terms with negative sign, putting $p = 1 - q$, we obtain

$$y_p = qy_0 + \frac{q(q^2-1^2)}{3!}\Delta^2 y_{-1} + \frac{q(q^2-1^2)(q^2-2^2)}{5!}\Delta^4 y_{-2} + \cdots$$

$$+ py_1 + \frac{p(p^2-1^2)}{3!}\Delta^2 y_0 + \frac{p(p^2-1^2)(p^2-2^2)}{5!}\Delta^4 y_{-2} + \cdots$$

This is known as *Laplace-Everett's formula.*

NOTE **Obs. 1.** *This formula is extensively used and involves only even differences on and below the central line as shown below:*

$$y_0 \qquad \Delta^2 y_{-1} \qquad \Delta^4 y_{-2} \qquad \Delta^6 y_{-3} \ \text{Central line}$$

$$y_1 \qquad \Delta^2 y_0 \qquad \Delta^4 y_{-1} \qquad \Delta^6 y_{-2}$$

Obs. 2. *There is a close relationship between Bessel's formula and Everett's formula and one can be deduced from the other by suitable rearrangements. It is also interesting to observe that Bessel's formula truncated after third differences is Everett's formula truncated after second differences.*

7.10 Choice of an Interpolation Formula

So far we have derived several interpolation formulae such as Newton's forward, Newton's backward, Gauss's forward, Gauss's backward, Stirling's, Bessel's and Everett's formulae for calculating y_p from equispaced values which are called **classical formulae**. Now, we have to see which formula yields most accurate results in a particular problem.

The coefficients in the central difference formulae are smaller and converge faster than those in Newton's formulae. After a few terms, the coefficients in the Stirling's formula decrease more rapidly than those of the Bessel's formula and the coefficients of Bessel's formula decrease more rapidly than those of Newton's formula. As such, whenever possible, central difference formulae should be used in preference to Newton's formulae.

The right choice of an interpolation formula however, depends on the position of the interpolated value in the given data.

The following rules will be found useful:

1. To find a tabulated value near the beginning of the table, use Newton's forward formula.

2. To find a value near the end of the table, use Newton's backward formula.

3. To find an interpolated value near the center of the table, use either Stirling's or Bessel's or Everett's formula.

If interpolation is required for p lying between $-\dfrac{1}{4}$ and $\dfrac{1}{4}$, prefer Stirling's formula

If interpolation is desired for p lying between $\dfrac{1}{4}$ and $\dfrac{3}{4}$, use Bessel's or Everett's formula.

EXAMPLE 7.7

Find $f(22)$ from the Gauss forward formula:

x:	20	25	30	35	40	45
$f(x)$:	354	332	291	260	231	204

Solution:

Taking $x_0 = 25$, $h = 5$, we have to find the value of $f(x)$ for $x = 22$.

i.e., for $p = \dfrac{x - x_0}{h} = \dfrac{22 - 25}{5} = -0.6$

The difference table is as follows:

x	p	y_p	Δy_p	$\Delta^2 y_p$	$\Delta^3 y_p$	$\Delta^4 y_p$	$\Delta^5 y_p$
20	-1	$354\ (= y_{-1})$	-22				
25	0	$332\ (= y_0)$	-41	-19	29		
30	1	$291\ (= y_1)$	-31	10	-8	-37	45
35	2	$260\ (= y_2)$	-29	2	0	8	
40	3	$231\ (= y_3)$	-27	2			
45	4	$204\ (= y_4)$					

Gauss forward formula is

$$y_p = y_0 + p\Delta y_0 + \frac{p(p-1)}{2!}\Delta^2 y_{-1} + \frac{(p+1)p(p-1)}{3!}\Delta^3 y_{-1}$$

$$+ \frac{(p+1)p(p-1)(p-2)}{4!}\Delta^4 y_{-2}$$

$$+ \frac{(p+1)(p-1)p(p-2)(p+2)}{5!}\Delta^5 y_{-2} + \dots..$$

$$\therefore f(22) = 332 + (0.6)(-41) + \frac{(-0.6)(-0.6-1)}{2!}(-19)$$

$$+ \frac{(-0.6+1)(-0.6)(-0.6-1)}{3!}(-8)$$

$$+ \frac{(-0.6+1)(-0.6)(-0.6-1)(-0.6-2)}{4!}(-37)$$

$$+ \frac{(-0.6+1)(-0.6)(-0.6-1)(-0.6-2)(-0.6+2)}{5!}(45)$$

$$= 332 + 24.6 - 9.12 - 0.512 + 1.5392 - 0.5241$$

Hence $f(22) = 347.983$.

EXAMPLE 7.8

Use Gauss's forward formula to evaluate y_{30}, given that $y_{21} = 18.4708$, $y_{25} = 17.8144$, $y_{29} = 17.1070$, $y_{33} = 16.3432$ and $y_{37} = 15.5154$.

Solution

Taking $x_0 = 29$, $h = 4$, we require the value of y for $x = 30$

i.e., for $p = \dfrac{x - x_0}{h} = \dfrac{30 - 29}{4} = 0.25$

The difference table is given below:

x	p	y_p	Δy_p	$\Delta^2 y_p$	$\Delta^3 y_p$	$\Delta^4 y_p$
21	– 2	18.4708				
			– 0.6564			
25	– 1	17.8144		– 0.0510		
			– 0.7074		– 0.7074	
29	0	17.1070		– 0.0564		– 0.0022
			– 0.7638		– 0.0076	
33	1	16.3432		– 0.0640		
			– 0.8278			
37	2	15.5154				

Gauss's forward formula is

$$y_p = y_0 + p\Delta y_0 + \frac{p(p+1)}{1.2}\Delta^2 y_{-1} + \frac{(p+1)p(p-1)}{1.2.3}\Delta^3 y_{-1}$$
$$+ \frac{(p+1)p(p-1)(p-2)}{1.2.3.4}\Delta^4 y_{-2} + \cdots$$

$$y_{30} = 17.1070 + (0.25)(-0.7638) + \frac{(0.25)(-0.75)}{2}(-0.0564)$$
$$+ \frac{(1.25)(0.25)(-0.75)}{6}(-0.0076) + \frac{(1.25)(0.25)(-0.75)(-1.75)}{24}$$
$$\times(-0.0022)$$

$$= 17.1070 - 0.19095 + 0.00529 + 0.0003 - 0.00004 = 16.9216 \text{ approx.}$$

EXAMPLE 7.9

Using Gauss backward difference formula, find y (8) from the following table.

x	0	5	10	15	20	25
y	7	11	14	18	24	32

Solution:

Taking $x_0 = 10, h = 5$, we have to find y for $x = 8$, *i.e.*, for

$$p = \frac{x - x_0}{h} = \frac{8 - 10}{5} = -0.4.$$

The difference table is as follows:

x	p	y_p	Δy_p	$\Delta^2 y_p$	$\Delta^3 y_p$	$\Delta^4 y_p$	$\Delta^5 y_p$
0	2	7					
			4				
5	1	11		− 1			
			3		2		
10	0	14		1		− 1	
			4		1		0
15	1	18		2		− 1	
			6		0		
20	2	24		2			
			8				
25	3	32					

Gauss backward formula is

$$y_p = y_0 + p\Delta y_{-1} + \frac{(p+1)p}{2!}\Delta^2 y_{-1} + \frac{(p+1)p(p-1)}{3!}\Delta^3 y_{-2}$$

$$+ \frac{(p+2)p(p+1)p(p-1)}{4!}\Delta^4 y_{-2} + \cdots$$

$$y(8) = 14 + (-0.4)(3) + \frac{(-0.4+1)(-0.4)}{2!}(1) + \frac{(-0.4+1)(-0.4)(-0.4-1)}{3!}(2)$$

$$+ \frac{(-0.4+2)(-0.4+1)(-0.4)(-0.4-1)}{4!}(-1)$$

$$= 14 - 1.2 - 0.12 + 0.112 + 0.034$$

Hence $y_{(8)} = 12.826$

EXAMPLE 7.10

Interpolate by means of Gauss's backward formula, the population of a town for the year 1974, given that:

Year:	1939	1949	1959	1969	1979	1989
Population: (in thousands)	12	15	20	27	39	52

Solution:

Taking $x_0 = 1969$, $h = 10$, the population of the town is to be found for

$$p = \frac{1974 - 1969}{10} = 0.5$$

The Central difference table is

x	p	y_p	Δy_p	$\Delta^2 y_p$	$\Delta^3 y_p$	$\Delta^4 y_p$	$\Delta^5 y_p$
1939	-3	12	3	2	0	3	-10
1949	-2	15					
			5				
1959	-1	20		2			
			7		3		
1969	0	27		5		-7	
			12		-4		
1979	1	39		1			
			13				
1989	2	52					

Gauss's backward formula is

$$y_p = y_0 + p\Delta y_{-1} + \frac{(p+1)p}{2!}\Delta^2 y_{-1} + \frac{(p+1)p(p-1)}{3!}\Delta^3 y_{-2}$$
$$+ \frac{(p+2)p(p+1)p(p-1)}{4!}\Delta^4 y_{-2}$$
$$+ \frac{(p+2)(p+1)p(p-1)(p-2)}{5!}\Delta^5 y_3 + \cdots$$

$$y_{0.5} = 27 + (0.5)(7) + \frac{(1.5)(0.5)}{2}(5) + \frac{(1.5)(0.5)(-0.5)}{6}$$
$$+ \frac{(2.5)(1.5)(-0.5)}{24}(-7) + \frac{(2.5)(1.5)(0.5)(-0.5)(1.5)}{120}(-10)$$

$$= 27 + 3.5 + 1.875 - 0.1875 + 0.2743 - 0.1172$$

$$= 32.532 \text{ thousands approx.}$$

EXAMPLE 7.11

Employ Stirling's formula to compute $y_{12.2}$ from the following table $(y_x = 1 + \log_{10}\sin x)$:

$x°$:	10	11	12	13	14
$10^5 y_x$:	23,967	28,060	31,788	35,209	38,368

Solution:

Taking the origin at $x_0 = 12°$, $h = 1$ and $p = x - 12$, we have the following central difference table:

p	y_x	Δy_x	$\Delta^2 y_x$	$\Delta^3 y_x$	$\Delta^4 y_x$
$-2 = x_{-2}$	$0.23967 = y_{-2}$				
		$0.04093 = \Delta y_{-2}$			
$-1 = x_{-1}$	$0.28060 = y_{-1}$		$-0.00365 = \Delta 2y_{-2}$		
		$0.03728 = \Delta y_{-1}$		$0.00058 = \Delta^3 y_{-2}$	
$0 = x_0$	$0.31788 = y_0$		$-0.00307 = \Delta^2 y_{-1}$		$-0.00013 = \Delta^4 y_{-2}$
		$0.03421 = \Delta y_0$		$-0.00045 = \Delta^2 y_{-1}$	
$1 = x_1$	$0.35209 = y_1$		$-0.00062 = \Delta^2 y_0$		
		$0.03159 = \Delta y_1$			
$2 = x_2$	$0.38368 = y_2$				

At $x = 12.2$, $p = 0.2$. (As p lies between $-\dfrac{1}{4}$ and $\dfrac{1}{4}$, the use of String's formula will be Quite suitable.)

Stirling's formula is

$$y_p = y_0 + \frac{p}{1}\frac{\Delta y_{-1} + \Delta y_{-0}}{2} + \frac{p^2}{2!}\Delta^2 y_{-1} + \frac{p(p^2-1)}{3!} \cdot \frac{\Delta^3 y_{-2} + \Delta^3 y_{-1}}{2}$$

$$+ \frac{p^2(p^2-1)}{4!}\Delta^4 y_{-2} + \cdots$$

When $p = 0.2$, we have

$$\therefore \quad y_{0.2} = 0.3178 + 0.2\left(\frac{0.03728 + 0.03421}{2}\right) + \frac{(0.2)^2}{2}(-0.00307)$$

$$+ \frac{(0.2)^2\left[(0.2)^2 - 1\right]}{6}\left(\frac{0.00058 + 0.00054}{2}\right) + \frac{(0.2)^2\left[(0.2)^2 - 1\right]}{24}(-0.00013)$$

$$= 0.31788 + 0.00715 - 0.00006 - 0.000002 + 0.0000002$$

$$= 0.32497.$$

EXAMPLE 7.12

Given

$\theta°$:	0	5	10	15	20	25	30
$\tan \theta$:	0	0.0875	0.1763	0.2679	0.3640	0.4663	0.5774

Using Stirling's formula, estimate the value of tan16°.

Solution:

Taking the origin at $\theta° = 15°$, $h = 5°$ and $p = \dfrac{\theta - 15}{5}$, we have the following central difference table:

p	$y = \tan\theta$	Δy	$\Delta^2 y$	$\Delta^3 y$	$\Delta^4 y$	$\Delta^5 y$
-3	0.0000					
		0.0875				
-2	0.0875		0.0013			
		0.0888		0.0015		
-1	0.1763		0.0028		0.0002	
		0.0916		0.0017		-0.0002
0	0.2679		0.0045		0.0000	
		0.0961		0.0017		0.0009
1	0.3640		0.0062		0.0009	
		0.1023		0.0026		
2	0.4663		0.0088			
		0.1111				
3	0.5774					

At $\theta = 16°$, $p = \dfrac{16 - 15}{5} = 0.2$

Stirling's formula is

$$y_p = y_o + \frac{p}{1}\cdot\frac{\Delta y_{-1} + \Delta y_0}{2} + \frac{p^2}{2!}\Delta^2 y_{-1} + \frac{p^2(p^2 - 1)}{3!}\cdot\frac{\Delta^2 y_{-2} + \Delta^3 y_{-1}}{2}$$
$$+ \frac{p^2(p^2 - 1)}{4!}\Delta^4 y_{-2} + \cdots$$

$$\therefore y_{0.2} = 0.2679 + 0.2\left(\frac{0.0916 + 0.0916}{2}\right) + \frac{(0.2)^2}{2}(0.0045) + \cdots$$
$$= 0.2679 + 0.01877 + 0.00009 + \cdots = 0.28676$$

Hence, $\tan 16° = 0.28676$.

EXAMPLE 7.13

Apply Bessel's formula to obtain y_{25}, given $y_{20} = 2854$, $y_{24} = 3162$, $y_{28} = 3544$, $y_{32} = 3992$.

Solution:

Taking the origin at $x_0 = 24$, $h = 4$, we have $p = (x - 24)$.

∴ The central difference table is

p	y	Δy	$\Delta^2 y$	$\Delta^3 y$
-1	2854			
		308		
0	3162		74	
		382		-8
1	3544		66	
		448		
2	3992			

At $x = 25, p = \dfrac{(25-24)}{4} = \dfrac{1}{4}$.. (As p lies between $\dfrac{1}{4}$ and $\dfrac{3}{4}$, the use of Bessel's formula will yield accurate results)

Bessel's formula is

$$y_p = y_0 + p\Delta y_0 + \frac{p(p-1)}{2!}\frac{\Delta^2 y_{-1} + \Delta^2 y_0}{2} + \frac{\left(p - \frac{1}{2}\right)p(p-1)}{2!}\Delta^3 y_{-1} + \cdots \qquad (1)$$

When $p = 0.25$, we have

$$y_p = 3162 + 0.25 \times 382 + \frac{0.25(-0.75)}{2!}\left(\frac{74+66}{2}\right) + \frac{(0.25)0.25(-0.75)}{2!} - 8$$

$$= 3162 + 95.5 - 6.5625 - 0.0625$$

$$= 3250.875 \text{ approx.}$$

EXAMPLE 7.14

Apply Bessel's formula to find the value of $f(27.5)$ from the table:

x:	25	26	27	28	29	30
$f(x)$:	4.000	3.846	3.704	3.571	3.448	3.333

Solution:

Taking the origin at $x_0 = 27$, $h = 1$, we have $p = x - 27$

The central difference table is

x	p	y	Δy	$\Delta^2 y$	$\Delta^3 y$	$\Delta^4 y$
25	−2	4.000				
			−0.154			
26	−1	3.846		0.012		
			−0.142		−0.003	
27	0	3.704		0.009		0.004
			−0.133		−0.001	
28	1	3.571		0.010		−0.001
			−0.123		−0.002	
29	2	3.448		0.008		
			−0.115			
30	3	3.333				

At $x = 27.5$, $p = 0.5$ (As p lies between 1/4 and 3/4, the use of Bessel's formula will yield an accurate result),

Bessel's formula is

$$y_p = y_0 + p\Delta y_0 + \frac{p(p-1)}{2!}\frac{\Delta^2 y_{-1} + \Delta^2 y_0}{2} + \frac{\left(p - \frac{1}{2}\right)p(p-1)}{3!}\Delta^3 y_{-1}$$

$$+ \frac{(p+1)p(p-1)(p-2)}{4!}\left(\frac{\Delta^4 y_{-2} + \Delta^4 y_{-1}}{2}\right) + \cdots$$

When $p = 0.5$, we have

$$y_p = 3.704 - \frac{(0.5)(0.5-1)}{2}\left(\frac{0.009 + 0.010}{2}\right) + 0$$

$$+ \frac{(0.5+1)(0.5)(0.5-1)(0.5-2)}{2}\frac{(-0.001-0.004)}{2}$$

$$= 3.704 - 0.11875 - 0.00006 = 3.585$$

Hence $f(27.5) = 3.585$.

EXAMPLE 7.15

Using Everett's formula, evaluate $f(30)$ if $f(20) = 2854$, $f(28) = 3162$, $f(36) = 7088$, $f(44) = 7984$

Solution:

Taking the origin at $x_0 = 28$, $h = 8$, we have $p = \dfrac{x - 28}{8}$. The central table is

x	p	y	Δy	$\Delta^2 y$	$\Delta^3 y$
20	-1	2854			
			308		
28	0	3162		3618	
			3926		-6648
36	1	7088		-3030	
			896		
44	2	7984			

At $\qquad x = 30, p = \dfrac{30 - 28}{8} = 0.25$ and $q = 1 - p = 0.75$

Everett's formula is

$$y_p = qy_0 + \frac{q(q^2 - 1^2)}{3!}\Delta^2 y_{-1} + \frac{q(q^2 - 1^2)(q^2 - 2^2)}{5!}\Delta^4 y_{-2} + \cdots$$

$$+ py_1 + \frac{p(p^2 - 1^2)}{3!}\Delta^2 y_0 + \frac{p(p^2 - 1^2)(p^2 - 2^2)}{5!}\Delta^4 y_{-2} + \cdots$$

$$= (0.75) + (3162) + \frac{0.75(0.75^2 - 1)}{6}(3618) + \cdots$$

$$+ 0.25 + (7080) + \frac{0.25\left(0.25^2 - 1\right)}{6}(-3030) + \cdots$$

$$= 2371.5 - 351.75 + 1770 + 94.69 = 3884.4$$

Hence $f(30) = 3884.4$

EXAMPLE 7.16

Given the table

x:	310	320	330	340	350	360
$\log x$:	2.49136	2.50515	2.51851	2.53148	2.54407	2.55630

find the value of log 337.5 by Everett's formula.

Solution:

Taking the origin at $x_0 = 330$ and $h = 10$, we have $p = \dfrac{x - 330}{10}$

∴ The central difference table is

p	y	Δy	$\Delta^2 y$	$\Delta^3 y$	$\Delta^4 y$	$\Delta^5 y$
–2	2.49136					
		0.01379				
–1	2.50515		– 0.00043			
		0.01336		0.00004		
0	2.51881		– 0.00039		– 0.00003	
		0.01297		0.00001		0.00004
1	2.53148		– 0.00038		0.00001	
		0.01259		0.00002		
2	2.54407		– 0.00036			
		0.01223				
3	2.55630					

To evaluate log 337.5, *i.e.*, for $x = 337.5$, $p = \dfrac{337.5 - 330}{10} = 0.75$

(As $p > 0.5$ and $= 0.75$, Everett's formula will be quite suitable)

Everett's formula is

$$y_p = qy_0 + \frac{q(q^2 - 1^2)}{3!}\Delta^2 y_{-1} + \frac{q(q^2 - 1^2)(q^2 - 2^2)}{5!}\Delta^4 y_{-2} + \cdots$$

$$+ py_1 + \frac{p(p^2 - 1^2)}{3!}\Delta^2 y_0 + \frac{p(p^2 - 1^2)(p^2 - 2^2)}{5!}\Delta^4 y_{-1} + \cdots$$

$$= 0.25 \times 2.51851 + \frac{0.25(0.0625 \text{-} 1)}{6} \times (-0.00039)$$

$$+ \frac{0.25(0.0625 \text{-} 1)(0.0625 \text{-} 4)}{120} \times (-0.00003)$$

$$+ 0.75 \times 2.53148 + \frac{0.75(0.5625 - 1)}{6} \times (-0.00038)$$

$$+ \frac{0.75(0.5625 - 1)(0.5625 - 4)}{6} \times (-0.00001)$$

$$= 0.62963 + 0.00002 - 0.0000002 + 1.89861 + 0.00002 + 0.0000001$$

$$= 2.52828 \text{ nearly.}$$

Exercises 7.2

1. Find the y (25), given that $y_{20} = 24$, $y_{24} = 32$, $y_{28} = 35$, $y_{32} = 40$, using Gauss for ward difference formula.

2. Using Gauss's forward formula, fin d a polynomial of degree four which takes the following values of the function $f(x)$:

x:	1	2	3	4	5
$f(x)$:	1	−1	1	−1	1

3. Using Gauss's forward formula, evaluate $f(3.75)$ from the table:

x:	2.5	3.0	3.5	4.0	4.5	5.0
Y:	24.145	22.043	20.225	18.644	17.262	16.047

4. From the following table:

x:	1.00	1.05	1.10	1.15	1.20	1.25	1.30
e^x:	2.7183	2.8577	3.0042	3.1582	3.3201	3.4903	3.6693

Find $e^{1.17}$, using Gauss forward formula.

5. Using Gauss's backward formula, estimate the number of persons earning wages between Rs. 60 and Rs. 70 from the following data:

Wages (Rs.):	Below 40	40—60	60—80	80—100	100—120
No. of persons: (in thousands)	250	120	100	70	50

6. Apply Gauss's backward formula to find $\sin 45°$ from the following table:

$\theta°$:	20	30	40	50	60	70	80
$\sin \theta$:	0.34202	0.502	0.64279	0.76604	0.86603	0.93969	0.98481

7. Using Stirling's formula find y_{35}, given $y_{20} = 512$, $y_{30} = 439$, $y_{40} = 346$, $y_{50} = 243$, where y_x represents the number of persons at age x years in a life table.

8. The pressure p of wind corresponding to velocity v is given by the following data. Estimate p when $v = 25$.

v:	v:	10	20	30	40
p:	1.1	2	4.4	7.9	

9. Use Stirling's formula to evaluate $f(1.22)$, given

x:	1.0	1.1	1.2	1.3	1.4
f(x):	0.841	0.891	0.932	0.963	0.985

10. Calculate the value of $f(1.5)$ using Bessels' interpolation formula, from the table

x:	0	1	2	3
f(x):	3	6	12	15

11. Use Bessel's formula to obtain y_{25}, given $y_{20} = 24$, $y_{24} = 32$, $y_{28} = 35$, $y_{32} = 40$.

12. Employ Bessel's formula to find the value of F at $x = 1.95$, given that

x:	1.7	1.8	1.9	2.0	2.1	2.2	2.3
F:	2.979	3.144	3.283	3.391	3.463	3.997	4.491

Which other interpolation formula can be used here? Which is more appropriate? Give reasons.

13. From the following table:

x:	20	25	30	35	40
f(x):	11.4699	12.7834	13.7648	14.4982	15.0463

Find $f(34)$ using Everett's formula.

14. Apply Everett's formula to obtain u_{25}, given $u_{20} = 2854$, $u_{24} = 3162$, $u_{28} = 3544$, $u_{32} = 3992$.

15. Given the table:

x:	310	320	330	340	350	360
log x:	2.4914	2.5052	2.5185	2.5315	2.5441	2.5563

16. Find the value of log 337.5 by Gauss, Stirling, Bessel, and Everett's formulae.

If $y_0, y_1, y_2, y_3, y_4, y_5$ (y_5 being constant) are given, prove that

$$y_{5/2} = \frac{3(a-c)+2.5(c-b)}{256} + \frac{c}{2} \text{ where } a = y_0 + y_5, b = y_1 + y_4, c = y_2 + y_3.$$

[**HINT:** Use Bessel's formula taking $p = 1/2$.]

7.11   Interpolation with Unequal Intervals

The various interpolation formulae derived so far possess the disadvantage of being applicable only to equally spaced values of the argument. It is, therefore, desirable to develop interpolation formulae for unequally spaced values of x. Now we shall study two such formulae:

(*i*) Lagrange's interpolation formula

(*ii*) Newton's general interpolation formula with divided differences.

7.12   Lagrange's Interpolation Formula

If $y = f(x)$ takes the value $y_0, y_1, \ldots, y_n$ corresponding to $x = x_0, x_1, \cdots, x_n$, then

$$f(x) = \frac{(x-x_1)(x-x_2)\cdots(x-x_n)}{(x_0-x_1)(x_0-x_2)\cdots(x_0-x_n)} y_0 + \frac{(x-x_0)(x-x_2)\cdots(x-x_n)}{(x_1-x_0)(x_1-x_2)\cdots(x_1-x_n)} y_1$$
$$+ \cdots + \frac{(x-x_0)(x-x_1)\cdots(x-x_{n-1})}{(x_n-x_0)(x_n-x_1)\cdots(x_n-x_{n-1})} y_n \tag{1}$$

This is known as *Lagrange's interpolation formula for unequal intervals.*

Proof: Let $y = f(x)$ be a function which takes the values (x_0, y_0), $(x_1, y_1), \cdots$, (x_n, y_n). Since there are $n + 1$ pairs of values of x and y, we can represent $f(x)$ by a polynomial in x of degree n. Let this polynomial be of the form

$$y = f(x) = a_0(x-x_1)(x-x_2)\ldots(x-x_n) + a_1(x-x_0)(x-x_2)\cdots(x-x_n)$$

$$+ a_2(x-x_0)(x-x_1)(x-x_3)\cdots(x-x_n) + \cdots + a_n(x-x_0)(x-x_1)\cdots(x-x_{n-1}) \tag{2}$$

Putting $x = x_0, y = y_0$, in (2), we get

$$y_0 = a_0(x_0-x_1)(x-x_2)\cdots(x-x_n)$$
$$a_0 = y_0 / [(x-x_1)(x-x_2)\cdots(x-x_n)]$$

Similarly putting $x = x_1, y = y_1$ in (2), we have

$$a_1 = y_1 / [(x_1-x_0)(x_1-x_2)\cdots(x_1-x_n)]$$

Proceeding the same way, we find $a_2, a_3 \ldots a_n$.

Substituting the values of $a_0, a_1, \cdots, a_n$ in (2), we get (1)

NOTE

Obs. *Lagrange's interpolation formula (1) for n points is a polynomial of degree (n – 1) which is known as the Lagrangian polynomial and is very simple to implement on a computer.*

This formula can also be used to split the given function into partial fractions.

For on dividing both sides of (1) by $(x - x_0)(x - x_1)\cdots(x - x_n)$, *we get*

$$\frac{f(x)}{(x - x_0)(x - x_1)\cdots(x - x_n)} = \frac{y_0}{(x_0 - x_1)(x_0 - x_2)\cdots(x_0 - x_n)} \cdot \frac{1}{(x - x_0)}$$

$$+ \frac{y_1}{(x_1 - x_0)(x_1 - x_2)\cdots(x_1 - x_n)} \cdot \frac{1}{(x - x_1)} + \cdots$$

$$+ \frac{y_n}{(x_n - x_0)(x_n - x_1)\cdots(x_n - x_{n-1})} \cdot \frac{1}{(x - x_n)}$$

EXAMPLE 7.17

Given the values

x:	5	7	11	13	17
$f(x)$:	150	392	1452	2366	5202

evaluate $f(9)$, using *Lagrange's formula*

Solution:

(*i*) Here $x_0 = 5$, $x_1 = 7$, $x_2 = 11$, $x_3 = 13$, $x_4 = 17$

and $\quad y_0 = 150$, $y_1 = 392$, $y_2 = 1452$, $y_3 = 2366$, $y_4 = 5202$.

Putting $x = 9$ and substituting the above values in Lagrange's formula, we get

$$f(9) = \frac{(9-7)(9-11)(9-13)(9-17)}{(5-7)(5-11)(5-13)(5-17)} \times 150 + \frac{(9-5)(9-11)(9-13)(9-17)}{(7-5)(7-11)(7-13)(7-17)} \times 392$$

$$+ \frac{(9-5)(9-7)(9-13)(9-17)}{(11-5)(11-7)(11-13)(11-17)} \times 1452$$

$$+ \frac{(9-5)(9-7)(9-11)(9-17)}{(13-5)(13-7)(13-11)(13-17)} \times 2366$$

$$+ \frac{(9-5)(9-7)(9-11)(9-13)}{(17-5)(17-7)(17-11)(17-13)} \times 5202$$

$$= -\frac{50}{3} + \frac{3136}{15} + \frac{3872}{3} + \frac{2366}{3} + \frac{578}{5} = 810$$

EXAMPLE 7.18

Find the polynomial $f(x)$ by using Lagrange's formula and hence find $f(3)$ for

x:	0	1	2	5
$f(x)$:	2	3	12	147

Solution:

Here $x_0 = 0$, $x_1 = 1$, $x_2 = 2$, $x_3 = 5$

and $y_0 = 2$, $y_1 = 3$, $y_2 = 12$, $y_3 = 147$.

Lagrange's formula is

$$y = \frac{(x-x_1)(x-x_2)(x-x_3)}{(x_0-x_1)(x_0-x_2)(x_0-x_3)}y_0 + \frac{(x-x_0)(x-x_2)(x-x_3)}{(x_1-x_0)(x_1-x_2)(x_1-x_3)}y_1$$

$$+ \frac{(x-x_0)(x-x_1)(x-x_3)}{(x_2-x_0)(x_2-x_1)(x_2-x_3)}y_2 + \frac{(x-x_0)(x-x_1)(x-x_2)}{(x_3-x_0)(x_3-x_1)(x_3-x_2)}y_3$$

$$= \frac{(x-1)(x-2)(x-5)}{(0-1)(0-2)(0-5)}(2) + \frac{(x-0)(x-2)(x-5)}{(1-0)(1-2)(1-5)}(3)$$

$$+ \frac{(x-0)(x-1)(x-5)}{(2-0)(2-1)(2-5)}(12) + \frac{(x-0)(x-1)(x-2)}{(5-0)(5-1)(5-2)}(147)$$

Hence $f(x) = x^3 + x^2 - x + 2$

$\therefore$ $\qquad f(3) = 27 + 9 - 3 + 2 = 35$

EXAMPLE 7.19

A curve passes through the points $(0, 18)$, $(1, 10)$, $(3, -18)$ and $(6, 90)$. Find the slope of the curve at $x = 2$.

Solution:

Here $x_0 = 0, x_1 = 1, x_2 = 3, x_3 = 6$ and $y_0 = 18, y_1 = 10, y_2 = -18, y_3 = 90$.

Since the values of x are unequally spaced, we use the Lagrange's formula:

$$y = \frac{(x-x_1)(x-x_2)(x-x_3)}{(x_0-x_1)(x_0-x_2)(x_0-x_3)}y_0 + \frac{(x-x_0)(x-x_2)(x-x_3)}{(x_1-x_0)(x_1-x_2)(x_1-x_3)}y_1$$

$$+ \frac{(x-x_0)(x-x_1)(x-x_3)}{(x_2-x_0)(x_2-x_1)(x_2-x_3)}y_2 + \frac{(x-x_0)(x-x_1)(x-x_2)}{(x_3-x_0)(x_3-x_1)(x_3-x_2)}y_3$$

$$= \frac{(x-1)(x-3)(x-6)}{(0-1)(0-3)(0-6)}(18) + \frac{(x-0)(x-3)(x-6)}{(1-0)(1-3)(1-6)}(10)$$

$$+ \frac{(x-0)(x-1)(x-6)}{(3-0)(3-1)(3-6)}(-18) + \frac{(x-0)(x-1)(x-3)}{(6-0)(6-1)(6-3)}(90)$$

$$= (-x^3 + 10x^2 - 27x + 18) + (x^3 - 9x^2 + 18x)$$

$$+ (x^3 - 7x^2 + 6x) + (x^3 - 4x^2 + 3x)$$

i.e., $y = 2x^3 - 10x^2 + 18$

Thus the slope of the curve at $x = 2 = \left(\dfrac{dy}{dx}\right)_{x-2}$

$$= (6x^2 - 20x)_{x=2} = -16$$

EXAMPLE 7.20

Using Lagrange's formula, express the function $\dfrac{3x^2 + x + 1}{(x-1)(x-2)(x-3)}$ as a sum of partial fractions.

Solution:

Let us evaluate $y = 3x^2 + x + 1$ for $x = 1$, $x = 2$ and $x = 3$

These values are

x:	$x_0 = 1$	$x_1 = 2$	$x_2 = 3$
y:	$y_0 = 5$	$y_1 = 15$	$y_2 = 31$

Lagrange's formula is

$$y = \frac{(x-x_1)(x-x_2)}{(x_0-x_1)(x_0-x_2)}y_0 + \frac{(x-x_0)(x-x_2)}{(x_1-x_0)(x_1-x_2)}y_1 + \frac{(x-x_0)(x-x_1)}{(x_2-x_0)(x_2-x_1)}y_2$$

Substituting the above values, we get

$$y = \frac{(x-2)(x-3)}{(1-2)(1-3)}(5) + \frac{(x-1)(x-3)}{(2-1)(2-3)}(15) + \frac{(x-1)(x-2)}{(3-1)(3-2)}(31)$$

$$= 2.5\,(x-2)\,(x-3) - 15\,(x-1)\,(x-3) + 15.5\,(x-1)\,(x-2)$$

Thus
$$\frac{3x^2 + x + 1}{(x-1)(x-2)(x-3)} = \frac{\begin{array}{c}2.5\,(x-2)\,(x-3) - 15(x-1)\,(x-3) + \\ 15.5\,(x-1)\,(x-2)\end{array}}{(x-1)(x-2)(x-3)}$$

$$= \frac{25}{x-1} - \frac{15}{x-2} + \frac{15.5}{x-3}$$

EXAMPLE 7.21

Find the missing term in the following table using interpolation:

x:	0	1	2	3	4
y:	1	3	9	...	81

Solution:

Since the given data is unevenly spaced, therefore we use Lagrange's interpolation formula:

$$= \frac{(x-x_1)(x-x_2)(x-x_3)}{(x_0-x_1)(x_0-x_2)(x_0-x_3)}\,_0 + \frac{(x-x_0)(x-x_2)(x-x_3}{(x_1-x_0)(x_1-x_2)(x_1-x_3)}\,_1$$

$$+ \frac{(x-x_0)(x-x_1)(x-x_3)}{(x_2-x_0)(x_2-x_1)(x_2-x_3)} \quad + \frac{(x-x_0)(x-x_1)(x-x_2)}{(x_3-x_0)(x_3-x_1)(x-_2)}$$

Here we have $x_0 = 0 \quad x_1 = 1 \quad x_2 = 2 \quad x_3 = 4$

$$y_0 = 1 \quad y_1 = 3 \quad y_2 = 9 \quad y_3 = 81$$

$$\therefore \quad y = \frac{(x-1)(x-2)(x-4)}{(0-1)(0-2)(0-4)}(1) + \frac{(x-0)(x-2)(x-4)}{(1-0)(1-2)(1-4)}(3)$$

$$+ \frac{(x-0)(x-1)(x-4)}{(2-0)(2-1)(2-4)}(9) + \frac{(x-0)(x-1)(x-2)}{(4-0)(4-1)(4-2)}(81)$$

When $x = 3$, then

$$\therefore \quad y = \frac{(3-1)(3-2)(3-4)}{-8} + 3(3-2)(3-4) + \frac{3(3-1)(3-4)(9)}{-4} +$$

$$+ \frac{3(3-1)(3-2)}{24}(81) = \frac{1}{4} - 3 + \frac{27}{2} + \frac{81}{24} = 31$$

Hence the missing term for $x = 3$ is $y = 31$.

EXAMPLE 7.22

Find the distance moved by a particle and its acceleration at the end of 4 seconds, if the time verses velocity data is as follows:

t:	0	1	3	4
v:	21	15	12	10

Solution:

Since the values of t are not equispaced, we use Lagrange's formula:

$$v = \frac{(t-t_1)(t-t_2)(t-t_3)}{(t_0-t_1)(t_0-t_2)(t_0-t_3)}v_0 + \frac{(t-t_0)(t-t_2)(t-t_3)}{(t_1-t_0)(t_1-t_2)(t_1-t_3)}v_1$$

$$+ \frac{(t-t_0)(t-t_1)(t-t_3)}{(t_1-t_0)(t_1-t_2)(t_1-t_3)}v_2 + \frac{(t-t_0)(t-t_1)(t-t_3)}{(t_1-t_0)(t_1-t_2)(t_1-t_3)}v_3$$

i.e. , $v = \dfrac{(t-1)(t-3)(t-4)}{(-1)(-2)(-4)}(21) + \dfrac{t(t-3)(t-4)}{(1)(-2)(-3)}(15)$

$$+ \frac{t(t-1)(t-4)}{(3)(2)(-1)}(12) + \frac{t(t-1)(t-3)}{(4)(3)(1)}(10)$$

i.e., $v = \dfrac{1}{12}(-5t^3 + 38t^2 - 105^t + 252)$

$\therefore$ Distance moved $s = \displaystyle\int_0^4 v\,dt = \int_0^4(-5t^3 + 38t^2 - 105^t + 252)$ $\quad\left[\because v = \dfrac{ds}{dt}\right]$

$$= \frac{1}{12}\left(-\frac{5t^4}{4} + \frac{38t^3}{3} - \frac{105t^2}{2} + 252t\right)\Bigg|_0^4$$

$$= \frac{1}{12}\left(-320 + \frac{2432}{3} - 840 + 1008\right) = 54.9$$

Also acceleration $\quad = \dfrac{dv}{dt} = \dfrac{1}{2}(-15t2 + 76t - 105 + 0)$

Hence acceleration at $(t=4) = \dfrac{1}{2}(-15\pm+76(4)-105) = -3.4$

Exercises 7.3

1. Use Lagrange's interpolation formula to find the value of y when $x = 10$, if the following values of x and y are given:

x:	5	6	9	11
y:	12	13	14	16

2. The following table gives the viscosity of oil as a function of temperature. Use Lagrange's formula to find the viscosity of oil at a temperature of 140°.

Temp°:	110	130	160	190
Viscosity:	10.8	8.1	5.5	4.8

3. Given $\log_{10} 654 = 2.8156$, $\log_{10} 658 = 2.8182$, $\log_{10} 659 = 2.8189$, $\log_{10} 661 = 2.8202$, find by using Lagrange's formula, the value of $\log_{10} 656$.

4. The following are the measurements T made on a curve recorded by oscilograph representing a change of current I due to a change in the conditions of an electric current.

T:	1.2	2.0	2.5	3.0
I:	1.36	0.58	0.34	0.20

Using Lagrange's formula, find I and $T = 1.6$.

5. Using Lagrange's interpolation, calculate the profit in the year 2000 from the following data:

Year:	1997	1999	2001	2002
Profit in Lakhs of Rs:	43	65	159	248

6. Use Lagrange's formula to find thee form of $f(x)$, given

x:	0	2	3	6
$f(x)$:	648	704	729	792

7. If $y(1) = -3$, $y(3) = 9$, $y(4) = 30$, $y(6) = 132$, fin d the Lagrange's interpolation polynomial that takes the same values as y at the given point s.

8. Given $f(0) = -18$, $f(1) = 0$, $f(3) = 0$, $f(5) = -248$, $f(6) = 0$, $f(9) = 13104$, find $f(x)$.

9. Find the missing term in the following table using interpolation

x:	1	2	4	5	6
y:	14	15	5	...	9

10. Using Lagrange's formula, express the function $\dfrac{x^2 + x - 3}{x^3 - 2x^2 - x + 2}$ as a sum of partial fractions.

11. Using Lagrange's formula, express the function $\dfrac{x^2 + 6x - 1}{(x^2 - 1)(x - 4)(x - 6)}$ as a sum of partial fractions.

[**Hint.** Tabulate the values of $f(x) = x^2 + 6x - 1$ for $\underline{x} = -1, 1, 4, 6$ and apply Lagrange's formula.]

12. Using **Lagrange's formula**, prove that

$$y_o = \frac{1}{2}(y_1 + y_{-1}) = \frac{1}{8}\left\{\frac{1}{2}(y_3 + y_1) - \frac{1}{2}(y_{-1} + y_{-3})\right\}.$$

[**Hint:** Here $x_0 = -3$, $x_1 = -1$, $x_2 = 1$, $x_3 = 3$.]

7.13 Divided Differences

The Lagrange's formula has the drawback that if another interpolation value were inserted, then the interpolation coefficients are required to be recalculated. This labor of recomputing the interpolation coefficients is saved by using Newton's general interpolation formula which employs what are called "**divided differences**." Before deriving this formula, we shall first define these differences.

If $(x_0, y_0), (x_1, y_1), (x_2, y_2), \cdots$ be given points, then the *first divided difference* for the arguments x_0, x_1 is defined by the relation $[x_0, x_1]$ or

$$\underset{x_1}{\Delta} y_0 = \frac{y_1 - y_0}{x_1 - x_0}$$

Similarly $[x_1, x_2]$ or $\underset{x_2}{\Delta} y_0 = \frac{y_2 - y_1}{x_2 - x_1}$ and $[x_2, x_3]$ or $\underset{x_3}{\Delta} y_0 = \frac{y_3 - y_2}{x_3 - x_2}$

The *second divided difference* for x_0, x_1, x_2 is defined as

$$[x_0, x_1, x_2] \text{ or } \underset{x_1, x_2}{\Delta^2} = \frac{[x_1, x_2] - [x_0, x_1]}{x_2 - x_0}$$

The *third divided difference* for x_0, x_1, x_2, x_3 is defined as

$$[x_0, x_1, x_2, x_3] \text{ or } \underset{x_1, x_2, x_3}{\Delta^3} y_0 = \frac{[x_1, x_2, x_3] - [x_0, x_1, x_2]}{x_2 - x_0}$$

Properties of Divided Differences

I. The divided differences are symmetrical in their arguments, i.e,. independent of the order of the arguments. For it is easy to write

$$[x_0, x_1] = \frac{y_0}{x_0 - x_1} + \frac{y_1}{x_1 - x_0} = [x_1, x_0], [x_0, x_1, x_2]$$

$$= \frac{y_0}{(x_0 - x_1)(x_0 - x_2)} + \frac{y_1}{(x_1 - x_0)(x_1 - x_2)} + \frac{y_2}{(x_2 - x_0)(x_2 - x_1)}$$

$$= [x_1, x_2, x_0] \text{ or } [x_2, x_0, x_1] \text{ and so on}$$

II. The nth divided differences of a polynomial of the nth degree are constant.

Let the arguments be equally spaced so that

$$x_1 - x_0 = x_2 - x_1 = \cdots = x_n - x_{n-1} = h. \text{ Then}$$

$$\left[x_0, x_1\right] = \frac{y_1 - y_0}{x_1 - x_0} = \frac{\Delta y_0}{h}$$

$$\left[x_0, x_1, x_2\right] = \frac{\left[x_1, x_2\right] - \left[x_0 - x_1\right]}{x_2 - x_0} = \frac{1}{2h}\left\{\frac{\Delta y_1}{h} - \frac{\Delta y_0}{h}\right\}$$

$$= \frac{1}{2!h^2}\Delta^2 y_0 \text{ and in general, } \left[x_0, x_1, x_2, \ldots\ldots, x_n\right] = \frac{1}{n!h^n}\Delta^n y_0$$

If the tabulated function is a nth degree polynomial, then $\Delta^n y_0$ will be constant. Hence the nth divided differences will also be constant

III. The divided difference operator Δ is linear

i.e., $$\Delta\left\{au_x + bv_x\right\} = a\Delta u_x + b\Delta v_x$$

We have $\Delta_{x_1}\left(au_{x_0} + bv_{x_0}\right) = \dfrac{\left(au_{x_1} + bv_{x_1}\right) - \left(au_{x_0} + bv_{x_0}\right)}{x_1 - x_0}$

$$= a\left\{\frac{u_{x_1} - u_{x_0}}{x_1 - x_0}\right\} + b\left\{\frac{v_{x_1} - v_{x_0}}{x_1 - x_0}\right\}$$

$$= a\,\Delta_{x_1} u_{x_0} + b\,\Delta_{x_0} v_{x_0}$$

In general $\Delta\left(au_x + bv_x\right) = a\Delta u_x + b\Delta v_x$ This property is also true for higher order differences.

7.14 Newton's Divided Difference Formula

Let $y_0, y_1, \cdots, y_n$ be the values of $y = f(x)$ corresponding to the arguments $x_0, x_1, \cdots, x_n$. Then from the definition of divided differences, we have

$$\left[x, x_0\right] = \frac{y - y_0}{x - x_0}$$

So that $$y = y_0 + \left(x - x_0\right)\left[x, x_0\right]$$

Again $$\left[x, x_0, x_1\right] = \frac{\left[x, x_0\right] - \left[x_0, x_1\right]}{x - x_1}$$

which gives $\left[x, x_0\right] = \left[x_0, x_1\right] + (x - x_1)\left[x, x_0, x_1\right]$

Substituting this value of $[x, x_0]$ in (1), we get

$$y = y_0 + (x - x_0)[x_0, x_1] + (x - x_0)(x - x_1)[x, x_0, x_1] \quad (2)$$

Also $\quad [x, x_0, x_1, x_2] = \dfrac{[x.x_0, x_1] - [x.x_0, x_2]}{x - x_2}$

which gives $[x, x_0, x_1] = [x_0, x_1, x_2] + (x - x_2)[x, x_0, x_1, x_2]$

Substituting this value of $[x, x_0, x_1]$ in (2), we obtain

$$y = y_0 + (x - x_0)[x_0, x_1] + (x - x_0)(x - x_1)[x_0, x_1, x_2]$$
$$+ (x - x_0)(x - x_1)(x - x_2)[x, x_0, x_1, x_2]$$

Proceeding in this manner, we get

$$y = y_0 + (x - x_0)[x_0, x_1] + (x - x_0)(x - x_1)[x_0, x_1, x_2]$$
$$+ (x - x_0)(x - x_1)\cdots(x - x_n)[x, x_0, x_1, \cdots x_n]$$
$$+ (x - x_0)(x - x_1)(x - x_2)[x, x_0, x_1, x_2] + \cdots \quad (3)$$

which is called *Newton's general interpolation formula with divided differences.*

7.15 Relation Between Divided and Forward Differences

If $(x_0, y_0), (x_1, y_1), (x_2, y_2), \cdots$ be the given points, then

$$[x_0, x_1] = \frac{y_1 - y_0}{x_1 - x_0}$$

Also $\quad \Delta y_0 = y_1 - y_0$

If $x_0, x_1, x_2, \cdots$ are equispaced, then $x_1 - x_0 = h$, so that

$$[x_0, x_1] = \frac{\Delta y_0}{h}$$

Similarly $\quad [x_1, x_2] = \dfrac{\Delta y_1}{h}$

Now $\quad [x_0, x_1, x_2] = \dfrac{[x_1, x_2] - [x_0, x_1]}{x_2 - x_0}$

$$= \frac{\Delta y_1 / h - \Delta y_0 / h}{2h} \qquad [\because x_2 - x_0 = 2h]$$

$$= \frac{\Delta y_1 - \Delta y_0}{2h^2}$$

Thus $\quad [x_0, x_1, x_2] = \dfrac{\Delta^2 y_0}{2! h^2}$

Similarly $[x_0, x_1, x_2] = \dfrac{\Delta^2 y_1}{2! h^2}$

$\therefore [x_0, x_1, x_2, x_3] = \dfrac{\Delta^2 y_1 / 2h^2 - \Delta^2 y_0 / 2h^2}{x_3 - x_0} = \dfrac{\Delta^2 y_1 - \Delta^2 y_0}{2h^2(3)}$ $\left[\because x_3 - x_0 = 3h \right]$

Thus $[x_0, x_1, x_2, x_3] = \dfrac{\Delta^3 y_0}{3! h^3}$

In general, $[x_0, x_1, \cdots x_n] = \dfrac{\Delta^n y_0}{n! h^n}$

This is the relation between divided and forward differences.

EXAMPLE 7.23

Given the values

x:	5	7	11	13	17
$f(x)$:	150	392	1452	2366	5202

evaluate $f(9)$, using Newton's divided difference formula

Solution:

The divided differences table is

x	y	Δy	$\Delta^2 y$	$\Delta^3 y$
5	150	$\dfrac{392 - 150}{7 - 5} = 121$		
7	392		$\dfrac{265 - 121}{11 - 5} = 24$	
		$\dfrac{1452 - 392}{11 - 7} = 265$		$\dfrac{32 - 24}{13 - 5} = 1$
11	1452		$\dfrac{457 - 265}{13 - 7} = 32$	
		$\dfrac{2366 - 1452}{13 - 11} = 457$		$\dfrac{42 - 32}{17 - 7} = 1$
13	2366		$\dfrac{709 - 457}{17 - 11} = 42$	
		$\dfrac{5202 - 2366}{17 - 13} = 709$		
17	5202			

Taking $x = 9$ in the Newton's divided difference formula, we obtain

$$f(9) = 150 + (9-5) \times 121 + (9-5)(9-7) \times 24 + (9-5)(9-7)(9-11) \times 1$$
$$= 150 + 484 + 192 - 16 = 810.$$

EXAMPLE 7.24

Using Newton's divided differences formula, evaluate $f(8)$ and $f(15)$ given:

x:	4	5	7	10	11	13
$y = f(x)$:	48	100	294	900	1210	2028

Solution:

The divided differences table is

x	$f(x)$	Δy	$\Delta^2 y$	$\Delta^3 y$	$\Delta^4 y$
4	48				0
		52			
5	100		15		
		97		1	
7	294		21		0
		202		1	
10	900		27		0
		310		1	
11	1210		33		
		409			
13	2028				

Taking $x = 8$ in the Newton's divided difference formula, we obtain

$$f(8) = 48 + (8-4)\,52 + (8-4)\,(8-5)\,15 + (8-4)\,(8-5)\,(8-7)\,1$$
$$= \mathbf{448.}$$

Similarly $f(15) = 3150$.

EXAMPLE 7.25

Determine $f(x)$ as a polynomial in x for the following data:

x:	-4	-1	0	2	5
$y = f(x)$:	1245	33	5	9	1335

Solution:

The divided differences table is

x	$f(x)$	Δy	$\Delta^2 y$	$\Delta^3 y$	$\Delta^4 y$
-4	1245				
		-404			
-1	33		94		
		-28		-14	
0	5		10		3
		2		13	
2	9		88		
		442			
5	1335				

Applying Newton's divided difference formula

$$f(x) = f(x_0) + (x - x_0)[x_0, x_1] + (x - x_0)(x - x_1)[x_0, x_1, x_2] + \cdots$$
$$= 1245 + (x + 4)(-404) + (x + 4)(x + 1)(94)$$
$$+ (x + 4)(x + 1)(x - 0)(-14) + (x + 4)(x + 1)x(x - 2)(3)$$
$$= 3x^4 - 5x^2 + 6x^2 - 14x + 5$$

EXAMPLE 7.26

Using Newton's divided difference formula, find the missing value from the table:

x:	1	2	4	5	6
y:	14	15	5	...	9

Solution:

The divided difference table is

x	y	Δy	$\Delta^2 y$	$\Delta^3 y$
1	14			
		$\dfrac{15-14}{2-1}=1$		
2	15		$\dfrac{-5-1}{4-1}=-2$	
		$\dfrac{5-15}{4-2}=-5$		$\dfrac{7/4+2}{6-1}=\dfrac{3}{4}$

x	y	Δy	$\Delta^2 y$	$\Delta^3 y$
4	5		$\dfrac{2+5}{6-2} = \dfrac{7}{4}$	
		$\dfrac{9-6}{6-4} = 2$		
6	9			

Newton's divided difference formula is

$$y = y_0 + (x - x_0)[x_0, x_1] + (x - x_0)(x - x_1)[x_0, x_1, x_2]$$
$$+ (x - x_0)(x - x_1)(x - x_2)[x_0, x_1, x_2, x_3] + \cdots$$
$$= 14 + (x - 1)(1) + (x - 1)(x - 2)(-2) + (x - 1)(x - 2)(x - 4)\frac{3}{4}$$

Putting $x = 5$, we get

$$y(5) = 14 + 4 + (4)(3)(-2) + (4)(3)(1)\frac{3}{4} = 3.$$

Hence missing value is 3

Exercises 7.4

1. Find the third divided difference with arguments 2, 4, 9, 10 of the function $f(x) = x^3 - 2x$.

2. Obtain the Newton's divided difference interpolating polynomial and hence find $f(6)$:

x:	3	7	9	10
$f(x)$:	160	120	72	63

3. Using Newton's divided differences interpolation, find $u(3)$, given that $u(1) = -26$, $u(2) = 12$, $u(4) = 256$, $u(6) = 844$.

4. A thermocouple gives the following output for rise in temperature

Tem p (°C)	0	10	20	30	40	50
Output (m V)	0.0	0.4	0.8	1.2	1.6	2.0

Find the output of thermocouple for 37°C temperature using Newton's divided difference formula.

5. Using Newton's divided difference interpolation, find the polynomial of the given data:

x:	−1	0	1	3
$f(x)$:	2	1	0	−1

6. For the following table, find $f(x)$ a s a polynomial in x using Newton's divided difference formula:

x:	5	6	9	11
$f(x)$:	12	13	14	16

7. Using the following data, find $f(x)$ a s a polynomial in x:

x:	-1	0	3	6	7
$f(x)$:	3	-6	39	822	1611

8. The observed values of a function are respectively 168, 120, 72, and 63 at the four positions 3, 7, 9, and 10 of the independent variable. What is the best estimate value of the function at the position 6?

9. Find the equation of the cubic curve which passes through the point s $(4, -43)$, $(7, 83)$, $(9, 327)$, and $(12, 1053)$.

10. Find the missing term in the following table using Newton's divided difference formula.

x:	0	1	2	3	4
y:	1	3	9	...	81

7.16 Hermite's Interpolation Formula

This formula is similar to the Lagrange's interpolation formula. In Lagrange's method, the interpolating polynomial $P(x)$ agrees with $y(x)$ at the points $x_0, x_1,......, x_n$, whereas in

Hermite's method $P(x)$ and $y(x)$ as well as $P'(x)$ and $y'(x)$ coincide at the $(n + 1)$ points, *i.e.*,

$$P(x_i) = y(x_i) \text{ and } P'(x_i) = y'(x_i); i = 0, 1,......, n \tag{1}$$

As there are $2(n + 1)$ conditions in (1), $(2n + 2)$ coefficients are to be determined.

Therefore $P(x)$ is a polynomial of degree $(2n + 1)$.

We assume that $P(x)$ is expressible in the form

$$p(x) = \sum_{i=0}^{n} U_i(x)y(x_i) + \sum_{i=0}^{n} V_i(x)y'(x_i) \tag{2}$$

where $U_i(x)$ and $V_i(x)$ are polynomials in x of degree $(2n + 1)$. These are to be determined. Using the conditions (1), we get

$$U_i(x_j) = \begin{cases} 0 \text{ when } i \neq j; \ V_i(x_j) = 0 \text{ for all } i \\ 1 \text{ when } i = j \end{cases}$$

$$U_i'(x_j) = 0 \text{ for all } i; V_i(x_j) = \begin{cases} 0 \text{ when } i \neq j; \\ 1 \text{ when } i = j \end{cases} \qquad (3)$$

We now write

$$U_i(x) = A_i(x)\left[L_i(x)\right]^2 ; V_i(x) = B_i(x)\left[L_i(x)\right]^2$$

where $\quad L_i(x) = \dfrac{(x-x_0)(x-x_1)\cdots(x-x_{i-1})(x-x_{i+1})\cdots(x-x_n)}{(x_i-x_0)(x_i-x_1)\cdots(x_i-x_{i-1})(x_i-x_{i+1})\cdots(x_i-x_n)}$

Since $[L_i(x)]^2$ is of degree $2n$ and $U_i(x)$, $V_i(x)$ are of degree $(2n + 1)$, therefore $A_i(x)$ and $Bi(x)$ are both linear functions

$\therefore$ We can write $\qquad U_i(x) = (a_i + b_i x)\ [L_i(x)]^2$

and $\qquad\qquad\qquad V_i(x) = (c_i + d_i x)[L_i(x)]^2 \qquad (4)$

Using conditions (3) in (4), we get $a_i + b_i x = 1,\ c_i\ + d_i x = 0$

and $\qquad\qquad\qquad\qquad\qquad b_i + 2L_i'(x_i) = 0,\ d_i = 1 \qquad (5)$

Solving these equations, we obtain

$$b_i = -\ 2L_i'(x_i), a_i = 1 + 2x_i L_i'(x_i)$$
$$d_i = 1 \text{ and } c_i = -\ x_i \qquad\qquad (6)$$

Now putting the above values in (4), we get

$$U_i(x) = [1 + 2x_i L_i'(x_i) - 2x L_i'(x_i)][L_i(x)]^2$$
$$= [1 - 2(x - x_i)L_i'(x_i)][L_i(x)]^2$$

and $V_i(x) = (x - x_i)\ [L_i(x)]^2$

Finally substituting $U_i(x)$ and $V_i(x)$ in (2), we obtain

$$p(x) = \sum_{i=0}^{n}\left[1 - 2(x - x_i)Li(x_i)\right][Li(x)]^2 y(x_i) + \sum_{i=0}^{n}(x - x_i)[Li(x)]^2 y'(x_i) \qquad (7)$$

This is the required *Hermite's interpolation formula* which is sometimes known as *osculating interpolation formula*.

NOTE   **Obs.** *In comparison to Lagrange's interpolation formula, the Hermite interpolation formula is computationally uneconomical*

EXAMPLE 7.27

For the following data:

x:	f(x)	f'(x)
0.5	4	− 16
1	1	− 2

Find the Hermite interpolating polynomial.

Solution:

We have $x_0 = 0.5$, $x_1 = 1$, $y(x_0) = 4$, $y(x_1) = 1$; $y'(x_0) = -16$, $y'(x_1) = -2$

Also $\quad L_i(x_0) = \dfrac{(x - x_0)}{(x_i - x_0)} = \dfrac{x-1}{-0.5} = -2(x-1); L_i'(x_0) = -2$

$$L_i(x_1) = \frac{(x - x_0)}{(x_i - x_0)} = \frac{x - 0.5}{1 - 0.5} = 2x - 1; L_i'(x_1) = 2$$

Hermite's interpolation formula in this case, is

$$P(x) = [1 - 2(x - x_0)L'(x_0)][L(x_0)]^2 y(x_0) + (x - x_0)[L(x_0)]^2 y'(x_0)$$

$$+ [1 - 2(x - x_1)L'(x_1)][L(x_1)]^2 y(x_1) + (x - x_1)[L(x_1)]^2 y'(x_1)$$

$$= [1 - 2(x - 0.5)(-2)][-2(x - 1)]^2 (4) + (x - 0.5)[-2(x - 1)]^2 (-16)$$

$$+ [1 - 2(x - 1)(2)](2x - 1)^2 (1) + (x - 1)(2x - 1)^2 (-2)$$

$$= 16[1 + 4(x - 0.5)](x^2 - 2x + 1) - 164(x - 0.5)(x^2 - 2x + 1)$$

$$+ [1 - 4(x - 1)](4x^2 - 4x + 1) - 2(x - 1)(4x^2 - 4x + 1)$$

Hence $P(x) = -24x^3 + 324x^2 - 130x + 23$

EXAMPLE 7.28

Apply Hermite's formula to interpolate for sin (1.05) from the following data:

x	sin x	cos x
1.00	0.84147	0.54030
1.10	0.89121	0.45360

Solution:

Here $y(x) = \sin x$ and $y'(x) = \cos x$

so that $y(x_0) = 0.84147$, $y'(x_0) = 0.54030$, $y(x_1) = 0.89121$,

$y'(x_1) = 0.45360$

Also $\quad L_i(x_0) = \dfrac{(x - x_0)}{(x_i - x_0)} = 11 - 10x, L_i'(x_0) = -10$

$\qquad L_i(x_1) = \dfrac{(x - x_0)}{(x_i - x_0)} = -10 + 10x, L_i'(x_1) = 10$

Hence the Hermite's interpolation formula in this case is

$P(x) = [1 - 2(x - x_0)L'(x_0)][L(x_0)]^2 y(x_0) + (x - x_0)[L(x_0)]^2 y'(x_0)$

$\qquad + [1 - 2(x - x_1)L'(x_1)][L(x_1)]^2 y(x_1) + (x - x_1)[L(x_1)]^2 y'(x_1)$

$\quad = [1 - 2(x - 1)(-10)](11 - 10x)^2 (0.84147) + (x - 1)(-11 + 10x)^2 (0.54030)$

$\qquad\qquad\qquad + [1 - 2(x - 1.1)(10)](-10 + 10x)^2 (0.89121)$

$\qquad\qquad\qquad + (x - 1.1)(-10 + 10x)^2 (0.4536)$

Putting $x = 1.05$ in $P(x)$, we get

$\sin(1.05) = 1 - 2(0.05)(-10)[-10(1.05) + 11]^2 (0.84147)$

$\qquad\qquad + (0.05)(-0.5)^2 (0.54030) + [1 - 2(0.05)(10)](0.5)^2 \times (0.89121)$

$\qquad\qquad + (-0.05)(0.5)^2 (0.4536) = 0.86742$

EXAMPLE 7.29

Determine the Hermite polynomial of degree 4 which fits the following data:

x:	0	1	2
$y(x)$:	0	1	0
$y'(x)$:	0	0	0

Solution:

Here $x_0 = 0$, $x_1 = 1$, $x_2 = 2$, $y(x_0) = 0$, $y(x_1) = 1$, $y(x_2) = 0$ and $y'(x_0) = 0$, $y'(x_1) = 0$, $y'(x2) = 0$.

Hermite's formula in this case is

$$P(x) = [1 - 2L_0'(x_0)(x - x_0)][L_0(x)]^2 y(x_0) + (x - x_0)[L_0(x)]^2 y'(x_0)$$
$$+ [1 - 2L_1'(x_1)(x - x_1)][L1(x)]^2 \times y(x_1) + (x - x_1)[L1(x)]^2 y'(x_1)$$
$$+ [1 - 2L_2'(x_2)(x - x_2)] \times [L_2(x)]^2 y(x_2) + (x - x_2)[L_2(x)]^2 y'(x_2)$$

Substituting the above values in $P(x)$, we get

$$P(x) = [1 - 2L_1'(x_1)(x - 1)][L_1(x)]^2$$

Where $L_1(x) = \dfrac{(x - x_0)(x - x_2)}{(x_1 - x_2)(x_1 - x_2)} = 2x - x^2$ and $L_1'(x_1) = (2 - 2x)_{x-1} = 0$

Hence $p(x) = [L_1(x)]^2 = (2x - x^2)^2$.

EXAMPLE 7.30

Using Hermite's intropolation, find the value of $f(-0.5)$ from the following

x:	-1	0	1
f(x):	1	1	3
f'(x):	-5	1	7

Solution:

Here $x_0 = -1$, $x_1 = 0$, $x_2 = 1$; $f(x_0) = 1$, $f(x_1) = 1$, $f(x_2) = 3$ and $f'(x_0) = -5$, $f'(x_1) = 1$, $f'(x_2) = 7$.

Hermite's formula in this case is

$$P(x) = U_0 f(x_0) + V_0 f'(x_0) + U_1 f(x_1) + V_1 f'(x_1) + U_2 f(x_2) + V_2 f'(x_2) \quad (i)$$

where $U_0 = [1 - 2L_0'(x_0)(x - x_0)][L_0(x)]^2, V_0 = (x - x_0)[L_0(x)]^2$

$U_1 = [1 - 2L_1'(x_1)(x - x_1)][L_1(x)]^2, V_1 = (x - x_1)[L_1(x)]^2$

$U_2 = [1 - 2L_2'(x_2)(x - x_2)][L_2(x_2)]^2, V_2 = (x - x_2)[L_2(x)]^2$

and $L_0(x) = \dfrac{(x - x_1)(x - x_2)}{(x_0 - x_1)(x_0 - x_2)} = \dfrac{x(x - 1)}{2}, L_0'(x) = x - \dfrac{1}{2}$

$L_1(x) = \dfrac{(x - x_0)(x - x_2)}{(x_1 - x_0)(x_1 - x_2)} = 1 - x^2 = L_1'(x) = -2x$

$L_2(x) = \dfrac{(x - x_0)(x - x_1)}{(x_2 - x_0)(x_2 - x_1)} = \dfrac{x(x + 1)}{2} = L_2'(x) = x + \dfrac{1}{2}$

Substituting the values of $L_0, L_0'; L_1, L_1'$ and L_2, L_2', we get

$$U_0 = \left[1 + 3(x+1)\right] \frac{x^2(x-1)^2}{4} = \frac{1}{4}\left(3x^5 - 2x^4 - 5x^3 + 4x^2\right)$$

$$V_0 = (x+1)\frac{x^2(x-1)^2}{4} = \frac{1}{4}\left(x^5 - x^4 - x^3 + x^2\right)$$

$$U_1 = x^4 - 2x^2 + 1, V1 = x^5 - 2x^3 + x$$

$$U_2 = \frac{1}{4}\left(3x^5 - 2x^4 - 5x^3 + 4x^2\right), V_2 = \frac{1}{4}\left(x^5 - x^4 - x^3 + x^2\right)$$

Substituting the values of $U_0, V_0, U_1, V_1; U_2, V_2$ in (i), we get

$$P(x) = \frac{1}{4}\left(3x^5 - 2x^4 - 5x^3 + 4x^2\right)(1) + \frac{1}{4}\left(x^5 - x^4 - x^3 + x^2\right)$$

$$+ (x^4 - 2x^2 + 1)(1) + (x^5 - 2x^3 + x)(1)$$

$$- \frac{1}{4}(3x^5 - 2x^4 - 5x^3 + 4x^2)(3) + \frac{1}{4}(x^5 - x^4 - x^3 + x^2)(7)$$

$$= 2x^4 - x^2 + x + 1$$

Hence $f(-0.5) = 2(-0.5)^4 - (-0.5)^2 + (-0.5) + 1 = 0.375$

Exercises 7.5

1. Find the Hermite's polynomial which fits the following data:

x:	0	1	2
f(x):	1	3	21
f'(x):	0	3	36

2. A switching path between parallel rail road tracks is to be a cubic polynomial joining positions (0, 0) and (4, 2) and tangents to the lines $y = 0$ and $y = 2$. Using Hermite's method, find the polynomial, given:

x	y	y'
0	0	0
4	2	0

3. Apply Hermite's formula estimate the values of log 3.2 from the following data:

x	$y = \log x$	$y' = 1/x$
3.0	1.0986	0.3333
3.5	1.2528	0.2857
4.0	1.3863	0.2500

7.17 Spline Interpolation

In the interpolation methods so far explained, a single polynomial has been fitted to the tabulated points. If the given set of points belong to the polynomial, then this method works well, otherwise the results are rough approximations only. If we draw lines through every two closest points, the resulting graph will not be smooth. Similarly we may draw a quadratic curve through points A_i, A_{i+1} and another quadratic curve through A_{i+1}, A_{i+2}, such that the slopes of the two quadratic curves match at A_{i+1} (Fig. 7.1). The resulting curve looks better but is not quite smooth. We can ensure this by drawing a cubic curve through A_i, A_{i+1} and another cubic through A_{i+1}, A_{i+2} such that the slopes and curvatures of the two curves match at A_{i+1}. Such a curve is called a **cubic spline.** We may use polynomials of higher order but the resulting graph is not better. As such, cubic splines are commonly used. This technique of "spline-fitting" is of recent origin and has important applications.

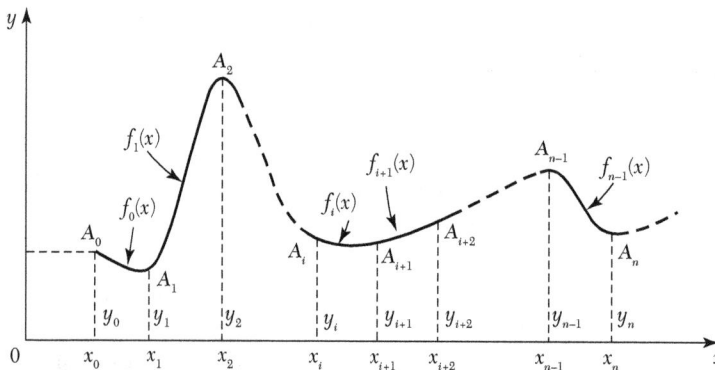

FIGURE 7.1

Cubic spline

Consider the problem of interpolating between the data points (x_0, y_0), $(x_1, y_1), \cdots (x_n, y_n)$ by means of spline fitting.

Then the cubic spline $f(x)$ is such that

(i) $f(x)$ is a linear polynomial outside the interval (x_0, x_n),

(ii) $f(x)$ is a cubic polynomial in each of the subintervals,

(iii) $f'(x)$ and $f''(x)$ are continuous at each point.

Since $f(x)$ is cubic in each of the subintervals $f''(x)$ shall be linear.

∴ Taking equally-spaced values of x so that $x_{i+1} - x_i = h$, we can write

$$f''(x) = \frac{1}{h}\left[(x_{i+1} - x)f''(x_i) + (x - x_i)f''(x_{i+1})\right]$$

Integrating twice, we have

$$f(x) = \frac{1}{h}\left[\frac{(x_{i+1} - x)}{3!}f''(x_i) + \frac{(x - x_i)}{3!}f''(x_{i+1})\right]a_i(x_{i+1} - x) + b_i(x - x_i) \qquad (1)$$

The constants of integration a_i, b_i are determined by substituting the values of $y = f(x)$ at x_i and x_{i+1}. Thus,

$$a_i = \frac{1}{h}\left[y_i - \frac{h^2}{3!}f''(x_i)\right] \text{ and } b_i = \frac{1}{h}\left[y_{i+1} - \frac{h^2}{3!}f''(x_{i+1})\right]$$

Substituting the values of ai, bi and writing $f''(x_i) = M_i$, (1) takes the form

$$f(x) = \frac{(x_{i+1} - x)^3}{6h}M_i + \frac{(x - x_i)^3}{6h}M_{i+1}$$

$$+ \frac{x_{i+1} - x}{h}\left(y_i - \frac{h^2}{6}M_i\right) + \frac{x - x_i}{h}\left(y_{i+1} - \frac{h^2}{6}M_{i+1}\right) \qquad (2)$$

$$\therefore f'(x) = -\frac{(x_{i+1} - x)^2}{2h}M_i + \frac{(x - x_i)^2}{6h}M_{i+1} - \frac{h}{6}(M_{i+1} - M_i) + \frac{1}{h}(y_{i+1} - y_i)$$

To impose the condition of continuity of $f'(x)$, we get
$f'(x - \varepsilon) = f'(x + \varepsilon)$ as $\varepsilon \to 0$

$$\therefore \frac{h}{6}(2M_i + M_{i-1}) + \frac{1}{h}(y_i - y_{i-1}) = -\frac{h}{6}(2M_i + M_{i+1}) + \frac{1}{h}(y_{i+1} - y_i)$$

$$M_{i-1} + 4M_i + M_{i+1} = \frac{6}{h^2}(y_{i-1} - 2y_i + y_{i+1}), i = 1 \text{ to } n - 1 \qquad (3)$$

Now since the graph is linear for $x < x_0$ and $x > x_n$, we have

$$M_0 = 0, M_n = 0 \qquad (4)$$

(3) and (4) give $(n + 1)$ equations in $(n + 1)$ unknowns M_i $(i = 0, 1, \cdots n)$ which can be solved. Substituting the value of M_i in (2) gives the concerned cubic spline.

EXAMPLE 7.31

Obtain the cubic spline for the following data

x:	0	1	2	3
y:	2	− 6	− 8	2

Solution:

Since the points are equispaced with $h = 1$ and $n = 3$, the cubic spline can be determined from $M_{i-1} + 4M_i + M_{i+1} = 6\,(y_{i-1} - 2y_i + y_{i+1})$, $i = 1, 2$.

$$\therefore \quad M_0 + 4M_1 + M_2 = 6\,(y_0 - 2y_1 + y_2)$$
$$M_1 + 4M_2 + M_3 = 6\,(y_1 - 2y_2 + y_3)$$

i.e., $\quad 4M_1 + M_2 = 36;\ M_1 + 4M_2 = 72 \qquad [\because M_0 = 0, M_3 = 0]$

Solving these, we get $M_1 = 4.8$, $M_2 = 16.8$.

Now the cubic spline in $(x_i \leq x \leq x_i + 1)$ is

$$f(x) = \frac{1}{6}(x_{i+1} - x)^3 M_i + \frac{1}{6}(x - x_i)^3 M_{i+1} + (x_{i+1} - x)\left(y_i - \frac{1}{6}M_i\right)$$
$$+ (x - x_i)\left(y_{i+1} - \frac{1}{6}M_{i+1}\right)$$

Taking $i = 0$ in (A) the cubic spline in $(0 \leq x \leq 1)$ is

$$f(x) = \frac{1}{6}(1-x)^3(0) + \frac{1}{6}(x-0)^3(4.8) + (1-x)(x-0) + x\left[-6 - \frac{1}{6}(4.8)\right]$$
$$= 0.8x^3 - 8.8x + 2 \qquad (0 \leq x \leq 1)$$

Taking $i = 1$ in (A), the cubic spline in $(1 \leq x \leq 2)$ is

$$f(x) = \frac{1}{6}(2-x)^3(4.8) + \frac{1}{6}(x-1)^3(16.8) + (2-x)\left[-6 - \frac{1}{6}(4.8)\right]$$
$$+ (x-1)[-8 - 1(16.8)]$$
$$= 2x^3 - 5.84x^2 - 1.68x + 0.8$$

Taking $i = 2$ in (A), the cubic spline in $(2 \leq x \leq 3)$ is

$$f(x) = \frac{1}{6}(3-x)^3(4.8) + \frac{1}{6}(x-2)^3(0) + (3-x)[-8 - 1(16.8)]$$
$$+ (x-2)[2 - 1(2)]$$
$$= -0.8x^3 + 2.64x^2 + 9.68x - 14.8$$

EXAMPLE 7.32

The following values of x and y are given:

x:	1	2	3	4
y:	1	2	5	11

Find the cubic splines and evaluate $y(1.5)$ and $y'(3)$.

Solution:

Since the points are equispaced with $h = 1$ and $n = 3$, the cubic splines can be obtained from

$$M_{i-1} + 4M_i + M_{i+1} = 6(y_{i-1} - 2y_i + y_{i+1}), \quad i = 1, 2.$$

$$\therefore \qquad M_0 + 4M_1 + M_2 = 6(y_0 - 2y_1 + y_2)$$

$$M_1 + 4M_2 + M_3 = 6(y_1 - 2y_2 + y_3)$$

i.e., $\qquad 4M_1 + M_2 = 12, \ M_1 + 4M_2 = 18 \qquad [\because M_0 = 0, M_3 = 0]$

which give, $\qquad M_1 = 2, M_2 = 4.$

Now the cubic spline in $(x_i \le x \le x_{i+1})$ is

$$f(x) = \frac{1}{6}(x_{i+1} - x)^3 M_i + \frac{1}{6}(x - x_i)^3 M_{i+1} + (x_{i+1} - x)\left(y_i - \frac{1}{6}M_i\right)$$

$$+ (x - x_i)\left(y_{i+1} - \frac{1}{6}M_{i+1}\right) \qquad \text{(A)}$$

Thus, taking $i = 0, i = 1, i = 2$ in (A), the cubic splines are

$$f(x) = \begin{cases} \dfrac{1}{3}\left(x^3 - 3x^2 + 5x\right) 1 \le x \le 2 \\[2mm] \dfrac{1}{3}\left(x^3 - 3x^2 + 5x\right) 2 \le x \le 3 \\[2mm] \dfrac{1}{3}\left(-2x^3 - 24x^2 - 76x + 81\right) 3 \le x \le 4 \end{cases}$$

$$\therefore \qquad y(1.5) = f(1.5) = 11/8$$

EXAMPLE 7.33

Find the cubic spline interpolation for the data:

x:	1	2	3	4	5
$f(x)$:	1	0	1	0	1

Solution:

Since the points are equispaced with $h = 1$, $n = 4$, the cubic spline can be found by means of

$$M_{i-1} + 4M_i + M_{i+1} = 6(y_{i-1} - 2y_i + y_{i+1}), i = 1,2,3$$

$\therefore \qquad M_0 + 4M_1 + M_2 = 6\left(y_0 - 2y_1 + y_2\right) = 12$

$$M_1 + 4M_2 + M_3 = 6(y_1 - 2y_2 + y_3) = -12$$

$$M_2 + 4M_3 + M_4 = 6(y_2 - 2y_3 + y_4) = 12$$

Since $\quad M_0 = y''_0 = 0$ and $M_4 = y''_4 = 0$

$\therefore \qquad 4M_1 + M_2 = 12; M_1 + 4M_2 + M_3 = -12; M_1 + 4M_3 = 12$

Solving these equations, we get $M_1 = 30/7$, $M_2 = -36/7$, $M_3 = 30/7$

Now the cubic spline in $(xi \le x \le xi+1)$ is

$$f(x) = \frac{1}{6}\left(x_{i+1} - x\right)^3 M_i + \frac{1}{6}\left(x - x_i\right)^3 M_{i+1} + \left(x_{i+1} - x\right)$$

$$\left(y_i - \frac{1}{6}M_i\right) + \left(x - x_i\right)\left(y_{i+1} - \frac{1}{6}M_{i+1}\right) \qquad (A)$$

Taking $i = 0$, in (A), the cubic spline in $(1 \le x \le 2)$ is

$$y = \frac{1}{6}\left[(x_1 - x)^3 M_0 + (x - x_0)^3 M_1\right] + (x_1 - x)\left(y_0 - \frac{1}{6}M_0\right)$$

$$+ (x - x_0)\left(y_1 - \frac{1}{6}M_1\right)$$

$$= \frac{1}{6}\left[(2 - x)^3 (0) + (x - x_0)^3 (30 / 7)\right] + (2 - x)\left(1 - \frac{1}{6}(0)\right)$$

$$+ (x - 1)\left(0 - \frac{1}{6}\left(\frac{30}{7}\right)\right)$$

i.e., $\quad y = 0.71x^3 - 2.14x^2 + 0.42x + 2 \quad (1 < x \le 2)$

Taking $i = 1$ in (A), the cubic spline in $(2 \le x \le 3)$ is

$$y = \frac{1}{6}\left[(3 - x)^3 \frac{30}{7} + (x - 2)^3 \left(-\frac{36}{7}\right)\right] + (3 - x)\left(0 - \frac{1}{6}\left(\frac{30}{7}\right)\right)$$

$$+ (x - 2)\left(1 - \frac{1}{6}\left(-\frac{36}{7}\right)\right)$$

i.e., $y = -1.57\,x^3 + 11.57x^2 - 27x + 20.28.$ $(2 \leq x \leq 3)$

Taking $i = 2$ in (A), the cubic spline in $(3 \leq x \leq 4)$ is

$$y = \frac{1}{6}(4-x)^3\left(-\frac{36}{7}\right) + \frac{1}{6}(x-3)^3\frac{30}{7} + (4-x)\left(1 - \frac{1}{6}\left(-\frac{36}{7}\right)\right) + (x-3)\left(0 - \frac{5}{7}\right)$$

i.e., $y = 1.57\,x^3 - 16.71\,x^2 + 57.86x - 64.57$ $(3 \leq x \leq 4)$

Taking $i = 3$ in (A), the cubic spline in $(4 \leq x \leq 5)$ is

$$y = \frac{1}{6}(1-x)^3\left(\frac{30}{7}\right) + (5-x)^3\left(-\frac{5}{7}\right) + (x-4)(1)$$

i.e., $y = -0.71x^3 + 2.14x^2 - 0.43x - 6.86.$ $(4 \leq x \leq 5)$

Exercises 7.6

1. Find the cubic splines for the following table of values:

x:	1	2	3
y:	− 6	− 1	16

Hence evaluate $y(1.5)$ and $y'(2)$.

2. The following values of x and y are given:

x:	1	2	3	4
y:	1	5	11	8

Usin g cubic spline, show that
(i) $y(1.5) = 2.575$ (ii) $y'(3) = 2.067$.

3. Find the cubic spline corresponding to the interval $[2,3]$ from the following table:

x:	1	2	3	4	5
y:	30	15	32	18	25

Hence compute (i) $y(2.5)$ and (ii) $y'(3)$.

7.18   Double Interpolation

So far, we have derived interpolation formulae to approximate a function of a single variable. In the case of functions, of two variables, we interpolate with respect to the first variable keeping the other variable constant. Then interpolate with respect to the second variable.

Similarly, we can extend the said procedure for functions of three variables.

7.19   Inverse Interpolation

So far, given a set of values of x and y, we have been finding the value of y corresponding to a certain value of x. On the other hand, the process of estimating the value of x for a value of y (which is not in the table) is called *inverse interpolation*. When the values of x are unequally spaced *Lagrange's method* is used and when the values of x are equally spaced, the *Iterative method* should be employed.

7.20   Lagrange's Method

This procedure is similar to Lagrange's interpolation formula (p. 207), the only difference being that x is assumed to be expressible as a polynomial in y.

Lagrange's formula is merely a relation between two variables either of which may be taken as the independent variable. Therefore, on interchanging x and y in the Lagrange's formula, we obtain

$$x = \frac{(y-y_1)(y-y_2)\cdots(y-y_n)}{(y-y_1)(y-y_2)\cdots(y-y_n)}x_0 + \frac{(y-y_0)(y-y_2)\cdots(y-y_n)}{(y_1-y_0)(y_1-y_2)\cdots(y1-y_n)}x_1$$
$$+ \frac{(y-y_0)(y-y_1)\cdots(y-y_{n-1})}{(y_n-y_0)(y_n-y_1)\cdots(y_n-y_{n-1})} \tag{1}$$

EXAMPLE 7.34

The following table gives the values of x and y:

x:	1.2	2.1	2.8	4.1	4.9	6.2
y:	4.2	6.8	9.8	13.4	15.5	19.6

Find the value of x corresponding to $y = 12$, using Lagrange's technique.

Solution:

Here $x_0 = 1.2$, $x_1 = 2.1$, $x_2 = 2.8$, $x_3 = 4.1$, $x_4 = 4.9$, $x_5 = 6.2$ and $y_0 = 4.2$, $y_1 = 6.8$, $y_2 = 9.8$, $y_3 = 13.4$, $y_4 = 15.5$, $y_5 = 19.6$.

Taking $y = 12$, the above formula (1) gives

$$x = \frac{(12-6.8)(12-9.8)(12-13.4)(12-15.5)(12-19.6)}{(4.2-6.8)(4.2-9.8)(4.2-13.4)(4.2-15.5)(4.2-19.6)} \times 1.2$$

$$+\frac{(12-4.2)(12-9.8)(12-13.4)(12-15.5)(12-19.6)}{(6.8-4.2)(6.8-9.8)(6.8-13.4)(6.8-15.5)(6.8-19.6)} \times 2.1$$

$$+\frac{(12-4.2)(12-6.8)(12-13.4)(12-15.5)(12-19.6)}{(9.8-4.2)(9.8-6.8)(9.8-13.4)(9.8-15.5)(9.8-19.6)} \times 2.8$$

$$+\frac{(12-4.2)(12-6.8)(12-9.8)(12-15.5)(12-19.6)}{(13.4-4.2)(13.4-6.8)(13.4-9.8)(13.4-15.5)(13.4-19.6)} \times 4.1$$

$$+\frac{(12-4.2)(12-6.8)(12-9.8)(12-13.4)(12-19.6)}{(15.5-4.2)(15.5-6.8)(15.5-9.8)(15.5-13.4)(15.5-19.6)} \times 4.9$$

$$+\frac{(12-4.2)(12-6.8)(12-9.8)(12-13.4)(12-15.5)}{(19.6-4.2)(19.6-6.8)(19.6-9.8)(19.6-13.4)(19.6-15.5)} \times 6.2$$

$$= 0.022 - 0.234 + 1.252 + 3.419 - 0.964 + 0.055 = 3.55$$

EXAMPLE 7.35

Apply Lagrange's formula inversely to obtain a root of the equation $f(x) = 0$, given that $f(30) = -30$, $f(34) = -13$, $f(38) = 3$, and $f'(42) = 18$.

Solution:

Here $x_0 = 30$, $x_1 = 34$, $x_2 = 38$, $x_3 = 42$

and $y_0 = -30$, $y_1 = -13$, $y_2 = 3$, $y_3 = 18$

It is required to find x corresponding to $y = f(x) = 0$.

Taking $y = 0$, Lagrange's formula gives

$$x = \frac{(y-y_1)(y-y_2)(y-y_3)}{(y_0-y_1)(y_0-y_2)(y_0-y_3)}x_0 + \frac{(y-y_0)(y-y_2)(y-y_3)}{(y_1-y_0)(y_1-y_2).(y1-y_3)}x_1$$

$$+\frac{(y-y_0)(y-y_1).(y-y_3)}{(y_2-y_0)(y_2-y_1)(y_2-y_3)}x_2 + \frac{(y-y_0)(y-y_1)(y-y_2)}{(y_3-y_0)(y_3-y_1)(y_3-y_2)}x_3$$

$$= \frac{13(-3)(-18)}{(-17)(-33)(-48)} \times 30 + \frac{30(-3)(-18)}{17(-16)(-31)} \times 34$$

$$+\frac{30(13)(-18)}{33(16)(-15)} \times 38 + \frac{30(13)(-3)}{48(31)(15)} \times 42$$

$$= -0.782 + 6.532 + 33.682 - 2.202 = 37.23.$$

Hence the desired root of $f(x) = 0$ is 37.23.

7.21   Iterative Method

Newton's forward interpolation formula (p. 274) is

$$y_p = y_0 + p\Delta y_0 + \frac{p(p-1)}{2!}\Delta^2 y_0 + \frac{p(p-1)(p-2)}{3!}\Delta^3 y_0 + \dots$$

From this, we get

$$p = \frac{1}{\Delta y_0}\left[y_p - y_0 + \frac{p(p-1)}{2!}\Delta^2 y_0 + \frac{p(p-1)(p-2)}{3!}\Delta^3 y_0 + \cdots \right] \qquad (1)$$

Neglecting the second and higher differences, we obtain the first approximation to p as

$$p_1 = (y_p - y_0)/\Delta y_0 \qquad (2)$$

To find the second approximation, retaining the term with second differences in (1) and replacing p by p_1, we get

$$p_2 = \frac{1}{\Delta y_0}\left[y_p - y_0 + \frac{p_1(p_1-1)}{2!}\Delta^2 y_0 \right] \qquad (3)$$

To find the third approximation, retaining the term with third differences in (1) and replacing every p by p_2, we have

$$p_3 = \frac{1}{\Delta y_0}\left[y_p - y_0 + \frac{p_2(p_2-1)}{2!}\Delta^2 y_0 + \frac{p_2(p_2-1)(p_2-2)}{3!}\Delta^3 y_0 \right]$$

and so on. This process is continued till two successive approximations of p agree with each other

NOTE   **Obs.** *This technique can be equally well be applied by starting with any other interpolation formula.*

This method is a powerful iterative procedure for finding the roots of an equation to a good degree of accuracy.

EXAMPLE 7.36

The following values of $y = f(x)$ are given

x:	10	15	20
y:	1754	2648	3564

Find the value of x for $y = 3000$ by iterative method.

Solution:

Taking $x_0 = 10$ and $h = 5$, the difference table is

x	y	Δy	$\Delta^2 y$
10	1754		
		894	
15	2648		22
		916	
20	3564		

Here $y_p = 3000$, $y_0 = 1754$, $\Delta y_0 = 894$ and $\Delta^2 y0 = 22$.

∴   The successive approximations to p are

$$p_1 = \frac{1}{894}(3000 - 1754) = 1.39$$

$$p_2 = \frac{1}{894}\left[3000 - 1754 - \frac{1.39(1.39-1)}{2} \times 22\right] = 1.387$$

$$p_3 = \frac{1}{894}\left[3000 - 1754 - \frac{1.387(1.387-1)}{2} \times 22\right] = 1.3871$$

We, therefore, take $p = 1.387$ correct to three decimal places. Hence the value of x (corresponding to $y = 3000$) = $x_0 + ph = 10 + 1.387 \times 5 = 16.935$.

EXAMPLE 7.37

Using inverse interpolation, find the real root of the equation $x3 + x - 3 = 0$, which is close to 1.2.

Solution:

The difference table is

x	y	$Y = x^3 + x - 3$	Δy	$\Delta^2 y$	$\Delta^3 y$	$\Delta^4 y$
1	-0.2	-1	0.431			
				0.066		
1.1	-0.1	-0.569	0.497		0.006	
				0.072		0
1.2	0	-0.072			0.006	

x	y	$Y = x^3 + x - 3$	Δy	$\Delta^2 y$	$\Delta^3 y$	$\Delta^4 y$
			0.569			
1.3	0.1	0.497		0.078		
			0.647			
1.4	0.2	1.144				

Clearly the root of the given equation lies between 1.2 and 1.3. Assuming the origin at $x = 1.2$ and using Stirling's formula

$$y = y_0 + x \frac{\Delta y_0 + \Delta y_{-1}}{2} + \frac{x^2}{2}\Delta^2 y_{-1} + \frac{x(x^2-1)}{6} \times \frac{\Delta^3 y_{-1} + \Delta^3 y_{-2}}{2} \text{, we get}$$

$$0 = -0.072 + x.\frac{0.569 + 0.467}{2} + \frac{x^2}{2}(0.072) + \frac{x(x^2-1)}{6}\left(\frac{0.006 + 0.006}{2}\right)$$

$$\left(\because y = 0 \right)$$

or $\qquad 0 = -0.072 + 0.532x + 0.036x^2 + 0.001x^3 \qquad\qquad (i)$

This equation can be written a s

$$x = \frac{0.072}{0.532} - \frac{0.036}{0.532}x^2 - \frac{0.001}{0.532}x^3$$

$\therefore$ First approximation $x^{(1)} = \dfrac{0.072}{0.532} = 0.1353$

Putting $x = x^{(1)}$ on R.H.S. of (i), we get

Second approximation

$$x^{(2)} = 0.1353 - 0.067(0.1353)^2 - 1.8797(0.1353)^3 = 0.134$$

Hence the desired root $= 1.2 + 0.1 \times 0.134 = 1.2134$.

Exercises 7.7

1. Apply Lagrange's method to find the value of x when f(x)=5 from the given data:

x:	5	6	9	11
f(x):	12	13	14	16

2. Obtain the value of t when A = 85 from the following table, using Lagrange's method:

t:	2	5	8	14
A:	94.8	87.9	81.3	68.7

3. Apply Lagrange's formula inversely to obtain the root of the equation $f(x) = 0$, given that $f(30) = -30, f(34) = -13, f(38) = 3$ and $f(42) = 18$.

4. From the following data:

x:	1.8	2.0	2.2	2.4	2.6
y:	2.9	3.6	4.4	5.5	6.7,

find x when $y = 5$ u sin g the iterative method.

5. The equation $x^3 - 15x + 4 = 0$ ha s a root close to 0.3. Obtain this root upto four decimal places using inverse interpolation.

6. Solve the equation $x = 10 \log x$, by iterative method given that

x:	1.35	1.36	1.37	1.38
log x:	0.1303	0.1355	0.1367	0.1392

7.22 Objective Type of Questions

Exercises 7.8

1. Select the correct answer or fill up the blanks in the following question:
Newton's back war d interpolation formula is.........

2. Bessel's formula is most appropriate when p lies between
(a) – 0. 25 an d 0.25   (b) 0.25 an d 0.75   (c) 0.75 an d 1.00

3. Form the divided difference table for the following data:

x:	5	15	22
y:	7	36	160

4. Interpolation is the technique of estimating the value of a function for any......

5. Bessel's formula for interpolation is......

6. The four divided differences for $x_0, x_1, x_2, x_3, x_4 = ...$

7. Stirling's formula is best suited for p lying between......

8. Newton's divided differences formula is.......

9. Given (x_0, y_0), (x_1, y_1), (x_2, y_2), Lagrange's interpolation formula is.......

10. If $f(0) = 1, f(2) = 5, f(3) = 10$ and $f(x) = 14$, then $x =$......

11. The difference between Lagrange's interpolating polynomial and Hermite's interpolating polynomial is.......

12. If $y(1) = 4$, $y(3) = 12$, $y(4) = 19$ and $y(x) = 7$, find x using Lagrange's formula.

13. Extrapolation is defined as.......

14. The second divided difference of $f(x) = 1/x$, with arguments a, b, c is......

15. The Gauss-forward interpolation formula is used to interpolate values of y for
(a) $0 < p < 1$ (b) $-1 < 1 < 0$
(c) $0 < p < -\alpha$ (d) $-\alpha < p < 0$

16. Given

x:	0	1	3	4
y:	−12	0	6	12

Using Lagrange's formula, a polynomial that can be fitted to the data is......

17. The nth divided difference of a polynomial of degree n is
(a) zero   (b) a constant
(c) a variable   (d) none of these.

18. The Gauss forward interpolation formula involves
(a) differences above the central line and odd differences on the central line
(b) even differences below the central line and odd differences on the central line
(c) odd differences below the central line and even differences on the central line
(d) odd differences above the central line and even differences on the central line.

19. Differentiate between interpolation polynomial and least square polynomial obtained for a set of data.

CHAPTER 8

NUMERICAL DIFFERENTIATION AND INTEGRATION

Chapter Objectives

- Numerical differentiation
- Formulae for derivatives
- Maxima and minima of a tabulated function
- Numerical integration
- Quadrature formulae
- Errors in quadrature formulae
- Romberg's method
- Euler-Maclaurin formula
- Method of undetermined coefficients
- Gaussian integration
- Numerical double integration
- Objective type of questions

8.1 Numerical Differentiation

It is the process of calculating the value of the derivative of a function at some assigned value of x from the given set of values (x_i, y_i). To compute dy/dx, we first replace the exact relation $y = f(x)$ by the best interpolating polynomial $y = \phi(x)$ and then differentiate the latter as many times as we desire. The choice of the interpolation formula to be used, will depend on the assigned value of x at which dy/dx is desired.

If the values of x are equispaced and dy/dx is required near the beginning of the table, we employ Newton's forward formula. If it is required near the end of the table, we use Newton's backward formula. For values near the middle of the table, dy/dx is calculated by means of Stirling's or Bessel's formula. If the values of x are not equispaced, we use Lagrange's formula or Newton's divided difference formula to represent the function.

Hence corresponding to each of the interpolation formulae, we can derive a formula for finding the derivative.

NOTE

Obs. *While using these formulae, it must be observed that the table of values defines the function at these points only and does not completely define the function and the function may not be differentiable at all. As such, the process of numerical differentiation should be used only if the tabulated values are such that the differences of some order are constants. Otherwise, errors are bound to creep in which go on increasing as derivatives of higher order are found. This is due to the fact that the difference between $f(x)$ and the approximating polynomial $\phi(x)$ may be small at the data points but $f'(x) - \phi'(x)$ may be large.*

8.2   Formulae for Derivatives

Consider the function $y = f(x)$ which is tabulated for the values $x_i (= x_0 + ih)$, $i = 0, 1, 2, \dots n$.

Derivatives using Newton's forward difference formula

Newton's forward interpolation formula (p. 274) is

$$y = y_0 + p\Delta y_0 + \frac{p(p-1)}{2!}\Delta^2 y_0 + \frac{p(p-1)(p-2)}{3!}\Delta^3 y_0 + \dots$$

Differentiating both sides w.r.t. p, we have

$$\frac{dy}{dp} = \Delta y_0 + \frac{2p-1}{2!}\Delta^2 y_0 + \frac{3p^2 - 6p + 2}{3!}\Delta^3 y_0 + \dots$$

Since $p = \dfrac{(x - x_0)}{h}$

Therefore $\dfrac{dp}{dx} = \dfrac{1}{h}$

Now $\dfrac{dy}{ds} = \dfrac{dy}{dp} \cdot \dfrac{dp}{dx} = \dfrac{1}{h}\left[\Delta y_0 + \dfrac{2p-1}{2!}\Delta^2 y_0 + \dfrac{3p^2 - 6p + 2}{3!}\Delta^3 y_0 \right.$

$$\left. + \dfrac{4p^3 - 18p^2 + 22p - 6}{4!}\Delta^4 y_0 + \cdots \right] \tag{1}$$

At $x = x_0$, $p = 0$. Hence putting $p = 0$,

$$\left(\dfrac{dy}{dx}\right)_{x_0} = \dfrac{1}{h}\left[\Delta y_0 - \dfrac{1}{2}\Delta^2 y_0 + \dfrac{1}{3}\Delta^3 y_0 - \dfrac{1}{4}\Delta^4 y_0 + \dfrac{1}{5}\Delta^5 y_0 - \dfrac{1}{6}\Delta^6 y_0 + \cdots \right] \tag{2}$$

Again differentiating (1) w.r.t. x, we get

$$\dfrac{d^2 y}{dx^2} = \dfrac{d}{dp}\left(\dfrac{dy}{dp}\right)\dfrac{dp}{dx}$$

$$= \dfrac{1}{h}\left[\dfrac{2}{2!}\Delta^2 y_0 + \dfrac{6p-6}{3!}\Delta^3 y_0 + \dfrac{12p^2 - 36p^2 - 36p + 22}{4!}\Delta^4 y_0 + \cdots \right]\dfrac{1}{h}$$

Putting $p = 0$, we obtain

$$\left(\dfrac{d^2 y}{dx^2}\right) = \dfrac{1}{h^2}\left[\Delta^2 y_0 - \Delta^3 y_0 + \dfrac{11}{12}\Delta^4 y_0 - \dfrac{5}{6}\Delta^5 y_0 + \dfrac{137}{180}\Delta^6 y_0 + \cdots \right] \tag{3}$$

Similarly $\qquad \left(\dfrac{d^3 y}{dx^3}\right) = \dfrac{1}{h^3}\left[\Delta^3 y_0 - \dfrac{3}{2}\Delta^4 y_0 + \cdots \right]$

Otherwise: We know that $1 + \Delta = E = e^{hD}$

$\therefore \qquad hD = \log(1 + \Delta) = \Delta - \dfrac{1}{2}\Delta^2 + \dfrac{1}{3}\Delta^3 - \dfrac{1}{4}\Delta^4 + \cdots$

or $\qquad D = \dfrac{1}{h}\left[\Delta - \dfrac{1}{2}\Delta^2 + \dfrac{1}{3}\Delta^3 - \dfrac{1}{4}\Delta^4 + \cdots \right]$

and $\qquad D^2 = \dfrac{1}{h^2}\left[\Delta - \dfrac{1}{2}\Delta^2 + \dfrac{1}{3}\Delta^3 - \dfrac{1}{4}\Delta^4 + \cdots \right]^2 = \dfrac{1}{h^2}\left[\Delta^2 - \Delta^3 + \dfrac{11}{12}\Delta^4 + \cdots \right]$

and $\qquad D^3 = \dfrac{1}{h^2}\left[\Delta^3 - \dfrac{3}{2}\Delta^4 + \cdots \right]$

Now applying the above identities to y_0, we get

Dy_0 i.e., $\left(\dfrac{dy}{dx}\right)_{x_0} = \dfrac{1}{h}\Delta y_0 - \dfrac{1}{2}\left[\Delta^2 y_0 \dfrac{1}{3}\Delta^3 y_0 - \dfrac{1}{4}\Delta^4 y_0 + \dfrac{1}{5}\Delta^5 y_0 - \dfrac{1}{6}\Delta^6 y_0 + \cdots \right]$

$$\left(\frac{d^2y}{dx^2}\right) = \frac{1}{h^2}\left[\Delta^2 y_0 - \Delta^3 y_0 + \frac{11}{12}\Delta^4 y_0 - \frac{5}{6}\Delta^5 y_0 + \frac{137}{180}\Delta^6 y_0 - \cdots\right]$$

and

$$\left(\frac{d^3y}{dx^3}\right) = \frac{1}{h^3}\left[\Delta^2 y_0 - \frac{3}{2}\Delta^4 y_0 + \cdots\right]$$

which are the same as (2), (3), and (4), respectively.

Derivatives using Newton's backward difference formula

Newton's backward interpolation formula (p. 274) is

$$y = y_n + p\nabla y_n + \frac{p(p+1)}{2!}\nabla^2 y_n + \frac{p(p+1)(p+2)}{3!}\nabla^3 y_n + \cdots$$

Differentiating both sides w.r.t. p, we get

$$\frac{dy}{dp} = \nabla y_n + \frac{2p+1}{2!}\nabla^2 y_n + \frac{3p^2 + 6p + 2}{3!}\nabla^3 y_n + \cdots$$

Since $p = \dfrac{x - x_n}{h}$, therefore. $\dfrac{dp}{dx} = \dfrac{1}{h}$

Now $\dfrac{dy}{dx} = \dfrac{dy}{dp}\cdot\dfrac{dp}{dx} = \dfrac{1}{h}\left[\nabla y_n + \dfrac{2p+1}{2!}\nabla^2 y_n + \dfrac{3p^3 + 6p + 2}{3!}\nabla^3 y_n + \cdots\right]$ (5)

At $x = x_n$, $p = 0$. Hence putting $p = 0$, we get

$$\left(\frac{dy}{dx}\right)_{x_n} = \frac{1}{h}\left[\nabla y_n + \frac{1}{2}\nabla^2 y_n + \frac{1}{3}\nabla^3 y_n + \frac{1}{4}\nabla^4 y_n + \frac{1}{5}\nabla^5 y_n + \frac{1}{6}\nabla^6 y_n + \cdots\right]$$ (6)

Again differentiating (5) w.r.t. x, we have

$$\frac{d^2y}{dx^2} = \frac{d}{dp}\left(\frac{dy}{dx}\right)\frac{dp}{dx3}$$

$$= \frac{1}{h^2}\left[\nabla^2 y_n + \frac{6p+6}{3!}\nabla^3 y_n + \frac{6p^2 + 18p + 11}{12}\nabla^4 y_n + \cdots\right]$$

Putting $p = 0$, we obtain

$$\left(\frac{d^2y}{dx^2}\right) = \frac{1}{h^2}\left[\nabla^2 y_n + \nabla^3 y_n + \frac{11}{12}\nabla^4 y_n + \frac{5}{6}\nabla^5 y_n + \frac{137}{180}\nabla^6 y_n + \cdots\right]$$ (7)

Similarly, $\left(\dfrac{d^3y}{dx^3}\right)_{x_n} = \dfrac{1}{h^3}\left[\nabla^3 y_n + \dfrac{3}{2}\nabla^4 y_n + \cdots\right]$ (8)

Otherwise: We know that $1 - \nabla = E^{-1} = e^{-hD}$

$$\therefore \qquad -hD = \log(1 - \nabla) = -[\nabla + \frac{1}{2}\nabla^2 + \frac{1}{3}\nabla^3 + \frac{1}{3}\nabla^4 + \cdots]$$

or

$$D = \frac{1}{h}\left[\nabla + \frac{1}{2}\nabla^2 + \frac{1}{3}\nabla^3 + \frac{1}{4}\nabla^4 + \cdots\right]$$

$$\therefore \qquad D^2 = \frac{1}{h^2}\left[\nabla + \frac{1}{2}\nabla^2 + \frac{1}{2}\nabla^3 + \cdots\right]^2 = \frac{1}{h^2}\left[\nabla^2 + \nabla^3 + \frac{11}{12}\nabla^4 + \cdots\right]$$

Similarly, $D^3 = \dfrac{1}{h^3}\left[\nabla^3 + \dfrac{3}{2}\nabla^4 + \cdots\right]$

Applying these identities to y_n, we get

Dy_n i.e., $\left(\dfrac{dy}{dx}\right)_{x_n} = \dfrac{1}{h}\left[\nabla y_n + \dfrac{1}{2}\nabla^2 y_n + \dfrac{1}{2}\nabla^3 y_n + \dfrac{1}{4}\nabla^4 y_n \dfrac{1}{5}\nabla^5 y_n + \dfrac{1}{6}\nabla^6 y_n + \cdots\right]$

$$\left(\frac{d^2 y}{dx^2}\right)_{x_n} = \frac{1}{h^2}\left[\nabla^2 y_n + \nabla^3 y_n + \frac{11}{12}\nabla^4 y_n + \frac{5}{6}\nabla^5 y_n + \frac{137}{180}\nabla^6 y_n + \cdots\right]$$

and

$$\left(\frac{d^3 y}{dx^3}\right)_{x_n} = \frac{1}{h^3}\left[\nabla^3 y_n + \frac{3}{2}\nabla^4 y_n + \cdots\right]$$

which are the same as (6), (7), and (8).

Derivatives using Stirling's central difference formula

Stirling's formula (p. 289) is

$$y_p = y_0 + \frac{p}{1!}\left(\frac{\Delta y_0 + \Delta y_{-1}}{2}\right) + \frac{p^2}{2!}\Delta^2 y_{-1}$$

$$+ \frac{p(p^2 - 1^2)}{3!}\left(\frac{\Delta^3 y_{-1} + \Delta^3 y_{-2}}{2}\right) + \frac{p^2(p^2 - 1^2)}{4!}\Delta^4 y_{-2} + \cdots$$

Differentiating both sides w.r.t. p, we get

$$\frac{dy}{dp} = \left(\frac{\Delta y_0 + \Delta y_{-1}}{2}\right) + \frac{2p}{2!}\Delta^2 y_{-1} + \frac{3p^2 - 1}{3!}\left(\frac{\Delta^3 y_{-1} + \Delta^3 y_{-2}}{2}\right)$$

$$+ \frac{4p^3 - 2p}{4!}\Delta^4 y_{-2} + \cdots$$

Since $p = \dfrac{x - x_0}{h}$, $\therefore \dfrac{dp}{dx} = \dfrac{1}{h}$.

Now $\dfrac{dy}{dx} = \dfrac{dy}{dp} \cdot \dfrac{dp}{dx} = \dfrac{1}{h}\left[\left(\dfrac{\Delta y_0 + \Delta y_{-1}}{26}\right) + p\Delta^2 y_{-1} + \dfrac{3p^2 - 1}{6}\left(\dfrac{\Delta^3 y_{-1} + \Delta^3 y_{-2}}{2}\right)\right.$

$$\left. + \dfrac{2p^3 - p}{12}\Delta^4 y_{-2} + \cdots\right]$$

At $x = x_0$, $p = 0$. Hence putting $p = 0$, we get

$$\left(\dfrac{dy}{dx}\right)_{x_0} = \dfrac{1}{h}\left[\dfrac{\Delta y_0 + \Delta y_{-1}}{2} - \dfrac{1}{6}\dfrac{\Delta^3 y_{-1} + \Delta^3 y_{-2}}{2} + \dfrac{1}{30}\dfrac{\Delta^5 y_{-2} + \Delta^5 y_{-3}}{2} + \cdots\right] \quad (9)$$

Similarly $\left(\dfrac{d^2 y}{dx^2}\right) = \dfrac{1}{h^2}\left[\Delta^2 y_{-1} - \dfrac{1}{12}\Delta^4 y_{-2} + \dfrac{1}{90}\Delta^6 y_{-3} - \cdots\right] \quad (10)$

Derivatives using Bessel's central difference formula

Bessel's formula (p. 290) is

$$y_p = y_0 + p\Delta y_0 + \dfrac{p(p-1)}{2!}\dfrac{\Delta^2 y_{-1} + \Delta^2 y_0}{2} + \left(p - \dfrac{1}{2}\right)\dfrac{p(p-1)}{3!}\Delta^3 y_{-1}$$

$$+ \dfrac{4p^3 - 6p^2 - 2p + 2}{4!}\dfrac{\Delta^4 y_{-2} + \Delta^4 y_{-1}}{2} + \cdots$$

Since $p = \dfrac{x - x_0}{h}$, $\therefore \dfrac{dp}{dx} = \dfrac{1}{h}$

Now $\dfrac{dy}{dx} = \dfrac{dy}{dp} \cdot \dfrac{dp}{dx} = \dfrac{1}{h}\left[\Delta y_0 + \dfrac{2p - 1}{2!}\dfrac{\Delta^2 y_{-1} + \Delta^2 y_0}{2}\right.$

$$\left. + \dfrac{3p^2 - 2p + \dfrac{1}{2}}{3!}\Delta^3 y_{-1} + \dfrac{4p^3 - 6p^2 - 2p + 2}{4!}\dfrac{\Delta^4 y_{-2} + \Delta^4 y_{-1}}{2} + \cdots\right]$$

At $x = x_0$, $p = 0$. Hence putting $p = 0$, we get

$$\left(\dfrac{dy}{dx}\right)_{x_0} = \dfrac{1}{h}\left[\Delta y_0 - \dfrac{1}{2}\left(\dfrac{\Delta^2 y_{-1} + \Delta^2 y_0}{2}\right) + \dfrac{1}{12}\Delta^3 y_{-1}\right. \quad (11)$$

$$\left. + \dfrac{1}{12}\left(\dfrac{\Delta^4 y_{-2} + \Delta^4 y_{-1}}{2}\right) + \cdots\right]$$

Similarly

$$\left(\dfrac{d^2 y}{dx^2}\right)_{x_0} = \dfrac{1}{h^2}\left[\left(\dfrac{\Delta^2 y_{-1} + \Delta^2 y_0}{2}\right) - \dfrac{1}{2}\Delta^3 y_{-1} - \dfrac{1}{12}\left(\dfrac{\Delta^4 y_{-2} + \Delta^4 y_{-1}}{2}\right) + \cdots\right] (12)$$

Derivatives using unequally spaced values of argument

(*i*) *Lagranges's interpolation formula is*

$$f(x) = \frac{(x-x_1)(x-x_2)\cdots(x-x_n)}{(x_0-x_1)(x_0-x_2)\cdots(x_0-x_n)} f(x_0)$$

$$+\frac{(x-x_0)(x-x_2)\cdots(x-x_n)}{(x_1-x_0)(x_1-x_2)\cdots(x_1-x_n)} f(x_1) + \cdots$$

Differentiating both sides w.r.t. *x*, we get $f(x)$.

(*ii*) *Newton's divided difference formula is*

$$f(x) = f(x_0) + (x-x_0)\Delta f(x_0) + (x-x_0)(x-x_1)\Delta^2 f(x_0)$$

$$+(x-x_0)(x-x_1)(x-x_2)\Delta^3 f(x_0) + \cdots$$

Differentiating both sides w.r.t. *x*, we obtain

$$f'(x) = \Delta f(x0) + [2x-(x_0+x_1)]\,\Delta^2 f(x_0) + [3x^2 - 2x(x_0+x_1+x_2)$$

$$+(x_0 x_1 + x_1 x_2 + x_2 x_3)]\,\Delta^3 f(x_0) + \cdots$$

EXAMPLE 8.1

Given that

x:	1.0	1.1	1.2	1.3	1.4	1.5	1.6
y:	7.989	8.403	8.781	9.129	9.451	9.750	10.031

find $\dfrac{dy}{dx}$ and $\dfrac{d^2 y}{dx^2}$ at (*a*) *x* = 1.1   (*b*) *x* = 1.6.

Solution:

(a) The difference table is:

x	y	Δ	Δ²	Δ³	Δ⁴	Δ⁵	Δ⁶
1.0	7.989						
		0.414					
1.1	8.403		− 0.036				
		0.378		0.006			
1.2	8.781		− 0.030		− 0.002		
		0.348		0.004		0.001	
1.3	9.129		− 0.026		− 0.001		0.002
		0.322		0.003		0.003	
1.4	9.451		− 0.023		0.002		

x	y	Δ	Δ^2	Δ^3	Δ^4	Δ^5	Δ^6
		0.299		0.005			
1.5	9.750		-0.018				
		0.281					
1.6	10.031						

We have

$$\left(\frac{dy}{dx}\right)_{x0} = \frac{1}{h}\left[\Delta y_0 - \frac{1}{2}\Delta^2 y_0 + \frac{1}{3}\Delta^3 y_0 + \frac{1}{4}\Delta^4 y_0 + \frac{1}{5}\Delta^5 y_0 - \frac{1}{6}\Delta^6 y_0 + \cdots\right] \quad (i)$$

and $$\left(\frac{d^2 y}{dx^2}\right)_{x0} = \frac{1}{h^2}\left[\Delta^2 y_0 - \Delta^3 y_0 + \frac{11}{12}\Delta^4 y_0 - \frac{5}{6}\Delta^5 y_0 + \frac{137}{180}\Delta^6 y_0 - \cdots\right] \quad (ii)$$

Here $h = 0.1$, $x_0 = 1.1$, $\Delta y_0 = 0.378$, $\Delta^2 y_0 = -0.03$ etc.

Substituting these values in (i) and (ii), we get

$$\left(\frac{dy}{dx}\right) = \frac{1}{0.1}\left[0.378 - \frac{1}{2}(-0.03) + \frac{1}{3}(0.004) - \frac{1}{4}(-0.001) + \frac{1}{5}(0.003)\right] = 3.952$$

$$\left(\frac{d^2 y}{dx^2}\right) = \frac{1}{(0.1)^2}\left[-0.03 - (0.004) + \frac{11}{12}(-0.001) - \frac{5}{6}(0.003)\right] = -3.74$$

(b) We use the above difference table and the backward difference operator ∇ instead of Δ.

$$\left(\frac{dy}{dx}\right)_{x_n} = \frac{1}{h}\left[\nabla y_n + \frac{1}{2}\nabla^2 y_n + \frac{1}{3}\nabla^3 y_n + \frac{1}{5}\nabla^5 y_n + \frac{1}{6}\nabla^6 y_n + \cdots\right] \quad (i)$$

and $$\left(\frac{d^2 y}{dx^2}\right)_{xn} = \frac{1}{h^2}\left[\nabla^2 y_n + \nabla^3 y_n + \frac{11}{12}\nabla^4 y_n + \frac{5}{6}\nabla^5 y_n + \frac{137}{180}\nabla^6 y_n + \cdots\right] \quad (ii)$$

Here $h = 0.1$, $x_n = 1.6$, $\nabla y_n = 0.281$, $\nabla^2 y_n = -0.018$ etc.

Putting these values in (i) and (ii), we get

$$\left(\frac{dy}{dx}\right)_{1.6} = \frac{1}{0.1}\left[0.281 + \frac{1}{2}(-0.018) + \frac{1}{3}(0.05) + \frac{1}{4}(0.002)\right.$$

$$\left. + \frac{1}{5}(0.003) + \frac{1}{6}(0.002)\right] = 2.75$$

$$\left(\frac{d^2 y}{dx^2}\right)_{1.6} = \frac{1}{(0.1)^2}\left[-0.018 + 0.005 + \frac{11}{12}(0.002) + \frac{5}{6}(0.003)\right.$$

$$\left. + \frac{137}{180}(0.002)\right] = -0.715.$$

EXAMPLE 8.2

The following data gives the velocity of a particle for twenty seconds at an interval of five seconds. Find the initial acceleration using the entire data:

Time t (sec):	0	5	10	15	20
Velocity v(m/sec): 0	3	14	69	228	

Solution:

The difference table is:

t	v	Δv	$\Delta^2 v$	$\Delta^3 v$	$\Delta^4 v$
0	0				
3					
5	3		8		
11		36			
10	14		44		24
55		60			
15	69		104		
159					
20	228				

An initial acceleration *i.e.,* $\left(\dfrac{dv}{dt}\right)$ at $t = 0$ is required, we use Newton's forward formula:

$$\left(\frac{dv}{dt}\right)_{t=0} = \frac{1}{h}\left(\Delta v_0 - \frac{1}{2}\Delta^2 v_0 + \frac{1}{3}\Delta^3 v_0 - \frac{1}{4}\Delta^4 v_0 + \cdots\right)$$

$$\therefore \qquad \left(\frac{dv}{dt}\right)_{t=0} = \frac{1}{5}\left[3 - \frac{1}{1}(8) + \frac{1}{3}(36) - \frac{1}{4}(24)\right]$$

$$= \frac{1}{5}(3 - 4 + 12 - 6) = 1$$

Hence the initial acceleration is 1 m/sec^2.

EXAMPLE 8.3

Find the value of cos (1.74) from the following table:

x:	1.7	1.74	1.78	1.82	1.86
sin x:	0.9916	0.9857	0.9781	0.9691	0.9584

Solution:

Let $y = f(x) = \sin x$. so that $f'(x) = \cos x$.

The difference table is

x	y	Δy	$\Delta^2 y$	$\Delta^3 y$	$\Delta^4 y$
1.7	0.9916				
		− 0.0059			
1.74	0.9857		− 0.0017		
		− 0.0076		0.0003	
1.78	0.9781		− 0.0014		− 0.0006
		− 0.0090		− 0.0003	
1.82	0.9691		− 0.0017		
		− 0.0107			
1.84	0.9584				

Since we require $f'(1.74)$, we use Newton's forward formula

$$\frac{dy}{dx} = \frac{1}{h}\left[\Delta y_0 = \frac{1}{2}\Delta^2 y_0 + \frac{1}{3}\Delta^3 y_0 - \frac{1}{4}\Delta^4 y_0 + \cdots \right] \qquad (i)$$

Here $h = 0.04$, $x_0 = 1.7$, $\Delta y_0 = -0.0059$, $\Delta^2 y_0 = -0.0017$ etc.

Substituting these values in (i), we get

$$\left(\frac{dy}{dx}\right)_{1.74} = \frac{1}{0.04}\left[0.0059 - \frac{1}{2}(-0.0017) + \frac{1}{3}(0.003) - \frac{1}{4}(-0.0006) \right]$$

$$= \frac{1}{0.04}(0.007) = 0.175$$

Hence cos (1.74) = 0.175

EXAMPLE 8.4

A slider in a machine moves along a fixed straight rod. Its distance x cm. along the rod is given below for various values of the time t seconds. Find the velocity of the slider and its acceleration when $t = 0.3$ second.

t =	0	0.1	0.2	0.3	0.4	0.5	0.6
x =	30.13	31.62	32.87	33.64	33.95	33.81	33.24

Solution:

The difference table is:

T	x	Δ	Δ²	Δ³	Δ⁴	Δ⁵	Δ⁶
0	30.13						
		1.49					
0.1	31.62		− 0.24				
		1.25		− 0.24			
0.2	32.87		− 0.48		0.26		
		0.77		0.02		− 0.27	
0.3	33.64		− 0.46		− 0.01		0.29
		0.31		0.01		0.02	
0.4	33.95		− 0.45		0.01		
		− 0.14		0.02			
0.5	33.81		− 0.43				
		− 0.57					
0.6	33.24						

As the derivatives are required near the middle of the table, we use Stirling's formulae:

$$\left(\frac{dx}{dt}\right) = \frac{1}{h}\left(\frac{\Delta x_0 + \Delta x_{-1}}{2}\right) - \frac{1}{6}\left(\frac{\Delta^3 x_{-1} + \Delta^3 x_{-2}}{2}\right) + \frac{1}{30}\left(\frac{\Delta^5 {}_{-2} + \Delta^5 x_{-3}}{2}\right) + \quad (i)$$

$$\left(\frac{d^2 x}{dt^2}\right) = \frac{1}{h^2}\left[\Delta^2 x_{-1} - \frac{1}{12}\Delta^4 x_{-2} + \frac{1}{90}\Delta^6 x_{-3} \cdots\right] \quad\quad\quad (ii)$$

Here $h = 0.1$, $t_0 = 0.3$, $\Delta x_0 = 0.31$, $\Delta x_{-1} = 0.77$, $\Delta^2 x_{-1} = -0.46$ etc.

Putting these values in (i) and (ii), we get

$$\left(\frac{dt}{dx}\right)_{0.3} = \frac{1}{0.1}\left[\frac{0.31 + 0.77}{2} - \frac{1}{6}\left(\frac{0.01 + 0.02}{2}\right) + \frac{1}{30}\left(\frac{0.02 - 0.27}{2}\right) - \cdots\right] = 5.33$$

$$\left(\frac{d^2 x}{dt^2}\right)_{0.3} = \frac{1}{(0.1)^2}\left[-0.46 - \frac{1}{12}(-0.01) + \frac{1}{90}(0.29) - \cdots\right] = -45.6$$

Hence the required velocity is 5.33 cm/sec and acceleration is − 45.6 cm/sec².

EXAMPLE 8.5

The elevation above a datum line of seven points of a road are given below:

x:	0	300	600	900	1200	1500	1800
y:	135	149	157	183	201	205	193

Find the gradient of the road at the middle point.

Solution:

Here $h = 300$, $x_0 = 0$, $y_0 = 135$, we require the gradient dy/dx at $x = 900$.

The difference table is

x	y	Δy	$\Delta^2 y$	$\Delta^3 y$	$\Delta^4 y$	$\Delta^5 y$
0	135					
		14				
300	149		−6			
		8		24		
600	157		18		−50	
		26		−26		70
900	183		−8		20	
		18		−6		−16
1200	201		−14		4	
		4		−2		
1500	205		−16			
		−12				
1800	193					

Using Stirling's formula for the first derivative [(9) p. 000], we get

$$y'(x_0) = \frac{1}{h}\left[\left(\frac{\Delta y_0 + \Delta y_{-1}}{26}\right) - \frac{1}{6}\left(\frac{\Delta^3 y_{-1} + \Delta^3 y_{-2}}{2}\right) + \frac{1}{30}\left(\frac{\Delta^5 y_{-2} + \Delta^5 y_{-3}}{2}\right)\right]$$

$$= \frac{1}{300}\left[\frac{1}{2}(18 + 26) - \frac{1}{12}(-6 - 26) + \frac{1}{60}(-16 + 70)\right]$$

$$= \frac{1}{300}(22 + 2.666 + 0.9) = 0.085$$

Hence the gradient of the road at the middle point is 0.085.

EXAMPLE 8.6

Using Bessel's formula, find $f'(7.5)$ from the following table:

x:	7.47	7.48	7.49	7.50	7.51	7.52	7.53
$f(x)$:	0.193	0.195	0.198	0.201	0.203	0.206	0.208

Solution:

Taking $x_0 = 7.50$, $h = 0.1$, we have $p = \dfrac{x - x_0}{h} = \dfrac{x - 7.50}{0.01}$

The difference table is

x	p	y_p	Δ	Δ^2	Δ^3	Δ^4	Δ^5	Δ^6
7.47	-3	0.193						
			0.002					
7.48	-2	0.195		0.001				
			0.003		-0.001			
7.49	-1	0.198		0.000		0.000		
			0.003		-0.001		0.003	
7.50	0	0.201		-0.001		0.003		-0.01
			0.002		0.002		-0.007	
7.51	1	0.203		0.001		-0.004		
			0.003		-0.002			
7.52	2	0.206		-0.001				
			0.002					
7.53	3	0.208						

Using Bessel's formula for the first derivative [(11) p. 000], we get

$$\left(\frac{dy}{dx}\right)_{x_0} = \frac{1}{h}\left[\Delta y_0 - \frac{1}{4}(\Delta^2 y_{-1} + \Delta^2 y_0) + \frac{1}{12}\Delta^3 y_{-1} + \frac{1}{24}(\Delta^4 y_{-2} + \Delta^4 y_{-1}) \right.$$

$$\left. - \frac{1}{120}\Delta^5 y_{-2} - \frac{1}{240}(\Delta^6 y_{-3} + \Delta^6 y_{-2}) \right]$$

$$\left(\frac{dy}{dx}\right)_{7.5} = \frac{1}{0.01}\left[0.002 - \frac{1}{4}(-0.001 + 0.001) + \frac{1}{12}(0.002) + \frac{1}{24}(-0.004 + 0.003) \right.$$

$$\left. - \frac{1}{120}(-0.007) - \frac{1}{240}(-0.010 + 0) \right]$$

$$[\because \Delta^6 y_{-2} = 0]$$

$$= 0.2 + 0 + 0.01666 - 0.0416 + 0.00583 + 0.00416 = 0.223.$$

EXAMPLE 8.7

Find $f'(10)$ from the following data:

x:	3	5	11	27	34
$f(x)$:	-13	23	899	17315	35606

Solution:

As the values of x are not equispaced, we shall use Newton's divided difference formula. The divided difference table is

x	$f(x)$	1st div. diff.	2nd div. diff.	3rd div. diff.	4th div. diff.
3	-13				
		18			
5	23		16		
		146		0.998	
11	899		39.96		0.0002
		1025		1.003	
27	17315		69.04		
		2613			
34	35606				

Fifth differences being zero, Newton's divided difference formula for the first derivative (p. 274), we get

$$f'(x) = f(x_0, x_1) + (2x - x_0 - x_1)f(x_0, x_1, x_2)$$
$$+ [3x^2 - 2x(x_0 + x_1 + x_2) + x_0x_1 + x_1x_2 + x_2x_0)] \times f(x_0, x_1, x_2, x_3)$$
$$+ [4x^3 - 3x^2(x_0 + x_1 + x_2 + x_3) + 2x(x_0x_1 + x_1x_2 + x_2x_3 + x_3x_0 + x_1x_3 + x_0x_2)$$
$$- (x_0x_1x_2 + x_1x_2x_3 + x_2x_3x_0 + x_0x_1x_3)] f(x_0, x_1, x_2, x_3, x_4)$$

Putting $x_0 = 3$, $x_1 = 5$, $x_2 = 11$, $x_3 = 27$ and $x = 10$, we obtain

$$f'(0) = 18 + 12 \times 16 + 23 \times 0.998 - 426 \times 0.0002 = 232.869.$$

8.3 Maxima and Minima of a Tabulated Function

Newton's forward interpolation formula is

$$y = y_0 + p\Delta y_0 + \frac{p(p-1)}{2!}\Delta^2 y_0 + \frac{p(p-1)(p-2)}{3!}\Delta^3 y_0 + \cdots$$

Differentiating it w.r.t. p, we get

$$\frac{dy}{dp} = \Delta y_0 + \frac{2p-1}{2}\Delta^2 y_0 + \frac{3p^2 - 6p + 2}{6}\Delta^3 y_0 + \cdots \qquad (1)$$

For maxima or minima, $dy/dp = 0$. Hence equating the right-hand side of (1) to zero and retaining only up to third differences, we obtain

$$\Delta y_0 + \frac{2p-1}{2}\Delta^2 y_0 + \frac{3p^2 - 6p + 2}{6}\Delta^3 y_0 = 0$$

i.e., $\left(\frac{1}{2}\Delta^3 y_0\right)p^2 + \left(\Delta^2 y_0 - \Delta^3 y_0\right)p + (\Delta y_0 - \frac{1}{2}\Delta^2 y_0 \frac{1}{3}\Delta^3 y_0) = 0.$

Substituting the values of Δy_0, $\Delta^2 y_0$, $\Delta^3 y_0$ from the difference table, we solve this quadratic for p. Then the corresponding values of x are given by $x = x_0 + p_h$ at which y is maximum or minimum.

EXAMPLE 8.8

From the table below, for what value of x, y is minimum? Also find this value of y.

x:	3	4	5	6	7	8
y:	0.205	0.240	0.259	0.262	0.250	0.224

Solution:

The difference table is

x	y	Δ	Δ^2	Δ^3
3	0.205			
		0.035		
4	0.240		-0.016	
		0.019		0.000
5	0.259		-0.016	
		0.003		0.001
6	0.262		-0.015	
		-0.012		0.001
7	0.250		-0.014	
		-0.026		
8	0.224			

Taking $x_0 = 3$, we have $y_0 = 0.205$, $\Delta y_0 = 0.035$, $\Delta^2 y_0 = -0.016$ and $\Delta^3 y_0 = 0$.

∴ Newton's forward difference formula gives

$$y = 0.205 + p(0.035) + \frac{p(p-1)}{2}(-0.016) \qquad (i)$$

Differentiating it w.r.t. p, we have

$$\frac{dy}{dp} = 0.035 + \frac{29-1}{2}(-0.016)$$

For y to be minimum, $dy/dp = 0$

∴ $0.035 - 0.008(2p - 1) = 0$

which gives $p = 2.6875$

∴ $x = x_0 + ph = 3 + 2.6875 \times 1 = 5.6875$.

Hence y is minimum when $x = 5.6875$.

Putting $p = 2.6875$ in (i), the minimum value of y

$$= 0.205 + 2.6875 \times 0.035 + \frac{1}{2}(2.6875 \times 1.6875)(-0.016) = 0.2628.$$

EXAMPLE 8.9

Find the maximum and minimum value of y from the following data:

x:	-2	-1	0	1	2	3	4
y:	2	-0.25	0	-0.25	2	15.75	56

Solution:

The difference table is

x	y	Δy	$\Delta^2 y$	$\Delta^3 y$	$\Delta^4 y$	$\Delta^5 y$
-2	2					
		-2.25				
-1	-0.25		2.5			
		0.25		-3		
0	0		-0.5		6	
		-0.25		3		0
1	-0.25		2.5		6	

x	y	Δy	$\Delta^2 y$	$\Delta^3 y$	$\Delta^4 y$	$\Delta^5 y$
		2.25		9		0
2	2		11.5		6	
		13.75		15		
3	15.75		26.5			
		40.25				
4	56					

Taking $x_0 = 0$, we have $y_0 = 0$, $\Delta y_0 = -0.25$, $\Delta^2 y_0 = 2.5$, $\Delta^3 y_0 = 9$, $\Delta^4 y_0 = 6$.

Newton's forward difference formula for the first derivative gives

$$\frac{dy}{dx} = \frac{1}{h}\left[\Delta y_0 - \frac{(2p-1)}{2!}\Delta^2 y_0 + \frac{3p^2 - 6p + 2}{3!}\Delta^3 y_0 - \frac{4p^3 - 18p^2 + 22p - 6}{4!}\Delta^4 y_0 - \cdots\right]$$

$$= \frac{1}{1} - 0.25 + \frac{2x-1}{2}(2.5) + \frac{1}{6}(3x^2 - 6x + 2)(9) + \frac{1}{24}(4x^3 - 18x^2 + 22x - 6)(6)$$

$$= \frac{1}{2}[-0.25 + 2.5x - 1.25 + 4.5x^2 - 9x + 3 + x^3 - 4.5x^2 + 5.5x - 1.5 = x^3 - x$$

For y to be maximum or minimum, $\dfrac{dy}{dx} = 0$ i.e., $x^3 - x = 0$

i.e., $\qquad x = 0, 1, -1$

Now $\qquad \dfrac{d^2 y}{dx^2} = 3x^2 - 1 = -\text{ve for } x = 0$

$$= +\text{ve for } x = 1$$
$$= +\text{ve for } x = -1.$$

Since $\qquad y = y_0 + x\Delta y_0 + \dfrac{x(x-1)}{2!}\Delta^2 y_0 + \cdots, y(0) = 0$

Thus y is maximum for $x = 0$, and maximum value $= y(0) = 0$.

Also y is minimum for $x = 1$ and minimum value $= y(0) = -0.25$

Exercises 8.1

1. Find $y'(0)$ and $y(0)$ from the following table:

x: 0		1	2	3	4	5
y: 4		8	15	7	6	2

2. Find the first, second and third derivatives of $f(x)$ at $x = 1.5$ if

x:	1.5	2.0	2.5	3.0	3.5	4.0
f(x):	3.375	7.000	13.625	24.000	38.875	59.000

3. Find the first and second derivatives of the function tabulated below, at the point $x = 1.1$:

x:	1.0	1.2	1.4	1.6	1.8	2.0
f(x):	0	0.128	0.544	1.296	2.432	4.00

4. Given the following table of values of x and y

x:	1.00	1.05	1.10	1.15	1.20	1.25	1.30
y:	1.000	1.025	1.049	1.072	1.095	1.118	1.140

find $\dfrac{dy}{dx}$ and $\dfrac{d^2y}{dx^2}$ at   (a) $x = 1.05$.   (b) $x = 1.25$ (c) $x = 1.15$.

5. For the following values of x and y, find the first derivative at $x = 4$.

x:	1	2	4	8	10
y:	0	1	5	21	27

6. Find the derivative of $f(x)$ at $x = 0.4$ from the following table:

x:	0.1	0.2	0.3	0.4
f(x):	1.10517	1.22140	1.34986	1.49182

7. From the following table, find the values of dy/dx and d^2y/dx^2 at $x = 2.03$.

x:	1.96	1.98	2.00	2.02	2.04
y:	0.7825	0.7739	0.7651 0.	7563	0.7473

8. Given sin $0° = 0.000$, sin $10° = 0.1736$, sin $20° = 0.3420$, sin $30° = 0.5000$, sin $40° = 0.6428$,

    (a) find the value of sin $23°$,

    (b) find the numerical value of cos x at $x = 10°$

    (c) find the numerical value of d^2y/dx^2 at $x = 20°$ for $y = \sin x$.

9. The population of a certain town is given below. Find the rate of growth of the population in 1961 from the following table

Year:	1931	1941	1951	1961	1971
Population: (in thousands)	40.62	60.80	71.95	103.56	132.68

Estimate the population in the years 1976 and 2003. Also find the rate of growth of population in 1991.

10. The following data gives corresponding values of pressure and specific volume of a superheated steam.

v:	2	4	6	8	10
p:	105	42.7	25.3	16.	7 13

Find the rate of change of
(i) pressure with respect to volume when $v = 2$,

(ii) volume with respect to pressure when $p = 105$.

11. The table below reveals the velocity v of a body during the specified time t find its acceleration at $t = 1.1$?

t:	1.0	1.1	1.2	1.3	1.4
v:	43.1	47.7	52.1	56.4	60.8

12. The following table gives the velocity v of a particle at time t. Find its acceleration at $t = 2$.

t:	0	2	4	6	8	10	12
v:	4	6	16	34	60	94	131

13. A rod is rotating in a plane. The following table gives the angle θ (radians) through which the rod has turned for various values of the time t second.

t:	0	0.2	0.4	0.6	0.8	1.0	1.2
θ:	0	0.12	0.49	1.12	2.02	3.20	4.67

Calculate the angular velocity and the angular acceleration of the rod, when $t = 0.6$ second.

14. Find dy/dx at $x = 1$ from the following table by constructing a central difference table:

x:	0.7	0.8	0.9	1.0	1.1	1.2	1.3
y:	0.644218	0.717356 0	0.783327	0.841471	0.891207	0.932039	0.963558

15. Find the value of $f'(x)$ at $x = 0.04$ from the following table using Bessel's formula.

x:	0.01	0.02	0.03	0.04	0.05	0.06
f(x):	0.1023	0.1047	0.1071	0.1096	0.1122	0.1148

16. If $y = f(x)$ and y_n denotes $f(x_0 + nh)$, prove that, if powers of h above h^6 are neglected.

$$\left(\frac{dy}{dx}\right)_{x_0} = \frac{3}{4h}\left[(y_1 - y_{-1}) - \frac{1}{5}(y_2 - y_{-2}) + \frac{1}{45}(y_3 - y_{-3})\right].$$

[**HINT:** Differentiate Stiling's formula w.r.t. x, and put $x = 0$]

17. Find the value of $f'(8)$ from the table given below:

x:	6	7	9	12
$f(x)$:	1.556	1.690	1.908	2.158

18. Given the following pairs of values of x and y:

x:	1	2	4	8	10
y:	0	1	5	21	27

Determine numerically dy/dx at $x = 4$.

19. Find $f'(6)$ from the following data:

x:	0	2	3	4	7	8
$f(x)$:	4	26	58	112	466	922

20. Find the maximum and minimum value of y from the following table:

x:	0	1	2	3	4	5
y:	0	0.25	0	2.25	16	56.25

21. Using the following data, find x for which y is minimum and find this value of y.

x:	0.60	0.65	0.70	0.75
y:	0.6221	0.6155	0.6138	0.6170

22. Find the value of x for which $f(x)$ is maximum, using the table

x:	9	10	11	12	13	14
$f(x)$:	1330	1340	1320	1250	1120	930

Also find the maximum value of $f(x)$.

8.4 Numerical Integration

The process of evaluating a definite integral from a set of tabulated values of the integrand $f(x)$ is called *numerical integration*. This process when applied to a function of a single variable, is known as *quadrature*.

The problem of numerical integration, like that of numerical differentiation, is solved by representing $f(x)$ by an interpolation formula and then integrating it between the given limits. In this way, we can derive quadrature formulae for approximate integration of a function defined by a set of numerical values only.

8.5   Newton-Cotes Quadrature Formula

Let $I = \int_a^b f(x)dx$

where $f(x)$ takes the values $y_0, y_1, y_2, \cdots y_n$ for $x = x_0, x_1, x_2, \cdots x_n$.

Let us divide the interval (a, b) into n sub-intervals of width h so that $x_0 = a, x_1 = x_0 + h, x_2 = x_0 + 2h, \cdots x_n = x_0 + nh = b$. Then

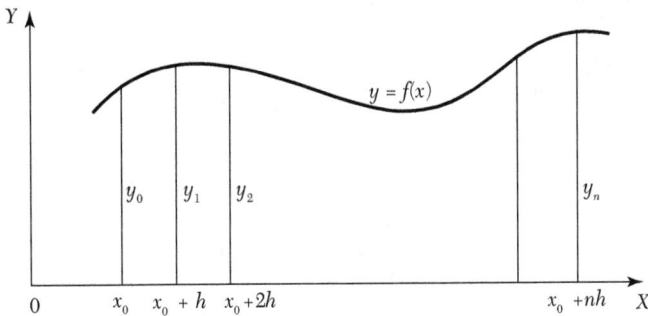

FIGURE 8.1

$$I = \int_{x_0}^{x_0+nh} f(x)dx = h\int_0^n f(x_0 + rh)dr, \text{ Putting } x = x_0 + rh, dx = hdr$$

$$= h\int_0^n [y_0 + r\Delta y_0 + \frac{r(r-1)}{2!}\Delta^2 y_0 + \frac{r(r-1)(r-2)}{3!}\Delta^3 y_0$$

$$+ \frac{r(r-1)(r-2)(r-3)}{4!}\Delta^4 y_0 + \frac{r(r-1)(r-20(r-3)(r-4)}{5!}\Delta^5 y_0$$

$$+ \frac{r(r-1)(r-2)(r-3)(r-4)(r-5)}{6!}\Delta^6 y_0 + ...]dr$$

[by Newton's forward interpolation formula]

Integrating term by term, we obtain

$$\int_{x_0}^{x_0+nh} f(x)dx = nh\left[y_0 + \frac{n}{2}\Delta y_0 + \frac{n(2n-3)}{12}\Delta^2 y_0 + \frac{n(n-2)^2}{24}\Delta^3 y_0\right.$$

$$+\left(\frac{n^4}{5}-\frac{3n^3}{2}+\frac{11n^2}{3}-3n\right)\frac{\Delta^4 y_0}{4!}$$

$$\left(+\frac{n^5}{6}-2n^4+\frac{34n^3}{4}-\frac{50n^2}{3}+12n\right)\frac{\Delta^5 y_0}{5!} \tag{1}$$

$$+\left(\frac{n^6}{7}-\frac{15n^5}{6}+17n^4-\frac{225n^3}{4}+\frac{274n^2}{3}-60n\right)\frac{\Delta^6 y_0}{6!}+\cdots\Bigg]$$

This is known as *Newton-Cotes quadrature formula*. From this general formula, we deduce the following important quadrature rules by taking $n = 1, 2, 3, \ldots$

I. Trapezoidal rule. Putting $n = 1$ in (1) and taking the curve through (x_0, y_0) and (x_1, y_1) as a straight line (Figure 8.2) *i.e.*, a polynomial of first order so that differences of order higher than first become zero, we get

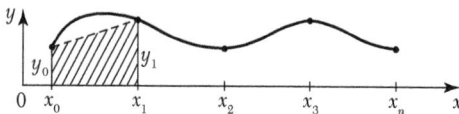

FIGURE 8.2

$$\int_{x_0}^{x_0+h} f(x)dx = h\left(y_0 + \frac{1}{2}\Delta y_0\right) = \frac{h}{2}(y_0 + y_1)$$

Similarly
$$\int_{x_0+h}^{x_0+2h} f(x)dx = h\left(y_1 + \frac{1}{2}\Delta y_1\right) = \frac{h}{2}(y_1 + y_2)$$

$$\cdots\cdots\cdots\cdots\cdots\cdots\cdots\cdots\cdots\cdots\cdots\cdots\cdots\cdots\cdots$$

$$\int_{x_0+(n-1)}^{x_0+nh} f(x)dx = \frac{h}{2}(y_{n-1} + yn)$$

Adding these n integrals, we obtain

$$\int_{x_0}^{x_0+nh} f(x)dx = \frac{h}{2}[(y_0 + y_n) + 2(y_1 + y_2 + \cdots + y_{n-1})] \tag{2}$$

This is known as the *trapezoidal rule*.

NOTE

Obs. *The area of each strip (trapezium) is found separately. Then the area under the curve and the ordinates at x_0 and x_n is approximately equal to the sum of the areas of the n trapeziums.*

II. Simpson's one-third rule. Putting $n = 2$ in (1) above and taking the curve through (x_0, y_0), (x_1, y_1), and (x_2, y_2) as a parabola (Figure 8.3), *i.e.*, a polynomial of the second order so that differences of order higher than the second vanish, we get

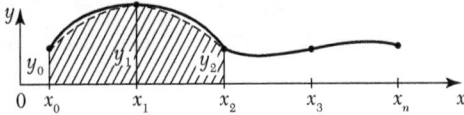

FIGURE 8.3

$$\int_{x_0}^{x_0+2h} f(x)dx = 2h\left(y_0 + \Delta y_0 + \frac{1}{6}\Delta^2 y_0\right) = \frac{h}{3}(y_0 + 4y_1 + y_2)$$

Similarly $\displaystyle\int_{x_0+2h}^{x_0+4h} f(x)dx = \frac{h}{3}(y_2 + 4y_3 + y_4)$

$$\cdots\cdots\cdots\cdots\cdots\cdots\cdots\cdots\cdots\cdots\cdots\cdots\cdots\cdots\cdots\cdots$$

$$\int_{x_0+(n-2)h}^{x_0+nh} f(x)dx = \frac{h}{3}(y_{n-2} + 4y_{n-1} + y_n), \ n \text{ being even.}$$

Adding all these integrals, we have when n is even

$$\int_{x_0}^{x_0+nh} f(x)dx = \frac{h}{3}[(y_0+y_n)+4(y_1+y_3+\ldots+y_{n-1})+2(y_2+y_4+\ldots y_{n-2})] \qquad (3)$$

This is known as the *Simpson's one-third rule* or simply *Simpson's rule* and is most commonly used.

NOTE **Obs.** *While applying (3), the given interval must be divided into an even number of equal subintervals, since we find the area of two strips at a time.*

III. Simpson's three-eighth rule. Putting $n = 3$ in (1) above and taking the curve through (x_i, y_i): $i = 0, 1, 2, 3$ as a polynomial of the third order (Figure 8.4) so that differences above the third order vanish, we get

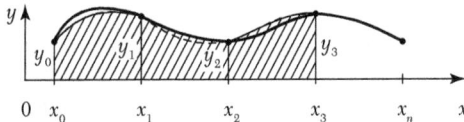

FIGURE 8.4

$$\int_{x_0}^{x_0+3h} f(x)dx = 3h\left(y_0 + \frac{3}{2}\Delta y_0 + \frac{3}{4}\Delta^2 y_0 + \frac{1}{8}\Delta^3 y_0\right)$$

$$= \frac{3h}{8}(y_0 + 3y_1 + 3y_2 + y_3)$$

Similarly,

$$\int_{x_0+3h}^{x_0+5h} f(x)dx = \frac{3h}{8}(y_3 + 3y_4 + 3y_5 + y_6) \text{ and so on.}$$

Adding all such expressions from x_0 to $x_0 + nh$, where n is a multiple of 3, we obtain

$$\int_{x_0}^{x_0+nh} f(x)dx = \frac{3h}{8}[(y_0 + y_n) + 3(y_1 + y_2 + y_4 + y_5 + \cdots + y_{n-1}) \qquad (4)$$
$$+ 2(y_3 + y_6 + \cdots + y_{n-3})]$$

NOTE **Obs.** *While applying (4), the number of sub-intervals should be taken as a multiple of 3.*

IV. Boole's rule. Putting $n = 4$ in (1) above and taking the curve (x_i, y_i), $i = 0, 1, 2, 3, 4$ as a polynomial of the fourth order (Figure 8.5) and neglecting all differences above the fourth, we obtain

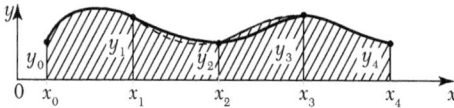

FIGURE 8.5

$$\int_{x_0}^{x_0+4h} f(x)dx = 4h\left(y_0 + 2\Delta y_0 + \frac{5}{3}\Delta^2 y_0 = \frac{2}{3}\Delta^3 y_0 + \frac{7}{90}\Delta^4 y_0\right)$$

$$= \frac{2h}{45}(7y_0 = 32y_1 + 12y_2 + 32y_{33} + 7y_4)$$

Similarly $\int_{x_0+4h}^{x_0+8h} f(x)dx = \frac{2h}{45}(7y_4 + 32y_5 + 12y_6 + 32y_7 + 7y_8)$ and so on.

Adding all these integrals from x_0 to $x_0 + nh$, where n is a multiple of 4, we get

$$\int_{x_0}^{x_0+nh} f(x)dx = \frac{2h}{45}(7y_0 + 32y_1 + 12y_2 + 32y_3 + 14y_4 + 32y_5 \qquad (5)$$
$$+ 12y_6 + 32y_7 + 14y_8 + \cdots$$

This is known as *Boole's rule.*

NOTE **Obs.** *While applying (5), the number of sub-intervals should be taken as a multiple of 4.*

V. Weddle's rule. Putting $n = 6$ in (1) above and neglecting all differences above the sixth, we obtain

$$\int_{x_0}^{x_0+6h} f(x)dx = 6h\left(y_0 + 3\Delta y_0 + \frac{9}{2}\Delta^2 y_0 + 4\Delta^3 y_0 + \frac{123}{60}\Delta^4 y_0\right.$$
$$\left. + \frac{11}{20}\Delta^5 x_0 + \frac{1}{6}\cdot\frac{41}{140}\Delta^6 y_0\right)$$

If we replace $\dfrac{41}{140}\Delta^6 y_0$ by $\dfrac{3}{10}\Delta^6 y_0$, the error made will be negligible.

$$\therefore \qquad \int_{x_0}^{x_0+6h} f(x)dx = \frac{3h}{10}(y_0 + 5y_1 + y_2 + 6y_3 + y_4 + 5y_5 + y_6)$$

Similarly

$$\int_{x_0+6h}^{x_0+12h} f(x)dx = \frac{3h}{10}(y_6 + 5y_7 + y_8 + 6y_9 + y_{10} + 5y_{11} + y_{12}) \text{ and so on.}$$

Adding all these integrals from x_0 to $x_0 + nh$, where n is a multiple of 6, we get

$$\int_{x_0}^{x_0+nh} f(x)dx = \frac{3h}{10}(y_0 + 5y_1 + y_2 + 6y_3 + y_4 + 5y_5 + 2y_6 + 5y_7 + y_8 + \cdots) \quad (6)$$

This is known as *Weddle's rule*.

NOTE **Obs.** *While applying (6), the number of sub-intervals should be taken as a multiple of 6. Weddle's rule is generally more accurate than any of the others. Of the two Simpson rules, the 1/3 rule is better.*

EXAMPLE 8.10

Evaluate $\displaystyle\int_0^6 \frac{dx}{1+x^2}$ by using

(*i*) Trapezoidal rule,

(*ii*) Simpson's 1/3 rule,

(*iii*) Simpson's 3/8 rule,

(*iv*) Weddle's rule and compare the results with its actual value.

Solution:

Divide the interval $(0, 6)$ into six parts each of width $h = 1$. The values of $f(x) = \dfrac{1}{1+x^2}$ are given below:

x	0	1	2	3	4	5	6
$f(x)$	1	0.5	0.2	0.1	0.0588	0.0385	0.027
$=y$	y_0	y_1	y_2	y_3	y_4	y_5	y_6

(*i*) By Trapezoidal rule,

$$\int_0^6 \frac{dx}{1+x^2} = \frac{h}{2}[(y_0 + y_6) + 2(y_1 + y_2 + y_3 + y_4 + y_5)]$$

$$= \frac{1}{2}[(1+0.027) + 2(0.5 + 0.2 + 0.1 + 0.0588 + 0.0385)] = 1.4108.$$

(*ii*) By Simpson's 1/3 rule,

$$\int_0^6 \frac{dx}{1+x^2} = \frac{h}{3}[(y_0 + y_6) + 4(y_1 + y_3 + y_5) + 2(y_2 + y_4)$$

$$= \frac{1}{3}[(1+0.027) + 4(0.5 + 0.1 + 0.0385) + 2(0.2 + 0.0588)] = 1.3662.$$

(*iii*) By Simpson's 3/8 rule,

$$\int_0^6 \frac{dx}{1+x^2} = \frac{3h}{8}[(y_0 + y_6) + 3(y_1 + y_2 + y_4 + y_5) + 2y_3]$$

$$= \frac{3}{8}[(1+0.027) + 3(0.5 + 0.2 + 0.0588 + 0.0385) + 2(0.1)] = 1.3571$$

(*iv*) By Weddle's rule,

$$\int_0^6 \frac{dx}{1+x^2} = \frac{3h}{10}[y_0 + 5y_1 + y_2 + 6y_3 + y_4 + 5y_5 + y_6]$$

$$= 0.3[1 + 5(0.5) + 0.2 + 6(0.1) + 0.0588 + 5(0.0385) + 0.027] = 1.3735$$

Also $\int_0^6 \frac{dx}{1+x^2} = \left|\tan^{-1} x\right|_0^6 = \tan^{-1} 6 = 1.4056$

This shows that the value of the integral found by Weddle's rule is the nearest to the actual value followed by its value given by Simpson's 1/3 rule.

EXAMPLE 8.11

Evaluate the integral $\int_0^1 \frac{x^2}{1+x^3}dx$ using Simpson's 1/3 rule. Compare the error with the exact value.

Solution:

Let us divide the interval (0, 1) into 4 equal parts so that $h = 0.25$. Taking $y = \frac{x^2}{(1+x^3)}$, we have

x:	0	0.25	0.50	0.75	1.00
y:	0	0.06153	0.22222	0.39560	0.5
	y_0	y_1	y_2	y_3	y_4

By Simpson's 1/3 rule, we have

$$\int_0^1 \frac{x^2}{1+x^3}dx = \frac{h}{3}[(y_0+y_4)+2(y_2)+4(y_1+y_3)]$$

$$= \frac{0.25}{3}[(0+0.5)+2(0.22222)+4(0.06153+0.3956)]$$

$$= \frac{0.25}{3}[0.5+0.44444+1.82852]=0.23108$$

Also $\int_0^1 \frac{x^2}{1+x^3}dx = \frac{1}{3}\left|\log(1+x^3)\right|_0^1 = \frac{1}{3}\log e2 = 0.23108$

Thus the error $= 0.23108 - 0.23105 = -0.00003$.

EXAMPLE 8.12

Use the Trapezoidal rule to estimate the integral $\int_0^2 ex^2 dx$ taking the number 10 intervals.

Solution:

Let $y = ex^2$, $h = 0.2$ and $n = 10$.

The values of x and y are as follows:

x:	0	0.2	0.4	0.6	0.8	1.0	1.2	1.4	1.6	1.8	2.0
y:	1	1.0408	1.1735	1.4333	1.8964	2.1782	4.2206	7.0993	12.9358	25.5337	54.5981
	y_0	y_1	y_2	y_3	y_4	y_5	y_6	y_7	y_8	y_9	y_{10}

By the Trapezoidal rule, we have

$$\int_0^1 e^{x2}dx = \frac{h}{2}[(y_0+y_{10})+2(y_1+y_2+y_3+y_4+y_5+y_6+y_7+y_8+y_9)]$$

$$= \frac{0.2}{2}[(1+54.5981)+2(1.0408+1.1735+1.8964+2.1782$$

$$+4.2206+7.0993+12.9358+25.5337)]$$

Hence $\int_0^2 e^{x^2}dx = 17.0621$.

EXAMPLE 8.13

Use Simpson's 1/3rd rule to find $\int_0^{0.6} e^{-x^2} dx$ by taking seven ordinates.

Solution:

Divide the interval $(0, 0.6)$ into six parts each of width $h = 0.1$. The values of $y = f(x) = e^{-x^2}$ are given below:

x	0	0.1	0.2	0.3	0.4	0.5	0.6
x^2	0	0.01	0.04	0.09	0.16	0.25	0.36
y	1	0.9900	0.9608	0.9139	0.8521	0.7788	0.6977
	y_0	y_1	y_2	y_3	y_4	y_5	y_6

By Simpson's 1/3rd rule, we have

$$e^{-x^2} dx = \frac{h}{3}[(y_0 + y_6) + 4(y_1 + y_3 + y_5) + 2(y_2 + y_4)]$$

$$\int_0^{0.6} = \frac{0.1}{3}[(1 + 0.6977) + 4(0.99 + 0.9139 + 0.7788) + 2(0.9608 + 0.8521)]$$

$$= \frac{0.1}{3}[1.6977 + 10.7308 + 3.6258] = \frac{0.1}{3}(16.0543) = 0.5351.$$

EXAMPLE 8.14

Compute the value of $\int_{0.2}^{1.4} (\sin x - \log x + e^x) dx$ using Simpson's 3/8 rule.

Solution:

Let $y = \sin x - \log_e x + e^x$ and $h = 0.2$, $n = 6$.

The values of y are as given below:

x:	0.2	0.4	0.6	0.8	1.0	1.2	1.4
y:	3.0295	2.7975	2.8976	3.1660	3.5597	4.0698	4.4042
	y_0	y_1	y_2	y_3	y_4	y_5	y_6

By Simpson's 3/8 rule, we have

$$\int_{0.2}^{1.4} y dx = \frac{3h}{8}[(y_0 + y_6) + 2(y_3) + 3(y_1 + y_2 + y_4 + y_5)]$$

$$= \frac{3}{8}(0.2)[7.7336 + 2(3.1660) + 3(13.3247)] = 4.053$$

Hence $\int_{0.2}^{1.4} (\sin x - \log e^x + e^x)dx = 4.053$.

NOTE **Obs. Applications of Simpson's rule.** *If the various ordinates in Section 8.5 represent equispaced cross-sectional areas, then Simpson's rule gives the volume of the solid. As such, Simpson's rule is very useful to civil engineers for calculating the amount of earth that must be moved to fill a depression or make a dam. Similarly if the ordinates denote velocities at equal intervals of time, the Simpson's rule gives the distance travelled. The following Examples illustrate these applications.*

EXAMPLE 8.15

The velocity v(km/min) of a moped which starts from rest, is given at fixed intervals of time t (min) as follows:

t:	2	4	6	8	10	12	14	16	18	20
v:	10	18	25	29	32	20	11	5	2	0

Estimate approximately the distance covered in twenty minutes.

Solution:

If s (km) be the distance covered in t (min), then $\dfrac{ds}{dt} = v$

$\therefore$ $|s|_{t=0}^{20} = \int_0^{20} v\,dt = \dfrac{h}{3}[X + 4.0 + 2.E]$, by Simpson's rule

Here $h = 2$, $v_0 = 0$, $v_1 = 10$, $v_2 = 18$, $v_3 = 25$ etc.

$\therefore$ $X = v_0 + v_{10} = 0 + 0 = 0$

$O = v_1 + v_3 + v_5 + v_7 + v_9 = 10 + 25 + 32 + 11 + 2 = 80$

$E = v_2 + v_4 + v_6 + v_8 = 18 + 29 + 20 + 5 = 72$

Hence the required distance $= |s|_{t=0}^{20} = \dfrac{2}{3}(0 + 4 \times 80 + 2 \times 72) = 309.33\,\text{km}$.

EXAMPLE 8.16

The velocity v of a particle at distance s from a point on its linear path is given by the following table:

s (m):	0	2.5	5.0	7.5	10.0	12.5	15.0	17.5	20.0
v (m/sec):	16	19	21	22	20	17	13	11	9

Estimate the time taken by the particle to traverse the distance of 20 meter, using Boole's rule.

Solution:

If t sec be the time taken to traverse a distance s (m) then $\dfrac{ds}{dt} = v$

or
$$\frac{dr}{ds} = \frac{1}{v} = y\,(\text{say}),$$

then
$$|t|_{s=0}^{s=20} = \int_0^{20} y\,ds$$

Here
$$h = 2.5 \text{ and } n = 8.$$

Also $y_0 = \dfrac{1}{16}, y_1 = \dfrac{1}{19}, y_2 = \dfrac{1}{4}, y_3 = \dfrac{1}{22}, y_4 = \dfrac{1}{20},$

$$y_5 = \frac{1}{17}, y_6 = \frac{1}{13}, y_7 = \frac{1}{11}, y_8 = \frac{1}{9}.$$

∴ By Boole's Rules, we have

$$|t|_{s=0}^{s=20} = \int_0^{20} y\,ds = \frac{2h}{45}[7y_0 + 32y_1 + 312y_2 + 14y_3 + 32y_5 + 12y_6 + 32y_7 + 14y_8]$$

$$= \frac{2(2.5)}{45}\left[7\left(\frac{1}{16}\right) + 32\left(\frac{1}{19}\right) + 12\left(\frac{1}{21}\right) + 32\left(\frac{1}{22}\right) + 14\left(\frac{1}{20}\right) + 32\left(\frac{1}{17}\right)\right.$$
$$\left. + 12\left(\frac{1}{3}\right) + 32\left(\frac{1}{11}\right) + 14\left(\frac{1}{9}\right)\right]$$

$$= \frac{1}{9}(12.11776) = 1.35$$

Hence the required time = 1.35 sec.

EXAMPLE 8.17

A solid of revolution is formed by rotating about the x-axis, the area between the x-axis, the lines $x = 0$ and $x = 1$ and a curve through the points with the following co-ordinates:

x:	0.00	0.25	0.50	0.75	1.00
y:	1.0000	0.9896	0.9589	0.9089	0.8415

Estimate the volume of the solid formed using Simpson's rule.

Solution:

Here $h = 0.25$, $y_0 = 1$, $y_1 = 0.9896$, $y_2 = 0.9589$ etc.

$\therefore$ Required volume of the solid generated

$$= \int_0^1 \pi y^2 dx = \pi \cdot \frac{h}{3} [(y_0^2 + y_4^2) + 4(y_1^2 + y_3^2) + 2y_2^2]$$

$$= \frac{0.25\pi}{3} [\{1 + 0.8415)^2\} + 4\{(0.9896)^2 + (0.9089)^2\} + 2(0.9589)^2]$$

$$= \frac{0.25 \times 3.1416}{3} [1.7081 + 7.2216 + 1.839]$$

$$= 0.2618(10.7687) = 2.8192.$$

Exercises 8.2

1. Use trapezoidal rule to evaluate $\int_0^1 x^3 dx$ considering five sub-intervals.

2. Evaluate $\int_0^1 \frac{dx}{1+x}$ applying

   (*i*) Trapezoidal rule
   (*ii*) Simpson's 1/3 rule
   (*iii*) Simpson's 3/8 rule.

3. Evaluate $\int_0^1 \frac{dx}{1+x^2}$ using

   (*i*) Trapezoidal rule taking $h = 1/4$.
   (*ii*) Simpson's 1/3rd rule taking $h = 1/4$.
   (*iii*) Simpson's 3/8th rule taking $h = 1/6$.
   (*iv*) Weddle's rule taking $h = 1/6$.
   Hence compute an approximate value of ϖ in each case.

4. Find an approximate value of log$e$ 5 by calculating to four decimal places, by Simpson's 1/3 rule, $\int_0^5 \frac{dx}{4x+5}$ dividing the range into ten equal parts.

5. Evaluate $\int_0^4 e^x dx$ by Simpson's rule, given that

   $e = 2.72$, $e^2 = 7.39$, $e^3 = 20.09$, $e^4 = 54.6$

   and compare it with the actual value.

6. Find $\int_0^6 \frac{e^x}{1+x} dx$ using Simpson's 1/3 rule.

7. Evaluate $\int_0^2 e^{-x^2} dx$, using Simpson's rule. (Take $h = 0.25$)

8. Evaluate using Simpson's 1/3 rule,

   (i) $\int_0^\pi \sin x \, dx$, taking eleven ordinates.

   (ii) $\int_0^{\pi/2} \sqrt{\cos\theta} \, d\theta$, taking nine ordinates.

9. Evaluate by Simpson's 3/8 rule:

   (i) $\int_0^9 \dfrac{dx}{1+x^3}$

   (ii) $\int_0^{\pi/2} \sin x \, dx$

   (iii) $\int_0^{\pi/2} e^{\sin x} dx$

   (iv) $\int_0^\pi \sqrt{(1 + 3\cos^2\theta)} \, d\theta$, using six ordinates

10. Given that

x:	4.0	4.2	4.4	4.6	4.8	5.0	5.2
$\log x$:	1.3863	1.4351	1.4816	1.5261	1.5686	1.6094	1.6487

   evaluate $\int_4^{5.2} \log x \, dx$ by

   (a) Trapezoidal rule

   (b) Simpson's 1/3 rule,

   (c) Simpson's 3/8 rule,

   (d) Weddle's rule.

   Also find the error in each case.

11. Use Boole's five point formula to compute $\int_0^{\pi/2} \sqrt{(\sin x)} \, / \, dx$.

12. The table below shows the temperature $f(t)$ as a function of time:

t:	1	2	3	4	5	6	7
$f(t)$:	81	75	80	83	78	70	60

   Using Simpson's $\dfrac{1}{3}$ rule to estimate $\int_1^7 f(t) dt$.

13. A curve is drawn to pass through the points given by the following table:

x:	1	1.5	2	2.5	3	3.5	4
y:	2	2.4	2.7	2.8	3	2.6	2.1

Estimate the area bounded by the curve, x-axis and the lines $x = 1$, $x = 4$.

14. A river is 80 feet wide. The depth d in feet at a distance x feet. from one bank is given by the following table:

x:	0	10	20	30	40	50	60	70	80
y:	0	4	7	9	12	15	14	8	3

15. Find approximately the area of the cross-section.

A curve is drawn to pass through the points given by the following table:

x:	1	1.5	2	2.5	3	3.5	4
y:	2	2.4	2.7	2.8	3	2.6	2.1

Using Weddle's rule, estimate the area bounded by the curve, the x-axis, and the lines $x = 1$, $x = 4$.

16. A curve is given by the table:

x:	0	1	2	3	4	5	6
y:	0	2	2.5	2.3	2 1.	7 1.	5

The x-coordinate of the C.G. of the area bounded by the curve, the end ordinates, and the x-axis is given by $A\bar{x} = \int_0^6 xydx$, where A is the area. Find $\bar{x}$ by using Simpson's rule.

17. A body is in the form of a solid of revolution. The diameter D in cms of its sections at distances x cm. from one end are given below. Estimate the volume of the solid.

x:	0	2.5	5.0	7.5	10.0	12.5	15.0
D:	5	5.5	6.0	6.75	6.25	5.5	4.0

18. The velocity v of a particle at distance s from a point on its path is given by the table:

s ft:	0	10	20	30	40	50	60
v ft/sec:	47	58	64	65	61	52	38

Estimate the time taken to travel sixty feet by using Simpson's 1/3 rule. Compare the result with Simpson's 3/8 rule.

19. The following table gives the velocity v of a particle at time t:

t (seconds):	0	2	4	6	8	10	12
v (m/sec.):	4	6	16	34	60	94	136

Find the distance moved by the particle in twelve seconds and also the acceleration at $t = 2$ sec.

20. A rocket is launched from the ground. Its acceleration is registered during the first eighty seconds and is given in the table below. Using Simpson's 1/3 rule, find the velocity of the rocket at $t = 80$ seconds.

t (sec):	0	10	20	30	40	50	60	70	80
f (cm/sec2):	30	31.63	33.34	35.47	37.75	40.33	43.25	46.69	50.67

21. A reservoir discharging water through sluices at a depth h below the water surface has a surface area A for various values of h as given below:

h (ft.):	10	11	12	13	14
A (sq. ft.):	950	1070	1200	1350	1530

If t denotes time in minutes, the rate of fall of the surface is given by $dh/dt = -48\sqrt{h}/A$.

Estimate the time taken for the water level to fall from fourteen to ten feet above the sluices.

8.6 Errors in Quadrature Formulae

The error in the quadrature formulae is given by

$$E = \int_a^b y\,dx - \int_a^b P(x)\,dx$$

where $P(x)$ is the polynomial representing the function $y = f(x)$, in the interval $[a, b]$.

Error in Trapezoidal rule. Expanding $y = f(x)$ around $x = x_0$ by Taylor's series, we get

$$y = y_0 + (x - x_0)y_0' + \frac{(x - x_0)^2}{2!}y_0'' + \dots \tag{1}$$

$$\therefore \int_{x_0}^{x_0+h} y\,dx = \int_{x_0}^{x_0+h}\left[y_0 + (x - x_0)y_0' + \frac{(x - x_0)^2}{2!}y_0'' + \dots\right]dx \tag{2}$$

$$= y_0 h + \frac{h^2}{2!}y_0' + \frac{h^3}{3!}y_0'' + \dots$$

Also A_1 = area of the first trapezium in the interval $[x_0, x_1] = \dfrac{1}{2}h(y_0 + y_1)$ (3)

Putting $x = x_0 + h$ and $y = y_1$ in (1), we get $y_1 = y_0 + hy_0' + \dfrac{h^2}{2!}y_0'' + \dots$

Substituting this value of y_1 in (3), we get

$$A_1 = \frac{1}{2}h\left[y_0 + y_0 + hy_0' + \frac{h^2}{2!}y_0'' + \dots\right] \tag{4}$$

$$= hy_0 + \frac{h^2}{2}y_0' + \frac{h^3}{2 \cdot 2!}y_0'' + \dots$$

$\therefore$ Error in the interval $[x_0, x1] = \displaystyle\int_{x_0}^{x_1} y\,dx - A_1$ $\quad$ [(2) – (4)]

$$= \frac{1}{3!} - \frac{1}{2.2!}h^3 y_0'' + \dots = -\frac{h^3}{12}y_0'' + \dots$$

i.e., Principal part of the error in $[x_0, x_1] = -\dfrac{h^3}{12}y_0''$

Hence the total error $E = -\dfrac{h^3}{12}[y_0'' + y_1'' + \dots + y_{n-1}'']$

Assuming that $y''(X)$ is the largest of the n quantities $y_0'', y_1'', \dots y_{n-1}''$ we obtain

$$E < -\frac{nh^3}{12}y''(X) = -\frac{(b-a)h^2}{12}y''(X) \qquad [\because nh = b - a \dots(5)$$

Hence the error in the trapezoidal rule is of the order h^2.

Error in Simpson's 1/3 rule. Expanding $y = f(x)$ around $x = x_0$ by Taylor's series, we get (1).

$\therefore$ Over the first doubt strip, we get

$$\int_{x_0}^{x_2} y\,dx = \int_{x_0}^{x_0+2h}\left[y_0 + (x - x_0)y_0' + \frac{(x - x_0)^2}{2!}y_0'' + \dots\right]dx \tag{6}$$

$$= 2hy_0 + \frac{4h^2}{2!}y_0' + \frac{8h^3}{3!}y_0'' + \frac{16h^4}{4!}y_0''' + \frac{32h^5}{5!}y_0^{iv} + \dots$$

Also A_1 = area over the first doubt strip by Simpson's 1/3 rule

$$= \frac{1}{3}h(y_0 + 4y_1 + y_2) \tag{7}$$

Putting $x = x_0 + h$ and $y = y_1$ in (1), we get

$$y_1 = y_0 + hy_0' + \frac{h^2}{2!}y'' + \frac{h^3}{3!}y_0''' + \cdots$$

Again putting $x = x_0 + 2h$ and $y = y_2$ in (1), we have

$$y_2 = y_0 + 2hy_0' + \frac{4h^2}{2!}y_0'' + \frac{8h^3}{3!}y_0''' + \cdots$$

Substituting these values of y_1 and y_2 in (7), we get

$$A_1 = \frac{h}{3}\left[y_0 + 4\left(y_0 + hy_0' + \frac{h^2}{2!}y_0''' + \cdots \right) + y_0 \right.$$

$$\left. + \left(2hy_0' + \frac{4h^2}{2!}y_0'' + \frac{8h^3}{3!}y_0''' + \cdots \right) \right]$$

$$= 2hy_0 + 2h^2 y_0' + \frac{4h^3}{3}y_0'' + \frac{2h^2}{3}y_0''' + \frac{5h^5}{18}y_0^{iv} + \cdots \qquad (8)$$

∴ Error in the interval $[x_0, x_2]$

$$= \int_{x_0}^{x_2} y\,dx - A_1 = \left(\frac{4}{5} - \frac{5}{18} \right) h^5 y_0^{iv} + \cdots \qquad [(6)-(8)]$$

i.e., Principal part of the error in $[x_0, x_2]$

$$= \left(\frac{4}{15} - \frac{5}{18} \right) h^5 y_0^{iv} = -\frac{h^5}{90}y_0^{iv}$$

Similarly principal part of the error in $[x_2, x_4] = -\dfrac{h^5}{90}y_2^{iv}$ and so on.

Hence the total error $E = -\dfrac{h^5}{90}[y_0^{iv} + y_2^{iv} + \cdots + y^{iv}2(n-1)]$

Assuming the $y^{iv}(X)$ is the largest of $y_0^{iv}, y_2^{iv}, \ldots, y^{iv}_{2n-2}$, we get

$$E < -\frac{nh^5}{90}y_0^{iv}(X) = -\frac{(b-a)h^4}{180}y^{iv}(X) \qquad [\because 2nh = b - a \ldots(9)]$$

i.e., the error in Simpson's 1/3 -rule is of the order h^4.

Error in Simpson's 3/8 rule. Proceeding as above, here the principal part of the error in the interval $[x_0, x_3]$

$$= -\frac{3h^5}{80}y^{iv} \qquad (10)$$

Error in Boole's rule. In this case, the principal part of the error in the interval

$$[x_0, x_4] = -\frac{8h^7}{945} y^{iv} \tag{11}$$

Error in Weddle's rule. In this case, principle part of the error in the interval

$$[x_0, x_6] = \frac{h^7}{140} y_0 iv \tag{12}$$

8.7 Romberg's Method

In Section 8.5, we have derived approximate quadrature formulae with the help of finite differences method. Romberg's method provides a simple modification to these quadrature formulae for finding their better approximations. As an illustration, let us improve upon the value of the integral

$$I = \int_a^b f(x)dx,$$

by the Trapezoidal rule. If I_1, I_2 are the values of I with sub-intervals of width h_1, h_2 and E_1, E_2 their corresponding errors, respectively, then

$$E_1 = -\frac{(b-a)h_1^2}{12} y''(X), E_2 = -\frac{(b-a)^2 h_2^2}{12} y''(\overline{X})$$

Since $y''(\overline{X})$ is also the largest value of $y''(x)$, we can reasonably assume that $y''(X)$ and $y''(\overline{X})$ are very nearly equal.

$$\therefore \qquad \frac{E_1}{E_2} = \frac{h_1^2}{h_2^2} \quad \text{or} \quad \frac{E_1}{E_2 - E_1} = \frac{h_1^2}{h_2^2 - h_1^2} \tag{1}$$

Now since $I = I_1 + E_1 = I_2 + E_2$,

$$\therefore \qquad E_2 - E_1 = I_1 - I_2 \tag{2}$$

From (1) and (2), we have

$$E_1 = \frac{h_1^2}{h_2^2 - h_1^2} (I_1 - I_2)$$

Hence $I = I_1 + E_1 = I_1 + \frac{h_1^2}{h_2^2 - h_1^2}(I_1 - I_2)$ i.e., $I = \frac{I_1 h_2^2 - I_2 h_1^2}{h_2^2 - h_1^2}$ (3)

which is a better approximation of I.

To evaluate I systematically, we take $h_1 = h$ and $h_2 = \dfrac{1}{2}h$

so that (3) gives $I = \dfrac{I_1(h/2)^2 - I_2 h_2^2}{(h/2)^2 - h^2} = \dfrac{4I_2 - I_1}{3}$

i.e., $\quad I(h, h/2) = \dfrac{1}{3}[4I(h/2) - I(h)]$ \hfill (4)

Now we use the trapezoidal rule several times successively halving h and apply (4) to each pair of values as per the following scheme:

$I(h)$
$\quad\quad\quad\quad I(h, h/2)$
$I(h/2)$
$\quad\quad\quad\quad\quad\quad\quad\quad\quad\quad I(h, h/2, h/4)$
$\quad\quad\quad\quad I(h/2, h/4)$
$\quad\quad\quad\quad\quad\quad\quad\quad\quad\quad\quad\quad\quad\quad\quad\quad\quad I(h, h/2, h/4, h/8)$
$I(h/4)$
$\quad\quad\quad\quad\quad\quad\quad\quad\quad\quad I(h/2, h/4, h/8)$
$\quad\quad\quad\quad I(h/4, h/8)$
$I(h/8)$

The computation is continued until successive values are close to each other. This method is called *Richardson's deferred approach to the limit* and its systematic refinement is called *Romberg's method.*

EXAMPLE 8.18

Evaluate $\displaystyle\int_0^1 \dfrac{dx}{1+x}$ correct to three decimal places using Romberg's method. Hence find the value of $\log_e 2$.

Solution:

Taking $h = 0.5$, 0.25, and 0.125 successively, let us evaluate the given integral by the Trapezoidal rule.

(i) When $h = 0.5$, the values of $y = (1 + x)^{-1}$ are:

x:$\quad\quad$ 0 $\quad\quad$ 0.5 $\quad\quad$ 1

y:$\quad\quad$ 1 $\quad\quad$ 0.6666 $\quad\quad$ 0.5

$\therefore\quad\quad I = \dfrac{0.5}{2}(1 + 0.5 + 2 \times 0.6666) = 0.7083.$

(ii) When $h = 0.25$, the values of $y = (1 + x)^{-1}$ are:

x:	0	0.25	0.5	0.75	1
y:	1	0.8	0.6666	0.5714	0.5

$$\therefore \qquad I = \frac{0.25}{2}[(1+0.5)+2(0.8+0.666+0.5714)] = 0.697$$

(*iii*) When $h = 0.125$, the values of $y = (1+x)^{-1}$ are:

x:	0	0.125	0.25	0.375	0.5	0.625	0.75	0.875	1
y:	1	0.8889	0.8 0.	7272	0.6667	0.6153	0.5714	0.5333	0.5

$$\therefore \qquad I = \frac{0.125}{2}[(1+0.5)+2(0.8889+0.8+0.7272+0.6667$$
$$+0.6513+0.5714+0.5333)]$$
$$= 0.6941$$

Using Romberg's formulae, we obtain

$$I(h,h/2) = \frac{1}{3}[4I(h/2) - I(h)] = \frac{1}{3}[4 \times 0.697 - 0.7083] = 0.6932$$

$$I(h/2,h/4) = \frac{1}{3}[4I(h/4) - I(h/2)] = \frac{1}{3}[4 \times 0.6941 - 0.697] = 0.6931$$

$$I(h,h/2,h/4) = \frac{1}{3}[4I(h/2,h/4) - I(h,h/2)] = 0.6931$$

Hence the value of the integral $\displaystyle\int_0^1 \frac{dx}{1+x} = 0.693$ (*i*)

Also $\displaystyle\int_0^1 \frac{dx}{1+x} = \left| \log(1+x) \right|_0^1 = \log 2$ (*ii*)

Hence from (*i*) and (*ii*), we have

$$\log_e 2 = 0.693.$$

EXAMPLE 8.19

Use Romberg's method to compute $\displaystyle\int_0^1 \frac{dx}{1+x^2}$ correct to four decimal places.

Solution:

We take $h = 0.5$, 0.25 and 0.125 successively and evaluate the given integral using the Trapezoidal rule.

(*i*) When $h = 0.5$, the values of $y = (1+x^2)^{-1}$ are

x:	0	0.5	1.0
y:	1	0.8	0.5

$$\therefore \quad I = \frac{0.5}{2}[1 + 2 \times 0.8 \times 0.5) = 0.775$$

(*ii*) When $h = 0.25$, the values of $y = (1 + x^2)^{-1}$ are

x:	0	0.25	0.5	0.75	1.0
y:	1	0.9412	0.8	0.64	0.5

$$\therefore I = \frac{0.25}{2}[1 + 2(0.9412 + 0.8 + 0.64) + 0.5] = 0.7828$$

(*iii*) When $h = 0.125$, we find that $I = 0.7848$

Thus we have

$$I(h) = 0.7750, \ I(h/2) = 0.7828, \ I(h/4) = 0.7848$$

Now using (4) above, we obtain

$$I(h, h/2) = \frac{1}{3}[4I(h/2) - I(h)] = \frac{1}{3}(3.1312 - 0.775) = 0.7854$$

$$I(h/2, h/4) = \frac{1}{3}[(4I(h/4) - I(h/2)] = \frac{1}{2}(3.1392 - 0.7828) = 0.7855$$

$$I(h, h/2, h/4) = \frac{1}{3}[4I(h/2, h/4) - I(h, h/2)] = \frac{1}{3}(3.142 - 0.7854) = 0.7855$$

$\therefore$ The table of these values is

```
0.7750
                0.7854
0.7828                          0.7855
                0.7855
0.7848
```

Hence the value of the integral = 0.7855.

EXAMPLE 8.20

Evaluate the integral $\int_0^{0.5} \left(\dfrac{x}{\sin x} \right) dx$ using Romberg's method, correct to three decimal places.

Solution:

Taking $h = 0.25, 0.125, 0.0625$ successively, let us evaluate the given integral by using Simpson's 1/3 rule.

(*i*) When $h = 0.25$, the values of $y = \dfrac{x}{\sin x}$ are

x:	0	0.25	0.5
y:	1	1.0105	1.0429
	y_0	y_1	y_2

∴ By Simpson's rule,

$$I = \frac{h}{3}[(y_0 + y_2) + 4y_1] = \frac{0.25}{3}[(1 + 1.0429) + 1.0105]$$

$$= 0.5071$$

(*ii*) When $h = 0.125$, the values of y are

x:	0	0.125	0.25	0.375	0.5
y:	1	1.0026	1.0105	1.1003	1.0429
	y_0	y_1	y_2	y_3	y_4

∴ By Simpson's rule

$$I = \frac{h}{3}[(y_0 + y_4) + 4(y_1 + y_3) + 2y_2]$$

$$= \frac{0.125}{3}[(1 + 1.0429) + 4(1.0026 + 1.1003) + 2(1.0105)]$$

$$= 0.5198$$

(*iii*) When $h = 0.0625$, the values of y are

x:	0	0.0625	0.125	0.1875	0.25	0.3125	0.1875	0.4375	0.5
y:	1	0.0006	1.0026	1.0059	1.0157	1.0165	1.1003	1.0326	1.0429
	y_0	y_1	y_2	y_3	y_4	y_5	y_6	y_7	y_8

∴ By Simpson's rule:

$$I = \frac{h}{3}[(y_0 + y_8) + 4(y_1 + y_3 + y_5 + y_7) + 2(y_2 + y_4 + y_6)]$$

$$= \frac{0.0625}{3}[(1 + 1.0429) + 4(1.0006 + 1.0059 + 1.0165 + 1.0326$$

$$+ 2(1.0026 + 1.0105 + 1.1003)]$$

$$= 0.510253$$

Using Romberg's formulae, we obtain

$$I(h,\frac{h}{2}) = \frac{1}{3}\left[4I\left(\frac{h}{2}\right) - I(h)\right] = 0.5241$$

$$I(\frac{h}{2},\frac{h}{4}) = \frac{1}{3}\left[4I\left(\frac{h}{4}\right) - I\left(\frac{h}{2}\right)\right] = 0.5070$$

$$I(h.\frac{h}{2},\frac{h}{4}) = \frac{1}{3}\left[4I\left(\frac{h}{2},\frac{h}{4}\right) - I\left(h,\frac{h}{2}\right)\right] = 0.5013$$

Hence $\int_0^{0.5}\left(\frac{x}{\sin x}\right)dx = 0.501$

8.8 Euler-Maclaurin Formula

Taking $\Delta F(x) = f(x)$, we define the inverse operator Δ^{-1} as

$$F(x) = \Delta^{-1} f(x) \quad (1)$$

Now $\qquad F(x_1) - F(x_0) = \Delta F(x_0) = f(x_0)$

Similarly, $\qquad F(x_2) - F(x_1) = f(x_1)$

$$\cdots\cdots\cdots\cdots\cdots\cdots\cdots\cdots$$

$$F(x_n) - F(x_{n-1}) = f(x_{n-1})$$

Adding all these, we get

$$F(x_n) - F(x_0) = \sum_{i=0}^{n=1} f(x_i) \qquad\qquad (2)$$

where $x_0, x_1, \ldots, x_n$ are the $(n + 1)$ equispaced values of x with difference h.

From (1), we have

$$F(x) = \Delta^{-1} f(x) = (E+1)^{-1} f(x) = (e^{hD} - 1)^{-1} f(x) \qquad [\because E = ehD]$$

$$= \left[\left(1 + hD + \frac{h^2 D^2}{2!} + \frac{h^3 D^3}{3!} + \cdots\right) - 1\right]^{-1} f(x)$$

$$= (hD)^{-1}\left[1 + \frac{hD}{2!} + \frac{h^2 D^2}{3!} + \cdots\right]^{-1} f(x) \qquad\qquad (3)$$

$$= \frac{1}{h}D^{-1}\left[1 - \frac{hD}{32} + \frac{h^2 D^2}{12} - \frac{h^4 D^4}{720} + \cdots\right] f(x)$$

$$= \frac{1}{h}\int f(x)dx - \frac{1}{2}f(x) + \frac{h}{12}f'(x) - \frac{h^3}{720}f'''(x) + \cdots$$

Putting $x = x_n$ and $x = x_0$ in (3) and then subtracting, we get

$$F(x_n) - F(x_0) = \frac{1}{h}\int_{x_0}^{x_n} f(x)dx - \frac{1}{2}[f(x_n) - f(x_0)] + \frac{h}{12}[f'(x_n)$$

$$-f'(x_0)] - \frac{h^3}{720}[f'''(x_n) - f'''(x_0)] + \cdots \tag{4}$$

∴ From (2) and (4), we have

$$\sum_{i=0}^{n-1} f(x_i) = \frac{1}{h}\int_{x_0}^{x_n} f(x)dx - \frac{1}{2}[f(x_n) - f(x_0)] + \frac{h}{12}[f'(x_n)$$

$$-f'(x_0)] - \frac{h^3}{720}[f'''(x_n) - f'''(x_0)] + \cdots$$

i.e., $\frac{1}{h}\int_{x_0}^{x_n} f(x)dx = \sum_{i=0}^{n=1} f(x_i) + \frac{1}{2}[f(x_n) - f(x_0)] - \frac{h}{12}[f'(x_n)$

$$-f'(x_0)] + \frac{h^3}{720}[f'''(x_n) - f'''(x_0)] + \cdots$$

$$= \frac{1}{2}[f(x_0) + 2f(x_1) + 2f(x_2) + \cdots + 2f(x_{n-1}) + f(x_n)]$$

$$-\frac{h}{12}[f'(x_n) - f'(x_0)] + \frac{h^3}{720}[f'''(x_n) - f'''(x_0)] + \cdots$$

Hence $\int_{x_0}^{x_0 - nh} y\,dx = \frac{h}{2}[y_0 + 2y_1 + 2y_2 + \cdots + 2y_{n-1} + y_n] \tag{5}$

$$-\frac{h^2}{12}(y_n' - y_0') + \frac{h^4}{720}(y_n''' - y_0''') + \cdots$$

which is called the *Euler-Maclaurin formula.*

NOTE **Obs.** *The first term on the right-hand side of (5) represents the approximate value of the integral obtained from trapezoidal rule and the other terms denote the successive corrections to this value. This formula is often used to find the sum of a series of the form*

$$y(x_0) + y(x_0 + h) + \cdots + y(x_0 + nh).$$

EXAMPLE 8.21

Using the Euler-Maclaurin formula, find the value of loge 2 from

$$\int_0^1 \frac{dx}{1+x}$$

Solution:

Taking $y = \dfrac{1}{(1+x)}, x_0 = 0, n = 10, h = 0.1,$ we have

$$y' = -\frac{1}{(1+x)^2} \quad \text{and} \quad y''' = \frac{-6}{(1+x)^4}.$$

Then the Euler-Maclaurin formula gives

$$\int_0^1 \frac{dx}{1+x} = \frac{0.1}{2}\left[\frac{1}{1+0} + \frac{2}{1+0.1} + \frac{2}{1+0.2} + \frac{2}{1+0.3} + \frac{2}{1+0.1}\right.$$

$$\left. + \frac{2}{1+0.5} + \frac{2}{1+0.6} + \frac{2}{1+0.7} + \frac{2}{1+0.8} + \frac{2}{1+0.9} + \frac{1}{1+1}\right]$$

$$- \frac{(0.1)^2}{12}\left[\frac{-1}{(1+1)^2} - \frac{-1}{(1+0)^2}\right] + \frac{(0.1)^4}{720}\left[\frac{-6}{(1+1)^4} - \frac{-6}{(1+0)^4}\right]$$

$$= 0.693773 - 0.000625 + 0.000002 = 0.693149$$

Also $\displaystyle\int_0^1 \frac{dx}{1+x} = \left|\log(1+x)\right|_0^1 = \log e^2$

Hence $\log_e 2 = 0.693149$ approx.

EXAMPLE 8.22

Apply the Euler-Maclaurin formula to evaluate

$$\frac{1}{51^2} + \frac{1}{53^2} + \frac{1}{55^2} + \ldots + \frac{1}{99^2}$$

Solution:

Taking $y = \dfrac{1}{x^2}, x_0 = 51, h = 2, n = 24,$ we have $y' = \dfrac{-2}{x^3}, y''' = \dfrac{-24}{x^5}$

Then the Euler-Maclaurin formula gives

$$\int_{51}^{99} \frac{dx}{x^2} = \frac{2}{2}\left[\frac{1}{51^2} + \frac{2}{53^2} + \frac{2}{55^2} + \ldots + \frac{2}{97^2} + \frac{1}{99^2}\right]$$

$$- \frac{(2)^2}{12}\left[\frac{-2}{99^3} - \frac{-2}{51^3}\right] + \frac{(2)^4}{720}\left[\frac{-24}{99^5} - \frac{-24}{51^5}\right]$$

$$\therefore \quad \frac{1}{51^2} + \frac{1}{53^2} + \frac{1}{55^2} + \ldots + \frac{1}{99^2} = \frac{1}{2}\int_{51}^{99} \frac{dx}{x^2}$$

$$+ \frac{1}{2}\left(\frac{1}{51^2} + \frac{1}{99^2}\right) + \frac{1}{3}\left(\frac{1}{51^3} - \frac{1}{99^3}\right) - \frac{8}{30}\left(\frac{1}{51^5} - \frac{1}{99^5}\right) + \ldots$$

$$= \frac{1}{2}\left|\frac{1}{x}\right|_{51}^{99} + 0.000243 + 0.0000022 - \ldots = 0.00499 \text{ approx.}$$

8.9   Method of Undetermined Coefficients

This method is based on imposing certain conditions on a preassigned formula involving certain unknown coefficients and then using these conditions for evaluating these unknown coefficients. Assuming the formula to be exact for the polynomials $1, x, \frac{1}{4}, x^n$ respectively and taking y_i for $y(x_i)$, we shall determine the unknown coefficients to derive the formulae.

Differentiation formulae. We first derive the two-term formula by assuming

$$y_0' = a_0 y_0 + a_1 y_1 \tag{1}$$

where the unknown constants a_0, a_1 are determined by making (1) exact for $y(x) = 1$ and x respectively.

So, putting $y(x) = 1$, x successively in (1), we get

$$0 = a_0 + a_1 \text{ and } 1 = a_0 x_0 + a_1 (x_0 + h)$$

where $a_1 = 1/h$ and $a_0 = -1/h$.

Hence $$y_0' = \frac{1}{h}(y_1 - y_0) \tag{2}$$

The three-term formula can be derived by taking

$$y_0' = a_{-1} y_{-1} + a_0 y_0 + a_1 y_1 \tag{3}$$

where the unknowns a_{-1}, a_0, a_1 are determined by making (3) exact for $y(x) = 1, x, x^2$, respectively.

∴ $$0 = a_{-1} + a_0 + a_1$$

$$1 = a_{-1}(x_0 - h) + a_0 x_0 + a_1(x_0 + h)$$

and $$2x_0 = a_{-1}(x_0 - h)^2 + a_0 x_0^2 + a_1(x_0 + h)^2.$$

To solve these equations, we shift the origin to x_0 *i.e.*, $x_0 = 0$. As such, y_0' being slope of the tangent to the curve $y = f(x)$ at $x = x_0$ remains unaltered. Thus the equations reduce to

$$a_{-1} + a_0 + a_1 = 0,$$

$$-a_{-1} + a_1 = 1/h \text{ and } a_{-1} + a_1 = 0,$$

giving $$a_{-1} = -1/2h, a_0 = 0, a_1 = 1/2h$$

Hence $$y_0' = \frac{1}{2h}(y_1 - y_{-1}), \tag{4}$$

Similarly for second order derivative, taking

$$y_0'' = a_{-1}y_{-1} + a_0y_0 + a_1y_1$$

and making it exact for $y(x) = 1, x, x^2$ and putting $x_0 = 0$, we get

$$y_0'' = \frac{1}{h^2}(y_1 - 2y_0 + y_{-1}) \tag{5}$$

Integration formulae. *The two-term formula is derived by assuming*

$$\int_{x_0}^{x_0+h} y\,dx = a_0y_0 + a_1y_1 \tag{6}$$

where the unknowns $a0, a1$ are determined by making (6) exact for $y(x) = 1, x$ respectively.

So putting $y(x) = 1, x$ successively in (6), we get

$$a_0 + a_1 = \int_{x_0}^{x_0+h} 1 \cdot dx = h$$

$$a_0x_0 + a_1(x_0 + h) = \int_{x_0}^{x_0+h} x \cdot dx = \frac{1}{2}[(x_0 + h)^2 - x_0^2]$$

To solve these, we shift the origin to x_0 and take $x_0 = 0$.

$\therefore$ The above equations reduce to

$$a_0 + a_1 = h \text{ and } a_1 = \frac{1}{2}h, \text{ whence } a_1 = \frac{1}{2}h, a_0 = \frac{1}{2}h$$

Hence $\int_{x_0}^{x_0+h} y\,dx = \frac{h}{2}(y_0 + y_1)$ which is *trapezoidal rule*. $\tag{7}$

The three-term formula is derived by assuming

$$\int_{x_0-h}^{x_0+h} y\,dx = a_{-1}y_{-1} + a_0y_0 + a_1y_1 \tag{8}$$

where the unknowns a_{-1}, a_0, a_1 are determined by making (8) exact for $y(x)$ = 1, x, x^2$ respectively.

So putting $y = 1, x, x^2$ successively in (8), we obtain

$$a_{-1} + a_0 + a_1 = \int_{x_0-h}^{x_0+h} 1 \cdot dx = 2h$$

$$a_{-1}(x_0 - h) + a_0x_0 + a(x_0 + h) = \int_{x_0-h}^{x_0+h} x\,dx = \frac{1}{2}[(x_0 + h)^2 - (x_0 - h)^2]$$

$$a_{-1}(x_0 - h)^2 + a_0x_0^2 + a_1(x_0 + h)^2 = \int_{x_0-h}^{x_0+h} x^2 dx = \frac{1}{3}[(x_0 + h)^3 - (x_0 - h)^3]$$

To solve these equations, we shift the origin to x_0 and take $x_0 = 0$.

$\therefore$ The above equations reduce to

$$a_{-1} + a_0 + a_1 = 2h, \; -a_{-1} + a_1 = 0 \text{ and } a_{-1} + a_1 = \frac{2}{3}h$$

Solving these, we get $a_{-1} = \frac{1}{3}h = a_1, a_0 = \frac{4}{3}h$

Hence $\displaystyle\int_{x_0-h}^{x_0+h} y\,dx = \frac{h}{3}(y_{-1} + 4y_0 + y_1)$ which is *Simpson's rule*. $\hspace{1cm}$ (9)

8.10 Gaussian Integration

So far the formulae derived for evaluation of $\displaystyle\int_a^b f(x)dx$, required the values of the function at equally spaced points of the interval. Gauss derived a formula which uses the same number of functional values but with different spacing and yields better accuracy.

Gauss formula is expressed as

$$\int_{-1}^{1} f(x)dx = w_1 f(x_1) + w_2 f(x_2) + \cdots + w_n f(x_n) = \sum_{i=1}^{n} w_i f(x_i) \hspace{1cm} (1)$$

where wi and xi are called the *weights* and *abscissae*, respectively. *The abscissae and weights are symmetrical with respect to the middle point of the interval.* There being $2n$ unknowns in (1), $2n$ relations between them are necessary so that the formula is exact for all polynomials of degree not exceeding $2n - 1$. Thus we consider

$$f(x) = c_0 + c_1 x + c_2 x + \cdots + c_{2n-1} x^{2n-1} \hspace{1cm} (2)$$

Then (1) gives

$$\int_{-1}^{1} f(x)dx = \int_{-1}^{1} (c_0 + c_1 x + c_2 x + \cdots + c_{2n-1} x^{2n-1})dx \hspace{1cm} (3)$$

$$= 2c_0 + \frac{2}{3}c_2 + \frac{2}{5}c_4 + \cdots$$

Putting $x = x_i$ in (2), we get

$$f(x_i) = c_0 + c_1 x_i + c_2 x_i^2 + c_3 x_i^3 + \ldots + c_{2n-1} x_i^{2n-1}$$

Substituting these values on the right hand side of (1), we obtain

$$\int_{-1}^{1} f(x)dx = w1(c_0 + c_1x_1 + c_2x_1^2 + c_3x_1^3 + \ldots c_{2n-1}x_1^{2n-1})$$

$$+ w_2(c_0 + c_1x_2 + c_2x_2^2 + c_3x_2^3 + \ldots + c_{2n-1}x_2^{2n-1})$$

$$+ w_3(c_0 + c_1x_3 + c_2x_3^2 + c_3x_3^3 + \ldots + c_{2n-1}x_3^{2n-1})$$

$$+ \ldots\ldots\ldots\ldots\ldots\ldots\ldots\ldots\ldots\ldots\ldots\ldots\ldots\ldots\ldots\ldots\ldots\ldots$$

$$+ wn(c_0 + c_1xn + c_2xn^2 + c_3xn^3 + \ldots + c_{2n-1}xn^{2n-1})$$

$$= c_0(w_1 + w_2 + w_3 + \cdots + w_n) + c_1(w_1x_1 + w_2x_2 + w_3x_3 + \cdots + w_nx_n)$$

$$+ c_2(w_1x_1^2 + w_2x_2^2 + w_3x_3^2 + \cdots + w_nx_n^2)$$

$$+ \ldots\ldots\ldots\ldots\ldots\ldots\ldots\ldots\ldots\ldots\ldots\ldots\ldots\ldots\ldots\ldots\ldots$$

$$+ c_{2n-1}(w_1x_1^{2n-1} + w_2x_2^{2n-1} + w_3x_3^{2n-1} + \cdots + w_nx_n^{2n-1}) \quad (4)$$

But the equations (3) and (4) are identical for all values of c_i, hence comparing coefficients of c_i, we obtain $2n$ equations in $2n$ unknowns w_i and x_i $(i = 1, 2, \ldots, n)$.

$$\left.\begin{array}{r}
w_1 + w_2 + w_3 + \cdots + w_n = 2 \\
w_1x_1 + w_2x_2 + w_3x_3 + \cdots + w_nx_n = 0 \\
w_1x_1^2 + w_2x_2^2 + w_3x_3^2 + \cdots + w_nx_n^2 = \dfrac{2}{3} \\
\ldots\ldots\ldots\ldots\ldots\ldots\ldots\ldots\ldots\ldots\ldots\ldots \\
w_1x_1^{2n-1} + w_2x_2^{2n-1} + w_3x_3^{2n-1} + \cdots + w_nx_n^{2n-1} = 0
\end{array}\right\} \quad (5)$$

The solution of the above equations is extremely complicated. It can however, be shown that x_i are the zeros of the $(n + 1)$th Legendre polynomial.

Gauss formula for n = 2 is

$$\int_{-1}^{1} f(x)dx = w_1 f(x_1) + w_2 f(x_2)$$

Then the equations (5) become

$$w_1 + w_2 = 2$$

$$w_1x_1 + w_2x_2 = 0$$

$$w_1x_1^2 + w_2x_2^2 = \frac{2}{3}$$

$$w_1x_1^3 + w_2x_2^3 = 0$$

Solving these equations, we obtain

$$w_1 = w_2 = 1, x_1 = -1/\sqrt{3} \text{ and } x_2 = 1/\sqrt{3}.$$

Thus *Gauss formula for n = 2* is

$$\int_{-1}^{1} f(x)dx = f(-1\sqrt{3}) + f(1/\sqrt{3}) \tag{6}$$

which gives the correct value of the integral of $f(x)$ in the range $(-1, 1)$ for any function up to third order. Equation (6) is also known as **Gauss-Legendre formula.**

Gauss formula for n = 3 is

$$\int_{-1}^{1} f(x)dx = \frac{8}{9} f(0) + \frac{5}{9}\left[f\left(-\sqrt{\frac{3}{5}}\right) + f\left(\sqrt{\frac{3}{5}}\right)\right] \tag{7}$$

which is exact for polynomials upto degree 5.

The abscissae xi and the weights wi in (1) are tabulated for different values of n. The following table lists the abscissae and weights for values of n from 2 to 5.

TABLE 8.1 Gauss integration: Abscissae and Weights

N	x_i	w_i
2	− 0.57735	1.0000
	0.57735	1.0000
3	− 0.7746	0 0.55555
	0.00000	0.88889
	0.77460	0.55555
4	− 0.86114	0.34785
	− 0.33998	0.65214
	0.33998	0.65214
	0.86114	0.34785
5	− 0.90618	0.23693
	− 0.53847	0.47863
	0.00000	0.56889
	0.53847	0.47863
	0.90618	0.23693

Gauss formula imposes a restriction on the limits of integration to be from − 1 to 1.

In general, the limits of the integral $\int_{a}^{b} f(x)dx$ are changed to − 1 to 1 by means of the transformation

$$x = \frac{1}{2}(b-1)u + \frac{1}{2}(b+a) \tag{8}$$

EXAMPLE 8.23

Evaluate $\int_{-1}^{1} \dfrac{dx}{1+x^2}$
using Gauss formula for $n = 2$ and $n = 3$.

Solution:

(i) Gauss formula for $n = 2$ is

$$I = \int_{-1}^{1} \frac{dx}{1+x^2} = f\left(-\frac{1}{\sqrt{3}}\right) + f\left(\frac{1}{\sqrt{3}}\right) \text{ where } f(x) = \frac{1}{1+x^2}$$

$$\therefore I = \frac{1}{1+(-1\sqrt{3})^2} + \frac{1}{1+(1/\sqrt{3})^2} = \frac{3}{4} + \frac{3}{4} = 1.5.$$

(ii) Gauss formula for $n = 3$ is

$$I = \frac{8}{9} f(0) + \frac{5}{9}\left[f-\left(\sqrt{\frac{3}{5}}\right) + f\left(\sqrt{\frac{3}{5}}\right)\right] \text{ where } f(x) = \frac{1}{1+x^2}$$

Thus $I = \dfrac{8}{9}(1) + \dfrac{5}{9}\left(\dfrac{5}{8} + \dfrac{5}{8}\right) = \dfrac{8}{9} + \dfrac{50}{72} = 1.5833$.

EXAMPLE 8.24

Using the three-point Gaussian quadrature formula, evaluate $\int_{0}^{1} \dfrac{dx}{1+x}$

Solution:

We first change the limits $(0, 1)$ to -1 to 1 by (8) above, so that

$$x = \frac{1}{2}(1-0)u\frac{1}{2}(1+0) = \frac{1}{2}(u+1).$$

$$\therefore \qquad I = \int_{0}^{1}\frac{dx}{1+x} = \int_{-1}^{1}\frac{\frac{1}{2}du}{1+\frac{1}{2}(u+1)} = \int_{-1}^{1}\frac{du}{u+3}$$

Gauss-formula for $n = 3$ is

$$I = \frac{8}{9} f(0) + \frac{5}{9} f\left(-\sqrt{\frac{3}{5}}\right) + f\left(\sqrt{\frac{3}{5}}\right) \text{ where } f(x) = \frac{1}{1+x^2}$$

Thus $\quad I = \dfrac{8}{9}\left(\dfrac{1}{3}\right) + \dfrac{5}{9}\left\{\dfrac{1}{\sqrt{(3/5)}+3} + \dfrac{1}{\sqrt{(3/5)}+3}\right\}$

$$= \frac{8}{27} + \frac{25}{63} = 0.29629 + 0.39682 = 0.6931$$

Otherwise (*using the table*):

$$I = w_1 f(u_1) + w_2 f(u_2) + w_3 f(u_3) \text{ where } f(u_i) = \frac{1}{u_i + 3}$$

Using the abscissae and weights corresponding to $n = 3$ in the above table, we obtain

$$I = \frac{1}{3 - 0.7746}(0.555) + \frac{1}{3 - 0}(0.8889) + \frac{1}{3 + 0.7746}(0.555)$$

$$= 0.4497 \times 0.5555 + \frac{1}{3}(0.8889) + 0.2649 \times 0.5555 = 0.6931.$$

EXAMPLE 8.25

Evaluate $\int_0^2 \frac{x^2 + 2x + 1}{1 + (x+1)^4} dx$ by the Gaussian three-point formula.

Solution:

Changing the limits of integration 0 to 2 to -1 to 1 by

$$x = \frac{1}{2}(b - a)u + \frac{1}{2}(b + a) = \frac{2 - 0}{2}u + \frac{2 + 0}{2} = u + 1$$

$$\therefore \quad I = \int_0^2 \frac{x^2 + 2x + 1}{1 + (x+1)^4} dx = \int_{-1}^1 \frac{(u+1)^2 + 2(u+1) + 1}{1 + (u+1+u)^4} du \quad [\because dx = du]$$

$$= \int_{-1}^1 \frac{u^2 + 4u + 4}{(u+2)^4 + 1} du = \int_{-1}^1 f(u) du$$

$$= w_1 f(u_1) + w_2 f(u_2) + w_3 f(u_3) \text{ where } f(u_i) = \frac{u_i^2 + 4u_i + 4}{(u_i + 2)^4 + 1}$$

Now $f(0) = \frac{4}{2^4 + 1} = \frac{4}{17}$

$$f\left(-\frac{3}{\sqrt{5}}\right) = \frac{(-\sqrt{(3/5)} + 2)^2}{[-\sqrt{(3/5)} + 2]^4 + 1} = \frac{15016}{3.2548} = 0.4614$$

$$f\left(\sqrt{\frac{3}{5}}\right) = \frac{\sqrt{(3/5)} + 2}{[\sqrt{(3/5)} + 2]^4 + 1} = \frac{7.6984}{60.2652} = 0.1277$$

Using the three-point Gaussian formula, we have

$$I = \int_{-1}^{1} f(u)du = \frac{8}{9}f(0) + \frac{5}{9}f\left[\left(-\sqrt{\frac{3}{5}}\right) + f\left(\sqrt{\frac{3}{5}}\right)\right]$$

$$= \frac{8}{9}\left(\frac{4}{17}\right) + \frac{5}{9}[0.4614 + 0.1277] = 0.5365$$

Solution:

Changing the limits of integration (0.2 to 1.5) to $(-1, 1)$ by

$$x = \frac{1}{2}(b-a)u + \frac{1}{2}(b+a) = \frac{1}{2}(1.5 - 0.2)u + \frac{1}{2}(1.5 + 0.2)$$

$$= 0.65u + 0.85$$

$$\therefore \qquad I = \int_{0.2}^{1.5} e^{-x^2} dx = 0.65 \int_{-1}^{1} e^{-(0.65u+0.85)2} du = 0.65 \int_{-1}^{1} f(u)du$$

so that $f(u) = e^{-(0.65u + 0.85)2}$

Now $f(0) = e^{-[0.65(0) + 0.85]2} = 0.4855$,

$$f(-\sqrt{3/5}) = f(-0.7746) = e^{-[0.65(-0.7746)+0.85]^2} = 0.8869$$

$$f(\sqrt{3/5}) = f(0.7746) = e^{-[0.65(0.7746)+0.85]^2} = 0.1601.$$

Using the Gauss three-point formula, we have

$$I = \int f(u)du = -f(0)[f(-\sqrt{3/5}) + f(\sqrt{3/5})]$$

$$= \frac{5}{9}(0.4855) + \frac{5}{9}[0.8869 + 0.1601] = 0.4316 + 0.5187 = 1.0133$$

Hence $\int_{0.2}^{1.5} e^{-x^2} dx = 0.65(1.0133) = 0.65865.$

Exercises 8.3

1. Obtain an estimate of the number of sub-intervals that should be chosen so as to guarantee that the error committed in evaluating $\int_{1}^{2} dx/x$ by trapezoidal rule is less than 0.001.

2. Evaluate $\int_{0}^{2} \frac{dx}{x^2 + 4}$ using the Romberg's method. Hence obtain an approximate value of π.

3. Apply Romberg's method to evaluate $\int_{4}^{5.2} \log x\, dx,$ given that

x:	4.0	4.2	4.4	4.6	4.8	5.0	5.2
$\log_e x$:	1.3863	1.4351	1.4816	1.526	1.5686	1.6094	1.6486.

4. Using the Euler-Maclaurin formula, find the value of $\int_{0}^{\pi/2} \sin x\, dx$ correct to five decimal places.

5. Using the Euler-Maclaurin formula, prove that

(a) $\sum_{1}^{n} x^2 = \dfrac{n(n+1)(2n+1)}{6}$ (b) $\sum_{1}^{n} x^3 = \left\{\dfrac{n(n+1)}{2}\right\}^2$

6. Apply the Euler-Maclaurin formula, to evaluate

(a) $\dfrac{1}{400} + \dfrac{1}{402} + \dfrac{1}{404} + \cdots + \dfrac{1}{500}$

(b) $\dfrac{1}{(201)^2} + \dfrac{1}{(203)^2} + \dfrac{1}{(205)^2} + \cdots + \dfrac{1}{(299)^2}$

7. Assuming that $\int_{0}^{h} y(x)dx = h(a_0 y_0 + a_1 y_1) + h^2(b_0 y_0' + b_1 y_1')$ derive the quadrature formula, using the method of undetermined coefficients.

8. Using the Gaussian two-point formula compute

(a) $\int_{-2}^{2} e^{-x2}dx$ (b) $\sum_{1}^{n} x^3 = \left\{\dfrac{n(n+1)}{2}\right\}^2$

9. Using three point Gaussian quadrature formula, evaluate:

(a) (i) $\int_{2}^{5} \dfrac{1}{x}dx$ (b) $\int_{1}^{2} \dfrac{1}{1+x^3}dx$.

10. Evaluate the following, integrals, using the Gauss three-point formula:

(a) $\int_{2}^{4} (1+x^4)dx$ (b) $\int_{3}^{5} \dfrac{4}{\left(2x^2\right)}dx$

11. Using the four point Gauss formula, compute $\int_{0}^{1} x dx$ correct to four decimal places.

8.11 Numerical Double Integration

The double integral

$$I = \int_c^d \int_a^b f(x,y)\,dx\,dy$$

is evaluated numerically by two successive integrations in x and y directions considering one variable at a time. Repeated application of trapezoidal rule (or Simpson's rule) yields formulae for evaluating I.

Trapezoidal rule. Dividing the interval (a, b) into n equal sub-intervals each of length h and the interval (c, d) into m equal sub-intervals each of length k, we have:

$$x_i = x_0 + ih,\ x_0 = a,\ x_n = b.$$

$$y_j = y_0 + jk,\ y_0 = c,\ y_m = d.$$

Using trapezoidal rule in both directions, we get

$$I = \frac{h}{2} \int_c^d [f(x_0,y) + f(x_n,y) + 2\{f(x_1,y) + f(x_2,y) + \ldots + f(x_{n-1},y)\}]\,dy$$

$$= \frac{hk}{4}[(f_{00} + f_{om}) + 2(f_{01} + f_{02} + \ldots + f_0, m-1)$$

$$+ (f_n0 + f_{nm}) + 2(f_{n1} + f_{n2} + \ldots + fn, m-1)$$

$$+ 2\sum_{i=1}^{n-1} \{(f_{i0} + f_{im}) + 2(f_{i1} + f_{i2} + \ldots + fi, m-1)\}]$$

where $f_{ij} = f(x_i, y_j)$.

Simpson's rule. We divide the interval (a, b) into $2n$ equal sub-intervals each of length h and the interval (c, d) into $2m$ equal sub-intervals each of length k. Then applying Simpson's rule in both directions, we get

$$\int_{y_{j-1}}^{y_{j+1}} \int_{x_{i-1}}^{x_{i+1}} f(x,y)\,dx\,dy = \frac{h}{3} \int_{y_{j-1}}^{y_{j+1}} [f(x_{i-1},y) + 4f(x_i,y) + f(x_{i+1},y)]\,dy$$

$$= \frac{hk}{9}[(f_{i-1,j-1} + 4f_{i-1},j + f_{i-1,j+1}) + 4(f_{i,j-1} + 4f_{i,j} + f_{i,j+1})$$

$$+ (f_{i+1,j-1} + 4f_{i+1},j + f_{i+1,j+1})]$$

Adding all such intervals, we obtain the value of I.

EXAMPLE 8.27

Using trapezoidal rule, evaluate $I = \int_1^2 \int_1^2 \dfrac{dxdy}{x\,y}$, taking four sub-intervals.

Solution:

Taking $h = k = 0.25$ so that $m = n = 4$, we obtain

$$I = \frac{1}{64}[f_{(1,1)} + f_{(1,2)} + 2(f_{(1,1.25)} + f(1,1.5) + f_{(1,1.75)})$$
$$+ f_{(2,1)} + f_{(2,2)} + 2(f_{(2,1.25)} + f_{(2,1.5)} + f(2,1.75)$$
$$+ 2\{f_{(1.25,1)} + f_{(1.25,2)} + 2f_{(1.25,1.25)} + f_{(1.25,1.5)} + f_{(1.25,1.75)})$$
$$+ f_{(1.5,1)} + f_{(1.5,2)} + 2(f_{(1.5,1.25)} + f_{(1.5,1.5)} + f_{(1.5,1.75)})$$
$$+ f_{(1.75,1)} + f_{(1.75,2)} + 2(f_{(1.75,1.25)} + f_{(1.75,1.5)} + f_{(1.75,1.75)})\}]$$
$$= 0.3407$$

EXAMPLE 8.28

Apply Simpson's rule to evaluate the integral

$$I = \int_2^{2.6} \int_4^{4.4} \frac{dxdy}{xy}$$

Solution:

Taking $h = 0.2$ and $k = 0.3$ so that $m = n = 2$, we get

$$I = \frac{hk}{91}[f(4,2) + 4f(4,2.3) + f(4,2.6)$$
$$+ 4\{f(4.2,2) + 4f(4.2,2.3) + f(4.2,2.6)\}$$
$$+ f(4.4,2) + 4f(4.4,2.3) + f(4.4,2.6)]$$
$$= \frac{0.06}{9}[0.6559 + 4(0.6246) + 0.5962]$$
$$= \frac{0.02}{3} \times 3.7505 = 0.025$$

Exercises 8.4

1. Evaluate $\int_0^1 \int_0^1 xe^y\,dxdy$ using the Trapezoidal rule $(h = k = 0.5)$.

2. Apply the Trapezoidal rule to evaluate

(a) $\int_1^5 \int_1^5 \dfrac{dxdy}{\sqrt{x^2 + y^2)}}$, taking two sub-intervals.

(b) $\int_0^1 \int_1^2 \frac{2xy\,dx\,dy}{(1+x^2)(1+y^2)}$, taking $h = k = 0.25$.

3. Evaluate $\int_0^2 \int_0^2 f(x,y)\,dx\,dy$ the Trapezoidal rule for the following data:

y/x	0	0.5	1	1.5	2
0	2	3	4	5	5
1	3	4	6	9	11
2	4	6	8	11	14

4. Using the Trapezoidal and Simpson's rules, evaluate

$\int_0^1 \int_0^1 e^{x+y}\,dx\,dy$ taking two sub-intervals.

5. Using Simpson's rule, evaluate

(a) $\int_1^{2.8} \int_2^{3.2} \frac{dx\,dy}{x+y}$ (b) $\int_0^1 \int_0^1 \frac{dx\,dy}{1+x+y}$, taking $h = k = 0.5$.

8.12   Objective Type of Questions

Exercises 8.5

Select the correct answer or fill up the blanks in the following questions:

1. The value of $\int_0^1 \frac{dx}{1+x}$ by Simpson's rule is

    (a) 0.96315   (b) 0.63915

    (c) 0.69315   (d) 0.69351.

2. Using forward differences, the formula for $f'(a) = $ ....

3. In application of Simpson's 1/3rd rule, the interval h for closer approximation should be ....

4. $f(x)$ is given by

x:	0	0.5	1
f(x):	1	0.8	0.5

then using Trapezoidal rule, the value of $\int_0^1 f(x)\,dx$ is...... .

5. If

x:	0	0.5	1	1.5	2
f(x):	0	0.25	1	2.25	4,

then the value of $\int_0^2 f(x)dx$ by Simpson's 1/3rd rule is .... .

6. Simpson's 3/8 rule states that ....

7. For the data:

t:	3	6	9	12
y(t):	−1	1	2	3

the value of $\int_3^{12} y(t)dt$ when computed by Simpson's 1/3 rule is

(a) 15   (b) 10   (c) 0   (d) 5.

8. While evaluating a definite integral by Trapezoidal rule, the accuracy can be increased by taking ....

9. The value of $\int_0^1 \dfrac{dx}{1+x^2}$ by Simpson's 1/3 rule (taking $n = 1/4$) is ....

10. For the data:

x:	2	4	6	8
f (x):	3	5	6	7,

$\int_2^8 f(x)dx$ when found by the Trapezoidal rule is

(a) 18   (b) 25   (c) 16   (d) 32.

11. Given $f_{00}, f_{01}, f_{02}, f_{10}, f_{11}, f_{12}, f_{20}, f_{21}, f_{22}$; then the Trapezoidal rule for evaluating $I = \int_{x_0}^{x_2} \int_{y_0}^{y_2} f(x,y)dxdy$ is

12. Gaussian two-point quadrature formula states that ....

13. The expression for $\left(\dfrac{dy}{dx}\right)_{x=x_0}$ using backward differences is ....

14. The number of strips required in Weddle's rule is ....

15. The error involved in Simpson's 1/3 rule is

(a) $-\dfrac{h^3}{12} f''(X)$ (b) $-\dfrac{h^5}{90} f^{iv}(X)$ (c) $-\dfrac{3h^5}{80} f^{iv}(X)$ (d) $-\dfrac{8h^7}{345} f^{vi}(X)$

16. The expression for Romberg integration is $I = \ldots\ldots$

17. The number of strips required in Simpson's 3/8 rule is a multiple of
   (a) 1   (b) 2   (c) 3   (d) 6.

18. Add two terms to the Euler–Maclaurin formula

$$\int_{x_0}^{x_0+nh} y\,dx = \frac{h}{2}(y_0 + 2y_1 + 2y_2 + \ldots + 2y_{n-1} + y_n) - \ldots$$

19. By the Gauss three-point formula, $\int_{-1}^{1} f(x)dx =$

20. The order of error in the Trapezoidal rule and Simpson's 1/3 rule is $\ldots..$ and $\ldots..$, respectively

21. If $y_0 = 1, y_1 = \frac{16}{17}, y_2 = \frac{4}{5}, y_3 = \frac{16}{25}, y_4 = \frac{1}{2}$ and $h = \frac{1}{4}$ then using the Trapezoidal rule, $\int_0^4 y\,dx = \ldots.$

22. The total error E in Trapezoidal rule = $\ldots\ldots$.

23. Using Simpson's 1/3 rule, $\int_0^1 \frac{dx}{x} = \ldots$ (taking $n = 4$)
   If $y_0 = 1, y_1 = 0.5, y_2 = 0.2, y_3 = 0.1, y_4 = 0.06, y_5 = 0.04$ and $y_6 = 0.03$, then

$\int_0^6 y\,dx$ by Simpson's 3/8 rule = $\ldots\ldots$

24. If $f(0) = 1, f(1) = 2.7, f(2) = 7.4, f(3) = 20.1, f(4) = 54.6$ and $h = 1$, then

$\int_0^4 f(x)dx$ by Simpson's 1/3 rule = $\ldots..$.

25. Simpson's 1/3 rule and direct integration give the same result if $\ldots\ldots$.

26. Whenever the Trapezoidal rule is applicable, Simpson's 1/3 rule can also be applied.   (True or False)

CHAPTER 9

*D*IFFERENCE *E*QUATIONS

Chapter Objectives

- Introduction
- Definitions
- Formation of difference equations
- Linear difference equations
- Rules for finding the complementary function
- Rules for finding the particular integral
- Difference equations reducible to linear form
- Simultaneous difference equations with constant coefficients
- Application to deflection of a loaded string
- Objective type of questions

9.1 Introduction

Difference calculus also forms the basis of Difference equations. These equations arise in all situations in which sequential relation exists at various discrete values of the independent variable. The need to work with discrete functions arises because there are physical phenomena which are inherently of a discrete nature. In control engineering, it often happens that the input is in the form of discrete pulses of short duration. The radar tracking devices receive such discrete pulses from the target which is being tracked. As such difference equations arise in the study of electrical networks, in the theory of probability, in statistical problems, and many other fields.

Just as the subject of Differential equations grew out of Differential calculus to become one of the most powerful instruments in the hands of a practical mathematician when dealing with continuous processes in nature, so the subject of Difference equations is forcing its way to the forefront for the treatment of discrete processes. Thus the difference equations may be thought of as the discrete counterparts of the differential equations.

9.2   Definition

A difference equation *is a relation between the differences of an unknown function at one or more general values of the argument.*

Thus
$$\Delta y_{(n+1)} + y(n) = 2 \tag{1}$$

and
$$\Delta y_{(n+1)} + \Delta^2 y_{(n-1)} = 1 \tag{2}$$

are difference equations.

An alternative way of writing a difference equation is as under:

Since $\Delta y_{(n+1)} = y_{(n+2)} - y_{(n+1)}$, therefore (1) may be written as

$$y_{(n+2)} - y(_{n+}1) + y_{(n)} = 2 \tag{3}$$

Also since, $\Delta^2 y_{(n-1)} = y_{(n+1)} - 2y_{(n)} + y_{(n-1)}$, therefore (2) takes the form:

$$y_{(n+2)} - 2y_{(n)} + y_{(n-1)} = 1 \tag{4}$$

Quite often, difference equations are met under the name of *recurrence relations*.

Order of a difference equation *is the difference between the largest and the smallest arguments occurring in the difference equation divided by the unit of increment.*

Thus (3) above is of the *second order*, for

$$\frac{\text{largest argument smallest argument}}{\text{unit of increment}} = \frac{(n+2)-n}{1} = 2,$$

and (4) is of the *third order*, for $\dfrac{(n+2)-(n-1)}{1} = 3$.

NOTE   **Obs.** *While finding the order of a difference equation, it must always be expressed in a form free of Δs, for the highest power of Δ does not give order of the difference equation.*

Solution *of a difference equation is an expression for y(n) which satisfies the given difference equation.*

The **general solution** *of a difference equation is that in which the number of arbitrary constants is equal to the order of the difference equation.*

A particular solution (or **particular integral**) *is that solution which is obtained from the general solution by giving particular values to the constants.*

9.3 Formation of Difference Equations

The following examples illustrate the way in which difference equations arise and are formed.

EXAMPLE 9.1

Form the difference equation corresponding to the family of curves

$$(1) \; y = ax + bx^2 \qquad\qquad (2) \; y_n = a \sin n\theta + b \cos n\theta \qquad\qquad (i)$$

Solution:

(*i*) We have $\Delta y = a \, \Delta(x) + b\Delta(x^2) = a \, (x + 1 - x) + b[(x + 1)^2 - x^2]$

$$= a + b(2x + 1) \qquad\qquad (ii)$$

and $\qquad\qquad \Delta^2 y = 2b[(x + 1) - x] = 2b \qquad\qquad (iii)$

To eliminate a and b, we have from (*iii*), $b = \dfrac{1}{2}\Delta^2 y$

and from (*ii*), $a = \Delta y - b(2x + 1) = \Delta y - \dfrac{1}{2}\Delta^2 y(2x + 1)$

Substituting these values of a and b in (*i*), we get

$$y = [\Delta y - \frac{1}{2}\Delta^2 y.(2x + 1)x + \frac{1}{2}\Delta2y.x^2$$

or $\qquad\qquad (x^2 + x)\Delta^2 y - 2x\Delta y + 2y = 0$

This is the desired difference equation which may equally well be written in terms of E as

$$(x^2 + x)y_{x+2} - (2x^2 + 4x)y_{x+}1 + (x^2 + 3x + 2)y_x = 0$$

(*ii*) $\qquad\qquad y_n = a \sin n\theta + b \cos n\theta$

∴ $\qquad\qquad y_{n+1} = a \sin (n + 1)\theta + b \cos (n + 1)\,\theta$

and $\qquad\qquad y_{n+2} = a \sin(n + 2)\,\theta + b \cos (n + 2)\,\theta$

Thus $y_{n+2} + y_n = a[\sin (n + 2)\, \theta + \sin n\theta] + b[\cos (n + 2)\, \theta + \cos n\theta]$

$$= 2a \sin (n + 1)\, \theta \cos \theta + 2b \cos (n + 1)\, \theta \cos \theta$$

$$= 2\cos \theta\, [a \sin (n + 1)\, \theta + b \cos (n + 1)\, \theta]$$

$$= 2 \cos \theta\, (yn + 1)$$

Hence $y_{n+2} - 2\, y_{n+1} \cos \theta + y_n = 0$.

EXAMPLE 9.2

From $y_n = A2^n + B(-3)^n$, derive a difference equation by eliminating the constants.

Solution:

We have $\qquad y_n = A.2^n + B(-3)^n, \ y_{n+}1 = 2A.2^n - 3B(-3)^n$

and $\qquad y_{n+2} = 4A.2^n + 9B(-3)^n.$

Eliminating A and B, we get

$$\begin{vmatrix} y_n & 1 & 1 \\ y_{n+1} & 2 & -3 \\ y_{n+2} & 4 & 9 \end{vmatrix} = 0 \ \text{ or } y_{n+2} + y_{n+1} - 6y_n = 0$$

which is the desired difference equation.

EXAMPLE 9.3

Show that n circles drawn in a plane so that each circle intersects all the others and no three circles meet in a point, divide the plane into $n^2 - n + 2$ parts.

Solution:

Let y_n denote the number of subregions into which the entire plane is divided by n circles. When $(n + 1)^{\text{th}}$ circle is drawn to intersect each of the previous n circles, $2n$ more subregions are added to y_n subregions.

i.e., $\qquad\qquad\qquad y_{n+1} = y_n + 2n$

$\therefore$ The diffference equation satisfied by y_n is

$y_{n+1} - y_n = 2n \quad i.e., \qquad \Delta_{yn} = 2[n]^1$

$$\therefore \qquad y_n = 2\Delta^{-1}[n]^1 = 2.\frac{[n]^2}{2} + c = n(n-1) + c$$

Obviously when $n = 1$, $y_n = 2$

Putting $n = 1$ in (i), we get $2 = 1\,(1-1) + c$ i.e., $c = 2$.

Hence $y_n = n(n-1) + 2$.

Exercises 9.1

1. Write the difference equation $\Delta^3 yx + \Delta^2 yx + \Delta y_x + y_x = 0$ in the subscript notation.

2. Assuming $\dfrac{\log(1-z)}{1+z} = y_0 + y_1 z + y_2 z^2 + \ldots + y_n z^n \ldots$, find the difference equations satisfied by y_n.

3. Form a difference equation by eliminating arbitrary constant from $u_n = a^{2n+1}$.

4. Find the difference equation satisfied by

   (i) $y = a/x + b$ $\qquad$ (ii) $y = ax^2 - bx$. $\qquad$ (iii) $y = (\sqrt{2})\left(a\sin\dfrac{n\pi}{4} + b\cos\dfrac{n\pi}{4}\right)$

5. Derive the difference equations in each of the following cases:

   (i) $y_n = A.3^n + B.5^n$. $\qquad$ (ii) $y_x = (A + Bx)2^x$.

6. Form the difference equations generated by

   (i) $yx = ax + b^2 x$ $\qquad\qquad$ (ii) $y_n = a2^n + b(-2)^n$

   (iii) $yx = a2^x + b3^x + c$.

7. Show that n straight lines, no two of which are parallel and no three of which meet in a point, divide the plane into $\dfrac{1}{2}(n^2 + n + 2)$ parts.

9.4 Linear Difference Equations

Def. *A* **linear difference equation** *is that in which* y_{n+1}, y_{n+2}, *etc. occur to the first degree only and are not multiplied together.*

A **linear difference equation with constant coefficients** *is of the form*

$$y_{n+r} + a_1 y_{n+r-1} + a_1 y_{n+r-2} + \cdots + a_r y_n = f(n) \qquad\qquad (1)$$

where $a_1, a_2, \cdots, a_r$ are constants.

Now we shall deal with linear difference equations with constant coefficients only. Their properties are analogous to those of linear differential equations with constant co-efficients.

Elementary properties. If $u_1(n), u_2(n), \cdots, u_r(n)$ are r independent solutions of the equation

$$y_{n+r} + a1y_{n+r-1} \cdots + a_r y_n = 0 \qquad (2)$$

then its complete solution is $U_n = c_1 u_1(n) + c_2 u_2(n) + \cdots + c_r u_r(n)$

where $c_1, c_2, \cdots, c_r$ are arbitrary constants.

If V_n is a particular solution of (1), then the complete solution of (1) is $y_n = U_n + V_n$.

The part Un is called the **complementary function (C.F.)** *and the part* V_n *is called the* **particular integral (P.I.)** *of* (1).

Thus the **complete solution (C.S.)** *of* (1) *is* yn = **C.F. + P.I.**

9.5   Rules for Finding the Complementary Function

(i.e., rules to solve a linear difference equation with constant coefficients having right hand side zero)

1. To begin with, consider the first order linear equation $y_{n+1} - \lambda_{yn} = 0$, where λ is a constant.

Rewriting it as $\dfrac{y_{n+1}}{\lambda^{n+1}} - \dfrac{y_n}{\lambda_n} = 0$, we have $\Delta\left(\dfrac{y_n}{\lambda_n}\right) = 0$, which gives $y_n/\lambda_n = c$, a constant.

Thus the solution of $(E - \lambda) y_n = 0$ is $y_n = \mathbf{c}.\lambda^n$.

2. Now consider the *second order linear equation* $y_{n+2} + ay_{n+1} + by_n = 0$ which in *symbolic form is*

$$(E^2 + aE + b)y_n = 0 \qquad (1)$$

Its symbolic co-efficient equated to zero *i.e.,* $E^2 + aE + b = 0$

is called the *auxiliary equation*. Let its roots be λ_1, λ_2.

Case I. *If these roots are real and distinct,* then (1) is equivalent to

$$(E - \lambda_1)(E - \lambda_2)y_n = 0 \qquad (2)$$

or
$$(E - \lambda_2)(E - \lambda_1)y_n = 0 \qquad (3)$$

If y_n satisfies the subsidiary equation $(E - \lambda_2)y_n = 0$, then it will also satisfy (3).

Similarly, if y_n satisfies the subsidiary equation $(E - \lambda_2)y_n = 0$, then it will also satisfy (2).

∴ It follows that we can derive two independent solutions of (1), by solving the two subsidiary equations

$$(E - \lambda_1)y_n = 0 \text{ and } (E - \lambda_2)y_n = 0$$

Their solutions are respectively, $y_n = c_1(\lambda_1)^n$ and $y_n = c_2(\lambda_2)^n$ where c_1, c_2 are arbitrary constants.

Thus the general solution of (1) is $y_n = c_1(\lambda_1)n + c_2(\lambda_2)n$.

Case II. *If the roots are real and equal (i.e., $\lambda_1 = \lambda_2$), then (2) becomes*

$$(E - \lambda_1)^2 y_n = 0 \tag{4}$$

Let $$y_n = (\lambda_1)^n z_n,$$

where z_n is a new dependent variable. Then (4) takes the form

$$(\lambda_1)^{n+2}z_{n+2} - 2\lambda_1(\lambda_1)^{n+1}z_{n+1} + \lambda_1^2 \cdot (\lambda_1)^n z_n = 0$$

or $\quad z_{n+2} - 2z_{n+1} + z_n = 0 \qquad i.e., \Delta^2 z_n = 0.$

∴ $\quad z_n = c_1 + c_2 n$, where c_1, c_2 are arbitrary constants.

Thus the solution of (1) becomes $y_n = (c_1 + c_2 n)(\lambda_1)n$.

Case III. If the roots are imaginary, (*i.e.* $\lambda_1 = \alpha + i\beta$, $\lambda_2 = \alpha - i\beta$), then the solution of (1)

is $\quad y_n = c_1(\alpha + i\beta)^n + c_2(\alpha - i\beta)^n \qquad$ [Put $\alpha = r \cos \theta$ and $\beta = r \sin \theta$]

$\qquad = r_n[c_1(\cos n\theta + i \sin n\theta) + c_2(\cos n\theta - i \sin n\theta)]$

$\qquad = r^n[A_1 \cos n\theta + A_2 \sin n\theta]$

where A_1, A_2 are arbitrary constants and $r = \sqrt{(\alpha^2 + \beta^2)}$, $\theta = \tan^{-1}(\beta/\alpha)$.

(3) *In general, to solve the equation* $y_{n+r} + a_1 y_{n+r-1} + a2 y_{n+r-2} + \cdots + a_r y_n = 0$ where a's are constants:

(*i*) Write the equation in the *symbolic form* $(E^r + a1E^{r-1} + \cdots + a_r)y_n = 0.$

(*ii*) Write down the *auxiliary equation i.e.*, $E^r + a1E^{r-1} \cdots + a_r = 0$ and solve it for E.

(*iii*) Write the solution as follows:

S.No.	Roots of A.E.	Solution, i.e. C.F.
1.	$\lambda_1, \lambda_2, \lambda_3, \cdots$ (real and distinct roots)	$c_1(\lambda_1)^n + c_2(\lambda_2)^n + c_3(\lambda_3)^n + \cdots$
2.	$\lambda_1, \lambda_1, \lambda_3, \cdots$ (2 real and equal roots)	$(c_1 + c_2 n)(\lambda_1)^n + c_3(\lambda_3)^n + \cdots$
3.	$\lambda_1, \lambda_1, \lambda_1, \cdots$ (3 real and equal roots)	$(c_1 + c_2 n + c_3 n_2)(y_1)^n + \cdots$
4.	$\alpha + i\beta, \alpha - i\beta, \cdots$ (a pair of imaginary roots)	$r^n(c_1 \cos n\theta + c_2 \sin n\theta)$ where $r = \sqrt{(\alpha^2 + \beta^2)}$ and $\theta = \tan^{-1}(\beta/\alpha)$.

EXAMPLE 9.4

Solve the difference equation $u_{n+3} - 2u_{n+2} - 5u_{n+1} + 6u_n = 0$.

Solution:

Given equation in *symbolic form* is $(E^3 - 2E^2 - 5E + 6)u_n = 0$

∴ Its auxiliary equation is $E^3 - 2E^2 - 5E + 6 = 0$

or $\qquad\qquad (E - 1)(E + 2)(E - 3) = 0.$ $\qquad$ ∴ $E = 1, -2, 3.$

Thus the complete solution is $u_n = c_1(1)^n + c_2(-2)^n + c_3(3)^n$.

EXAMPLE 9.5

Solve $u_{n+2} - 2u_{n+1} + u_n = 0$.

Solution:

Given difference equation in *symbolic form* is $(E^2 - 2E + 1)u_n = 0$.

∴ Its auxiliary equation is $E^2 - 2E + 1 = 0$

or $(E - 1)^2 = 0.$ $\qquad$ ∴ $E = 1, 1$

Thus the required solution is $un = (c_1 + c_2 n)(1)^n$, *i.e.*, $u_n = c_1 + c_2 n$.

EXAMPLE 9.6

Solve $y_{n+1} - 2y_n \cos \alpha + y_{n-1} = 0$.

Solution:

This is a second order difference equation in y_{n-1}; which in *symbolic form* is

$$(E^2 - 2E \cos \alpha + 1) \, y_n = 0.$$

The *auxiliary equation* is $E^2 - 2E \cos \alpha + 1 = 0$

$$\therefore \qquad E = \frac{2 \cos \alpha \pm (4 \cos^2 \alpha - 4)}{4} = \cos \alpha \pm i \sin \alpha.$$

Thus, the *solution* is $y_{n-1} = (1)^{n-1}[c1 \cos (n-1) \alpha + c_2 \sin (n-1)\alpha]$

or $\qquad y_n = c_1 \cos n\alpha + c_2 \sin n\alpha.$

EXAMPLE 9.7

The integers 0, 1, 1, 2, 3, 5, 8, 13, 21,⋯ are said to form a Fibonacci sequence. Form the Fibonacci difference equation and solve it.

Solution:

In this sequence, each number beyond the second, is the sum of its two previous numbers. If y_n be the nth number then $y_n = y_{n-1} + y_{n-2}$ for $n > 2$.

or $\qquad y_{n+2} - y_{n+1} - y_n = 0$ (for $n > 0$)

or $\qquad (E^2 - E - 1)y_n = 0$ is the difference equation.

Its A.E. is $E^2 - E - 1 = 0$ which gives $E = \frac{1}{2}(1 \pm \sqrt{5}).$

Thus the solution is $y_n = c_1 \left(\dfrac{1+\sqrt{5}}{2}\right)^n + c_2 \left(\dfrac{1-\sqrt{5}}{2}\right)^n$, for $n > 0$

When $n = 1, y_1 = 0$

$$\therefore \qquad c_1 \left(\frac{1+\sqrt{5}}{2}\right) + c_2 \left(\frac{1-\sqrt{5}}{2}\right) = 0 \qquad\qquad (i)$$

When $n = 2, y_2 = 1$

$$\therefore \qquad c_1 \left(\frac{1+\sqrt{5}}{2}\right) + c_2 \left(\frac{1-\sqrt{5}}{2}\right)^2 = 1 \qquad\qquad (ii)$$

Solving (i) and (ii), we get

$$c_1 = \frac{5-\sqrt{5}}{10} \quad \text{and} \quad c_2 = \frac{5+\sqrt{5}}{10}$$

Hence the complete solution is

$$y_n = \frac{5-\sqrt{5}}{10}\left(\frac{1+\sqrt{5}}{2}\right) + \frac{5+\sqrt{5}}{10}\left(\frac{1-\sqrt{5}}{2}\right)^n$$

Exercises 9.2

Solve the following difference equations:

1. $u_{x+2} - 4u_{x+1} + 4U_x = 0$, given $u_0 = 1$, $u_1 = 0$.

2. $y_{n+2} + y_{n+1} + 2y_n = 0$.

3. $\Delta^2 u_n + 2\Delta u_n + u_n = 0$.

4. $(\Delta^2 - 3\Delta + 2)y_n = 0$

5. $4y_n - y_{n+2} = 0$ given that $y_0 = 0$, $y_1 = 2$.

6. $u_{k+3} - 3u_{k+2} + 4u_k = 0$.

7. $f(x + 3) - 3f(x + 1) - 2f(x) = 0$.

8. $u_{n+}3 - 3_{un+}1 + 2_{un} = 0$, given $u1 = 0$, $u_2 = 8$ and $u_3 = -2$.

9. $(E^3 - 5E^2 + 8E - 4)y_n = 0$, given that $y_0 = 3$, $y_1 = 2$, $y_4 = 22$.

10. $u_{n+1} - 2u_n + 2u_n - 1 = 0$.

11. $y_{m+}3 + 16_{ym} - 1 = 0$.
  [**HINT.** $E^4 = -16 = 16[\cos (2n + 1)\pi + i \sin (2n + 1)\pi]$; use De Moivre's theorem.]

12. Show that the difference equation $I_{m+1} - (2 + r_0/r) I_m + I_{m-1} = 0$ has the solution $I_m = I_0 \sinh (n - m)\alpha/\sinh (n - 1)\alpha$, if $I = I_0$ and $I_n = 0$, α being $= 2 \sinh^{-1}\dfrac{1}{2}(r_0 / r)^{1/2}$.

13. A series of values of yn satisfy the relation $y_{n+2} + a y_{n+1} + by_n = 0$. Given that $y_0 = 0$, $y_1 = 1$, $y_2 = y_3 = 2$. Show that $y_n = 2^{n/2} \sin n\pi/4$.

14. A particle is moving in a horizontal direction. In each second, it travels a distance which is twice the distance moved in the previous second. If the distance moved in the rth second is x_r and $x_0 = 3$, $x_1 = 4$, then show that $x_r = 2^n + 2$.

15. A plant is such that each of its seeds when one year old produces eight-fold and produces eighteen-fold when two years old or more. A seed is planted and as soon as a new seed is produced it is planted. Taking y_n to be the number of seeds produced at the end of the nth year, show that
$y_{n+1} = 8y_n + 18 (y_1 + y_2 + \cdots + y_{n-1})$.
Hence show that $y_{n+2} - 9y_{n+1} - 10y_n = 0$ and find y_n.

9.6 Rules for Finding the Particular Integral

Consider the equation $y_{n+r} + a_1 y_{n+r-1} + \cdots + a_r y_n = f(n)$

which in *symbolic form is* $\phi(E)_{yn} = f(n)$ $\qquad$ (1)

where $\phi(E) = E^r + a_1 E^{r-1} + \ldots + a_r$

Then the particular integral is given by $\text{P.I} = \dfrac{1}{\phi(E)} f(n)$

Case I. *When* $f(n) = a^n$ $\qquad\qquad$ (*Power function*)

$$\text{P.I} = \frac{1}{\phi(E)} a^n, \text{ put } E = a$$

$$= \frac{1}{\phi(a)} a^n, \text{provided } \phi(a) \neq 0.$$

If $\phi(a) = 0$, *then for the equation*

(*i*) $(E-a)y_n = a^n$ $\text{ P.I} = \dfrac{1}{E-a} a^n = na^{n-1}$

(*ii*) $(E-a)^2 y_n = a^n$ $\text{ P.I.} = \dfrac{1}{(E-a)^2} a^n = \dfrac{n(n-1)}{2!} a^{n-2}$

(*iii*) $(E-a)^3 y^n = a^n$ $\text{ P.I} = \dfrac{1}{(E-a)^3} a^n = \dfrac{n(n-1)(n-2)}{3!} a^{n-3}$

and so on.

EXAMPLE 9.8

Solve $y_{n+2} - 4y_{n+1} + 3y_n = 5^n$.

Solution:

Given equation in *symbolic form* is $(E^2 - 4E + 3)y_n = 5^n$

$\therefore$ The auxiliary equation is $E^2 - 4E + 3 = 0$

or $(E-1)(E-3) = 0$ $\qquad\qquad \therefore E = 1, 3$

$\therefore$ C.F. $= c_1(1)_n + c_2(3)^n = c_1 + c_2.3^n$

and $\qquad\qquad\qquad\qquad \text{P.I} = \dfrac{1}{E^2 - 4E + 3} 5^n \qquad\qquad$ [Put $E = 5$]

$$= \frac{1}{25 - 4.5 + 3} 5^n = \frac{1}{8} \cdot 5^n$$

Thus the complete solution is $y_n = c1 + c_2.3^n + 5^{n/8}$.

EXAMPLE 9.9

Solve $u_{n+2} - 4u_{n+1} + 4u_n = 2n$.

Solution:

Given equation in *symbolic form* is $(E^2 - 4E + 4)u_n = 2n$.

The auxiliary equation is $E^2 - 4E + 4 = 0$. $\therefore E = 2, 2$.

$\therefore$ C.F. $= (c_1 + c_2 n)2^n$

$$\text{P.I} = \frac{1}{(E-2)^2} \cdot 2^n = \frac{n(n-1)}{2!} \cdot 2^{n-2} = n(n-1)2^{n-3}.$$

Hence the complete solution is $u_n = (c1 + c2n)\,2^n + n(n-1)\,2^{n-3}$.

Case II. (1) *When $f(n) = \sin kn$.*   (*trigonometric function*)

$$P.I = \frac{1}{\phi(E)}\sin kn = \frac{1}{\phi(E)}\left(\frac{e^{ikn} - e^{-ikn}}{2i}\right) = \frac{1}{2i}\left[\frac{1}{\phi(E)}a^n - \frac{1}{\phi(E)}b^n\right]$$

where $a = e^{ik}$ and $b = e^{-ik}$.

Now proceed as in case I.

(2) *When $f(n) = \cos kn$*

$$P.I. = \frac{1}{\phi(E)}\cos kn = \frac{1}{\phi(E)}\left(\frac{e^{ikn} + e^{-ikn}}{2}\right)$$

$$= \frac{1}{2}\left[\frac{1}{\phi(E)}a^n + \frac{1}{\phi(E)}b^n\right]\text{as before.}$$

Now proceed as in case I

EXAMPLE 9.10

Solve $y_{n+2} - 2\cos\alpha.y_{n+1} + y_n = \cos\alpha n$.

Solution:

Given equation in *symbolic form* is $(E^2 - 2\cos\alpha.\,E + 1)\,y_n = \cos\alpha n$

The *auxiliary equation is* $E^2 - 2\cos\alpha.\,E + 1 = 0$.

$\therefore$ $E = \dfrac{2\cos\alpha \pm \sqrt{(4\cos^2\alpha - 4)}}{2} = \cos\alpha \pm i\sin\alpha.$

$\therefore$ C.F. $= (1)n\,[c_1 \cos\alpha n + c_2 \sin\alpha n]$, *i.e.*, $c_1 \cos\alpha n + c_2 \sin\alpha n$.

$$P.I = \frac{1}{E^2 - 2E\cos\alpha + 1}\cos\alpha n$$

$$= \frac{1}{E^2 - E(e^{i\alpha} + e^{-i\alpha}) + 1}\left(\frac{e^{i\alpha n} + e^{-i\alpha n}}{2}\right)$$

$$= \frac{1}{2}\left[\frac{1}{(E - e^{i\alpha})(E - e^{-i\alpha})}e^{i\alpha n} + \frac{1}{(E - e^{i\alpha})(E - e^{-i\alpha})}e^{-i\alpha n}\right]$$

$$[\text{Put } E = e^{i\alpha}] \qquad\qquad\qquad\qquad [\text{Put } E = e^{-i\alpha}]$$

$$= \frac{1}{2}\left[\frac{1}{(E - e^{i\alpha})}\cdot\frac{1}{e^{i\alpha} - e^{-i\alpha}}e^{i\alpha n} + \frac{1}{E - e^{-i\alpha}}\cdot\frac{1}{e^{-i\alpha} - e^{i\alpha}}e^{-i\alpha n}\right]$$

$$= \frac{1}{4i\sin\alpha}\left[\frac{1}{E - e^{i\alpha}}e^{i\alpha n} - \frac{1}{E - e^{-i\alpha}}e^{-i\alpha n}\right]$$

$$= \frac{1}{4i\sin\alpha}[n\cdot e^{i\alpha(n-1)} - n\cdot e^{-i\alpha(n-1)}]$$

$$= \frac{n}{2\sin\alpha}\left[\frac{e^{i\alpha(n-1)} - e^{-i\alpha(n-1)}}{2i}\right] = \frac{n\sin(n-1)\alpha}{2\sin\alpha}$$

Hence the complete solution is

$$y_n = c_1\cos\alpha n + c_2\sin\alpha n + \frac{n\sin(n-1)}{2\sin\alpha}.$$

Case III. When $f(n) = np$. \hfill *(Polynomial function)*

$$P.I = \frac{1}{\phi(E)}n^p = \frac{1}{\phi(1 + \Delta)}n^p$$

1. Expand $[\phi(1 + \Delta)]^{-1}$ in ascending powers of Δ by the binomial theorem as far as the term in Δp.

2. Express n^p in the factorial form and operate on it with each term of the expansion.

EXAMPLE 9.11

Solve $y_{n+2} - 4y_n = n^2 + n - 1$.

Solution: Given equation is $(E^2 - 4)y_n = n^2 + n - 1$.

The *auxiliary equation is* $E^2 - 4 = 0$, $\therefore E = \pm 2$.

$$\therefore \text{C.F.} = c_1(2)^n + c_2(-2)^n$$

$$\therefore \quad P.I = \frac{1}{E^2 - 4}(n^2 + n - 1) = \frac{1}{(1+\Delta)^2 - 4}[n(n-1) + 2n - 1]$$

$$= \frac{1}{\Delta^2 + 2\Delta - 3}([n]^2 + 2[n] - 1) = -\frac{1}{3}\left[1 - \left(\frac{2}{3}\Delta + \frac{\Delta^2}{3}\right)\right]^{-1}\{[n]^2 + 2[n] - 1\}$$

$$= -\frac{1}{3}\left[1 + \left(\frac{2}{3}\Delta + \frac{\Delta^2}{3}\right) + \left(\frac{2}{3}\Delta + \frac{\Delta^2}{3}\right)^2 + \ldots\right] \times \{[n]^2 + 2[n] - 1\}$$

$$= -\frac{1}{3}[1 + \frac{2}{3}\Delta + \frac{7}{9}\Delta^2 + \ldots]\{[n]^2 + 2[n] - 1\}$$

$$= -\frac{1}{3}\{[n]^2 + 2[n] - 1\frac{2}{3}(2\lfloor n \rfloor) + 2 + \frac{7}{9}n - \frac{17}{27}$$

Hence the *complete solution* is $y_n = c_1 2^n + c_2(-2)^n - \dfrac{n^2}{3} - \dfrac{7n}{9} - \dfrac{17}{27}$.

Case IV. *When* $f(n) = a^n F(n)$, $F(n)$ *being a polynomial of finite degree in* n.

$$\text{P.I} = \frac{1}{\phi(E)} a^n F(n) = a^n \cdot \frac{1}{\phi(aE)} F(n).$$

Now $F(n)$ being a polynomial in n, proceed as in case III.

EXAMPLE 9.12

Solve $y_{n+2} - 2y_{n+1} + y_n = n^2 . 2^n$.

Solution:

Given equation is $(E^2 - 2E + 1)y_n = n^2 . 2n$.

Its   C.F. $= c_1 + c_2 n$

and   P.I $= \dfrac{1}{(E-1)^2} 2^n . n^2 = 2^n \dfrac{1}{(2E-1)^2} n^2 = 2^n \dfrac{1}{(1+2\Delta)^2} n^2$

$$= 2^n (1 + 2\Delta)^{-2}\{n(n-1) + n\} = 2^n(1 - 4\Delta + 12\Delta^2 - \cdots)([n]^2 + [n])$$

$$= 2^n\{[n]^2 + [n] - 4(2[n] + 1) + 12 \times 2\}$$

$$= 2^n([n]^2 - 7[n] + 20) = 2^n(n^2 - 8n + 20).$$

Hence the *complete solution* is $y_n = c_1 + c_2 n + 2^n(n^2 - 8n + 20)$.

Exercises 9.3

Solve the following difference equations:

1. $y_{n+2} - 5y_{n+1} - 6y_n = 4_n$, $y_0 = 0$, $y_1 = 1$.

2. $y_{n+2} + 6y_{n+1} + 9y_n = 2_n$, $y_0 = y_1 = 0$.

3. $y_{p+3} - 3y_{p+2} + 3y_{p+1} - yp = 1$.

4. $(E^2 - 4E + 3)y = 3x$.

5. $u_{x+2} - 7u_{x+1} + 10ux = 12.4x$.

6. $y_{x+2} - 4y_{x+1} + 4yx = 3.2x + 5.4x$.

7. $u_{n+2} - u_n = \cos n/2$.

8. $y_{p+2} - (2 \cos \dfrac{1}{2})y_{p+1} + y_p = \sin p/2$.

9. $y_{n+2} - 2y_{n+1} + 4y_n = 6$, given that $y_0 = 0$ and $y_1 = 2$.

10. $(E^2 - 4)y_x = x^2 - 1$.

11. $y_{n+3} + y_n = n^2 + 1$, $y0 = y_1 = y_2 = 0$.

12. $y_{n+3} - 5y_{n+2} + 3y_{n+1} + 9y_n = 2^n + 3n$.

13. $(4E^2 - 4E + 1)\, y = 2^n + 2^{-n}$.

14. $y_{n+2} + 5y_{n+1} + 6y_n = n + 2^n$.

15. $u_{x+2} + 6u_{x+1} + 9u_x = x2^x + 3x + 7$.

16. $y_{n+3} + 8yn = (2n + 3)\, 2n$.

17. $u_{n+2} - 4u_{n+1} + 4un = n^2 2n$.

18. $(E^2 - 5E + 6)\, y_k = 4k(k^2 - k + 5)$.

19. $(E^2 - 2E + 4)y_n = -2^n \left\{ 6\cos\dfrac{n\pi}{3} + 2\sqrt{3}\sin\dfrac{n\pi}{3} \right\}$

20. A beam of length l, supported at n points carries a uniform load w per unit length The bending moments $M_1, M_2 \ldots M_n$ at the supports satisfy the Clapeyron's equation:
$$M_{r+2} + 4M_{r+1} + M_r = -\dfrac{1}{2}wl^2.$$

21. If a beam weighing 30 kg is supported at its ends and at two other supports dividing the beam into three equal parts of 1 meter length, show that the bending moment at each of the two middle supports is 1 kg meter.

9.7 Difference Equations Reducible to Linear Form

At times non-linear difference equations can be reduced to the linear form by a suitable substitution. We shall consider the following types of such equations:

I. Homogeneous equation of the type $F\{y_{x+1}/y_x, x\} = 0$.

Putting $y_{x+1}/y_x = u_x$, this equation takes the linear form $F(u_x, x) = 0$.

EXAMPLE 9.13

Solve $y_{x+1}{}^2 - 3y_{x+1} y_x + 2yx^2 = 0$.

Solution:

Dividing throughout by $y_x{}^2$, it becomes $\left(\dfrac{y_{x+1}}{y_x}\right)^2 - 3\left(\dfrac{y_{x+1}}{y_x}\right) + 2 = 0$

Putting $y_{x+1}/y_x = u_x$, we get $u_x{}^2 - 3u_x + 2 = 0$

or *Case I.* When $ux = 1$ *i.e.*, $y_{x+1} - yx = 0$.

Its A.E. is $E - 1 = 0$ or $E = 1$.

∴ Solution is $y_x = c_1.(1)x = c_1$.

Case II. When $u_x = 2$ *i.e.*, $y_{x+1} - 2yx = 0$.

Its A.E. is $E - 2 = 0$ or $E = 2$.

∴   Solution is $y_x = c_2(2)^x$.

II. Equation of the type $p(x)\, y_x y_{x+1} + q(x)y_{x+1} + r(x)y_x = 0$

Dividing throughout by $y_x\, y_{x+1}$, it reduces to $p(x) + \dfrac{q(x)}{y_x} + \dfrac{r(x)}{y_{x+1}} = 0$

Putting $1/y_x = u_x$, we get $p(x) + q(x)\, u_x + r(x)\, u_{x+1} = 0$

which is a linear equation.

EXAMPLE 9.14

Solve $y_{x+1} - y_x + xy_{x+1}y_x = 0$ given $y_1 = 2$.

Solution:

Dividing throughout by $y_x y_{x+1}$, the given equation becomes

$$\frac{1}{y_{x+1}} - \frac{1}{y_x} = x$$

Putting $1/y_x = u_x$, we get $u_{x+1} - u_x = x$ where $u_1 = \frac{1}{2}$ as $y_1 = 2$.

or $\qquad \Delta u_x = x$ or $u_x = \Delta^{-1}[x] = [x]^2/2 + c_1$

But $\qquad u_1 = 1/2.$ $\qquad\qquad \therefore c_1 = 1/2.$

Thus $\qquad u_x = \frac{1}{2}x(x-1) + \frac{1}{2}$ i.e. $y_x = \frac{1}{u_x} = \frac{2}{x^2 - x + 1}$

III. Equation of the type $y_x y_{x+1} + p(x)\, y_{x+1} + q(x)\, y_x = r(x)$.

We have $y_{x+1}[y_x + p(x)] + q(x)\, y_x = r(x)$ $\qquad\qquad\qquad\qquad$ (1)

Putting $y_x + p(x) = u_{x+1}/u_x$ or $y_x = (u_{x+1}/u_x) - p(x)$, (1) reduces to

$$\left\{\frac{u_{x+2}}{u_{x+1}} - p(x+1)\right\}\frac{u_{x+1}}{u_x} + q(x)\left\{\frac{u_{x+1}}{u_x} - p(x)\right\} = r(x)$$

or $\qquad u_{x+2} + [q(x) - p(x+1)]u_{x+1} - [p(x)\,q(x) + r(x)]u_x = 0$

which is a linear equation.

EXAMPLE 9.15

Solve $y_{x+1}\, y_x + (x+2)y_{x+1} + xy_x + x^2 + 2x + 2 = 0$.

Solution:

We have $y_{x+1}[y_x + x + 2] + xy_x = -x^2 - 2x - 2$

Putting $y_x = \frac{u_{x+1}}{u_x}(x+2)$,

it reduces to $\frac{u_{x+2}}{u_x} = -2$ or $u_{x+2} - 3u_{x+1} + 2u_x = 0$.

Its A.E. is $E^2 - 3E + 2 = 0$, $\qquad\qquad \therefore E = 1, 2$.

Thus the solution is $u_x = c_1 + c_2\, 2^x$.

Hence the solution of the given equation is $y_x = \dfrac{c_1 = c_2\, 2^{x+1}}{c_1 + c_2\, 2^x} - x - 2$.

Exercises 9.4

Solve the following difference equations:

1. $y_x\, y_{x+2} = y_{x+1}^{\,2}$.

2. $y_{x+2}\, y_x^{\,2} = y_{x+1}^{\,3}$, if $y_1 = 1$, $y_2 = 2$.

3. $2y_{x+1}^{\,2} + y_{x+1}\, yx - yx^2 = 0$.

4. $y_{n+1} = \sqrt{y_n} \cdot y_{n+1} =$.

5. $y_x\, y_{x+1} - 3y_x + 2 = 0$.

6. $y_{x+1}\, y_x + 5y_{x+1} + y_x + 9 = 0$.

9.8 Simultaneous Difference Equations with Constant Coefficients

The method used for solving simultaneous differential equations with constant coefficients also applies to simultaneous difference equations with constant coefficients. The following example illustrates the technique.

EXAMPLE 9.16

Solve the simultaneous difference equations

$$u_{x+1} + v_x - 3u_x = x,\ 3u_x + v_{x+1} - 5v_x = 4^x$$

subject to the conditions $u_1 = 2$, $v_1 = 0$.

Solution:

Given equations in symbolic form, are

$$(E - 3)u_x + v_x = x \qquad (i)$$
$$3u_x + (E - 5)v_x = 4^x \qquad (ii)$$

Operating the first equation with $E - 5$ and subtracting the second from it, we get

$$[(E - 5)(E - 3) - 3]u_x = (E - 5)x - 4^x$$

or
$$(E^2 - 8E + 12)u_x = 1 - 4x - 4^x$$

Its solution is

$$u_x = c_1 2^x + c_2 6^x - \frac{4}{5}x - \frac{19}{25} + \frac{4^x}{4}$$

Substituting the value of u_x from (iii) in (i), we get

$$v_x = c_1 2^x - 3c_2 6^x - \frac{3x}{5} - \frac{34}{25} - \frac{4^x}{4}$$

Taking $u_1 = 2$, $v_1 = 0$, in (iii) and (iv), we obtain

$$2c_1 + 6c_2 = \frac{64}{25}, 2c_1 - 18c_2 = \frac{74}{25}$$

Where $\qquad c_1 = 1.33, c_2 = -0.0167$

Hence $\qquad u_x = 1.332^x - 0.01676.6^x - 0.8x - 0.76 + 4^{x-1}$

$$v_x = 1.33.2^x - 0.05.6^x - 1.36 - 4^{x-1}.$$

Exercises 9.5

Solve the following simultaneous difference equations:

1. $y_{x+1} - z_x = 2(x + 1)$, $z_{x+1} - y_x = -2(x + 1)$.

2. $y_{n+1} - y_n + 2z_{n+1} = 0$, $z_{n+1} - z_n - 2y_n = 2^n$.

3. $u_{n+1} + n = 3u_n + 2v_n$, $v_{n+1} - n = u_n + 2v_n$, given $u_0 = 0$, $v_0 = 3$.

4. $u_{x+1} + v_x + w_x = 1$, $u_x + v_{x+1} + w_x = x$, $u_x + v_x + w_{x+1} = 2x$.

9.9 Application to Deflection of a Loaded String

Consider a light string of length l stretched tightly between A and B. Let the forces P_i be acting at its equispaced points x_i ($i = 1, 2,..., n - 1$) and perpendicular to AB resulting in small transverse displacements y_i at these points (Figure 9.1). Assuming the angle θ_i made by the portion between xi and $xi+1$ with the horizontal, to be small, we have

$$\sin \theta_i = \tan \theta_i = \theta_i \text{ and } \cos \theta_i = 1.$$

If T is the tension of the string at x_i, then $T \cos \theta_i = T$ i.e., the tension may be taken as uniform.

Taking $x_{i+1} - x_i = h$, we have

$$y_{i+1} - y_i = h \tan \theta_i = h\theta_i. \qquad (1)$$
$$y_i - y_{i-1} = h \tan \theta_{i-1} = h\theta_{i-1} \qquad (2)$$

Also resolving the forces in equilibrium at $(xi, yi) \perp$ to AB, we get

$$T \sin \theta_i - T \sin \theta_i{-}1 + P_i = 0 \ i.e. \ T(\theta_i - \theta_i{-}1) + P_i = 0 \tag{3}$$

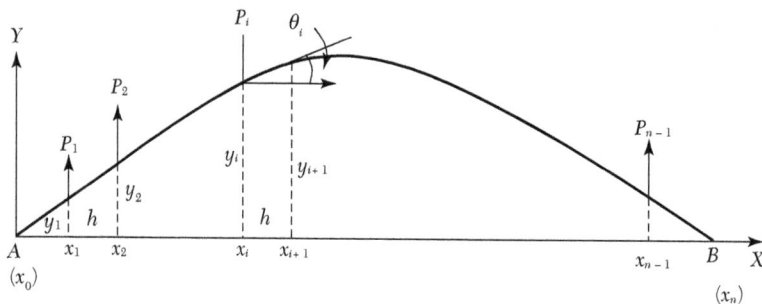

FIGURE 9.1

Eliminating θ_i and $\theta_i{-}1$ from (1), (2) and (3), we obtain

$$y_{t+1} - 2y_1 + y_{i-1} = \frac{hp_i}{T} \tag{4}$$

which is a difference equation and its solution gives the displacements yi. To obtain the arbitrary constants in the solution, we take $y_0 = y_n = 0$ as the boundary conditions, since the ends A and B of the string are fixed.

EXAMPLE 9.17

A light elastic string stretched between two fixed nails is 120 cm apart, carries 11 loads of weight at 5 gm each at equal intervals and the resulting tension is 500 gm weight. Show that the sag at the mid-point is 1.8 cm.

Solution:

Taking $h = 10$ cm, $P_i = 5$ gm and $T = 500$ gm weight.,

the above equation (4) becomes $y_{i+1} - 2y_i + y_{i-1} = -1/10$

i.e., $\quad y_{i+2} - 2y_{i+1} + y_1 = -\dfrac{1}{10}.$

Its A.E. is $(E - 1)^2 = 0 \quad i.e., E = 1, 1.$ $\qquad \therefore$ C.F. $= c_1 + c_2 i$

and P.I $= \dfrac{1}{(E-1)^2}\left(-\dfrac{1}{10}\right) = -\dfrac{1}{10}\dfrac{1}{(E-1)^2}(1)^i = -\dfrac{1}{10}\dfrac{i(i-1)}{2} = \dfrac{1}{20}$

Thus the C.S. is $y_i = c_1 + c_2 i + \dfrac{1}{20}(i - i^2)$

Since $y_0 = 0$, $\therefore c_1 = 0$

and $y_{12} = 0$, $\therefore c_2 = \dfrac{11}{20}$

Hence $y_i \approx \dfrac{}{20} i \quad \dfrac{}{20}(i \quad i^2)$

At the mid-point $i = 6$, we get $y_6 = 1.8$ cm.

Exercises 9.6

A light string of length $(n + 1)l$ is stretched between two fixed points with a force P. It is loaded with n equal masses m at distances l. If the system starts rotating with angular velocity ω, find the displacement y_i of the ith mass.

9.10   Objective Type of Questions

Exercises 9.7

Select the correct answer or fill up the blanks in the following questions:

1. $y_n = A \, 2^n + B \, 3^n$, is the solution of the difference equation $\cdots$

2. The solution of $(E - 1)^3 \, u_n = 0$ is $\cdots$ ..

3. The solution of the difference equation $u_{n+3} - 2u_{n+2} - 5u_{n+1} + 6u_n = 0$ is $\cdots$

4. The solution of $y_{n+1} - y_n = 2^n$ is $\cdots$ given that $y_0 = 2$.

5. The difference equation $y_{n+1} - 2y_n = n$ given that $y_0 = 2$ has $y_n = \cdots$ as its solution.

6. The difference equation corresponding to the family of curves $y = ax^2 + bx$ is $\cdots$

7. The particular integral of the equation $(E - 2) \, y_n = 1$.

8. The solution of $4y_n = y_{n+} 2$ such that $y_0 = 0$, $y_1 = 2$, is $\cdots$.

9. The equation $\Delta^2 u_{n+1} + \dfrac{1}{2}\Delta^2 u_n = 0$ is of order $\cdots$ ..

10. The difference equation satisfied by $y = a + b/x$ is $\cdots$.

11. The order of the difference equation $y_{n+2} - 2y_{n+1} + y_n = 0$ is $\cdots$.

12. The solution of $y_{n+2} - 4y_{n+1} + 4y_n = 0$ is $\cdots$.

13. The particular integral of $u_{x+2} - 6u_{x+1} + 9u_x = 3$ is $\cdots$.

14. The difference equation generated by $u_n = (a + bn)\, 3^n$ is $\cdots$

15. Solution of $6y_{n+2} + 5y_{n+1} - 6y_n = 2^n$ is $y_n = A(2/3)^n + B(-3/2)^n + 2^n/28$.
(True or False)

CHAPTER 10

NUMERICAL SOLUTION OF ORDINARY DIFFERENTIAL EQUATIONS

Chapter Objectives

- Introduction
- Picard's method
- Taylor's series method
- Euler's method
- Modified Euler's method
- Runge's method
- Runge-Kutta method
- Predictor-corrector methods.
- Milne's method
- Adams-Bashforth method
- Simultaneous first order differential equations
- Second order differential equations.
- Error analysis
- Convergence of a method
- Stability analysis
- Boundary-value problems
- Finite-difference method
- Shooting method
- Objective type of questions

10.1   Introduction

A number of problems in science and technology can be formulated into differential equations. The analytical methods of solving differential equations are applicable only to a limited class of equations. Quite often differential equations appearing in physical problems do not belong to any of these familiar types and one is obliged to resort to numerical methods. These methods are of even greater importance when we realize that computing machines are now readily available which reduce numerical work considerably.

Solution of a differential equation. The solution of an ordinary differential equation means finding an explicit expression for y in terms of a finite number of elementary functions of x. Such a solution of a differential equation is known as the *closed* or *finite form of solution*. In the absence of such a solution, we have recourse to numerical methods of solution.

Let us consider the first order differential equation

$$dy/dx = f(x, y), \text{ given } y(x_0) = y_0 \qquad (1)$$

to study the various numerical methods of solving such equations. In most of these methods, we replace the differential equation by a difference equation and then solve it. These methods yield solutions *either* as a power series in x from which the values of y can be found by direct substitution, *or* a set of values of x and y. The methods of Picard and Taylor series belong to the former class of solutions. In these methods, y in (1) is approximated by a truncated series, each term of which is a function of x. *The information about the curve at one point is utilized and the solution is not iterated. As such, these are referred to as* **single-step methods.**

The methods of Euler, Runge-Kutta, Milne, Adams-Bashforth, etc. belong to the latter class of solutions. *In these methods, the next point on the curve is evaluated in short steps ahead, by performing iterations until sufficient accuracy is achieved. As such, these methods are called* **step-by-step methods**.

Euler and Runga-Kutta methods are used for computing y over a limited range of x- values whereas Milne and Adams methods may be applied for finding y over a wider range of x-values. Therefore Milne and Adams methods require starting values which are found by Picard's Taylor series or Runge-Kutta methods.

Initial and boundary conditions. An ordinary differential equation of the nth order is of the form

$$F(x, y, dy/dx, d^2y/dx^2, \cdots, d^ny/dx^n) = 0 \qquad (2)$$

Its general solution contains n arbitrary constants and is of the form

$$\phi(x, y, c_1, c_2, \cdots, c_n) = 0 \qquad (3)$$

To obtain its particular solution, n conditions must be given so that the constants $c_1, c_2 \ldots, c_n$ can be determined.

If these conditions are prescribed at one point only (say:x_0), then the differential equation together with the conditions constitute an **initial value problem** *of the nth order.*

If the conditions are prescribed at two or more points, then the problem is termed as **boundary value problem.**

In this chapter, we shall first describe methods for solving initial value problems and then explain the **finite difference method** and **shooting method** for solving boundary value problems.

10.2 Picard's Method

Consider the first order equation $\dfrac{dy}{dx} = f(x, y)$ \qquad (1)

It is required to find that particular solution of (1) which assumes the value y_0 when $x = x_0$. Integrating (1) between limits, we get

$$\int_{y_0}^{y} dy = \int_{x_0}^{x} f(x, y) dx \text{ or } y = y_0 + \int_{x_0}^{x} f(x, y) dx \qquad (2)$$

This is an integral equation equivalent to (1), for it contains the unknown y under the integral sign.

As a first approximation y_1 to the solution, we put $y = y_0$ in $f(x, y)$ and integrate (2), giving

$$y_1 = y_0 + \int_{x_0}^{x} f(x, y_0) dx$$

For a second approximation y_2, we put $y = y_1$ in $f(x, y)$ and integrate (2), giving

$$y_2 = y_0 + \int_{x_0}^{x} f(x, y_1) dx$$

Similarly, *a third approximation is*

$$y_3 = y_0 + \int_{x_0}^{x} f(x, y_2)dx$$

Continuing this process, we obtain y_4, y_5, $\cdots$ y_n where

$$y_n = y_0 + \int_{x_0}^{x} f(x, y_{n-1})dx$$

Hence this method gives a sequence of approximations y_1, y_2, y_3 $\cdots$ each giving a better result than the preceding one.

NOTE **Obs.** *Picard's method is of considerable theoretical value, but can be applied only to a limited class of equations in which the successive integrations can be performed easily. The method can be extended to simultaneous equations and equations of higher order (See Sections 10.11 and 10.12).*

EXAMPLE 10.1

Using Picard's process of successive approximations, obtain a solution up to the fifth approximation of the equation $dy/dx = y + x$, such that $y = 1$ when $x = 0$. Check your answer by finding the exact particular solution.

Solution:

(*i*) We have $y = 1 + \int_{x_0}^{x} (x + y)dx$

First approximation. Put $y = 1$ in $y + x$, giving

$$y_1 = 1 + \int_{x_0}^{x} (1 + x)dx = 1 + x + x^2/2$$

Second approximation. Put $y = 1 + x + x^2/2$ in $y + x$, giving

$$y_1 = 1 + \int_{x_0}^{x} (1 + x + x^2/2)dx = 1 + x + x^2 + x^3/6$$

Third approximation. Put $y = 1 + x + x^2 + x^3/6$ in $y + x$, giving

$$y_3 = 1 + \int_{x_0}^{x} (1 + x + x^2 + x^3/6)dx = 1 + 2x + x^2 + \frac{x^3}{3} + \frac{x^4}{24}$$

Fourth approximation. Put $y = y_3$ in $y + x$, giving

$$y_4 = 1 + \int_{0}^{x} \left(1 + 2x + x^2 + \frac{x^3}{3} + \frac{x^4}{24}\right)dx$$

$$= 1 + x + x^2 + \frac{x^3}{3} + \frac{x^4}{12} + \frac{x^5}{120}$$

Fifth approximation, Put $y = y_4$ in $y + x$, giving

$$y_5 = 1 + \int_0^x \left(1 + 2x + x^2 + \frac{x^3}{3} + \frac{x^4}{12} + \frac{x^5}{120}\right) dx \tag{1}$$

$$= 1 + x + x^2 + \frac{x^3}{3} + \frac{x^4}{12} + \frac{x^5}{60} + \frac{x^6}{720}$$

(*ii*) Given equation

$$\frac{dy}{dx} - y = x \quad \text{is a Leibnitzs linear in } x$$

Its, I.F. being e^{-x} the solution is

$$ye^{-x} = \int xe^{-x} dx + c$$

$$= -xe^{-x} - \int \left(-e^{-x}\right) dx + c = -xe^{-x} - e^{-x} + c$$

∴ $\qquad y = ce^x - x - 1$

Since $y = 1$, when $x = 0$, $\qquad\qquad$ ∴ $c = 2$.

Thus the desired particular solution is

$$y = 2e^x - x - 1 \tag{2}$$

Or using the series: $e^x = 1 + x + \frac{x^2}{2!} + \frac{x^3}{3!} + \frac{x^4}{4!} + \cdots$

We get $\qquad\qquad y = 1 + x + x^2 + \frac{x^3}{3} + \frac{x^4}{12} + \frac{x^5}{60} + \frac{x^6}{360} + \cdots \infty \tag{3}$

Comparing (1) and (3), it is clear that (1), approximates to the exact particular solution (3) upto the term in x^5.

NOTE **Obs.** *At* $x = 1$, *the fourth approximation* $y_4 = 3.433$ *and the fifth approximation* $y_5 = 3.434$ *whereas the exact value is* 3.44.

EXAMPLE 10.2

Find the value of y for $x = 0.1$ by Picard's method, given that

$$\frac{dy}{dx} = \frac{y - x}{y + x}, y(0) = 1.$$

Solution:

We have $y = 1 + \int_0^x \frac{y - x}{y + x} dx$

First approximation. Put $y = 1$ in the integrand, giving

$$y_1 = 1 + \int_0^x \frac{y-x}{y+x} dx = 1 + \int_0^x \left(-1 + \frac{2}{1+x}\right) dx$$

$$= 1 + \left[-x + 2\log(1+x)\right]_0^x = 1 - x + 2\log(1+x)$$

Second approximation. Put $y = 1 - x + 2\log(1 + x)$ in the integrand, giving

$$y_2 = 1 + \int_0^x \frac{1 - x + 2\log(1+x) - x}{1 - x + 2\log(1+x) + x} dx$$

$$= 1 + \int_0^x \left[1 - \frac{2x}{1 + 2\log(1+x)}\right] dx$$

which is very difficult to integrate.

Hence we use the first approximation and taking $x = 0.1$ in (i) we obtain

$$y(0.1) = 1 - (0.1) + 2\log 1.1 = 0.9828.$$

10.3 Taylor's Series Method

Consider the first order equation $\dfrac{dy}{dx} = f(x,y)$ $\hfill$ (1)

Differentiating (1), we have $\dfrac{d^2y}{dx^2} = \dfrac{\partial f}{\partial x} + \dfrac{\partial f}{\partial y}\dfrac{dy}{dx}$ *i.e.* $y'' = f_x + f_y f'$ $\hfill$ (2)

Differentiating this successively, we can get y''', y^{iv} etc. Putting $x = x_0$ and $y = 0$, the

Values of $(y')_0, (y'')_0, (y''')_0$ can be obtained. Hence the Taylor's series

$$y = y_0 + (x - x_0)(y')_0 + \frac{(x - x_0)^2}{2!}(y'')_0 + \frac{(x - x_0)^3}{3!}(y''')_0 + \dots$$ $\hfill$ (3)

gives the values of y for every value of x for which (3) converges.

On finding the value y_1 for $x = x_1$ from (3), $y' \cdot y''$ etc. can be evaluated at $x = x_1$ by means of (1), (2) etc. Then y can be expanded about $x = x_1$. In this way, the solution can be extended beyond the range of convergence of series (3).

Obs. *This is a single step method and works well so long as the successive derivatives can be calculated easily. If (x, y) is somewhat complicated and the calculation of higher order derivatives becomes tedious, then Taylor's method cannot be used gainfully. This is the main drawback of this method and therefore, has little application for computer programs. However, it is useful for finding starting values for the application of powerful methods like Runga-Kutta, Milne and Adams- Bashforth which will be described in the subsequent sections.*

EXAMPLE 10.3

Solve $y' = x + y$, $y(0) = 1$ by Taylor's series method. Hence find the values of y at $x = 0.1$ and $x = 0.2$.

Solution:

Differentiating successively, we get

$$y' = x + y \qquad y'(0) = 1 \qquad [\because y(0) = 1]$$
$$y'' = 1 + y' \qquad y''(0) = 2$$
$$y''' = y'' \qquad y'''(0) = 2$$
$$y''' = y''' \qquad y'''(0) = 2, \text{ etc.}$$

Taylor's series is

$$y = y_0 + (x - x_0)(y')_0 + \frac{(x - x_0)^2}{2!}(y'')_0 + \frac{(x - x_0)^3}{3!}(y''')_0 + \cdots$$

Here $x_0 = 0$, $y_0 = 1$

$$\therefore \qquad y = 1 + x(1) + \frac{x^2}{2}(2) + \frac{(x)^3}{3!}(2) + \frac{(x)^4}{4!}(4)\cdots$$

Thus $\quad y(0.1) = 1 + 0.1 + (0.1)^2 + \frac{(0.1)^3}{3!} + \frac{(0.1)^4}{4!}\cdots$

$$= 1.1103$$

and $\quad y(0.2) = 1 + 0.2 + (0.2)^2 + \frac{(0.2)^3}{3} + \frac{(0.2)^4}{6} + \cdots$

$$= 1.2427$$

EXAMPLE 10.4

Find by Taylor's series method, the values of y at $x = 0.1$ and $x = 0.2$ to five places of decimals from $dy/dx = x^2 y - 1$, $y(0) = 1$.

Solution:

Differentiating successively, we get

$$y' = x^2 y - 1, \qquad\qquad (y')_0 = -1 \qquad [\because y(0) = 1]$$

$$y'' = 2xy + x^2 y', \qquad\qquad (y'')_0 = 0$$

$$y''' = 2y + 4xy' + x2y'', \qquad (y''')_0 = 2$$

$$y^{iv} = 6y' + 6xy'' + x2y''', \qquad (y^{iv})_0 = -6, \text{ etc.}$$

Putting these values in the Taylor's series, we have

$$y = 1 + x(-1) + \frac{x^2}{2}(0) + \frac{(x)^3}{3!}(2) + \frac{(x)^4}{4!}(-6) + \cdots$$

$$= 1 + -x + \frac{x^3}{3} - \frac{x^4}{4} + \cdots$$

Hence $y(0.1) = 0.90033$ and $y(0.21) = 0.80227$

EXAMPLE 10.5

Employ Taylor's method to obtain approximate value of y at $x = 0.2$ for the differential equation $dy/dx = 2y + 3e^x$, $y(0) = 0$. Compare the numerical solution obtained with the exact solution.

Solution:

(a) We have $y' = 2y + 3e^x$; $y'(0) = 2y(0) + 3e^0 = 3$.

Differentiating successively and substituting $x = 0$, $y = 0$ we get

$$y'' = 2y' + 3e^x, \qquad\qquad y''(0) = 2y'(0) + 3 = 9$$

$$y''' = 2y'' + 3e^x, \qquad\qquad y'''(0) = 2y''(0) + 3 = 21$$

$$y^{iv} = 2y''' + 3e^x, \qquad\qquad y^{iv}(0) = 2y'''(0) + 3 = 45 \text{ etc.}$$

Putting these values in the Taylor's series, we have

$$y(x) = y(0) + xy'(0) + \frac{x^2}{2!}y''(0) + \frac{x^3}{3!}y'''(0) + \frac{x^4}{4!}y^{iv}(0) + \cdots$$

$$= 0 + 3x + \frac{9}{2}x^2 + \frac{21}{6}x^3 + \frac{45}{24}x^4 + \cdots$$

$$= 3x + \frac{9}{2}x^2 + \frac{21}{6}x^3 + \frac{15}{8}x^4 + \cdots$$

Hence $y(0.2) = 3(0.2) + 4.5(0.2)^2 + 3.5(0.2)^3 + 1.875(0.2)^4 + \cdots = 0.8110$ $\quad (i)$

(b) Now $\dfrac{dy}{dx} - 2y = 3e^x$ is a Leibnitz's linear in x

Its I.F. being e_{-2x}, the solution is

$$ye^{-2x} = \int 3e^x \cdot e^{-2x}\, dx + c = -3e^{-x} + c \quad \text{or} \quad y = -3e^x + ce^{2x}$$

Since $y = 0$ when $x = 0$, $\qquad\qquad\qquad \therefore c = 3.$

Thus the exact solution is $y = 3(e^{2x} - e^x)$

When $x = 0.2$, $y = 3(e^{0.4} - e^{0.2}) = 0.8112$ $\hfill (ii)$

Comparing (i) and (ii), it is clear that (i) approximates to the exact value up to three decimal places

EXAMPLE 10.6

Solve by Taylor series method of third order the equation $\dfrac{dy}{dx} = \dfrac{x^3 + xy^2}{e^x}$, $y(0) = 1$ for y at $x = 0.1$, $x = 0.2$ and $x = 0.3$

Solution:

We have $\quad y' = \left(x^3 + xy^2\right)e^{-x}; \quad y'(0) = 0$

Differentiating successively and substituting $x = 0$, $y = 1$.

$$y'' = (x^3 + xy^2)(-e^{-x}) + (3x^2 + y^2 + x.2y.y')e^{-x}$$
$$= (-x^3 - xy^2 + 3x^2 + y^2 + 2xyy')e^{-x}; \quad y''(0) = 1$$
$$y''' = (-x^3 - xy^2 + 3x^2 + y^2 + 2xyy')(-e^{-x})$$
$$+\{-3x^2 - (y^2 + x.2y.y') + 6x + 2yy'$$
$$+2[yy' + x(y'^2 + yy'')]\}e^{-x} \quad y'''(0) = -2$$

Substituting these values in the Taylor's series, we have

$$y(x) = y(0) + xy'(0) + \frac{x^2}{2!}y''(0) + \frac{x^3}{3!}y'''(0) + \cdots$$

$$= 1 + x(0) + \frac{x^2}{2}(1) + \frac{x^3}{6}(-2) + \cdots$$

$$= 1 + \frac{x^2}{2} - \frac{x^3}{6} + \cdots$$

Hence $y(0.1) = 1 + \dfrac{1}{2}(0.1)^2 - \dfrac{1}{3}(0.1)^3 = 1.005$

$$y(0.2) = 1 + \dfrac{1}{2}(0.2)^2 - \dfrac{1}{3}(0.2)^3 = 1.017$$

$$y(0.3) = 1 + \dfrac{1}{2}(0.3)^2 - \dfrac{1}{3}(0.3)^3 = 1.036$$

EXAMPLE 10.7

Solve by Taylor's series method the equation $\dfrac{dy}{dx} = \log(xy)$ for $y(1.1)$ and $y(1.2)$, given $y(1) = 2$.

Solution:

We have $y' = \log x + \log y$; $y'(1) = \log 2$

Differentiating w.r.t., x and substituting $x = 1$, $y = 2$, we get

$$y'' = \frac{1}{x} + \frac{1}{y}y' \quad y'' = 1 + \frac{1}{2}\log 2$$

$$y''' = \frac{1}{x^2} + \frac{1}{y}y'' + y'\left(-\frac{1}{y^2}\right)y';$$

$$y''' = 1 + \frac{1}{2}\left(1 + \frac{1}{2}\log 2\right) - \frac{1}{4}\left(\log 2\right)^2$$

Substituting these values in the Taylor's series about $s = 1$, we have

$$y(x) = y(1) + (x-1)y'(1) + \frac{(x-1)^2}{2!}y''(1) + \frac{(x-1)^3}{3!}y'''(1) + \cdots$$

$$= 2 + (x-1)\log 2 + \frac{1}{2}(x-1)^2\left(1 + \frac{1}{2}\log 2\right)$$

$$+ \frac{1}{6}(x-1)^3\left[-\frac{1}{2} + \frac{1}{4}\log 2 - \frac{1}{4}\left(\log 2\right)^2\right]$$

$$\therefore \quad y(1.1) = 2 + (0.1)\log 2 + \frac{(0.1)^2}{2}\left(1 + \frac{1}{2}\log 2\right) + \frac{(0.1)^3}{6}\left[-\frac{1}{2} + \frac{1}{4}\log 2 - \frac{1}{4}\left(\log 2\right)^2\right]$$

$$= 2.036$$

$$y(1.2) = 2 + (0.2)\log 2 + \frac{(0.2)^2}{2}\left(1 + \frac{1}{2}\log 2\right) + \frac{(0.2)^3}{6}\left[-\frac{1}{2} + \frac{1}{4}\log 2 - \frac{1}{4}\left(\log 2\right)^2\right]$$

$$= 2.081$$

Exercises 10.1

1. Using Picard's method, solve $dy/dx = -xy$ with $x_0 = 0$, $y_0 = 1$ up to the third approximation.

2. Employ Picard's method to obtain, correct to four places of decimals the, solution of the differential equation $dy/dx = x^2 + y^2$ for $x = 0.4$, given that $y = 0$ when $x = 0$.

3. Obtain Picard's second approximate solution of the initial value problem
$$y' = x^2/(y^2 + 1), \ y(0)=0.$$

4. Find an approximate value of y when $x = 0.1$, if $dy/dx = x - y^2$ and $y = 1$ at $x = 0$, using
(a) Picard's method   (b) Taylor's series.

5. Solve $y' = x + y$ given $y(1) = 0$. Find $y(1.1)$ and $y(1.2)$ by Taylor's method. Compare the result with its exact value.

6. Using Taylor's series method, compute $y(0.2)$ to three places of decimals from $\dfrac{dy}{dx} = 1 - 2xy$ given that $y(0) = 0$.

7. Evaluate $y(0.1)$ correct to six places of decimals by Taylor's series method if $y\,(x)$ satisfies
$$y' = xy + 1, \ y(0) = 1.$$

8. Solve $y' = y^2 + x$, $y(0) = 1$ using Taylor's series method and compute $y(0.1)$ and $y(0.2)$.

9. Evaluate $y(0.1)$ correct to four decimal places using Taylor's series methods if $dy/dx = x^2 + y^2$, $y(0) = 1$.

10. Using Taylor series method, find $y(0.1)$ correct to three decimal places given that $dy/dx = e^x - y^2$, $y(0) = 1$

10.4   Euler's Method

Consider the equation $\dfrac{dy}{dx} = f(x,y)$ (1)

given that $y(x_0) = y_0$. Its curve of solution through $P(x_0, y_0)$ is shown dotted in Figure.10.1. Now we have to find the ordinate of any other point Q on this curve.

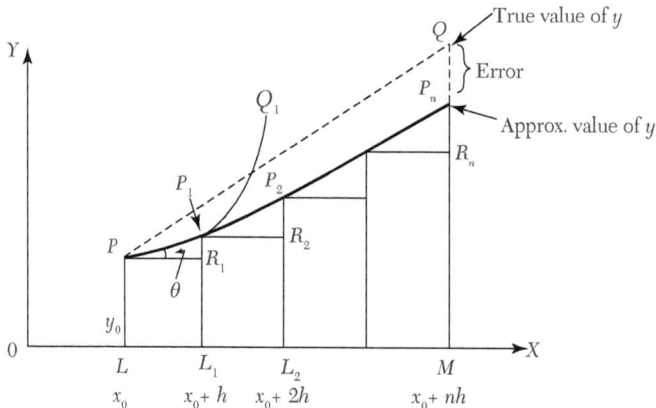

FIGURE 10.1

Let us divide LM into n sub-intervals each of width h at L_1, $L_2 \cdots$so that h is quite small

In the interval LL_1, we approximate the curve by the tangent at P. If the ordinate through L_1 meets this tangent in $P_1(x_0 + h, y_1)$, then

$$y_1 = L_1 P_1 = LP + R_1 P_1 = y_0 + PR_1 \tan \theta$$

$$= y_0 + h \left(\frac{dy}{dx}\right)_p = y_0 + hf(x_0, y_0)$$

Let $P_1 Q_1$ be the curve of solution of (1) through P_1 and let its tangent at P_1 meet the ordinate through L_2 in $P_2(x_0 + 2h, y_2)$. Then

$$y_2 = y_1 + hf(x_0 + h, y_1) \tag{1}$$

Repeating this process n times, we finally reach on an approximation MP_n of MQ given by

$$y_n = y_{n-1} + hf(x_0 + \overline{n-1}h, y_{n-1})$$

This is *Euler's method* of finding an approximate solution of (1).

NOTE **Obs.** *In Euler's method, we approximate the curve of solution by the tangent in each interval, i.e., by a sequence of short lines. Unless h is small, the error is bound to be quite significant. This sequence of lines may also deviate considerably from the curve of solution. As such, the method is very slow and hence there is a modification of this method which is given in the next section.*

EXAMPLE 10.8

Using Euler's method, find an approximate value of y corresponding to $x = 1$, given that $dy/dx = x + y$ and $y = 1$ when $x = 0$.

Solution:

We take $n = 10$ and $h = 0.1$ which is sufficiently small. The various calculations are arranged as follows:

x	y	$x + y = dy/dx$	Old y + 0.1 $(dy/dx) = new\ y$
0.0	1.00	1.00	1.00 + 0.1 (1.00) = 1.10
0.1	1.10	1.20	1.10 + 0.1 (1.20) = 1.22
0.2	1.22	1.42	1.22 + 0.1 (1.42) = 1.36
0.3	1.36	1.66	1.36 + 0.1 (1.66) = 1.53
0.4	1.53	1.93	1.53 + 0.1 (1.93) = 1.72
0.5	1.72	2.22	1.72 + 0.1 (2.22) = 1.94
0.6	1.94	2.54	1.94 + 0.1 (2.54) = 2.19
0.7	2.19	2.89	2.19 + 0.1 (2.89) = 2.48
0.8	2.48	3.29	2.48 + 0.1 (3.29) = 2.81
0.9	2.81	3.71	2.81 + 0.1 (3.71) = 3.18
1.0	3.18		

Thus the required approximate value of $y = 3.18$.

NOTE **Obs.** *In Example 10.1(Obs.), we obtained the true values of y from its exact solution to be 3.44 where as by Euler's method y = 3.18 and by Picard's method y = 3.434. In the above solution, had we chosen n = 20, the accuracy would have been considerably increased but at the expense of double the labor of computation. Euler's method is no doubt very simple but cannot be considered as one of the best.*

EXAMPLE 10.9

Given $\dfrac{dy}{dx} = \dfrac{y - x}{y + x}$ with initial condition $y = 1$ at $x = 0$; find y for $x = 0.1$ by Euler's method.

Solution:

We divide the interval $(0, 0.1)$ in to five steps, *i.e.*, we take $n = 5$ and $h = 0.02$. The various calculations are arranged as follows:

x	y	dy/dx	$Old y + 0.02\ (dy/dx) = new\ y$
0.00	1.0000	1.0000	$1.0000 + 0.02(1.0000) = 1.0200$
0.02	1.0200	0.9615	$1.0200 + 0.02(0.9615) = 1.0392$
0.04	1.0392	0.926	$1.0392 + 0.02(0.926) = 1.0577$
0.06	1.0577	0.893	$1.0577 + 0.02(0.893) = 1.0756$
0.08	1.0756	0.862	$1.0756 + 0.02(0.862) = 1.0928$
0.10	1.0928		

Hence the required approximate value of $y = 1.0928$.

10.5 Modified Euler's Method

In Euler's method, the curve of solution in the interval LL_1 is approximated by the tangent at P (Figure 10.1) such that at P_1, we have

$$y_1 = y_0 + h f(x_0, y_0) \tag{1}$$

Then the slope of the curve of solution through P_1

$$[\text{i.e., } (dy/dx)P_1 = f(x_0 + h, y_1)]$$

is computed and the tangent at P_1 to P_1Q_1 is drawn meeting the ordinate through L_2 in

$$P_2(x_0 + 2h, y_2).$$

Now we find a better approximation $y_1^{(1)}$ of $y(x_0 + h)$ by taking the slope of the curve as the mean of the slopes of the tangents at P and P_1, i.e.,

$$y_1^{(1)} = y_0 + \frac{h}{2}[f(x_0, y_0) + f(x_0 + h, y_1)]$$

As the slope of the tangent at $P1$ is not known, we take y_1 as found in (1) by Euler's method and insert it on R.H.S. of (2) to obtain the first modified value $y_1(1)$

Again (2) is applied and we find a still better value $y_{1(2)}$ corresponding to L_1 as

$$y_1^{(2)} = y_0 + \frac{h}{2}[f(x_0, y_0) + f(x_0 + h, y_1^{(1)})]$$

We repeat this step, until two consecutive values of y agree. This is then taken as the starting point for the next interval L_1L_2.

Once y_1 is obtained to a desired degree of accuracy, y corresponding to L_2 is found from (1).

$$y_2 = y_1 + hf(x_0 + h, y_1)$$

and a better approximation $y_2^{(1)}$ is obtained from (2)

$$y_2^{(1)} = y_1 + \frac{h}{2}[f(x_0 + h, y_1) + f(x_0 + 2h, y_2)]$$

We repeat this step until y_2 becomes stationary. Then we proceed to calculate y_3 as above and so on.

This is the *modified Euler's method* which gives great improvement in accuracy over the original method.

EXAMPLE 10.10

Using modified Euler's method, find an approximate value of y when $x = 0.3$, given that $dy/dx = x + y$ and $y = 1$ when $x = 0$.

Solution:

The various calculations are arranged as follows taking $h = 0.1$:

x	$x + y = y'$	*Mean slope*	*Old y* + 0.1 (*mean slope*) = *new y*
0.0	$0 + 1$	—	$1.00 + 0.1\,(1.00) = 1.10$
0.1	$0.1 + 1.1$	$\frac{1}{2}(1 + 1.2)$	$1.00 + 0.1\,(1.1) = 1.11$
0.1	$0.1 + 1.11$	$-(1 \quad 1.21)$	$1.00 + 0.1\,(1.105) = 1.1105$
0.1	$0.1 + 1.1105$	$\frac{1}{2}(1 + 1.2105)$	$1.00 + 0.1\,(1.1052) = 1.1105$
Since the last two values are equal, we take $y(0.1) = 1.1105$.			
0.1	1.2105	—	$1.1105 + 0.1\,(1.2105) = 1.2316$
0.2	$0.2 + 1.2316$	$\frac{1}{2}(1.12105 + 1.4316)$	$1.1105 + 0.1\,(1.3211) = 1.2426$
0.2	$0.2 + 1.2426$	$\frac{1}{2}(1.2105 + 1.4426)$	$1.1105 + 0.1\,(1.3266) = 1.2432$
0.2	$0.2 + 1.2432$	$\frac{1}{2}(1.2105 + 1.4432)$	$1.1105 + 0.1\,(1.3268) = 1.2432$
Since the last two values are equal, we take $y(0.2) = 1.2432$.			
0.2	1.4432	—	$1.2432 + 0.1\,(1.4432) = 1.3875$
0.3	$0.3 + 1.3875$	$\frac{1}{2}(1.4432 + 1.6875)$	$1.2432 + 0.1\,(1.5654) = 1.3997$
0.3	$0.3 + 1.3997$	$\frac{1}{2}(1.4432 + 1.6997)$	$1.2432 + 0.1\,(1.5715) = 1.4003$
0.3	$0.3 + 1.4003$	$\frac{1}{2}(1.4432 + 1.7003)$	$1.2432 + 0.1\,(1.5718) = 1.4004$
0.3	$0.3 + 1.4004$	$\frac{1}{2}(1.4432 + 1.7004)$	$1.2432 + 0.1\,(1.5718) = 1.4004$
Since the last two values are equal, we take $y(0.3) = 1.4004$.			

Hence $y(0.3) = 1.4004$ approximately.

NOTE **Obs.** *In Example 10.8, we obtained the approximate value of y for x = 0.3 to be 1.53 whereas by the modified Euler's method the corresponding value is 1.4003 which is nearer its true value 1.3997, obtained from its exact solution y = 2ex − x − 1 by putting x = 0.3.*

EXAMPLE 10.11

Using the modified Euler's method, find $y(0.2)$ and $y(0.4)$ given

$$y' = y + e^x, y(0) = 0.$$

Solution:

We have $y' = y + ex = f(x, y)$; $x = 0$, $y = 0$ and $h = 0.2$

The various calculations are arranged as under:

To calculate $y(0.2)$:

x	$y + ex = y'$	Mean slope	Old $y + h$ (Mean slope) $=$ new y
0.0	1	—	$0 + 0.2\,(1) = 0.2$
0.2	$0.2 + e^{0.2} = 1.4214$	$\frac{1}{2}(1 + 1.4214) = 1.2107$	$0 + 0.2\,(1.2107) = 0.2421$
0.2	$0.2421 + e^{0.2} = 1.4635$	$\frac{1}{2}(1 + 1.4635) = 1.2317$	$0 + 0.2\,(1.2317) = 0.2463$
0.2	$0.2463 + e^{0.2} = 1.4677$	$\frac{1}{2}(1 + 1.4677) = 1.2338$	$0 + 0.2\,(1.2338) = 0.2468$
0.2	$0.2468 + e^{0.2} = 1.4682$	$\frac{1}{2}(1 + 1.4682) = 1.2341$	$0 + 0.2\,(1.2341) = 0.2468$

Since the last two values of y are equal, we take $y(0.2) = 0.2468$.

To calculate $y(0.4)$:

x	$y + ex$	Mean slope	Old$y + 0.2$ (mean slope) new y
0.2	$0.2468 + e^{0.2} = 1.4682$	—	$0.2468 + 0.2\,(1.4682) = 0.5404$
0.4	$0.5404 + e^{0.4} = 2.0322$	$\frac{1}{2}(1.4682 + 2.0322)$ $= 1.7502$	$0.2468 + 0.2\,(1.7502) = 0.5968$

x	$y + ex$	Mean slope	Old y + 0.2 (mean slope) new y
0.4	$0.5968 + e^{0.4} = 2.0887$	$\frac{1}{2}(1.4682 + 2.0887)$ $= 1.7784$	$0.2468 + 0.2\,(1.7784) = 0.6025$
0.4	$0.6025 + e^{0.4} = 2.0943$	$\frac{1}{2}(1.4682 + 2.0943)$ $= 1.78125$	$0.2468 + 0.2\,(1.78125) = 0.6030$
0.4	$0.6030 + e^{0.4} = 2.0949$	$\frac{1}{2}(1.4682 + 2.0949)$ $= 1.7815$	$0.2468 + 0.2\,(1.7815) = 0.6031$
0.4	$0.6031 + e^{0.4} = 2.0949$	$\frac{1}{2}(1.4682 + 2.0949)$ $= 1.7816$	$0.2468 + 0.2\,(1.7815) = 0.6031$

Since the last two value of y are equal, we take $y(0.4) = 0.6031$

Hence $y(0.2) = 0.2468$ an d $y(0.4) = 0.6031$ approximately.

EXAMPLE 10.12

Solve the following by Euler's modified method:

$$\frac{dy}{dx} = \log(x + y), y(0) = 2$$

at $x = 1.2$ and 1.4 with $h = 0.2$.

Solution:

The various calculations are arranged as follows:

x	$\log (x + y) = y'$	Mean slope	Old y + 0.2 (mean slope) = new y
0.0	$\log (0 + 2)$	—	$2 + 0.2(0.301) = 2.0602$
0.2	$\log (0.2 + 2.0602)$	$\frac{1}{2}(0.310 = 0.3541)$	$2 + 0.2\,(0.3276) = 2.0655$
0.2	$\log (0.2 + 2.0655)$	$\frac{1}{2}(0.301 + 0.3552)$	$2 + 0.2\,(0.3281) = 2.0656$
0.2	0.3552	—	$2.0656 + 0.2\,(0.3552) = 2.1366$
0.4	$\log (0.4 + 2.1366)$	$\frac{1}{2}(0.3552 + 0.4042)$	$2.0656 + 0.2\,(0.3797) = 2.1415$
0.4	$\log (0.4 + 2.1415)$	$\frac{1}{2}(0.3552 + 0.4051)$	$2.0656 + 0.2\,(0.3801) = 2.1416$
0.4	0.4051	—	$2.1416 + 0.2\,(0.4051) = 2.2226$
0.6	$\log (0.6 + 2.2226)$	$\frac{1}{2}(0.4051 + 0.4506)$	$2.1416 + 0.2\,(0.4279) = 2.2272$
0.6	$\log (0.6 + 2.2272)$	$\frac{1}{2}(0.4051 + 0.4514)$	$2.1416 + 0.2\,(0.4282) = 2.2272$

x	$\log(x+y) = y'$	Mean slope	Old $y + 0.2$ (mean slope) = new y
0.6	0.4514	—	$2.2272 + 0.2(0.4514) = 2.3175$
0.8	$\log(0.8 + 2.3175)$	$\frac{1}{2}(0.4514 + 0.4938)$	$2.2272 + 0.2(0.4726) = 2.3217$
0.8	$\log(0.8 + 2.3217)$	$\frac{1}{2}(0.4514 + 0.4943)$	$2.2272 + 0.2(0.4727) = 2.3217$
0.8	0.4943	—	$2.3217 + 0.2(0.4943) = 2.4206$
1.0	$\log(1 + 2.4206)$	$\frac{1}{2}0.4943 + 0.5341)$	$2.3217 + 0.2(0.5142) = 2.4245$
1.0	$\log(1 + 2.4245)$	$\frac{1}{2}(0.4943 + 0.5346)$	$2.3217 + 0.2(0.5144) = 2.4245$
1.0	0.5346	—	$2.4245 + 0.2(0.5346) = 2.5314$
1.2	$\log(1.2 + 2.5314)$	$\frac{1}{2}(0.5346 + 0.5719)$	$2.4245 + 0.2(0.5532) - 2.5351$
1.2	$\log(1.2 + 2.5351)$	$\frac{1}{2}(0.5346 + 0.5723)$	$2.4245 + 0.2(0.5534) = 2.5351$
1.2	0.5723	—	$2.5351 + 0.2(0.5723) = 2.6496$
1.4	$\log(1.4 + 2.6496)$	$\frac{1}{2}(0.5723 + 0.6074)$	$2.5351 + 0.2(0.5898) = 2.6531$
1.4	$\log(1.4 + 2.6531)$	$\frac{1}{2}(0.5723 + 0.6078)$	$2.5351 + 0.2(0.5900) = 2.6531$

Hence $y(1.2) = 2.5351$ an d $y(1.4) = 2.6531$ approximately.

EXAMPLE 10.13

Using Euler's modified method, obtain a solution of the equation

$$dy/dx = x + \left|\sqrt{y}\right|$$

with initial conditions $y = 1$ at $x = 0$, for the range $0 \pounds x \pounds 0.6$ in steps of 0.2.

Solution:

The various calculations are arranged as follows:

x	$x + \left|\sqrt{y}\right| = y'$	Mean slope	Old $y + 0.2$ (mean slope) = new y
0.0	$0 + 1 = 1$	—	$1 + 0.2(1) = 1.2$
0.2	$0.2 + \left|\sqrt{(1.2)}\right|$ $= 1.2954$	$\frac{1}{2}(1 + 1.2954)$ $= 1.1477$	$1 + 0.2(1.1477) = 1.2295$

x	$x+\left\lvert\sqrt{y}\right\rvert=y'$	Mean slope	Old $y+0.2$ (mean slope) = new y
0.2	$0.2+\left\lvert\sqrt{(1.2295)}\right\rvert$ $=1.3088$	$\frac{1}{2}(1+1.3088)$ $=1.1544$	$1+0.2\,(1.1544)=1.2309$
0.2	$0.2+\left\lvert\sqrt{(1.2309)}\right\rvert$ $=1.3094$	$\frac{1}{2}(1+1.3094)$ $=1.1547$	$1+0.2\,(1.1547)=1.2309$
0.2	1.3094	—	$1.2309+0.2\,(1.3094)=1.4927$
0.4	$0.4+\left\lvert\sqrt{(1.4927)}\right\rvert$ $=1.6218$	$\frac{1}{2}(1.3094+1.6218)$ $=1.4654$	$1.2309+0.2\,(1.4654)=1.5240$
0.4	$0.4+\left\lvert\sqrt{(1.524)}\right\rvert$ $=1.6345$	$\frac{1}{2}(1.3094+1.6345)$ $=1.4718$	$1.2309+0.2\,(1.4718)=1.5253$
0.4	$0.4+\left\lvert\sqrt{(1.5253)}\right\rvert$ $=1.6350$	$\frac{1}{2}(1.3094+1.6350)$ $=1.4721$	$1.2309+0.2\,(1.4721)=1.5253$
0.4	1.6350	—	$1.5253+0.2\,(1.635)=1.8523$
0.6	$0.6+\left\lvert\sqrt{(1.8523)}\right\rvert$ $=1.9610$	$\frac{1}{2}(1.635+1.961)$ $=1.798$	$1.5253+0.2\,(1.798)=1.8849$
0.6	$0.6+\left\lvert\sqrt{(1.8849)}\right\rvert$ $=1.9729$	$\frac{1}{2}(1.635+1.9729)$ $=1.8040$	$1.5253+0.2\,(1.804)=1.8861$
0.6	$0.6+\left\lvert\sqrt{(1.8861)}\right\rvert$ $=1.9734$	$\frac{1}{2}(1.635+1.9734)$ $=1.8042$	$1.5253+0.2\,(1.8042)=1.8861$

Hence $y(0.6) = 1.8861$ approximately.

Exercises 10.2

1. Apply Euler's method to solve $y' = x + y$, $y(0) = 0$, choosing the step length = 0.2. (Carry out six steps).

2. Using Euler's method, find the approximate value of y when $x = 0.6$ of $dy/dx = 1 - 2xy$, given that $y = 0$ when $x = 0$ (take $h = 0.2$).

3. Using the simple Euler's method solve for y at $x = 0.1$ from $dy/dx = x + y + xy$, $y(0) = 1$, taking step size $h = 0.025$.

4. Solve $y' = 1 - y, y(0) = 0$

by the modified Euler's method and obtain y at $x = 0.1, 0.2, 0.3$

5. Given that $dy/dx = x^2 + y$ and $y(0) = 1$. Find an approximate value of $y(0.1)$, taking $h = 0.05$ by the modified Euler's method.

6. Given $y' = x + \sin y, y(0) = 1$. Compute $y(0.2)$ and $y(0.4)$ with $h = 0.2$ using Euler's modified method.

7. Given $\dfrac{dy}{dx} = \dfrac{y-x}{y+x}$ with boundary conditions $y = 1$ when $x = 0$, find

approximately y for $x = 0.1$, by Euler's modified method (five steps)

8. Given that $dy/dx = 2 + \sqrt{(xy)}$ and $y = 1$ when $x = 1$. Find approximate value of y at $x = 2$ in steps of 0.2, using Euler's modified method.

10.6   Runge's Method*

Consider the differential equation, $\dfrac{dy}{dx} = f(x,y), y(x_0) = y_0$ \hfill (1)

Clearly the slope of the curve through $P(x_0, y_0)$ is $f(x_0, y_0)$ (Figure 10.2).

Integrating both sides of (1) from (x_0, y_0) to $(x_0 + h, y_0 + k)$, we have

$$\int_{y_0}^{y_0+k} dy = \int_{x_0}^{x_0+h} f(x,y)dx \hfill (2)$$

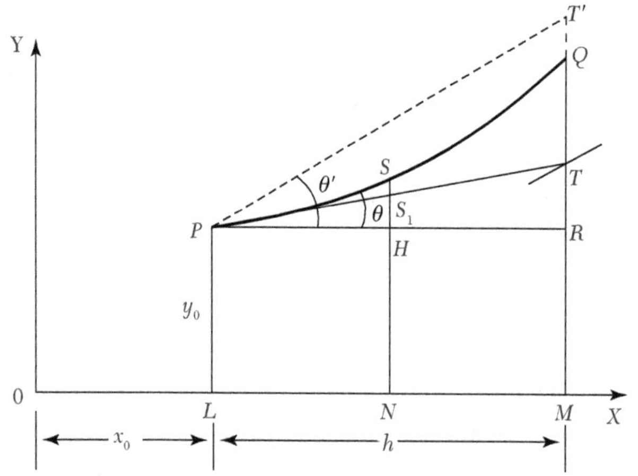

FIGURE 10.2

---

*Called after the German mathematician *Carl Runge* (1856-1927).

To evaluate the integral on the right, we take N as the mid-point of LM and find the values of $f(x, y)$ (*i.e., dy/dx*) at the points $x_0, x_0 + h/2, x_0 + h$. For this purpose, we first determine the values of y at these points.

Let the ordinate through N cut the curve PQ in S and the tangent PT in S_1. The value of y_S is given by the point S_1

$$\therefore \qquad y_s = NS_1 = LP + HS_1 = y_0 + PH. \tan\theta$$

$$= y_0 + h(dy/dx)_p = y_0 + \frac{h}{2} f(x_0, y_0) \qquad (3)$$

Also $\qquad y_T = MT = LP + RT = y_0 + PR. \tan\theta = y_0 + hf(x_0 + y_0).$

Now the value of y_Q at $x_0 + h$ is given by the point T'' where the line through P draw with slope at $T(x_0 + h, y_T)$ meets MQ.

$\therefore$ Slope at $T = \tan\theta' = f(x_0 + h, y_T) = f[x_0 + h, y_0 + hf(x_0, y_0)]$

$$\therefore \qquad y_Q = R + RT = y_0 + PR. \tan\theta' = y_0 + hf[x_0 + h, y_0 + hf(x_0, y_0)] \quad (4)$$

Thus the value of $f(x, y)$ at $P = f(x_0, y_0)$,

the value of $f(x, y)$ at $S = f(x_0 + h/2, y_S)$

and the value of $f(x, y)$ at $Q = (x_0 + h, y_Q)$

where y_S and y_Q are given by (3) and (4).

Hence from (2), we obtain

$$k = \int_{x_0}^{x_0+h} f(x,y)dx = \frac{h}{6}\left[f_P + 4f_S + f_Q\right] \qquad \text{by Simpson's rule}$$

$$= \frac{h}{6}\left[f(x_0 + y_0) + f(x_0 + h/2, y_S) + (x_0 + h, y_Q)\right]$$

Which gives a sufficiently accurate value of k and also $y = y_0 + k$

The repeated application of (5) gives the values of y for equi-spaced points.

Working rule *to solve* (1) *by Runge's method*:

Calculate successively

$$k_1 = hf(x_0, y_0),$$

$$k_2 = hf\left(x_0 + \frac{1}{2}hy_0 + \frac{1}{2}k_1\right)$$

$$k' = hf\left(x_0 + h, y_0 + k_1\right)$$

and
$$k_3 = hf\left(x_0 + h, y_0 + k'\right)$$

Finally compute, $k = \dfrac{1}{6}\left(k_1 + 4k_2 + k_3\right)$

which gives the required approximate value as $y_1 = y_0 + k$.

(Note that k is the weighted mean of k_1, k_2, and k_3).

EXAMPLE 10.14

Apply Runge's method to find an approximate value of y when $x = 0.2$, given that $dy/dx = x + y$ and $y = 1$ when $x = 0$.

Solution:

Here we have $x_0 = 0$, $y_0 = 1$, $h = 0.2$, $f(x_0, y_0) = 1$

$\therefore \qquad k_1 = hf(x_0, y_0) = 0.2(1) = 0.200$

$$k_2 = hf\left(x_0 + \frac{1}{2}hy_0 + \frac{1}{2}k_1\right) = 0.2f(0.1, 1.1) = 0.240$$

$$k' = hf\left(x_0 + h, y_0 + k_1\right) = 0.2f(0.2, 1.2) = 0.280$$

and $\qquad k_3 = hf\left(x_0 + h, y_0 + k'\right) = 0.2f(0.1, 1.28) = 0.296$

$\therefore \qquad k = \dfrac{1}{6}\left(k_1 + 4k_2 + k_3\right) = \dfrac{1}{6}(0.200 + 0.960 + 0.296) = 0.2426$

Hence the required approximate value of y is 1.2426.

10.7   Runge-Kutta Method*

The Taylor's series method of solving differential equations numerically is restricted by the labor involved in finding the higher order derivatives. However, there is a class of methods known as Runge-Kutta methods which do not require the calculations of higher order derivatives and give greater accuracy. The Runge-Kutta formulae possess the advantage of requiring only the function values at some selected points. These methods agree with Taylor's series solution up to the term in h^r where r differs from method to method and is called the *order of that method*.

First order R-K method. We have seen that Euler's method (Section 10.4) gives

$$y_1 = y_0 + hf(x_0, y_0) = y_0 + hy_0' \qquad\qquad [\because y' = f(x, y)]$$

Expanding by Taylor's series

$$y_1 = y(x_0 + h) = y_0 + hy_0' + \frac{h^2}{2} y_0'' + \cdots$$

It follows that the Euler's method agrees with the Taylor's series solution upto the term in h.

Hence, *Euler's method is the Runge-Kutta method of the first order.*

Second order R-K method. The modified Euler's method gives

$$y_1 = y + \frac{h}{2} \left[f(x_0, y_0) + f(x_0 + h, y_1) \right] \qquad (1)$$

Substituting $y_1 = y_0 + hf(x_0, y_0)$ on the right-hand side of (1), we obtain

$$y_1 = y_0 + \frac{h}{2} \left[f_0 + f(x_0 + h), y_0 + hf_0 \right] \quad \text{where} \quad f_0 = (x_0, y_0) \qquad (2)$$

Expanding L.H.S. by Taylor's series, we get

$$y_1 = y(x_0 + h) = y_0 + hy_0' + \frac{h^2}{2!} y_0'' + \frac{h^3}{3!} y_0''' + \cdots \qquad (3)$$

Expanding $f(x_0 + h, y_0 + hf_0)$ by Taylor's series for a function of two variables, (2) gives

$$y_1 = y_0 + \frac{h}{2} \left[f_0 + \left\{ f_0 = (x_0, y_0) + h \left(\frac{\partial f}{\partial x} \right)_0 + hf_0 \left(\frac{\partial f}{\partial y} \right)_0 + O(h^2)^{\circ\circ} \right\} \right]$$

$$= y_0 + \frac{1}{2} \left[hf_0 + hf_0 + h^2 \left\{ \left(\frac{\partial f}{\partial x} \right)_0 + \left(\frac{\partial f}{\partial y} \right)_0 \right\} + O(h^3) \right]$$

$$= y_0 + hf_0 + \frac{h^2}{2} f_0' + O(h^3) \qquad \left[\because \frac{df(x,y)}{dx} = \frac{\partial f}{\partial x} + f \frac{\partial f}{\partial y} \right]$$

$$= y_0 + hy_0' + \frac{h^2}{2!} y_0'' + O(h^3) \qquad (4)$$

Comparing (3) and (4), it follows that the modified Euler's method agrees with the Taylor's series solution upto the term in h^2.

Hence the modified Euler's method is the Runge-Kutta method of the second order.

$\circ\circ O(h^2)$ means "terms containing second and higher powers of h" and is read as *order of* h^2.

∴ *The second order Runge-Kutta formula is*

$$y_1 = y_0 + \frac{1}{2}(k_1 + k_2)$$

Where $k_1 = hf(x_0, y_0)$ and $k_2 = hf(x_0 + h, y_0 + k)$

(*iii*) *Third order R-K method.* Similarly, it can be seen that Runge's method (Section 10.6) agrees with the Taylor's series solution upto the term in h^3.

As such, *Runge's method is the Runge-Kutta method of the third order.*

∴ *The third order Runge-Kutta formula is*

$$y_1 = y_0 + \frac{1}{6}(k_1 + 4k_2 + k_3)$$

Where, $k_1 = hf(x_0, y_0)$, $k_2 = hf\left(x_0 + \frac{1}{2}h, y_0 + \frac{1}{2}k_1\right)$

And $k_3 = hf(x_0 + h, y_0 + k')$, where $k' = k_3 = hf(x_0 + h, y_0 + k_1)$.

(*iv*) *Fourth order R-K method.* This method is most commonly used and is often referred to as the *Runge-Kutta method* only.

Working rule *for finding the increment k of y corresponding to an increment h of x by Runge-Kutta method from*

$$\frac{dy}{dx} = f(x, y), y(x_0)$$

is as follows:

Calculate successively $k_1 = hf(x_0, y_0)$,

$$k_2 = hf\left(x_0 + \frac{1}{2}h, y_0 + \frac{1}{2}k_1\right)$$

$$k_3 = hf\left(x_0 + \frac{1}{2}h, y_0 + \frac{1}{2}k_2\right)$$

and $$k_4 = hf(x_0 + h, y_0 + k_3)$$

Finally compute $$k = \frac{1}{6}(k_1 + 2k_2 + 2k_3 + k_4)$$

which gives the required approximate value as $y_1 = y_0 + k$.

(Note that k is the weighted mean of k_1, k_2, k_3, and k_4).

NOTE **Obs.** *One of the advantages of these methods is that the operation is identical whether the differential equation is linear or non-linear.*

EXAMPLE 10.15

Apply the Runge-Kutta fourth order method to find an approximate value of y when $x = 0.2$ given that $dy/dx = x + y$ and $y = 1$ when $x = 0$.

Solution:

Here $x_0 = 0, y_0 = 1, h = 0.2, f(x_0, y_0) = 1$

$\therefore \quad k_1 = hf(x_0, y_0) = 0.2 \times 1 = 0.2000$

$k_2 = hf\left(x_0 + \frac{1}{2}h, y_0 + \frac{1}{2}k_1\right) = 0.2 \times f(0.1, 1.1) = 0.2400$

$k_3 = hf\left(x_0 + \frac{1}{2}h, y_0 + \frac{1}{2}k_2\right) = 0.2 \times f(0.1, 1.12) = 0.2440$

and $\quad k_4 = hf(x_0 + h, y_0 + k_3) = 0.2 \times f(0.2, 1.244) = 0.2888$

$\therefore \quad k = \frac{1}{6}(k_1 + 2k_2 + 2k_3 + k_4)$

$\quad = \frac{1}{6}(0.2000 + 0.4800 + 0.4880 + 0.2888)$

$\quad = \frac{1}{6} \times (1.4568) = 0.2428$

Hence the required approximate value of y is 1.2428.

EXAMPLE 10.16

Using the Runge-Kutta method of fourth order, solve $\dfrac{dy}{dx} = \dfrac{y^2 - x^2}{y^2 + x^2}$ with $y(0) = 1$ at $x = 0.2, 0.4$.

Solution:

We have $f(x, y) = \dfrac{y^2 - x^2}{y^2 + x^2}$

To find $y(0.2)$

Hence $x_0 = 0, y_0 = 1, h = 0.2$

$k_1 = hf(x_0, y_0) = 0.2f(0, 1) = 0.2000$

$k_2 = hf\left(x_0 + \frac{1}{2}h, y_0 + \frac{1}{2}k_1\right) = 0.2 \times f(0.1, 1.1) = 0.19672$

$$k_3 = hf\left(x_0 + \frac{1}{2}h, y_0 + \frac{1}{2}k_2\right) = 0.2f(0.1, 1.09836) = 0.1967$$

$$k_4 = hf\left(x_0 + h, y_0 + k_3\right) = 0.2f(0.2, 1.1967) = 0.1891$$

$$k = \frac{1}{6}\left(k_1 + 2k_2 + 2k_3 + k_4\right)$$

$$= \frac{1}{6}[0.2 + 2(0.19672) + 2(0.1967) + 0.1891] = 0.19599$$

Hence $y(0.2) = y_0 + k = 1.196$.

To find $y(0.4)$:

Here $x_1 = 0.2$, $y_1 = 1.196$, $h = 0.2$.

$$k_1 = hf(x_1, y_1) = 0.1891$$

$$k_2 = hf\left(x_1 + \frac{1}{2}h, y_1 + \frac{1}{2}k_1\right) = 0.2f(0.3, 1.2906) = 0.1795$$

$$k_3 = hf\left(x_1 + \frac{1}{2}h, y_1 + \frac{1}{2}k_2\right) = 0.2f(0.3, 1.2858) = 0.1793$$

$$k_4 = hf\left(x_1 + h, y_1 + k_3\right) = 0.2f(0.4, 1.3753) = 0.1688$$

$$k = \frac{1}{6}\left(k_1 + 2k_2 + 2k_3 + k_4\right)$$

$$= \frac{1}{6}[0.1891 + 2(0.1795) + 2(0.1793) + 0.1688] = 0.1792$$

Hence $y(0.4) = y_1 + k = 1.196 + 0.1792 = 1.3752$.

EXAMPLE 10.17

Apply the Runge-Kutta method to find the approximate value of y for $x = 0.2$, in steps of 0.1, if $dy/dx = x + y^2$, $y = 1$ where $x = 0$.

Solution:

Given $f(x, y) = x + y^2$.

Here we take $h = 0.1$ and carry out the calculations in two steps.

Step I. $x0 = 0$, $y0 = 1$, $h = 0.1$

$\therefore \qquad k_1 = hf(x_0, y_0) = 0.1f(0, 1) = 0.1000$

$$k_2 = hf\left(x_0 + \frac{1}{2}h, y_0 + \frac{1}{2}k_1\right) = 0.1f(0.05, 1.1) = 0.1152$$

$$k_3 = hf\left(x_0 + \frac{1}{2}h, y_0 + \frac{1}{2}k_2\right) = 0.1f(0.05, 1.1152) = 0.1168$$

$$k_4 = hf\left(x_0 + h, y_0 + k_3\right) = 0.1f(0.1, 1.1168) = 0.1347$$

$$\therefore \qquad k = \frac{1}{6}\left(k_1 + 2k_2 + 2k_3 + k_4\right)$$

$$= \frac{1}{6}(0.1000 + 0.2304 + 0.2336 + 0.1347) = 0.1165$$

giving $y(0.1) = y_0 + k = 1.1165$

Step II. $x_1 = x_0 + h = 0.1$, $y_1 = 1.1165$, $h = 0.1$

$$\therefore \qquad k_1 = hf(x_1, y_1) = 0.1f(0.1, 1.1165) = 0.1347$$

$$k_2 = hf\left(x_1 + \frac{1}{2}h, y_1 + \frac{1}{2}k_1\right) = 0.1f(0.15, 1.1838) = 0.1551$$

$$k_3 = hf\left(x_1 + \frac{1}{2}h, y_1 + \frac{1}{2}k_2\right) = 0.1f(0.15, 1.194) = 0.1576$$

$$k_4 = hf(x_1 + h, y_2 + k_3) = 0.1f(0.2, 1.1576) = 0.1823$$

$$\therefore \qquad k = \frac{1}{6}\left(k_1 + 2k_2 + 2k_3 + k_4\right) = 0.1571$$

Hence $y(0.2) = y_1 + k = 1.2736$

EXAMPLE 10.18

Using the Runge-Kutta method of fourth order, solve for y at $x = 1.2$, 1.4

From $\dfrac{dy}{dx} = \dfrac{2xy + e^x}{x^2 + xe^x}$ given $x_0 = 1$, $y_0 = 0$

Solution:

We have $f(x, y) = \dfrac{2xy + e^x}{x^2 + xe^x}$

To find $y(1.2)$:

Here $x_0=1$, $y_0=0$, $h=0.2$

$$k_1 = hf(x_0, y_0) = 0.2 \frac{0+e'}{1+e'} = 0.1462$$

$$k_2 = hf\left(x_0 + \frac{h}{2}, y_0 + \frac{k_1}{2}\right) = 0.2\left\{\frac{2(1+0.1)(0+0.073)e^{1+0.1}}{(1+0.1)^2 + (1+0.1)e^{1+0.1}}\right\}$$

$$= 0.1402$$

$$k_3 = hf\left(x_0 + \frac{1}{2}h, y_0 + \frac{1}{2}k_2\right) = 0.2\left\{\frac{2(1+0.1)(0+0.07)e^{1.1}}{(1+0.1)^2 + (1+0.1)e^{1.1}}\right\}$$

$$= 0.1399$$

$$k_4 = hf\left(x_0 + h, y_0 + k_3\right) = 0.2\left\{\frac{2(1.2)(0.1399)e^{1.2}}{(1.2)^2 + (1.2)e^{1.2}}\right\}$$

$$= 0.1348$$

and $\quad k = \frac{1}{6}\left(k_1 + 2k_2 + 2k_3 + k_4\right) = \frac{1}{6}[0.1462 + 0.2804 + 0.2798 + 0.1348]$

$$= 0.1402$$

Hence $y(1.2) = y0 + k = 0 + 0.1402 = 0.1402$.

To find $y(1.4)$:

Here $\quad x_1 = 1.2, y_1 = 0.1402, h = 0.2$

$$k_1 = hf(x_1, y_1) = 0.2f(1.2, 0) = 0.1348$$

$$k_2 = hf\left(x_1 + h/2, y_1 + k_1/2\right) = 0.2f(1.3, 0.2076) = 0.1303$$

$$k_3 = hf\left(x_1 + h/2, y_1 + k_2/2\right) = 0.2f(1.3, 0.2053) = 0.1301$$

$$k_4 = hf(x_1 + h, y_1 + k_3) = 0.2f(1.3, 0.2703) = 0.1260$$

$\therefore \qquad k = \frac{1}{6}\left(k_1 + 2k_2 + 2k_3 + k_4\right)$

$$= \frac{1}{6}[0.1348 + 0.2606 + 0.2602 + 0.1260]$$

$$= 0.1303$$

Hence $\quad y(1.4) = y_1 + k = 0.1402 + 0.1303 = 0.2705$.

Exercises 10.3

1. Use Runge's method to approximate y when $x = 1.1$, given that $y = 1.2$ when $x = 1$ and $dy/dx = 3x + y^2$.

2. Using the Runge-Kutta method of order 4, find $y(0.2)$ given that $dy/dx = 3x + y^2$, $y(0) = 1$ taking $h = 0.1$.

3. Using the Runge-Kutta method of order 4, compute $y(0.2)$ and $y(0.4)$ from $10 \dfrac{dy}{dx} = x^2 + y^2$ $y(0) = 1$, taking $h = 0.1$.

4. Use the Runge Kutta method to find y when $x = 1.2$ in steps of 0.1, given that $dy/dx = x^2 + y^2$ and $y(1) = 1.5$.

5. Given $dy/dx = x^3 + y$, $y(0) = 2$. Compute $y(0.2)$, $y(0.4)$, and $y(0.6)$ by the Runge-Kutta method of fourth order.

6. Find $y(0.1)$ and $y(0.2)$ using the Runge-Kutta fourth order formula, given that $y' = x^2 - y$ and $y(0) = 1$.

7. Using fourth order Runge-Kutta method, solve the following equation, taking each step of $h = 0.1$, given $y(0) = 3$. dy/dx $(4x/y - xy)$. Calculate y for $x = 0.1$ and 0.2.

8. Find by the Runge-Kutta method an approximate value of y for $x = 0.6$, given that $y = 0.41$ when $x = 0.4$ and $dy/dx = \sqrt{(x+y)}$

9. Using the Runge-Kutta method of order 4, find $y(0.2)$ for the equation $\dfrac{dy}{dx} = \dfrac{y-x}{y+x}$, , $y(0) = 1$. Take $h = 0.2$.

10. Using fourth order Runge-Kutta method, integrate $y' = -2x^3 + 12x^2 - 20x + 8.5$, using a step size of 0.5 and initial condition of $y = 1$ at $x = 0$.

11. Using the fourth order Runge-Kutta method, find y at $x = 0.1$ given that $dy/dx = 3e^x + 2y$, $y(0) = 0$ and $h = 0.1$.

12. Given that $dy/dx = (y^2 - 2x)/(y^2 + x)$ and $y = 1$ at $x = 0$, find y for $x = 0.1$, 0.2, 0.3, 0.4, and 0.5.

10.8   Predictor-Corrector Methods

If x_{i-1} and x_i are two consecutive mesh points, we have $x_i = x_{i-1} + h$. In Euler's method (Section 10.4), we have

$$y_i = y_{i-1} + hf(x_0 + \overline{i-1}h, y_{i-1}); \quad i = 1, 2, 3 \cdots \tag{1}$$

The modified Euler's method (Section 10.5), gives

$$y_i = y_{i-1} + \frac{h}{2}\left[f(x_{i-1}, y_{i-1}) + f(x_i, y_i)\right]$$

The value of y_i is first estimated by using (1), then this value is inserted on the right side of (2), giving a better approximation of y_i. This value of y_i is again substituted in (2) to find a still better approximation of y_i. This step is repeated until two consecutive values of y_i agree. *This technique of refining an initially crude estimate of y_i by means of a more accurate formula is known as* **predictor-corrector method.** The equation (1) is therefore called the *predictor* while (2) serves as a *corrector* of y_i.

In the methods so far described to solve a differential equation over an interval, only the value of y at the beginning of the interval was required. In the predictor-corrector methods, four prior values are needed for finding the value of y at x_i. Though slightly complex, these methods have the advantage of giving an estimate of error from successive approximations to y_i.

We now describe two such methods, namely: Milne's method and Adams-Bashforth method.

10.9   Milne's Method

Given $dy/dx = f(x, y)$ and $y = y_0$, $x = x_0$; to find an approximate value of y for $x = x_0 + nh$ by Milne's method, we proceed as follows:

The value $y_0 = y(x_0)$ being given, we compute

$$y_1 = y(x_0 + h), y_2 = y(x_0 + 2h), y_3 = y(x_0 + 3h),$$

by Picard's or Taylor's series method.

Next we calculate,

$$f_0 = f(x_0, y_0), f_1 = f(x_0 + h, y_1), f_2 = f(x_0 + 2h, y_2), f_3 = f(x_0 + 3h, y_3)$$

Then to find $y_4 = y(x_0 + 4h)$, we substitute Newton's forward interpolation formula

$$f(x, y) = f_0 + n\Delta f_0 + \frac{n(n-1)}{2}\Delta^2 f_0 + \frac{n(n-1)(n-2)}{6}\Delta^3 f_0 + \cdots$$

In the relation

$$y_4 = y_0 + \int_{x_0}^{x_0+4h} f(x,y)dx$$

$$y_4 = y_0 + \int_{x_0}^{x_0+4h} \left(f_0 + n\Delta f_0 + \frac{n(n-1)}{2}\Delta^2 f_0 + ... \right) dx$$

$$[\text{Put } x = x_0 + nh, \, dx = hdn]$$

$$= y_0 + \int_0^4 \left(f_0 + n\Delta f_0 + \frac{n(n-1)}{2}\Delta^2 f_0 + ... \right) dn$$

$$= y_0 + h\left(4f_0 + 8\Delta f_0 + \frac{20}{3}\Delta^2 f_0 + ... \right)$$

Neglecting fourth and higher order differences and expressing $\Delta f_0, \Delta^2 f_0$ and $\Delta^3 f_0$ and in terms of the function values, we get

$$y_4^{(p)} = y_0 + \frac{4h}{3}(2f_1 - f_2 + 2f_3)$$

which is called a *predictor*.

Having found y_4, we obtain a first approximation to

$$f_4 = f(x_0 + 4h, y_4)$$

Then a better value of y_4 is found by Simpson's rule as

$$y_4^{(c)} = y_2 + \frac{h}{3}(f_2 + 4f_3 + f_4)$$

which is called a *corrector*.

Then an improved value of f_4 is computed and again the corrector is applied to find a still better value of y_4. We repeat this step until y_4 remains unchanged. Once y_4 and f_4 are obtained to desired degree of accuracy, $y_5 = y(x_0 + 5h)$ is found from the *predictor* as

$$y_5^{(p)} = y_1 + \frac{4h}{3}(2f_2 - f_3 + 2f_4)$$

and $f_5 = f(x_0 + 5h, y_5)$ is calculated. Then a better approximation to the value of y_5 is obtained from the *corrector* as

$$y_5^{(c)} = y_3 + \frac{h}{3}(f_3 + 4f_4 + f_5)$$

We repeat this step until y_5 becomes stationary and, then proceed to calculate y_6 as before.

This is *Milne's predictor-corrector method*. To insure greater accuracy, we must first improve the accuracy of the starting values and then subdivide the intervals.

EXAMPLE 10.19

Apply Milne's method, to find a solution of the differential equation $y' = x - y^2$ in the range $0 \leq x \leq 1$ for the boundary condition $y = 0$ at $x = 0$.

Solution:

Using Picard's method, we have

$$y = y(0) + \int_0^x f(x,y)dx, \text{ where } f(x,y) = x - y^2$$

To get the first approximation, we put $y = 0$ in $f(x, y)$,

Giving $y_1 = 0 + \int_0^x x dx = \dfrac{x^2}{2}$

To find the second approximation, we put

Giving $y_2 = \int_0^x \left(x - \dfrac{x^4}{4} \right) dx = \dfrac{x^2}{2} - \dfrac{x^5}{20}$

Similarly, the third approximation is

$$y_3 = \int_0^x \left[x - \left(\dfrac{x^2}{2} - \dfrac{x^5}{20} \right)^2 \right] dx = \dfrac{x^2}{2} - \dfrac{x^5}{20} + \dfrac{x^8}{160} - \dfrac{x^{11}}{4400} \qquad (i)$$

Now let us determine the starting values of the Milne's method from (*i*), by choosing $h = 0.2$.

$$x_0 = 0.0, \qquad y_0 = 0.0000, \qquad f_0 = 0.0000$$
$$x_1 = 0.2, \qquad y_1 = 0.020, \qquad f_1 = 0.1996$$
$$x_2 = 0.4, \qquad y_2 = 0.0795 \qquad f_2 = 0.3937$$
$$x_3 = 0.5, \qquad y_3 = 0.1762, \qquad f_3 = 0.5689$$

Using the predictor, $y_4^{(p)} = y_0 + \dfrac{4h}{3}(2f_1 - f_2 + 2f_3)$

$$x = 0.8 \qquad y_4^{(p)} = 0.3049, \qquad f_4 = 0.7070$$

and the corrector, $y_4^{(c)} = y_2 + \dfrac{h}{3}(f_2 + 4f_3 + f_4)$, yields

$$y_4^{(c)} = 0.3046 \qquad f_4 = 0.7072 \qquad (ii)$$

Again using the *corrector*,
$$y_4^{(c)} = 0.3046, \text{, which is the same as in (ii)}$$

Now using the predictor,
$$y_4^{(p)} = y_1 + \frac{4h}{3}(2f_2 - f_3 + 2f_4)$$

$$x = 0.1, \qquad y_5^{(p)} = 0.4554 \qquad f_5 = 0.7926$$

and the corrector $y_5^{(c)} = y_3 + \frac{h}{3}(f_3 + 4f_4 + f_5)$ gives

$$y_5^{(c)} = 0.4555 \qquad f_5 = 0.7925$$

Again using the corrector,
$$y_5^{(c)} = 0.4555, \text{ a value which is the same as before.}$$

Hence $y(1) = 0.4555$.

EXAMPLE 10.20

Using Milne's method find $y(4.5)$ given $5xy' + y^2 - 2 = 0$ given $y(4) = 1$, $y(4.1) = 1.0049$, $y(4.2) = 1.0097$, $y(4.3) = 1.0143$; $y(4.4) = 1.0187$.

Solution:

We have $y' = (2 - y^2)/5x = f(x)$ [say]

Then the starting values of the Milne's method are

$$x_0 = 0, \qquad y_0 = 1, \qquad f_0 = \frac{2-12}{5\times4} = 0.05$$

$$x_1 = 4.1, \qquad y_1 = 1.0049, \qquad f_1 = 0.0485$$

$$x_2 = 4.2, \qquad y_2 = 1.0097, \qquad f_2 = 0.0467$$

$$x_3 = 4.3, \qquad y_3 = 1.0143, \qquad f_3 = 0.0452$$

$$x_4 = 4.4, \qquad y_4 = 1.0187, \qquad f_4 = 0.0437$$

Since y_5 is required, we use the predictor

$$y_5^{(p)} = y_1 + \frac{4h}{3}(2f_2 - f_3 + 2f_4') \qquad\qquad (h = 0.1)$$

$$x = 4.5, \; y_5^{(p)} = 1.0049 + \frac{4(0.1)}{3}(2\times2.0467 - 0.0452 + 2\times0.0437) = 1.023$$

$$f_5 = \frac{2 - y_5^2}{5x_5} = \frac{2 - (1.023)^2}{5 \times 4.5} = 0.0424$$

Now using the corrector $y_5^{(c)} = y_3 + \dfrac{h}{3}(f_3 + 4f_4 + f_5)$, we get

$$y_5^{(c)} = 1.0143 + \frac{0.1}{3}(0.0452 + 4 \times 0.0437 + 0.0424) = 1.023$$

Hence $y(4.5) = 1.023$

EXAMPLE 10.21

Given $y' = x(x^2 + y^2)\,e^{-x}$, $y(0) = 1$, find y at $x = 0.1, 0.2$, and 0.3 by Taylor's series method and compute $y(0.4)$ by Milne's method.

Solution:

Given $y(0) = 1$ and $h = 0.1$

We have $\qquad y'(x) = x(x^2 + y^2)e^{-x} \qquad y'(0) = 0$

$$\therefore \qquad y''(x) = \left[(x^3 + xy^2)(e^{-x}) + (3x^2 + y^2 + x(2y)y')\right]e^{-x}$$

$$= e^{-x}\left[-x^3 - xy^2 + 3x^2 + y^2 + 2xyy'\right]; \qquad y''(0) = 1$$

$$y'''(x) = e^{-x}\left[-x^3 - xy^2 + 3x^2 + y^2 + 2xyy' - 6x - 2yy' - 2xy'^2 - 2xyy'\right]$$

$$y'''(0) = 2$$

Substituting these values in the Taylor's series,

$$y(x) = y(0) + \frac{x}{1!}y'(0) + \frac{x^2}{2!}y''(0) + \frac{x^3}{3!}y'''(0) + \cdots$$

$$y(0.1) = 1 + (0.1)(0) + \frac{1}{2}(0.1)^2(1) + \frac{1}{6}(0.1)^3(-2) + \cdots$$

$$= 1 + 0.005 - 0.0003 = 1.0047, \textit{ i.e., } 1.005$$

Now taking $\qquad x = 0.1, y(0.1) = 1.005, h = 0.1$

$$y'(0.1) = 0.092, y''(0.1) = 0.849, y'''(0.1) = -1.247$$

Substituting these values in the Taylor's series about $x = 0.1$,

$$y(0.2) = y(0.1) + \frac{0.1}{1!}y'(0.1) + \frac{(0.1)^2}{2!}y''(0.1) + \frac{(0.1)^3}{3!}y'''(0.1) + \cdots$$

$$= 1.005 + (0.1)(0.092) + \frac{(0.1)^2}{2}(0.849) + \frac{(0.1)^3}{6}(-1247) + \cdots$$

$$= 1.018$$

Now taking $x = 0.2, y(0.2) = 1.018, h = 0.1$

$$y'(0.2) = 0.176, y''(0.2) = 0.77, y'''(0.2) = 0.819$$

Substituting these values in the Taylor's series

$$y(0.2) = y(0.2) + \frac{0.1}{1!}y'(0.2) + \frac{(0.1)^2}{2!}y''(0.2) + \frac{(0.1)^3}{3!}y'''(0.2) + \cdots$$

$$= 1.018 + 0.0176 + 0.0039 + 0.0001$$

$$= 1.04$$

Thus the starting values of the Milne's method with $h = 0.1$ are

$$x_0 = 0.0, \qquad y_0 = 1 \qquad\qquad f_0 = y_0 = 0$$
$$x_1 = 0.1, \qquad y_1 = 1.005 \qquad f_1 = 0.092$$
$$x_2 = 0.2, \qquad y_2 = 1.018 \qquad f_2 = 0.176$$
$$x_3 = 0.3, \qquad y_3 = 1.04 \qquad f_3 = 0.26$$

Using the predictor, $y_4^{(p)} = y_0 + \dfrac{4h}{3}\left(2f_1 - f_2 + 2f_3\right)$

$$= 1 + \frac{4(0.1)}{3}[2(0.092) - 0.176 + 2(0.26)]$$

$$= 1.09.$$

$x = 0.4 \quad y_4^{(p)} = 1.09, \qquad f_4 = y'(0.4) = 0.362$

Using the corrector, $y_4^{(c)} = y_2 + \dfrac{h}{3}\left(f_2 + 4f_3 + f_4\right)$, yields

$$y_4^{(c)} = 0.018 + \frac{0.1}{3}(0.176 + 4(0.26) + 0.362) = 1.071$$

Hence $y(0.4) = 1.071$

EXAMPLE 10.22

Using the Runge-Kutta method of order 4, find y for $x = 0.1, 0.2, 0.3$ given that $dy/dx = xy + y^2$, $y(0) = 1$. Continue the solution at $x = 0.4$ using Milne's method.

Solution:

We have $\qquad f(x, y) = xy + y^2.$

To find $y(0.1)$:

Here $x_0 = 0, y_0 = 1, h = 0.1.$

$\therefore \qquad k_1 = hf(x_0, y_0) = (0.1) \times f(0,1) = 0.1000$

$$k_2 = hf\left(x_0 + \frac{1}{2}h, y_0 + \frac{1}{2}k_1\right) = (0.1) \times f(0.05, 1.05) = 0.1155$$

$$k_3 = hf\left(x_0 + \frac{1}{2}h, y_0 + \frac{1}{2}k_2\right) = (0.1) \times f(0.05, 1.0577) = 0.1172$$

$$k_4 = hf\left(x_0 + h, y_0 + k_3\right) = (0.1) \times f(0.1, 1.1172) = 0.13598$$

$$k = \frac{1}{6}\left(k_1 + 2k_2 + 2k_3 + k_4\right)$$

$$= \frac{1}{6}(0.1 + 0.231 + 0.2343 + 0.13598) = 0.11687$$

Thus $y(0.1) = y_1 = y_0 + k = 1.1169$

To find $y(0.2)$:

Here $x_1 = 0.1$, $y_1 = 1.1169$, h=0.1

$$k_1 = hf\left(x_1, y_1\right) = (0.1) \times f(0.1, 1.1169) = 0.1359$$

$$k_2 = hf\left(x_1 + \frac{1}{2}h, y_1 + \frac{1}{2}k_1\right) = (0.1) \times f(0.15, 1.1848) \quad 0.1581$$

$$k_3 = hf\left(x_1 + \frac{1}{2}h, y_1 + \frac{1}{2}k_2\right) = (0.1) \times f(0.15, 1.1959) = 0.1609$$

$$k_4 = hf\left(x_1 + h, y_1 + k_3\right) = (0.1) \times f(0.2, 1.2778) = 0.1888$$

$$k = \frac{1}{6}\left(k_1 + 2k_2 + 2k_3 + k_4\right) = 0.1605$$

Thus $y(0.2) = y_2 = y_1 + k = 1.2773$.

To find $y(0.3)$:

Here $x_2 = 0.2$, $y_2 = 1.2773$, $h = 0.1$.

$$k_1 = hf(x_2, y_2) = (0.1) \times f(0.2, 1.2773) = 0.1887$$

$$k_2 = hf\left(x_2 + \frac{1}{2}h, y_2 + \frac{1}{2}k_1\right) = (0.1) \times f(0.25, 1.3716) = 0.2224$$

$$k_3 = hf\left(x_2 + \frac{1}{2}h, y_2 + \frac{1}{2}k_2\right) = (0.1) f(0.25, 1.3885) = 0.2275$$

$$k_4 = hf\left(x_2 + h, y_2 + k_3\right) = (0.1) f(0.3, 1.5048) = 0.2716$$

$$k = \frac{1}{6}(k_1 + 2k_2 + 2k_3 + k_4) = 0.2267$$

$$y(0.3) = y_3 = y_2 + k = 1.504$$

Now the starting values for the Milne's method are:

$x_0 = 0.0$	$y_0 = 1.0000$	$f_0 = 1.0000$
$x_1 = 0.1$	$y_1 = 1.1169$	$f_1 = 1.3591$
$x_2 = 0.2$	$y_2 = 1.2773$	$f_2 = 1.8869$
$x_3 = 0.3$	$y_3 = 1.5049$	$f_3 = 2.7132$

Using the *predictor*

$$y_4^{(p)} = y_0 + \frac{4h}{3}(2f_1 - f_2 + 2f_3)$$

$$x_4 = 0.4 \quad y_4^{(p)} = 1.8344 \quad f_4 = 4.0988$$

and the *corrector*,

$$y_4^{(c)} = y_2 + \frac{h}{3}(f_2 + 4f_3 + f_4)$$

$$y_4^{(c)} = 1.2773 + \frac{0.1}{3}[1.8869 + 4(2.7132) + 4.098]$$

$$= 1.8397 \quad f_4 = 4.1159.$$

Again using the *corrector*,

$$y_4^{(c)} = 1.2773 + \frac{0.1}{3}[1.8869 + 4(2.7132) + 4.1159]$$

$$= 1.8391 \quad f_4 = 4.1182 \tag{i}$$

Again using the *corrector*,

$$y_4^{(c)} = 1.2773 + \frac{0.1}{3}[1.8869 + 4(2.7132) + 4.1182]$$

$$= 1.8392 \text{ which is same as (i)}$$

Hence $y(0.4) = 1.8392.$

Exercises 10.4

1. Given $\frac{dy}{dx} = x^3 + y$, $y(0) = 2$. The values of $y(0.2) = 2.073$, $y(0.4) = 2.452$,

and $y(0.6) = 3.023$ are gotten by the R.K. method of the order. Find $y(0.8)$ by Milne's predictor-corrector method taking $h = 0.2$

2. Given $2\,dy/dx = (1+x^2)y^2$ and $y(0) = 1$, $y(0.1) = 1.06$, $y(0.2) = 1.12$, $y(0.3) = 1.21$, evaluate $y(0.4)$ by Milne's predictor corrector method.

3. Solve that initial value problem

$$\frac{dy}{dx} = 1 + xy^2, y(0) = 1$$

for $x = 0.4$ by using Milne's method, when it is given that

x:	0.1	0.2	0.3
y:	1.105	1.223	1.355

4. From the data given below, find y at $x = 1.4$, using Milne's predictor-corrector formula: $dy/dx = x^2 + y/2$:

x = 1	1.1	1.2	1.3
y = 2	2.2156	2.4549	2.7514

5. Using Taylor's series method, solve $\dfrac{dy}{dx} = xy + x^2$, $y(0) = 1$; at $x = 0.1$, $0.2, 0.3$. Continue the solution at $x = 0.4$ by Milne's predictor-corrector method.

6. If $y = 2e^x - y$, $y(0) = 2$, $y(0.1) = 2.01$, $y(0.2) = 2.04$, and $y = 2.09$, find $y(0.4)$ using Milne's predictor-corrector method.

7. Using the Runge-Kutta method, calculate $y\,(0.1)$, $y(0.2)$, and $y(0.3)$ given that $\dfrac{dy}{dx} = \dfrac{2xy}{1+x^2} = 1$. $y(0) = 0$. Taking these values as starting values, find $y(0.4)$ by Milne's method.

10.10   Adams-Bashforth Method

Given $\dfrac{dy}{dx} = f(x,y)$ and $y_0 = y(x_0)$, we compute

$$y_{-1} = y(x_0 - h), y_{-2} = y(x_0 - 2h), y_{-3} = y(x_0 - 3h)$$

by Taylor's series or Euler's method or the Runge-Kutta method.

Next we calculate

$$f_{-1} = f(x_0 - h, y_{-1}), f_{-2} = f(x_0 - 2h, y_{-2}), f_{-3} = f(x_0 - 3h, y_{-3})$$

Then to find y_1, we substitute Newton's backward interpolation formula

$$f(x,y) = f_0 + n\nabla f_0 + \frac{n(n+1)}{2}\nabla^2 f_0 + \frac{n(n+1)(n+2)}{6}\nabla^3 f_0 + \cdots$$

in $\qquad y_1 = y_0 + \int_{x_0}^{x_0+h} f(x,y)$ (1)

$\therefore \qquad y_1 = y_0 + \int_{x_0}^{x_1} \left(f_0 + n\nabla f_0 + \frac{n(n+1)}{2}\nabla^2 f_0 + \cdots \right) dx$

$[\text{Put } x = x_0 + nh, \, dx = hdn]$

$= y_0 + h\int_0^1 \left(f_0 + n\nabla f_0 + \frac{n(n+1)}{2}\nabla^2 f_0 + \cdots \right) dn$

$= y_0 + h\left(f_0 + \frac{1}{2}\nabla f_0 + \frac{5}{12}\nabla^2 f_0 + \frac{3}{8}\nabla^3 f_0 + \cdots \right)$

Neglecting fourth and higher order differences and expressing $\nabla f_0, \nabla^2 f_0$ and $\nabla^3 f_0$ in terms of function values, we get

$$y_1 = y_0 + \frac{h}{24}\left(55f_0 - 59f_{-1} + 37f_{-2} - 9f_{-3} \right) \qquad (2)$$

This is called the *Adams-Bashforth predictor formula.*

Having found y_1, we find $f_1 = f(x_0 + h_1, y_1)$.

Then to find a better value of $y1$, we derive a *corrector formula* by substituting Newton's backward formula at $f1$, *i.e.,*

$$f(x,y) = f_1 + n\nabla f_1 + \frac{n(n+1)}{2}\nabla^2 f_1 + \frac{n(n+1)(n+2)}{6}\nabla^3 f_1 + \cdots$$

in (1)

$\therefore \qquad y_1 = y_0 + \int_{x_0}^{x_1} \left(f_1 + n\nabla f_1 + \frac{n(n+1)}{2}\nabla^2 f_1 + \cdots \right) dx$

$[\text{Put } x = x_1 + nh, \, dx = h \, dn]$

$= y_0 + \int_{-1}^0 \left(f_1 + n\nabla f_1 + \frac{n(n+1)}{2}\nabla^2 f_1 + \cdots \right) dn$

$= y_0 + h\left(f_1 - \frac{1}{2}\nabla f_1 - \frac{1}{12}\nabla^2 f_0 - \frac{1}{24}\nabla^3 f_1 + \cdots \right)$

Neglecting fourth and higher order differences and expressing $\nabla f_1, \nabla^2 f_1$ and $\nabla^3 f_1$ and in terms of function values, we obtain

$$y_1^{(c)} = y_0 + \frac{h}{24}\left(9f_1 + 19f_0 - 5f_{-1} + 9f_{-2} \right)$$

which is called the *Adams-Moulton corrector formula.*

Then an improved value of $f1$ is calculated and again the corrector (3) is applied to find a still better value $y1$. This step is repeated until y_1 remains unchanged and then we proceed to calculate y_2 as above.

NOTE

Obs. *To apply both Milne and Adams-Bashforth methods, we require four starting values of y which are calculated by means of Picard's method or Taylor's series method or Euler's method or the Runge-Kutta method. In practice, the Adams formulae (2) and (3) above together with the fourth order Runge-Kutta formulae have been found to be the most useful.*

EXAMPLE 10.23

Given $\dfrac{dy}{dx} = x^2\left(1+y\right)$ and $y(1) = 1$, $y(1.1) = 1.233$, $y(1.2) = 1.548$, $y(1.3) = 1.979$, evaluate $y(1.4)$ by the Adams-Bashforth method.

Solution:

Here $f(x, y) = x^2(1 + y)$

Starting values of the Adams-Bashforth method with $h = 0.1$ are

$x = 1.0$, $y_{-3} = 1.000$, $f_{-3} = (1.0)^2(1 + 1.000) = 2.000$

$x = 1.1$, $y_{-2} = 1.233$, $f_{-2} = 2.702$

$x = 1.2$, $y_{-1} = 1.548$, $f_{-1} = 3.669$

$x = 1.3$, $y_0 = 1.979$, $f_0 = 5.035$

Using the *predictor*,

$$y_1^{(p)} = y_0 + \frac{h}{24}\left(55f_0 - 59f_{-1} + 37f_{-2} - 9f_{-3}\right)$$

$$x_4 = 1.4, \quad y_1^{(p)} = 2.573 \quad f_1 = 7.004$$

Using the *corrector*

$$y_1^{(c)} = y_0 + \frac{h}{24}\left(9f_1 + 19f_0 - 5f_{-1} + f_{-2}\right)$$

$$y_1^{(c)} = 1.979 + \frac{0.1}{24}\left(9 \times 7.004 + 19 \times 5.035 - 5 \times 3.669 + 2.702\right) = 2.575$$

Hence $y(1.4) = 2.575$

EXAMPLE 10.24

If $\dfrac{dy}{dx} = 2e^x y$, $y(0) = 2$, find $y(4)$ using the Adams predictor corrector formula by calculating $y(1)$, $y(2)$, and $y(3)$ using Euler's modified formula.

Solution:

We have $f(x, y) = 2e^x y$

x	$2e^x y$	Mean slope	Old $y + h$ (mean slop) = new y
0	4		$2 + 0.1(4) = 2.4$
0.1	$2e^{0.1}(2.4) = 5.305$	$\frac{1}{2}(4 + 5.305) = 4.6524$	$2 + 0.1\,(4.6524) = 2.465$
0.1	$2e^{0.1}(2.465) = 5.449$	$\frac{5}{1}(4 + 2.462) = 4.2544$	$2 + 0.1\,(4.7244) = 2.472$
0.1	$2e^{0.1}(2.4724) = 5.465$	$\frac{1}{2}(4 + 5.465) = 4.7324$	$2 + 0.1\,(4.7324) = 2.473$
0.1	$2e^{0.1}(2.478) = 5.467$	$\frac{1}{2}(4 + 5.467) = 4.7333$	$2 + 0.1\,(4.7333) = 2.473$
0.1	5.467	—	$2 + 0.1\,(5.467) = 3.0199$
0.2	$2e^{0.2}(3.0199) = 7.377$	$\frac{1}{2}(5.467 + 7.377) = 6.422$	$2.473 + 0.1\,(6.422) = 3.1155$
0.2	7.611	$\frac{1}{2}(5.467 + 7.611) = 6.539$	$2.473 + 0.1\,(6.539) = 3.127$
0.2	7.639	$\frac{1}{2}(5.467 + 7.639) = 6.553$	$2.473 + 0.1\,(6.553) = 3.129$
0.2	7.643	$\frac{1}{2}(5.467 + 7.643) = 6.555$	$2.473 + 0.1\,(6.555) = 3.129$
0.2	7.463	—	$3.129 + 0.1\,(7.643) = 3.893$
0.3	$2e^{0.3}(3.893) = 10.51$	$\frac{1}{2}(7.643 + 10.51) = 9.076$	$3.129 + 0.1\,(9.076) = 4.036$
0.3	10.897	$\frac{1}{2}(7.643 + 10.897) = 9.266$	$3.129 + 0.1\,(9.2696) = 4.056$
0.3	10.949	$\frac{1}{2}(7.643 + 10.949) = 9.296$	$3.129 + 0.1\,(9.296) = 4.058$
0.3	10.956	$\frac{1}{2}(7.643 + 10.956) = 9.299$	$3.129 + 0.1\,(9.299) = 4.0586$

To find $y(0.4)$ *by Adam's method*, the starting values with $h = 0.1$ are

$$x = 0.0 \qquad y_{-3} = 2.4 \qquad f_{-3} = 4$$
$$x = 0.1 \qquad y_{-2} = 2.473 \qquad f_{-2} = 5.467$$
$$x = 0.2 \qquad y_{-1} = 3.129 \qquad f_{-1} = 7.643$$
$$x = 0.3 \qquad y_{0} = 4.059 \qquad f_{0} = 10.956$$

Using the predictor formula

$$y_1^{(p)} = y_0 + \frac{h}{24}\left(55f_0 - 59f_{-1} + 37f_{-2} - 9f_{-3}\right)$$

$$= 4.059 + \frac{0.1}{24}\left(55 \times 10.957 - 59 \times 7.643 + 37 \times 5.467 - 9 \times 4\right)$$

$$= 5.383$$

Now $x = 0.4$ $y_1 = 5.383$ $f_1 = 2e^{0.4}(5.383) = 16.061$

Using the corrector formula

$$y_1^{(c)} = y_0 + \frac{h}{24}\left(9f_1 + 19f_0 - 5f_{-1} + f_{-2}\right)$$

$$= 4.0586 + \frac{0.1}{24}\left(9 \times 16.061 + 19 \times 10.956 - 5 \times 7.643 + 5.467\right)$$

$$= 5.392$$

Hence $y(0.4) = 5.392$

EXAMPLE 10.25

Solve the initial value problem $dy/dx = x - y^2$, $y(0) = 1$ to find $y(0.4)$ by Adam's method. Starting solutions required are to be obtained using the Runge-Kutta method of the fourth order using step value $h = 0.1$

Solution:

We have $f(x, y) = x - y^2$.

To find $y(0.1)$:

Here $x_0 = 0$, $y_0 = 1$, $h = 0.1$.

$$\therefore \qquad k_1 = hf(x_0, y_0) = (0.1)f(0,1) = -0.1000$$

$$k_2 = hf\left(x_0 + \frac{1}{2}h, y_0 + \frac{1}{2}k_1\right) = (0.1)f(0.05, 0.95) = -0.08525$$

$$k_3 = hf\left(x_0 + \frac{1}{2}h, y_0 + \frac{1}{2}k_2\right) = (0.1)f(0.05, 0.9574) = -0.0867$$

$$k_4 = hf\left(x_0 + h, y_0 + k_3\right) = (0.1)f(0.1, 0.9137) = -0.07341$$

$$k = \frac{1}{6}(k_1 + 2k_2 + 2k_3 + k_4)0 = -0.0883$$

Thus $\quad y(0.1) = y_1 = y_0 + k = 1 - 0.0883 = 0.9117$

To find $y(0.2)$:

Here $x_1 = 0.1$, $y_1 = 0.9117$, $h = 0.1$

$$k_1 = hf\left(x_1, y_1\right) = (0.1) \times f(0.1, 0.9117) = -0.0731$$

$$k_2 = hf\left(x_1 + \frac{1}{2}h, y_1 + \frac{1}{2}k_1\right) = (0.1)f(0.15, 0.8751) \quad 0.0616$$

$$k_3 = hf\left(x_1 + \frac{1}{2}h, y_1 + \frac{1}{2}k_2\right) = (0.1)f(0.15, 0.8809) = 0.0626$$

$$k_4 = hf\left(x_1 + h, y_1 + k_3\right) = (0.1) \times f(0.2, 0.8491) = 0.0521$$

$$k = -\left(k_1 + 2k_2 + 2k_3 + k_4\right) \quad 0.06$$

Thus $\quad y(0.2) = y_2 = y_1 + k = 0.8494.$

To find $y(0.3)$:

Here $x_2 = 0.2$, $y_2 = 0.8494$, $h = 0.1$.

$$k_1 = hf(x_2, y_2) = (0.1) \times f(0.25, 0.8494) = 0.0521$$

$$k_2 = hf\left(x_2 + \frac{1}{2}h, y_2 + \frac{1}{2}k_1\right) = (0.1)f(0.25, 0.8233) = 0.0428$$

$$k_3 = hf\left(x_2 + \frac{1}{2}h, y_2 + \frac{1}{2}k_2\right) = (0.1)f(0.25, 0.828) = 0.0436$$

$$k_4 = hf\left(x_2 + h, y_2 + k_3\right) = (0.1)f(0.3, 0.058) = 0.0349$$

$$k = \frac{1}{6}\left(k_1 + 2k_2 + 2k_3 + k_4\right) = 0.0438$$

Thus $\quad y(0.3) = y_3 = y_2 + k = 0.8061$

Now the starting values for the Milne's method are:

$x_0 = 0.0$	$y_0 = 1.0000$	$f_0 = 0.0 - (0.1)^2 = 1.0000$
$x_1 = 0.1$	$y_1 = 0.9117$	$f_1 = 0.1 - (0.9117)^2 = -0.7312$
$x_2 = 0.2$	$y_2 = 0.8494$	$f_2 = 0.2 - (0.8494)^2 = -0.5215$
$x_3 = 0.3$	$y_3 = 0.8061$	$f_3 = 0.3 - (0.8061)^2 = -0.3498$

Using the predictor,

$$y_1^{(p)} = y_0 + \frac{h}{24}\left(55f_0 - 59f_{-1} + 37f_{-2} - 9f_{-3}\right)$$

$x = 0.4$

$$y_1^{(p)} = 0.8061 + \frac{0.1}{24}\left(55(-0.3498) - 59(-0.5215) + 37(-0.7312) - 9(-1)\right)$$

$$= 0.7789 \quad f_1 = -0.2.67$$

Using the *corrector*,

$$y_1^{(c)} = y_0 + \frac{h}{24}\left(9f_1 + 19f_0 - 5f_{-1} + f_{-2}\right)$$

$$y_1^{(c)} = 0.8061 + \frac{0.1}{24}\left[9(-0.2067) + 19 \times (-0.3498) - 5(-0.5215) - 0.7312\right]$$

$$= 0.7785$$

Hence $y(0.4) = 0.7785$

Exercises 10.5

1. Using the Adams-Bashforth method, obtain the solution of $dy/dx = x - y^2$ at $x = 0.8$, given the values

x:	0	0.2	0.4	0.6
y:	0	0.0200	0.0795	0.1762

2. Using the Adams-Bashforth formulae, determine $y(0.4)$ given the differential equation $dy/dx = \frac{1}{2}xy$ and the data:

x:	0	0.1	0.2	0.3
y:	1	1.002	5 1.0101	1.0228

3. Given $y' = x^2 - y$, $y(0) = 1$ and the starting values $y(0.1) = 0.90516$, $y(0.2) = 0.82127$, $y(0.3) = 0.74918$, evaluate $y(0.4)$ using the Adams-Bashforth method.

4. Using the Adams-Bashforth method, find $y(4.4)$ given $5xy' + y^2 = 2$, $y(4) = 1$, $y(4.1) = 1.0049$, $y(4.2) = 1.0097$ and $y(4.3) = 1.0143$.

5. Given the differential equation $dy/dx = x^2y + x^2$ and the data:

x:	1	1.1	1.2	1.3
y:	1	1.233	1.548488	1.978921

determine $y(1.4)$ by any numerical method.

6. Using the Adams-Bashforth method, evaluate $y(1.4)$; if y satisfies $dy/dx + y/x = 1/x^2$ and $y(1) = 1$, $y(1.1) = 0.996$, $y(1.2) = 0.986$, $y(1.3) = 0.972$.

10.11 Simultaneous First Order Differential Equations

The simultaneous differential equations of the type

$$\frac{dy}{dx} = f(x,y,z) \tag{1}$$

and

$$\frac{dz}{dx} = \phi(x,y,z) \tag{2}$$

with initial conditions $y(x_0) = y_0$ and $z(x_0) = z_0$ can be solved by the methods discussed in the preceding sections, especially Picard's or Runge-Kutta methods.

Picard's method gives

$$y_1 = y_0 + \int f(x,y_0,z_0)dx, \ z_1 = z_0 + \int \phi(x,y_0,z_0)dx$$

$$y_2 = y_0 + \int f(x,y_1,z_1)dx, \ z_2 = z_0 + \int \phi(x,y_1,z_1)dx$$

$$y_3 = y_0 + \int f(x,y_2,z_2)dx, \ z_3 = z_0 + \int \phi(x,y_2,z_2)dx$$

and so on.

(*ii*) *Taylor's series method* is used as follows:

If h be the step-size, $y_1 = y(x_0 + h)$ and $z_1 = z(x_0 + h)$. Then Taylor's algorithm for (1) and (2) gives

$$y_1 = y_0 + hy_0' + \frac{h^2}{2!}y_0'' + \frac{h^3}{3!}y_0''' + \cdots \tag{3}$$

$$z_1 = z_0 + hz_0' + \frac{h^2}{2!}z_0'' + \frac{h^3}{3!}z_0''' + \cdots \tag{4}$$

Differentiating (1) and (2) successively, we get y'', z'', etc. So the values $y_0', y_0'', y_0''' \cdots$ and $z_0, z_0'', z_0''' \cdots$ are known. Substituting these in (3) and (4), we obtain y_1, z_1 for the next step.

Similarly, we have the algorithms

$$y_2 = y_1 + hy_1' + \frac{h^2}{2!}y_1'' + \frac{h^3}{3!}y_1'' + \cdots \tag{5}$$

$$z_2 = z_1 + hz_1' + \frac{h^2}{2!}z_1'' + \frac{h^3}{3!}z_1''' + \cdots \tag{6}$$

Since y_1 and z_1 are known, we can calculate y_1', y_1'', and $z_1', z_1'', \cdots$. Substituting these in (5) and (6), we get y_2 and z_2.

Proceeding further, we can calculate the other values of y and z step by step.

(iii) Runge-Kutta method is applied as follows:

Starting at (x_0, y_0, z_0) and taking the step-sizes for x, y, z to be h, k, l respectively, the Runge-Kutta method gives,

$$k_1 = hf(x_0, y_0, z_0)$$

$$l_1 = h\phi(x_0, y_0, z_0)$$

$$k_2 = hf\left(x_0 + \frac{1}{2}h, y_0 + \frac{1}{2}k_1, z_0 + \frac{1}{2}l_1\right)$$

$$l_2 = h\phi\left(x_0 + \frac{1}{2}h, y_0 + \frac{1}{2}k_1, z_0 + \frac{1}{2}l_1\right)$$

$$k_3 = hf\left(x_0 + \frac{1}{2}h, y_0 + \frac{1}{2}k_2, z_0 + \frac{1}{2}l_2\right)$$

$$l_3 = h\phi\left(x_0 + \frac{1}{2}h, y_0 + \frac{1}{2}k_2, z_0 + \frac{1}{2}l_2\right)$$

$$k_4 = hf\left(x_0 + \frac{1}{2}h, y_0 + \frac{1}{2}k_3, z_0 + \frac{1}{2}l_3\right)$$

$$l_4 = h\phi\left(x_0 + \frac{1}{2}h, y_0 + \frac{1}{2}k_3, z_0 + \frac{1}{2}l_3\right)$$

Hence
$$y_1 = y_0 + \frac{1}{6}\left(k_1 + 2k_2 + 2k_3 + k_4\right)$$

and
$$z_1 = y_0 + \frac{1}{6}\left(l_1 + 2l_2 + 2l_3 + l_4\right)$$

To compute y_2 and z_2, we simply replace x_0, y_0, z_0 by x_1, y_1, z_1 in the above formulae.

EXAMPLE 10.26

Using Picard's method, find approximate values of y and z corresponding to $x = 0.1$, given that $y(0) = 2$, $z(0) = 1$ and $dy/dx = x + z$, $dz/dx = x - y^2$.

Solution:

Here $x_0 = 0$, $y_0 = 2$, $z_0 = 1$,

and $\dfrac{dy}{dx} = f(x, y, z) = x + z$

$\dfrac{dz}{dx} = \phi(x, y, z) = x - y^2$

$\therefore \quad y = y_0 + \displaystyle\int_{x_0}^{x} f(x, y, z)dx$ and $z = z_0 + \displaystyle\int_{x_0}^{x} \phi(x, y, z)dx$

First approximations

$$y_1 = y_0 + \int_{x_0}^{x} f(x, y_0, z_0)dx = 2 + \int_{x_0}^{x} (x+1)dx = 2 + x + \frac{1}{2}x^2$$

$$z_1 = z_0 + \int_{x_0}^{x} \phi(x, y_0, z_0)dx = 1 + \int_{x_0}^{x} (x-4)dx = 1 - 4x + \frac{1}{2}x^2$$

Second approximations

$$y_2 = y_0 + \int_{x_0}^{x} f(x, y_1, z_1)dx = 2 + \int_{0}^{x}(1 - 4x + \frac{1}{2}x^2)dx$$

$$= 2 + x - \frac{3}{2}x^2 + \frac{x^3}{6}$$

$$z_2 = z_0 + \int_{x_0}^{x} \phi(x, y_1, z_1)dx = 1 + \int_{x_0}^{x}\left[x - \left(2 + x + \frac{1}{2}x^2\right)^2\right]dx$$

$$= 1 - 4x + \frac{3}{2}x^2 - x^3 - \frac{x^4}{4} - \frac{x^5}{20}$$

Third approximations

$$y_3 = y_0 + \int_{x_0}^{x} f(x, y_2, z_2)dx = 2 + x - \frac{3}{2}x^2 - \frac{1}{2}x^3 - \frac{1}{4}x^4 - \frac{1}{20}x^5 - \frac{1}{120}x^6$$

$$z_3 = z_0 + \int_{x_0}^{x} \phi(x, y_2, z_2)dx$$

$$= 1 - 4x + \frac{3}{2}x^2 + \frac{5}{3}x^3 + \frac{7}{12}x^4 - \frac{31}{60}x^5 + \frac{1}{12}x^6 - \frac{1}{252}x^7$$

and so on.

When $x = 0.1$

$$y_1 = 2.105, \qquad y_2 = 2.08517, \qquad y_3 = 2.08447$$
$$z_1 = 0.605, \qquad z_2 = 0.58397, \qquad z_3 = 0.58672.$$

Hence $y(0.1) = 2.0845,$ $z(0.1) = 0.5867$

correct to four decimal places.

EXAMPLE 10.27

Find an approximate series solution of the simultaneous equations $dx/dt = xy + 2t$, $dy/dt = 2ty + x$ subject to the initial conditions $x = 1$, $y = -1$, $t = 0$.

Solution:

x and y both being functions of t, Taylor's series gives

and
$$\left. \begin{aligned} x(t) &= x_0 + tx_0' + \frac{t^2}{2!}x_0'' + \frac{t^3}{3!}x_0''' + \cdots \\ y(t) &= y_0 + ty_0' + \frac{t^2}{2!}y_0'' + \frac{t^3}{3!}y_0''' + \cdots \end{aligned} \right\} \qquad (i)$$

Differentiating the given equations
$$x' = xy + 2t \qquad\qquad (ii)$$
$$y' = 2ty + x \qquad\qquad (iii)$$

w.r.t. t, we get

$$\left. \begin{aligned} x'' &= xy' + x'y + 2 & y'' &= 2ty' + 2y + x' \\ x''' &= (xy'' + x'y') + x''y + x'y' & y''' &= 2ty'' + 2y' + 2y' + x'' \end{aligned} \right\} \qquad (iv)$$

Putting $x_0 = 1$, $y_0 = -1$, $t_0 = 0$ in (ii), (iii), and (iv), we obtain

$$x_0 = -1 + 2(0) = -1 \qquad\qquad y_0' = 1$$
$$x_0'' = x_0 y_0' + x_0' y_0 + 2 \qquad\qquad y_0''' = 0 + 2y_0 + x_0'$$
$$= 1.1 + (-1)(-1) + 2 = 4 \qquad = 2(-1) + (-1) = -3$$
$$x_0''' = -3 + (-1)(1) + 4(-1) - 1 = -9 \qquad y_0''' = 2 + 2 + 4 = 8 \text{ etc}$$

Substituting these values in (i), we get
$$x(t) = 1 - t + 4\frac{t^2}{2!} + (-9)\frac{t^3}{3!} + \cdots = 1 - t + 2t^2 - \frac{3}{2}t^3 + \cdots$$

$$x(t) = 1 + t + 3\frac{t^2}{2!} + 8\frac{t^3}{3!} + \cdots = 1 + t - \frac{3}{2}t^2 + \frac{4}{3}t^3 + \cdots$$

EXAMPLE 10.28

Solve the differential equations

$$\frac{dy}{dx} = 1 + xz, \frac{dz}{dx} = -xy \text{ for } x = 0.3$$

using the fourth order Runge-Kuta method. Intial values are $x = 0$, $y = 0$, $z = 1$.

Solution:

Here $f(x, y, z) = 1 + xz, \phi(x, y, z) = -xy$

$x_0 = 0, y_0 = 0, z_0 = 1$. Let us take $h = 0.3$.

$\therefore$ $k_1 = hf(x_0, y_0, z_0) = 0.3f(0, 0, 1) = 0.3(1 + 0) = 0.3$.

$l_1 = h\phi(x_0, y_0, z_0) = 0.3(-0 \times 0) = 0$

$k_2 = hf\left(x_0 + \frac{1}{2}h, y_0 + \frac{1}{2}k_1, z_0 + \frac{1}{2}l_1\right)$

$= (0.3) f(0.15, 0.15, 1) = 0.3(1 + 0.15) = 0.345$

$l_2 = h\phi\left(x_0 + \frac{1}{2}h, y_0 + \frac{1}{2}k_1, z_0 + \frac{1}{2}l_1\right)$

$= (0.3)[-(0.15)(0.15)] = -0.00675$

$k_3 = hf\left(x_0 + \frac{1}{2}h, y_0 + \frac{k_2}{2}, z_0 + \frac{l_2}{2}\right)$

$= (0.3) f(0.15, 0.1725, 0.996625)$

$= 0.3[1 + 0.996625 \times 0.15] = 0.34485$

$l_3 = h\phi\left(x_0 + \frac{1}{2}h, y_0 + \frac{k_2}{2}, z_0 + \frac{l_2}{2}\right)$

$= 0.3[-(0.15)(0.1725)] = -0.007762$

$k_4 = hf\left(x_0 + h, y_0 + k_3, z_0 + l_3\right)$

$= 0.3f(0.3, 0.34485, 0.99224) = 0.3893$

$l_4 = h\phi\left(x_0 + h, y_0 + k_3, z_0 + l_3\right)$

$= 0.3[-(0.3)(0.34485)] = -0.03104$

Hence $y(x_0 + h) = y_0 + \frac{1}{6}(k_1 + 2k_2 + 2k_3 + k_4)$

i.e., $\qquad y(0.3) = 0 + \dfrac{1}{6}[0.3 + 2(0.345) + 2(0.34485) + 0.3893] = 0.34483$

and $\qquad z(x+h) = y_0 + \dfrac{1}{6}(l_1 + 2l_2 + 2l_3 + l_4)$

i.e. $\qquad z(0.3) = 1 + \dfrac{1}{6}[0 + 2(-0.00675) + 2(0.0077625) + (-0.03104)]$

$\qquad\qquad\qquad = 0.98999$

10.12 Second Order Differential Equations

Consider the second order differential equation

$$\frac{d^2 y}{dx_2} = f\left(x, y, \frac{dy}{dx}\right)$$

By writing $dy/dx = z$, it can be reduced to two first order simultaneous differential Equations

$$\frac{dy}{dx} = z, \frac{dz}{dx} f(x, y, z)$$

These equations can be solved as explained above.

EXAMPLE 10.29

Find the value of $y(1.1)$ and $y(1.2)$ from $y'' + y^2 y' = x3; y(1) = 1, y'(1) = 1$, using the Taylor series method

Solution:

Let $\qquad y' = z$ so that $y'' = z'$

Then the given equation becomes $z' + y^2 z = z^3$

$\therefore \qquad\qquad y' = z$

$\qquad\qquad z' = x^3 - y^2 z \qquad\qquad\qquad\qquad\qquad (i)$

such that $\qquad y(1) = 1, z(1) = 1, h = 0.1. \qquad\qquad (ii)$

Now from (i) $\qquad y' = z, y'' = z', y''' = z'' \qquad\qquad (iii)$

$$z' = x^3 - y^2 z, z'' = 3x^2 - y^2 z' - 2yz^2 \;(\because y' = z)$$

and from (ii) $\qquad z''' = 6x - (y^2 z'' + 2yy'z') - 2(y'z^2 + y^2 zz') \qquad (iv)$

$$= 6x - (y^2 z'' + 2yz'^2) - 2(z^3 + 2yzz')$$

Taylor's series for $y(1.1)$ is

$$y(1.1) = y(1) + hy'(1) + \frac{h^2}{2!}y''(1) + \frac{h^3}{3!}y'''(1) + ...$$

Also $\qquad y(1) = 1, y'(1) = 1, y''(1) = z'(1) = 0, y'''(1) = z''(1) = 1$

$\therefore \qquad y(1.1) = (1) + 0.1(1) + \frac{(0.1)^2}{2}(0) + \frac{(0.1)^3}{6}(0) = 1.1002.$

Taylor's series for $z(1.1)$ is

$$z(1.1) = z(1) + hz'(1) + \frac{h^2}{2!}z''(1) + \frac{h^3}{3!}z'''(1) + ...$$

Here $\qquad z(1) = 1, z'(1) = 0, z''(1) = 1, z'''(1) = 3$

$$z(1.1) = (1) + 0.1(0) + \frac{(0.1)^2}{2}(1) + \frac{(0.1)^3}{6}(3) = 1.0055$$

Hence $y(1.1) = 1.1002$ and $z(1.1) = 1.0055$.

EXAMPLE 10.30

Using the Runge-Kutta method, solve $y'' = xy'^2 - y^2$ for $x = 0.2$ correct to 4 decimal places. Initial conditions are $x = 0$, $y = 1$, $y' = 0$.

Solution:

Let $dy/dx = z = f(x, y, z)$

Then $\qquad \dfrac{dy}{dx} = xz^2 - y^2 = \phi(x, y, z)$

We have $x_0 = 0$, $y_0 = 1$, $z_0 = 0$, $h = 0.2$

$\therefore$ Runge-Kutta formulae become

$k_1 = hf(x_0, y_0, z_0) = 0.2(0) = 0$

$k_2 = hf\left(x_o + \dfrac{1}{2}h, y_0 + \dfrac{1}{2}k_1, z_0 + \dfrac{1}{2}l_1\right)$

$\qquad = 0.2(-0.1) = -0.02$

$k_3 = hf\left(x_o + \dfrac{1}{2}h, y_0 + \dfrac{1}{2}k_2, z_0 + \dfrac{1}{2}l_2\right)$

$\qquad = 0.2(-0.0999) = -0.02$

$k_4 = hf\left(x_o + h, y_0 + k_3, z_0 + l_3\right)$

$\qquad = 0.2(-0.1958) = -0.0392$

$$\therefore \qquad k = \frac{1}{6}\left(k_1 + 2k_2 + 2k_3 + k_4\right) = 0.0199$$

$$l_1 = hf(x_0, y_0, z0) = 0.2(-1) = -0.2$$

$$l_2 = h\phi\left(x_0 + \frac{1}{2}h, y_0 + \frac{1}{2}k_1, z_0 + \frac{1}{2}l_1\right)$$
$$= 0.2(-0.999) = -0.1998$$

$$l_3 = h\phi\left(x_0 + \frac{1}{2}h, y_0 + \frac{1}{2}k_2, z_0 + \frac{1}{2}l_2\right)$$
$$= 0.2(-0.9791) = -0.1958$$

$$l_4 = h\phi\left(x_0 + h, y_0 + k_3, z_0 + l_3\right)$$
$$= 0.2(0.9527) = -0.1905$$

$$l = \frac{1}{6}\left(l_1 + 2l_2 + 2l_3 + l_4\right) = -0.1970$$

Hence at $x = 0.2$,

$$y = y_0 + k = 1 - 0.0199 = 0.9801$$

and $\qquad y' = z = z0 + l = 0 - 0.1970 = -0.1970.$

EXAMPLE 10.31

Given $y'' + xy' + y = 0$, $y(0) = 1$, $y'(0) = 0$, obtain y for $x = 0(0.1)\,0.3$ by any method. Further, continue the solution by Milne's method to calculate $y(0.4)$.

Solution:

Putting $y' = z$, the given equation reduces to the simultaneous equations

$$z' + xz + y = 0, \; y' = z \qquad\qquad (1)$$

We employ Taylor's series method to find y.

Differentiating the given equation n times, we get

$$y_{n+2} + x_{n+1} + ny_n + y_n = 0$$

At $\qquad x = 0$, $(y_{n+2})_0 = -(n+1)(y_n)_0$

$\therefore \qquad y(0) = 1$, gives $y_2(0) = -1$, $y_4(0) = 3$, $y_6(0) = -5 \times 3$, ...

and $\qquad y_1(0) = 0$ yields $y_3(0) = y_5(0) = \ldots\ldots = 0.$

Expanding $y(x)$ by Taylor's series, we have

$$y(x) = y(0) + xy_1(0) + \frac{x^2}{2!}y_2(0) + \frac{x^3}{3!}y_3(0) + \cdots$$

$\therefore \qquad y(x) = 1 - \frac{x^2}{2!} + \frac{3}{4!}x^4 - \frac{5 \times 3}{6!}x^6 + \cdots$ \hfill (2)

and $\qquad z(x) = y'(x) = -x + \frac{1}{2}x^3 - \frac{1}{8}x^5 + \cdots = -xy,$ \hfill (3)

From (2), we have

$$y(0.1) = 1 - \frac{(0.1)^2}{2} + \frac{1}{8}(0.1)^4 - \cdots = 0.995$$

$$y(0.2) = 1 - \frac{(0.2)^2}{2} + \frac{(0.2)^4}{8} - \cdots = 0.9802$$

$$y(0.3) = 1 - \frac{(0.3)^2}{2} + \frac{(0.3)^4}{8} - \frac{(0.3)^6}{48} \cdots = 0.956$$

From (3), we have

$$z(0.1) = -0.0995, \; z(0.2) = -0.196, \; z(0.3) = -0.2863.$$

Also from (1), $z'(x) = -(xz + y)$

$\therefore \qquad z'(0.1) = 0.985, \; z'(0.2) = -0.941, \; z'(0.3) = -0.87.$

Applying Milne's predictor formula, first to z and then to y, we obtain

$$z(0.4) = z(0) + \frac{4}{3}(0.1)\{2z'(0.1) - z'(0.2) + 2z'(0.3)\}$$

$$= 0 + \left(\frac{0.4}{3}\right)\{-1.79 + 0.941 - 1.74\} = -0.3692$$

and $\qquad y(0.4) = y(0) + \frac{4}{3}(0.1)\{2y'(0.1) - y'(0.2) + 2y'(0.3)\} \quad [\because y' = z]$

$$= 0 + \left(\frac{0.4}{3}\right)\{-0.199 + 0.196 - 0.5736\} = 0.9231$$

Also $\quad z'(0.4) = -\{x(0.4)\,z(0.4) + y(0.4)\}$

$$= -\{0.4(-0.3692) + 0.9231\} = -0.7754.$$

Now applying Milne's corrector formula, we get

$$z(0.4) = z(0.2) + \frac{h}{3}\{z'(0.2) + 4z'(0.3) + z'(0.4)\}$$

$$= -0.196 + \left(\frac{0.1}{3}\right)\{-0.941 - 3.48 - 0.7754\} = -0.3692$$

and $\qquad y(0.4) = y(0.2) + \dfrac{h}{3}\{y'(0.2) + 4y'(0.3) + y'(0.4)\}$

$$= 0.9802 + \left(\dfrac{0.1}{3}\right)\{-0.196 - 1.1452 - 0.3692\} = 0.9232$$

Hence $y(0.4) = 0.9232$ and $z(0.4) = -0.3692$.

Exercises 10.6

1. Apply Picard's method to find the third approximations to the values of y and z, given that
   $$dy/dx = z,\ dz/dx = x^3(y + z),\ \text{given } y = 1,\ z = 1/2 \text{ when } x = 0.$$

2. Using Taylor's series method, find the values of x and y for $t = 0.4$, satisfying the differential equations
   $dx/dt = x + y + t,\ d^2y/dt^2 = x - t$ with initial conditions $x = 0,\ y = 1,$
   $dy/dt = -1$ at $t = 0$.

3. Solve the following simultaneous differential equations, using Taylor series method of the fourth order, for $x = 0.1$ and 0.2:
   $$\dfrac{dy}{dx} = xz + 1;\ \dfrac{dz}{dy} == xy;\ y(0) = 1.$$

4. Find $y(0.1),\ z(0.1),\ y(0.2),$ and $z(0.2)$ from the system of equations: $y' = x + z,\ z' = x - y^2$ given $y(0) = 0,\ z(0) = 1$ using Runge-Kutta method of the fourth order.

5. Using Picard's method, obtain the second approximation to the solution of
   $$\dfrac{d^2y}{dx^2} = x^3\dfrac{dy}{dx} + x^3y \text{ so that } y(0) = 1, y'(0) = \dfrac{1}{2}.$$

6. Use Picard's method to approximate y when $x = 0.1$, given that
   $$\dfrac{d2y}{dx2} + 2x\dfrac{d^2y}{dx^2} + 2x\dfrac{dy}{dx} + y = 0 \text{ and } y = 0.5, \dfrac{dy}{dx} = 0.1 \text{ when } x = 0.$$

7. Find $y(0.2)$ from the differential equation $y'' + 3xy' - 6y = 0$ where $y(0) = 1,\ y'(0) = 0.1$, using the Taylor series method.

8. Using the Runge-Kutta method of the fourth, solve $y'' = y + xy',\ y(0) = 1,$ $y'(0) = 0$ to find $y(0.2)$ and $y'(0.2)$.

9. Consider the second order initial value problem $y'' - 2y' + 2y = e^{2t} \sin t$ with $y(0) = -0.4$ and $y'(0) = -0.6$. Using the fourth order Runga-Kutta method, find $y(0.2)$.

10. The angular displacement θ of a simple pendulum is given by the equation

$$\frac{d^2\phi}{dt^2} + \frac{g}{1}\sin\phi = 0$$

where $l = 98$ cm and $g = 980$ cm/sec². If $\theta = 0$ and $d\theta/dt = 4.472$ at $t = 0$, use the Runge-Kutta method to find θ and $d\theta/dt$ when $t = 0.2$ sec.

11. In a L-R-C circuit the voltage $v(t)$ across the capacitor is given by the equation

$$LC\frac{d^2v}{dt^2} + RC\frac{dv}{dt} + v = 0$$

subject to the conditions $t = 0$, $v = v_0$, $dv/dt = 0$.
Taking $h = 0.02$ sec, use the Runge-Kutta method to calculate v and dv/dt when $t = 0.02$, for the data $v_0 = 10$ volts, $C = 0.1$ farad, $L = 0.5$ henry and $R = 10$ ohms.

10.13   Error Analysis

The numerical solutions of differential equations certainly differ from their exact solutions. *The difference between the computed value yi and the true value y(xi) at any stage is known as the* **total error.** *The total error at any stage is comprised of* **truncation error** *and* **round-off error.**

The most important aspect of numerical methods is to minimize the errors and obtain the solutions with the least errors. It is usually not possible to follow error development quite closely. We can make only rough estimates. That is why, our treatment of error analysis at times, has to be somewhat intuitive.

In any method, the truncation error can be reduced by taking smaller sub-intervals. The round-off error cannot be controlled easily unless the computer used has the double precision arithmetic facility. In fact, this error has proved to be more elusive than the truncation error.

The truncation error in Euler's method is $\frac{1}{2}h^2 yn'''$, i.e., of (h^2) while

that of modified Euler's method is $\frac{1}{2}h^3 y_{n'''}$, .i.e., of (h^3)

Similarly in the fourth order of the Runge-Kutta method, the truncation error is of $O(h^5)$.

In the Milne's method, the truncation error

due to predictor formula $= \dfrac{14}{45} y_n^v h^5$

and due to corrector formula $= -\dfrac{1}{90} y_n^v h^5$.

i.e., the truncation error in Milne's method is also of $O(h^5)$.

Similarly the error in the Adams-Bashforth method is of the fifth order. Also the predictor error T_p and the corrector error T_c are so related that $19 T_p \approx -251\, T_c$.

The **relative error** *of an approximate solution is the ratio of the total error to the exact value.* It is of greater importance than the error itself for if the true value becomes larger, then a larger error may be acceptable. If the true value diminishes, then the error must also diminish otherwise the computed results may be absurd.

EXAMPLE 10.32

Does applying Euler's method to the differential equation

$dy/dx = f(x, y)$, $y(x_0) = y_0$, estimate the total error?

When $f(x, y) = -y$, $y(0) = 1$, compute this error neglecting the round-off error.

Solution:

We know that Euler's solution of the given differential equation is

$$y_{n+1} = y_n + hf(x_n, y_n) \text{ where } x_n = x_0 + nh.$$

i.e., $\qquad y_{n+1} = y_n + hy_n' \qquad\qquad\qquad (1)$

Denoting the exact solution of the given equation at $x = x_n$ by $y(x_n)$ and expanding $y(x_{n+1})$ by Taylor's series, we obtain

$$y(x_{n+1}) = y(x_n) + hy'(x_n) + \frac{h^2}{2!} y''(\xi_n), x_n \le \xi_n \le x_n + 1 \qquad (2)$$

$\therefore \qquad$ The truncation error $T_{n+1} = y(x_{n+1}) - y_{n+1} = (1/2)h^2\, y''(\zeta_n)$

Thus the truncation error is of $O(h^2)$ as $h \to 0$.

To include the effect of round-off error R_n, we introduce a new approximation y_n which is defined by the same procedure allowing for the round-off error.

$$\bar{y}_{n+1} = \bar{y}_n + hf(x_n, \bar{y}_n) - R_{n+1} \qquad (3)$$

∴ The total error is defined by

$$E_{n+1} = y(x_{n+1}) - \bar{y}_{n+1} \qquad [(2) - (3)]$$

$$= y(x_n) + hy'(x_n) + \frac{h^2}{2!}y''(\xi_n) - \left\{\bar{y}_n + hf(x_n, y_n) - R_{n+1}\right\}$$

$$= \left[y(x_n) - \bar{y}_n\right] + h\left[h'(x_n) - f(x_n, y_n)\right] + T_{n+1} + R_{n+1} \qquad (4)$$

Assuming continuity of $\partial f/\partial y$ and using Mean-Value theorem, we have

$f[x_n, y(x_n)] - f(x_n, y_n) = [y(x_n) - y_n] fy(x_n, \zeta_n)$, where ζ_n lies between $y(x_n)$ and y_n.

∴ (4) takes the form

$$E_{n+1} = [y(x_n) - \bar{y}_n\left[1 + hf_y(x_n, \xi_n)\right] + T_{n+1} + R_{n+1}$$

or $$E_{n+1} = E_n[1 + hf_y(x_n, \zeta_n)] + T_{n+1} + R_{n+1} \qquad (5)$$

This is the *recurrence formula* for finding the total error. The first terms on the right-hand side is the *inherited error*, i.e., the propagation of the error from the previous step y_n to y_{n+1}.

(*b*) We have $dy/dx = -y$, $y(0) = 1$.

Taking $h = 0.01$ and applying (1) successively, we obtain

$$y(0.01) = 1 + 0.01(-1) = 0.99$$

$$y(0.02) = 0.99 + 0.01\ (-0.99) = 0.9801$$

$$y(0.03) = 0.9703, y(0.04) = 0.9606$$

∴ The truncation error

$$T_{n+1} = (1/2)h_2 y''(\xi) = 0.00005 y\xi) \le 5 \times 10^{-5}\ y(x_n) \qquad [\because dy/dx \text{ is } -ve]$$

i.e., $$T_1 \le 5 \times 10^{-5}\ y(0) = 5 \times 10^{-5}$$

$$T_2 \le 5 \times 10^{-5}\ y(0.01) = 5 \times 10^{-5}\ (0.99) < 5 \times 10{-}5$$

$$T_3 \le 5 \times 10^{-5}\ y(0.02) = 5 \times 10^{-5}\ (0.9801) < 5 \times 10{-}5$$

$$T_4 \le 5 \times 10^{-5}\ y(0.03) = 5 \times 10^{-5}\ (0.9703) < 5 \times 10{-}5 \text{ etc.}$$

Also $1 + hf_0(x_n, y_n) = 1 + 0.01(-1) = 0.99$.

Neglecting the round-off error and using the above results, (5) gives

$$E_0 = 0, \ E_1 = E0(0.99) + T_1 \le 5 \times 10^{-5} = 0.00005$$
$$E_2 = E_1(0.99) + T_2 < 5 \times 10{-5} + 5 \times 10^{-5} = 0.0001$$
$$E_3 = E_2(0.99) + T_3 < 10{-4} + 5 \times 10^{-5} = 0.00015$$
$$E_4 = E_3(0.99) + T_4 < 1.5 \times 10{-4} + 5 \times 10^{-5} = 0.0002 \text{ etc.}$$

NOTE
Obs. *The exact solution is $y = e^{-x}$.*
$\therefore$ *Actual error in $y(0.03) = e^{-0.03} - 0.9703 = 0.00014$*
and actual error in $y(0.04) = e^{-0.04} - 0.9606 = 0.00019$.
Clearly the total error E_4 agrees with the actual error in $y(0.04)$.

10.14 Convergence of a Method

Any *numerical method for solving a differential equation is said to be convergent if the approximate solution y_n approaches the exact solution $y(x_n)$ as h tends to zero provided the rounding errors arise from the initial conditions approach zero.* This means that as a method is continually re-fined by taking smaller and smaller step-sizes, the sequence of approximate solutions must converge to the exact solution.

Taylor's series method is convergent provided $f(x, y)$ possesses enough continuous derivatives. The Runge-Kutta methods are also convergent under similar conditions. Predictor corrector methods are convergent if $f(x, y)$ satisfies the *Lipschitz condition, i.e.*,

$$| f(x,y) - f(x,\overline{y}) | \le k|y - \overline{y} |,$$

k being a constant, then the sequence of approximations to the numerical solution converges to the exact solution.

10.15 Stability Analysis

There is a limit to which the step-size h can be reduced for controlling the truncation error, beyond which a further reduction in h will result in the increase of round-off error and hence increase in the total error. This behavior of the error bound is shown in Figure 10.3.

In such situations, we have to use stable methods so that an error introduced at any stage does not get magnified.

A method is said to be **stable** *if it produces a bounded solution which imitates the exact solution. Otherwise it is said to be* **unstable.** If a method is stable for all values of the parameter, it is said to be *absolutely* or *unconditionally stable*. If it is stable for some values of the parameter, it is said to be *conditionally stable*.

The Taylor's method and Adams-Bashforth method prove to be relatively stable. Euler's method and the Runge-Kutta method are conditionally stable as will be seen from Example 10.23.

The Milne's method is however, unstable since when the parameter is negative, each of the errors is magnified while the exact solution decays.

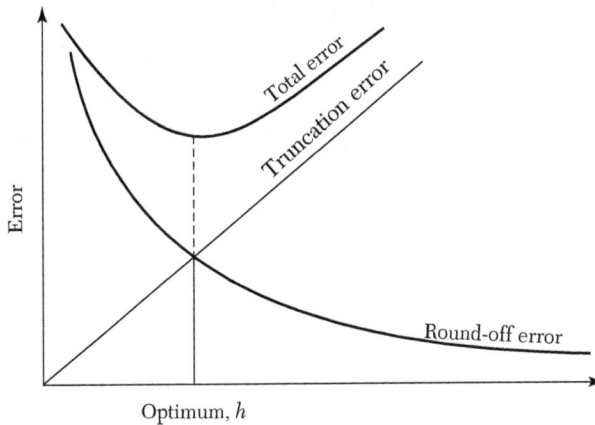

FIGURE 10.3

EXAMPLE 10.33

Does applying Euler's method to the equation

$dy/dx = \lambda y$, given $y(x_0) = y_0$,

determine its stability zone? What would be the range of stability when $\lambda = -1$?

Solution:

We have $\qquad y' = \lambda y, y(x_0) = y_0$ $\hfill$ (1)

By Euler's method,

$$y_n = y_{n-1} + hy'_{n-1} = y_{n-1} + \lambda h y_{n-1} = (1 + \lambda h)y_{n-1} \qquad \text{[by (1)}$$

$$\therefore \quad y_{n-1} = (1 + \lambda h)\, y_{n-2}$$

$$\cdots\cdots\cdots\cdots\cdots\cdots\cdots\cdots\cdots$$

$$y_2 = (1 + \lambda h)\, y_1$$
$$y_1 = (1 + \lambda h)\, y_0$$

Multiplying all these equations, we obtain

$$y_n = (1 + \lambda h)^n\, y_0 \qquad (2)$$

Integrating (1), we get $y = ce^{\lambda x}$

Using $\ y(x_0) = y_0, y_0 = ce^{\lambda x0} \qquad \therefore y = y_0 e^{l(x-x0)}$

In particular, the exact solution through (x_n, y_n) is

$$y_n = y_0 e^{\lambda(xn-x0)} = y_0 e^{\lambda nh} \qquad\qquad [\because x_n = x_0 + nh]$$

or $$y_n = y_0(e^{\lambda h})^n = y_0 \left[1 + \lambda h + \frac{(\lambda h)^2}{2} + \cdots\right]^n \qquad (3)$$

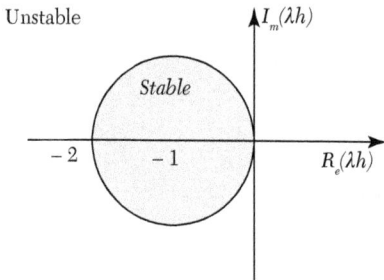

FIGURE 10.4

Clearly the numerical solution (2) agrees with exact solution (3) for small values of h. The solution (2) increases if $|1 + \lambda h| > 1$.

Hence $|1 + \lambda h| < 1$ defines a stable zone.

When λ is real, then the method is stable if $|1 + \lambda h| < 1$ i.e. $-2 < \lambda h < 0$

When λ is complex $(= a + ib)$, then it is stable if

$|1 + (a + ib)\, h\,| < 1$ i.e. $(1 + ah)^2 + (bh)^2 < 1$

i.e., $(x + 1)2 + y_2 < 1,$ \hfill [where $x = ah, y = bh.$]

i.e., λh lies within the unit circle shown in Figure 10.4.

When λ is imaginary $(= ib), |1 + \lambda h| = 1$, then we have a *periodic-stability*.

Hence Euler's method is absolutely stable if and only if

(*i*) real λ: $-2 < \lambda h = 0$.

(*ii*) complex λ: λh lies within the unit circle (Figure 10.4), *i.e.*, Euler's method is conditionally convergent.

When $\lambda = -1$, the solution is stable in the range $-2 < -h < 0$ *i.e.* $0 < h < 2$.

Exercises 10.7

1. Show that the approximate values y_i, obtained from $y' = y$ with $y(0) = 1$ by Taylor's series method, converge to the exact solution for h tending to zero.

2. Show that the modified Euler's method is convergent.

3. Starting with the equation $y' = \lambda y$, show that the modified Euler's method is relatively stable.

4. Apply the fourth order Runge-Kutta method to the equation $dy/dx = \mu y$, $y(x_0) = y_0$ and show that the range of absolute stability is
$$-2.78 < \mu h < 0.$$

5. Find the range of absolute stability of the equation
$$y' + 10y = 0, \, y(0) = 1, \text{ using}$$

(*a*) Euler's method, (*b*) Runge-Kutta method.

6. Show that the local truncation errors in the Milne's predictor and corrector formulae are

$$\frac{14}{45}h^5 y'' \text{ and } -\frac{1}{90}h^5 y'', \text{ respectively.}$$

10.16 Boundary Value Problems

Such a problem requires the solution of a differential equation in a region R subject to the various conditions on the boundary of R. Practical applications give rise to many such problems. We shall discuss two-point linear boundary value problems of the following types:

(*i*) $\dfrac{d\,y}{dx} + \lambda(x)\dfrac{dy}{dx} + \mu(x)y = \gamma(x)$ with the conditions $y(x_0) = a$,

$$y(x_n) = b.$$

(ii) $\dfrac{d^4 y}{dx^4} + \lambda(x)y = \mu(x)$ with the conditions $y(x_0) = y'(x_0) = a$ and

$$y(x_n) = y'(x_n) = b.$$

There exist two numerical methods for solving such boundary value problems. The first one is known as the *finite difference method* which makes use of finite difference equivalents of derivatives. The second one is called the *shooting method* which makes use of the techniques for solving initial value problems.

10.17   Finite-Difference Method

In this method, the derivatives appearing in the differential equation and the boundary conditions are replaced by their finite-difference approximations and the resulting linear system of equations are solved by any standard procedure. These roots are the values of the required solution at the pivotal points.

The *finite-difference approximations to the various derivatives are derived as under:*

If $y(x)$ and its derivatives are single-valued continuous functions of x then by Taylor's expansion, we have

$$y(x+h) = y(x) + hy'(x) + \frac{h^2}{2!}y''(x) + \frac{h^3}{3!}y'''(x) + \cdots \tag{1}$$

and $\qquad y(x-h) = y(x) + hy'(x) + \dfrac{h^2}{2!}y''(x) + \dfrac{h^3}{3!}y'''(x) + \cdots \tag{2}$

Equation (1) gives

$$y'(x) = \frac{1}{h}\left[y(x+h) - y(x)\right] - \frac{h}{2}y''(x) - \cdots$$

i.e., $\qquad y'(x) = \dfrac{1}{h}\left[y(x+h) - y(x)\right] + O(h)$

which is the *forward difference approximation of* $y'(x)$ with an error of the order h.

Similarly (2) gives

$$y'(x) = \frac{1}{h}\left[y(x) - y(x-h)\right] + O(h)$$

which is the *backward difference approximation of* $y'(x)$ with an error of the order h.

Subtracting (2) from (1), we obtain

$$y'(x) = \frac{1}{2h}\left[y(x+h) - y(x-h)\right] + O(h^2)$$

which is the *central-difference approximation of* $y'(x)$ with an error of the order h^2. Clearly this central difference approximation to $y'(x)$ is better than the forward or backward difference approximations and hence should be preferred.

Adding (1) and (2), we get

$$y''(x) = \frac{1}{h^2}\left[y(x+h) - 2y(x) + y(x-h)\right] + O(h^2)$$

which is the *central difference approximation of* $y''(x)$. Similarly we can derive central difference approximations to higher derivatives.

Hence *the working expressions for the central difference approximations to the first four derivatives of* y_i *are as under:*

$$y_i' = \frac{1}{2h}(y_{i+1} - y_{i-1}) \tag{3}$$

$$y_i'' = \frac{1}{h^2}(y_{i+1} - 2y_i + y_{i-1}) \tag{4}$$

$$y_i''' = \frac{1}{2h^3}(y_{i+2} - 2y_{i+1} + 2y_{i-1} - y_{i-2}) \tag{5}$$

$$y_i^{iv} = \frac{1}{h^4}(y_{i+2} - 4y_{i+1} + 6y_i - 4y_{i-1} + y_{i-2}) \tag{6}$$

NOTE **Obs.** *The accuracy of this method depends on the size of the sub-interval h and also on the order of approximation. As we reduce h, the accuracy improves but the number of equations to be solved also increases.*

EXAMPLE 10.34

Solve the equation $y'' = x + y$ with the boundary conditions $y(0) = y(1) = 0$.

Solution:

We divide the interval $(0, 1)$ into four sub-intervals so that $h = 1/4$ and the pivot points are at $x_0 = 0$, $x_1 = 1/4$, $x_2 = 1/2$, $x_3 = 3/4$, and $x_4 = 1$.

Then the differential equation is approximated as

$$\frac{1}{h^2}\left[y_{i+1} - 2y_i + y_{i-1}\right] = x_i + y_i$$

or $\qquad 16y_{i+1} - 33y_i + 16_{i-1} = x_i, i = 1, 2, 3.$

Using $y_0 = y_4 = 0$, we get the system of equations

$$16y_2 - 33y_1 = \frac{1}{4}; 16y_3 - 33y_2 + 16y_1 = \frac{1}{2}; -33y_3 + 16y_2 = \frac{3}{4}$$

Their solution gives

$$y_1 = -0.03488, \; y_2 = -0.05632, \; y_3 = -0.05003.$$

<u>**NOTE**</u>   **Obs.** *The exact solution being* $y(x) = \dfrac{\sinh x}{\sinh 1} - x$, *the error at each nodal point is given in the table below:*

x	Computed value $y(x)$	Exact value $y(x)$	Error
0.25	– 0.03488	– 0.03505	0.00017
0.5	– 0.05632	– 0.05659	0.00027
0.75	– 0.05003	– 0.05028	0.00025

EXAMPLE 10.35

Using the finite difference method, find $y(0.25)$, $y(0.5)$, and $y(0.75)$ satisfying the differential equation $\dfrac{d^2y}{dx^2} + y = x$, subject to the boundary conditions $y(0) = 0$, $y(1) = 2$.

Solution:

Dividing the interval $(0, 1)$ into four sub-intervals so that $h = 0.25$ and the pivot points are at $x_0 = 0$, $x_1 = 0.25$, $x_2 = 0.5$, $x_3 = 0.75$, and $x_4 = 1$.

The given equation $y''(x) + y(x) = x$, is approximated as

$$\frac{1}{h^2}\left[y_{i+1} - 2y_i + y_{i-1}\right] + y_i = x_i$$

or $\qquad 16y_{i+1} - 31y_i + 16y_{i-1} = x_i \qquad\qquad\qquad (i)$

Using $\; y_0 = 0$ and $y_4 = 2$, (i) gives the system of equation,

$(i = 1)\; 16y_2 - 31y_1 = 0.25; \qquad\qquad\qquad\qquad\qquad (ii)$

$(i = 2)\; 16y_3 - 31y_2 + 16y_1 = 0.5 \qquad\qquad\qquad\qquad (iii)$

$(i = 3)\; 32 - 31y_3 + 16y_2 = 0.75, \; i.e., -31y_3 + 16y_2 = -31.25 \;(iv)$

Solving the equations (*ii*), (*iii*), and (*iv*), we get
$$y_1 = 0.5443, y_2 = 1.0701, y_3 = 1.5604$$
Hence y (0.25) = 0.5443, y(0.5) = 1.0701, y(0.75) = 1.5604

EXAMPLE 10.36

Determine values of y at the pivotal points of the interval (0, 1) if y satisfies the boundary value problem $y^{iv} + 81y = 81x^2$, $y(0) = y(1) = y''(0) = y''(1) = 0$. (Take $n = 3$).

Solution:

Here $h = 1/3$ and the pivotal points are $x_0 = 0$, $x_1 = 1/3$, $x_2 = 2/3$, $x_3 = 1$. The corresponding y-values are $y_0(= 0)$, y_1, y_2, $y_3(= 0)$.

Replacing y^{iv} by its central difference approximation, the differential equation becomes

$$\frac{1}{h^4}\left(y_{i+2} - 4y_{i+1} + 6y_i - 4y_{i-1} + y_{i-2}\right) + 81y_i = 81x_i^2$$

or $\quad y_{i+2} - 4y_{i+1} + 7y_i - 4y_{i-1} + y_{i-2} = x_i^2, i = 1, 2$

At $\quad i = 1, y_3 - 4y_2 + 7y_1 - 4y_0 + y{-1} = 1/9$

At $\quad i = 2, y_4 - 4y_3 + 7y_2 - 4y_1 + y_0 = 4/9$

Using $y_0 = y_3 = 0$, we get $- 4y_2 + 7y_1 + y{-1} = 1/9$ $\quad (i)$

$$y_4 + 7y_2 - 4y_1 = 4/9 \qquad\qquad (ii)$$

Regarding the conditions $y_0'' = y_3'' = 0$, we know that

$$yi'' = \frac{1}{h^2}(y_{i+1} - 2y_i + y_{i-1})$$

At $i = 0$, $\quad y_0'' = 9 (y_1 - 2y_0 + y{-1})$ or $y{-1} = - y_1 [\because y_0 = y_0'' = 0]$ $\quad (iii)$

At $i = 3$, $\quad y_3'' = 9(y_4 - 2y_3 + y_2)$ or $y_4 = - y_2$ $\quad [\because y_3 = y_3'' = 0]$ $\quad (iv)$

Using (*iii*), the equation (*i*) becomes

$$- 4y_2 + 6y_1 = 1/9 \qquad\qquad (v)$$

Using (*iv*), the equation (*ii*) reduces to

$$6y_2 - 4y_1 = 4/9 \qquad\qquad (vi)$$

Solving (*v*) and (*vi*), we obtain
$$y_1 = 11/90 \text{ and } y_2 = 7/45.$$

Hence $y(1/3) = 0.1222$ and $y(2/3) = 0.1556$.

EXAMPLE 10.37

The deflection of a beam is governed by the equation $\dfrac{d^4y}{dx^4} + 81y = \phi(x)$, where $f(x)$ is given by the table

x	1/3	2/3	1
$\phi(x)$	81	162	243

and boundary condition $y(0) = y'(0) = y''(1) = y'''(1) = 0$. Evaluate the deflection at the pivotal points of the beam using three sub-intervals.

Solution:

Here $h = 1/3$ and the pivotal points are $x_0 = 0$, $x_1 = 1/3$, $x_2 = 2/3$, $x3 = 1$. The corresponding y-values are $y_0(= 0)$, y_1, y_2, y_3.

The given differential equation is approximated to

$$\frac{1}{h^4}\left(y_{i+2} - 4y_{i+1} + 6y_i - 4y_{i-1} + y_{i-2}\right) + 81y_i = \phi(x_i)$$

At $i = 1$, $y_3 - 4y_2 + 7y_1 - 4y_0 + y{-1} = 1$ (*i*)

At $i = 2$, $y_4 - 4y_3 + 7y_2 - 4y_1 + y_0 = 2$ (*ii*)

At $i = 3$, $y_5 - 4y_4 + 7y_3 - 4y_2 + y_1 = 3$ (*iii*)

We have $y_0 = 0$ (*iv*)

Since $yi' = \dfrac{1}{2h}\left(y_{i+1} - y_{i-1}\right)$

$\therefore$ for $i = 0$, $0 = y_0' = \dfrac{1}{2h}\left(y_1 - y{-1}\right)$ i.e., $y{-1} = y_1$ (*v*)

Since $yi'' = \dfrac{1}{h^2}\left(y_{i+1} - 2y_i + y_{i-1}\right)$

$\therefore$ for $i = 3$, $0 = y_3'' = \dfrac{1}{h^2}\left(y_4 - 2y_3 + y_2\right)$, i.e., $y_4 = 2y_3 - y_2$ (*vi*)

Also $y_i''' = \dfrac{1}{2h^3}\left(y_{i+2} - 2y_{i+1} + 2y_{i-1} - y_{i-2}\right)$

$\therefore$ for $i = 3$, $0 = y_3''' = \dfrac{1}{2h^3}\left(y_5 - 2y_4 + 2y_2 - y_1\right)$

i.e., $y_5 = 2y_4 - 2y_2 + y_1$ (*vii*)

Using (*iv*) and (*v*), the equation (*i*) reduces to

$$y_3 - 4y_2 + 8y_1 = 1$$ (*viii*)

Using (iv) and (vi), the equation (ii) becomes

$$-y_3 + 3y_2 - 2y_1 = 1 \qquad (ix)$$

Using (vi) and (vii), the equation (iii) reduces to

$$3y_3 - 4y_2 + 2y_1 = 3 \qquad (x)$$

Solving ($viii$), (ix), and (x), we get

$$y_1 = 8/13,\ y_2 = 22/13,\ y_3 = 37/13.$$

Hence $y(1/3) = 0.6154$, $y(2/3) = 1.6923$, $y(1) = 2.8462$.

10.18 Shooting Method

In this method, the given boundary value problem is first transformed to an initial value problem. Then this initial value problem is solved by Taylor's series method or Runge-Kutta method, etc. Finally the given boundary value problem is solved. The approach in this method is quite simple.

Consider the boundary value problem

$$y''(x) = y(x),\ y(x) = A,\ y(b) = B \qquad (1)$$

One condition is $y(a) = A$ and let us assume that $y'(a) = m$ which represents the slope. We start with two initial guesses for m, then find the corresponding value of $y(b)$ using any initial value method.

Let the two guesses be m_0, m_1 so that the corresponding values of $y(b)$ are $y(m_0, b)$ and $y(m_1, b)$. Assuming that the values of m and $y(b)$ are linearly related, we obtain a better approximation m_2 for m from the relation:

$$\frac{m_2 - m_1}{y(b) - y(m_1, b)} = \frac{m_1 - m_0}{y(m_1, b) - y(m_0, b)}$$

This gives $\quad m_2 = m_1 - (m_1 - m_0)\dfrac{y(m_1, b) - y(b)}{y(m_1, b) - y(m_0, b)} \qquad (2)$

We now solve the initial value problem

$$y''(x) = y(x),\ y(a) = A,\ y'(a) = m_2$$

and obtain the solution $y(m_2, b)$.

To obtain a better approximation m_3 for m, we again use the linear relation (2) with $[m_1, y(m_1, b)]$ and $[m_2, y(m_2, b)]$. This process is repeated until the value of $y(m_i, b)$ agrees with $y(b)$ to desired accuracy.

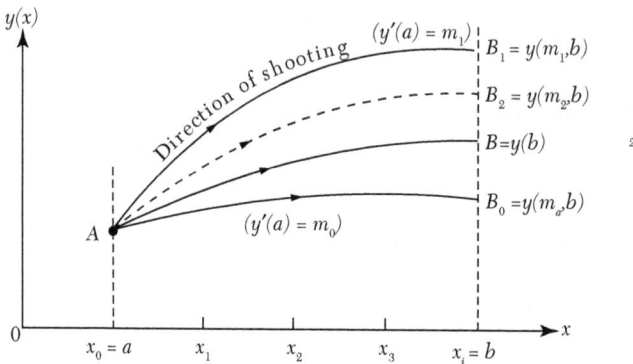

FIGURE 10.5

NOTE   **Obs.** *This method resembles an artillery problem and as such is called the shooting method (Figure 10.5). The speed of convergence in this method depends on our initial choice of two guesses for m. However, the shooting method is quite slow in practice. Also this method is quite tedius to apply to higher order boundary value problems.*

EXAMPLE 10.38

Using the shooting method, solve the boundary value problem:

$$y''(x) = y(x),\ y(0) = 0 \text{ and } y(1) = 1.17.$$

Solution:

Let the initial guesses for $y'(0) = m$ be $m_0 = 0.8$ and $m_1 = 0.9$. Then $y''(x) = y(x)$, $y(0) = 0$ gives

$$y'(0) = m \qquad\qquad y''(0) = y(0) = 0$$
$$y'''(0) = y'(0) = m, \qquad y^{iv}(0) = y''(0) = 0$$
$$yv(0) = y'''(0) = m, \qquad yvi(0) = yiv(0) = 0$$

and so on.

Putting these values in the Taylor's series, we have

$$y(x) = y(0) + xy'(0) + \frac{x^2}{2!}y''(0) + \frac{x^3}{3!}y'''(0) + \cdots$$

$$= m\left(x + \frac{x^3}{6} + \frac{x^5}{120} + \frac{x^7}{5040} + \cdots\right)$$

$\therefore \qquad y(1) = m(1 + 0.1667 + 0.0083 + 0.0002 + \cdots) = m \ (1.175)$

For $m_0 = 0.8$, $y(m_0, 1) = 0.8 \times 1.175 = 0.94$

For $m_1 = 0.9$, $y(m_1, 1) = 0.9 \times 1.175 = 1.057$

Hence a better approximation for m, i.e., m_2 is given by

$$m_2 = m_1 - (m_1 - m_0) \frac{y(m_1, 1) - y(1)}{y(m_1, 1) - y(m_0, 1)}$$

$$= 0.9 - (0.1) \frac{1.057 - 1.175}{1.057 - 0.94} = 0.9 + 0.10085 = 1.00085$$

which is closer to the exact value of $y'(0) = 0.996$

We now solve the initial value problem

$$y''(x) = y(x), \ y(0) = 0, \ y'(0) = m_2.$$

Taylor's series solution is given by

$$y(m_2, 1) = m_2 \ (1.175) = 1.1759$$

Hence the solution at $x = 1$ is $y = 1.176$ which is close to the exact value of $y(1) = 1.17$.

Exercises 10.8

1. Solve the boundary value problem for $x = 0.5$:

$$\frac{d^2 y}{dx^2} + y + 1 = 0, y(0) = y(1) = 0. \qquad \text{(Take } n = 4\text{)}$$

2. Find an approximate solution of the boundary value problem:

$y'' + 8(\sin^2 \pi y) \, y = 0, \ 0 \le x \le 1, \ y(0) = y(1) = 1.$ \qquad (Take $n = 4$)

3. Solve the boundary value problem:

$xy'' + y = 0, \ y(1) = 1, \ y(2) = 2.$ \qquad (Take $n = 4$)

4. Solve the equation $y'' - 4y' + 4y = e^{3x}$, with the conditions $y(0) = 0$, $y(1) = -2$, taking $n = 4$.

5. Solve the boundary value problem $y'' - 64y + 10 = 0$ with $y(0) = y(1) = 0$ by the finite difference method. Compute the value of $y(0.5)$ and compare with the true value.

6. Solve the boundary value problem
$$y'' + xy' + y = 3x^2 + 2, \ y(0) = 0, \ y(1) = 1.$$

7. The boundary value problem governing the deflection of a beam of length three meters is given by

$$\frac{d^4y}{dx^4} + 2y = \frac{1}{9}x^2 + \frac{2}{3}x + 4, y(0) = y'(0) = y(3) = y'(3) = 0.$$

The beam is built-in at the left end $(x = 0)$ and simply supported at the right end $(x = 3)$.
Determine y at the pivotal points $x = 1$ and $x = 2$.

8. Solve the boundary value problem,

$$\frac{d^4y}{dx^4} + 81y = 729x^2 y(0) = y'(0) = y''(1) = y'''(1) = 0. \text{ Use } n = 3$$

9. Solve the equation $y^{iv} - y''' + y = x^2$, subject to the boundary conditions
$$y(0) = y'(0) = 0 \text{ and } y(1) = 2, \ y'(1) = 0. \hspace{2cm} \text{(Take } n = 5).$$

10. Apply shooting method to solve the boundary value problem

$$\frac{d^2y}{dx2} = y, \ y(0) = 0 \text{ and } y(1) = 1.1752.$$

11. Using shooting method, solve the boundary value problem

$$\frac{d^2y}{dx2} = 6y^2, \ y(0) = 1, \ y(0.5) = 0.44$$

10.19   Objective Type of Questions

Exercises 10.9

Select the correct answer or fill up the blanks in the following questions:

1. Which of the following is a step by step method:
(a) Taylor's   (b) Adams-Bashforth
(c) Picard's   (d) None.

2. The finite difference scheme for the equation $2y'' + y = 5$ is ....

3. If $y' = x + y, y(0) = 1$ and $y^{(1)} = 1 + x + x^2/2$, then by Picard's method, the value of $y^{(2)}(x)$ is ......

4. The iterative formula of Euler's method for solving $y' = f(x, y)$ with $y(x_0) = y_0$, is ....

5. Taylor's series for solution of first order ordinary differential equations is ....

6. The disadvantage of Picard's method is ....

7. Given y_0, y_1, y_2, y_3, Milne's corrector formula to find y_4 for $dy/dx = f(x, y)$, is ....

8. The second order Runge-Kutta formula is ....

9. Adams-Bashforth predictor formula to solve $y' = f(x, y)$, given $y_0 = y(x_0)$ is .... .

10. The Runge-Kutta method is better than Taylor's series method because ....

11. To predict Adam's method atleast ...... values of y, prior to the desired value, are required.

12. Taylor's series solution of $y' - xy = 0$, $y(0) = 1$ upto x^4 is ....

13. If dy/dx is a function of x alone, the fourth order Runge-Kutta method reduces to ....

14. Milne's Predictor formula is ....

15. Adam's Corrector formula is ....

16. Using Euler's method, $dy/dx = (y - 2x)/y$, $y(0) = 1$; gives $y(0.1) = $ ....

17. $\dfrac{d^2 y}{dx^2} + y^2 \dfrac{dy}{dx} + y = 0$ is equivalent to a set of two first order differential equations ...... and ....

18. The formula for the fourth order Runge-Kutta method is ....

19. Taylor's series method will be useful to give some ...... of Milne's method.

20. The names of two self-starting methods to solve $y' = f(x, y)$ given $y(x_0) = y_0$ are ....

21. In the derivation of the fourth order Runge-Kutta formula, it is called fourth order because .....

22. If $y' = x - y$, $y(0) = 1$ then by Picard's method, the value of $y^{(1)}(1)$ is ....
    (a) 0.915 (b) 0.905 (c) 1.091 (d) none.

23. The finite difference formulae for $y'(x)$ and $y''(x)$ are ....

24. If $y' = -y$, $y(0) = 1$, then by Euler's method, the value of $y(1)$ is
    (a) 0.99 (b) 0.999 (c) 0.981 (d) none.

25. Write down the difference between initial value problem and boundary value problem ..... .

26. Which of the following methods is the best for solving initial value problems:
    (a) Taylor's series method
    (b) Euler's method
    (c) Runge-Kutta method of the fourth order
    (d) Modified Euler's method.

27. The finite difference scheme of the differential equation $y'' + 2y = 0$ is .....

28. Using the modified Euler's method, the value of $y(0.1)$ for
$$\frac{dy}{dx} = x - y, y(0) = 1 \text{ is}$$
    (a) 0.809 (b) 0.909 (c) 0.0809 (c) none.

29. The multi-step methods available for solving ordinary differential equations are ....

30. Using the Runge Kutta method, the value of $y(0.1)$ for $y' = x - 2y$, $y(0) = 1$, taking $h = 0.1$, is ......
    (a) 0.813 (b) 0.825 (c) 0.0825 (c) none.

31. In Euler's method, if h is small the method is too slow, if h is large, it gives inaccurate value. (True or False)

32. Runge-Kutta method is a self-starting method.   (True or False)

33. Predictor-corrector methods are self-starting methods.   (True or False)

11

NUMERICAL SOLUTION OF PARTIAL DIFFERENTIAL EQUATIONS

Chapter Objectives

- Introduction
- Classification of second order equations
- Finite-difference approximations
- Elliptic equations to partial derivatives
- Solution of Laplace equation
- Solution of Poisson's equation
- Solution of elliptic equations by relaxation
- Parabolic equations method
- Solution of one-dimensional heat equation
- Solution of two-dimensional heat equation
- Hyperbolic equations
- Solution of wave equation

11.1   Introduction

Partial differential equations arise in the study of many branches of applied mathematics, e.g., in fluid dynamics, heat transfer, boundary layer flow, elasticity, quantum mechanics, and electromagnetic theory. Only a few of these equations can be solved by analytical methods which are also complicated by requiring use of advanced mathematical techniques. In most of the cases, it is easier to develop approximate solutions by numerical methods. Of all the numerical methods available for the solution of partial differential equations, the method of finite differences is most commonly used. In this method, the derivatives appearing in the equation and the boundary conditions are replaced by their finite difference approximations. Then the given equation is changed to a system of linear equations which are solved by iterative procedures. This process is slow but produces good results in many boundary value problems. An added advantage of this method is that the computation can be carried by electronic computers. To accelerate the solution, sometimes the method of relaxation proves quite effective.

Besides discussing the finite difference method, we shall briefly describe the relaxation method also in this chapter.

11.2   Classification of Second Order Equations

The general linear partial differential equation of the second order in two independent variables is of the form

$$A(x,y)\frac{\partial^2 u}{dx^2} + B(x,y)\frac{\partial^2 u}{\partial x \partial y} + C(x,y)\frac{\partial^2 u}{\partial y^2} + \left(x,y,u\frac{\partial u}{\partial x},\frac{\partial u}{\partial y}\right) = 0 \qquad (1)$$

Such a partial differential equation is said to be

(*i*) **elliptic** *if* $B^2 - 4AC < 0$,   (*ii*) **parabolic** *if* $B^2 - 4AC = 0$, and
(*iii*) **hyperbolic** *if* $B^2 - 4AC > 0$.

NOTE   **Obs.** *A partial equation is classified according to the region in which it is desired to be solved. For instance, the partial differential equation* $f_{xx} + f_{yy} = 0$ *is elliptic if* $y > 0$, *parabolic if* $y = 0$, *and hyperbolic if* $y < 0$.

EXAMPLE 11.1

Classify the following equations:

(i) $\dfrac{\partial^2 u}{\partial x^2} + 4\dfrac{\partial^2 u}{\partial x \partial y} + 4\dfrac{\partial^2 u}{\partial y^2} - \dfrac{\partial u}{\partial x} + 2\dfrac{\partial u}{\partial y} = 0$

(ii) $x^2 \dfrac{\partial^2 u}{\partial x^2} + \left(1 - y^2\right)\dfrac{\partial^2 u}{\partial y^2} = 0, -\infty < x < \infty, - < y < 1$

(iii) $\left(1+x^2\right)\dfrac{\partial^2 u}{\partial x^2}+\left(5+2x^2\right)\dfrac{\partial^2 u}{\partial x\partial t}+\left(4+x^2\right)\dfrac{\partial^2 u}{\partial t^2}=0.$

Solution:

(*i*) Comparing this equation with (1) above, we find that t^2

$$A = 1,\ B = 4,\ C = 4$$

$\therefore$ $\qquad B^2 - 4AC = (4)^2 - 4\times 1\times 4 = 0$

So the equation is parabolic.

(*ii*) Here $A = x^2,\ B = 0,\ C = 1 - y^2$

$$B^2 - 4AC = 0 - 4x^2\ (1 - y^2) = 4x^2(y^2 - 1)$$

For all x between $-\infty$ and ∞, x^2 is positive

For all y between -1 and 1, $y^2 < 1$

$$B^2 - 4AC < 0$$

Hence the equation is elliptic

(*iii*) Here $A = 1 + x^2,\ B = 5 + 2x^2,\ C = 4 + x^2$

$\therefore$ $\qquad B^2 - 4AC = (5 + 2x^2)^2 - 4(1 + x^2)(4 + x2) = 9\ i.e.\ > 0$

So the equation is hyperbolic

Exercises 11.1

1. What is the classification of the equation $f_{xx} + 2f_{xy} + f_{yy} = 0.$

2. Determine whether the following equation is elliptic or hyperbolic?

$(x + 1)u_{xx} - 2(x + 2)u_{xy} + (x + 3)u_{yy} = 0.$

3. Classify the equation

(*i*) $y^2 u_{xx} - 2xy u_{xy} + x^2 u_{yy} + 2u_x - 3u = 0.$

(*ii*) $x^2\dfrac{\partial^2 u}{\partial x^2} + y^2\dfrac{\partial^2 u}{\partial y^2} = x\dfrac{\partial u}{\partial x} - y\dfrac{\partial u}{\partial y}$

(*iii*) $3\dfrac{\partial^2 u}{\partial x^2} + 4\dfrac{\partial^2 u}{\partial x\partial y} + 6\dfrac{\partial^2 u}{\partial y^2} - 2\dfrac{\partial u}{\partial x} + \dfrac{\partial u}{\partial y} - u = 0$

4. In which parts of the (x, y) plane is the following equation elliptic?

$$\dfrac{\partial^2 u}{\partial x^2} + \dfrac{\partial^2 u}{\partial x\partial y} + \left(x^2 + 4y^2\right)\dfrac{\partial^2 u}{\partial y^2} = 2\sin\left(xy\right).$$

11.3   Finite Difference Approximations to Partial Derivatives

Consider a rectangular region R in the x, y plane. Divide this region into a rectangular network of sides $\Delta x = h$ and $\Delta y = k$ as shown in Figure 11.1. The points of intersection of the dividing lines are called *mesh points, nodal points, or grid points*

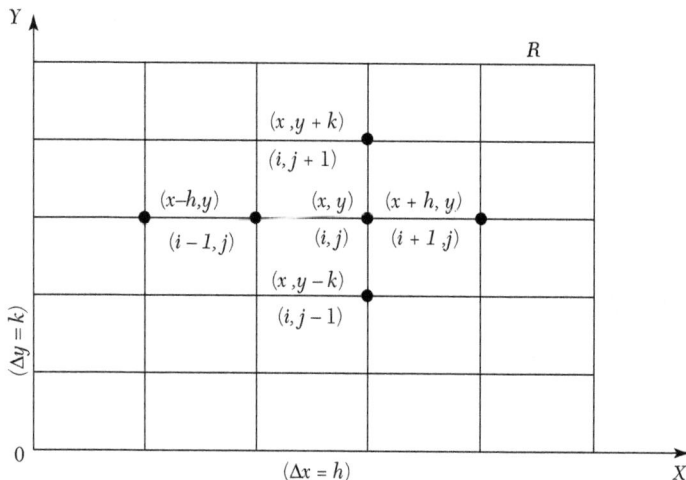

FIGURE11.1

Then we have the finite difference approximations for the partial derivatives in x-direction (Section 10.17):

$$\frac{\partial u}{\partial x} = \frac{u(x+h,y)-u(x,y)}{h}+O(h) = \frac{u(x,y)-u(x-h,y)}{h}+O(h)$$

$$= \frac{u(x+h,y)-u(x-h,y)}{2h}+O(h^{2)}$$

And $$\frac{\partial^2 u}{\partial x^2} = \frac{(x-h,y)-2u(x,y)+u(x+h,y)}{h^2}+O(h^2)$$

Writing $u(x, y) = u(ih, jk)$ as simply $u_{i,j}$, the above approximations become

$$ux = \frac{u_{i+1},j - U_{i,j}}{h}+(h) \tag{1}$$

$$= \frac{u_{i,j} - u_{i-1},j}{h}+O(h) \tag{2}$$

$$= \frac{u_{i+1,j} - u_{i-1,j}}{2h} + O(h^2) \tag{3}$$

$$u_{xx} = \frac{u_{i-1,j} - 2u_{i,j} + u_{i+1,j}}{h^2} + O(h^2) \tag{4}$$

Similarly we have the approximations for the derivatives w.r.t. y:

$$u_y = \frac{u_{i,j+1} - u_{i,j}}{k} + O(k) \tag{5}$$

$$= \frac{u_{i,j} - u_{i,j-1}}{k} + O(k) \tag{6}$$

$$= \frac{u_{i,j} + 1 - u_{i,j} - 1}{2k} + O(k^2) \tag{7}$$

and $\qquad u_{yy} = \frac{u_{i,j-1} - 2u_{i,j} + u_{i,j+1}}{k^2} + O(k^2) \tag{8}$

Replacing the derivatives in any partial differential equation by their corresponding difference approximations (1) to (8), we obtain the finite-difference analogues of the given equation.

11.4 Elliptic Equations

The Laplace equation $\nabla^2 u = \dfrac{\partial^2 u}{\partial x^2} + \dfrac{\partial^2 u}{\partial y^2} = 0$ and the Poisson's equation

$\dfrac{\partial^2 u}{\partial x^2} + \dfrac{\partial^2 u}{\partial y^2} = f(x,y)$ are Example s of elliptic partial differential equations.

The Laplace equation arises in steady-state flow and potential problems. Poisson's equation arises in fluid mechanics, electricity and magnetism and torsion problems.

The solution of these equations is a function $u(x, y)$ which is satisfied at every point of a region R subject to certain boundary conditions specified on the closed curve C (Figure 11.2).

In general, problems concerning steady viscous flow, equilibrium stresses in elastic structures etc., lead to elliptic type of equations.

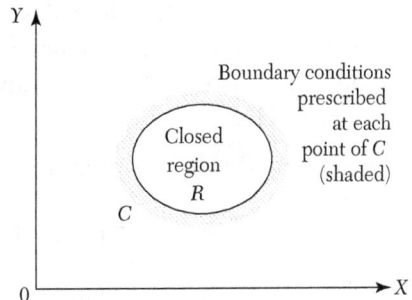

FIGURE 11.2

11.5   Solution of Laplace's Equation

$$\frac{\partial^2 u}{\partial x^2} + \frac{\partial^2 u}{\partial y^2} = 0 \qquad (1)$$

Consider a rectangular region R for which $u(x, y)$ is known at the boundary. Divide this region into a network of square mesh of side h, as shown in Figure 11.3 (assuming that an exact sub-division of R is possible). Replacing the derivatives in (1) by their difference approximations, we have

$$\frac{1}{h^2}\Big[u_{i-1,j} - 2u_{i,j} + u_{i+1,j}\Big] + \frac{1}{h^2}\Big[u_{i,j-1} - 2_{ui,j} + u_{i,j+1}\Big] = 0$$

or $\qquad\qquad u_{i,j} = \dfrac{1}{4}\Big[u_{i-1,j} + u_{i+1,j} + u_{ij+1} + u_{i,j-1}\Big] \qquad (2)$

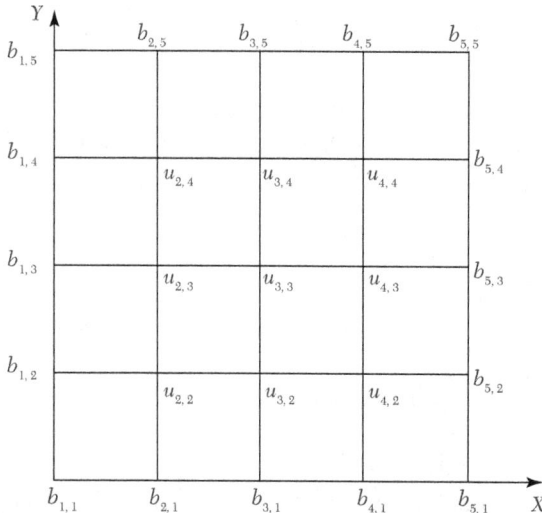

FIGURE 11.3

This shows that the value of u at any interior mesh point is the average of its values at four neighboring points to the left, right, above and below. (2) is called the **standard 5-point formula** which is exhibited in Figure 11.4.

Sometimes a formula similar to (2) is used which is given by

$$u_{i,j} = \frac{1}{4}(u_{i-1,j+1} + u_{i+1,j-1} + u_{i+1,j+1} + u_{i-1,j-1}) \qquad (3)$$

This shows that the value of $u_{i,j}$ is the average of its values at the four neighboring diagonal mesh points. (3) is called the **diagonal five-point**

formula which is represented in Figure 11.5. Although (3) is less accurate than (2), yet it serves as a reasonably good approximation for obtaining the starting values at the mesh points.

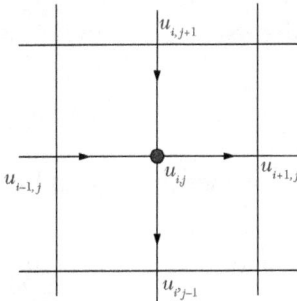

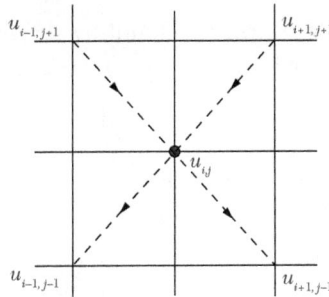

FIGURE 11.4   **FIGURE 11.5**

Now to find the initial values of u at the interior mesh points, we first use the diagonal five-point formula (3) and compute $u_{3,3}, u_{2,4}, u_{4,4}, u_{4,2}$ and $u_{2,2}$, in this order. Thus we get,

$$u_{3,3} = \frac{1}{4}\,(b_{1,5} + b_{5,1} + b_{5,5} + b_{1,1});$$

$$u_{2,4} = \frac{1}{4}\,(b_{1,5} + u_{3,3} + b_{3,5} + b_{1,3})$$

$$u_{4,4} = \frac{1}{4}\,(b_{3,5} + b_{5,3} + b_{3,5} + u_{3,3}); u_{4,2}$$

$$= \frac{1}{4}\,(u_{3,3} + b_{5,1} + b_{3,1} + b_{5,3})$$

$$u_{2,2} = \frac{1}{4}\,(b_{1,3} + b_{3,1} + u_{3,3} + b_{1,1})$$

The values at the remaining interior points, i.e., $u_{2,3}, u_{3,4}, u_{4,3}$ and $u_{3,2}$ are computed by the standard five-point formula (2). Thus, we obtain

$$u_{2,3} = \frac{1}{4}\,(b_{1,3} + u_{3,3} + u_{2,4} + u_{2,2}), u_{3,4}$$

$$= \frac{1}{4}\,(u_{2,4} + u_{4,4} + b_{3,5} + u_{3,3})$$

$$u_{4,3} = \frac{1}{4}\,(u_{3,3} + b_{5,3} + u_{4,4} + u_{4,2}), u_{3,2}$$

$$= \frac{1}{4}\,(u_{2,2} + u_{4,2} + u_{3,3} + u_{3,1})$$

Having found all the nine values of $u_{i,j}$ once, their accuracy is improved by either of the following iterative methods. In each case, the method is repeated until the difference between two consecutive iterates becomes negligible.

(i) **Jacobi's method.** Denoting the nth iterative value of $u_{i,j}$ by $u^{(n)}_{i,j}$, the iterative formula to solve (2) is

$$u^{(n+1)}_{i,j} = \frac{1}{4}\left[u^{(n)}_{i-1,j} + u^{(n)}_{i+1,j} + u^{(n)}_{i,j+1} + u^{(n)}_{i,j-1}\right] \qquad (4)$$

It gives improved values of $u_{i,j}$ at the interior mesh points and is called the *point Jacobi's formula*.

(ii) **Gauss-Seidal method.** In this method, the iteration formula is

$$u^{(n+1)}_{i,j} = \frac{1}{4}\left[u^{(n+1)}_{i-1,j} + u^{(n)}_{i+1,j} + u^{(n+1)}_{i,j+1} + u^{(n)}_{i,j-1}\right]$$

It utilizes the latest iterative value available and scans the mesh points symmetrically from left to right along successive rows.

NOTE **Obs.** *The Gauss-Seidal method is simple and can be adapted to computer calculations. Its convergence being slow, the working is somewhat lengthy. It can however, be shown that the Gauss-Seidal scheme converges twice as fast as Jacobi's scheme.*

The accuracy of calculations depends on the mesh-size, *i.e.*, smaller the h, the better the accuracy. But if h is too small, it may increase rounding-off errors and also increases the labor of computation.

EXAMPLE 11.2

Solve the elliptic equation $u_{xx} + u_{yy} = 0$ for the following square mesh with boundary values as shown in Figure 11.6.

Solution:

Let $u_1, u_2, \cdots, u_9$ be the values of u at the interior mesh-points. Since the boundary values of u are symmetrical about AB,

$\therefore\ u_7 = u_1, u_8 = u_2, u_9 = u_3$.

Also the values of u being symmetrical about CD. $u3 = u_1, u_6 = u_4$, $u_9 = u_7$.

Thus it is sufficient to find the values u_1, u_2, u_4, and u_5.

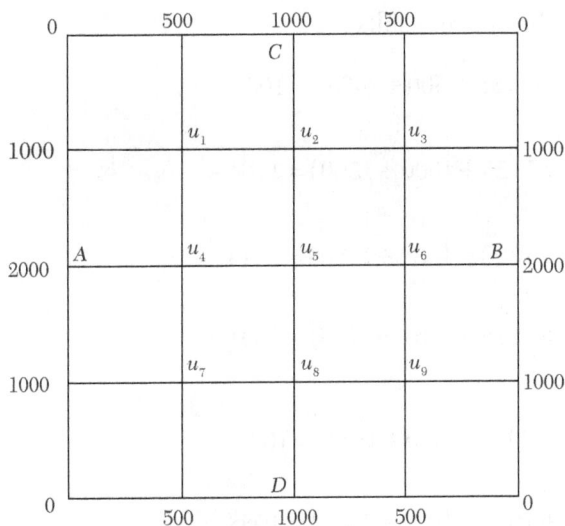

FIGURE 11.6

Now we find their initial values in the following order:

$$u_5 = \frac{1}{4}(2000 + 2000 + 1000 + 1000) = 1500 \qquad \text{(Std. formula)}$$

$$u_1 = \frac{1}{4}(0 + 1500 + 1000 + 2000) = 1125 \qquad \text{(Diag.formula)}$$

$$u_2 = \frac{1}{4}(1125 + 1125 + 1000 + 1500) \approx 1188 \qquad \text{(Std. formula)}$$

$$u_4 = \frac{1}{4}14(2000 + 1500 + 1125 + 1125) \approx 1438 \qquad \text{(Std. formula)}$$

Now we carry out the iteration process using the standard formulae:

$$u_1^{(n+1)} = \frac{1}{4}\left[1000 + u_2^{(n)} + 500 + u_4^{(n)}\right]$$

$$u_2^{(n+1)} = \frac{1}{4}\left[u_1^{(n+1)} + u_1^{(n)} + 1000 + u_5^{(n)}\right]$$

$$u_4^{(n+1)} = \frac{1}{4}\left[2000 + u_5^{(n)} + u_1^{(n+1)} + u_1^{(n)}\right]$$

$$u_5^{(n+1)} = \frac{1}{4}\left[u_4^{(n+1)} + u_4^{(n)} + u_2^{(n+1)} + u_2^{(n)}\right]$$

First iteration: (put $n = 0$ in the above results)

$$u_1^{(1)} = \frac{1}{4}(1000 + 1188 + 500 + 1438) \approx 1032$$

$$u_2^{(1)} = \frac{1}{4}(1032 + 1125 + 1000 + 1500) = 1164$$

$$u_4^{(1)} = \frac{1}{4}(2000 + 1500 + 1032 + 1125) = 1414$$

$$u_5^{(1)} = \frac{1}{4}(1414 + 1438 + 1164 + 1188) = 1301$$

Second iteration: (put $n = 1$)

$$u_1^{(2)} = \frac{1}{4}(1000 + 1164 + 500 + 1414) = 1020$$

$$u_2^{(2)} = \frac{1}{4}(1020 + 1032 + 1000 + 1301) = 1088$$

$$u_4^{(2)} = \frac{1}{4}(2000 + 1301 + 1020 + 1032) = 1338$$

$$u_5^{(2)} = \frac{1}{4}(1338 + 1414 + 1088 + 1164) = 1251$$

Third iteration:

$$u_1^{(3)} = \frac{1}{4}(1000 + 1088 + 500 + 1338) = 982$$

$$u_2^{(3)} = \frac{1}{4}(982 + 1020 + 1000 + 1251) = 1063$$

$$u_4^{(3)} = \frac{1}{4}(2000 + 1251 + 982 + 1020) = 1313$$

$$u_5^{(3)} = \frac{1}{4}(1313 + 1338 + 1063 + 1088) = 1201$$

Fourth iteration:

$$u_1^{(4)} = \frac{1}{4}(1000 + 1063 + 500 + 1313) \approx 969$$

$$u_2^{(4)} = \frac{1}{4}(969 + 982 + 1000 + 1201) = 1038$$

$$u_4^{(4)} = \frac{1}{4}(2000 + 1201 + 969 + 982) = 1288$$

$$u_5^{(4)} = \frac{1}{4}(1288 + 1313 + 1038 + 1063) = 1176$$

Fifth iteration:

$$u_1^{(5)} = \frac{1}{4}(1000 + 1038 + 500 + 1288) = 957$$

$$u_2^{(5)} = \frac{1}{4}(957 + 969 + 1000 + 1176) \approx 1026$$

$$u_4^{(5)} = \frac{1}{4}(2000 + 1176 + 957 + 969) \approx 1276$$

$$u_5^{(5)} = \frac{1}{4}(1276 + 1288 + 1026 + 1038) = 1157$$

Similarly,

$$u_1^{(6)} = 951, u_2^{(6)} = 1016, u_4^{(6)} = 1266, u_5^{(6)} = 1146$$

$$u_1^{(7)} = 946, u_2^{(7)} = 1011, u_4^{(7)} = 1260, u_5^{(7)} = 1138$$

$$u_1^{(8)} = 943, u_2^{(8)} = 1007, u_4^{(8)} = 1257, u_5^{(8)} = 1134$$

$$u_1^{(9)} = 941, u_2^{(9)} = 1005, u_4^{(9)} = 1255, u_5^{(9)} = 1131$$

$$u_1^{(10)} = 940, u_2^{(10)} = 1003, u_4^{(10)} = 1253, u_5^{(10)} = 1129$$

$$u_1^{(11)} = 939, u_2^{(11)} = 1002, u_4^{(11)} = 1252, u_5^{(11)} = 1128$$

$$u_1^{(12)} \approx 939, u_2^{(12)} \approx 1001, u_4^{(12)} \approx 1251, u_5^{(12)} = 1126$$

There is a negligible difference between the values obtained in the eleventh and twelfth iterations.

Hence $u_1 = 939$, $u_2 = 1001$, $u_4 = 1251$ and $u_5 = 1126$.

EXAMPLE 11.3

Given the values of $u(x, y)$ on the boundary of the square in the Figure 11.7, evaluate the function $u(x, y)$ satisfying the Laplace equation $\nabla^2 u = 0$ at the pivotal points of this figure by

(a) *Jacobi's method*   (b) *Gauss-Seidal method*

Solution:

To get the initial values of u_1, u_2, u_3, u_4, we assume that $u_4 = 0$. Then

$$u_1 = \frac{1}{4}\ (1000 + 0 + 1000 + 2000) = 1000 \qquad \text{(Diag. formula)}$$

$$u_2 = \frac{1}{4}\ (1000 + 500 + 1000 + 0) = 625 \qquad \text{(Std. formula)}$$

$$u_3 = \frac{1}{4}\ (2000 + 0 + 1000 + 500) = 875 \qquad \text{(Std. formula)}$$

$$u_4 = \frac{1}{4}(875 + 0 + 625 + 0) = 375 \qquad \text{(Std. formula)}$$

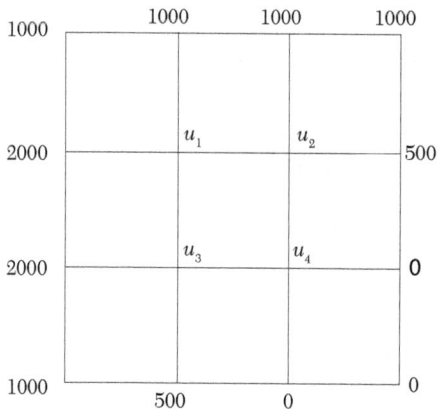

FIGURE 11.7

(*a*) We carry out the successive iterations, using Jacobi's formulae:

$$u_1^{(n+1)} = \frac{1}{4}\Big[2000\ + u_2^{(n)} + \ 1000\ + u_3^{(n)}\Big]$$

$$u_2^{(n+1)} = \frac{1}{4}\Big[u_1^{(n)} + 500 + 1000 + u_4^{(n)}\Big]$$

$$u_3^{(n+1)} = \frac{1}{4}\Big[2000 + u_4^{(n)} + \ u_1^{(n)} + 500\Big]$$

$$u_4^{(n+1)} = \frac{1}{4}\Big[u_3^{(n)} + 0 + u_2^{(n)} + 0\Big]$$

First iteration: (put $n = 0$ in the above results)

$$u_1^{(1)} = \frac{1}{4}(2000 + 625 + 1000 + 875) = 1125$$

$$u_2^{(1)} = \frac{1}{4}(1000 + 500 + 1000 + 375) \approx 719$$

$$u_3^{(1)} = \frac{1}{4}(2000 + 375 + 1000 + 500) \approx 969$$

$$u_4^{(1)} = \frac{1}{4}(875 + 0 + 625 + 0) \approx 375$$

Second iteration: (put $n = 1$)

$$u_1^{(2)} = \frac{1}{4}(2000 + 719 + 1000 + 969) = 1172$$

$$u_2^{(2)} = \frac{1}{4}(1125 + 500 + 1000 + 375) = 750$$

$$u_3^{(2)} = \frac{1}{4}(2000 + 375 + 1125 + 500) = 1000$$

$$u_4^{(2)} = \frac{1}{4}(969 + 0 + 719 + 0) = 422$$

Similarly, $\quad u_1^{(3)} \approx 1188, u_2^{(3)} \approx 774, u_3^{(3)} \approx 1024, u_4^{(3)} \approx 438$

$$u_1^{(4)} \approx 1200, u_2^{(4)} \approx 782, u_3^{(4)} \approx 1032, u_4^{(4)} \approx 450$$

$$u_1^{(5)} \approx 1204, u_2^{(5)} \approx 788, u_3^{(5)} \approx 1038, u_4^{(5)} \approx 454$$

$$u_1^{(6)} \approx 1206.5, u_2^{(6)} \approx 790, u_3^{(6)} \approx 1040, u_4^{(6)} \approx 456.5$$

$$u_1^{(7)} \approx 1208, u_2^{(7)} \approx 791, u_3^{(7)} \approx 1041, u_4^{(7)} \approx 458$$

and $\quad u_1^{(8)} \approx 1208, u_2^{(8)} \approx 791.5, u_3^{(8)} \approx 1041.5, u_4^{(8)} \approx 458$.

There is no significant difference between the seventh and eighth iteration values.

Hence $u_1 = 1208$, $u_2 = 792$, $u_3 = 1042$ and $u_4 = 458$.

(*b*) We carry out the successive iterations, using Gauss-Seidal formulae

$$u_1^{(n+1)} = \frac{1}{4}\left[2000 + u_2^{(n)} + 1000 + u_3^{(n)}\right]$$

$$u_2^{(n+1)} = \frac{1}{4}\left[u_1^{(n+1)} = 500 + 1000 + u_4^{(n)}\right]$$

$$u_3^{(n+1)} = \frac{1}{4}\left[2000 + u_4^{(n)} + u_1^{(n+1)} = 500\right]$$

$$u_4^{(n+1)} = \frac{1}{4}\left[u_3^{(n+1)} + 0 + u_2^{(n+1)} + 0\right]$$

First iteration:(put $n = 0$ in the above results)

$$u_1^{(1)} = \frac{1}{4}(2000 + 625 + 1000 + 875) = 1125$$

$$u_2^{(1)} = \frac{1}{4}(1125 + 500 + 1000 + 375) = 750$$

$$u_3^{(1)} = \frac{1}{4}(2000 + 375 + 1125 + 500) = 1000$$

$$u_4^{(1)} = \frac{1}{4}(1000 + 0 + 750 + 0) \approx 438$$

Second iteration: (put $n = 1$)

$$u_1^{(2)} = \frac{1}{4}(2000 + 750 + 1000 + 1000) \approx 1188$$

$$u_2^{(2)} = \frac{1}{4}(1188 + 500 + 1000 + 438) \approx 782$$

$$u_3^{(2)} = \frac{1}{4}(2000 + 438 + 1188 + 500) \approx 1032$$

$$u_4^{(2)} = \frac{1}{4}(1032 + 0 + 782 + 0) \approx 454$$

Similarly $\quad u_1^{(3)} \approx 1204, u_2^{(3)} \approx 789, u_3^{(3)} \approx 1040, u_4^{(3)} \approx 458$

$$u_1^{(4)} \approx 1207, u_2^{(4)} \approx 791, u_3^{(4)} \approx 1041, u_4^{(4)} = 458$$

and $\quad u_1^{(5)} = 1208, u_2^{(5)} \approx 791.5, u_3^{(5)} \approx 1041.5, u_4^{(5)} \approx 458.25$

Thus there is no significant difference between the fourth and fifth iteration values.

Hence $u_1 = 1208, u_2 = 792, u_3 = 1042$ and $u_4 = 458$.

EXAMPLE 11.4

Solve the Laplace equation $u_{xx} + u_{yy} = 0$ given that

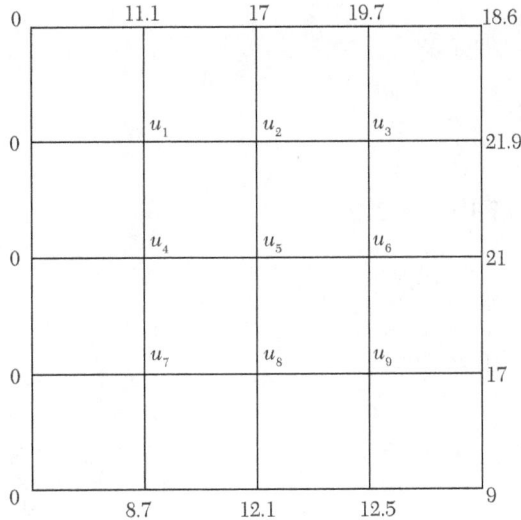

FIGURE 11.8

Solution:

We first find the initial values in the following order:

$$u_5 = \frac{1}{4}(0 + 17 + 21 + 12.1) = 12.5 \qquad \text{(Std. formula)}$$

$$u_1 = \frac{1}{4}(0 + 12.5 + 0 + 17) = 7.4 \qquad \text{(Diag. formula)}$$

$$u_3 = \frac{1}{4}(12.5 + 18.6 + 17 + 21) = 17.28 \qquad \text{(Diag. formula)}$$

$$u_7 = \frac{1}{4}(12.5 + 0 + 0 + 12.1) = 6.15 \qquad \text{(Diag. formula)}$$

$$u_9 = \frac{1}{4}(12.5 + 9 + 21 + 12.1) = 13.65 \qquad \text{(Diag. formula)}$$

$$u_2 = \frac{1}{4}(17 + 12.5 + 7.4 + 17.3) = 13.55 \qquad \text{(Std. formula)}$$

$$u_4 = \frac{1}{4}(7.4 + 6.2 + 0 + 12.5) = 6.52 \qquad \text{(Std. formula)}$$

$$u_6 = \frac{1}{4}(17.3 + 13.7 + 12.5 + 21) = 16.12 \qquad \text{(Std. formula)}$$

$$u_8 = \frac{1}{4}(12.5 + 12.1 + 6.2 + 13.7) = 11.12 \qquad \text{(Std. formula)}$$

Now we carry out the iteration process using the standard formula:

$$u_1^{(n+1)} = \frac{1}{4}[0 + 11.1 + u_4^{(n)} + u_2^{(n)}]$$

$$u_2^{(n+1)} = \frac{1}{4}[u_1^{(n+1)} + 17 + u_5^{(n)} + u_3^{(n)}]$$

$$u_3^{(n+1)} = \frac{1}{4}[u_1^{(n+1)} + 19.7 + u_6^{(n)} + 219]$$

$$u_4^{(n+1)} = \frac{1}{4}[0 + u_1^{(n+1)} + u_7^{(n)} + u_5^{(n)}]$$

$$u_5^{(n+1)} = \frac{1}{4}[u_4^{(n+1)} + u_2^{(n+1)} + u_8^{(n)} + u_6^{(n)}]$$

$$u_6^{(n+1)} = \frac{1}{4}[u_5^{(n+1)} + u_3^{(n+1)} + u_9^{(n)} + 21]$$

$$u_7^{(n+1)} = \frac{1}{4}[0 + u_4^{(n+1)} + 8.7 + u_8^{(n)}]$$

$$u_8^{(n+1)} = \frac{1}{4}[u_7^{(n+1)} + u_5^{(n+1)} + 12.1 + u_9^{(n)}]$$

$$u_9^{(n+1)} = \frac{1}{4}[u_8^{(n+1)} + u_6^{(n+1)} + 12.8 + 17]$$

First iteration: (put $n = 0$, in the above results)

$$u_1^{(1)} = \frac{1}{4}(0 + 11.1 + u_4^{(0)} + u_2^{(0)})$$

$$= \frac{1}{4}(0 + 11.1 + 6.52 + 13.55) = 7.79$$

$$u_2^{(1)} = \frac{1}{4}(7.79 + 17 + 12.5 + 17.28) = 13.64$$

$$u_3^{(1)} = \frac{1}{4}(13.64 + 19.7 + 16.12 + 21.9) = 12.84$$

$$u_4^{(1)} = \frac{1}{4}(0 + 7.79 + 6.15 + 12.5) = 6.61$$

$$u_5^{(1)} = \frac{1}{4}(6.61 + 13.64 + 11.12 + 16.12) = 11.88$$

$$u_6^{(1)} = \frac{1}{4}(11.88 + 17.84 + 13.65 + 21) = 16.09$$

$$u_7^{(1)} = \frac{1}{4}(0 + 6.61 + 8.7 + 11.12) = 6.61$$

$$u_8^{(1)} = \frac{1}{4}(6.61 + 11.88 + 12.1 + 13.65) = 11.06$$

$$u_9^{(1)} = \frac{1}{4}(11.06 + 16.09 + 12.8 + 17) = 12.238$$

Second iteration: (put $n = 1$)

$$u_1^{(2)} = \frac{1}{4}(0 + 11.1 + 6.61 + 13.64) = 7.84$$

$$u_2^{(2)} = \frac{1}{4}(7.84 + 17 + 11.88 + 17.84) = 16.64$$

$$u_3^{(2)} = \frac{1}{4}(13.64 + 19.7 + 16.09 + 21.9) = 17.83$$

$$u_4^{(2)} = \frac{1}{4}(0 + 7.84 + 6.61 + 11.88) = 6.58$$

$$u_5^{(2)} = \frac{1}{4}(6.58 + 13.64 + 11.06 + 16.09) = 11.84$$

$$u_6^{(2)} = \frac{1}{4}(11.84 + 17.83 + 14.24 + 21) = 16.23$$

$$u_7^{(2)} = \frac{1}{4}(0 + 6.58 + 8.7 + 11.06) = 6.58$$

$$u_8^{(2)} = \frac{1}{4}(6.58 + 11.84 + 12.1 + 14.24) = 11.19$$

$$u_9^{(2)} = \frac{1}{4}(11.19 + 16.23 + 12.8 + 17) = 14.30$$

Third iteration: (put $n = 2$)

$$u_1^{(3)} = \frac{1}{4}(0 + 11.1 + 6.58 + 13.64) = 7.83$$

$$u_2^{(3)} = \frac{1}{4}(7.83 + 17 + 11.84 + 17.83) = 13.637$$

$$u_3^{(3)} = \frac{1}{4}(13.63 + 19.7 + 16.23 + 21.9) = 17.86$$

$$u_4^{(3)} = \frac{1}{4}(0 + 7.83 + 6.58 + 11.84) = 6.56$$

$$u_5^{(3)} = \frac{1}{4}(6.56 + 13.63 + 11.19 + 16.23) = 11.90$$

$$u_6^{(3)} = \frac{1}{4}(11.90 + 17.86 + 14.30 + 21) = 16.27$$

$$u_7^{(3)} = \frac{1}{4}(0 + 6.56 + 8.7 + 11.19) = 6.61$$

$$u_8^{(3)} = \frac{1}{4}(6.61 + 11.90 + 12.1 + 14.30) = 11.23$$

$$u_9^{(3)} = \frac{1}{4}(11.23 + 16.27 + 12.8 + 17) = 14.32$$

Similarly

$$u_1^{(4)} = 7.82, u_2^{(4)} = 13.65, u_3^{(4)} = 17.88, u_4^{(4)} = 6.58, u_5^{(4)} = 11.94,$$
$$u_6^{(4)} = 16.28, u_7^{(4)} = 6.63, u_8^{(4)} = 11.25, u_9^{(4)} = 14.33$$
$$u_1^{(5)} = 7.83, u_2^{(5)} = 13.66, u_3^{(5)} = 17.89, u_4^{(5)} = 6.50, u_5^{(5)} = 11.95,$$
$$u_6^{(5)} = 16.29, u_7^{(5)} = 6.64, u_8^{(5)} = 11.25, u_9^{(5)} = 14.34$$

There is no significant difference between the fourth and fifth iteration values.

Hence $u_1 = 7.83$, $u_2 = 13.66$, $u_3 = 17.89$, $u_4 = 6.6$, $u_5 = 11.95$, $u_6 = 16.29$, $u_7 = 6.64, u_8 = 11.25$, $u_9 = 14.34$.

11.6   Solution of Poisson's Equation

$$\frac{\partial^2 u}{\partial x^2} + \frac{\partial^2 u}{\partial y^2} = f(x,y) \tag{1}$$

Its method of solution is similar to that of the Laplace equation. Here the standard five-point formula for (1) takes the form

$$u_{i-1,j} + u_{i+1,j} + u_{i,j+1} + u_{i,j-1} - 4u_{i,j} = h^2 f(ih, jh) \tag{2}$$

By applying (2) at each interior mesh point, we arrive at linear equations in the nodal values $u_{i,j}$. These equations can be solved by the Gauss-Seidal method.

NOTE   **Obs.** *The error in replacing u_{xx} by the finite difference approximation is of the order $O(h^2)$. Since $k = h$, the error in replacing u_{yy} by the difference approximation is also of the order $O(h^2)$. Hence the error in solving Laplace and Poisson's equations by finite difference method is of the order $O(h^2)$.*

EXAMPLE 11.5

Solve the Poisson equation $u_{xx} + u_{yy} = -81xy$, $0 < x < 1$, $0 < y < 1$ given that $u(0, y) = 0$, $u(x, 0) = 0$, $u(1, y) = 100$, $u(x, 1) = 100$ and $h = 1/3$.

Solution:

Here $h = 1/3$.

The standard five-point formula for the given equation is

$$u_{i-1,j} + u_{i+1,j} + u_{i,j+1} + u_{i,j-1} - 4_{ui,j} = h^2 f(ih, jh)$$
$$= h^2 \left[-81(ih \cdot jh) \right] = h^4 (-81) \, ij = -ij \qquad (i)$$

For u_1 $(i = 1, j = 2)$, (i) gives $0 + u_2 + u_3 + 100 - 4u_1 = -2$

i.e., $\qquad\qquad -4u_1 + u_2 + u_3 = -102 \qquad\qquad (ii)$

For u_2 $(i = 2, j = 2)$, (i) gives $u_1 + 100 + u_4 + 100 - 4u_2 = -4$

i.e., $\qquad\qquad u_1 - 4u_2 + u_4 = -204 \qquad\qquad (iii)$

For u_3 $(i = 1, j = 1)$, (i) gives $0 + u_4 + 0 + u_1 - 4u_3 = -1$

i.e., $\qquad\qquad u_1 - 4u_3 + u_4 = -1 \qquad\qquad (iv)$

For u_4 $(i = 2, j = 1)$ gives $u_3 + 100 + u_2 - 4u_4 = -2$

i.e., $\qquad\qquad u_2 + u_3 - 4u_4 = -102 \qquad\qquad (v)$

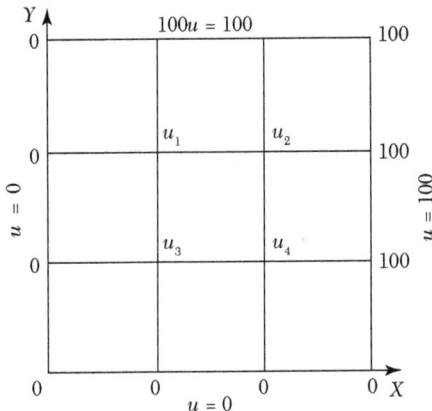

FIGURE 11.9

Subtracting (v) from (ii), $-4u_1 + 4u_4 = 0$, i.e., $u_1 = u_4$

Then (iii) becomes $2u_1 - 4u_2 = -204 \qquad\qquad (vi)$

and (iv) becomes $2u_1 - 4u_3 = -1 \qquad\qquad (vii)$

Now $(4) \times (ii) + (vi)$ gives $-14u_1 + 4u_3 = -612$ $\qquad$ $(viii)$

$(vii) + (viii)$ gives $-12u_1 = -613$

Thus $\qquad u_1 = 613/12 = 51.0833 = u_4.$

From (vi), $\qquad u_2 = \dfrac{1}{2}(u_1 + 102) = 76.5477$

From (vii), $\qquad u_3 = \dfrac{1}{2}\left(u_1 + \dfrac{1}{2}\right) = 25.7916$

EXAMPLE 11.6

Solve the equation $\nabla^2 u - 10(x^2 + y^2 + 10)$ over the square with sides $x = 0 = y$, $x = 3 = y$ with $u = 0$ on the boundary and mesh length $= 1$.

Solution:

Here $h = 1$.

$\therefore$ The standard five-point formula for the given equation is

$$u_{i-1,j} + u_{i+1,j} + u_{i,j+1} + u_{i,j-1} - 4u_{i,j} = -10(i^2 + j^2 + 10) \qquad (i)$$

For u_1 $(i = 1, j = 2)$, (i) gives $0 + u_2 + 0 + u_3 - 4u_1 = -10(1 + 4 + 10)$

i.e., $\qquad u_1 = \dfrac{1}{4}(u_2 + u_3 + 150) \qquad (ii)$

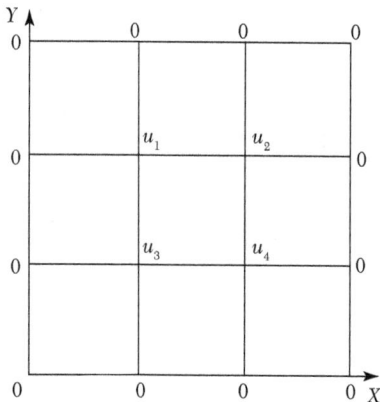

FIGURE 11.10

For u_2 $(i = 2, j = 2)$, (i) gives $\qquad u_2 = \dfrac{1}{4}(u_1 + u_4 + 180) \qquad (iii)$

For u_3 $(i = 1, j = 1)$, we have $\qquad u_3 = \dfrac{1}{4}(u_1 + u_4 + 120) \qquad (iv)$

For u_4 $(i = 2, j = 1)$, we have $\qquad u_4 = \dfrac{1}{4}(u_2 + u_3 + 150)$ $\qquad\qquad$ (v)

Equations (ii) and (v) show that $u_4 = u_1$. Thus the above equations reduce to

$$u_1 = \frac{1}{4}(u_2 + u_3 + 150), \ u_2 = \frac{1}{4}(u_2 + 90), \ u_3 = \frac{1}{4}(u_1 + 60)$$

Now let us solve these equations by the Gauss-Seidal iteration method.

First iteration: Starting from the approximations $u_2 = 0$, $u_3 = 0$, we obtain $u_1^{(1)} = 37.5$

Then $\qquad\qquad\qquad u_2^{(1)} = \dfrac{1}{2}(37.5 + 90) \approx 64$

$$u_2^{(1)} = \frac{1}{2}(37.5 + 60) \approx 49$$

Second iteration: $\quad u_1^{(2)} = \dfrac{1}{4}(64 + 49 + 150) \approx 66, u_2^{(2)} = \dfrac{1}{2}(66 + 90) = 78$

$$u_3^{(2)} = \frac{1}{2}(66 + 60) = 63$$

Third iteration: $\quad u_1^{(3)} = \dfrac{1}{4}(78 + 63 + 150) \approx 73, u_2^{(3)} = \dfrac{1}{2}(73 + 90) \approx 82,$

$$u_3^{(3)} = (73 + 60) \approx 67$$

Fourth iteration: $\quad u_1^{(4)} = (82 + 67 + 150) \approx 75, u_2^{(4)} = (75 + 90) = 82.5,$

$$u_3^{(4)} = (75 + 60) = 67.5$$

Fifth iteration: $\quad u_1^{(5)} = (82.5 + 67.5 + 150) = 75, u_2^{(5)} = (75 + 90) = 82.5,$

$$u_3^{(5)} = (75 + 90) = 67.5$$

Since these values are the same as those of fourth iteration, we have $u_1 = 75$, $u_2 = 82.5, u_3 = 67.5$ and $u_4 = 75$.

Exercises 11.2

1. Solve the equation $u_{xx} + u_{yy} = 0$ for the square mesh with the boundary values as shown in Figure 11.11.

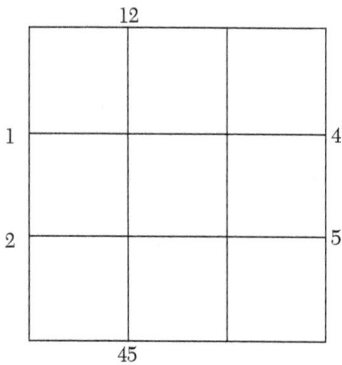

FIGURE 11.11

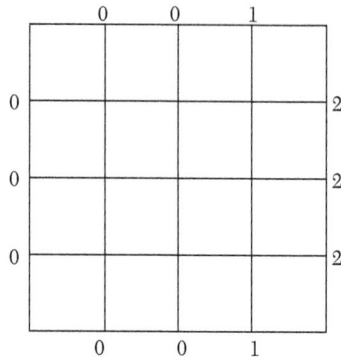

FIGURE 11.12

2. Solve $u_{xx} + u_{yy} = 0$ over the square mesh of side four units satisfying the following boundary conditions: $u(0, y) = 0$ for $0 \le y \le 4$, $u(4, y) = 12 + y$ for $0 \le y \le 4$; $u(x, 0) = 3x$ for $0 \le x \le 4$, $u(x, 4) = x^2$ for $0 \le x \le 4$.

3. Solve the elliptic equation $u_{xx} + u_{yy} = 0$ for the square mesh with boundary values as shown in Figure 11.12. Iterate until the maximum difference between successive values at any point is less than 0.005.

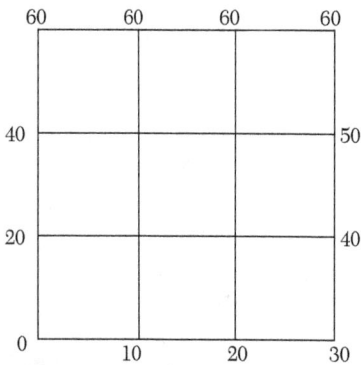

FIGURE 11.13

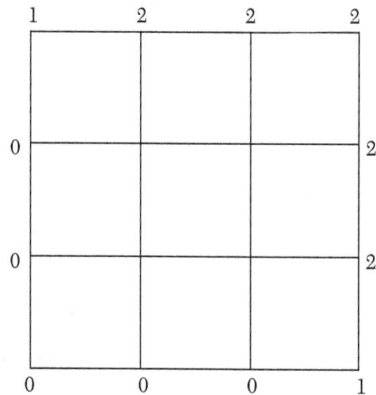

FIGURE 11.14

4. Using central-difference approximation solve $\nabla^2 u = 0$ at the nodal points of the square grid of Figure 11.13 using the boundary values indicated.

5. Solve $u_{xx} + u_{yy} = 0$ for the square mesh with boundary values as shown in Figure 11.14. Iterate till the mesh values are correct to two decimal places.

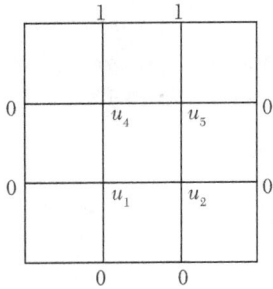

FIGURE 11.15

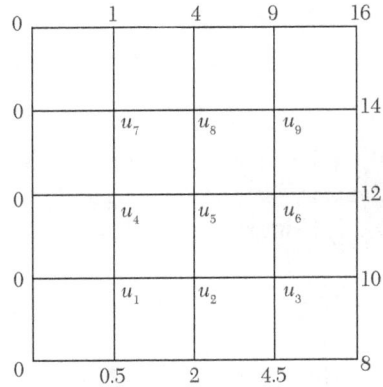

FIGURE 11.16

6. Solve the Laplace's equation $u_{xx} + u_{yy} = 0$ in the domain of Figure 11.15 by (a) Jacobi's method, (b) Gauss-Seidal method.

7. Solve the Laplace's equation $\nabla^2 u = 0$ in the domain of the Figure 11.16.

8. Solve the Poisson's equation $\nabla^2 u = 8x^2 y^2$ for the square mesh of Figure 11.17 with $u(x, y) = 0$ on the boundary and mesh length $= 1$.

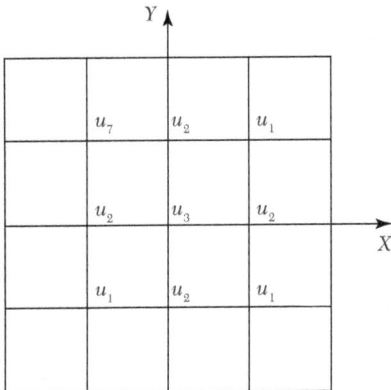

FIGURE 11.17

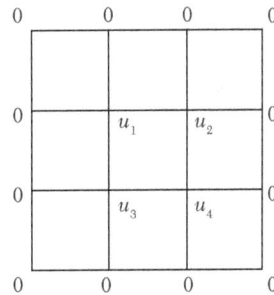

FIGURE 11.18

11.7   Solution of Elliptic Equations by Relaxation Method

If the equations for all the mesh points are written using (2) of Section 11.6, we get a system of equations which can be solved by any method. For this purpose, the method of relaxation is particularly well-suited. Here we shall describe this method in relation to elliptic equations.

Consider the Laplace equation

$$\frac{\partial^2 u}{\partial x^2} + \frac{\partial^2 u}{\partial y^2} = 0 \qquad (1)$$

We take a square region and divide it into a square net of mesh size h. Let the value of u at A be u_0 and its values at the four adjacent points be u_1, u_2, u_3, u_4 (Figure 11.19). Then

$$\frac{\partial^2 u}{\partial x^2} \approx \frac{u_1 + u_3 - 2u_0}{h^2} \text{ and } \frac{\partial^2 u}{\partial y^2} \approx \frac{u_2 + u_4 - 2u_0}{h^2}$$

If (1) is satisfied at A, then

$$\frac{u_1 - u_3 - 2u_0}{h^2} + \frac{u_2 + u_4 - 2u_0}{h^2} \approx 0$$

or
$$u_1 + u_2 + u_3 + u_4 - 4u_0 \approx 0$$

If r_0 be the residual (discrepancy) at the mesh point A,

then
$$r_0 = u_1 + u_2 + u_3 + u_4 - 4u_0 \qquad (2)$$

Similarly the residual at the point B, is given by

$$r_1 = u_0 + u_5 + u_6 + u_7 - 4u_1 \text{ and so on} \qquad (3)$$

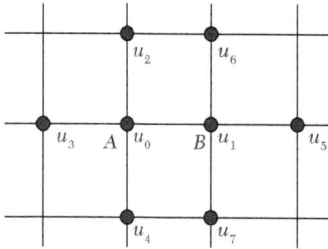

FIGURE 11.19

The main aim of the relaxation process is to reduce all the residuals to zero by making them as small as possible step by step. We, therefore, try to adjust the value of u at an internal mesh point so as to make the residual thereat zero. But when the value of u is changing at a mesh point, the values of the residuals at the neighboring interior points will also be changed. If u_0 is given an increment 1, then

(i) (2) shows that r_0 is changed by -4.

(ii) (3) shows that r_1 is changed by 1.

i.e., if the value of the function is increased by 1 *at a mesh point* (shown by a double ring), *then the residual at that point is decreased by* 4 *while the residuals at the adjacent interior points* (shown by a single ring), *get increased each by* 1. This relaxation pattern is shown in Figure 11.20.

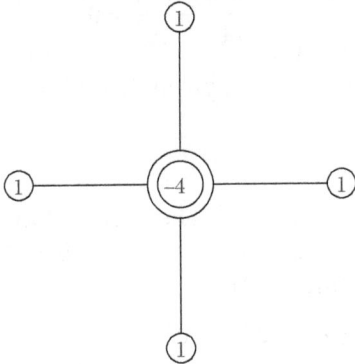

FIGURE 11.20

Working procedure *to solve an equation by the relaxation method:*

I. *Write down by trial, the initial values of u at the interior mesh points by diagonal averaging or cross-averaging.*

II. *Calculate the residuals at each of these points by* (2) *above. If we apply this formula at a point near the boundary, one or more end points get chopped off since there are no residuals at the boundary.*

III. *Write the residuals at a mesh-point on the right of this point and the value of u on its left.*

IV. *Obtain the solution by reducing the residuals to zero, one by one, by giving suitable increments to u and using Figure* 11.20. *At each step, we reduce the numerically largest residual to zero and record the increment of u on the left* (below the earlier value thereat) *and the modified residual on the right* (below the earlier residual).

V. *When a round of relaxation is completed, the value of u and its increments are added at each point. Using these values, calculate all the residuals afresh. If some of there calculated residuals are large, liquidate these again.*

VI. *Stop the relaxation process, when the current values of the residuals are quite small. The solution will be the current value of u at each of the nodes.*

NOTE **Obs.** *Relaxation method combines simplicity with the speed of convergence. Its only drawback is its unsuitability for computer calculations.*

EXAMPLE 11.7

Solve by relaxation method, the Laplace equation $\dfrac{\partial^2 u}{\partial x^2} + \dfrac{\partial^2 u}{\partial y^2} = 0$ inside the square bounded by the lines $x = 0$, $x = 4$, $y = 0$, $y = 4$, given that $u = x^2 y^2$ on the boundary.

Solution:

Taking $h = 1$, we find u on the boundary from $u = x^2 y^2$. The initial values of u at the nine mesh points are estimated to be 24, 56, 104; 16, 32, 56; 8, 16, 24 as shown on the left of the points in Figure 11.21.

∴ Residual at A, *i.e.*, $r_A = 0 + 56 + 16 + 16 - 4 \times 24 = -8$

Similarly $r_B = 0$, $r_C = -16$, $r_D = 0$, $r_E = 16$, $r_F = 0$, $r_G = 0$, $r_H = 0$, $r_I = -8$.

(*i*) The numerically largest residual is 16 at E. To liquidate it, we increase u by 4 so that the residual becomes zero and the residuals at neighboring nodes get increased by 4.

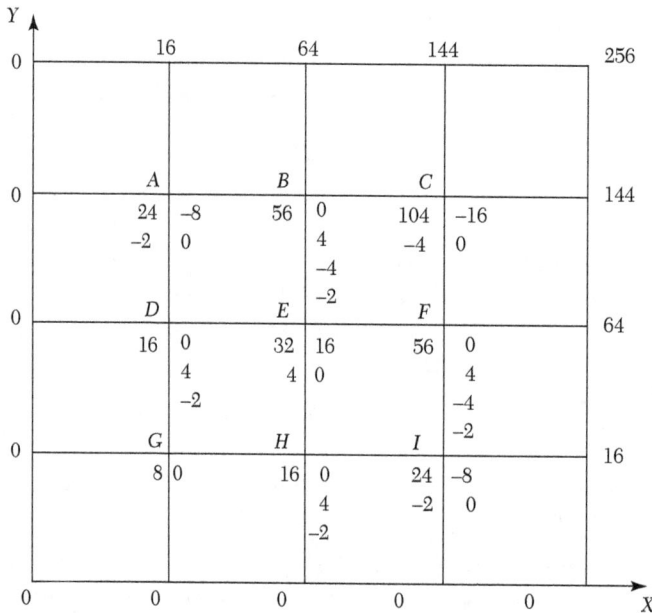

FIGURE 11.21

(*ii*) Next, the numerically largest residual is – 16 at *C*. To reduce it to zero, we increase *u* by – 4 so that the residuals at the adjacent nodes are increased by – 4.

(*iii*) Now, the numerically largest residual is – 8 at *A*. To liquidate it, we increase *u* by– 2 so that the residuals at the adjacent nodes are increased by – 2.

(*iv*) Finally, the largest residual is – 8 at *I*. To liquidate it, we increase *u* by – 2 so that the residuals at the adjacent points are increased by – 2.

(*v*) The numerically largest current residual being 2, we stop the relaxation process. Hence the final values of *u* are:

$$u_A = 22, \qquad u_B = 56, \qquad u_C = 100,$$
$$u_D = 16, \qquad u_E = 36, \qquad u_F = 56,$$
$$u_G = 8, \qquad u_H = 16, \qquad u_I = 22.$$

EXAMPLE 11.8

Solve by relaxation method Example 11.3.

Solution:

(*i*) The initial values of *u* at *A*, *B*, *C*, and *D* are estimated to be 1000, 625, 875, and 375 [Figure 11.22 (*i*)].

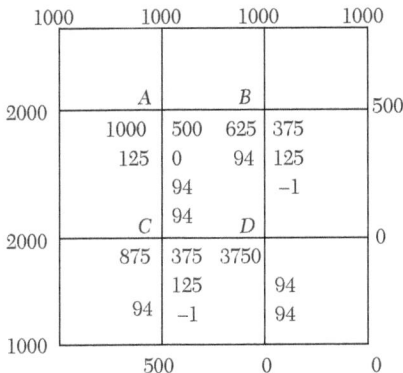

FIGURE 11.22 (I)

∴ $\qquad r_A = 500, r_B = 375, r_C = 375, r_D = 0$

To liquidate r_A, increase *u* by 125

To liquidate r_B, increase u by 94

To liquidate r_C, increase u by 94

(ii) Modified values of u are 1125, 719, 969, 375 [Figure (ii)]

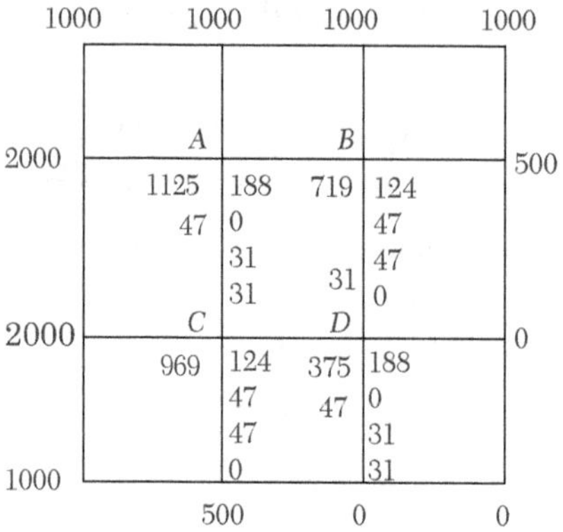

FIGURE 11.22 (II)

$\therefore \qquad r_A = 188, r_B = 124, r_C = 124, r_D = 188.$

To liquidate r_A, r_D, r_B, r_C increase u by 47, 47, 31, 31 in turn.

(iii) Revised values of u are 1172, 750, 1000, 422 [Figure (iii)]

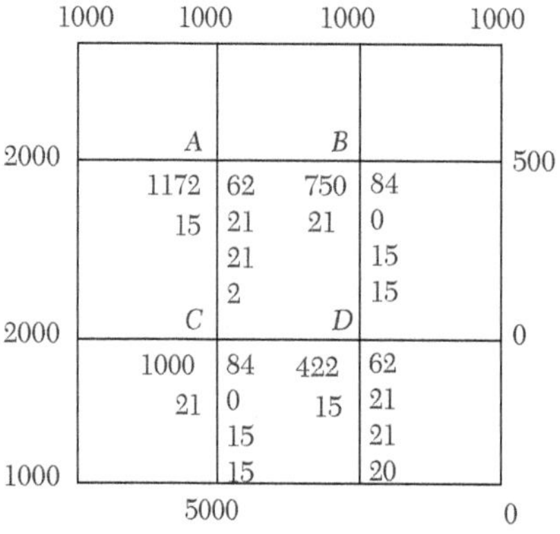

FIGURE 11.22 (III)

$\therefore \qquad r_A = 62, r_B = 84, r_C = 84, r_D = 62$

To liquidate r_B, r_C, r_A, r_D increase u by 21, 21, 15, 15, respectively.

(iv) Improved values of u are 1187, 771, 1021, 437 [Figure (iv)]

$\therefore \qquad r_A = 44, r_B = 40, r_C = 40, r_D = 44.$

To liquidate r_A, r_D, r_B, r_C increase u by 11, 11, 10, 10, respectively

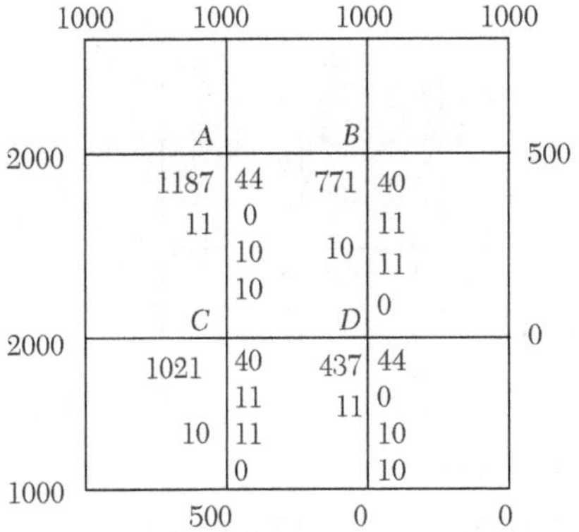

FIGURE 11.22 (IV)

(v) Modified values of u are 1198, 781, 1031, 448 [Figure (v)]

$\therefore \qquad r_A = 20, r_B = 22, r_C = 22, r_D = 20.$

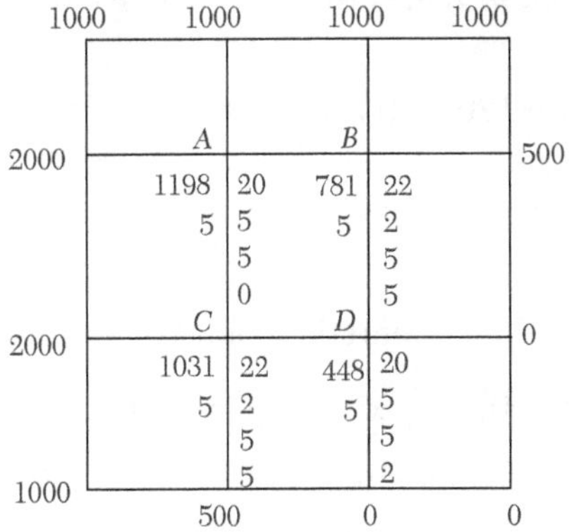

FIGURE 11.22 (V)

To liquidate r_B, r_C, r_A, r_D increase u by 5, 5, 5, 5, respectively.

(vi) Revised values of u are 1203, 786, 1036, 453 [Figure (vi)]

$\therefore \qquad r_A = 10, r_B = 12, r_C = 12, r_D = 10$

To liquidate r_B, r_C, r_A, r_D increase u by 3, 3, 2, 2, respectively.

```
  1000      1000      1000      1000          1000      1000      1000      1000

        A         B                                 A         B
 2000                            500        2000                           500
        1203 |10   786 |12                         1205 |8   789 |4
           2 |3       3 |0                            2 |0      1 |2
             |3         |2                              |1        |2
             |2         |2                              |1        |0
        C         D                                 C         D
 2000                              0        2000                             0
        1036 |12   453 |10                         1039 |4   455 |8
           3 |0       2 |3                            2 |      2 |0
             |2         |3                            1 |2        |1
             |2         |2                              |0        |1
  1000                                        1000
            500       0         0                        500       0
  FIGURE 11.22 (VI)                            FIGURE 11.22 (VII)
```

(*vii*) Improved values of u are 1205, 789, 1039, 455 [Figure (*vii*)]

∴ $r_A = 8, r_B = 4, r_C = 4, r_D = 8$.

To liquidate r_A, r_D, r_B, r_C increase u by 2, 2, 1, 1.

(*viii*) Finally the current residuals being 1, 0, 0, 1, we stop the relaxation process.

Hence the values of u at A, B, C, D are 1207, 790, 1040, 457.

Exercises 11.3

1. Given that $u(x, y)$ satisfies the equation $\nabla^2 u = 0$ and the boundary conditions are $u(0, y) = 0$, $u(4, y) = 8 + 2y$, $u(x, 0) = \dfrac{1}{2} x^2$, $u(x, 4) = x^2$, find the values $u(i, j)$, $i = 1, 2, 3; j = 1, 2, 3$ by the relaxation method.

2. Apply the relaxation method to solve the equation $\nabla^2 u = -400$, when the region of u is the square bounded by $x = 0, y = 0, x = 4$, and $y = 4$ and u is zero on the boundary of the square.

3. Solve by relaxation method, the equation $\nabla^2 u = 0$ in the square region with square meshes(Figure 11.23) starting with the initial values $u_1 = u_2 = u_3 = u_4 = 1$.

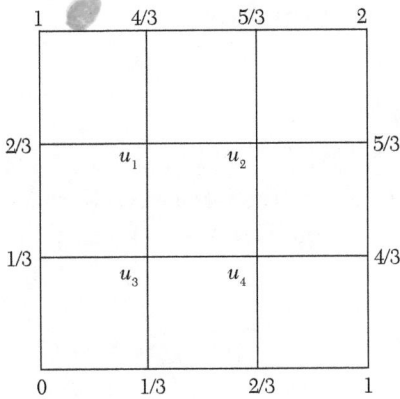

FIGURE 11.23

11.8 Parabolic Equations

The one-dimensional heat conduction equation $\dfrac{\partial u}{\partial t} = c^2 \dfrac{\partial^2 u}{\partial x^2}$ is a well-known Example of parabolic partial differential equations. The solution of this equation is a temperature function u(x, t) which is defined for values of x from 0 to l and for values of time t from 0 to ∞. The solution is not defined in a closed domain but advances in an open-ended region from initial values, satisfying the prescribed boundary conditions (Figure 11.24).

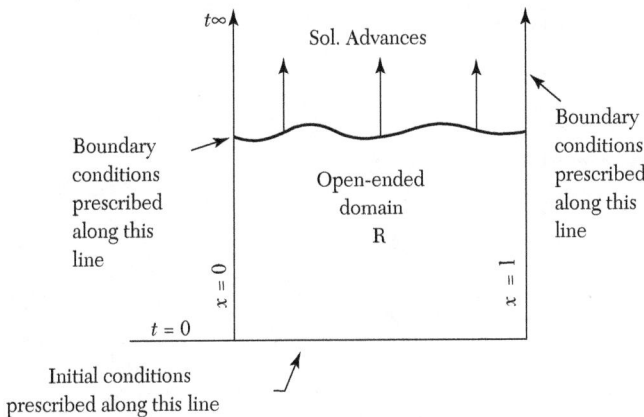

FIGURE 11.24

In general, the study of pressure waves in a fluid, propagation of heat and unsteady state problems lead to parabolic type of equations.

11.9 Solution of One Dimensional Heat Equation

$$\frac{\partial u}{\partial t} = c^2 \frac{\partial^2 u}{\partial x^2} \tag{i}$$

where $c^2 = k/s\rho$ is the diffusivity of the substance (cm²/sec.)

Schmidt method. Consider a rectangular mesh in the x-t plane with spacing h along x direction and k along time t direction. Denoting a mesh point $(x, t) = (ih, jk)$ as simply i, j, we have

$$\frac{\partial u}{\partial t} = \frac{u_{i,j+1} - u_{i,j}}{k} \qquad \text{[by (5) Section 11.3.}$$

and

$$\frac{\partial^2 u}{\partial x^2} = \frac{u_{i-1,j} - 2u_{i,j} + u_{i+1,j}}{h^2} \qquad \text{[by (4) Section 11.3.}$$

Substituting these in (1), we obtain $u_{i,j+1} - u_{i,j} = \dfrac{kc^2}{h^2}\left[u_{i-1,j} - 2u_{i,j} + u_{i+1,j}\right]$

or $\qquad u_{i,j+1} = \alpha u_{i-1,j} + (1 - 2\alpha)\, u_{i,j} + \alpha u_{i+1,j} \tag{2}$

where $\alpha = kc^2/h^2$ is the mesh ratio parameter.

This formula enables us to determine the value of u at the $(i, j + 1)$th mesh point in terms of the known function values at the points x_{i-1}, x_i, and x_{i+1} at the instant t_j. It is a relation between the function values at the two time levels $j + 1$ and j and is therefore, called a *two-level formula*. In schematic form (2) is shown in Figure 11.25.

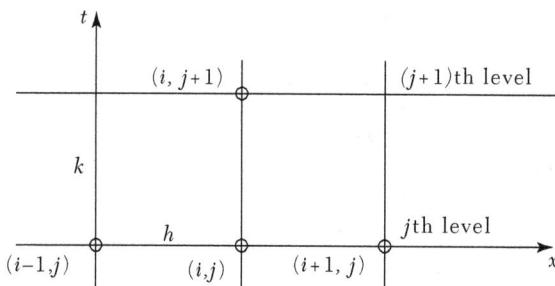

FIGURE 11.25

Hence (2) is called the *Schmidt explicit formula* which is valid only for $0 < \alpha \le 12$.

NOTE

Obs. *In particular when α = 1/2, (2) reduces to*

$$u_{i,j+1} = 1/2, (u_{i-1,j} + u_{i+1,j}) \tag{3}$$

which shows that the value of u at x_i at time t_{j+1} is the mean of the u-values at x_{i-1} and x_{i+1} at time t_j. This relation, known as Bendre-Schmidt recurrence relation, gives the values of u at the internal mesh points with the help of boundary conditions.

Crank-Nicolson method. We have seen that the Schmidt scheme is computationally simple and for convergent results $α ≤ 12$ *i.e.*, $k ≤ h^2/2c^2$. To obtain more accurate results, h should be small *i.e.* k is necessarily very small. This makes the computations exceptionally lengthy as more time levels would be required to cover the region. A method that does not restrict $α$ and also reduces the volume of calculations was proposed by Crank and Nicolson in 1947.

According to this method, $\partial^2 u/\partial x^2$ is replaced by the average of its central-difference approximations on the jth and $(j + 1)$th time rows. Thus (1) is reduced to

$$\frac{u_{i,j+1} - u_{i,j}}{h} = c^2 \frac{1}{2} \left\{ \frac{u_{i-1,j} - 2u_{i,j} + u_{i+1,j}}{h^2} \right\} + \left\{ \frac{u_{i-1,j+1} - 2u_{i,j+1} + u_{i+1,j+1}}{h^2} \right\}$$

or $\quad -αu_{i-1,j+1} + (2 + 2α)u_{i,j+1} - αu_{i+1,j+1} = αu_{i-1,j} + (2 - 2α)u_{i,j} + αu_{i+1,j} \tag{4}$

where $α = kc^2/h^2$.

Clearly the left side of (4) contains three unknown values of u at the $(j + 1)$th level while all the three values on the right are known values at the jth level. Thus (4) is a two *level implicit relation* and is known as *Crank-Nicolson formula*. It is convergent for all finite values of $α$. Its computational model is given in Figure 11.26.

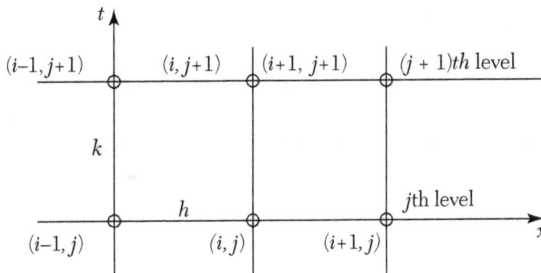

FIGURE 11.26

If there are n internal mesh points on each row, then the relation (4) gives n simultaneous equations for the n unknown values in terms of the known boundary values. These equations can be solved to obtain the values at these mesh points. Similarly, the values at the internal mesh points on all rows can be found. A method such as this in which the calculation of an unknown mesh value necessitates the solution of a set of simultaneous equations, is known as an *implicit scheme*.

Iterative methods of solution *for an implicit scheme.*

From (4), we have

$$(1+\alpha)\,u_{i,j+1} = \frac{1}{2}\,\alpha(u_{i-1,j+1+ui+1,j+1}) + u_{i,j} + \frac{1}{2}\,\alpha(u_{i-1,j} - 2_{ui,j} + u_{i+1,j}) \qquad (5)$$

Here only $u_{i,j+1}$, $u_{i-1,j+1}$ and $u_{i+1,j+1}$ are unknown while all others are known since these were already computed in the jth step.

Writing
$$b_i = u_{i,j} + \frac{\alpha}{2}\Big(u_{i-1,j} - 2u_{i,j} + u_{i+1,j}\Big)$$

and dropping j's (5) becomes $u_i = \dfrac{\alpha}{2(1+\alpha)}\big(u_{i-1} + u_{i+1}\big) + \dfrac{b}{1+\alpha}$

This gives the iteration formula

$$u_i^{(n+1)} = \frac{\alpha}{2(1+\alpha)}\big\{u_{i-1}^{(n)} + u_{i+1}^{(n)}\big\} + \frac{b_i}{1+\alpha} \qquad (6)$$

which expresses the $(n + 1)$th iterates in terms of the nth iterates only. This is known as the *Jacobi's iteration formula*.

As the latest value of u_{i-1} i.e., $u_{i-1}^{(n+1)}$ is already available, the convergence of the iteration formula (6) can be improved by replacing $u_{i-1}^{(n)}$ by $u_{i-1}^{(n+1)}$. Accordingly (6) may be written as

$$u_i^{(n+1)} = \frac{\alpha}{2(1+\alpha)}\big\{u_{i-1}^{(n+1)} + u_{i+1}^{(n)}\big\} + \frac{b_i}{1+\alpha} \qquad (7)$$

which is known as the *Gauss-Seidal iteration formula*.

NOTE   **Obs.** *Gauss-Seidal iteration scheme is valid for all finite values of α and converges twice as fast as Jacobi's scheme.*

Du Fort and Frankel method. If we replace the derivatives in (1) by the central difference approximations

$$\frac{\partial u}{\partial t} = \frac{u_{i,j+1} - u_{i,j-1}}{2k} \qquad \text{[From (7) Section 11.3]}$$

and $$\frac{\partial^2 u}{\partial x^2} = \frac{u_{i-1,j} - 2u_{i,j-1} + u_{i+1,j}}{h^2}$$ [From (4) Section 11.3]

we obtain $$u_{i,j+1} - u_{i,j-1} = \frac{2kc^2}{h^2} \left[u_{i-1,j} - 2u_{i,j} + u_{i+1,j}\right]$$

i.e., $$u_{i,j+1} = u_{i,j-1} + 2\alpha \left[u_{i-1,j} - 2u_{i,j} + u_{i+1,j}\right] \tag{8}$$

where $\alpha = kc^2/h^2$. This difference equation is called the *Richardson scheme* which is a *three-level method*.

If we replace $u_{i,j}$ by the mean of the values $u_{i,j-1}$ and $u_{i,j+1}$

i.e., $u_{i,j} = (u_{i,j-1} + \frac{1}{2} u_{i,j+1})$ in (8), then we get

$$u_{i,j+1} = u_{i,j-1} + 2\alpha[u_{i-1,j} - (u_{i,j-1} + u_{i,j+1}) + u_{i+1,j}]$$

On simplification, it can be written as

$$u_{i,j+1} = \frac{1-2\alpha}{1+2\alpha}u_{i,j-1} + \frac{2\alpha}{1+2\alpha}\left\{u_{i-1,j} + u_{i+1,j}\right\} \tag{9}$$

This difference scheme is called *Du Fort-Frankel method* which is a *three level explicit method*. Its computational model is given in Figure 11.27

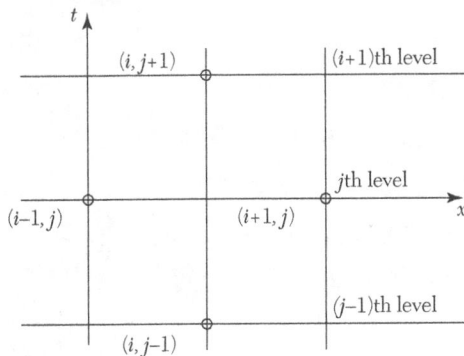

FIGURE 11.27

EXAMPLE 11.9

Solve $\dfrac{\partial u}{\partial u} = \dfrac{\partial^2 u}{\partial^2 x^2}$ in $0 < x < 5$, $t \geq 0$ given that $u(x, 0) = 20$, $u(0, t) = 0$, $u(5, t) = 100$. Compute u for the time-step with $h = 1$ by the Crank-Nicholson method.

Solution:

Here $c^2 = 1$ and $h = 1$.

Taking α (*i.e.*, $c^2 k/h$) = 1, we get $k = 1$.

Also we have

J \ I	0	1	2	3	4	5
0	0	20	20	20	20	100
1	0	u_1	u_2	u_3	u_4	100

Then Crank-Nicholson formula becomes

$$4u_{i,j+1} = u_{i-1,j+1} + u_{i+1,j+1} + u_{i-1,j} + u_{i+1,j}$$

$$\therefore \quad 4u_1 = 0 + 20 + 0 + u_2 \; i.e., \; 4u_1 - u_2 = 20 \tag{1}$$

$$4u_2 = 20 + 20 + u_1 + u_3 \; i.e., \; u_1 - 4u_2 + u_3 = -40 \tag{2}$$

$$4u_3 = 20 + 20 + u_2 + u_4 \; i.e., \; u_2 - 4u_3 + u_4 = -40 \tag{3}$$

$$4u_4 = 20 + 100 + u_3 + 100 \; i.e., \; u_3 - 4u_4 = -220 \tag{4}$$

Now (1) – 4(2) gives $15u_2 - 4u_3 = 180$ \hfill (5)

4(3) + (4) gives $4u_2 - 15u_3 = -380$ \hfill (6)

Then 15(5) – 4(6) gives $209\, u_2 = 4220 \; i.e., \; u_2 = 20.2$

From (5), we get $\quad 4u_3 = 15 \times 20.2 - 180 \; i.e., \; u_3 = 30.75$

From (1), $\quad 4u_1 = 20 + 20.2 \; i.e., \; u1 = 10.05$

From (4), $\quad 4u_4 = 220 + 30.75 \; i.e., \; u4 = 62.69$

Thus the required values are 10.05, 20.2, 30.75 and 62.68.

EXAMPLE 11.10

Solve the boundary value problem $u_t = u_{xx}$ under the conditions $u(0, t) = u(1, t) = 0$ and $u(x, 0) = \sin px$, $0 \le x \le 1$ using the Schmidt method (Take $h = 0.2$ and $\alpha = 1/2$).

Solution:

Since $\quad h = 0.2$ and $\alpha = \frac{1}{2}$

$$\therefore \quad \alpha = \frac{k}{h^2} \text{ gives } k = 0.02$$

Since $\alpha = 1/2$, we use the Bendre-Schmidt relation

$$u_{i,j+1} = \frac{1}{2}(i_{i-1,j} + u_{i+1,j}) \qquad (i)$$

We have $u(0, 0) = 0$, $u(0.2, 0) = \sin \pi/5 = 0.5875$

$$u(0.4, 0) = \sin 2\pi/5 = 0.9511, u(0.6, 0) = \sin 3\pi/5 = 0.9511$$

$$u(0.8, 0) = \sin 4\pi/5 = 0.5875, u(1, 0) = \sin \pi = 0$$

The values of u at the mesh points can be obtained by using the recurrence relation (i) as shown in the table below:

$x \rightarrow$		i	0	0.2	0.4	0.6	0.8	1.0
$t \downarrow$ 0	j		0	1	2	3	4	5
	0		0	0.5878	0.9511	0.9511	0.5878	0
0.02	1		0	0.4756	0.7695	0.7695	0.4756	0
0.04	2		0	0.3848	0.6225	0.6225	0.3848	0
0.06	3		0	0.3113	0.5036	0.5036	0.3113	0
0.08	4		0	0.2518	0.4074	0.4074	0.2518	0
0.1	5		0	0.2037	0.3296	0.3296	0.2037	0

EXAMPLE 11.11

Find the values of $u(x, t)$ satisfying the parabolic equation $\dfrac{\partial u}{\partial t} = 4\dfrac{\partial^2 u}{\partial x^2}$
and the boundary conditions $u(0, t) = 0 = u(8, t)$ and $u(x, 0) = 4x - (1/2) x^2$ at the points $x = i{:}i = 0, 1, 2, \cdots, 7$ and $t = 1/8 \ j{:}j = 0, 1, 2, \cdots, 5$

Solution:

Here $c^2 = 4$, $h = 1$ and $k = 1/8$. Then $\alpha = c^2k/h^2 = 1/2$.

∴ We have Bendre-Schmidt's recurrence relation

$$u_{i,j+1} = 1/2 \ (u_{i-1,j} + u_=) \qquad (i)$$

Now since $\qquad u(0, t) = 0 = u(8, t)$

∴ $u_{0,i} = 0$ and $u_{8,j} = 0$ for all values of j, $i.e.$, the entries in the first and last column are zero.

Since $u(x, 0) = 4x - (1/2) x^2$

$\therefore$ $u_{i,0} = 4i - (1/2) i^2$

$= 0, 3.5, 6, 7.5, 8, 7.5, 6, 3.5$ for $i = 0, 1, 2, 3, 4, 5, 6, 7$

at $t = 0$

These are the entries of the first row.

Putting $j = 0$ in (i), we have $u_{i,1} = (1/2) (u_{i-1,0} + u_{i+1,0})$

Taking $i = 1, 2, \cdots, 7$ successively, we get

$$u_{1,1} = \frac{1}{2}(u_{0,0} + u_{2,0}) = \frac{1}{2}(0 + 6) = 3$$

$$u_{2,1} = \frac{1}{2}(u_{1,0} + u_{3,0}) = \frac{1}{2}(3.5 + 7.5) = 5.5$$

$$u_{3,1} = \frac{1}{2}(u_{2,0} + u_{4,0}) = \frac{1}{3}(6 + 8) = 7$$

$u_{4,1} = 7.5, u_{5,1} = 7, u_{6,1} = 5.5, u_{7,1} = 3.$

These are the entries in the second row.

Putting $j = 1$ in (i), the entries of the third row are given by

$$u_{1,2} = \frac{1}{2}(u_{i-1,1} + u_{i+1,1})$$

Similarly putting $j = 2, 3, 4$ successively in (i), the entries of the fourth, fifth, and sixth rows are obtained.

Hence the values of $u_{i,j}$ are as given in the following table:

i / j	0	1	2	3	4	5	6	7	8
0	0	3.5	6	7.5	8	7.5	6	3.5	0
1	0	3	5.5	7	7.5	7	5.5	2	0
2	0	2.75	5	6.5	7	6.5	5	2.75	0
3	0	2.5	4.625	6	6.5	6	4.625	2.5	0
4	0	2.3125	4.25	5.5625	6	5.5625	4.25	2.3125	0
5	0	2.125	3.9375	5.125	5.5625	5.125	3.9375	2.125	0

EXAMPLE 11.12

Solve the equation $\dfrac{\partial u}{\partial y} = \dfrac{\partial^2 u}{\partial x^2}$ subject to the conditions $u(x,\ 0) = \sin \pi x$,

$0 \le x \le 1; u(0, t) = u(1, t) = 0$, using (a) Schmidt method, (b) Crank-Nicolson method, (c) Du Fort-Frankel method. Carryout computations for two levels, taking $h = 1/3, k = 1/36$.

Solution:

Here $c^2 = 1, h = 1/3, k = 1/36$ so that $\alpha = kc^2/h^2 = 1/4$.

Also $u_{1,0} = \sin \pi/3 = \sqrt{3}/2, u_{2,0} = \sin 2\pi/3 = \sqrt{3}/2$ and all boundary values are zero as shown in Figure 11.28.

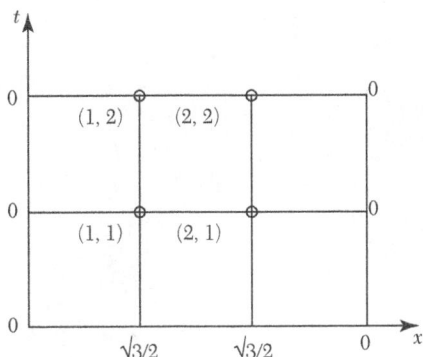

FIGURE 11.28

(a) *Schmidt's formula* [(2) of Section 11.9]

$$u_{i,j+1} = \alpha u_{i-1,j} + (1 - 2\alpha) u_{i,j} + \alpha u_{i+1,j}$$

becomes $\qquad u_{i,j+1} = \dfrac{1}{4}[u_{i-1,j} + 2u_{i,j} + u_{i+1,j}]$

For $i = 1, 2; j = 0$:

$$u_{1,1} = \frac{1}{4}\ [u_{0,0} + 2u_{1,0} + u_{2,0}] = \frac{1}{4}\ (0 + 2 \times \sqrt{3}/2 + \sqrt{3}/2) = 0.65$$

$$u_{2,1} = \frac{1}{2}\ [u_{1,0} + 2u_{2,0} + u_{3,0}] = \frac{1}{4}\ (\sqrt{3}/2 + 2 \times \sqrt{3}/2 + 0) = 0.65$$

For $i = 1, 2; j = 1$:

$$u_{1,2} = \frac{1}{4}\ (u_{0,1} + 2u_{1,1} + u_{2,1}) = 0.49$$

$$u_{2,2} = \frac{1}{4}\ (u_{1,1} + 2u_{2,1} + u_{3,1}) = 0.49$$

(b) *Crank-Nicolson formula* [(4) of Section 11.9] becomes

$$-\frac{1}{4}\,u_{i-1,j+1} + \frac{5}{2}\,u_{i,j+1} - \frac{1}{4}\,u_{i+1,j+1} = \frac{1}{4}\,u_{i-1,j} + -u_{i,j} + \frac{1}{4}\,u_{i+1,j}$$

For $i = 1, 2; j = 0$:

$$-u_{0,1} + 10u_{1,1} - u_{2,1} = u_{0,0} + 6u_{1,0} + u_{2,0}$$

i.e.,
$$10u_{1,1} - u_{2,1} = 7\sqrt{3}/2$$

$$-u_{1,1} + 10u_{2,1} - u_{3,1} = u_{1,0} + 6u_{2,0} + u_{3,0}$$

i.e.,
$$-u_{1,1} + 10u_{2,1} = 7\sqrt{3}/2$$

Solving these equations, we find
$$u_{1,1} = u_{2,1} = 0.67$$

For $i = 1, 2; j = 1$:

$$-u_{0,2} + 10u_{1,2} - u_{2,2} = u_{0,1} + 6u_{1,1} + u_{2,1}$$

i.e.,
$$10u_{1,2} - u_{2,2} = 4.69$$

$$-u_{1,2} + 10u_{2,2} - u_{3,2} = u_{1,1} + 6u_{2,1} + u_{3,1}$$

i.e,
$$-u_{1,2} + 10_{u2,2} = 4.69$$

Solving these equations, we get $u_{1,2} = u_{2,2} = 0.52$.

(c) *Du Fort-Frankel formula* [(8) of Section 11.9] becomes $u_{i,j+1} = \frac{1}{3}\,(u_{i,j-1} + u_{i-1,j} + u_{i+1,j})$

To start the calculations, we need $u_{1,1}$ and $u_{2,1}$.

We may take $u_{1,1} = u_{2,1} = 0.65$ from Schmidt method.

For $i = 1, 2; j = 1$:

$$u_{1,2} = \frac{1}{3}\,(u_{1,0} + u_{0,1} + u_{2,1}) = \frac{1}{3}\,(\sqrt{3}/2 + 0 + 0.65) = 0.5$$

$$u_{2,2} = \frac{1}{3}\,(u_{2,0} + u_{1,1} + u_{3,1}) = \frac{1}{3}\,(\sqrt{3}/2 + 0.65 + 0) = 0.5.$$

11.10   Solution of Two Dimensional Heat Equation

$$\frac{\partial u}{\partial t} = c^2 \left(\frac{\partial^2 u}{\partial x^2} + \frac{\partial^2 u}{\partial y^2} \right) \tag{1}$$

The methods employed for the solution of one dimensional heat equation can be readily extended to the solution of (1).

Consider a square region $0 \le x \le y \le a$ and assume that u is known at all points within and on the boundary of this square.

If h is the step-size then a mesh point $(x, y, t) = (ih, jh, nl)$ may be denoted as simply (i, j, n).

Replacing the derivatives in (1) by their finite difference approximations, we get

$$\frac{u_{li,j,n+1} - u_{i,j,n}}{l} = \frac{c^2}{h^2}\{(u_{i-1,j,n} - 2u_{i,j,n} + u_{i+1,j,n}) + (u_{i,j-1,n} - 2u_{i,j,n} + u_{i,j+1,n})\}$$

i.e.,
$$u_{i,j,n+1} = u_{i,j,n} + \alpha(u_{i-1,j,n} + u_{i+1,j,n} + u + + u_{i,j-1,n} - 4u_{i,j,n}) \qquad (2)$$

where $\alpha = lc^2/h^2$. This equation needs the five points available on the nth plane (Figure 11.29).

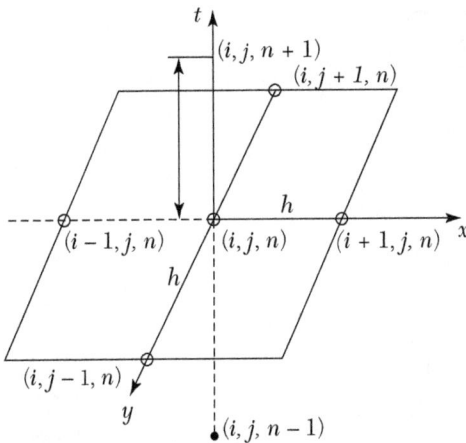

FIGURE 11.29

The computation process consists of point-by-point evaluation in the $(n + 1)$th plane using the points on the nth plane. It is followed by plane-by-plane evaluation. This method is known as *ADE* (*Alternating Direction Explicit*) *method*.

EXAMPLE 11.13

Solve the equation $\dfrac{\partial u}{\partial u} = \dfrac{\partial^2 u}{\partial x^2} + \dfrac{\partial^2 u}{\partial y^2}$ subject to the initial conditions

$u(x,y, 0) = \sin 2\pi x \sin 2\pi y$, $0 \le x, y \le 1$, and the conditions $u(x, y, t) = 0$,

$t > 0$ on the boundaries, using ADE method with $h = 1/3$ and $\alpha = 1/8$. (Calculate the results for one time level).

Solution:

The equation (2) above becomes

$$u_{i,j,n+1} = u_{i,j,n} + \frac{1}{8}\left(u_{i-1,j,n} + u_{i+1,j,n} + u_{i,j+1,n} + u_{i,j-1,n} - 4u_{i,j,n}\right)$$

$$i.e., \quad u_{i,j,n+1} = \frac{1}{2}u_{i,j,n} + \frac{1}{8}\left(u_{i-1,j,n} + u_{i+1,j,n} + u_{i,j+1,n} + u_{i,j-1,n}\right) \tag{1}$$

The mesh points and the computational model are given in Figure 11.30.

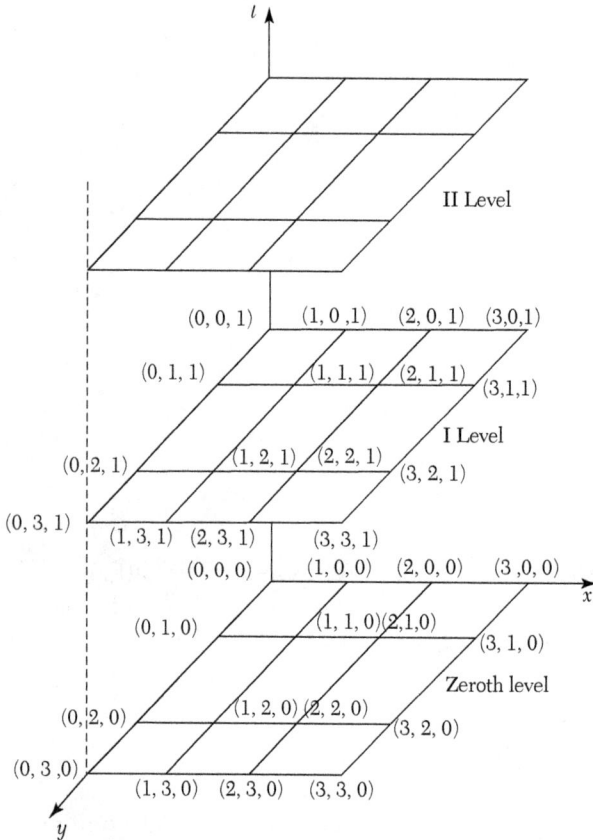

FIGURE 11.30

At the zero level ($n = 0$), the initial and boundary conditions are

$$u_{i,j,0} = \sin\frac{2\pi i}{3}\sin\frac{2\pi i}{3}$$

and $\quad u_i, 0, 0 = u_0, j, 0 = u_3, j, 0 = u_i, 3, 0 = 0; i, j = 0, 1, 2, 3.$

Now we calculate the mesh values at the first level:

For n = 0, (1) gives

$$u_{i,j,1} = \frac{1}{2}\ u_{i,j,0} + \frac{1}{8}\ (u_{i-1,j,0} + u_{i+1,j,0} + u_{i,j+1,0} + u_{i,j-1,0}) \qquad (2)$$

(*i*) Put $i = j = 1$ in (2):

$$u_{1,1,1} = \frac{1}{2}\ u_{1,1,0} + \frac{1}{8}\ (u_{0,1,0} + u_{2,1,0} + u_{1,2,0} + u_{1,0,0})$$

$$= \frac{1}{2}\left(\sin\frac{2\pi}{3}\right)^2 + \frac{1}{8}\left(0 + \sin\frac{4\pi}{3}\sin\frac{2\pi}{3} + \sin\frac{2\pi}{3}\sin\frac{4\pi}{3} + 0\right)$$

$$= \frac{3}{8} + \frac{1}{8}\left(-\frac{\sqrt{3}}{2}\times\frac{\sqrt{3}}{2}\times\frac{\sqrt{3}}{2}\times\frac{\sqrt{3}}{2}\times\right) = \frac{3}{16}$$

(*ii*) Put $i = 2, j = 1$ in (2)

$$u_{2,1,1} = \frac{1}{2}u_{2,1,0} + \frac{1}{8}\ (u_{1,1,0} + u_{3,1,0} + u_{2,2,0} + u_{2,0,0})$$

$$= \frac{1}{2}\sin\frac{4\pi}{3}\sin\frac{2\pi}{3} + \frac{1}{8}\left\{\left(\sin\frac{2\pi}{3}\right)^2 + 0 + \left(\sin\frac{4\pi}{3}\right)^2 + 0\right.$$

$$= -\frac{1}{2}\left(\frac{\sqrt{3}}{2}\right)^2 + \frac{1}{8}\left\{\left(\frac{\sqrt{3}}{2}\right)^2 + \left(-\frac{\sqrt{3}}{2}\right)^2\right\} = -\frac{3}{16}$$

(*iii*) Put $i = 1, j = 2$ in (2):

$$u_{1,2,1} = \frac{1}{2}\ u_{1,2,0} + \frac{1}{8}\ (u_{0,2,0} + u_{2,2,0} + u_{1,1,0})$$

$$= \frac{1}{2}\sin\frac{2\pi}{3}\sin\frac{4\pi}{3} + \frac{1}{8}\left\{0 + \left(\sin\frac{4\pi}{3}\right)^2 + 0 + \left(\sin\frac{2\pi}{3}\right)^2\right\}$$

$$= -\frac{3}{8} + \frac{1}{8}\left(\frac{3}{4} + \frac{3}{4}\right) = -\frac{3}{16}$$

(*iv*) Put $i = 2, j = 2$ in (2):

$$u_{2,2,1} = \frac{1}{2} u_{2,2,0} + \frac{1}{8} (u_{1,2,0} + u_{3,2,0} + u_{2,3,0} + u_{2,1,0})$$

$$= \frac{1}{2} \left(\sin \frac{4\pi}{3} \right)^2 + \frac{1}{8} \left(\sin \frac{2\pi}{3} \sin \frac{4\pi}{3} + 0 + 0 + \sin \frac{4\pi}{3} \sin \frac{2\pi}{3} \right)$$

$$= \frac{3}{8} + \frac{1}{8} \left(-\frac{3}{4} - \frac{3}{4} \right) = -\frac{3}{16}$$

Similarly the mesh values at the second and higher levels can be calculated.

Exercises 11.4

1. Find the solution of the parabolic equation $u_{xx} = 2u_t$ when $u(0, t) = u(4, t) = 0$ and $u(x, 0) = x(4 - x)$, taking $h = 1$. Find the values up to $t = 5$.

2. Solve the equation $\dfrac{\partial^2 u}{\partial x^2} = \dfrac{\partial u}{\partial t}$ with the conditions $u(0, t) = 0$, $u(x, 0) = x(1 - x)$, and $u(1, t) = 0$. Assume $h = 0.1$. Tabulate u for $t = k$, $2k$ and $3k$ choosing an appropriate value of k.

3. Given $\dfrac{\partial^2 f}{\partial x^2} - \dfrac{\partial f}{\partial f} = 0$; $f(0, t) = f(5, t) = 0$, $f(x, 0) = x^2(25 - x^2)$; find the values of f for $x = ih$ ($i = 0, 1, ..., 5$) and $t = jk$ ($j = 0, 1, ..., 6$) with $h = 1$ and $k = 1/2$, using the explicit method.

4. Given $\partial u/\partial t = \partial^2 u/\partial t^2$, $u(0, t) = 0$, $u(4, t) = 0$ and $u(x, 0) = x/3(16 - x^2)$. Obtain $u_{i,j}$ for $10 = 1, 2, 3, 4$ and $j = 1, 2$ using Crank-Nicholson's method.

5. Solve the heat equation $\dfrac{\partial u}{\partial t} = \dfrac{\partial^2 u}{\partial x^2}$ subject to the conditions $u(0, t) = u(1, t) = 0$ and

$$u(x,0) = \begin{cases} 2x \text{ for } 0 \le x \le 1/2 \\ 2(1 - x) \text{ for } 1/2 \le x \le 1 \end{cases}$$

Take $h = 1/4$ and k according to the Bandre-Schmidt equation.

6. Solve the two dimensional heat equation $\dfrac{\partial u}{\partial t} = \dfrac{\partial^2 u}{\partial x^2} + \dfrac{\partial^2 u}{\partial y^2}$ satisfying the initial condition: $u(x, y, 0) = \sin \pi x \sin \pi y$, $0 \le x, y \le 1$ and the boundary conditions: $u = 0$ at $x = 0$ and $x = 1$ for $t > 0$. Obtain the solution up to two time levels with $h = 1/3$ and $\alpha = 18$.

11.11   Hyperbolic Equations

The wave equation $\dfrac{\partial^2 u}{\partial t^2} = c^2 \dfrac{\partial^2 u}{\partial x^2}$ is the simplest Example of hyperbolic partial differential equations. Its solution is the displacement function $u(x, t)$ defined for values of x from 0 to 1 and for t from 0 to ∞, satisfying the initial and boundary conditions. The solution, as for parabolic equations, advances in an open-ended region (Figure 11.24). In the case of hyperbolic equations however, we have two initial conditions and two boundary conditions.

Such equations arise from convective type of problems in vibrations, wave mechanics, and gas dynamics.

11.12   Solution of Wave Equation

$$\frac{\partial^2 u}{\partial t^2} = c^2 \frac{\partial^2 u}{\partial x^2} \tag{1}$$

subject to the initial conditions: $u = f(x)$, $\partial u/\partial t = g(x)$, $0 \le x \le 1$ at $t = 0$ (2)

and the boundary conditions: $u(0, t) = \phi(t)$, $u(1, t) = \psi(t)$ (3)

Consider a rectangular mesh in the x-t plane spacing h along x direction and k along time direction. Denoting a mesh point $(x, t) = (ih, jk)$ as simply i, j, we have

$$\frac{\partial^2 u}{\partial x^2} = \frac{u_{i-1,j} - 2u_{i+1,j}}{h^2} \quad \text{and} \quad \frac{\partial^2 u}{\partial t^2} = \frac{u_{i,j-1} - 2u_{i,j} + u_{i,j+1}}{h^2}$$

Replacing the derivatives in (1) by their above approximations, we obtain

$$u_{i,j-1} - 2u_{i,j} + u_{i,j+1} = \frac{c^2 k^2}{h^2}\left(u_{i-1,j} - 2u_{i,j} + u_{i,j+1}\right)$$

or
$$u_{i,j+1} = 2(1 - \alpha^2 c^2)\,u_{i,j} + \alpha^2 c^2 (u_{i-1,j} + u_{i+1,j}) - u_{i,j-1} \tag{4}$$

where $\alpha = k/h$.

Now replacing the derivative in (2) by its central difference approximation, we get

$$\frac{ui, j+1 - ui, j-1}{2k} = \frac{\partial u}{\partial t} = g(x)$$

or
$$u_{i,j+1} = u_{i,j-1} + 2kg(x) \text{ at } t = 0$$

i.e.,
$$u_{i,1} = u_{i,-1} + 2kg(x) \text{ for } j = 0 \tag{5}$$

Also initial condition $u = f(x)$ at $t = 0$ becomes $ui, -1 = f(x)$ (6)

Combining (5) and (6), we have $u_{i,1} = f(x) + 2kg(x)$ (7)

Also (3) gives $u_{0,j} = \phi(t)$ and $u_{1,j} = \psi(t)$.

Hence the explicit form (4) gives the values of $u_{i,j+1}$ at the $(j + 1)$th level when the nodal values at $(j - 1)$th and jth levels are known from (6) and (7) as shown in Figure 11.31. Thus (4)*gives an* **implicit scheme** *for the solution of the wave equation.*

A special case. The coefficient of $u_{i,j}$ in (4) will vanish if $\alpha c = 1$ or $k = h/c$. Then (4) reduces to the simple form

$$u_{i,j+1} = u_{i-1,j} + u_{i+1,j} - u_{i,j-1}$$ (8)

NOTE   **Obs. 1.** *This provides an* **explicit scheme** *for the solution of the wave equation.*
For $\alpha = 1/c$, the solution of (4) is stable and coincides with the solution of (1).
For $\alpha < 1/c$, the solution is stable but inaccurate.
For $\alpha > 1/c$, the solution is unstable.

NOTE   **Obs. 2.** *The formula (4) converges for $\alpha \leq 1$ i.e., $k \leq h$.*

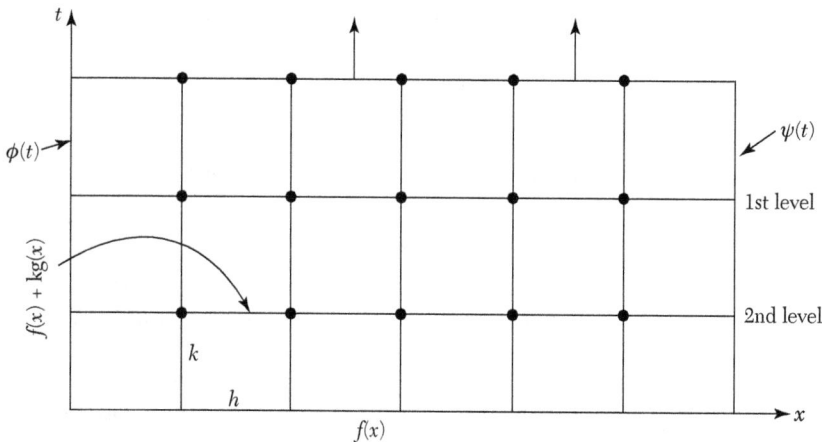

FIGURE 11.31

EXAMPLE 11.14

Evaluate the pivotal values of the equation $u_{tt} = 16u_{xx}$, taking $\Delta x \equiv 1up$ to $t = 1.25$. The boundary conditions are $u(0, t) = u(5, t) = 0$, $ut(x, 0) = 0$ and $u(x, 0) = x^2(5 - x)$.

Solution:

Here $c^2 = 16$.

$\therefore$ The difference equation for the given equation is

$$u_{i,j+1} = 2(1 - 16\alpha^2) u_{i,j} + 16\alpha^2 (u_{i-1,j} + u_{i+1,j}) - u_{i,j-1} \qquad (i)$$

where $\alpha = k/h$.

Taking $h = 1$ and choosing k so that the coefficient of $u_{i,j}$ vanishes, we have $16\alpha^2 = 1$, i.e.,$k = h/4 = 1/4$.

$\therefore$ (1) reduces to $u_{i,j+1} = u_{i-1,j} + u_{i+1,j} - u_{i,j-1} \qquad (ii)$

which gives a convergent solution (since $k/h < 1$). Its solution coincides with the solution of the given differential equation.

Now since $u(0, t) = u(5, t) = 0$, $\therefore$ $u_{0,j} = 0$ and $u_{5,j} = 0$ for all values of j

i.e., the entries in the first and last columns are zero.

Since $\qquad u_{(x, 0)} = x^2 (5 - x)$

$\therefore \qquad u_{i, 0} = i^2(5 - i) = 4, 12, 18, 16$ for $i = 1, 2, 3, 4$ at $t = 0$.

These are the entries for the first row.

Finally since $u_{t(x, 0)} = 0$ becomes

$\therefore \qquad \dfrac{u_{i,j+1} - u_{i,j-1}}{2k} = 0$, when $j = 0$, giving $u_{i, 1} = u_{i, -1} \qquad (iii)$

Thus the entries of the second row are the same as those of the first row.

Putting $j = 0$ in (ii),

$$u_{i, 1} = u_{i-1, 0} + u_{i+1}, 0 - u_{i, -1} = u_{i-1, 0} + u_{i+1, 0} - u_{i, 1}, \text{ using } (iii)$$

or $\qquad u_{i, 1} = 1/2 (u_{i-1, 0} + u_{i+1, 0}) \qquad (iv)$

Taking $i = 1, 2, 3, 4$ successively, we obtain

$$u_{1,1} = (u_{0, 0} + u_{2, 0}) = (0 + 12) = 6$$

$$u_{1,1} = \frac{1}{2}(u_{0,0} + u_{2,0}) = \frac{1}{2}(0 + 12) = 6$$

$$u_{2,1} = \frac{1}{2}(u_{1,0} + u_{3,0}) = \frac{1}{2}(4 + 18) = 11$$

$$u_{3,1} = \frac{1}{2}(u_{2,0} + u_{4,0}) = \frac{1}{2}(12 + 16) = 14$$

$$u_{4,1} = \frac{1}{2}(u_{3,0} + u_{5,0}) = \frac{1}{2}(18 + 0) = 9$$

These are the entries of the *second row*.

Putting $j = 1$ in (*ii*), we get $u_{i,2} = u_{i-1,1} + u_{i+1,1} - u_{i,0}$

Taking $i = 1, 2, 3, 4$ successively, we obtain

$$u_{1,2} = u_{0,1} + u_{2,1} - u_{1,0} = 0 + 11 - 4 = 7$$
$$u_{2,2} = u_{1,1} + u_{3,1} - u_{2,0} = 6 + 14 - 12 = 8$$
$$u_{3,2} = u_{2,1} + u_{4,1} - u_{3,0} = 11 + 9 - 18 = 2$$
$$u_{4,2} = u_{3,1} + u_{5,1} - u_{4,0} = 14 + 0 - 16 = -2$$

These are the entries of the *third row*.

Similarly putting $j = 2, 3, 4$ successively in (*ii*), the entries of the fourth, fifth, and six throws are obtained.

Hence the values of $u_{i,j}$ are as shown in the table below:

i / j	0	1	2	3	4	5
0	0	4	12	18	16	0
1	0	6	11	14	9	0
2	0	7	8	2	-2	0
3	0	2	-2	-8	-7	0
4	0	-9	-14	-11	-6	0
5	0	-16	-18	-12	-4	0

EXAMPLE 11.15

Solve $y_{tt} = y_{xx}$ up to $t = 0.5$ with a spacing of 0.1 subject to $y(0, t) = 0$, $y(1, t) = 0$, $yt\,(x, 0) = 0$ and $y(x, 0) = 10 + x(1 - x)$.

Solution:

As $c^2 = 1$, $h = 0.1$, $k = (h/c) = 0.1$; we use the formula

$$u_{i,j+1} = y_{i-1,j} + y_{i+1,j} - y_{i,j-1} \qquad (i)$$

Since $y(0, t) = 0$, $y(1, t) = 0$,

$\therefore \qquad y_{0,j} = 0$, $y_{1,j} = 0$ for all values of i.

i.e., all the entries in the first and last columns are zero.

Since $y(x, 0) = 10 + x\,(1 - x)$, $\therefore y_{i,0} = 10 + i\,(1 - i)$

$\therefore \qquad y_{0.1,0} = 10.09$, $y_{0.2,0} = 10.16$, $y_{0.3,0} = 10.21$, $y_{0.4,0} = 10.24$

$\therefore \qquad y_{0.5,0} = 10.25$, $y_{0.6,0} = 10.24$, $y_{0.7,0} = 10.21$, $y_{0.8,0} = 10.16$,

$\qquad \qquad y_{0.9,0} = 10.09$

These are the entries of the first row.

Since $y_t\,(x, 0) = 0$, we have $1/2(y_{i,j+1} - y_{i,j-1}) = 0 \qquad (ii)$

When $\qquad j = 0$, $y_{i,1} = y_{i,-1}$

Putting $\qquad j = 0$ in (i), $y_{i,1} = y_{i-1,0} + y_{i+1,0} - y_{i,-1}$

Using (ii) $\quad y_{i,1} = 1/2(y_{i-1,0} + y_{i+1,0})$

Taking $i = 1, 2, 3 \cdots, 9$ successively, we obtain the entries of the second row.

Putting $j = 1$ in (i), $y_{i,2} = y_{i-1,1} + y_{i+1,1} - y_{i,0}$

Taking $i = 1, 2, 3, \cdots, 9$ successively, we get the entries of the third row.

Similarly putting $j = 2, 3, \ldots, 7$ successively in (i), the entries of the fourth to ninth row are obtained. Hence the values of $u_{i,j}$ are as given in the table below:

j \ i	0	1	2	3	4	5	6	7	8	9	10
0	0	10.19	10.16	10.21	10.24	10.25	10.24	10.21	10.16	10.09	0
1	0	5.08	10.15	10.20	10.23	10.24	10.23	10.20	10.15	5.08	0
2	0	0.06	5.12	10.17	10.20	10.21	10.20	10.17	10.12	0.06	0
3	0	0.04	0.08	5.12	10.15	10.16	10.15	10.12	10.08	0.04	0
4	0	0.02	0.04	0.06	5.08	10.09	10.08	10.16	10.04	0.02	0
5	0	0	0	0	0	0	0	0	0	-0.02	0

EXAMPLE 11.16

The transverse displacement u of a point at a distance x from one end and at any time t of a vibrating string satisfies the equation $\partial^2 u/\partial t^2 = 4\partial^2 u/\partial x^2$, with boundary conditions $u = 0$ at $x = 0$, $t > 0$ and $u = 0$ at $x = 4$, $t > 0$ and initial conditions $u = x(4 - x)$ and $\partial u/\partial t = 0$, $0 \le x \le 4$. Solve this equation numerically for one-half period of vibration, taking $h = 1$ and $k = 1/2$.

Solution:

Here, $h/k = 2 = c$.

$\therefore$ The difference equation for the given equation is

$$u_{i,j+1} = u_{i-1,j} \, u_{i+1,j} - u_{i,j-1} \qquad (i)$$

which gives a convergent solution (since $k < h$).

Now since $u(0, t) = u(4, t) = 0$,

$\therefore \qquad u_{0,j} = 0$ and $u4, j = 0$ for all values of j.

i.e., the entries in the first and last columns are zero.

Since $\qquad u_{(x, 0)} = x(4 - x)$,

$\therefore \qquad u_{i, 0} = i(4 - i) = 3, 4, 3$ for $i = 1, 2, 3$ at $t = 0$.

These are the entries of the *first row*.

Also $u_t(x, 0) = 0$ becomes

$$\frac{u_{i,j+1} - u_{i,j-1}}{2k} = 0 \text{ when } j = 0, \text{ giving } u_{i, 1} = u_{i, -1} \qquad (ii)$$

Putting $j = 0$ in (i), $u_{i, 1} = u_{i-1, 0} + u_{i+1, 0} - u_{i, -1} = u_{i-1, 0} + u_{i+1, 0} - u_{i, 1}$, using (ii)

or $\qquad u_{i, 1} = 1/2 \, (u_{i-1, 0} + u_{i+1, 0}) \qquad (iii)$

Taking $i = 1, 2, 3$ successively, we obtain

$$u_{1, 1} = 1/2 \, (u_{0,0} + u_{1, 0}) = 2; \, u_{2, 1} = 1/2 \, (u_{1, 0} + u_{3, 0})$$
$$= 3, \, u_{3, 1} = 1/2 \, (u_{2, 0} + u_{4, 0}) = 2$$

These are the entries of the *second row*.

Putting $j = 1$ in (i), $u_{i, 2} = u_{i-1}, 1 + u_{i+1}, 1 - u_{i, 0}$

Taking $i = 1, 2, 3$, successively, we get

$$u_{1, 2} = u_{0, 1} + u_{2, 1} - u_{1, 0} = 0 + 3 - 3 = 0$$
$$u_{2, 2} = u_{1, 1} + u_{3, 1} - u_{2, 0} = 2 + 2 - 4 = 0$$
$$u_{3, 2} = u_{2, 1} + u_{4, 1} - u_{3, 0} = 3 + 0 - 3 = 0$$

These are the entries of the *third row* and so on.

Now the equation of the vibrating string of length l is $u_{tt} = c^2 u_{xx}$.

∴ Its period of vibration $\dfrac{2l}{c} = \dfrac{2 \times 4}{2} = 4\,\text{sec}$ $\qquad\qquad [\because l = 4 \text{ and } c = 2]$

This shows that we have to compute $u_{(x,\,t)}$ up to $t = 2$

i.e. Similarly we obtain the values of $u_{i,\,2}$ (fourth row) and $u_{i,\,3}$ (fifth row).

Hence the values of $u_{i,\,j}$ are as shown in the next table:

$\overset{i}{\underset{j}{\diagdown}}$	0	1	2	3	4
0	0	3	4	3	0
1	0	2	3	2	0
2	0	0	0	0	0
3	0	-2	-3	-2	0
4	0	-3	-4	-3	0

EXAMPLE 11.17

Find the solution of the initial boundary value problem:

$\dfrac{\partial^2 u}{\partial t^2} = \dfrac{\partial^2 u}{\partial x^2}$, $0 \le x \le 1$; subject to the initial conditions $u(x, 0) = \sin \pi x$,

$0 \le x \le 1$, $\left(\dfrac{\partial u}{\partial t}\right)(x, 0) = 0$, $0 \le x \le 1$ and the boundary conditions $u(0, t) = 0$, $u(1, t) = 0$, $t > 0$; by using in the (*a*) the explicit scheme (*b*) the implicit scheme.

Solution:

(*a*) Explicit scheme

Take $h = 0.2$, $k = h/c = 0.2$ $\qquad\qquad\qquad [\because c = 1]$

∴ We use the formula $u_{i,\,j+1} = u_{i-1,\,j} + u_{i+1,\,j} - u_{i,\,j-1}$ $\qquad\qquad$ (*i*)

Since $u(0, t) = 0$, $u(1, t) = 0$, $u_{0,\,j} = 0$, $u_{1,\,j} = 0$ for all values of j

i.e., the entries in the first and last columns are zero.

Since $u(x, 0) = \sin \pi x$, $u_{i,\,0} = \sin \pi x$

∴ $\qquad u_{1,\,0} = 0$, $u_{2,\,0} = \sin (.2\pi) = 0.5878$, $u_{3,\,0} = \sin (.4\pi) = 0.9511$,

$u_{4,0} = \sin(.6\pi) = 0.5878$.

These are the entries of the first row.

Since $u_t(x, 0) = 0$ we have $1/2(u_{i,j+1} - u_{i,j-1}) = 0$, when $j = 0$

i.e., $\hspace{4cm} u_{i,1} = u_{i,-1}$ $\hspace{4cm}$ (ii)

Putting $j = 0$ in (i), $u_{i,1} = u_{i-1,0} + u_{i+1,0} - u_{i,-1}$

Using (ii) $u_{i,1} = 1/2(u_{i-1,0} + u_{i+1,0})$

Taking $i = 1, 2, 3, 4$ successively, we obtain the entries of the second row.

Putting $\hspace{2cm} j = 1$ in (i), $u_{i,2} = u_{i-1,1} + u_{i+1,1} - u_{i,0}$

Now taking $i = 1, 2, 3, 4$ successively, we get the entries of the third row.

Similarly taking $j = 2, j = 3, j = 4$ successively, we obtain the entries of the fourth, fifth, and sixth rows, respectively.

Hence the values of $u_{i,j}$ are as given in the table below:

j \ i	0	1	2	3	4	5
0	0	0.5878	0.9511	0.9511	0.5878	0
1	0	0.4756	0.7695	0.9511	0.7695	0
2	0	0.1817	0.4756	0.5878	0.3633	0
3	0	0	0.0001	− 0.1122	− 0.1816	0
4	0	− 0.1816	− 0.5878	− 0.7694	0.4755	0
5	0	− 0.5878	− 0.9511	− 0.9511	− 0.5878	0

(b) Implicit scheme

We have the formula:

$u_{i,j+1} = 2(1 - \alpha^2 c^2) u_{i,j} + \alpha^2 c^2 (u_{i-1,j} + u_{i+1,j}) - u_{i,j-1}$, where $\alpha = k/h$. $\hspace{1cm}$ (i)

Here $c^2 = 1$, Take $h = 0.25$ and $k = 0.5$ so that $\alpha = k/h = 2$.

$\therefore$ (i) reduces to

$$u_{i,j+1} = -6u_{i,j} + 4(u_{i-1,j} + u_{i+1,j}) - u_{i,j-1} \hspace{2cm} (ii)$$

Since $\hspace{1cm} u(i,0) = \sin \pi x$

$\therefore \hspace{1cm} u_{(1,0)} = 0.7071, u_{(2,0)} = 0.5, u_{(3,0)} = 0.7071$

There are the entries of the first row.

Since $u_t(x, 0) = 0$, we have $1/2(y_{i, i+1} - y_{i, i-1}) = 0$, where $j = 0$

$\therefore$ $$y_{i, 1} = y_{i, -1} \qquad (ii)$$

Putting $j = 0$ and using (iii), (ii) reduces to

$$u_{i, 1} = -3 u_{i, 0} + 2 (u_{i-1, 0} + u_{i+1, 0})$$

Now taking $\quad i = 1, u_{1, 1} = -3 u_{1, 0} + 2 (u_{0, 0} + u_{2, 0}) = -0.1213$

$\qquad\qquad i = 2, u_{2, 1} = -3 u_{2, 0} + 2 (u_{1, 0} + u_{3, 0}) = -0.1716$

$\qquad\qquad i = 3, u_{3, 1} = -3 u_{3, 0} + 2 (u_{2, 0} + u_{4, 0}) = -0.1213$

These are the entries of the second row.

Putting $j = 1$, (ii) reduces to

$$u_{i, 2} = -6u_{i, 1} + 4 (u_{i-1, 1} + u_{i+1, 1})$$

Now taking $\quad i = 1, u_{1, 2} = -6u_{1, 1} + 4 (u_{0, 1} + u_{2, 1}) = 0.414$

$\qquad\qquad i = 2, u_{2, 2} = -6u_{2, 1} + 4 (u_{1, 1} + u_{3, 1}) = 0.0592$

$\qquad\qquad i = 3, u_{3, 2} = -6u_{3, 1} + 4 (u_{2, 1} + u_{4, 1}) = 0.0414$

These are the entries of the third row.

Putting $j = 2$, (ii) reduces to

$$u_{i, 3} = -6u_{i, 2+4} (u_{i-1, 2} + u_{i+1, 2}) - ui,$$

Now taking $\quad i = 1, u_{1, 3} = -6u_{1, 2+4} (u_{0, 2} + u_{2, 2}) - u_{1, 1} = 0.1097$

$\qquad\qquad i = 2, u_{2, 3} = -6u_{2, 2+4} (u_{1, 2} + u_{3, 2}) - u_{2, 1} = 0.1476$

$\qquad\qquad i = 3, u_{3, 3} = -6u_{3, 2+4}(u_{2, 2} + u_{4, 2}) - u_{3, 1} = 0.1097$

These are the entries of the fourth row.

Hence the values of $u_{i,j}$ are as tabulated below:

i \ j	0	1	2	3	4
0	0	0.7071	0.5	0.7071	0
1	0	−0.1213	−0.1716	−0.1213	
2	0	0.0414	0.0592	0.0414	0

EXERCISES 11.5

1. Solve the boundary value problem $u_{tt} = u_{xx}$ with the conditions $u(0, t) = u(1, t) = 0$, $u(x, 0) = 1/2 \; x(1 - x)$ and $ui(x, 0) = 0$, taking $h = k = 0.1$ for $0 \le t \le 0.4$. Compare your solution with the exact solution at $x = 0.5$ and $t = 0.3$.

2. The transverse displacement of a point at a distance x from one end and at any time t of a vibrating string satisfies the equation $\dfrac{\partial^2 u}{\partial t^2} = 25 \dfrac{\partial^2 u}{\partial x^2}$ with the boundary conditions $u(0,t) = u(5, t) = 0$ and the initial conditions

$$u(x, 0) = \begin{cases} 20x \text{ for } 0 \le x < 1 \\ 5(5 - x) \text{ for } 1 \le x < 5 \end{cases} \text{ and } u_t(x, 0) = 0. \text{ Solve this equation nu-}$$

merically for one-half period of vibration, taking $h = 1$, $k = 0.2$.

3. The function u satisfies the equation $\dfrac{\partial^2 u}{\partial t^2} = \dfrac{\partial^2 u}{\partial x^2}$ and the conditions: $u(x, 0) = 1/8 \sin \pi x, u_t(x, 0) = 0$ for $0 \le x \le 1$, $u(0, t) = u(1, t) = 0$ for $t \ge 0$.
Use the explicit scheme to calculate u for $x = 0(0.1) \; 1$ and $t = 0(0.1) \; 0.5$.

4. Solve $\dfrac{\partial^2 u}{\partial t^2} = \dfrac{\partial^2 u}{\partial x^2}$, $0 < x < 1$, $t > 0$, given $u(x, 0) = u_t (x, 0) = u(0, 1) = 0$ and $u (1, t) = 100 \sin \pi t$. Compute u for four times with $h = 0.25$.

EXERCISES 11.6

1. Which of the following equations is parabolic:
(a) $f_{xy} - f_x = 0$ (b) $f_{xx} + 2f_{xy} + f_{yy} = 0$ (c) $f_{xx} + 2f_{xy} + 4f_{yy} = 0.$

2. $u_{ij} = 1/4(u_{i+1,j} - u_{i-1,j} + u_{i,j+1} - u_{i,j-1})$ is Leibmann's five-point formula.
(True or False)

3. $u_{xx} + 3u_{xy} + u_{yy} = 0$ is classified as $\cdots$.

4. $\nabla^2 u = f(x, y)$ is known as $\cdots$.

5. The simplest formula to solve $u_{tt} = \alpha^2 u_{xx}$ is $\cdots..$.

6. The finite difference form of $\partial^2 u/\partial x^2$ is $\cdots..$.

7. Schmidt's finite difference scheme to solve $u_t = c^2 u_{xx}$ is $\cdots..$.

8. The five point diagonal formula gives $u_{ij} = ...$.

9. The partial differential equation $(x + 1)\, u_{xx} - 2(x + 2)\, u_{xy} + (x + 3)\, u_{yy} = 0$ is classified as····.

10. $u_{i,j+1} = 1/2(u_{i+1,j} + u_{i-1,j})$ is called ····· recurrence relation.

11. In terms of difference quotients $4u_{xx} = u_{tt}$ is ···.

12. The Bendre-Schmidt recurrence relation for one dimensional heat equation is ....

13. The diagonal five point formula to solve the Laplace equation $u_{xx} + u_{yy} = 0$ is ....

14. The Crank-Nicholson formula to solve $u_{xx} = au_{t}$ when $k = ah^2$, is ....

15. In the parabolic equation $u_t = \alpha^2\, u_{xx}$ if $\lambda = k\alpha^2/h^2$, where $k = \Delta t$, and $h = \Delta x$, then explicit method is stable if $\lambda = $ ....

16. The Bendre-Schmidt recurrence scheme is useful to solve ..... equation.

17. The two methods of solving one-dimensional diffusion (heat) equation are ....

18. $2\dfrac{\partial^2 u}{\partial x^2} + 4\dfrac{\partial^2 u}{\partial x \partial y} + 3\dfrac{\partial^2 u}{\partial y^2} = 0$ is classified as.... .

19. The order of error in solving Laplace and Poisson's equations by finite difference method is ···.

20. The difference scheme for solving the Poisson equation $\nabla^2 u = f(x, y)$ is....

21. The explicit formula for one-dimensional wave equation with $1 - \lambda^2\alpha^2 = 0$ and $\lambda = k/h$ is ···.

22. The general form of Poisson's equation in partial derivatives is ···.

23. If u satisfies Laplace equation and $u = 100$ on the boundary of a square, the value of u at an interior grid point is ···.

24. The Laplace equation $u_{xx} + u_{yy} = 0$ in difference quotients is ···.

25. The equation $yu_{xx} + u_{yy} = 0$ is hyperbolic in the region ···.

26. To solve $\dfrac{\partial u}{\partial u} = \dfrac{1}{2}\dfrac{\partial^2 u}{\partial x^2}$ by the Bendre-Schmidt method with $h = 1$, the value of k is ···.

27. Crank Nicholson's scheme is called an implicit scheme because ···.

CHAPTER 12

LINEAR PROGRAMMING

Chapter Objectives

- Introduction
- Formulation of the problem
- Graphical method
- Elliptic equations to partial derivatives
- General linear programming problem
- Canonical and standard forms of L.P.P.
- Simplex method
- Working procedure of the simplex method
- Artificial variable techniques—M method, Two-phase method
- Exceptional cases—Degeneracy
- Duality concept
- Duality theorem
- Dual simplex method
- Transportation problem
- Working procedure for transportation problems
- Degeneracy in transportation problems
- Assignment problem
- Objective type of questions

12.1 Introduction

We often face situations where decision making is a problem of planning activity. The problem generally, is of utilizing the scarce resources in an efficient manner so as to maximize the profit or to minimize the cost or to yield the maximum production. Such problems are called optimization problems. Linear programming in particular, deals with the optimization (maximization or minimization) of linear functions subject to linear constraints. This technique was propounded by George B. Dantzig in 1947 while working on a project for the U.S. Air Force. He also developed a powerful iterative process known as the "simplex method" for solving linear programming problems in 1951.

Linear programming is widely used to tackle a number of industrial, economic, marketing, and distribution problems. This technique has found its applications to important areas of product mix, blending problems, and diet problems. Oil refineries, chemical industries, steel industries, and food processing industry are also using linear programming with considerable success. In defense, this technique is being employed in inspection, optimal bombing patterns, design of weapons, etc. In fact, linear programming may be applied to any situation where a linear function of variables has to be optimized subject to a set of linear equations or inequalities.

In this chapter, our purpose is to present the principles of linear programming and the techniques of its application in a manner that will suit both engineers and scientists who are increasingly using this technique to solve their problems. Beginning with the graphical method which provides a great deal of insight into the basic concepts, the simplex method of solving linear programming problems is developed. Then the reader is introduced to the duality concept. Finally a special class of linear programming problems namely: transportation and assignment problems, is taken up.

12.2 Formulation of the Problem

To begin with, a problem is to be presented in a linear programming form which requires defining the variables involved, establishing relationships between them, and formulating the objective function and the constraints. We illustrate this through a few examples, wherein the stress will be on the analysis of the problem and formulation of the linear programming model.

EXAMPLE 12.1

A manufacturer produces two types of models M_1 and M_2. Each M_1 model requires 4 hours of grinding and 2 hours of polishing; whereas each M_2 model requires 2 hours of grinding and 5 hours of polishing. The manufacturer has 2 grinders and 3 polishers. Each grinder works for 40 hours a week and each polisher works for 60 hours a week. Profit on an M_1 model is $ 3 and on an M_2 model is $ 4. Whatever is produced in a week is sold in the market. How should the manufacturer allocate his production capacity to the two types of models so that he may make the maximum profit in a week

Solution:

Let x_1 be the number of M_1 models and x_2, the number of M_2 models produced per week. Then the weekly profit (in $) is

$$Z = 3x_1 + 4x_2 \qquad\qquad (i)$$

To produce these number of models, the total number of grinding hours needed per week

$$= 4x_1 + 2x_2$$

and the total number of polishing hours required per week

$$= 2x_1 + 5x_2$$

Since the number of grinding hours available is not more than 80 and the number of polishing hours is not more than 180, therefore

$$4x_1 + 2x_2 \le 80 \qquad\qquad (ii)$$
$$2x_1 + 5x_2 \le 180 \qquad\qquad (iii)$$

Also since the negative number of models are not produced, obviously we must have

$$x_1 \ge 0 \text{ and } x_2 \ge 0 \qquad\qquad (iv)$$

Hence this allocation problem is to find x_1, x_2 which

Maximize $Z = 3x_1 + 4x_2$

subject to $4x_1 + 2x_2 \le 80, 2x_1 + 5x_2 \le 180, x_1, x_2 \ge 0.$

NOTE

Obs. *The variables that enter into the problem are called* **decision variables.**

The expression (i) showing the relationship between the manufacturer's goal and the decision variables, is called the **objective function.**

The inequalities (ii), (iii), and (iv) are called the **constraints.**

The objective function and the constraints being all linear, it is a *linear programming problem(L.P.P.).* This is an example of a real situation from industry.

EXAMPLE 12.2

Consider the following problem faced by a production planner in a soft-drink plant. He has two bottling machines A and B. A is designed for 8-ounce bottles and B for 16 ounce bottles. However, each can be used on both types with some loss of efficiency. The following is available:

Machine	8-ounce bottles	16-ounce bottles
A	100/minute	40/minute
B	60/minute	75/minute

The machines can be run 8 hours per day, 5 days per week. Profit in a 8-ounce bottle is 15 paise and on a 16-ounce bottle is 25 paise. Weekly production of drink cannot exceed 300,000 ounces and the market can absorb 25,000 8-ounce bottles and 7,000 16-ounce bottles per week. The planner wishes to maximize his profit subject, of course, to all the production and marketing restrictions. Formulate this as a linear programming problem.

Solution:

Let x_1 units of 8-ounce bottle and x_2 units of 16-ounce bottle be produced per week. Than the weekly profit (in $) of the production planner is

$$Z = 0.15x1 + 0.25x_2 \qquad (i)$$

Since an 8-ounce bottle takes 1/100 minutes and a 16-ounce bottle 1/40 minutes on machine A and the machine can run 8 hours per day, 5 days per week, *i.e.,* 2400 minutes per week, therefore we have

$$\frac{1}{100}x_1 + \frac{1}{40}x_2 \leq 2400 \qquad (ii)$$

Also since an 8-ounce bottle takes 1/60 minutes and a 16-ounce bottle takes 1/75 minutes on machine B which can run for 2400 minutes per week, therefore we have

$$\frac{1}{60}x_1 + \frac{1}{75}x_2 \leq 2400 \qquad (iii)$$

As the total weekly production cannot exceed 300,000 ounces, therefore,

$$8x_1 + 16x_2 \le 300,000 \qquad\qquad (iv)$$

As the market can absorb at the most 25,000, 8-ounce bottles and 7,000, 16-ounce bottles per week, therefore,

$$0 \le x_1 \le 25,000 \text{ and } 0 \le x_2 \le 7,000 \qquad\qquad (v)$$

Hence this allocation problem of the production planner is to find x_1, x_2 which

Maximize $Z = 0.15x_1 + 0.25x_2$

subject to $2x_1 + 5x_2 \le 480,000,\ 5x_1 + 4x_2 \le 720,000,\ x_1 + 2x_2 \le 37,500$

$0 \le x1 \le 25,000 \text{ and } 0 \le x2 \le 7,000.$

EXAMPLE 12.3

A firm making castings uses electric furnace to melt iron with the following specifications:

	Minimum	Maximum
Carbon	3.20%	3.40%
Silicon	2.25%	2.35%

Specifications and costs of various raw materials used for this purpose are given below:

Material	Carbon%	Silicon%	Cost ($)
Steel scrap	0.4	0.15	850/metric ton
Cast iron scrap	3.80	2.40	900/metric ton
Remelt from foundary	3.50	2.30	500/metric ton

If the total charge of iron metal required is 4 metric tons, find the weight in kg of each raw material that must be used in the optimal mix at minimum cost.

Solution:

Let x_1, x_2, x_3 be the amounts (in kg) of these raw materials. The objective is to minimize the cost *i.e.*,

Minimize $Z = \dfrac{850}{1000}x_1 + \dfrac{900}{1000}x_2 + \dfrac{500}{1000}x_3$ \qquad (i)

For iron melt to have a minimum of 3.2% carbon,

$$0.4\,x_1 + 3.8\,x_2 + 3.5\,x_3 \ge 3.2 \times 4,000 \qquad\qquad (ii)$$

For iron melt to have a maximum of 3.4% carbon,

$$0.4\,x_1 + 3.8\,x_2 + 3.5\,x_3 \le 3.4 \times 4{,}000 \qquad (iii)$$

For iron melt to have a minimum of 2.25% silicon,

$$0.15\,x_1 + 2.41\,x_2 + 2.35\,x_3 \ge 2.25 \times 4{,}000 \qquad (iv)$$

For iron melt to have a maximum of 2.35% silicon,

$$0.15\,x_1 + 2.41\,x_2 + 2.35\,x_3 \le 2.35 \times 4{,}000 \qquad (v)$$

Also, since the materials added up must be equal to the full charge weight of 4 metric tons,

$$\therefore \qquad x_1 + x_2 + x_3 = 4{,}000 \qquad (vi)$$

Finally since the amounts of raw material cannot be negative

$$x_1 \ge 0,\, x_2 \ge 0,\, x_3 \ge 0 \qquad (vii)$$

Thus the linear programming problem is to find x_1, x_2, x_3 which

Minimize $Z = 0.85\,x_1 + 0.9\,x_2 + 0.5\,x_3$

subject to $0.4\,x_1 + 3.8\,x_2 + 3.5\,x_3 \ge 12{,}800,\ 0.4\,x_1 + 3.8\,x_2 + 3.5\,x_3 \le 13{,}600$

$0.15\,x_1 + 2.41\,x_2 + 2.35\,x_3 \ge 9{,}000,\ 0.15\,x_1 + 2.41\,x_2 + 2.35\,x_3 \le 9{,}400$

$x_1 + x_2 + x_3 = 4{,}000,\ x_1,\, x_2,\, x_3 \ge 0.$

Exercises 12.11

1. A firm manufactures two items. It purchases castings which are then machined, bored, and polished. Castings for items A and B cost \$ 3 and \$ 4 each and are sold at \$ 6 and \$ 7 each, respectively. Running costs of these machines are \$ 20, \$ 14, and \$17.50 per hour, respectively. Formulate the problem so that the product mix maximizes the profit. Capacities of the machines are

	Part A	Part B
Machining capacity	25 per hr.	40 per hr.
Boring capacity	28 per hr.	35 per hr.
Polishing capacity	35 per hr.	25 per hr.

2. A firm manufactures 3 products A, B, and C. The profits are \$ 3, \$ 2, and \$ 4, respectively. The firm has two machines M_1 and M_2 and below

is the required capacity processing time in minutes for each machine on each product.

Machine	Product		
	A	B	C
M_1	4	3	5
M_2	2	2	4

Machines M_1 and M_2 have 2000 and 2500 machine-minutes respectively. The firm must manufacture 100 A's, 200 B's and 50 C's but not more than 150 A's. Set up an *L.P.P.* to maximize profit.

3. Three products are processed through three different operations. The time (in minutes) required per unit of each product, the daily capacity of the operations (in minutes per day), and the profit per unit sold for each product (in Dollars) are as follows:

Operation	Time per unit			Operation capacity
	Product I	Product II	Product III	
1	3	4	3	42
2	5	0	3	45
3	3	6	2	41
Profit ($)	3	2	1	

The zero time indicates that the product does not require the given operation. The problem is to determine the optimum daily production for three products that maximize the profit. Formulate this production planning problem as a linear programming problem assuming that all units produced are sold.

4. An aeroplane can carry a maximum of 200 passengers. A profit of $ 400 is made on each first class ticket and a profit of $ 300 is made on each economy class ticket. The airline reserves at least twenty seats for first class. However, at least four times as many passengers prefer to travel by economy class than by the first class. How many tickets of each class must be sold in order to maximize profit for the airline? Formulate the problem as an L.P. model.

5. A firm manufactures headache pills in two sizes A and B. Size A contains 2 grains of asprin, 5 grains of bicarbonate, and 1 grain of codeine. Size B contains 1 grain of asprin, 8grains of bicarbonate and 6 grains of codeine. It is found by users that it requires at least 12 grains of asprin,

74 grains of bicarbonate, and 24 grains of codeine for providing immediate effect. It is required to determine the least number of pills a patient should take to get immediate relief. Formulate the problem as a standard L.P.P.

6. A dairy feed company may purchase and mix one or more of three types of grains containing different amounts of nutritional elements. The data is given in the table below. The production manager specifies that any feed mix for his live stock must meet at least minimum nutritional requirements and seeks the least costly among all three mixes.

Item	One unit weight of			Minimum requirement
	Grain 1	Grain 2	Grain 3	
Nutritional Ingredients	A	2	3	7	1,250
	B	1	1	0	250
	C	5	3	0	900
	D	6	25	1	232.5
Cost per weight of		41	35	96	

Formulate the problem as a L.P. model.

7. A firm produces an alloy with the following specifications:
   (*i*) specific gravity ≤ 0.97
  (*ii*) chromium content $\geq 15\%$
 (*iii*) melting temperature $\geq 494°C$

The alloy requires three raw materials A, B, and C whose properties are as follows:

Property	Properties of raw material		
	A	B	C
Sp. gravity	0.94	1.00	1.05
Chromium	10%	15%	17%
Melting pt.	470°C	500°C	520°C

Find the values of A, B, C to be used to make 1 meric ton of alloy of desired properties, keeping the raw material costs at the minimum when they are $ 105/metric ton for A, $ 245/metric ton for B and $ 165/ metric ton for C. Formulate an *L.P.* model for the problem.

8. The owner of Metro sports wishes to determine how many advertisements to place in the selected three monthly magazines A, B, and C. His objective is to advertise in such a way that total exposure to principal buyers of expensive sports goods is maximized. Percentages of readers for magazine are known. Exposure in any particular magazine is the number of advertisements placed multiplied by the number of principal buyers. The following data may be used:

	Magazine		
	A	B	C
Readers	1 lakh	0.6 lakh	0.4 lakh
Principal buyers	10%	15%	7%
Cost per advertisement ($)	5000	4500	4250

The budgeted amount is at most $100,000 for advertisements. The owner has already decided that magazine A should have no more than six advertisements and that B and C each have at least two advertisements. Formulate an $L.P.$ model for the problem.

12.3 Graphical Method

Linear programming problems involving only two variables can be effectively solved by a graphical technique. In actual practice, we rarely come across such problems. Even then, the graphical method provides a pictorial representation of the solution and one gets ample insight into the basic concepts used in solving large $L.P.P.$

Working procedure *to solve a linear programming problem graphically*:

Step 1. Formulate the given problem as a linear programming problem.

Step 2. Plot the given constraints as equalities on $x_1 x_2$-coordinate plane and determine the convex region* formed by them.

Step 3. Determine the vertices of the convex region and find the value of the objective function at each vertex. The vertex which gives the optimal

*A **region** or a **set** of points is said to be convex if the line joining any two of its points lies completely in the region (or the set). Figures 12.1 and 12.2 represent convex regions while Figures 12.3 and 12.4 do not form convex sets.

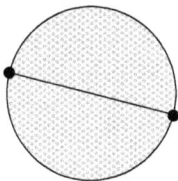 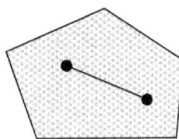 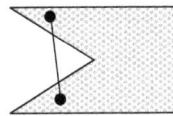 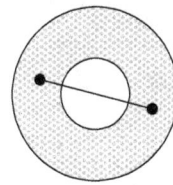

FIGURE 12.1   *FIGURE 12.2*   *FIGURE 12.3*   *FIGURE 12.4*

(maximum or minimum) value of the objective function gives the desired optimal solution to the problem.

Otherwise. Draw the dotted line through the origin representing the objective function with $Z = 0$. As Z is increased from zero, this line moves to the right remaining parallel to itself. We go on sliding this line (parallel to itself), till it is *farthest* away from the origin and passes through only one vertex of the convex region. This is the vertex where maximum value of Z is attained.

When it is required to minimize Z, the value of Z is increased until the dotted line passes through the *nearest* vertex of the convex region.

EXAMPLE 12.4

Solve the L.P.P. of Example 12.1 graphically.

Solution:

The problem is:

Maximize $\qquad\qquad Z = 3x_1 + 4_{x2}$ $\qquad\qquad\qquad$ (i)

subject to $\qquad\qquad 4x_1 + 2x_2 \le 80$ $\qquad\qquad\qquad$ (ii)

$\qquad\qquad\qquad\qquad 2x_1 + 5x_2 \le 180$ $\qquad\qquad\qquad$ (iii)

$\qquad\qquad\qquad\qquad x_1, x_2 \ge 0$ $\qquad\qquad\qquad\qquad$ (iv)

Consider the x_1x_2-coordinate system as shown in Figure 12.5. The non-negativity restrictions (iv) imply that the values of x_1, x_2 lie in the first quadrant only.

We plot the lines $4x_1 + 2x_2 = 80$ and $2x_1 + 5x_2 = 180$.

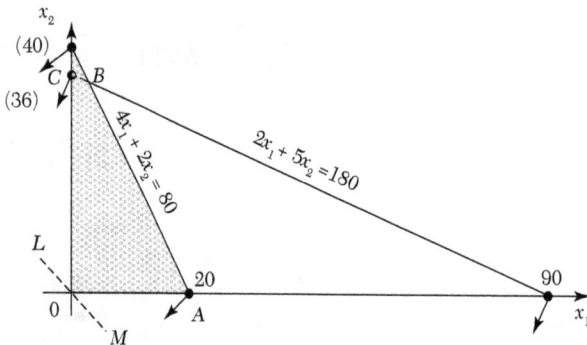

FIGURE 12.5

Then any point on or below $4x_1 + 2x_2 = 80$ satisfies (ii) and any point on or below $2x_1 + 5x_2 = 180$ satisfies (iii). This shows that the desired point $(x_1,$

x_2) must be somewhere in the shaded convex region $OABC$. This region is called the *solution space* or *region of feasible solutions* for the given problem. Its vertices are $O(0,0)$, $A(20, 0)$, $B(2.5, 35)$, and $C(0, 36)$.

The values of the objective function (i) at these points are

$$Z(O) = 0,\ Z(A) = 60,\ Z(B) = 147.5,\ Z(C) = 144.$$

Thus the maximum value of Z is 147.5 and it occurs at B. Hence the optimal solution to the problem is

$$x_1 = 2.5,\ x_2 = 35 \text{ and } Z_{max} = 147.5.$$

Otherwise. Our aim is to find the point in the solution space which maximizes the profit function Z. To do this, we observe that on making $Z = 0$, (i) becomes $3x_1 + 4x_2 = 0$ which is represented by the dotted line LM through O. As the value of Z is increased, the line LM starts moving parallel to itself towards the right. larger the value of Z, more will be the company's profit. In this way, we go on sliding LM until it is farthest away from the origin and passes through one of the corners of the convex region. This is the point where the maximum value of Z is attained. Just possibly, such a line may be one of the edges of the solution space. In that case every point on that edge gives the same maximum value of Z.

Here Z_{max} is attained at $B(2.5, 35)$. Hence the optimal solution is $x_1 = 2.5$, $x_2 = 35$ and $Z_{max} = 147.5$.

EXAMPLE 12.5

Find the maximum value of $Z = 2x + 3y$

Subject to the constraints: $x + y \leq 30$, $y \geq 3$, $0 \leq y \leq 12$, $x - y \geq 0$, and $0 \leq x \leq 20$.

Solution:

Any point (x, y) satisfying the conditions $x \geq 0$, $y \geq 0$ lies in the first quadrant only. Also since,

$x + y \leq 30$, $y \geq 3$, $y \leq 12$, $x \geq y$ and $x \leq 20$, the desired point (x, y) lies within the convex region $ABCDE$ (shown shaded in Figure 12.6). Its vertices are $A(3, 3)$, $B(20, 3)$, $C(20, 10)$, $D(18, 12)$ and $E(12, 12)$.

The values of Z at these five vertices are $Z(A) = 15$, $Z(B) = 49$, $Z(C) = 70$, $Z(D) = 72$, and $Z(E) = 60$.

Since the maximum value of Z is 72 which occurs at the vertex D, the solution to the L.P.P. is

$x = 18$, $y = 12$ and maximum $Z = 72$.

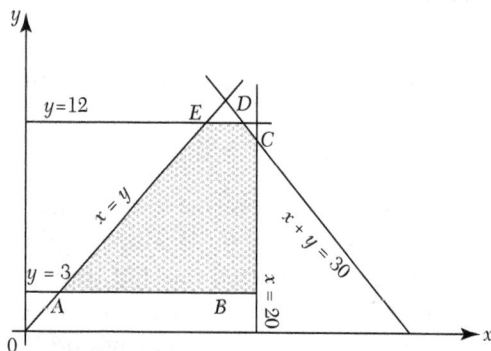

FIGURE 12.6

EXAMPLE 12.6

A company manufactures two types of cloth, using three different colours of wool. One yard length of type A cloth requires 4 oz of red wool, 5 oz of green wool and 3 oz of yellow wool. One yard length of type B cloth requires 5 oz of red wool, 2 oz of green wool and 8 oz of yellow wool. The wool available for manufacture is 1000 oz of red wool, 1000 oz of green wool and 1200 oz of yellow wool. The manufacturer can make a profit of \$ 5 on one yard of type A cloth and \$ 3 on one yard of type B cloth. Find the best combination of the quantities of type A and type B cloth which gives him maximum profit by solving the L.P.P. graphically.

Solution:

Let the manufacturer decide to produce x_1 yards of type A cloth and x_2 yards of type B cloth. Then the total income in dollars, from these units of cloth is given by

$$Z = 5x_1 + 3x_2 \qquad (i)$$

To produce these units of two types of cloth, he requires

red wool $= 4x_1 + 5x_2$ oz, green wool $= 5x_1 + 2x_2$ oz,

and   yellow wool $= 3x_1 + 8x_2$ oz.

Since the manufacturer does not have more than 1000 oz of red wool, 1000 oz of green wool and 1200 oz of yellow wool, therefore

$$4x_1 + 5x_2 \le 1000 \qquad (ii)$$
$$5x_1 + 2x_2 \le 1000 \qquad (iii)$$
$$3x_1 + 8x_2 \le 1200 \qquad (iv)$$

Also $\qquad x_1 \ge 0, x_2 \ge 0 \qquad (v)$

Thus the given problem is to maximize Z subject to the constraints (ii) to (v).

Any point satisfying the condition (v) lies in the first quadrant only. Also the desired point satisfying the constraints (ii) to (iv) lies in the convex region $OABCD$ (Figure 12.7). Its vertices are $O(0, 0)$, $A(200, 0)$, $B(3000/17, 1000/17)$, $C(2000/17, 1800/17)$, and $D(0,150)$.

The values of Z at these vertices are given by $Z(O) = 0$, $Z(A) = 1000$, $Z(B) = 1057.6$, $Z(C) = 905.8$ and $Z(D) = 450$.

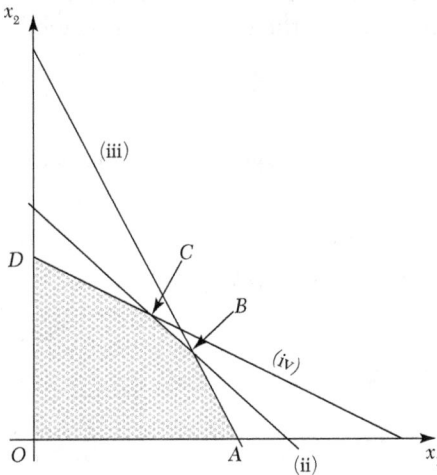

FIGURE 12.7

Since the maximum value of Z is 1058.8 which occurs at the vertex B, the solution to the given problem is

$$x_1 = 3000/17, x_2 = 1000/17 \text{ and max.}$$
$$Z = 1058.8.$$

Hence the manufacturer should produce 176.5 yards of type A cloth 58.8 yards of type B cloth, so as to get the maximum profit of \$ 1058.8.

EXAMPLE 12.7

A company making cold drinks has two bottling plants located at towns T_1 and T_2. Each plant produces three drinks A, B, and C and their production capacity per day is shown below:

Cold drinks	Plant at	
	T_1	T_2
A	6,000	2,000
B	1,000	2,500
C	3,000	3,000

The marketing department of the company forecasts a demand of 80,000 bottles of A, 22,000 bottles of B and 40,000 bottles of C during the month of June. The operating costs per day of plants at T_1 and T_2 are \$ 6,000 and \$ 4,000 respectively. Find (graphically) the number of days for which each plant must be run in June so as to minimize the operating costs while meeting the market demand.

Solution:

Let the plants at T_1 and T_2 be run for x_1 and x_2 days. Then the objective is to minimize the operation costs, *i.e.*,

$$\text{min. } Z = 6000\, x_1 + 4000 x_2 \qquad (i)$$

Constraints on the demand for the three cold drinks are:

for A, $6,000\, x_1 + 2,000 x_2 \geq 80,000$ or $3\, x_1 + x_2 \geq 40$ $\qquad (ii)$

for B, $1,000\, x_1 + 2,500 x_2 \geq 22,000$ or $x_1 + 2.5 x_2 \geq 22$ $\qquad (iii)$

for C, $3,000\, x_1 + 3,000 x_2 \geq 40,000$ or $x_1 + x_2 \geq 40/3$ $\qquad (iv)$

Also $\qquad\qquad\qquad x_1, x_2 \geq 0.$ $\qquad\qquad\qquad\qquad (v)$

Thus *the L.P.P. is to minimize (i) subject to constraints (ii) to (v).*

The solution space satisfying the constraints (ii) to (v) is shown shaded in Figure 12.8. As seen from the direction of the arrows, the solution space is unbounded. The constraint (iv) is dominated by the constraints (ii) and (iii) and hence does not affect the solution space. Such a constraint as (iv) is called the *redundant constraint*.

The vertices of the convex region ABC are A(22, 0), B(12, 4), and C(0, 40).

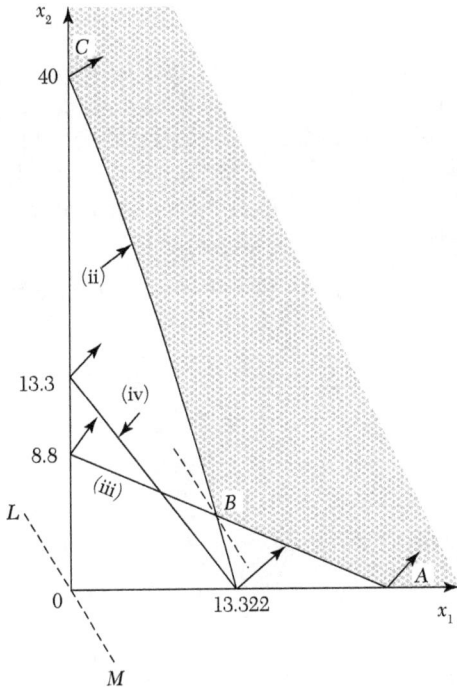

FIGURE 12.8

Values of the objective function (i) at these vertices are

$$Z(A) = 132{,}000, \ Z(B) = 88{,}000, \ Z(C) = 160{,}000.$$

Thus the minimum value of Z is \$ 88,000 and it occurs at B. Hence the solution to the problem is

$$x_1 = 12 \text{ days}, \ x_2 = 4 \text{ days}, \ Z_{min} = \$ \ 88{,}000.$$

Otherwise. Making $Z = 0$, (i) becomes $3\,x_1 + 2x_2 = 0$ which is represented by the dotted line LM through O. As Z is increased, the line LM moves parallel to itself, to the right. Since we are interested in finding the minimum value of Z, value of Z is increased until LM passes through the vertex nearest to the origin of the shaded region, *i.e.*, $B(12, 4)$.

Thus the operating cost will be minimum for $x_1 = 12$ days, $x_2 = 4$ days, and $Z_{min} = 6000 \times 12 + 4000 \times 4 = \$ \ 88{,}000.$

NOTE **Obs.** *The dotted line parallel to the line LM is called the iso-cost line since it represents all possible combinations of x_1, x_2 which produce the same total cost.*

12.4   Some Exceptional Cases

The constraints generally, give a region of feasible solution which may be bounded or unbounded. In problems involving two variables and having a finite solution, it was observed that the optimal solution existed at a vertex of the feasible region. In fact, this is true for all *L.P.* problems for which solutions exist. Thus it may be stated that *if there exists an optimal solution of an L.P.P., it will be at one of the vertices of the solution space.*

In each of the above examples, the optimal solution was unique. But it is not always so. In fact, *L.P.P. may have*

> (*i*) *a unique optimal solution,*

or (*ii*) *an infinite number of optimal solutions,*

or (*iii*) *an unbounded solution,*

or (*iv*) *no solution.*

Below are a few examples to illustrate the *exceptional cases (ii) to (iv).*

EXAMPLE 12.8

A firm uses milling machines, grinding machines, and lathes to produce two motor parts. The machining times required for each part, the machining times available on different machines and the profit on each motor part are given below:

Type of machine	Machining time reqd. for the motor part (mts)		Max. time available per week (minutes)
	I	*II*	
Milling machines	10	4	2,000
Grinding machines	3	2	900
Lathes	6	12	3,000
Profit/unit ($)	100	40	

Determine the number of parts I and II to be manufactured per week to maximize the profit.

Solution:

Let x_1, x_2 be the number of parts I and II manufactured per week. Then *objective* being to maximize the profit, we have

$$\text{maximize } Z = 100x_1 + 40x_2 \qquad (i)$$

Constraints being on the time available on each machine, we obtain

for milling machines,	$10x_1 + 4x_2 \leq 2,000$	(ii)
for grinding machines,	$3x_1 + 2x_2 \leq 900$	(iii)
for lathes,	$6x_1 + 12x_2 \leq 3,000$	(iv)
Also	$x_1, x_2 \geq 0$	(v)

Thus the problem is to determine x_1, x_2 which maximize (i) subject to the constraints (ii) to (v).

The solution space satisfying (ii), (iii), (iv) and meeting the non-negativity restrictions (v) is shown shaded in Figure 12.9.

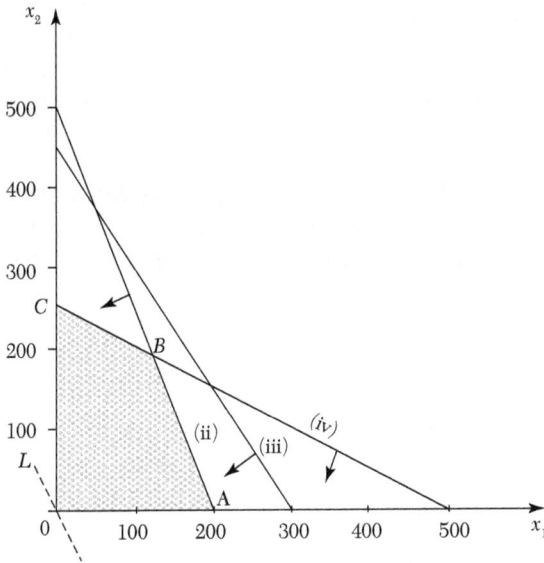

FIGURE 12.9

Note that (iii) is a redundant constraint as it does not affect the solution space. The vertices of the convex region $OABC$ are

$$O(0, 0), A(200, 0), B(125, 187.5), C(0, 250).$$

Values of the objective function (i) at these vertices are

$$Z(O) = 0, Z(A) = 20,000, Z(B) = 20,000 \text{ and } Z(C) = 10,000$$

Thus the maximum value of Z occurs at two vertices A and B.

∴ Any point on the line joining A and B will also give the same maximum value of Z *i.e.*, there are an infinite number of feasible solutions which yield the same maximum value of Z.

Thus there is no unique optimal solution to the problem and any point on the line AB can be taken to give the profit of $ 20,000.

Obs. *An L.P.P. having more than one optimal solution, is said to have alternative or multiple optimal solutions. It implies that the resources can be combined in more than one way to maximize the profit.*

EXAMPLE 12.9

Using graphical method, solve the following L.P.P:

Maximize	$Z = 2x_1 + 3x_2$	(i)
subject to	$x_1 - x_2 \leq 2$	(ii)
	$x_1 + x_2 \geq 4$	(iii)
	$x_1, x_2 \geq 0$	(iv)

Solution:

Consider $x_1 x_2$ coordinate system. Any point (x_1, x_2) satisfying the restrictions (iv) lies in the first quadrant only. The solution space satisfying the constraints (ii) and (iii) is the convex region shown shaded in Figure 12.10.

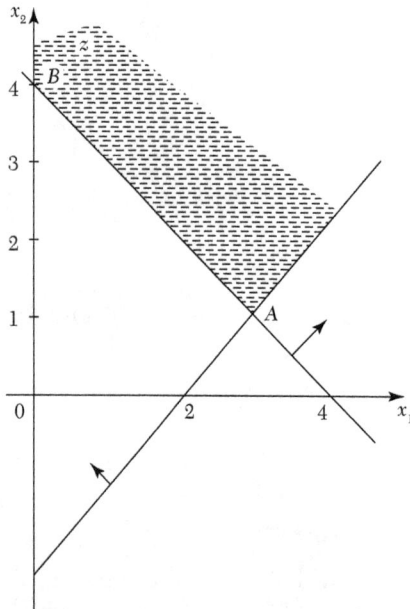

FIGURE 12.10

Here the solution space is unbounded. The vertices of the feasible region (in the finite plane) are $A(3, 1)$ and $B(0, 4)$.

Values of the objective function (i) at these vertices are $Z(A) = 9$ and $Z(B) = 12$.

But there are points in this convex region for which Z will have much higher values. For instance, the point $(5, 5)$ lies in the shaded region and the value of Z thereafter is 12.5. In fact, the maximum value of Z occurs at infinity. Thus the problem has an unbounded solution.

EXAMPLE 12.10

Solve graphically the following L.P.P:

Maximize	$Z = 4x_1 + 3x_2$	(i)
subject to	$x_1 - x_2 \leq -1,$	(ii)
$-x_1 + x_2 \leq 0,$		(iii)
And	$x_1, x_2 \geq 0.$	(iv)

Solution:

Consider $x_1 x_2$-coordinate system. Any point (x_1, x_2) satisfying (iv) lies in the first quadrant only. The two solution spaces, one satisfying (ii) and the other satisfying (iii) are shown in Figure 12.11.

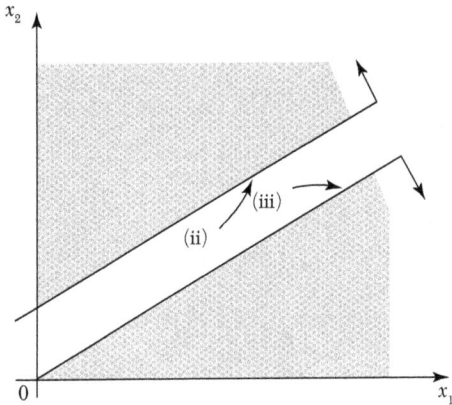

FIGURE12.11

There being no point (x_1, x_2) common to both the shaded regions, the problem cannot be solved. Hence the solution does not exist since the constraints are inconsistent.

NOTE **Obs.** *The above problem had no solution because the constraints were incompatible. There may be cases in which the constraints are compatible but the problem may still have no feasible solution.*

This is an example of insoluble programming problems. At times, management sets such goals which are unattainable within the available resources for a number of reasons. Such exceptional management problems are solved with the help of "Goal Programming Technique" which has recently been developed.

Exercises 12.2

Using the graphical method, solve the following L.P. problems:

1. Max. $Z = 5x_1 + 3x_2$
subject to $3x_1 + 5x_2 \leq 15$
$5x_1 + 2x_2 \leq 10$
$x_1, x_2 \geq 0$

2. Max. $Z = 5x_1 + 7x_2$
subject to $x_1 + x_2 \leq 4$,
$5x_1 + 8x_2 \leq 24$,
$10x_1 + 7x_2 \leq 35$ and $x_1, x_2 \geq 0$.

3. Min. $Z = 20x_1 + 10x_2$
subject to $x_1 + 2x_2 \leq 40$
$3x_1 + x_2 \geq 30$
$4x_1 + 3x_2 \geq 60$ and $x_1, x_2 \geq 0$

4. Max. $Z = 120x_1 + 100x_2$
subject to $10x_1 + 5x_2 \leq 80$
$6x_1 + 6x_2 \leq 66$
$4x_1 + 8x_2 \geq 24$
$5x_1 + 6x_2 \leq 90$ and $x_1, x_2 \geq 0$.

5. If x_1, x_2 are real, show that the set $S = \left\{ (x_1, x_2) \middle| \begin{array}{l} x_1 + x_2 \leq 50 \\ x_1 + 2x_2 \leq 80 \\ 2x_1 + x_2 \geq 20 \\ x_1, x_2 \geq 0 \end{array} \right\}$ is a convex

set. Find the extreme points of this set. Hence solve L.P.P. (graphically):
Maximize $Z = 4x_1 + 3x_2$ subject to constraints given in S.

6. A firm manufactures two products A and B on which the profits earned per unit are $ 3 and $ 4, respectively. Each product is processed on two machines M_1 and M_2. Product A requires one minute of processing time on M_1 and 2 minutes on M_2 while B requires one minute on M_1 and one minute on M_2. Machine M_1 is available for not more than 7 hours and 30 minutes while M_2 is available for 10 hours during any working day. Find the number of units of products A and B to be manufactured to get maximum profit.

7. Two spare parts X and Y are to be produced in a batch. Each one has to go through two processes A and B. The time required in hours per unit and total time available are given below:

	X	Y	Total hours available
Process A	3	4	24
Process B	9	4	36

Profit per unit of X and Y are $ 5 and $ 6 respectively. Find how many number of spare parts of X and Y are to be produced in this batch to maximize the profit. (Each batch is complete in all respects and one cannot produce fractional units and stop the batch).

8. A manufacturer has two products I and II both of which are made in steps by machines A and B. The process times per hundred for the two products on the two machines are:

Product	M/c. A	M/c. B
I	4 hrs.	5 hrs.
II	5 hrs.	2 hrs.

Set-up times are negligible. For the coming period machine A has 100 hrs. and B has 80 hrs. The contribution for product I is $ 10 per 100 units and for product II is $ 5 per 100 units. The manufacturer is in a market which can absorb both products as much as he can produce for the immediate period ahead. Determine graphically, how much of products I and II, he should produce to maximize his contribution.

9. Two grades of paper M and N are produced on a paper machine. Because of raw material restrictions not more than 400 metric tons of grade M and 300 metric tons of grade N can be produced in a week. It requires 0.2 and 0.4 hours to produce a metric ton of products M and N respectively, with corresponding profits of $ 20 and $ 50 per metric ton.

It is given that there are 160 hours in a week. Formulate the problem as an L.P.P. and determine the optimum product mix.

10. A production manager wants to determine the quantity to be produced per month of products A and B manufactured by his firm. The data on resources required and availability of resources are given below:

Resources	Requirements		Available per month
	Product A	Product B	
Raw material (kg)	60	120	12,000
Machine hrs/piece	8	5	600
Assembly man hrs.	3	4	500
Sale price/piece	$ 30	$ 40	

Formulate the problem as a standard L.P.P. Find product mix that would give maximum profit by graphical technique.

11. A pineapple firm produces two products: canned pineapple and canned juice. The specific amounts of material, labor, and equipment required to produce each product and the availability of each of these resources are shown in the table given below:

	Canned Juice	Pine-apple	Available resources
Labor (man hrs.)	3	2.0	12.0
Equipment (m/c hrs)	1	2.3	6.9
Material (units)		1.4	4.9

Assuming one unit each of canned juice and canned pineapple has profit margins of $2 and $1, respectively. Formulate it as L.P. problem and solve it graphically.

12. The sales manager of a company has budgeted $ 120,000 for an advertising program for one of the firm's products. The selected advertising program consists of running advertisements in two different magazines. The advertisement for magazine A costs $ 2,000 per run while the advertisement for magazine B costs $ 5,000 per run. Past experience has indicated that at least 20 runs in magazine A and at least 10 runs in magazine B are necessary to penetrate the market with any appreciable effect. Also, experience has indicated that there is no reason to make more than 50 runs in either of the two magazines. How many runs in magazine A and how many in magazine B should be made?

Solve the following L.P.P. graphically:

13. Maximize $\quad Z = 3x + 2y$

subject to $\quad -2x + 3y \le 9$, $x - 5y \ge -20$ and $x, y \ge 0$.

14. Maximize $\quad Z = x_1 + 8x_2$

subject to $\quad x_1 + 8x_2 \le 8$, $x_1 + 2x_2 \le 6$, $2x_1 + 3x_2 \le 6$,

$\quad 6x_1 + x_2 \le 8$, $x_1 \ge 0$, $x_2 \ge 0$.

15. Minimize $\quad Z = 8x_1 + 12x_2$

subject to $\quad 60x_1 + 30x_2 \ge 240$, $30x_1 + 60x_2 \ge 300$,

$\quad 30x_1 + 180x_2 \ge 540$, and $x_1, x_2 \ge 0$.

16. G.J. Breveries Ltd. have two bottling plants one located at "G" and other "J". Each plant produces three drinks: whiskey, beer, and brandy. The number of bottles produced per day are as follows:

Drink	Plant at "G"	Plant at "J"
Whiskey	1,500	1,500
Beer	3,000	1,000
Brandy	2,000	5,000

A market survey indicates that during the month of July, there will be a demand of 20,000 bottles of whiskey, 40,000 bottles of beer, and 44,000 bottles of brandy. The operating cost per day for plants at G and J are $ 600 and $ 400. For how many days each plant be run in July so as to minimize the production cost, while still meeting the market demand. Solve graphically.

12.5 General Linear Programming Problem

Any L.P. problem involving more than two variables may be expressed as follows: Find the values of the variables $x_1, x_2, \cdots, x_n$ which maximize (or minimize) the objective function

$$Z = c_1 x_1 + c_2 x_2 + \cdots + c_n x_n \qquad (i)$$

subject to the constraints

$$\left.\begin{aligned}
a_{11}x_1 + a_{12}x_2 + \cdots + a_{1n}x_n &\le b_1 \\
a_{21}x_1 + a_{22}x_2 + \cdots + a_{2n}x_n &\le b_2 \\
\cdots\cdots\cdots\cdots\cdots\cdots\cdots\cdots \\
a_{m1}x_1 + a_{m2}x_2 + \cdots + a_{mn}x_n &\le b_m
\end{aligned}\right\} \qquad (ii)$$

and meet the non-negative restrictions.

$$x_1, x_2, \cdots, x_n \geq 0 \qquad\qquad (iii)$$

Def. 1. *A set of values x_1, x_2, $\cdots\cdots$, x_n which satisfies the constraints of the L.P.P. is called its* **solution.**

Def. 2. *Any solution to a L.P.P. which satisfies the non-negativity restrictions of the problem is called its* **feasible solution.**

Def. 3. *Any feasible solution which maximizes (or minimizes) the objective function of the L.P.P. is called its* **optimal solution.**

Some of the constraints in (ii) may be equalities, some others may be inequalities of $(\leq)$ type and remaining ones inequalities of $(\geq)$ type. The inequality constraints are changed to equalities by adding (or subtracting) non-negative variables to (from) the left-hand side of such constraints.

Def. 4. *If the constraints of a general L.P.P. be*

$$\sum_{j=1}^{n} a_{ij} x_i \leq b_i \,(i = 1, 2, \cdots\cdots k)$$

then the non-negative variables s_i which satisfy

$$\sum_{j=1}^{n} a_{ij} x_i + s_i = b_i \,(i = 1, 2, \cdots\cdots k)$$

are called **slack variables.**

Def. 5. *If the constraints of a general L.P.P. be*

$$\sum_{j=1}^{n} a_{ij} x_i \geq b_i \,(i = k, k+1, \cdots\cdots)$$

then the non-negative variables s_i which satisfy

$$\sum_{j=1}^{n} a_{ij} x_i - s_i = b_i, = (i = k, k+1, \cdots\cdots)$$

are called **surplus variables.**

12.6 Canonical and Standard Forms of L.P.P.

After the formulation of L.P.P., the next step is to obtain its solution. But before any method is used to find its solution, the problem must be presented in a suitable form. As such, we explain its following two forms:

Canonical form. The general *L.P.P.* can always be expressed in the following form:

Maximize $\quad Z = c_1 x_1 + c_2 x_2 + \cdots + c_n x_n$

subject to the constraints

$$a_{i1} x_1 + a_{i2} x_2 + \cdots + a_{in} x_n \le b_i \,; i = 1, 2, \cdots m$$

$$x_1, x_2, \cdots xn \ge 0,$$

by making some elementary transformations. This form of the *L.P.P.* is called its **canonical form** and has the following characteristics:

(*i*) Objective function is of maximization type,

(*ii*) All constraints are of ($\le$) type,

(*iii*) All variables x_i are non-negative.

The canonical form is a format for a *L.P.P.* which finds its use in the Duality theory.

Standard form. The general *L.P.P.* can also be put in the following form:

Maximize $Z = c_1 x_1 + c_2 x_2 + \cdots + c_n x_n$

subject to the constraints

$$a_{i1} x_1 + a_{i2} x_2 + \cdots + a_{in} x_n = b_i \,; i = 1, 2, \cdots m$$

$$x_1, x_2, \cdots xn \ge 0,$$

This form of the *L.P.P.* is called its **standard form** and has the following characteristics:

(*i*) Objective function is of maximization type,

(*ii*) All constraints are expressed as equations,

(*iii*) Right hand side of each constraint is non-negative,

(*iv*) All variables are non-negative.

NOTE

Obs. Any L.P.P. can be expressed in the standard form.

As minimize $Z = c_1 x_1 + c_2 x_2 + \cdots + c_n x_n$
is equivalent to maximize $\quad Z' (= -Z) = -c_1 x_1 - c_2 x_2 \ldots - c_n x_n,$
the objective function can always be expressed in the maximization form.

The inequality constraints can always be converted to equalities by adding (or subtracting) the slack (or surplus) variables to the left hand sides of such constraints.

So far, the decision variables $x_1, x_2, \cdots, x_n$ have been assumed to be all non-negative. In actual practice, these variables could also be zero or negative. If a variable is negative, it can always be expressed as the difference of two non-negative variables, e.g., a variable x_i can be written as

$$x_i = x_i' - x_i''$$

where $\qquad x_i' \geq 0, x_i'' \geq 0.$

EXAMPLE 12.11

Convert the following L.P.P. to the standard form:

Maximize $\quad Z = 3x_1 + 5x_2 + 7x_3$,
subject to $\qquad 6x_1 - 4x_2 \leq 5, \qquad 3x_1 + 2x_2 + 5x3 \geq 11,$
$\qquad\qquad 4x_1 + 3_{x3} \leq 2, \qquad x_1, x_2 \geq 0.$

Solution:

As x_3 is unrestricted, let $x_3 = x_3' - x_3''$ where $x_3', x_3'' \geq 0$. Now the given constraints can be expressed as

$$6x_1 - 4x_2 \leq 5, \qquad 3x_1 + 2x_2 + 5x_3' - 5x_3'' \geq 11,$$
$$4x_1 + 3x_3' - 3x_3'' \leq 2, \qquad x_1, x_2, x_3', x_3'' \geq 0$$

Introducing the slack/surplus variables, the problem in standard form becomes:

Maximize $\quad Z = 3x_1 + 5x_2 + 7x_3' - 7x_3''$
subject to $\quad 6x_1 - 4x_2 + s_1 = 5, \qquad\qquad 3x_1 + 2x_2 + 5x_3' - 5x_3'' - s_2 = 11,$
$\qquad\qquad 4x_1 + 3x3' - 3x3'' + s_3 = 2, \quad x_1, x_2, x_3', x_3'', s_1, s_2, s_3 \leq 0.$

EXAMPLE 12.12

Express the following problem in the standard form:

Minimize $Z = 3x_1 + 4x_2$
subject to $\qquad 2x_1 - x_2 - 3x_3 = -4, \qquad 3x_1 + 5x_2 + x_4 = 10,$
$\qquad\qquad x_1 - 4x_2 = 12, \qquad\qquad x_1, x_3, x_4 \geq 0$

Solution:

Here x_3, x_4 are the slack/surplus variables and x_1, x_2 are the decision variables. As x_2 is unrestricted, let $x_2 = x_2' - x_2''$ where $x_2', x_2'' \geq 0$.

∴ The problem is standard form is

Maximize $Z'\ (=-Z)=-3\,x_1-4x_2'+4x_2''$

subject to $\quad -2\,x_1+x_2'-x_2''+3x_3=4,\ 3x_1+5x_2'-5x_2''+x_4=10,$

$$x_1-4x_2'+4x_2''=12, \qquad x_1, x_2', x_2'', x_3, x_4 \geq 0.$$

12.7   Simplex Method

While solving an *L.P.P.* graphically, the region of feasible solutions was found to be convex, bounded by vertices and edges joining them. The optimal solution occurred at some vertex. If the optimal solution was not unique, the optimal points were on an edge. These observations also hold true for the general *L.P.P.* Essentially the problem is that of finding the particular vertex of the convex region which corresponds to the optimal solution. The most commonly used method for locating the optimal vertex is the **simplex method.** This method consists in moving step by step from one vertex to the adjacent one. Of all the adjacent vertices, the one giving better value of the objective function over that of the preceding vertex, is chosen. This method of jumping from one vertex to the other is then repeated. Since the number of vertices is finite, the simplex method leads to an optimal vertex in a finite number of steps.

In simple method, an infinite number of solutions is reduced to a finite number of promising solutions by using the following facts:

(*i*) When there are m constraints and $m+n$ (decision and slack) variables (m being $\leq n$), the starting solution is found by setting n variables equal to zero and then solving the remaining m equations, provided the solution exists and is unique. *The n zero variables are known as* **non-basic variables** *while the remaining m variables are called* **basic variables** *and they form a* **basic solution.** This reduces the number of alternatives (basic solutions) for obtaining the optimal solution to $^{m+n}C_m$ only.

(*ii*) In an *L.P.P.*, the variables must always be non-negative. Some of the basic solutions may contains negative variables. Such solutions are called *basic infeasible solutions* and should not be considered. To achieve this, we start with a basic solution which is non-negative. The next basic solution must always be non-negative. This is ensured by the feasibility condition. Such a solution is known as **basic feasible solution.**

*If all the variables in the basic feasible solution are non-zero, then it is called **non-degenerate solution** and if some of the variables are zero, it is called **degenerate solution.***

(*iii*) A new basic feasible solution may be obtained from the previous one by equating one of the basic variables to zero and replacing it by a new non-basic variables. *The eliminated variable is called the **leaving** or **outgoing variable** while the new variable is known as the **entering** or **incoming variable.***

The incoming variable must improve the value of the objective function which is ensured by the optimality condition. *This process is repeated until no further improvement is possible. This process is repeated until no further improvement is possible. The resulting solution is called the **optimal basic feasible solution** or simply **optimal solution.***

The simplex method is, therefore, based on the following two conditions:

I. Feasibility condition. It ensures that if the starting solution is basic feasible, the subsequent solutions will also be basic feasible.

II. Optimality condition. It ensures that only improved solutions will be obtained.

Now, we shall elaborate the above terms in relation to the general linear programming problem in standard form, i.e.,

$$\text{Maximize} \quad Z = c_1 x_1 + c_2 x_2 + \ldots + c_n x_n \tag{1}$$

$$\text{subject to} \quad \sum_{j=1}^{n} a_{ij} x_i - s_i = b_i, = 1, 2, \cdots\cdots, m \tag{2}$$

$$\text{and} \quad x_j \geq 0, s_i \geq 0, j = 1, 2, \ldots n \tag{3}$$

(*i*) *Solution.* $x_1, x_2, \cdots . x_n$ is a solution of the general *L.P.P.* if it satisfies the constraints (2).

(*ii*) *Feasible solution.* $x_1, x_2, \cdots . x_n$ is a feasible solution of the general *L.P.P.* if it satisfies both the constraints (2) and the non-negativity restrictions (3). The set S of all feasible solutions is called the feasible region. A linear program is said to be *infeasible* when the set S is empty.

(*iii*) *Basic solution* is the solution of the m basic variable when each of the n non-basic variables is equated to zero.

(*iv*) *Basic feasible solution* is that *basic solution* which also satisfies the non-negativity restriction (3).

(v) *Optimal solution* is that basic feasible solution which also optimizes the objective function (1) while satisfying the conditions (2) and (3).

(iv) *Non-degenerate basic feasible solution* is that basic feasible solution which contains exactly m non-zero basic variables. If any of the basic variables becomes zero, it is called a *degenerate basic feasible solution.*

EXAMPLE 12.13

Find all the basic solutions of the following system of equations identifying in each case the basic and non-basic variables:

$$2x_1 + x_2 + 4x_3 = 11, \; 3x_1 + x_2 + 5x_3 = 14.$$

Investigate whether the basic solutions are degenerate basic solutions or not. Hence find the basic-feasible solution of the system.

Solution:

Since there are $m + n = 3$ variables and there are $m = 2$ constraints in this problem, a basic solution can be obtained by setting any one variable equal to zero and then solving the resulting equations. Also the total number of basic solutions $= {}^{m+n}C_m = {}^3C_2 = 3$.

The characteristics of the various basic solutions are given below:

No. of basic solution	Basic variables	Non-basic variables	Values of basic variables	Is the Solution: feasible? (Are all $x_j > 0$?)	Is the Solution: degenerate?
1.	x_1, x_2	x_3	$2x_1 + x_2 = 11$ $3x_1 + x_2 = 14$ $\therefore x_1 = 3, x_2 = 5$	Yes	No
2.	x_2, x_3	x_1	$x_2 + 4x_3 = 11$ $x_2 + 5x_3 = 14$ $\therefore x_2 = 3, x_3 = -1$	No	Yes
3.	x_1, x_3	x_2	$2x_1 + 4x_3 = 11$ $3x_1 + 5x_3 = 14$ $\therefore x_1 = 1/2, x_3 = 5/2$	Yes	No

The basic feasible solutions are:

(i) $x_1 = 3, x_2 = 5, x_3 = 0$ (ii) $x_1 = 1\,2, x_2 = 0, x3 = 5/2$

which are also non-degenerate basic solutions.

EXAMPLE 12.14

Find an optimal solution to the following L.P.P. be computing all basic solutions and then finding one that maximizes the objective function:

$$2x_1 + 3x_2 - x_3 + 4x_4 = 8, \ x_1 - 2x_2 + 6x_3 - 7x_4 = -3$$
$$x_1, x_2, x_3, x_4 \geq 0 \qquad Max. \ Z = 2 \ x_1 + 3x_2 + 4x_3 + 7x_4.$$

Solution:

Since there are four variables and two constraints, a basic solution can be obtained by setting any two variables equal to zero and then solving the resulting equations. Also the total number of basic solutions = $^4C_2 = 6$.

The characteristics of the various basic solutions are given below:

No. of basic solution	Basic variables	Non-basic variables	Values of basic variables	Is the solution feasible? (Are all $x_j \geq 0$?)	Value of Z	Is the solution optimal?
1.	x_1, x_2	$x_3, x_4 = 0$	$2x_1 + 3x_2 = 8$ $x_1 - 2x_2 = -3$ $\therefore x_1 = 1, x2 = 2$	Yes	8	No
2.	x_1, x_3	$x_2, x_4 = 0$	$2x_1 - x_3 = 8$ $x_1 + 6x_3 = -3$ $\therefore x_1 = -14/13,$ $x_3 = -67/13$	No	-	-
3.	x_1, x_4	$x_2, x_3 = 0$	$2x1 + 4x_4 = 8$ $x_1 - 7x_4 = -3$ $\therefore x_1 = 22/9,$ $x_4 = 7/9$	Yes	10.3	No
4.	x_2, x_3	$x_1, x_4 = 0$	$3x_2 - x_3 = 8$ $-2x_2 + 6x_3 = -3$ $\therefore x_2 = 45/16,$ $x_3 = 7/16$	Yes	10.2	No
5.	x_2, x_4	$x1, x3 = 0$	$3x_2 + 4x_4 = 8$ $-2x_2 - 7x_4 = -3$ $\therefore x_2 = 132/39$ $x_4 = -7/13$	No	-	-
6.	x_3, x_4	$x_1, x_2 = 0$	$-x_3 + 4x_4 = 8$ $6x_3 - 7x_4 = -3$ $\therefore x_3 = 44/17,$ $x_4 = 45/17$	Yes	28.9	Yes

Hence the optimal basic feasible solution is

$$x_1 = 0, x_2 = 0, x_3 = 44/17, x_4 = 45/17$$

and the maximum value of $Z = 28.9$.

Exercises 12.3

1. Reduce the following problem to the standard form:

Determine $x_1 \geq 0, x_2 \geq 0, x_3 \geq 0$ so as to

Maximize $Z = 3x_1 + 5x_2 + 8x_3$

subject to the constraints $2x_1 - 5x_2 \leq 6, 3x_1 + 2x_2 + x3 \geq 5, 3x_1 + 4x_3 \leq 3$.

2. Express the following *L.P.P.* in the standard form:

Minimize $Z = 3x_1 + 2x_2 + 5x_3$

subject to $\quad -5x_1 + 2x_2 \leq 5, \quad 2x_1 + 3x_2 + 4x_3 \geq 7,$

$\quad\quad\quad\quad 2x_1 + 5x3 \leq 3, \quad\quad x_1, x_2, x3 \geq 0.$

3. Convert the following *L.P.P.* to standard form:

Maximize $Z = 3x_1 - 2x_2 + 4x_3$

Subject to $\quad x_1 + 2x_2 + x3 \leq 8, 2x_1 - x_2 + x3 \geq 2,$

$\quad\quad\quad\quad 4x_1 - 2x_2 - 3x_3 = -6, x_1, x_2 \geq 0.$

4. Obtain all the basic solutions to the following system of linear equations:

$$x_1 + 2x_2 + x_3 = 4, 2x_1 + x_2 + 5x_3 = 5.$$

5. Show that the following system of linear equations has two degenerate feasible basic solutions and the non-degenerate basic solution is not feasible:

$$2x_1 + x_2 - x_3 = 2, 3x_1 + 2x_2 + x_3 = 3.$$

6. Find all the basic feasible solutions of the equations:

$$2x_1 + 6x_2 + 2x_3 + x_4 = 3, 6x_1 + 4x_2 + 4x_3 + 6x_4 = 2.$$

7. Find all the basic solutions to the following problem:

Maximize $Z = x_1 + 3x_2 + 3x_3$,

subject to $x_1 + 2x_2 + 3x_3 = 4, 2x_1 + 3x_2 + 5x_3 = 7$.

Which of the basic solutions are

(*a*) non-degenerate basic feasible, (*b*) optimal basic feasible?

8. Show that the feasible solution

$$x_1 = 1, x_2 = 0, x_3 = 1; z = 6$$

to the system of equations

$$x_1 + x_2 + x_3 = 2; x_1 - x_2 + x_3 = 2$$

with maximum $Z = 2x_1 + 3x_2 + 4x_3$ is not basic.

12.8   Working Procedure of the Simplex Method

Assuming the existence of an initial basic feasible solution, an optimal solution to any L.P.P. by simplex method is found as follows:

Step 1. (*i*) *Check whether the objective function is to be maximized or minimized.*

If $Z = c1\, x_1 + c_2 x_2 + c_3 x_3 + \cdots + c_n x_n$

is to be minimized, then convert it into a problem of maximization, by writing.

Minimize Z = Maximize $(-Z)$

(*ii*) *Check whether all b's are positive.*

If any of the b_i's is negative, multiply both sides of that constraint by -1 so as to make its right hand side positive.

Step 2. *Express the problem in the standard form.*

Convert all inequalities of constraints into equations by introducing slack/surplus variables in the constraints giving equations of the form.

$$a_{11}x_1 + a_{12}x_2 + a_{13}x_3 + \cdots + s_1 + os_2 + os_3 + \cdots = b_1.$$

Step 3. *Find an initial basic feasible solution.*

If there are m equation involving n unknowns, then assign zero values to any $(n-m)$ of the variables for finding a solution. Starting with a basic solution for which $x_j : j = 1, 2, \cdots, (n-m)$ are each zero, find all s_i. If all s_i are ≥ 0, the basic solution is feasible and non-degenerate. If one or more of the s_i values are zero, then the solution is degenerate.

The above information is conveniently expressed in the following simplex table:

c_B	c_j Basis	c_1 x_1	c_2 x_2	$c_3.....0$ $x_3.....s_1$	0 s_2	$0......$ $a_{13}.....1$
0	s_1	a_{11}	a_{12}	$a_{13}.....1$	0	$0.....b_1$
0	s_2	a_{21}	a_{22}	$a_{23}.....0$	1	$0.....b_2$
0	s_2	a_{31}	a_{32}	$a_{33}.....0$	0	$1....b_3$
:	:	:	:	:	:	:	:

Body matrix   Unit matrix

[The variables s_1, s_2, s_3 etc. are called *basic variables* and variables $x_1, x_2,$ x_3 etc. are called non-basic variables. *Basis* refers to the basic variables $s_1, s_2,$ s_3 ···., c_j row denotes the coefficients of the variables in the objective function, while c_B–column denotes the coefficients of the basic variables only in the objective function. b-column denotes the values of the basic variables while remaining variables will always be zero. The coefficients of x's (decision variables) in the constraint equations constitute the *body matrix* while coefficients of slack variables constitute the *unit matrix*].

Step 4. Apply optimality test.

Compute $C_j = c_j - Z_j$ where $Z_j = Sc_B \, a_{ij}$

[C_j-row is called *net evaluation row* and indicates the per unit increase in the objective functions if the variable heading the column is brought into the solution.]

If all C_j are negative, then the initial basic feasible solution is *optimal*. If even one C_j is positive, then the current feasible solution is not optimal (*i.e.,* can be improved) and proceed to the next step.

Step 5. (i) Identify the incoming and outgoing variables.

If there are more than one positive C_j, then the *incoming variable* is the one that heads the column containing maximum C_j. The column containing it is known as the *key column* which is shown marked with an arrow at the bottom. If more than one variable has the same maximum C_j, any of these variables may be selected arbitrarily as the incoming variable.

Now divide the elements under b-column by the corresponding elements of key column and choose the row containing the minimum positive ratio θ. Then replace the corresponding basic variable (by making its value zero). It is termed as the *outgoing variable*. The corresponding row is called the *key row* which is shown marked with an arrow on its right end. The element at the intersection of the key row and key column is called the *key element* which is shown bracketed. If all these ratios are ≤ 0, the incoming variable can be made as large as we please without violating the feasibility condition. Hence the problem has an *unbounded solution* and no further iteration is required.

(ii) Iterate towards an optimal solution.

Drop the outgoing variable and introduce and incoming variable along-with its associated value under cB column. Convert the key element to unity by dividing the key row by the key element. Then make all other elements of the key column zero by subtracting proper multiples of key row from the other rows.

[This is nothing but the sweep-out process used to solve the linear equations. The operations performed are called *elementary row operations*.]

Step 6. Go to step 4 and repeat the computational procedure until either an optimal (or an unbounded) solution is obtained.

EXAMPLE 12.15

Using simplex method

Maximize $Z = 5x_1 + 3x_2$

subject to $x_1 + x_2 \le 2, 5x_1 + 2x_2 \le 10,$

$$3x_1 + 8x_2 \le 12, x_1, x_2 \ge 0.$$

Solution:

Consists of the following steps:

Step 1. Check whether the objective function is to be maximized and all b's are positive.

The problem consists of maximization type and all b's are ≥ 0, so this step is not necessary.

Step 2. Express the problem in the standard form.

By introducing the slack variables s_1, s_2, s_3, the problem in standard form becomes

Maximize. $Z = 5 x_1 + 3x_2 + os_1 + os_2 + os_3$

subject to
$$x_1 + x_2 + s_1 + os_2 + os_3 = 2 \qquad (i)$$

$$5x_1 + 2x_2 + os_1 + s_2 + os_3 = 10 \qquad (ii)$$

$$3x_1 + 8x_2 + os_1 + os_2 + os_3 = 12 \qquad (iii)$$

$$x_1, x_2, s_1, s_2, s_3 \ge 0$$

Step 3. Find an initial basic feasible solution.

There are three equations involving five unknowns and for obtaining a solution, we assign zero values to any two of the variables. We start with a basic solution for which we set $x_1 = 0$ and $x_2 = 0$. (This basic solution corresponds to the origin in the graphical method.) Substituting $x_1 = x_2 = 0$ in (*i*), (*ii*), and (*iii*), we get the basic solution

$$s_1 = 2, s_2 = 10, s_3 = 12.$$

Since all s_1, s_2, s_3 are positive, the basic solution is also feasible and non-degenerate.

∴ The basic feasible solution is

$x_1 = x_2 = 0$ (non-basic) and $s_1 = 2, s_2 = 10, s_3 = 12$ (basic)

∴ *Initial basic feasible solution* is given by the following table:

	c_j	5	3	0	0	0		
c_B	Basis	x_1	x_2	$s_1 1$	s_2	s_2	b	θ
0	s_1	(1)	1	1	0	0	2	2/1←
0	s_2	5	2	0	1	0	10	10/5
0	s_3	3	8	0	0	1	12	12/3
	$Z_j = \Sigma c_B a_{ij}$	0	0	0	0	0	0	
	$C_j = c_j - Z_j$	5	3	0	0	0		
		↑						

[For x_1-column ($j = 1$), $Z_j = \Sigma c_B\, a_{i1} = 0(1) + 0(5) + 0(3) = 0$

and for x_2-column ($j = 2$), $Z_j = \Sigma c_B\, a_{i2} = 0(1) + 0(2) + 0(8) = 0$

Similarly $Z_j(b) = 0(2) + 0(10) + 0(12) = 0.$]

Step 4. Apply optimality test.

As C_j is positive under some columns, the initial basic feasible solution is not optimal (*i.e.*, can be improved) and we proceed to the next step.

Step 5. (i) Identify the incoming and outgoing variables.

The previous table showed that x_1 is the *incoming variable* as its incremental contribution C_j (= 5) is maximum and the column in which it appears is the *key column* (shown marked by an arrow at the bottom).

Dividing the elements under the *b*-column by the corresponding elements of key-column, we find a minimum positive ratio θ is 2 in two row. We, therefore, arbitrarily choose the row containing s_1 as the *key row* (shown marked by an arrow on its right end). The element at the intersection of the key row and the key column *i.e.*, (1), is the *key element* s_1 therefore, the *outgoing basic variable* will now become non-basic.

Having decided that x_1 is to enter the solution, we have tried to find as to what maximum value x_1 could have without violating the constraints. So removing s_1, the new basis will contain x_1, s_2, and s_3 as the basic variables.

(*ii*) *Iterate towards the optimal solution.*

To transform the initial set of equations with a basic feasible solution into an equivalent set of equations with a different basic feasible solution, we make the key element unity. Here the key element being unity, we retain the key row as it is. Then to make all other elements in key column zero, we subtract proper multiples to key row from the other rows. Here we subtract five times the elements of key row from the second row and three times the elements of key row from the third row. These become the second and third rows of the next table. We also change the corresponding value under c_B column from 0 to 5, while replacing s_1 by x_1 under the basis. Thus the *second basic feasible solution* is given by the following table:

	c_j	5	3	0	0	0		
c_B	Basis	x_1	x_2	s_1	s_2	s_3	b	θ
5	x_1	1	1	1	0	0	2	
0	s_2	0	-3	-5	1	0	0	
0	s_3	0	5	-3	0	1	6	
	$Z_j = \Sigma\, c_B\, a_{ij}$	5	5	5	0	0	10	
	$C_j = c_j - Z_j$	0	-2	-5	0	0		

As C_j is either zero or negative under all columns, the above table gives the optimal basic feasible solution. This optimal solution is $x_1 = 2$, $x_2 = 0$ and maximum $Z = 10$.

EXAMPLE 12.16

A firm produces three products which are processed on three machines. The relevent data is given next:

Machine	Time per unit (minutes)			Machine capacity (minutes/day)
	Product A	Product B	Product C	
M_1	2	3	2	440
M_2	4	—	3	470
M_3	2	5	—	430

The profit per unit for products A, B, and C is $ 4, $ 3 and $ 6, respectively. Determine the daily number of units to be manufactured for each product. Assume that all the units produced are consumed in the market.

Solution:

Let the firm decide to produce x_1, x_2, x_3 units of products A, B, C respectively. Then the L.P. model for this problem is:

Max. $Z = 4 x_1 + 3x_2 + 6x_3$

subject to $\qquad 2x_1 + 3x_2 + 2x_3 \le 440, \qquad 4x_1 + 3x_2 \le 470$

$\qquad\qquad 2x_1 + 5x_2 \le 430, \qquad\qquad x_1, x_2, x3 \ge 0.$

Step 1. *Check whether the objective function is to be maximized and all b's are non-negative.*

The problem consists of maximization type and b's are ≥ 0, so this step is not necessary.

Step 2. *Express the problem in the standard form.*

By introducing the slack variables s_1, s_2, s_3, the problem in standard form becomes:

Max. $Z = 4x_1 + 3x_2 + 6x_3 + 0s_1 + 0s_2 + 0s_3$

subject to $\quad 2x_1 + 3x_2 + 2x_3 + s_{10} + s_2 + 0s_3 = 440$

$\qquad\qquad 4x_1 + 0x_2 + 3x_3 + 0s_1 + s_2 + 0s_3 = 470$

$\qquad\qquad 2x_1 + 5x_2 + 0x_3 + 0s_1 + 0s_2 + s_3 = 430$

Step 3. *Find an initial basic feasible solution.*

The basic (non-degenerate) feasible solution is

$\qquad x_1 = x_2 = x_3 = 0$ (non-basic)

$\qquad s_1 = 440, s_2 = 470, s_3 = 430$ (basic)

$\therefore$ *Initial basic feasible solution* is given by the following table:

	c_j	4	3	6	0	0	0		
c_B	Basis	x_1	x_2	x_3	s_1	s_2	s_3	b	θ
0	s_1	2	3	2	1	0	0	440	440/2
0	s_2	4	0	(3)	0	1	0	470	470/3←
0	s_3	2	5	0	0	0	1	430	430/0
$Z_j = \Sigma c_B a_{ij}$		0	0	0	0	0	0		
$C_j = c_j - Z_j$		4	3	6	0	0	0		
				↑					

Step 4. Apply optimality test.

As Cj is positive under some columns, the initial basic feasible solution is not optimal and we proceed to the next step.

Step 5. (i) Identify the incoming and outgoing variables.

The above table shows that x_3 is the incoming variable while s_2 is the outgoing variable and (3) is the key element.

(ii) Iterate towards the optimal solution.

Drop s_2 and introduce x_3 with its associated value 6 under c_B column. Convert the key element to unity and make all other elements of key column zero. Then the *second feasible solution* is given by the table below:

c_j		4	3	6	0	0	0		
c_B	Basis	x_1	x_2	x_3	s_1	s_2	s_3	b	θ
0	s_1	−2/3	(3)	0	1	−2/3	0	380/3	380/9←
6	x_3	4/3	0	1	0	1/3	0	470/3	∞
0	s_3	2	5	0	0	0	1	430	86
	Z_j	8	0	6	0	2	0	940	
	C_{ji}	−4	3	0	0	−2	0		
			↑						

Step 6. As C_j is positive under the second column, the solution is not optimal and we proceed further. Now x_2 is the incoming variable and s_1 is the outgoing variable and (3) is the key element for the next iteration.

Drop s_1 and introduce x_2 with its associated value 3 under c_B column. Convert the key element to unity and make all other elements of the key column zero. Then the *third basic feasible solution* is given by the following table:

c_j		4	3	6	0	0	0		
c_B	Basis	x_1	x_2	x_3	s_1	s_2	s_3	b	θ
3	x_2	−2/9	1	0	1/3	−2/9	0	380/9	
6	x_3	4/3	0	1	0	1/3	0	470/3	
0	s_3	28/9	0	0	−5/3	10/9	0	1970/9	
	Z_j	22/3	3	6	1	4/3	0	3200/3	
	C_{ji}	−10/3	0	0	−1	−4/3	0		

Now since each $C_j \leq 0$, therefore it gives the optimal solution

$$x_1 = 0, x_2 = 380/9, x_3 = 470/3$$

and $\quad Z_{max} = 3200/3$ *i.e.,* 1066.67 Dollars.

EXAMPLE 12.17

Solve the following L.P.P. the by simplex method:

Minimize $Z = x_1 - 3x_2 + 3x_3$,

subject to $\quad 3x_1 - x_2 + 2x_3 \leq 7, \qquad\qquad 2x_1 + 4x_2 \geq -12,$

$\qquad\qquad -4\,x_1 + 3x_2 + 8x_3 \leq 10, \qquad x_1, x_2, x3 \geq 0.$

Solution:

Consists of the following steps:

Step 1. Check whether objective function is to be maximized and all b's are non-negative.

As the problem is that of minimizing the objective function, converting it to the maximization type, we have

Max. $Z' = -x_1 + 3x_2 - 3x_3$

As the right-hand side of the second constraint is negative, we write it as

$$-2x_1 - 4x_2 \leq 12$$

Step 2. Express the problem in the standard form.

By introducing the slack variables s_1, s_2, s_3, the problem in the standard form becomes

Max. $Z' = -x_1 + 3x_2 - 3x_3 + 0s_1 + 0s_2 + 0s_3$

subject to $3x_1 - x_2 + 2x_3 + s_1 + 0s_2 + 0s_3 = 7$

$\qquad\qquad -2x_1 - 4x_2 + 0x_3 + 0s_1 + s_2 + 0s_3 = 12$

$\qquad\qquad -4x_1 + 3x_2 + 8x_3 + 0s_1 + 0s_2 + s_3 = 10$

$\qquad\qquad\qquad x_1, x_2, x_3, s_1, s_2, s_3 \geq 0.$

Step 3. Find initial basic feasible solution.

The basic (non-degenerate) feasible solution is

$\quad x_1 = x_2 = x_3 = 0$ (non-basic), $s_1 = 7, s_2 = 12, s_3 = 10$ (basic)

∴ *Initial basic feasible solution* is given by the table below:

	c_j	−1	3	−3	0	0	0		
c_B	Basis	x_1	x_2	x_3	s_1	s_2	s_3	b	θ
0	s_1	3	−1	2	1	0	0	7	7(−4)
0	s_2	−2	−4	0	0	1	0	12	12(−4)
0	s_3	−4	(3)	8	0	0	1	10	10/3←
	$Z_j = \Sigma\, c_B a_{ij}$	0	0	0	0	0	0	0	
	$C_j = c_j - Z_j$	−1	3	−3	0	0	0		
			↑						

Step 4. Apply optimality test.

As C_j is positive under second column, the initial basic feasible solution is not optimal and we proceed further.

Step 5. (i) Identify the incoming and outgoing variables.

The above table shows that x_2 is the incoming variable, s_3 is the outgoing variable and (3) is the key element.

(ii) Iterate towards the optimal solution.

∴ Drop s_3 and introduce x_2 with its associated value 3 under c_B column. Convert the key element to unity and make all other elements of the key column zero. Then the *second basic feasible solution* is given by the following table:

	c_j	−1	3	−3	0	0	0		
c_B	Basis	x_1	x_2	x_3	s_1	s_2	s_3	b	θ
0	s_1	(5/3)	0	14/3	1	0	1/3	31/3	31/5←
0	s_2	−22/3	0	32/3	0	1	4/3	76/3	−38/11
3	x_2	−4/3	1	8/3	0	0	1/3	10/3	−5/2
	Z_j	−4	3	8	0	0	1	10	
	C_j	3	0	−11	0	0	−1		
		↑							

Step 6. As C_j is positive under first column, the solution is not optimal and we proceed further. x_1 is the incoming variable, s_1 is the outgoing variable and (5/3) is the key element.

$\therefore$ Drop s_1 and introduce x_1 with its associated value -1 under c_B column. Convert the key element to unity and make all other elements of the key column zero. Then the *third basic feasible solution* is given by the table below:

	c_j	-1	3	-3	0	0	0	
c_B	Basis	x_1	x_2	x_3	s_1	s_2	s_3	b
-1	x_1	1	0	$14/5$	$3/5$	0	$1/5$	$31/5$
0	s_2	0	0	$156/5$	$22/5$	1	$14/5$	$354/5$
3	x_2	0	1	$32/5$	$4/5$	0	$3/5$	$58/5$
	Z_j	-1	3	$82/5$	$9/5$	0	$8/5$	$143/5$
	C_j	0	0	$-97/5$	$-9/5$	0	$-8/5$	

Now since each $C_j \leq 0$, therefore it gives the optimal solution

$x_1 = 31/5$, $x_2 = 58/5$, $x_3 = 0$ (non-basic) and $Z'_{max} = 143/5$.

Hence $Z_{min} = -143/5$.

EXAMPLE 12.18

Maximize $Z = 107x_1 + x_2 + 2x_3$,

subject to the constraints: $14x_1 + x_2 - 6x_3 + 3x_4 = 7$, $\quad 16x_1 + \dfrac{1}{2}x_2 - 6x_3 \leq 5$

$$3x_1 - x_2 - x_3 \leq 0, \qquad x_1, x_2, x_3, x_4 \geq 0$$

Solution:

Consists of the following steps:

Step 1. Check whether objective function is to be maximized and all b's are non-negative.

This step is not necessary.

Step 2. Express the problem in the standard form.

Here x_4 is a slack variable. By introducing other slack variables s_1 and s_2 the problem in standard form becomes

Max. $Z = 107x_1 + x_2 + 2x_3 + 0x_4 + 0s_1 + 0s_2$

subject to $\dfrac{14}{3}x_1 + \dfrac{1}{3}x_2 - 2x_3 + x_4 + 0s_1 + 0s_2 = \dfrac{7}{3}$,

$$16x_1 + \dfrac{1}{2}x_2 - 6x_3 + 0x_4 + s_1 + 0s_2 = 5$$

$3x_1 - x_2 - x_3 + 0x_4 + 0s_1 + s_2 = 0$, $x_1, x_2, x_3, x_4, s_1, s_2 \geq 0$.

Step 3. Find initial basic feasible solution.

The basic feasible solution is

$$x_1 = x_2 = x_3 = 0 \text{ (non-basic)}$$
$$x_4 = 7/3, s_1 = 5, s_2 = 0 \text{ (basic)}$$

∴ *Initial basic feasible solution* is given in the table below:

	c_j	107	1	2	0	0	0		
c_B	Basis	x_1	x_2	x_3	x_4	s_1	s_2	b	θ
0	x_4	14/3	1/3	−2	1	0	0	7/3	$\dfrac{7}{3}\Big/\dfrac{14}{3}$
0	s_1	16	1/2	−6	0	1	0	5	5/16
0	s_2	(3)	−1	−1	0	0	1	0	0/3←
	$Z_j = \Sigma c_B a_{ij}$	0	0	0	0	0	0		
	$C_j = c_j - Z_j$	107	1	2	0	0	0		
		↑							

Step 4. Apply optimality test.

As C_j is positive under some columns, the initial basic feasible solution is not optimal and we proceed further.

Step 5. (i) Identify the incoming and outgoing variables

The above table shows that x_1 is the incoming variable, s_2 is the outgoing variable, and (3) is the key element.

(ii) Iterate towards the optimal solution.

Drop s_2 and introduce x_1 with its associated value 107 under c_B column. Convert, key element to unity and make all other elements of the key column zeros. Then the *second basic feasible solution* is given by the following table:

	c_j	107	1	2	0	0	0		
c_B	Basis	x_1	x_2	x_3	x_4	s_1	s_2	b	θ
0	x_4	0	17/9	−4/9	1	0	−14/9	7/3	−21/4
0	s_1	0	35/6	−2/3	0	1	−16/3	5	−15/2
107	x_1	1	−1/3	−1/3	0	0	1/3	0	0
	Z_j	107	−107/3	−107/3	0	0	107/3		
	C_j	0	110/3	113/3	0	0	−107/3		
				↑					

As C_j is positive under some columns, the solution is not optimal. Here $113/3$ is the largest positive value of C_j, and x_3 is the incoming variable. But all the values of θ being ≤ 0, x_3 will not enter the basis. This indicates that the solution to the problem is unbounded.

[**Remember that** (*i*) *the incoming variable is the non-basic variable corresponding to the largest positive value of C_j and (ii) the outgoing variable is the basic-variable corresponding to the least positive ratio θ, obtained by dividing the b-column elements by the corresponding key-column elements.*]

Exercises 12.4

Using simplex method, solve the following *L.P.P.* $(1 - 9)$:

1. Maximize $Z = x_1 + 3x_2$
subject to $x_1 + 2x_2 \leq 10$, $0 \leq x_1 \leq 5$, $0 \leq x_2 \leq 4$.

2. Maximize $Z = 4x_1 + 10x_2$
subject to $2x_1 + x_2 \leq 50$, $2x_1 + 5x_2 \leq 100$, $2x_1 + 3x_2 \leq 90$, $x_1, x_2 \geq 0$.

3. Maximize $Z = 4x_1 + 5x_2$,
subject to $x_1 - 2x_2 \leq 2$, $2x_1 + x_2 \leq 6$, $x_1 + 2x_2 \leq 5$, $-x_1 + x_2 \leq 2$, $x_1, x_2 \geq 0$.

4. Maximize $Z = 10x_1 + x_2 + 2x_3$,
subject to $x_1 + x_2 - 2x_3 \leq 10$, $4x_1 + x_2 + x_3 \leq 20$, $x_1, x_2, x_3 \geq 0$.

5. Maximize $Z = x_1 + x_2 + 3x_3$,
subject to $3x_1 + 2x_2 + x_3 \leq 3$, $2x_1 + x_2 + 2x_3 \leq 2$, $x_1, x_2, x_3 \geq 0$.

6. Maximize $Z = x_1 - x_2 + 3x_3$
subject to $x_1 + x_2 + x_3 \leq 10$, $2x_1 - x_2 \leq 2$, $2x_1 - 2x_2 + 3x_3 \leq 0$, $x_1, x_2, x_3 \geq 0$.

7. Minimize $Z = 3x_1 + 5x_2 + 4x_3$
subject to $2x_1 + 3x_2 \leq 8$, $2x_2 + 5x_3 \leq 10$,
$3x_1 + 2x_2 + 4x_3 \leq 15$, $x_1, x_2, x_3 \geq 0$.

8. Minimize $Z = x_1 - 3x_2 + 2x_3$,
subject to $3x_1 - x_2 + 2x_3 \leq 7$, $-2x_1 + 4x_2 \leq 12$,
$-4x_1 + 3x_2 + 8x_3 \leq 10$, $x_1, x_2, x_3 \geq 0$.

9. Maximize $Z = 4x_1 + 3x_2 + 4x_3 + 6x_4$
subject to $x_1 + 2x_2 + 2x_3 + 4x_4 \leq 80$, $2x_1 + 2x_3 + x_4 \leq 60$,
$3x_1 + 3x_2 + x_3 + x_4 \leq 80$, $x_1, x_2, x_3, x_4 \geq 0$.

10. A firm produces products A and B and sells them at a profit of $ 2 and $ 3 each respectively. Each product is processed on machines G and H. Product A requires 1 minute on G and 2 minutes on H whereas product B requires 1 minute on each of the machines. Machine G is not available for more than 6 hours 40 min/day whereas the time constraint for machine H is 10 hours. Solve this problem via the simplex method for maximizing the profit.

11. A company makes two types of products. Each product of the first type requires twice as much labor time as the second type. If all products are of the second type only, the company can produce a total of 500 units a day. The market limits daily sales of the first and the second type to 150 and 250 units, respectively. Assuming that the profits per unit are $ 8 for type I and $ 5 for type II, determine the number of units of each type to be produced to maximize profit.

12. The owner of a dairy is trying to determine the correct blend of two types of feed. Both contain various percentages of four essential ingredients. With the following data determine the least cost blend?

Ingredient	% per kg of feed		Min. requirement in kg.
	Feed 1	Feed 2	
1	40	20	4
2	10	30	2
3	20	40	3
4	30	10	6
Cost ($/kg.)	5	3	

13. A manufacturing firm has discontinued production of a certain unprofitable product line. This created considerable excess production capacity. Management is considering to devote their excess capacity to one or more of three products 1, 2, and 3. The available capacity on machines and the number of machine hours required for each unit of the respective product, is given below:

Machine Type	Available Time (hrs/week)	Productivity (hrs/unit)		
		Product I	Product II	Product III
Milling Machine	250	8	2	3
Lathe	150	4	3	0
Grinder	50	2	–	1

The unit profit would be $ 20, $ 6 and $ 8, respectively for products 1, 2, and 3. Find how much of each product the firm should produce in order to maximize profit.

14. The following table gives the various vitamin contents of three types of food and daily requirements of vitamins along with cost per unit. Find the combination of food for minimum cost.

Vitamin (mg)	Food F	Food G	Food	Mimimum daily requirement (mg)
A	1	1	10	1
C	100	10	10	50
D	10	100	10	10
Cost/unit ($)	10	15	5	

15. A farmer has 1,000 acres of land on which he can grow corn, wheat, or soyabeans. Each acre of corn costs $ 100 for preparation, requires s man-days of work and yiellds a profit of $ 30. An acre of wheat costs $ 120 to prepare, requires ten man-days of work and yields a profit of $ 40. An acre of soyabeans costs $ 70 to prepare, requires eight man-days of work and yields a profit of $ 20. If the farmer has $ 1,00,000 for preparation and can count on 8,000 man-days of work, how many acres should be allocated to each crop to maximize profits ?

12.9   Artificial Variable Techniques

So far we have seen that the introduction of slack/surplus variables provided the initial basic feasible solution. But there are many problems wherein at least one of the constraints is of ($\geq$) or ($=$) type and slack variables fail to give such a solution. There are two similar methods for solving such problems which we explain below

M-method or **Method of penalties.** This method is due to A. *Charnes* and consists of the following steps:

Step 1. Express the problem in standard form.

Step 2. Add non-negative variables to the left hand side of all those constraints which are of ($\geq$) or ($=$) type. Such new variables are called *artificial variables* and the purpose of introducing these is just to obtain an initial basic feasible solution. But their addition causes violation of the corresponding constraints. As such, we would like to get rid of these variables and would not allow them to appear in the final solution. For this purpose, we assign a very large penality $(-M)$ to these artificial variables in the objective function.

Step 3. Solve the modified *L.P.P.* by simplex method.

At any iteration of the simplex method, one of the following three cases may arise:

(*i*) There remains no artificial variable in the basis and the optimality condition is satisfied. Then the solution is an optimal basic feasible solution to the problem.

(*ii*) There is atleast one artificial variable in the basis at zero level (with zero value in b-column) and the optimality condition is satisfied. Then the solution is a degenerate optimal basic feasible solution

(*iii*) There is at least one artificial variable in the basis at the non-zero level (with positive value in b-column) and the optimality condition is satisfied. Then the problem has no feasible solution. The final solution is not optimal, since the objective function contains an unknown quantity M. Such a solution satisfies the constraints but does not optimize the objective function and is therefore, called *pseudo optimal solution*.

Step 4. Continue the simplex method until either an optimal basic feasible solution is obtained or an unbounded solution is indicated.

NOTE **Obs.** *The artificial variables are only a computational device for getting a starting solution. Once an artificial variable leaves the basis, it has served its purpose and we forget about it, i.e., the column for this variable is omitted from the next simplex table.*

EXAMPLE 12.19

Use Charne's penalty method to Minimize $Z = 2x_1 + x_2$

subject to $\quad 3x_1 + x_2 = 3,\ 4x_1 + 3x_2 \geq 6,$

$\quad\quad\quad x_1 + 2x_2 \leq 3,\ x_1,\ x_2 \geq 0.$

Solution:

Consists of the following steps:

Step 1. *Express the problem in standard form.*

The second and third inequalities are converted into equations by introducing the surplus and slack variables s_1, s_2, respectively.

Also the first and second constraints being of (=) and (≥) type, we introduce two artificial variables A_1, A_2.

Converting the minimization problem to the maximization form for the *L.P.P.* can be rewritten as

Max. $Z' = -2x_1 - x_2 + 0s_1 + 0s_2 - MA1 - MA2$

subject to

$$3x_1 + x_2 + 0s_1 + 0s_2 + A_1 + 0A_2 = 3$$

$$4x_1 + 3x_2 - s_1 + 0s_2 + 0A_1 + A_2 = 6$$

$$x_1 + 2x_2 + 0s_1 + s_2 + 0A_1 + 0A_2 = 3$$

$$x_1, x_2, s_1, s_2, A_1, A_2 \geq 0.$$

Step 2. Obtain an initial basic feasible solution.

Surplus variable s_1 is not a basic variable since its value is – 6. As negative quantities are not feasible, s_1 must be prevented from appearing in the initial solution. This is done by taking $s_1 = 0$. By setting the other non-basic variables x_1, x_2 each = 0, we obtain the initial basic feasible solution as

$$x_1 = x_2 = 0, s_1 = 0;$$
$$A_1 = 3, A_2 = 6, s_2 = 3$$

Thus the initial simplex table is

	c_j	–2	–1	0	0	–M	–M		
c_B	Basis	x_1	x_2	s_1	s_2	A_1	A_2	b	θ
–M	A_1	(3)	1	0	0	1	0	3	3/3 ←
–M	A_2	4	3	–1	0	0	1	6	6/4
0	s_2	1	2	0	1	0	0	3	3/1
	$Z_j = \Sigma c_B a_{ij}$	–7M	–4M	M	0	–M	–M	–9M	
	$C_j = c_j - Z_j$	7M – 2	4M – 1	–M	0	0	0		
		↑							

Since C_j is positive under x_1 and x_2 columns, this is not an optimal solution.

Step 3. Iterate towards optimal solution.

Introduce x_1 and drop A_1 from basis.

∴ The new simplex table is

	c_j	-2	-1	0	0	-M		
c_B	Basis	x_1	x_2	s_1	s_2	A_2	b	θ
-2	x_1	1	1/3	0	0	0	1	3
-M	A_2	0	(5/3)	-1	0	1	2	6/5 ←
0	s_2	0	5/3	0	1	0	2	6/5
	Z_j	-2	$-\dfrac{2}{3}-\dfrac{5M}{3}$	M	0	$-M$	$-2-2M$	
	C_j	0	$-\dfrac{1}{3}+\dfrac{5M}{3}$	$-M$	0	0		
			↑					

Since C_j is positive under x_2 column, this is not an optimal solution.

∴ Introduce x_2 and drop A_2.

Then the revised simplex table is

	c_j	-2	-1	0	0	
c_B	Basis	x_1	x_2	s_1	s_2	b
-2	x_1	1	0	1/5	0	3/5
-1	x_2	0	1	-3/5	0	6/5
0	s_2	0	0	1	1	0
	Z_j	-2	-1	1/5	0	-12/5
	C_j	0	0	-1/5	0	

Since none of C_j is positive, this is an optimal solution. Thus, an optimal basic feasible solution to the problem is

$x_1 = 3/5$, $x_2 = 6/5$, Max. $Z' = -12/5$.

Hence the optimal value of the objective function is

Min. $Z = -$ Max. $Z' = -(-12/5) = 12/5$.

EXAMPLE 12.20

Maximize $Z = 3x_1 + 2x_2$

subject to the constraints: $\quad 2x_1 + x_2 \le 2$, $3x_1 + 4x_2 \ge 12$, $x_1, x_2 \ge 0$.

Solution:

Consists of the following steps:

Step 1. Express the problem in standard form.

The inequalities are converted into equations by introducing the slack and surplus variables s_1, s_2, respectively. Also the second constraint being of ($\geq$) type, we introduce the artificial variable A. Thus the L.P.P. can be rewritten as

$$\text{Max. } Z = 3x_1 + 2x_2 + 0s_1 + 0s_2 - MA$$

$$\text{subject to} \quad 2x_1 + x_2 + s_1 + 0s_2 + 0A = 2,$$

$$3x_1 + 4x_2 + 0s_1 - s_2 + A = 12,$$

$$x_1, x_2, s_1, s_2, A \geq 0.$$

Step 2. Find an initial basic feasible solution.

Surplus variable s_2 is not a basic variable since its value is -12. Since a negative quantity is not feasible, s_2 must be prevented from appearing in the initial solution. This is done by letting $s_2 = 0$. By taking the other non-basic variables x_1 and x_2 each $= 0$, we obtain the initial basic feasible solution as

$$x_1 = x_2 = s_2 = 0, s_1 = 2, A = 12$$

$\therefore$ The initial simplex table is

c_B	Basis	c_j 3	2	0	0	$-M$		
		x_1	x_2	s_1	s_2	A	b	θ
0	s_1	2	(1)	1	0	0	2	2←
$-M$	A	3	4	0	-1	1	12	3
	s_2	0	5/3	0	1	0	2	6/5
	$Z_j = \Sigma c_B a_{ij}$	$-3M$	$-4M$	0	M	$-M$	$-12M$	
	$C_j = c_j - Z_j$	$3+3M$	$2+4M$	0	$-M$	0		
			↑					

Since C_j is positive under some columns, this is not an optimal solution.

Step 3. Iterate towards optimal solution.

Introduce x_2 and drop s_1.

∴ The new simplex table is

c_j		3	2	0	0	$-M$	
c_B	Basis	x_1	x_2	s_1	s_2	A	b
2	x_2	2	1	1	0	0	2
$-M$	A	-5	0	-4	-1	1	4
Z_j		$4+5M$	2	$2+4M$	M	$-M$	$4{-}4M$
C_j		$-(1+5M)$	0	$-(2+4M)$	$-M$	0	

Here each C_j is negative and an artificial variable appears in the basis at the non-zero level. Thus there exists a *pseudo optimal solution* to the problem.

Two-phase method. This is another method to deal with the artificial variables wherein the L.P.P. is solved in two phases.

Phase I. *Step* 1. Express the given problem in the standard form by introducing slack, surplus, and artificial variables.

Step 2. Formulate an artificial objective function
$$Z^* = -A_1 - A_2, \cdots. -A_m$$
by assigning (-1) cost to each of the artificial variables A_i and zero cost to all other variables.

Step 3. Maximize Z^* subject to the constraints of the original problem using the simplex method. Then three cases arise:

(*a*) *Max.* $Z^* < 0$ *and at least one artificial variable appears in the optimal basis at a positive level.*

In this case, the original problem does not possess any feasible solution and the procedure comes to an end.

(*b*) *Max.* $Z^* = 0$ *and no artificial variable appears in the optimal basis.*

In this case, a basic feasible solution is obtained and we proceed to phase II for finding the optimal basic feasible solution to the original problem.

(*c*) *Max.* $Z^* = 0$ *and at least one artificial variable appears in the optimal basis at zero level.*

Here a feasible solution to the auxiliary L.P.P. is also a feasible solution to the original problem with all artificial variables set $= 0$.

To obtain a basic feasible solution, we prolong phase I for pushing all the artificial variables out of the basis (without proceeding on to phase II).

Phase II. The basic feasible solution found at the end of phase I is used as the starting solution for the original problem in this phase, *i.e.*, the final simplex table of phase I is taken as the initial simplex table of phase II and the artificial objective function is replaced by the original objective function. Then we find the optimal solution.

EXAMPLE 12.21

Use a two-phase method to

Minimize $Z = 7.5x_1 - 3x_2$

subject to the constraints $3x_1 - x_2 - x_3 \geq 3$, $x_1 - x_2 + x_3 \geq 2$, $x_1, x_2, x_3 \geq 0$.

Solution:

Phase I. *Step* 1. *Express the problem in standard form.*

Introducing surplus variables s_1, s_2 and artificial variables A_1, A_2. The phase I problem in standard form becomes

Max. $Z^* = 0x_1 + 0x_2 + 0x_3 + 0s_1 + 0s_2 - A_1 - A_2$

subject to $3x_1 - x_2 - x_3 - s_1 + 0s_2 + A_1 + 0A_2 = 3$

$$x_1 - x_2 + x_3 + 0s_1 - s_2 + 0A_1 + A_2 = 2$$

$$x_1, x_2, x_3, s_1, s_2, A_1, A_2 \geq 0$$

Step 2. *Find an initial basic feasible solution.*

Setting $x_1 = x_2 = x_3 = s_1 = s_2 = 0$,

we have $A_1 = 3$, $A_2 = 2$ and $Z^* = -5$

∴ Initial simplex table is

c_B	Basis	c_j	0	0	0	0	0	-1	-1		
			x_1	x_2	x_3	s_1	s_2	A_1	A_2	b	θ
-1	A_1		(3)	-1	-1	-1	0	1	0	3	1←
-1	A_2		1	-1	1	0	-1	0	1	2	2
	$Z_j^* = \Sigma c_B a_{ij}$		-4	2	0	1	1	-1	-1	-5	
	$C_j = c_j - Z_j^*$		4	-2	0	-1	-1	0	0		
			↑								

As C_j is positive under x_1 column, this solution is not optimal.

Step 3. *Iterate towards an optimal solution.*

Making key element (3) unity and replacing A_1 by x_1, we have the new simplex table:

c_j		0	0	0	0	0	−1	−1		
c_B	Basis	x_1	x_2	x_3	s_1	s_2	A_1	A_2	b	θ
0	x_1	1	−1/3	−1/3	−1/3	0	1/3	0	1	−3
−1	A_2	0	−2/3	4/3	1/3	−1	−1/3	1	1	3/4←
	$Z_j°$	0	2/3	−4/3	−1/3	1	1/3	−1	−1	
	C_j	0	−2/3	4/3	1/3	−1	−1/3	0		
				↑						

Since C_j is positive under x_3 and s_1 columns, this solution is not optimal.

Making key element (4/3) unity and replacing A_2 by x_3, we obtain the revised simplex table:

c_j		0	0	0	0	0	−1	−1	
c_B	Basis	x_1	x_2	x_3	s_1	s_2	A_1	A_2	b
0	x_1	1	−1/2	0	−1/4	−1/4	1/4	1/4	5/4
0	x_2	0	−1/2	1	1/4	−3/4	−1/4	3/4	3/4
$Z_j°$		0	0	0	0	0	0	0	0
C_j		0	0	0	0	0	−1	−1	

Since all $C_j \leq 0$, this table gives the optimal solution. Also $Z^*_{max} = 0$ and no artificial variable appears in the basis. Thus an optimal basic feasible solution to the auxiliary problem and therefore to the original problem, has been attained.

Phase II. Considering the actual costs associated with the original variables, the objective function is

$$\text{Max. } Z' = -15/2x_1 + 3x_2 + 0x_3 + 0s_1 + 0s_2 - 0A_1 - 0A_2$$

$$\text{subject to} \quad 3x_1 - x_2 - x_3 - s_1 + 0s_2 + A_1 + 0A_2 = 3,$$

$$x_1 - x_2 + x_3 + 0s_1 - s_2 + 0A_1 + A_2 = 2,$$

$$x_1, x_2, x_3, s_1, s_2, A_1, A_2 \geq 0$$

The optimal initial feasible solution thus obtained, will be an optimal basic feasible solution to the original *L.P.P.*

Using final table of phase I, the initial simplex table of phase II is as follows:

	c_j	$-15/2$	3	0	0	0	
c_B	Basis	x_1	x_2	x_3	s_1	s_2	b
$-15/2$	x_1	1	$-1/2$	0	$-1/4$	$-1/4$	$5/4$
0	x_3	0	$-1/2$	1	$1/4$	$-3/4$	$3/4$
Z_j^*		$-15/2$	$15/4$	0	$15/5$	$15/8$	$-75/8$
C_j		0	$-3/4$	0	$-15/8$	$-15/8$	

Since all $C_j \leq 0$, this solution is optimal.

Hence an optimal basic feasible solution to the given problem is

$$x_1 = 5/4, x_2 = 0, x_3 = 3/4 \text{ and min. } Z = 75/8.$$

12.10 Exceptional Cases

Tie for the incoming variable. When more than one variable has the same largest positive value in C_j row (in maximization problem), a tie for the choice of incoming variable occurs. As there is no method to break this tie, we choose any one of the prospective incoming variables arbitrarily. Such an arbitrary choice does not in any way affect the optimal solution.

Tie for the outgoing variable. When more than one variable has the same least positive ratio under the θ-column, a tie for the choice of outgoing variable occurs. If the equal values of said ratio are > 1, choose any one of the prospective leaving variables arbitrarily. Such an arbitrary choice does not affect the optimal solution.

If the equal values of ratios are zero, the simplex method fails and we make use of the following degeneracy technique.

Degeneracy. We know that a basic feasible solution is said to be degenerate if any of the basic variables vanishes. This phenomenon of getting a degenerate basic feasible solution is called *degeneracy* which may arise

(*i*) *at the initial stage*, when at least one basic variable is zero in the initial basic feasible solution or

(*ii*) *at any subsequent stage*, when the least positive ratios under θ the-column are equal for two or more rows.

In this case, an arbitrary choice of one of these basic variables may result in one or more basic variables becoming zero in the next iteration. At times, the same sequence of simplex iterations is repeated endlessly without improving the solution. These are termed as *cycling* type of problems. Cycling occurs very rarely. Intact, cycling has seldom occurred in practical problems.

To avoid cycling, we apply the following **perturbation procedure:**

(*i*) Divide each element in the tied rows by the *positive coefficients* of the key column in that row.

(*ii*) Compare the resulting ratios (from left to right) first of unit matrix and then of the body matrix, column by column.

(*iii*) The outgoing variable lies in that row which first contains the smallest algebraic ratio.

EXAMPLE 12.22

Maximize $Z = 5x_1 + 3x_2$

subject to $x_1 + x_2 \le 2$, $5x_1 + 2x_2 \le 10$, $3x_1 + 8x_2 \le 12$; $x_1, x_2 \ge 0$.

Solution:

Consists of the following steps:

Step 1. Express the problem in the standard form.

Introducing the slack variables s_1, s_2, s_3, the problem in the standard form is

Max. $Z = 5x_1 + 3x_2 + 0s_1 + 0s_2 + 0s_3$

$x_1 + x_2 + s_1 + 0s_2 + 0s_3 = 2$, $5x_1 + 2x_2 + 0s_1 + s_2 + 0s_3 = 10$

$3x_1 + 8x_2 + 0s_1 + 0s_2 + s_3 = 12$, $x_1, x_2, s_1, s_2, s_3 \ge 0$.

Step 2. Find the initial basic feasible solution.

The initial basic feasible solution is

$x_1 = x_2 = 0$ (non-basic)

$s_1 = 2, s_2 = 10, s_3 = 12$ (basic) and $Z = 0$.

∴ Initial simplex table is

c_B	c_j Basis	5 x_1	3 x_2	0 s_1	0 s_2	0 s_3	b	θ
0	s_1	1	1	1	0	0	2	2/1
0	s_2	(5)	2	0	1	0	10	10/5←
0	s_3	3	8	0	0	1	12	12/3
	$Z_j = \Sigma\, c_B a_{ij}$	0	0	0	0	0	0	
	$C_j = c_j - Z_j$	5	3	0	0	0		
		↑						

As C_j is positive under some columns, this solution is not optimal.

Step 3. Iterate towards optimal solution.

x_1 is the incoming variable. But the first two rows have the same ratio under θ-column. Therefore we apply *perturbation* method.

First column of the unit matrix has 1 and 0 in the tied rows. Dividing these by the corresponding elements of the key column, we get 1/1 and 0/5. s_2-row gives the smaller ratio and therefore s_2 is the outgoing variable and (5) is the key element.

Thus the new simplex table is

c_B	c_j Basis	5 x_1	3 x_2	0 s_1	0 s_2	0 s_3	b	θ
0	s_1	0	(3/5)	1	–1/5	0	0	0 ←
5	s_2	1	2/5	0	1/5	0	2	5
0	s_3	0	34/5	0	–3/5	1	6	15/17
	Z_j	5	2	0	1	0	10	
	C_j	0	1	0	–1	0		
			↑					

As C_j is positive under x_2 column, this solution is not optimal.

Making key element (3/5) unity and replacing s_1 by x_2, we obtain the revised simplex table:

c_B	c_j Basis	5 x_1	3 x_2	0 s_1	0 s_2	0 s_3	b
3	x_1	0	1	5/3	–1/3	0	0
5	x_2	1	0	–2/3	1/3	0	2
0	s_3	0	0	–34/3	5/3	1	6
	Z_j	5	3	5/3	2/3	0	10
	C_j	0	0	–5/3	–2/3	0	

As $C_j \leq 0$ under all columns, this table gives the optimal solution. Hence an optimal basic feasible solution is $x_1 = 2$, $x_2 = 0$ and $Z_{max} = 10$.

Exercises 12.5

Solve the following L.P. problems using the M-method:

1. Maximize $Z = 3x_1 + 2x_2 + 3x_3$
subject to: $2x_1 + x_2 + x_3 \leq 2$, $3x_1 + 4x_2 + 2x_3 \geq 8$, $x_1, x_2, x_3 \geq 0$.

2. Maximize $Z = 2x_1 + x_2 + 3x_3$
subject to: $x_1 + x_2 + 2x_3 \leq 5$, $2x_1 + 3x_2 + 4x_3 = 12$, $x_1, x_2, x_3 \geq 0$.

3. Maximize $Z = 8x_2$,
subject to: $x_1 - x_2 \geq 0$, $2x_1 + 3x_2 \leq -6$, x_1, x_2 unrestricted.

4. Minimize $Z = 4x_1 + 3x_2 + x_3$
subject to: $x_1 + 2x_2 + 4x_3 \geq 12$, $3x_1 + 2x_2 + x_3 \geq 8$, $x_1, x_2, x_3 \geq 0$.

5. Maximize $Z = x_1 + 2x_2 + 3x_3 - x_4$
subject to: $x_1 + 2x_2 + 3x_3 = 15$, $2x_1 + x_2 + 5x_3 = 20$,
$x_1 + 2x_2 + x_3 + x_4 = 10$, $x_1, x_2, x_3, x_4 \geq 0$.
Use two phase method to solve the following L.P. problems:

6. Minimize $Z = x_1 + x_2$
subject to: $2x_1 + x_2 \geq 4$, $x_1 + 7x_2 \geq 7$, $x_1, x_2 \geq 0$.

7. Maximize $Z = 5x_1 + 3x_2$
subject to: $2x_1 + x_2 \leq 1$, $x_1 + 4x_2 \geq 6$, $x_1, x_2 \geq 0$.

8. Maximize $Z = 5x_1 - 2x_2 + 3x_3$,
subject to: $2x_1 + 2x_2 - x_3 \geq 2$,
$3x_1 - 4x_2 \leq 3$, $x_2 + x_3 \leq 5$, $x_1, x_2, x_3 \geq 0$.

9. Maximize $Z = 5x_1 - 4x_2 + 3x_3$
subject to: $2x_1 + x_2 - 6x_3 = 20$, $6x_1 + 5x_2 + 10x_3 \leq 76$,
$8x_1 - 3x_2 + 6x_3 \leq 50$, $x_1, x_2, x_3 \geq 0$.
Solve the following degenerate L.P. problems:

10. Maximize $Z = 9x_1 + 3x_2$
subject to: $4x_1 + x_2 \leq 8$, $2x_1 + x_2 \leq 4$, $x_1, x_2 \geq 0$.

11. Maximize $Z = 2x_1 + 3x_2 + 10x_3$
subject to: $x_1 + 2x_3 = 0$, $x_2 + x_3 = 1$, $x_1, x_2, x_3 \geq 0$.

12. Maximize $Z = 0.5x_1 + 6x_2 + 5x_3$

subject to: $4x_1 + 6x_2 + 3x_3 \leq 24$, $x_1 + 1.5x_2 + 3x_3 \leq 12$, $3x_1 + x_2 \leq 12$,

$x_1, x_2, x_3 \geq 0$.

12.11 Duality Concept

One of the most interesting concepts in linear programming is the *duality* theory. Every linear programming problem has associated with it, another linear programming problem involving the same data and closely related optimal solutions. Such two problems are said to be *duals* of each other. While one of these is called the *primal*, the other the *dual*.

The importance of the duality concept is due to two main reasons. First, if the primal contains a large number of constraints and a smaller of variables, the labor of computation can be considerably reduced by converting it into the dual problem and then solving it. Secondly, the interpretation of the dual variables from the cost or economic point of view proves extremely useful in making future decisions in the activities being programmed.

Formulation of dual problem. Consider the following L.P.P:

Maximize $Z = c1x_1 + c2x_2 + \cdots + c_n x_n$,

subject to the constraints $a_{11}x_1 + a_{12}x_2 + \cdots + a_{1n}x_n \leq b_1$,

$$a_{21}x_1 + a_{22}x_2 + \cdots + a_{2n}x_n \leq b2,$$

....

$$a_{m1}x_1 + a_{m2}x_2 + \cdots + a_{mn}x_n \leq b_m,$$

$$x_1, x_2, \cdots, xn \geq 0.$$

To construct the dual problem, we adopt the following guide-lines:

(*i*) The maximization problem in the primal becomes the minimization problem in the dual and *vice versa*.

(*ii*) ($\leq$) type of constraints in the primal become ($\geq$) type of constraints in the dual and *vice versa*.

(*iii*) The coefficients $c_1, c_2, \cdots, c_n$ in the objective function of the primal become $b_1, b_2, \cdots, b_n$ in the objective function of the dual.

(*iv*) The constants $b_1, b_2, \cdots, b_n$ in the constraints of the primal become $c_1, c_2, \cdots, c_n$ in the constraints of the dual.

(v) If the primal has n variables and m constraints, the dual will have m variables and n constraints, i.e., the transpose of the body matrix of the primal problem gives the body matrix of the dual.

(vi) The variables in both the primal and dual are non-negative.

Then the dual problem will be

Minimize $W = b_1 y_1 + b_2 y_2 + \cdots + b_m y_m$

subject to the constraints $\quad a_{11} y_1 + a_{12} y_2 + \cdots + a_{m1} y_m \geq c_1,$

$$a_{21} y_1 + a_{22} y_2 + \cdots + a_{m2} y_m \geq c_2,$$

$$\cdots\cdots\cdots\cdots\cdots\cdots\cdots\cdots\cdots\cdots\cdots\cdots$$

$$a_{1n} y_1 + a_{2n} y_2 + \cdots + a_{mn} y_m \geq c_n,$$

$$y_1, y_2, \cdots, y_m \geq 0.$$

EXAMPLE 12.23

Write the dual of the following L.P.P:

Minimize $Z = 3x_1 - 2x_2 + 4x_3,$

subject to $\quad 3x_1 + 5x_2 + 4x_3 \geq 7, \; 6x_1 + x_2 + 3x_3 \geq 4,$

$$7x_1 - 2x_2 - x_3 \leq 10, \; x_1 - 2x_2 + 5x_3 \geq 3,$$

$$4x_1 + 7x_2 - 2x_3 \geq 2, \; x_1, x_2, x_3 \geq 0.$$

Solution:

Since the problem is of minimization, all constraints should be of $\geq$ type. We multiply the third constraint throughout by -1 so that

$$-7x_1 + 2x_2 + x_3 \geq -10$$

Let y_1, y_2, y_3, y_4 and y_5 be the dual variables associated with the above five constraints. Then the dual problem is given by

Maximize $W = 7y_1 + 4y_2 - 10y_3 + 3y_4 + 2y_5$

subject to $3y_1 + 6y_2 - 7y_3 + y_4 + 4y_5 \leq 3, \; 5y_1 + y_2 + 2y_3 - 2y_4 + 7y_5 \leq -2$

$4y_1 + 3y_2 + y_3 + 5y_4 - 2y_5 \leq 4, \; y_1, y_2, y_3, y_4, y_5 \geq 0.$

Formulation of dual problem when the primal has equality constraints. Consider the problem.

Maximize $\quad Z = c_1 x_1 + c_2 x_2$

subject to $\quad a_{11} x_1 + a_{12} x_2 = b_1 \, ; \, a_{21} x_1 + a_{22} x_2 \le b_2 \, ; \, x_1, x_2 \ge 0.$

The equality constraint can be written as

$$a_{11} x_1 + a_{12} x_2 \le b_1 \text{ and } a_{11} x1 + a_{12} x_2 \ge b_1$$

or $\quad a_{11} x_1 + a_{12} x_2 \le b_1 \text{ and } - a_{11} x_1 - a_{12} x_2 \le - b_1$

Then the above problem can be restated as

Maximize $Z = c_1 x_1 + c_2 x_2$

subject to $\quad a_{11} x_1 + a_{12} x_2 \le b_1, \, - a_{11} x_1 - a_{12} x_2 \le - b_1,$

$$a_{21} x_1 + a_{22} x_2 \le b_2, x_1, x_2 \ge 0.$$

Now we form the dual using y_1', y_1'', y_2 as the dual variables.

Then the dual problem is

Minimize $W = b_1 (y_1' - y_1'') + b_2 y_2,$

subject to $\quad a_{11}(y_1' - y_1'') + a_{21} y_2 \ge c_1, \, a_{12}(y_1' - y_1'') + a_{22} y_2 \ge c2, \, y_1', y_1'',$
$y_2 \ge 0.$

The term $(y_1' - y_1'')$ appears in both the objective function and all the constraints of the dual. This will always happen whenever there is an equality constraint in the primal. Then the new variable $y_1' - y_1'' \, (= y_1)$ becomes unrestricted in sign being the difference of two non-negative variables and the above dual problem takes the form.

Minimize $W = b_1 y_1 + b_2 y_2,$

subject to $\quad a_{11} b_1 + a_{21} y_2 \ge c1, \, a_{12} y_1 + a_{22} y_2 \ge c_2 ,$

y_1 unrestricted in sign, $y_2 \ge 0.$

In general, if the primal problem is

Maximize $Z = c_1 x_1 + c_2 x_2 + \cdots + c_n x_n ,$

subject to $\quad a_{11} x_1 + a_{12} x_2 + \cdots + a_{1n} x_n = b_1$

$$a_{21} x_1 + a_{22} x_2 + ... + a_{2n} x_n = b_2$$

$$\cdots\cdots\cdots\cdots\cdots\cdots\cdots\cdots$$

$$a_{m1} x_1 + a_{m2} x_2 + \cdots + a_{mn} x_n = b_m$$

$$x_1, x_2, \cdots, x_n \ge 0;$$

then the dual problem is

Minimize $W = b_1 y_1 + b_2 y_2 + \cdots + b_m y_m$

subject to $\quad a_{11} y_1 + a_{21} y_2 + \dots + a_{m1} y_m \geq c_1,$

$\qquad\qquad a_{12} y_1 + a_{22} y_2 + \dots + a_{m2} y_m \geq c_2,$

$\qquad\qquad \dots\dots\dots\dots\dots\dots\dots\dots\dots\dots$

$\qquad\qquad a_{1n} y_1 + a_{2n} y_2 + \dots + a_{mn} y_m \geq c_n,$

$\qquad\qquad y_1, y_2, \cdots, y_m$ all unrestricted in sign.

Thus *the dual variables corresponding to equality constraints are unrestricted in sign. Conversely when the primal variables are unrestricted in sign, the corresponding dual constraints are equalities.*

EXAMPLE 12.24

Construct the dual of the L.P.P:

Maximize $Z = 4x_1 + 9x_2 + 2x_3,$

subject to $\qquad 2x_1 + 3x_2 + 2x_3 \leq 7,\ 3x_1 - 2x_2 + 4x_3 = 5,\ x_1, x_2, x_3 \geq 0.$

Solution:

Let y_1 and y_2 be the dual variables associated with the first and second constraints. Then the dual problem is

Minimize $W = 7y_1 + 5y_2,$

subject to $2y_1 + 3y_2 \leq 4,\ 3y_1 - 2y_2 \leq 9,\ 2y_1 + 4y_2 \leq 2,\ y_1 \geq 0,\ y_2$ is unrestricted in sign.

Exercises 12.6

Write the duals of the following problems:

1. Maximize $Z = 10x_1 + 13x_2 + 19x_3$

subject to $\qquad 6x_1 + 5x_2 + 3x_3 \leq 26,\ 4x_1 + 2x_2 + 5x_3 \leq 7,\ x_1, x_2, x_3 \geq 0.$

2. Minimize $Z = 2x_1 + 4x_2 + 3x_3$

subject to $\qquad 3x_1 + 4x_2 + x_3 \geq 11,\ -2x_1 - 3x_2 + 2x_3 \leq -7,$

$\qquad\qquad x_1 - 2x_2 - 3x_3 \leq -1,\ 3x_1 + 2x_2 + 2x_3 \geq 5,\ x_1, x_2, x_3 \geq 0.$

3. Maximize $Z = 3x_1 + x_2 + 4x_3 + x_4 + 9x_5$,
$$4x_1 - 5x_2 - 9x_3 + x_4 - 2x5 \leq 6,\ 2x_1 + 3x_2 + 4x_3 - 5x_4 + x5 \leq 9,$$
$$x_1 + x_2 - 5x_3 - 7x_4 + 11x5 \leq 10,\ x_1, x_2, x_3, x_4, x5 \geq 0.$$

4. Maximize $Z = 3x_1 + 16x_2 + 7x_3$
   subject to $\quad x_1 - x_2 + 3x_3 \geq 3,\ -3x_1 + 2x_3 \leq 1,$
$$2x_1 + x_2 - x_3 = 4,\ x_1, x_2, x_3 \geq 0.$$

5. Maximize $Z = 3x_1 + x_2 + 2x_3$
   subject to $\quad x_1 + x_2 + x_3 \geq 6,\ 3x_1 - 2x_2 + 3x_3 = 3,$
$$-4x_1 + 3x_2 - 6x_3 = 4,\ x_1, x_2, x_3 \geq 0.$$

6. Minimize $Z = 2x_1 + 3x_2 + 4x_3$
   subject to $\quad 2x_1 + 3x_2 + 5x_3 \geq 2,\ 3x_1 + x_2 + 7x_3 = 3,$
$$x_1 + 4x_2 + 6x_3 \leq 5,\ x_1, x_2 \geq 0 \text{ and } x_3 \text{ is unrestricted.}$$

7. Obtain the dual problem of the following L.P.P:
   Maximize $f(x) = 2x_1 + 5x_2 + 6x_3$
   subject to the constraints:
$$5x_1 + 6x_2 - x_3 \leq 3,\ -2x_1 + x_2 + 4x_3 \leq 4,$$
$$x_1 - 5x_2 + 3x_3 \leq 1,\ -3x_1 - 3x_2 + 7x_3 \leq 6,\ x_1, x_2, x_3 \geq 0.$$

Also verify that the dual of the dual problem is the primal problem.

12.12   Duality Principle

If the primal and the dual problems have feasible solutions then both have optimal solutions and the optimal value of the primal objective function is equal to the optimal value of the dual objective function, i.e.,

Max. Z = Min. W

This is the fundamental theorem of duality. It suggests that an optimal solution to the primal can directly be obtained from that of the dual problem and *vice-versa*.

Working rules for obtaining an optimal solution to the primal (dual) problem from that of the dual (primal):

Suppose we have already found an optimal solution to the dual (primal) problem by the simplex method.

Rule I. If the primal variable corresponds to a slack starting variable in the dual problem, then its optimal value is directly given by the coefficient of the slack variable with a changed sign, in the C_j row of the optimal dual simplex table and *vice-versa*.

Rule II. If the primal variable corresponds to an artificial starting variable in the dual problem, then its optimal value is directly given by the coefficient of the artificial variable, with a changed sign, in the C_j row of the optimal dual simplex table, after deleting the constant M and *vice-versa*.

On the other hand, if the primal has an unbounded solution, then the dual problem will not have a feasible solution and *vice-versa*.

Now we shall work out two examples to demonstrate the primal dual relationships.

EXAMPLE 12.25

Construct the dual of the following problem and solve the primal and the dual:

Maximize $Z = 2x_1 + x_2$,

subject to $-x_1 + 2x_2 \leq 2,\ x_1 + x_2 \leq 4,\ x_1 \leq 3,\ x_1, x_2 \geq 0.$

Solution:

Using the primal problem. Since only two variables are involved, it is convenient to solve the problem graphically.

In the $x_1 x_2$-plane, the five constraints show that the point (x_1, x_2) lies within the shaded region *OABCD* of Figure 12.12.

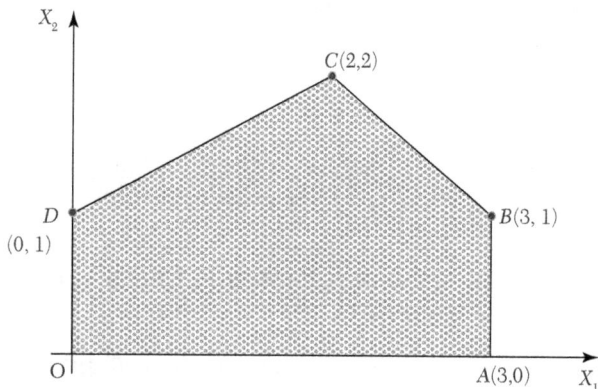

FIGURE 12.12

Values of the objective function $Z = 2x_1 + x_2$ at these corners are $Z(0) = 0$, $Z(A) = 6$, $Z(B) = 7$, $Z(C) = 6$, and $Z(D) = 1$.

Hence the optimal solution is $x_1 = 3$, $x_2 = 1$ and max. $Z = 7$.

Solution:

Using the dual problem. The dual problem of the given primal is:
Minimize $W = 2y_1 + 4y_2 + 3y_3$
subject to $-y_1 + y_2 + y_3 \geq 2, 2y_1 + y_2 \geq 1, y_1, y_2 \geq 0.$

Step 1. Express the problem in the standard form.

Introducing the slack and the artificial variables, the dual problem in the standard form is

Max. $W' = -2y_1 - 4y_2 - 3y_3 + 0s_1 + 0s_2 - MA_1 - MA_2$
subject to $-y_1 + y_2 + y_3 - s_1 + 0s_2 + A_1 + 0A_2 = 2,$
$2y_1 + y_2 + 0y_3 + 0s_1 - s_2 + 0A_1 + A_2 = 1$

Step 2. Find an initial basic feasible solution.

Setting the non-basic variables y_1, y_2, y_3, s_1, s_2 each equal to zero, we get the initial basic feasible solution as

$y_1 = y_2 = y_3 = s_1 = s_2 = 0$ (non-basic), $A_1 = 2, A_2 = 1.$ (basic)

$\therefore$ Initial simplex table is

c_B	c_j Basis	-2 y_1	-4 y_2	-3 y_3	0 s_1	0 s_2	$-M$ A_1	$-M$ A_2	b	θ
$-M$	A_1	-1	1	1	-1	0	1	0	2	2/1
$-M$	A_2	2	(1)	0	0	-1	0	1	1	1/1←
	Z_j	$-M$	$-2M$	$-M$	M	M	$-M$	$-M$	$-3M$	
	C_j	$M-2$	$2M-4$	$M-3$	$-M$	$-M$	0	0		
			↑							

As C_j is positive under some columns, the initial solution is not optimal.

Step 3. Iterate toward an optimal solution.

(i) Introduce y_2 and drop A_2. Then the new simplex table is

c_B	c_j Basis	-2 y_1	-4 y_2	-3 y_3	0 s_1	0 s_2	$-M$ A_1	$-M$ A_2	b	θ
$-M$	A_1	-3	0	(1)	-1	1	1	-1	1	1/1←
-4	y_2	2	1	0	0	-1	0	1	1	1/0
	Z_j	$3M-8$	-4	$-M$	M	$4-M$	$-M$	$M-4$	$-M-4$	
	C_j	$6-3M$	0	$M-3$	$-M$	$M-4$	0	0		
				↑						

As C_j is positive under some columns, this solution is not optimal.

(*ii*) Now introduce y_3 and drop A_1. Then the revised simplex table is

	c_j	-2	-4	-3	0	0	$-M$	$-M$	
c_B	Basis	y_1	y_2	y_3	s_1	s_2	A_1	A_2	b
-3	Y_3	-3	0	1	-1	1	1	-1	1
-4	y_2	2	1	0	0	-1	0	1	1
	Z_j	1	-4	-3	3	1	-3	1	-7
	C_j	-3	0	0	-3	-1	$3-M$	$1-M$	
				$\uparrow$					

As all $C_j \le 0$, the optimal solution is attained.

Thus an optimal solution to the dual problem is

$$y_1 = 0, y_2 = 1, y_3 = 1, \text{ Min. } W = -\text{ Max. } (W') = 7.$$

To derive the optimal basic feasible solution to the primal problem, we note that the primal variables x_1, x_2 correspond to the artificial starting dual variables A_1, A_2, respectively. In the final simplex table of the dual problem, C_j corresponding to A_1 and A_2 are 3 and 1, respectively after ignoring M. Thus by rule II, we get opt. $x_1 = 3$ and opt. $x_2 = 1$.

Hence an optimal basic feasible solution to the given primal is

$$x_1 = 3, x_2 = 1; \text{ max. } Z = 7.$$

NOTE   **Obs.** *The validity of the duality theorem is therefore, checked since max. Z = min. W = 7 from both the methods.*

EXAMPLE 12.26

Using duality solve the following problem:

Minimize $Z = 0.7 x_1 + 0.5x_2$

subject to $x_1 \ge 4, x_2 \ge 6, x_1 + 2x_2 \ge 20, 2x_1 + x_2 \ge 18, x_1, x_2 \ge 0.$

Solution:

The dual of the given problem is

Max. $W = 4y_1 + 6y_2 + 20y_3 + 18y_4$,

subject to $y_1 + y_3 + 2y_4 \le 0.7, y_2 + 2y_3 + y_4 \le 0.5, y_1, y_2, y_3, y_4 \ge 0$

Step 1. Express the problem in the standard form.

Introducing slack variables, the dual problem in the standard form becomes

$$\text{Max. } W = 4y_1 + 6y_2 + 20y_3 + 18y_4 + 0s_1 + 0s_2,$$

$$\text{subject to} \quad y_1 + 0y_2 + y_3 + 2y_4 + s_1 + 0s_2 = 0.7,$$

$$0y_1 + y_2 + 2y_3 + y_4 + 0s_1 + s_2 = 0.5, \, y_1, y_2, y_3, y_4 \geq 0.$$

Step 2. Find an initial basic feasible solution.

Setting non-basic variables y_1, y_2, y_3, y_4 each equal to zero, the basic solution is

$$y_1 = y_2 = y_3 = y_4 = 0 \text{ (non-basic)}, s_1 = 0.7, s_2 = 0.5 \text{ (basic)}$$

Since the basic variables s_1, $s_2 > 0$, the initial basic solution is feasible and non-degenerate.

$\therefore$ Initial simplex table is

c_B	Basis	c_j 4	6	20	18	0	0	b	θ
		y_1	y_2	y_3	y_4	s_1	s_2		
0	s_1	1	0	1	2	1	0	0.7	0.7/1
0	s_2	0	1	(2)	1	0	1	0.5	0.5/2←
	Z_j	0	0	0	0	0	0	0	
	C_j	4	6	20	18	0	0		
				↑					

As C_j is positive in some columns, the initial basic solution is not optimal.

Step 3. Iterate towards an optimal solution.

(*i*) Introduce y_3 and drop s_2. Then the new simplex table is

c_B	Basis	c_j 4	6	20	18	0	0	b	θ
		y_1	y_2	y_3	y_4	s_1	s_2		
0	s_1	1	−1/2	0	(3/2)	1	−1/2	9/20	3/10 ←
2	y_3	0	1/2	1	1/2	0	1/2	1/4	1/2
	Z_j	0	10	20	10	0	10	5	
	C_j	4	−4	0	8	0	−10		
					↑				

As C_j is positive under some of the columns, this solution is not optimal.

(*ii*) Introduce y_4 and drop s_1. Then the revised simplex table is

	c_j	4	6	20	18	0	0	
c_B	*Basis*	y_1	y_2	y_3	y_4	s_1	s_2	b
18	y_1	2/3	−1/3	0	1	2/3	−1/3	3/10
20	y_3	−1/3	2/3	1	0	−1/3	2/3	1/10
	Z_j	16/3	22/3	18	18	16/3	22/3	74/10
	C_j	−4/3	−4/3	0	0	−16/3	−22/3	

As all $C_j \leq 0$, the table gives the optimal solution.

Thus the optimal basic feasible solution is

$$y_1 = 0, y_2 = 0, y_3 = 20, y_4 = 18 \text{ max. } W = 7.4$$

Step 4. Derive optimal solution to the primal.

We note that the primal variable x_1, x_2 corresponds to the slack starting dual variables s_1, s_2 respectively. In the final simplex table of the dual problem. C_j values corresponding to s_1 and s_2 are − 16/3 and − 22/3, respectively.

Thus, by rule I, we conclude that

opt. x_1 = 16/3 and opt. x_2 = 22/3.

Hence an optimal basic feasible solution to the given primal is

$$x_1 = 16/3, x_2 = 22/3 \text{ ; min. } Z = 7.4.$$

NOTE **Obs.** *To check the validity of the duality theorem, the student is advised to solve the given L.P.P. directly by simplex method and see that*

$$\text{min. } Z = \text{max. } W = 7.4.$$

Exercises 12.7

Using duality solve the following problems (1—3):

1. Minimize $Z = 2x_1 + 9x_2 + x_3$,
subject to $x_1 + 4x_2 + 2x_3 \geq 5$, $3x_1 + x_2 + 2x_3 \geq 4$, $x_1, x_2, x_3 \geq 0$

2. Maximize $Z = 2x_1 + x_2$,
subject to $x_1 + 2x_2 \leq 10$, $x_1 + x_2 \leq 6$, $x_1 - x_2 \leq 2$, $x_1 - 2x_2 \leq 1$, $x_1, x_2 \geq 0$.

3. Maximize $Z = 3x_1 + 2x_2$,
subject to $x_1 + x_2 \geq 1$, $x_1 + x_2 \leq 7$, $x_1 + 2x_2 \leq 10$, $x_2 \leq 3$, $x_1, x_2 \geq 0$.

4. Maximize $Z = 3x_1 + 2x_2 + 5x_3$

subject to $x_1 + 2x_2 + x_3 \leq 430$, $3x_1 + 2x_3 \leq 460$, $x_1 + 4x_2 \leq 420$, $x_1, x_2, x_3 \geq 0$.

5. Write the dual of the following problem and solve the dual.

Maximize $Z = -2x_1 - 2x_2 - 4x_3$,

subject to $2x_1 + 3x_2 + 5x_3 \geq 2$, $3x_1 + x_2 + 7x_3 \geq 3$,

$x_1 + 4x_2 + 6x_3 \leq 5$, $x_1, x_2, x_3 \geq 0$.

12.13   Dual Simplex Method

In Section 12.9, we have seen that a set of basic variables giving a feasible solution can be found by introducing artificial variables and using the M-method or Two-phase method. Using the primal-dual relationships for a problem, we have another method (known as *Dual simplex method*) for finding an initial feasible solution. Whereas the regular simplex method starts with a basic feasible (but non-optimal) solution and works towards optimality, the dual simplex method starts with a basic infeasible (but optimal) solution and works towards feasibility. The dual simplex method is quite similar to the regular simplex method, the only difference lies in the criteria used for selecting the incoming and outgoing variables, In the dual simplex method, we first determine the outgoing variable and then the incoming variable while in the case of regular simplex method the reverse is done.

Working procedure for dual simplex method:

Step 1. (*i*) *Convert the problem to maximization form,* if it is not so.

(*ii*) *Convert* (≥) *type constraints, if any to* (≤) *type* by multiplying such constraints by – 1.

(*iii*) *Express the problem in standard form* by introducing slack variables.

Step 2. *Find the initial basic solution* and express this information in the form of *dual simplex table.*

Step 3. *Test the nature of* $C_j = c_j - Z_j$:

(*a*) If all $C_j \leq 0$ and all $bi \geq 0$, then optimal basic feasible solution has been attained.

(*b*) If all $C_j \leq 0$ and at least one $bi < 0$, then go to step 4.

(*c*) If any $C_j \geq 0$, the method fails.

Step 4. *Mark the outgoing variable.* Select the row that contains the most negative bi. This will be the key row and the corresponding basic variable is the outgoing variable.

Step 5. *Test the nature of key row elements:*

(*a*) If all these elements are ≥ 0, the problem does not have a feasible solution.

(*b*) If at least one element < 0, find the ratios of the corresponding elements of C_j-row to these elements. Choose the smallest of these ratios. The corresponding column is the key column and the associated variable is the *incoming variable.*

Step 6. *Iterate towards optimal feasible solution.* Make the key element unity. Perform row operations as in the regular simplex method and repeat iterations until either an optimal feasible solution is attained or there is an indication of non-existence of a feasible solution.

EXAMPLE 12.27

Using dual simplex method:

maximize $-3x_1 - 2x_2$,

subject to $x_1 + x_2 \geq 1$, $x_1 + x_2 \leq 7$, $x_1 + 2x_2 \geq 10$, $x_2 \leq 3$, $x_1 \geq 0$, $x_2 \geq 0$.

Solution:

Consists of the following steps:

Step 1. (*i*) *Convert the first and third constraints into* ($\leq$) *type.*

These constraints become

$$-x_1 - x_2 \leq -1, -x_1 - 2x_2 \leq -10.$$

(*ii*) *Express the problem in standard form*

Introducing slack variables s_1, s_2, s_3, s_4 the given problem takes the form

Max. $Z = -3x_1 - 2x_2 + 0s_1 + 0s_2 + 0s_3 + 0s_4$

subject to $-x_1 - x_2 + s_1 = -1$, $x_1 + x_2 + s_2 = 7$,

$$-x_1 - 2x_2 + s_3 = -10, x_2 + s_4 = 3, x_1, x_2, s_1, s_2, s_3, s_4 \geq 0.$$

Step 2. Find the initial basic solution

Setting the decision variables x_1, x_2 each equal to zero, we get the basic solution

$$x_1 = x_2 = 0, s_1 = -1, s_2 = 7, s_3 = -10, s_4 = 3 \text{ and } Z = 0.$$

∴ Initial solution is given by the table below:

	c_j	−3	−2	0	0	0	0	
c_B	Basis	x_1	x_2	s_1	s_2	s_3	s_4	b
0	s_1	−1	−1	1	0	0	0	−1
0	s_2	1	1	0	1	0	0	7
0	s_3	−1	(−2)	0	0	1	0	−10 ←
	s_4	0	1	0	0	0	1	3
	$Z_j = \Sigma c_B a_{ij}$	0	0	0	0	0	0	0
	$C_j = c_j - Z_j$	−3	−2	0	0	0	0	
			↑					

Step 3. Test nature of C_j.

Since all C_j values are ≤ 0 and $b_1 = -1$, $b3 = -10$, the initial solution is optimal but infeasible. We therefore, proceed further.

Step 4. Mark the outgoing variable.

Since b_3 is negative and numerically largest, the third row is the key row and s_3 is the outgoing variable.

Step 5. Calculate ratios of elements in C_j-row to the corresponding negative elements of the key row.

These ratios are $-3/-1 = 3$, $-2/-2 = 1$ (neglecting ratios corresponding to + ve or zero elements of key row). Since the smaller ratio is 1, therefore, x_2-column is the key column and (−2) is the key element.

Step 6. Iterate towards optimal feasible solution.

(*i*) Drop s_3 and introduce x_2 alongwith its associated value −2 under c_B column. Convert the key element to unity and make all other elements of the key column zero. Then the second solution is given by the table below:

c_j		-3	-2	0	0	0	0	
c_B	Basis	x_1	x_2	s_1	s_2	s_3	s_4	b
0	s_1	$-1/2$	0	1	0	$-1/2$	0	4
0	s_2	$1/2$	0	0	1	$1/2$	0	2
-2	x_2	$1/2$	1	0	0	$-1/2$	0	5
0	s_4	$-1/2$	0	0	0	$1/2$	1	$-2 \leftarrow$
	$Z_j = \Sigma\, c_B a_{ij}$	-1	-2	0	0	1	0	-10
	$C_j = c_j - Z_j$	-2	0	0	0	-1	0	
		$\uparrow$						

Since all C_j values are ≤ 0 and $b4 = -2$, this solution is optimal but infeasible. We therefore proceed further.

(*ii*) *Mark the outgoing variable*

Since b_4 is negative, the fourth row is the key row and s_4 is the outgoing variable. (*iii*) *Calculate ratios of elements in C_j-row to the corresponding negative elements of the key row.*

This ratio is $\dfrac{-2}{-\frac{1}{2}} = 4$ (neglecting other ratios corresponding to + ve or 0 elements of key row).

∴ x_1-column is the key column and $\left(-\dfrac{1}{2}\right)$ is the key element.

(*iv*) *Drop s_4 and introduce x_1* with its associated value -3 under the c_B column. Convert the key element to unity and make all other elements of the key column zero. Then the third solution is given by the table below:

c_j		-3	-2	0	0	0	0	
c_B	Basis	x_1	x_2	s_1	s_2	s_3	s_4	b
0	s_1	0	0	1	0	-1	-1	6
0	s_2	0	0	0	1	1	1	0
-2	x_2	0	1	0	0	0	1	3
-3	x_1	1	0	0	0	-10	-2	4
Z_j		-3	-2	0	0	3	4	-18
C_j		0	0	0	0	-3	-4	

Since all C_j values are ≤ 0 and all b's are ≥ 0, therefore this solution is optimal and feasible. Thus the *optimal solution* is $x_1 = 4$, $x_2 = 3$ and $Z_{max} = -18$.

EXAMPLE 12.28

Using dual simplex method, solve the following problem:

Minimize $Z = 2x_1 + 2x_2 + 4x_3$

subject to $2x_1 + 3x_2 + 5x_3 \geq 2,\ 3x_1 + x_2 + 7x_3 \leq 3,$

$$x_1 + 4x_2 + 6x_3 \leq 5,\ x_1,\ x_2,\ x_3 \geq 0$$

Solution:

Consists of the following steps:

Step 1. (*i*) *Convert the given problem to maximization form* by writing

Maximize $Z' = -2x_1 - 2x_2 - 4x_3$.

(*ii*) *Convert the first constraint into* ($\leq$) *type.* Thus it is equivalent to

$-2x_1 - 3x_2 - 5x_3 \leq -2$

(*iii*) *Express the problem in standard form.*

Introducing slack variables s_1, s_2, s_3, the given problem becomes

max. $Z' = -2x_1 - 2x_2 - 4x_3 + 0s_1 + 0s_2 + 0s_3$

subject to $-2x_1 - 3x_2 - 5x_3 + s_1 + 0s_2 + 0s_3 = -2,\ 3x_1 + x_2 + 7x_3 + 0s_1 + s_2 + 0s_3 = 3,\ x_1 + 4x_2 + 6x_3 + 0s_1 + 0s_2 + s_3 = 5,\ x_1, x_2, x_3, s_1, s_2, s_3 \geq 0.$

Step 2. *Find the initial basic solution.*

Setting the decision variables x_1, x_2, x_3 each equal to zero, we get the basic solution

$$x_1 = x_2 = x_3 = 0,\ s_1 = -2,\ s_2 = 3,\ s_3 = 5 \text{ and } Z' = 0.$$

∴ Initial solution is given by the table below:

c_B	c_j Basis	-2 x_1	-2 x_2	-4 x_3	0 s_1	0 s_2	0 s_3	b
0	s_1	-2	(-3)	-5	1	0	0	$-2 \leftarrow$
0	s_2	3	1	7	0	1	0	3
0	s_3	1	4	6	0	0	1	5
	Z_j	0	0	0	0	0	0	0
	C_j	-2	-2	4	0	0	0	
			↑					

Step 3. Test nature of C_j.

Since all C_j values are ≤ 0 and $b_1 = -2$, the initial solution is optimal but infeasible.

Step 4. Mark the outgoing variable.

Since $b_1 < 0$, the first row is the key row and s_1 is the outgoing variable.

Step 5. Calculate the ratio of elements of C_j-row to the corresponding negative elements of the key row.

These ratios are $-2/-2 = 1$, $-2/-3 = 0.67$, $-4/-5 = 0.8$.

Since 0.67 is the smallest ratio, x_2-column is the key column and (-3) is the key element. *Step 6. Iterate towards optimal feasible solution.*

Drop s_1 and introduce x_2 with its associated value -2 under c_B column. Then the revised dual simplex table is

c_B	Basis	c_j	-2	-2	-4	0	0	0	
			x_1	x_2	x_3	s_1	s_2	s_3	b
0	x_2		$2/3$	1	$5/3$	$-1/3$	0	0	$2/3$
0	s_2		$7/3$	0	$16/3$	$1/3$	1	0	$7/3$
0	S_3		$-5/3$	0	$-2/3$	$4/3$	0	1	$7/3$
	Z_j		$-4/3$	-2	$-10/3$	$2/3$	0	0	$-4/3$
	C_j		$-2/3$	0	$-2/3$	$-2/3$	0	0	

Since all $C_j \leq 0$ and all bi are > 0, this solution is optimal and feasible. Thus the optimal solution is $x_1 = 0$, $x_2 = 2/3$, $x_3 = 0$ and max. $Z' = -4/3$

i.e.,   min. $Z = 4/3$.

Exercises 12.8

Using dual simplex method, solve the following problems:

1. Maximize $Z = -3x_1 - x_2$
subject to $x_1 + x_2 \geq 1$, $2x_1 + 3x_2 \geq 2$; $x_1, x_2 \geq 0$.

2. Minimize $Z = 2x_1 + x_2$,
subject to $3x_1 + x_2 \geq 3$, $4x_1 + 3x_2 \geq 6$, $x_1 + 2x_2 \leq 3$, $x_1, x_2 \geq 0$.

3. Minimize $Z = x_1 + 2x_2 + 3x_3$,
subject to $2x_1 - x_2 + x_3 \geq 4$, $x_1 + x_2 + 2x_3 \leq 8$, $x_2 - x_3 \geq 2$; $x_1, x_2, x_3 \geq 0$.

4. Minimize $Z = 6x_1 + 7x_2 + 3x_3 + 5x_4$,

subject to $5x_1 + 6x_2 - 3x_3 + 4x_4 \geq 12$, $x_2 + 5x_3 - 6x_4 \geq 10$,

$2x_1 + 5x_2 + x_3 + x_4 \geq 8$, $x_1, x_2, x_3, x_4 \geq 0$.

5. Minimize $Z = 3x_1 + 2x_2 + x_3 + 4x_4$

subject to $2x1 + 4x2 + 5x3 + x4 \geq 10$, $3x1 - x2 + 7x3 - 2x4 \geq 2$,

$5x_1 + 2x_2 + x_3 + 6x_4 \geq 15$, $x_1, x_2, x_3, x_4 \geq 0$.

12.14 Transportation Problem

This is a special class of linear programming problems in which the objective is to transport a single commodity from various origins to different destinations at a minimum cost.

Formulation of a transportation problem. There are m plant locations (origins) and n distribution center (destinations). The production capacity of the ith plant is ai and the number of units required at the jth destilnation is b_j. The transportation cost of one unit from the ith plant to the jth destination is c_{ij}. Our objective is to determine the number of units to be transported from the ith plant to jth destination so that the total transportation cost is minimum.

Let x_{ij} be the number of units shipped from ith plant to jth destination, then *the general transportation problem is:*

$$\text{Minimize } Z = \sum_{i=1}^{m} \sum_{j=1}^{n} c_{ij} x_{ij}$$

subject to the constraints

$x_{i1} + x_{i2} + \cdots + x_{in} = a_i$, for ith origin $(i = 1, 2, \cdots m)$

$x_{1j} + x_{2j} + \cdots + x_{mj} = b_j$, for destination $(j = 1, 2, \cdots n)$

$x_{ij} \geq 0$.

Def. 1. The two sets of constraints will be consistent if

$$\sum_{i=1}^{m} a_1 = \sum_{j=1}^{n} b_j$$

which is the condition for a transportation problem to have a *feasible solution*. Problems satisfying this condition are called *balanced transportation problems*.

2. A feasible solution to a transportation problem is said to be a *basic feasible solution* if it contains at the most $(m + n - 1)$ strictly positive allocations, otherwise the solution will *degenerate*. If the total number of positive (non-zero) allocations is exactly $(m + n - 1)$, then the basic feasible solution is said to be *non-degenerate*.

3. A feasible solution which minimizes the transportation cost is called an *optimal solution*.

This problem is explicitly represented in the following *transportation table*:

Distribution centers (Destinations)

The mn squares are called *cells*. The per unit cost c_{ij} of transporting from the ith origin to the jth destination is displayed in the *lower right side of the (i, j)th cell*. Any feasible solution is shown in the table by entering the value of x_{ij} in the *small square at the upper left side of the (i, j)th cell*. The various a's and b's are called *rim requirements*. The feasibility of a solution can be verified by summing the values of x_{ij} along the rows and down the columns.

NOTE **Obs. 1.** *The special features of a transportation problem are that*
(i) the coefficients of all x_{ij} in the constraints are unity, and
(ii) the total supply $\Sigma\, a_i =$ total demand Σb_j.

Obs. 2. *The objective function and the constraints being all linear, the problem can be solved be the simplex method. But the number of variables being large, there will be too many* *calculations. However, the coefficients of all x_{ij} in the constraints being unity, we can look for some technique which would be simpler than the simplex method.*

12.15   Working Procedure for Transportation Problems

Step 1. *Construct transportation table.* Express the supply from the origins a_i, demand at destinations b_j and the unit shipping cost c_{ij} in the form of a matrix, know as transportation table. If the supply and demand are equal, the problem is *balanced.*

Step 2. *Find the initial basic feasible solution.* We find an initial allocation which satisfies the demand at each project site without violating the capacities of the plants (origins) and also meeting the non-negativity restrictions. There are several methods for initial allocations *e.g.,* North-West corner rule, Row minima method, Least cost method, and Vogel's approximation method. *The Vogel's approximation method (VAM) takes into account not only the least cost c_{ij} but also the costs that just exceed the least cost c_{ij} and therefore yields a better initial solution than obtained from other methods. As such we shall confine ourselves to VAM only which consists of the following steps:*

(*i*) Display the difference between the least and the next to least costs in each row, by enclosing them in brackets to the right of the row. Similarly display the differences for each column within brackets below that column.

(*ii*) Identify the row or column with the largest difference among all the rows and columns and allocate as much as possible under the rim requirements, to the lowest cost cell in that row or column. In case of a tie allocate to the cell associated with the lower cost.

If the greatest difference corresponds to ith row and c_{ij} is the lowest cost in the ith row, allocate as much as possible, *i.e.,* min (a_i, b_j) in the (i, j)th cell and cross off the ith row or the jth column.

(*iii*) Recalculate the row and column differences for the reduced table and go to the previous step.

(*iv*) Repeat the procedure till all the rim requirements are satisfied. Note the solution in the upper left corner small squares of the basic cells.

Step 3. Apply optimality check.

In the above solution, the number of allocations must be "$m + n - 1$", otherwise the basic solution degenerates.

Now to test for optimality, we apply the modified distribution (MODI) method and examine each unoccupied cell to determine whether making an allocation in it reduces the total transportation cost and then repeat this procedure until the lowest possible transportation cost is obtained. This method consists of the following steps:

(*i*) Note the numbers *ui* along the left and *vj* along the top of the cost matrix such that their sums equal to the original costs of occupied cells, *i.e.*, solve the equations $[u_i + v_j = c_{ij}]$ starting initially with some $u_i = 0$.

(*ii*) *Compute the net evaluations* $w_{ij} = u_i + v_j - c_{ij}$ for all the empty cells and enter them in upper right hand corners of the corresponding cells.

(*iii*) *Examine the sign of each* w_{ij}. If all $w_{ij} \leq 0$, then the current basic feasible solution is optimal. If even one $w_{ij} > 0$, this solution is not optimal and we proceed further.

Step 4. Iterate towards an optimal solution

(*i*) Choose the unoccupied cell with the largest w_{ij} and mark θ in it.

(*ii*) Draw a closed path consisting of horizontal and vertical lines beginning and ending at θ-cell and having its other corners at the allocated cells.

(*iii*) Add and subtract θ alternately to and from the transition cells of the loop subject to rim requirements. Assign a maximum value to θ so that one basic variable becomes zero and the other basic variables remain non-negative. Now the basic cell whose allocation has been reduced to zero leaves the basis.

Step 5. Return to step 3 and repeat the process until an optimal basic feasible solution is obtained.

EXAMPLE 12.29

Solve the following transportation problem:

		A	B	C	D		
	I	21	16	25	13	11	
Source	II	17	18	14	23	13	Availability
	III	32	27	18	41	19	
	Requirement	6	10	12	15	43	

Solution *consists of the following steps:*

Step 1. *Transportation table.* Here the total availability and the total requirement being the same, *i.e.*, 43, the problem is balanced.

Step 2. *Find the initial basic feasible solution.* Following VAM, the differences between the smallest and next to the smallest costs in each row and each column are computed and displayed within brackets against the respective rows and columns (table 1). The largest of these differences is (10) which is associated with the fourth column.

Table 1

21	16	25	11⌐ 13	11 (3)
17	18	14	23	13 (3)
32	27	18	41	19 (9)
6	10	12	15	
(4)	(2)	(4)	(10)	

Table 2

17	18	14 4⌐	23	13 (3)
32	27	18	41	19 (9)
6	10	12	4	
(15)	(9)	(4)	(18)	

Since c_{14} (= 13) is the minimum cost, we allocate x_{14} = min (11, 15) = 11. This exhausts the availability of first row and therefore we cross it.

Table 3

6⌐ 17	18	14	9 (3)
32	27	18	19 (9)
6	10	12	
(15)	(9)	(4)	

Table 4

3⌐ 18	14	3 (4)
27	18	19 (9)
10	12	
(9)	(4)	

Table 5

7⌐	12⌐	19
27	18	
7	12	

The row and column differences are now computed for reduced Table 2 and displayed within brackets. The largest of these is (18) which is against the fourth column. Since c_{14} (= 23) is the minimum cost, we allocate x_{14} = min(13, 4) = 4.

This exhausts the availability of the fourth column which we cross off. Proceeding in this way, the subsequent reduced transportation tables and differences for the remaining rows and columns are shown in Tables 3, 4, and 5.

Finally the initial basic feasible solution is as shown in Table 6.

Table 6

21	16	25	13 [11]
6 [3] 17	18	14	[4] 23
[7] 32	[12] 27	18	41

Table 7

v_j \ u_i	17	18	9	23
-10	(−) 21	(−) 16	(−) 25	[11] 13
0	6 17	3 18	(−) 14	[4] 23
9	(−) 32	[7] 27	12 18	(−) 41

Step 3. Apply optimality check

As the number of allocations $= m + n - 1$ (*i.e.*, 6), we can apply the MODI method.

(*i*) We have $u_2 + v_1 = 17$, $u_2 + v_2 = 18$, $u_3 + v_2 = 27$

$$u_3 + v_3 = 18, u_1 + v_4 = 13, u_2 + v_4 = 23$$

Let $u_2 = 0$, then $v_1 = 17$, $v_2 = 18$, $u_3 = 9$, $v_3 = 9$, $v_4 = 23$, $u_1 = -10$.

(*ii*) Net evaluations $w_{ij} = (u_i + v_j) - c_{ij}$ for all empty cells are

$$w_{11} = -14, w_{12} = -8, w_{13} = -26, w_{23} = -5, w_{31} = -6, w_{34} = -9.$$

(*iii*) Since all the net evaluations are negative, the current solution is optimal. Hence the optimal allocation is given by

$$x_{14} = 11, x_{21} = 6, x_{22} = 3, x_{24} = 4, x_{32} = 7 \text{ and } x_{33} = 12.$$

∴ The optimal (minimum) transportation cost

$$= 11 \times 13 + 6 \times 17 + 3 \times 18 + 4 \times 23 + 7 \times 27 + 12 \times 18 = \$ 796.$$

EXAMPLE 12.30

A company has three cement factories located in cities 1, 2, and 3 which supply cement to four projects located in towns 1, 2, 3, and 4. Each plant can supply 6, 1, and 10 truck loads of cement daily respectively and the daily cement requirements of the projects are respectively 7, 5, 3, and 2 truck loads. The transportation costs per truck load of cement (in hundreds of Dollars) from each plant to each project site are as follows:

		Project sites			
		1	2	3	4
	1	2	3	11	7
Factories	2	1	0	6	1
	3	5	8	15	9

Determine the optimal distribution for the company so as to minimize the total transportation cost.

Solution *consists of the following steps:*

Step 1. Construct the transportation table. Express the supply from the factories, demands at sites, and the unit shipping cost in the form of the following transportation table (Table 1). Here the supply being equal to the demand, the problem is balanced.

Table 1

		Project sites				
		1	2	3	4	Supply
	1	2	3	11	7	6
Factories	2	1	0	6	1	1
	3	5	8	15	9	10
Demand		7	5	3	2	17

Step 2. Find the initial basic feasible solution.

Using VAM, the initial basic feasible solution is as shown in Table 2. The transportation cost according to this route is given by

$$Z = \$ (1 \times 2 + 5 \times 3 + 1 \times 1 + 6 \times 5 + 3 \times 15 + 1 \times 9) \times 100 = \$ 102,00.$$

Step 3. Apply optimality check.

As the numbers of allocations $= (m + n - 1)$, *i.e.,* 6, we can apply the MODI method.

We now compute the net evaluations $w_{ij} = (u_i + v_j) - c_{ij}$ which are exhibited in Table 3. Since the net evaluations in two cells are positive, a better solution can be found.

Table 2

1 [5]				
2	3	11	7	6
1	0	6 [1]	1	1
6			[1]	
5	8 [3]	15	9	10
7	5	3	2	

Table 3

v_j	2	3	12	6
u_i 0	1	[5]	(+)	(−)
	2	3	11	7
− 5	(−)	(−)	(+) [1]	
	1	0	6	1
3	[6]	(−) [3]	[1]	
	5	8	15	9

Step 4. Iterate towards optimal solution.

First iteration:

(a) Next basic feasible solution.

(i) Choose the unoccupied cell with the maximum w_{ij}. In case of a tie, select the one with lower original cost. In Table 3, cells (1, 3) and (2, 3) each have $w_{ij} = 1$ and out of these cell (2, 3) has the lower original cost 6, therefore we take this as the next basic cell and note θ in it.

(ii) Draw a closed path beginning and ending at θ-cell. Add and subtract θ, alternately to and from the transition cells of the loop subject the rim requirements. Assign a maximum value to θ so that one basic variable becomes zero and the other basic variables remain ≥ 0. Now the basic cell whose allocation has been reduced to zero leaves the basis. This gives the *second basic feasible solution* (Table 5).

	Table 4			
[1]	[5]		
2	3	11	7
	[1]	θ	[1] $-\theta$
1	0	6	1
[6]		[3] $-\theta$	[1] $+\theta$
5	8	15	9

	Table 5			
[1]	[5]		
2	3	11	7
		$\theta = 1$	$1-1$
1	0	6	1
[6]		$3-1$	$1+1$
5	8	15	9

∴ Total transportation cost of this revised solution.

$$= \$ \left[1 \times 2 + 5 \times 3 + 1 \times 6 + 6 \times 5 + 2 \times 15 + 2 \times 9 \right] \times 100 = \$ \, 101,00.$$

(b) Optimality check. As the number of allocations in table $5 = m + n - 1$ (*i.e.,* 6), we can apply the MODI method. We compute the net evaluations which are shown in Table 6. Since the cell (1, 3) has a positive value, the second basic feasible solution is not optimal.

Table 6

u_i \ v_j	2	3	12	6
0	[1] / 2	[5] / 3	(+) / 11	(−) / 7
−6	(−) / 1	(−) / 0	[1] / 6	(−) / 1
3	[6] / 5	(−) / 8	[2] / 15	[2] / 9

Table 7

1−1 / 2	[5] / 3	θ = 1 / 11	/ 7
/ 1	/ 0	[1] / 6	/ 1
6+1 / 5	/ 8	2−1 / 15	[2] / 9

Second iteration:

(*a*) *Next basic feasible solution.* In the second basic feasible solution introduce the cell (1, 3) taking $\theta = 1$ and drop the cell (1, 1) giving Table 7. Thus we obtain the third basic feasible solution (Table 8).

Table 8

/ 2	[5] / 3	[1] / 11	/ 7
/ 1	/ 0	[1] / 6	/ 1
[7] / 5	/ 8	[1] / 15	[2] / 9

Table 9

u_i \ v_j	1	3	11	5
0	(−) / 2	[5] / 3	[1] / 11	(−) / 7
−5	(−) / 1	(−) / 0	[1] / 6	(−) / 1
1	[7] / 5	(−) / 8	[1] / 15	[2] / 9

Optimality check. As the number of allocations in Table 8 = $m + n - 1$ (*i.e.*, 6), we can apply the MODI method.

We compute the net evaluations which are shown in Table 9. Since all the net evaluations are ≤ 0, this basic feasible solution is optimal.

Thus the optimal transportation policy is as shown in Table 9 and the optimal transportation cost

$$= \$\,[5 \times 3 + 1 \times 11 + 1 \times 6 + 7 \times 5 + 1 \times 15 + 2 \times 9] \times 100 = \$\,10{,}000$$

12.16 Degeneracy in Transportation Problems'

When the number of basic cells in a mn-transportation table, is less than "m + n – 1" the basic solution degenerates. To remove the degeneracy, we assign a small positive value ε to as many zero-valued variables as may be necessary to complete "$m + n - 1$" basic variables. The cells containing ε

are then treated like other basic cells and the problem is solved in the usual way. The ε's are kept till the optimum solution is attained. Then we let each $\varepsilon \to 0$.

EXAMPLE 12.31

Solve the following transportation problem:

			To			
9	12	9	6	9	10	5
7	3	7	7	5	5	6
6	5	9	11	3	11	2
6	8	11	2	2	10	9
4	4	6	2	4	2	22

(*From* label at left, *To* label at top)

Solution:

Consists of the following steps:

Step 1. Transportation table. The total supply and total demand being equal, the transportation problem is balanced.

Step 2. Find the initial basic feasible solution.

Using VAM, the initial basic feasible solution is as shown in table 1.

Step 3. Apply optimality check. Since the number of basic cells is 8 which is less than $m + n - 1 = 9$, the basic solution degenerates. In order to complete the basis and thereby remove degeneracy, we require only one more positive basic variable. We select the variable x_{23} and allocate a small positive quantity ε to the cell $(2, 3)$.

Table 1

						Supply
9	12	9 [5]	6	9	10	5
7	3 [4]	7 [ε]	7	5	5 [2]	$6 + \varepsilon = 6$
6 [1]	5	9 [1]	11	3	11	2
6 [3]	8	11	2 [2]	2 [4]	10	9
4	4	$6 + \varepsilon = 6$	2	4	2	

We now compute the net evaluations $w_{ij} = (u_i + v_j) - c_{ij}$ which are exhibited in Table 2. Since all the net evaluations are ≤ 0, the current solution is optimal. Hence the optimal allocation is

$$x_{13} = 5,\ x_{22} = 4,\ x_{26} = 2,\ x_{31} = 1,\ x_{33} = 1,\ x_{41} = 3,\ x_{44} = 2 \text{ and } x_{45} = 4.$$

Table 2

$v_j \to$	4	3	7	0	0	5
$u_i = 2$	(–)	(–)	[5]	(–)	(–)	(–)
	9	12	9	6	9	10
$u_i = 0$	(–)	[4]	[ε]	(–)	(–)	[2]
	7	3	7	7	5	5
$u_i = 2$	[1]	(0)	[1]	(–)	(–)	(–)
	6	5	9	11	3	11
$u_i = 2$	[3]	(–)	(–)	[2]	[4]	(–)
	6	8	11	2	2	10

$\therefore$ The minimum (optimal) transportation cost

$$= 5 \times 9 + 4 \times 3 + \varepsilon \times 7 + 2 \times 5 + 1 \times 6 + 1 \times 9 + 3 \times 6 + 2 \times 2 + 4 \times 2$$

$$= 112 + 7\varepsilon = \$\,112 \text{ as } \varepsilon \to 0.$$

Exercises 12.9

1. Obtain an initial basic feasible solution to the following transportation problem:

		To				
		D	E	F	G	
	A	11	13	17	14	250
From	B	16	18	14	10	300
	C	21	24	13	10	400
		200	225	275	250	

2. Solve the following transportation problem:

Consumers / Suppliers	A	B	C	Available
I	6	8	4	14
II	4	9	8	12
III	1	2	6	5
Required	6	10	15	31

3. Consider four bases of operations B_i and three targets T_j. The tons of bombs per aircraft from any base that can be delivered to any target are given in the following table:

B_i \ T_j	1	2	3
1	8	6	5
2	6	6	6
3	10	8	4
4	8	6	4

The daily sortie capability of each of the four bases is 150 sorties per day. The daily requirement in sorties over each target is 200. Find the allocation of sorties from each base to each target which maximizes the total tonnage over all the three targets.

4. Solve the following transportation problem:

Destination

Origin	D_1	D_2	D_3	D_4	
O_1	1	2	1	4	30
O_2	3	3	2	1	50 Availability
O_3	4	2	5	9	20
	20	40	30	10	100

5. A company has factories F_1, F_2, F_3 which supply ware-houses at W_1, W_2, and W_3. Weekly factory capacities, weekly ware-house requirements and unit shipping costs (in Dollars) are as follows:

Factories	Warehouses			Supply
	W_1	W_2	W_3	
F_1	16	20	12	200
F_2	14	8	18	160
F_3	26	24	16	90
Demand	180	120	150	450

Determine the optimal distribution for this company to minimize shipping costs.

6. A company is spending $ 1,000 on transportation of its units from plants to four distribution centers. The supply and demand of units, with unit cost of transportation are given below:

Plants	Distribution centers				Availabilities
	D_1	D_2	D_3	D_4	
P_1	19	30	50	12	7
P_2	70	30	40	60	10
P_3	40	10	60	20	18
Requirements	5	8	7	15	

What can be the maximum saving by optimal scheduling.?

7. A departmental store wishes to stock the following quantities of a popular product in three types of containers:

Container type:	1	2	3
Quantity:	170	200	180

Tenders are submitted by four dealers who undertake to supply not more than the quantities shown below:

Dealer:	1	2	3	4
Quantity:	150	160	110	130

The store estimates that profit per unit will vary with the dealer as shown below:

Dealers →	1	2	3	4
Container type ↓				
1	8	9	6	3
2	6	11	5	10
3	3	8	7	9

Find the maximum profit of the store.

8. Obtain an optimum basic feasible solution to the following transportation problem:

From	To				Available
	7	3	4	2	
	2	1	3	3	
	3	4	6	5	
	4	1	5	10	
	Demand				

9. A company has three plants A, B, and C and three warehouses X, Y,and Z. The number of units available at the plant is 60, 70, and 80, respectively. The demands at X, Y,and Z are 50, 80, 80, respectively. The unit costs of transportation are as follows:

	X	Y	Z
A	8	7	3
B	3	8	9
C	11	3	5

Find the allocation so that the total transportation cost is minimum.

10. A company has three plants at locations A, B, and C which supply to warehouses located as D, E, F, G, and H. Monthly plant capacities are 800, 500, and 900 units, respectively. Monthly warehouse requirements are 400, 400, 500, 400, and 800 units, respectively. Unit transportation costs in dollars are given below:

		To				
		D	E	F	G	H
	A	5	8	6	6	3
From	B	4	7	7	6	6
	C	8	4	6	6	3

Determine an optimum distribution for the company in order to minimize the total transportation cost.

12.17 Assignment Problem

An assignment problem is a special type of transportation problem in which the objective is to assign a number of origins to an *equal* number of destinations at a minimum cost (or maximum profit).

Formulation of an assignment problem. There are n new machines M_i $(i = 1, 2, \cdots n)$ which are to be installed in a machine shop. There are n vacant spaces S_j $(j = 1, 2, \cdots n)$ available. The cost of installing the machine M_i at space S_j is c_{ij} Dollars. Let us formulate the problem of assigning machines to spaces so as to minimize the overall cost.

Let x_{ij} be the assignment of machine M_i to space S_j, i.e., let x_{ij} be a variable such that

$$x_{ij} = \begin{cases} 1, \text{ if th machine is installed at th space} \\ 0, \text{ otherwise} \end{cases}$$

Since one machine can only be installed at each space, we have

$x_{i1} + x_{i2} + \cdots.. + x_{in} = 1$, for machine M_i $(i = 1, 2, ...n)$

$x_{1j} + x_{2j} + \cdots.. + x_{nj} = 1$, for space S_j $(j = 1, 2, ...n)$

Also the total installation cost is $\displaystyle\sum_{i=1}^{n}\sum_{j=1}^{n} c_{ij} x_{ij}$

Thus the assignment problem can be stated as follows:

Determine $x_{ij} \geq 0$ $(i, j = 1, 2, ...n)$ so as to

minimize $Z = \displaystyle\sum_{i=1}^{n}\sum_{j=1}^{n} c_{ij} x_{ij}$

subject to the constraints

$$\sum_{i=1}^{n} x_{ij} = 1, j = 1, 2, \cdots\cdots n, \text{and} \sum_{i=1}^{n} x_{ij} = 1, i = 1, 2, \cdots\cdots n.$$

This problem is explicitly represented by the following $n \times n$ cost matrix:

		Spaces				
		S_1	S_2	S_3	......	S_n
	M_1	c_{11}	c_{12}	c_{13}	......	c_{1n}
	M_2	c_{21}	c_{22}	c_{23}	......	c_{2n}
Machines	M_3	c_{31}	c_{32}	c_{33}	......	c_{3n}
	:	:	:	:		
	:	:	:	:		
	M_n	c_{n1}	c_{n2}	c_{n3}		c_{nn}

NOTE **Obs.** *This assignment problem constitutes n ! possible ways of installing n machines at n spaces. If we enumerate all these n ! alternatives and evaluate the cost of each one of them and select the one with the minimum cost, the problem would be solved. But this method would be very slow and time consuming, even for small value of n and hence it is not at all suitable. However, a much more efficient method of solving such problems is available. This is the* **Hungarian method** *for solution of assignment problems which we describe below.*

Working procedure to solve an assignment problem: *Step* 1. *Reduce the matrix.* Subtract the smallest element of each row (of the given cost matrix) from all elements of that row. See if each row contains at least one zero. If not, subtract the smallest element of each column (not containing zero) from all the elements of that column. This gives the *reduced matrix.*

Step 2. *Assign the zeros*

(*a*) Examine rows (of the reduced matrix) successively until a row with exactly one unmarked zero is found. Make an assignment to this single zero by encircling it. Cross all other zeros in the column of this encircled zero, as these will not be considered for any future assignment. Continue in this way until all the rows have been examined.

(*b*) Now examine columns successively until a column with exactly one unmarked zero is found. Encircle this zero and make an assignment there. Then cross any other zero in its row. Continue in this way until all the columns have been examined.

In case, some rows or columns contain more than one unmarked zeros, encircle any unmarked zero arbitrarily and cross all other zeros in its row or column. Proceed in this way, until no zero is left unmarked.

Step 3. *Apply optimality check.*

Repeat step 2 (*a*) and (*b*) until one of the following occurs:

(*i*) If no row or no column is without assignment (encircled zero), then the current assignment is optimal.

(*ii*) If there is some row and/or column without an assignment, then the current assignment is not optimal and we go to next step.

Step 4. Find the minimum number of lines crossing all zeros.

(*a*) Tick (√) the rows which do not have assignments.

(*b*) Tick (√) the columns (not already marked) which have zeros in the ticked row.

(*c*) Tick (√) the rows (not already marked) which have assignments in ticked columns.

Repeat (*b*) and (*c*) until no more marking is required.

(*d*) Draw lines through all unticked rows and ticked columns. If the number of these lines is equal to the order of the matrix then it is an optimal solution otherwise not.

Step 5. Iterate towards an optimal solution.

Select the smallest element and subtract it from all uncovered elements. Add this smallest element to every element lying at the intersection of two lines. The resulting matrix is the second basic feasible solution.

Step 6. Go to Step 2 and repeat the procedure until the optimal solution is attained.

EXAMPLE 12.32

Four jobs are to be done on four different machines. The cost (in dollars) of producing *i*th job on the *j*th machine is given below:

		M_1	M_2	M_3	M_4
	J_1	15	11	13	15
Jobs	J_2	17	12	12	13
	J_3	14	15	10	14
	J_4	16	13	11	17

Assign the jobs to different machines so as to minimize the total cost.

Solution:

Consists of the following steps:

Step 1. *Reduce the matrix.* Subtract the smallest element 11 of row 1 from all its elements. Similarly subtract 12, 10, and 11 from rows 2, 3, and 4, respectively. The resulting matrix is as shown in Table 1. Columns 1 and 4 do not have any zero element. Subtract the smallest element 4 of column

1 from all its elements and element 1 from all elements of column 4. The *reduced matrix* is as given in Table 1.

<table>
<tr><td colspan="5" align="center">*Table 1*</td></tr>
<tr><td></td><td>M_1</td><td>M_2</td><td>M_3</td><td>M_4</td></tr>
<tr><td>J_1</td><td>4</td><td>0</td><td>2</td><td>4</td></tr>
<tr><td>J_2</td><td>5</td><td>0</td><td>0</td><td>1</td></tr>
<tr><td>J_3</td><td>4</td><td>5</td><td>0</td><td>4</td></tr>
<tr><td>J_4</td><td>5</td><td>2</td><td>0</td><td>6</td></tr>
</table>

<table>
<tr><td colspan="5" align="center">*Table 2*</td></tr>
<tr><td></td><td>M_1</td><td>M_2</td><td>M_3</td><td>M_4</td></tr>
<tr><td>J_1</td><td>⊠</td><td>⓪</td><td>2</td><td>3</td></tr>
<tr><td>J_2</td><td>1</td><td>⊠</td><td>⊠</td><td>⓪</td></tr>
<tr><td>J_3</td><td>⓪</td><td>5</td><td>⊠</td><td>3</td></tr>
<tr><td>J_4</td><td>1</td><td>2</td><td>⓪</td><td>5</td></tr>
</table>

Step 2. *Assign the zeros.* Row 4 has a single unmarked zero in column 3. Encircle it and cross all other zeros in column 3. Row 3 has a single unmarked zero in column 1. Encircle it and cross the other zero in column 1. Row 1 has a single unmarked zero in column 2. Encircle it and cross the other zero in column 2. Finally row 2 has a single unmarked zero in column 4. Encircle it (Table 2).

Step 3. *Apply optimality check.* Since we have one encircled zero in each row and in each column, this gives the optimal solution.

∴ The optimal assignment policy is

Job 1 to machine 2, Job 2 to machine 4,

Job 3 to machine 1, Job 4 to machine 3,

and the minimum assignment cost = $(11 + 13 + 14 + 11) = $ 49.

EXAMPLE 12.33

A marketing manager has 5 salesmen and 5 sales districts. Considering the capabilities of the salesmen and the nature of districts, the marketing manager estimates that sales per month (in hundred Dollars) for each salesman in each district would be as follows:

	Sales districts				
	A	B	C	D	E
Salesmen 1	32	38	40	28	40
2	40	24	28	21	36
3	41	27	33	30	37
4	22	38	41	36	36
5	29	33	40	35	39

Find the assignment of salesmen to districts that will result in maximum sales.

Solution:

Consists of the following steps:

Step 1. Reduce the matrix. Convert the given maximization problem into a minimization problem, by making all the profits negative, since max. $Z = $ min. $(-Z)$. Then subtract the smallest element of each row from the elements of that row. Now subtract the smallest element of each column (not containing zero) from the elements of that column. This gives the *reduced matrix* (Table 1).

Table 1

8	(0)	✗	7	(0)
(0)	14	12	14	4
(0)	12	8	6	4
19	1	(0)	✗	5
11	5	(0)	✗	1

Table 2

12	0	0	7	0
0	10	8	10	0
0	8	4	2	0
23	1	0	0	5
15	5	0	0	1

Step 2. Assign the zeros. Rows 2 and 3 have each a single unmarked zero in column 1. Encircle these. Columns 2 and 5 have each a single unmarked zero in row 1. Encircle these and cross the zero in row 1. Columns 3 and 4 have each unmarked zeros. Encircle the zeros in each of the rows 4 and 5 as shown in Table 1 and cross other zeros.

Step 3. Apply optimality check. As column 4 is without assignment, this solution is not optimal. Therefore we go to next step.

Step 4. Find minimum number of lines crossing all zeros. Draw the least number of horizontal and vertical (dotted) lines which cover all the zeros. Since there are four dotted lines which are less than the order of the cost matrix (= 5), we go to Step 5.

Step 5. Iterate toward an optimal solution. Select the smallest element in the Table 1, not covered by the dotted lines. Such an element is 4 which lies at two different positions. Selecting the element that lies at position (3, 5) arbitrarily, subtract it from all the uncovered elements of the cost matrix (Table 1) and add the same to the elements lying at the intersection of two dotted lines. Now draw more minimum number of dotted lines so as to cover the new zero. Here we draw such a line in column 5 (Table 2).

Now, since the number of dotted lines is equal to the order of the cost matrix, the optimal solution is attained.

Finally, to determine this optimal assignment, we consider only the zero elements (Table 3):

Table 3

	A	B	C	D	E
1		⓪	⦻		
2	⓪				⦻
3	⦻				⓪
4			⓪	⦻	
5			⦻	⓪	

(*i*) Examine successively the rows with exactly one zero. There is no such row.

(*ii*) Examine successively the columns with exactly one zero. Column 2 has one zero, encircle it and cross all zeros of row 1.

(*iii*) Encircle arbitrarily the zero in position (2, 1) and cross all zeros in row 2 and column 1. Then encircle the unmarked zero in row 3. Now encircle arbitrarily the zero in position (4, 3) and cross all zeros in row 4 and column 3. Finally encircle the remaining unmarked zero in row 5.

Now each row and each column has one encircled zero, therefore the optimal assignment policy is:

Salesman 1 to district B, 2 to A, 3 to E, 4 to C and 5 to D.

Hence the maximum sales

$= \$ (38 + 40 + 37 + 41 + 35) \times 100 = \$ 19100.$

Exercises 12.10

1. A firm plans to begin production of three new products on its three plants. The unit cost of producing i at plant j is as given below. Find the assignment that minimizes the total unit cost.

		Plant		
		1	2	3
Product	1	10	8	12
	2	18	6	14
	3	6	4	2

2. Solve the following assignment problem:

	1	2	3	4
A	10	12	19	11
B	5	10	7	8
C	12	14	13	11
D	8	15	11	9

3. A machine tool company decides to make four sub-assemblies through four contractors. Each contractor is to receive only one sub assembly. The cost of each sub-assembly is determined by the bids submitted by each contractor and is shown in the table below (in hundreds of Dollars). Assign different assemblies to contractors so as to minimize the total cost.

		Contractor			
		A	B	C	D
Sub-assembly	I	15	13	14	17
	II	11	12	15	13
	III	18	12	10	11
	IV	15	17	14	16

4. Four professors are each capable of teaching any one of the four different courses. Class preparations time in hours for different topics varies from professor to professor and is given in the table below. Each professor is assigned only one course. Find the assignment policy schedule so as to minimize the total course preparation time for all courses.

Prof.	L.P.	Queuing Theory	Dynamic Programming	Regression Analysis
A	2	10	9	7
B	15	4	14	8
C	13	14	16	11
D	3	15	13	8

5. Consider the problem of assigning five jobs to five persons. The assignment costs are given below:

		Jobs				
		1	2	3	4	5
	A	8	4	2	6	1
	B	0	9	5	5	4
Persons	C	3	8	9	2	6
	D	4	3	1	0	3
	E	9	5	8	9	5

Determine the assignment schedule.

6. The head of the department has five jobs A, B, C, D, E and five subordinates V, W, X, Y and Z. The number of hours each man would take to perform each job is as follows:

	V	W	X	Y	Z
A	3	5	10	15	8
B	4	7	15	18	8
C	8	12	20	20	12
D	5	5	8	10	6
E	10	10	15	25	10

How should the jobs be allocated to minimize the total time?

7. A company has six jobs to be processed by six mechanics. The following table gives the return in Dollars when the ith job is assigned to the jth mechanic. How should the jobs be assigned to the mechanics so as to maximize the over all return?

Mechanic ↓	Job					
	I	II	III	IV	V	VI
1	9	22	58	11	19	27
2	43	78	72	50	63	48
3	41	28	91	37	45	33
4	74	42	27	49	39	32
5	36	11	57	22	25	18
6	13	56	53	31	17	28

8. A company has four machines on which to do three jobs. Each job can be assigned to one and only one machine. The cost of each job on each machine is given in the following table:

Job	Machine			
	A	B	C	D
1	18	24	28	32
2	8	13	17	19
3	10	15	19	22

Determine the optimum assignment.

HINT. Whenever the cost matrix of an assignment problem is not a square matrix, the problem is called an *unbalanced assignment problem*. In such problems, we add dummy rows (or columns) so as to form a square matrix. Then we solve the resulting balanced problem in the usual way. In this problem, we add a dummy fourth row so as to get the following balanced assignment problem:

Job	Machine			
	A	B	C	D
1	18	24	28	32
2	8	13	17	19
3	10	15	19	22
4	0	0	0	0

9. Determine an optimum assignment schedule for the following assignment problem. The cost matrix is given:

Job ↓	Machines					
	1	2	3	4	5	6
A	11	17	8	16	20	15
B	9	7	12	6	15	13
C	13	16	15	12	16	8
D	21	24	17	28	26	15

E	14	10	12	11	15	6

12.18   Objective Type of Questions

Exercises 12.11

Fill up the blanks in the following questions:

1. Infeasibility in a linear programming problem means ⋯⋯ ·

2. The significance of the $(Z_j - C_j)$ row in the simplex solution procedure is that ⋯⋯ ·

3. The duality principle states that ⋯⋯ ·

4. The difference between the transportation problem and the assignment problem is ⋯⋯ ·

5. The special features of a transportation problem are ⋯⋯ ·.

6. The canonical form of an L.P.P. is such that ⋯⋯ ·

7. The dual problem of the L.P.P:

Max. $Z = 4x_1 + 9x_2 + 2x_3$,

subject to $2x_1 + 3x_2 + 2x_3 \leq 7$, $3x_1 - 2x_2 + 4x_3 = 5$, $x_1, x_2, x_3 \geq 0$, is ⋯⋯ ·

8. The optimality and feasibility conditions related with Dual simplex method are ⋯⋯ ·

9. Feasible and basic solutions related with a transportation problem are ⋯⋯ ·

10. A transportation problem is

					Supply
	2	3	11	4	15
	5	6	8	7	20
Demand	10	5	12	8	

Its linear programming problem is ⋯⋯

11. The basic feasible solutions of $2x_1 + x_2 + 4x_3 = 11$, $3x_1 + x_2 + 5x_3 = 14$ are ⋯⋯ ·

12. A slack variable is defined as······ ·

13. The advantage of the dual simplex method is ······ ·

14. If the total availability is equal to the total requirements, the transportation problem is called ······ ·

15. An artificial variable is that ······ ·

16. Two conditions on which the simplex method is based are ······ ·

17. A feasible solution which minimizes the transportation cost is called an ······ · solution.

18. The dual problem of: Max.$5x_1 + 6x_2$ subject to $x_1 + 2x_2 = 5, -x_1 + 5x_2 \geq 3$, x_1 unrestricted and $x_2 \geq 0$, is ······ ·

19. For a balanced transportation problem with 3 rows and 3 columns, the number of basic variables will be ······ ·

20. Using graphical method, Max. $Z = 5x_1 + 3x_2$ subject to $5x_1 + 2x_2 \leq 10, 3x_1 + 5x_2 \leq 15, x_1, x_2 \geq 0$, is ······ ·

21. In an L.P. problem, unbounded solution is that ······ ·

22. Degeneracy in a transportation problem is resolved by ······ ·

23. A basic solution is said to be non-degenerate in L.P.P. when ······ ·

24. The dual of the problem Max. $Z = 2x_1 + x_2$ subject to $-x_1 + 2x_2 \leq 2, x_1 + x_2 \leq 4, x_1 \leq 3, x_1, x_2 \geq 0$, is ······ ·

25. The two methods used to find the initial solution of a transportation problem are ······ ·

26. Constraints involving "equal to sign" do not require use of ······ or ······ variables.

13

A Brief Review of Computers

Chapter Objectives

- Introduction
- Structure of a computer
- Computer representation of numbers
- Floating point representation of numbers
- Computer calculations: algorithm, flowchart
- Program writing

13.1 Introduction

Many problems in modern science and engineering involve so much of computation that they would require years of labor for their solution even using the best of calculators. With the advent of high speed computers, the picture has changed completely. Such complicated problems can now be solved very quickly by using electronic computers. For instance, a system of thirty linear equations with thirty unknowns is a monstrous problem for a human being but it is just a routine job for a digital computer. In fact, modern numerical techniques can best be appreciated within the context of some basic knowledge of computers. As such, a concise introduction to the digital computer is given in this chapter.

13.2   Structure of a Computer

A digital computer has the following interconnected units each of which performs a specific task:

 (*i*)   **Input unit** accepts (or reads) the data and instructions which are typed using a keyboard of *video display unit*. It consists of a magnetic tape (or disk).

 (*ii*) **Memory unit** stores the data and procedures. It consists of magnetic cores or semi-conductor storage.

 (*iii*) **Central Processing Unit (C.P.U.)** is the vital component that makes the computer work. It takes care of all arithmetic and logical operations. It comprises of the following two parts:

   (*a*) **Control unit** interprets and carries out the instructions stored in the memory. It has electronic circuitry to decode instructions and activates other units.

   (*b*) **Arithmetic unit** carries out the required calculations. It consists of electronic registers, accumulators.

 (*iv*) **Output unit** presents the results of the calculations. It consists of a video display terminal and a printer. Various discs as **input** as well as **output** devices.

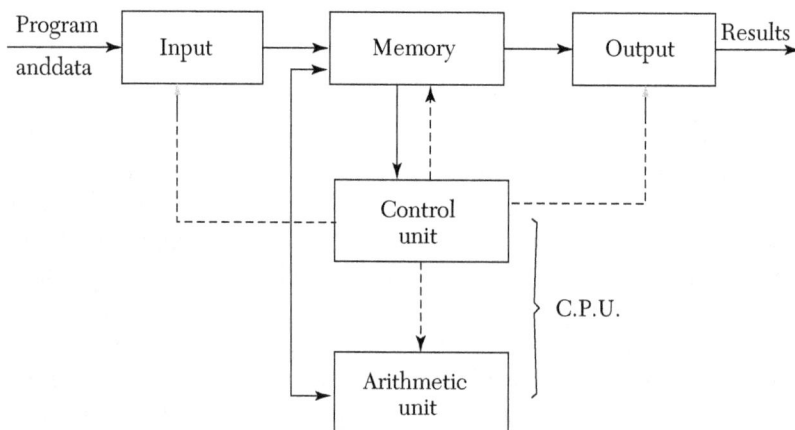

FIGURE 13.1

The memory and control units deal with numbers. So they must have a method of recording numbers. Such a recording is achieved by the magnetic cores or semi-conductor storage which require the numbers to be made up of zeros and ones only.

13.3   Computer Representation of Numbers

The numbers are first converted to machine numbers consisting of 0 and 1 with a base depending on a computer. Most of the computers have a base 2 which is called a **binary system** of numbers. The decimal system has the base 10. We, therefore, convert the decimal numbers into the binary system for the **input** and reconvert the binary numbers to the decimal form for the **output**.

Binary numbers. Any number is written in binary notation as

$$b_{n-1}b_{n-2} \cdots b_1 b_0 . b_{-1} b_{-2} \cdots b_{-m} \tag{A}$$

where b's are binary bits 0 or 1 and the point is the binary point.

Rule I. To convert the binary number (A) *to the decimal form, use the formula:*

$$b_{n-1} 2^{n-1} + b_{n-2} 2^{n-2} + \cdots + b_1 2^1 + b_0 2^0 + b_{-1} 2^{-1} + b_{-2} 2^{-2} + \cdots + b_{-m} 2^{-m}$$

Rule II. To convert an integer to a binary number:

(*i*) divide it by 2 and write the remainder,

(*ii*) continue the process until the quotient is zero,

(*iii*) write the remainders from bottom to top. This will give the required binary equivalent.

Rule III. To convert a decimal number to a binary fraction:

(*i*) multiply the given number by 2 and separate the integral part,

(*ii*) multiply the fractional part again by 2 and separate the integral part,

(*iii*) continue this process, until the fractional part reduces to zero,

(*iv*) write the integral parts and prefix the binary point. This will be the desired binary fraction.

EXAMPLE 13.1

Find the decimal number corresponding to the binary number 1101001.1110011.

Solution:

$$(1101001. 1110011)_2 = 1 \times 2^6 + 1 \times 2^5 + 0 \times 2^4 + 1 \times 2^3 + 0 \times 2^2$$
$$+ 0 \times 2^1 + 1 \times 2^0 + 1 \times 2^{-1} + 1 \times 2^{-2} + 1 \times 2^{-3} + 0 \times 2^{-4}$$

$$+0 \times 2^{-5} + 1 \times 2^{-6} + 1 \times 2^{-7} \qquad\qquad\qquad [\text{Rule I}]$$

$$= 64 + 32 + 8 + 1 + 0.5 + 0.25 + 0.125 + 0.015625 + 0.0078125$$

$$= (105.8984375)_{10}.$$

EXAMPLE 13.2

Convert 78.59375 to the binary system.

Solution:

We first convert 78 to binary form using Rule II.

$$78 = 1001110$$

Then we convert .59375 to the binary fraction using Rule III.

$$\therefore \qquad\qquad 0.59375 = 0.10011$$

Hence $(78.59375)_{10} = (1001110.10011)_2$

Verification: Using Rule I,

$(1001110.10011)_2$

$$= 1 \times 2^6 + 0 \times 2^5 + 0 \times 2^4 + 1 \times 2^3$$
$$+ 1 \times 2^2 + 1 \times 2^1 + 0 \times 2^0$$
$$+ 1 \times 2^{-1} + 0 \times 2^{-2} + 0 \times 2^{-3}$$
$$+ 1 \times 2^{-4} + 1 \times 2^{-5}$$
$$= 78.59375$$

2	78
2	39 – 0
2	19 – 1
2	9 – 1
2	4 – 1
2	2 - 0
2	1 – 0
	0 – 1

```
      0.59375
         ×2
  1 0.18750
         ×2
  0 0.037500
         ×2
  0 00.75000
         ×2
  1 0.50000
         ×2
  1 0.00000
```

Some of the computers have a base 8 which is called the **octal system** and uses the symbols 0, 1, 2, 3, 4, 5, 6, 7. As $8 = 2^3$, a group of three binary bits can be represented by an equivalent octal digit. Equivalence between the octal and binary systems is given below:

Octal:	0	1	2	3	4	5	6	7
Binary:	000	001	010	011	100	101	110	111

Another base commonly used is 16 which is known as **hexadecimal**. The symbols used are 0 to 9 and A, B, C, D, E, F. As $16 = 2^4$, a group of four binary bits can be represented by an equivalent hexadecimal symbol. Equivalence between hexadecimal and binary numbers is as follows:

Hexa:	0	1	2	3	4	5	6	7
Binary: 0000	0001	0010	0011	0100	0101	0110	0111	
Hexa:	8	9	A	B	C	D	E	F
Binary: 1000	1001	1010	1011	1100	1101	1110	1111	

EXAMPLE 13.3

Convert the binary number 1011101.1100101 to the octal and hexa-decimal systems.

Solution:

(i) Given number $= (001 \quad 011 \quad 101. 110 \quad 010 \quad 100)_2$

$= (135.624)_8$ *i.e.*, octal equivalent.

(ii) Given number $= (0101 \quad 1101 \, . \, 1100 \quad 1010)_2$

$= (5D.CA)_{16}$ *i.e.*, hexa. equivalent

$= 5 \times 16^1 + 13 \times 16^0 + 12 \times 16^{-1} + 10 \times 16^{-2}$

$= (93.7890625)_{10}$ *i.e.*, decimal equivalent.

13.4 Floating Point Representation of Numbers

The memory of a digital computer has separate cells called "words." Each word contains the same number of binary digits called "bits." The number of digits which can be stored in a computer is known as its *word length*. The numbers are stored in a computer in two forms: *fixed-point* and *floating-point* forms. The fixed point mode is used to represent integers while the floating-point mode is used to represent real numbers.

A floating-point number is of the form

$$.d_1 d_2 \cdots d_n \times b^m$$

where $d_1, d_2, \cdots$, dn are all digits in the base "b" of the number system used and lie between o and b. The exponent m is such that $M_1 \leq m \leq M_2$ where M_1 and M_2 vary with the computer. The fractional part $.d_1 d_2 \cdots d_n$ is called the *mantissa* which lies between ±1 and restricts the size of a number. If a floating-point number has a fixed mantissa of k digits, we say that the word length of the computer is k.

For instance, the number 33.74×106 is represented as .3374E8 (E8 is used to represent 108). Hence the mantissa is .3374 and the exponent is 8. While storing numbers, the leading digit in the mantissa is always made non-zero by suitably shifting it and adjusting the value of the exponent accordingly. This process is called *normalization*. Therefore the number.003374 in normalized floating point mode would be stored as .3374E-2.

Thus the shifting of the mantissa to the left until its most significant digit is non-zero is called **normalization**.

<table>
<tr><td>

NOTE

</td><td>

Obs. *To perform arithmetic operations with numbers in normalized floating point modes, we assume a hypothetical computer with a four decimal digit mantissa.*

</td></tr>
</table>

Arithmetic operations

(i) To add two numbers represented in normalized floating point notation, we make their exponents equal by shifting the mantissa appropriately.

The operation of subtraction is nothing but addition of a negative number. After addition/subtraction of the mantissas, the resulting mantissa is normalized and the exponent is suitably adjusted.

EXAMPLE 13.4

Evaluate (a).6756E4 + .7644E6 (b).4546E-4 – .8524E-5.

Solution:

(a) The exponent of the number with the smaller exponent is increased by 2 so that .6756E4 becomes .0067E6.

Thus .6756E4 +.7644E6 =.0067E6 +.7644E6 = 0.7711E6.

(b) Increasing the exponent of .8524E-5 by 1, it becomes .0852E-4.

Thus .4546E-4 – .8524E-5 = .4546E-4 – .0852E-4 = .3694E-4.

(ii) To multiply two numbers given in the normalized floating point mode, we multiply their mantissas and add their exponents.

After multiplication of the mantissas, the resulting mantissa is normalized and the exponent is suitably adjusted. The mantissa is only four digits of the resulting mantissa which is retained by dropping the rest of the digits.

EXAMPLE 13.5

Evaluate (a).6543E11 × .5123E-14 (b).1234E12 ×.1111E9.

Solution

(a).6543E11 × .5123E-14 = .33519789E-3 = .3351E-3.

(b).1234E12 ×.1111E9 = .0137097E21 = .1370E20.

(iii) To divide a number by another, the mantissa of the numerator is divided by the mantissa of the denominator and the denominator exponent is subtracted from the exponent of the numerator. The quotient mantissa is then normalized retaining four digits and the exponent adjusted suitably.

EXAMPLE 13.6

Divide.1000E5 by.8889E3.

Solution:

∴ .1000E5 ÷ .8889E3 = .1124E2.

NOTE

Obs. *While performing the arithmetic operations with numbers in normalized floating-point mode, the numbers have to be truncated to fit the four digit mantissa of our hypothetical computer. This leads to results with wide disparity. In fact, the associative and distributive laws do not yield valid results in floating point representation, i.e.,*

$$l + m - n \neq (l - n) + m \text{ and } l(m - n) \neq lm - ln.$$

For instance, if l = .6776E1, m = .6667E – 1 and n = .6755E1, then

$$(l + m) - n = (.6776E1 + .0067E1) - .6755E1$$

$$= .6842E1 - .6755E1 = .0087E1 = .8700E-1.$$

But $(l - n) + m = (.6776E1 - .6755E1) + .6667E-1$

$$= .0021E1 + .6667E-1 = .2100E-1 + .6667E-1$$

$$= .8767E-1$$

∴ We see that $(l + m) - n \neq (l - n) + m.$

In fact, the correct answer is.8767E-1 because no number has been truncated. *Thus inaccuracies creep in floating point arithmetic due to truncation of numbers. As such utmost care should be taken before accepting the validity of a computer solution.* Moreover, we can never ensure exact equality of a number to zero because most of the numbers in floating-point mode are only approximations.

13.5 Computer Calculations

Algorithm. Once the method of calculation has been decided, we must describe clearly the computational steps to be followed in a particular sequence. These steps constitute the *algorithm* of the method.

Flow chart. A pictorial representation of a specific sequence of steps to be used by a computer is called a *flow chart*. It is essentially a convenient way of planning the order of operations involved in an algorithm and helps in writing a program. The programmer, then knows clearly where to start, what information to use, what operations to be carried out, and in which order and where to stop. As such a flow-chart is additional help for writing the program in any language.

A flow-chart contains certain symbols to represent the various operations. These symbols are connected by arrows to indicate the flow of information. The commonly used symbols and their meanings are given below:

A symbol used to indicate "Start" or "Stop/End" of a program. It is also used to mark end of a "Subprogram." In that case "Return" is written in it.

[In FORTRAN the subprograms are functions and subroutines while in "C" these are functions only.]

A *parallelogram* is used to indicate an "Input" or "Output" of data.

A *rectangle* is a processing symbol, *e.g.*, addition, subtraction and movement of data to computer memory.

A *diamond* is a decision-making symbol. A particular path is chosen depending on the "Yes" or "No" answer.

A *small circle* with any number or letter in it, is used as a *connector symbol*. It connects various parts of a flow-chart until that are far apart or spread over pages.

A *rectangle with double vertical sides* is used to denote a *subprocess* which is given else-where as indicated by the connector symbol. When this box is encountered the flow goes to the subroutine and it continues until a

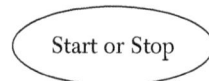

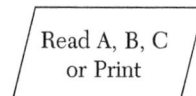

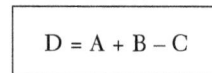

"Return" statement is encountered. Then it goes back to the main flow chart and flow resumes onward processing.

The flow chart can be translated into any computer language and can be executed on the computer.

The flow-chart can be translated into any computer language and can be executed on the computer.

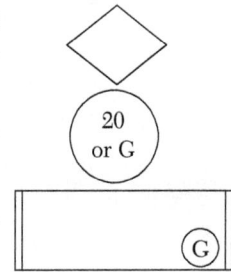

EXAMPLE 13.7

Develop a flow chart to select the largest number of a given set of 500 numbers.

Solution:

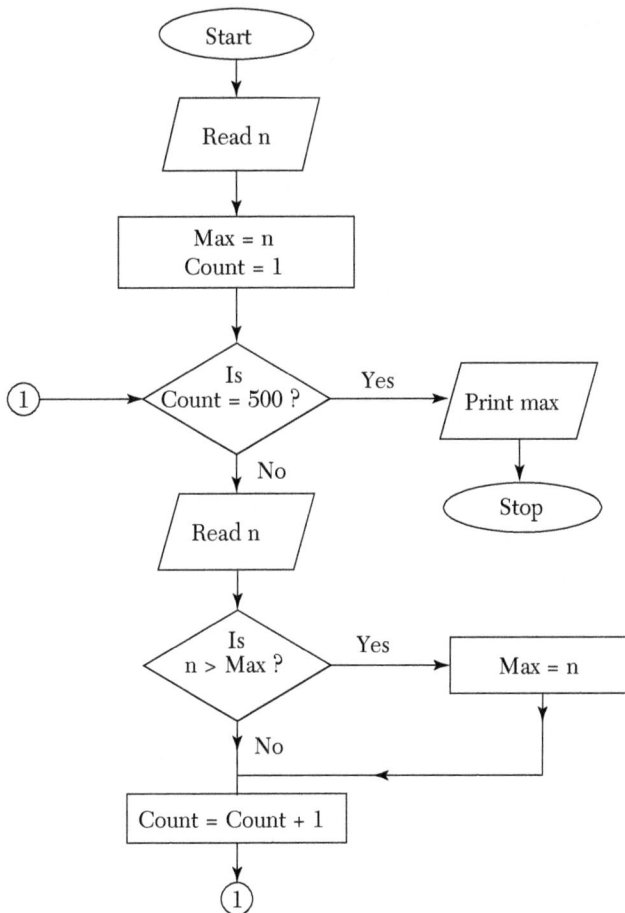

This technique of finding the largest number, i.e., assuming the first element in the list is the maximum and then scanning the rest of the list for anything greater is used in Section 14.12, and Section 15.12.

EXAMPLE 13.8

Draw a flow chart for computing the roots of the quadratic equation $ax^2 + bx + c = 0$.

Solution:

We know that its roots are given by

$$x_1 = \frac{-b + \sqrt{d}}{2a}, x_2 = \frac{-b - \sqrt{d}}{2a} \qquad \text{where } d = b^2 - 4ac.$$

Flow-chart:

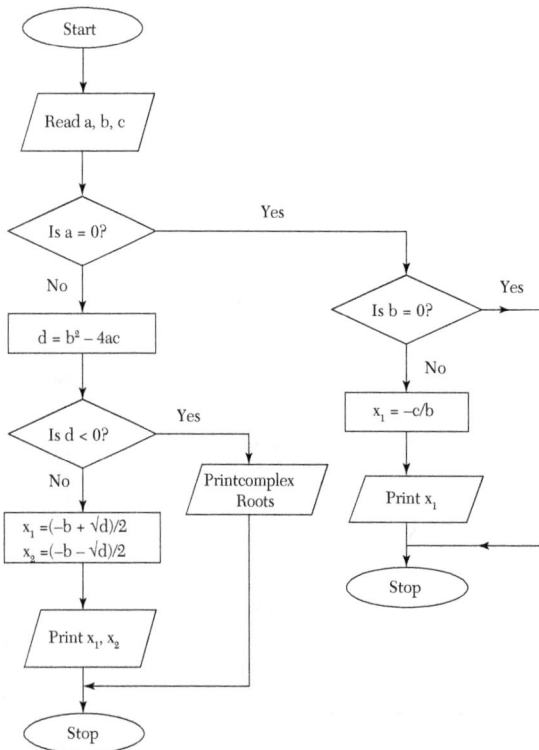

This method of finding the roots is used in Section 14.6.

Exercises 13.1

1. Convert the following binary numbers to decimal form:
   (a) 1101101   (b) 0.11011
   (c) 0.1010101   (d) 1.0110101.

2. Convert the following decimal numbers to binary form:
   (a) 22.625   (b) – 10.125.

3. Show that
   (a) $(176)_8 = (126)_{10}$ (b) $(17AB)_{16} = (6059)_{10}$.

4. If $A = (111010)_2$ and $B = (1011)_2$, evaluate $A + B$ and $A - B$.

5. Find the product of the binary numbers
   (a) 10101 and 110   (b) 11.1101 and 101.101.

6. Add the numbers 83.72 and 1.529 in a decimal computer with a fixed word length of four. Find the absolute and relative errors involved.

7. Draw a flow chart to evaluate
   $1 + 4 + 7 + \cdots + 1003$.

8. Draw a flow-chart to pick up the largest of three given distinct numbers.

9. Draw a flow-chart to arrange a given set of N numbers in an ascending order.

13.6   Program Writing

Based on the flow chart, we write the instructions in a code that the computer can understand. A series of such instructions is called a *program*. If there are any errors in the program these will be pointed out by the computer during compilation. After correcting the compilation errors, the program is executed with the input data to check for logical errors which may be due to misinterpretation of the algorithm or due to incorrect usage of computer language. The process of finding the errors and correcting them is termed *debugging*.

While writing a program, our aim should be that the same program is able to run on any machine with the minimum number of modifications.

CHAPTER 14

NUMERICAL METHODS USING C LANGUAGE

Chapter Objectives

- Introduction
- An overview of "C" features
- 3–30 Programs of standard methods in "C" language.

14.1   Introduction

C is a general purpose programming language, originally designed by Dennis Ritchie in 1972 at Bell laboratories. In 1988, it was standardized by the American National Standards Institute (ANSI) and named as ANSI C.

C, a powerful language, is used for many purposes like writing operating systems, business and scientific applications and even the C compiler itself. It is not tied to any particular hardware and programs written in C are portable across any system. The programs written in C are efficient and fast. It is one of the most popular computer languages today.

An overview of C features is given below for ready reference. It is followed by Programs of Standard Numerical methods in C language alongwith input/output of numerous examples solved in Chapters 1 to 12.

14.2 An Overview of "C" Features

C constants are numbers, which do not change during execution of a program. These may be of three types:

Type	Example
Integer	27, 10897 etc.
Floating point (Real)	2.723, – 0.123 etc.
String	"Enter the value"

NOTE *The string constants are enclosed in double quotes (").*

C variables can contain different C constants during the execution of the pro- gram. These are declared in a C program by first specifying the type *int* for an integer and *float* for the floating point and then the variable names separated by commas. The general format is

    type   *list of variables*

For *e.g.,* to declare integer variables the statement is

$$\text{int } a, b, c;$$

and to declare a floating point variables it is

$$\text{float } a, b, c;$$

Variables can be initialized at the same time as they are declared. For example,

$$\text{float } a = 1.5;$$

declares *a* as a float variable having a value 1.5.

Rules for naming C variables:

(*i*) A variable name may contain only alphabets, digits, and the underscore (_).

(*ii*) It must begin with a alphabet or an underscore.

(*iii*) It can be as long as you wish, but on some C systems only the first thirty-one 31 characters are considered.

NOTE *Lower-case and upper case alphabets are treated as different in C. For e.g. Num and num are two entirely different variable names. As a matter of convention, lower- case alphabets are used.*

Arrays. An array is an aggregate of variables of the same type. These variables are called *elements* of the array. The following statements declare arrays in C:

int $b[10]$;

float $c[2][2]$;

The first statement creates a one dimensional array named b having ten elements, each element being referred by an appropriate subscript in rectangle brackets, *i.e.*, $b[0], b[1], ..., b[9]$.

The second statement creates a 2-dimensional array named c having four elements $c[0][0], c[0][1], c[1][0], c[1][1]$.

Rules for the naming of arrays are same as those for variable names.

NOTE *Subscripts always start from zero in C.*

User defined types. Apart from the built in types *int* and *float* C allows users to define an identifier that can represent an existing data type.

The syntax is

> *typedef type identifier*

For *e.g.*,

> typedef int number;

> typedef float matrix [2][2];

The first statement defines *number* to mean the same as *int*. The second defines *ma- trix* to be mean the same as 2 × 2 array of float.

The above two statements enable declarations of the form number a, $b, c;$

which declares three integers a, b and c, and

> matrix $x;$

which declares a two-dimensional array x having four elements $x[0][0], x[0][1], x[1][0]$ and $x[1][1]$.

Initialization of arrays at the time of declaration

The syntax is

$$type \ array\text{-}name \ [size] = \{list \ of \ values\}$$

for *e.g.*, int $a[2] = \{2, 1\}$;

initializes $a[0]$ to 2 and $a[1]$ to 1.

$$int \ a[2][2] = \{\{0, 1\}, \{3, 5\}\};$$

initializes $a[0][0]$ to 0, $a[0][1]$ to 1, $a[1][0]$ to 3 and $a[1][1]$ to 5.

Arithmetic operators. These are as follows:

Symbol	Use
+	Addition
−	Subtraction
*	Multiplication
/	Division

while using the operators, the following order of precedence is adopted

(*i*) *, /   (*ii*) +, −

In this case, the order of operators is that different circular brackets are used.

There is no exponentiation operator in C, but there are various *library functions* avail- able for the same.

For *e.g.*, to calculate the square root *sqrt* function is used.

Further details on functions are presented later.

Mathematical expressions consist of a sequence of arithmetic operators and variable names. For *e.g.*,

(*i*) $a + b$ is written as $a + b$.   (*ii*) $\dfrac{a}{b} + c$

(*iii*) $\dfrac{a}{b+c}$ is written as $a/a + b$

(*iv*) $\sqrt{b^2 - 4ac}$ is written as **sqrt** $(b*b - 4*a*c)$.

(*v*) $\alpha(\beta + \gamma)$ is written as alpha * (beta + gamma).

(*vi*) a^b is written as **exp** $(b*\mathbf{ln} \ (a))$

*The multiplication operator * has to be written explicitly. In C, its presence is never assumed.*

exp and **ln** are **library functions**.

Arithmetic statements are of the form

$$var = exp;$$

where *var* is an integer or a floating point variable.

exp is a mathematical expression written in C format.

The = sign has a special meaning. It tells C to calculate the value of *exp*. and *assign* it to **var**.

For *e.g.*, $n = i * i;$

calculates the value of $i * i$ and assigns the result to the variable *n*. If $i = 10$, then *n* gets the value 100.

A C statement is always terminated by a semi colon (;).

C also permits statements of the type $k = n = i * i;$

This is equivalent to the following statements $n = i * i; k = n;$

NOTE *To test the equality of two expressions C uses "= =".*

Shorthand assignment operators

Apart from the assignment operator =, C can also support certain short hand assignment operators (+ +, − −, + =, − =, * =, / =). Their use is illustrated by the following examples.

Statement using the assignment operator	*Statement using the shorthand assignment operator*
$a = a + 1$	$a + +$ or $+ + a$
$a = a − 1$	$a − −$ or $− − a$
$a = a + 4$	$a + = 4$
$a = a − 4$	$a − = 4$
$a = a * 4$	$a * = 4$
$a = a / 4$	$a / = 4$

NOTE *The use of shorthand assignment operators not only results in concise programs but more efficient programs also.*

Comments in C start with "/ *" and end with "* /".

For *e.g.* / * Euler's Method * /.

Input statement is scanf.

syntax scanf *("control string", & variable 1, & variable 2,.......);*

The **control string** contains the format of the data being input by the user. It contains

$\%d$ for an integer, and

$\%f$ for a floating point type.

The ampersand (&) symbol is necessary before the integer or floating-point type variable. Its significance is discussed under **functions**.

An example of the scanf statement:

Assuming the declarations

int $c;$

float $a[3][3];$

the statement

scanf *("%d %f"*, & c, & $a[1][2]$);

takes input from the user and stores it in the corresponding variables

$\%d$ corresponds to c,

$\%f$ corresponds to *a[1][2]*.

Output statement is printf.

Syntax printf *("control string", argument 1, argument 2,.......);*

The arguments can be C constants or variables. The *control string* can consist of

(*i*) The characters that will be printed as such.

(*ii*) Format specifications for variables.

(*iii*) Escape sequences.

Format specifications

for *integers* % *w* d

where *w* specifies the minimum width (*i.e.*, number of digits) for output for

floating point % *w.p* f

where *p* is number of digits to be displayed after the decimal point, after rounding off if necessary. In this *w* includes the decimal point too.

If width of the number is less than the specified width, the number is right justified in that width. However if width of the number is greater than the specified width, it will be printed in full.

Escape sequences. These are sequence of two characters meant for performing special tasks. The first character is always a back slash (\).

The most commonly used escape sequence is \n. It causes the output to start from next line.

An example of the printf statement:

For *e.g.*, Assuming the declarations

int *i* = 1;

float *a* = 27.23;

the statement

printf ("%3d\n %8.2f \n", *i*, *a*);

will output

ƀƀ1

ƀƀƀ27.23

where ƀ is a blank space.

Relational operators available in C are:

Mathematical symbol	C symbol
>	>
≥	> =
<	<
≤	< =
=	= =
≠	! =

Logical expressions are mathematical expressions connected by relational operators. Their value is either true or false.

Examples of logical expressions: Assuming $i = 2, j = 3$

$i < j$ is true

$i = = j$ is false

$(i * j) > (i + j)$ is true.

In C the result of a logical expression is an integer. 0 is taken as false, any non-zero integer is taken as true.

Logical operators are used to test more than one conditions *i.e.*, to combine more than one logical expressions.

Logical operator	C symbol
AND	&&
OR	||
NOT	!

The following tables illustrates their use.

AND

Logical expression 1	Logical expression 2	Result
True	True	True
True	False	False
False	True	False
False	False	False

OR

Logical expression 1	Logical expression 2	Result
True	True	True
True	False	True
False	True	True
False	False	False

NOT

Logical expression	Result
True	False
False	True

Decision making statement—If

Syntax

> **if** (Lexp)
>> {Tstatements}
>
> **else**
>> {Fstatements}

where *Lexp* is a logical expression.

> *Tstatements* are C statements executed when value of Lexp is true.

> *Fstatements* are C statements executed when value of Lexp is false.

> The *else* part is optional.

Loops

> (*i*) *While Loop*

> Syntax

> (*a*)   while *(Lexp)*
>> {statements}

> (*b*)   **do**
>> {statements}
>
>> **while** *(Lexp)*

Both of these forms of the while loop cause execution of *statements* while value of *Lexp* is true. The difference between the two forms is that in the latter, the statements are executed at least once irrespective of the value of *Lexp*.

> (*ii*) *For loop*

> *syntax*

> **for** *(initialization statement; Lexp; increment statement)*
>> {statements}

The loop is best explained by the following flow chart:

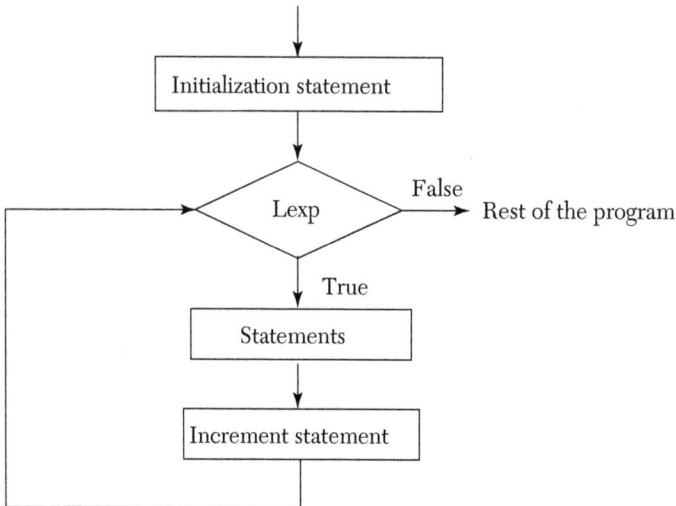

Break statement. When a break statement is encountered inside a loop, that loop is exited, *irrespective* of the value of *Lexp*, and the program continues with the statement immediately following the loop.

Functions. These are the basic block of a C program. *Functions* contain *State- ments* that specify what is to be done.

Every (program *has to* contain a function named *main*. The program begins executing at the first statement of *main*. Apart from *main* the C functions are classified into

— *Library functions*

— *User defined functions*

These functions are called from *main* to accomplish various tasks.

(*i*) *Library functions* are already available and we just have to use them. *e.g.*, printf, scanf, sqrt, cos, sin, fabs (used to get the absolute value of a floating point variable) etc.

(*ii*) *User defined functions* have to be written by the user in the program.

Syntax

return-type function-name (Argument-list)

{

statements

}

Program for understanding the various terms and concepts related with functions:

1. / * Sample program * /

2. # include < stdio.*h*>

3. float add (float *a*, float *x*);

4. void half (float *°x*)

5. {

6. *°x* / = 2;

7. return;

8. }

9. main ( )

10. {

11. float *a* = 2, *b* = 2, *c;*

12. *c* = add (*a, b*);

13. print ("%f % f %f\n", *a, b, c*);

14. half (&*a*); half (&*b*);

15. printf ("&f &f &f\n", *a, b, c*);

16. }

17. float add (float *a*, float *x*)

18. {

19. float sum;

20. sum = *a + x;*

21. *a* = 20, *x* = 20; /*changing the formal arguments*/

22. return sum;

23. }

[Line numbers have been added for reference purpose and are not part of the program.]

NOTES:

(a) Declaration and Definition

Line number 3 is the *declaration* of the function named *add*. It indicates that there is a function *add* which takes two arguments *a* and *b* both of type float and returns a float. *i.e.*, the *Argument list* is *float a, float b* and *return-type* is *float* (which can be *void* if function doesn't return anything— see line 4. It also indicates that the function is defined later in the program).

Lines 17–23 are the *definition* of the function *add i.e.*, they define how the function will make use of the *arguments* it received and return the required sum.

Lines 4–8 constitute both the *declaration* and *definition* of the function *half*.

(b) Calling a function

Line number 12 calls the function *add* with *a* and *b* as arguments and stores the value returned by it into the variable *c*.

(c) Actual and Formal arguments

The variables *a* and *x* in the *declaration* of function *add* are called the *formal arguments* (line 3).

The variables *a* and *b* in the call to the function (line 12) are the *actual arguments*.

(d) Call by value/Call by reference

In C language the values of actual arguments are *always copied* to the formal arguments when a function is called. This way of passing arguments is called *call by value*. In this any change made to the formal arguments in the function does *not* affect the value of actual arguments.

But in other languages, notably FORTRAN any change made to the formal arguments *is reflected* in the actual arguments. This is called *call by reference*, as the formal argument is treated as just another name for the actual argument. Both of them *refer* to the same location in the computer's memory.

(e) Simulating a call by reference in C

The following thumb rule can be followed to make the formal argument *refer* to the actual argument (and not just receive a copy of it)

"Precede the actual argument with an ampersand & (line 14) and precede the formal argument with an asterisk * (line 4)."

The actual working of this involves the concepts of *pointers*, which are another data type in C. The reader can refer to any standard C book for a complete understanding.

(f) Return statement in a function passes the control back to the calling function along with the calculated value (line 22)

Syntax   *return*   *expression;*

The *expression* can be omitted, in this case the return statement causes the function to just terminate then and there and pass control back to the calling function (Line 7).

In case no return statement is present in a function, an implicit return takes place on encountering the right curly brace } (For e.g., in the function *main*, the control passes back to the caller, *i.e.*, the operating system in this case after line 16).

Preprocessor directives. The lines in a C program that begin with a hash (#) sign are called *preprocessor directives*. The two most commonly used are # define and # include.

NOTE   *There is no semi colon (;) after the directive.*

(*i*) **# define**

syntax

> **# define** *name replacement*

It instructs the computer to replace all occurences of *name* with the *replacement* even before the program is processed, *i.e.*, checked for syntax.

For *e.g.*, consider the following statements

> # define N 2
>
> int $a[N]$;

Before the program is processed by the compiler, the second line *i.e.*, int $a[N]$; is changed to int $a[2]$; and the first line is removed.

∴ The resulting statement that is processed is int $a[2]$.

(*ii*) # include

> *syntax*
>
> **# include** < *header-file-name* >

This instructs the computer to insert the contents of the mentioned header file at the place where the directive appeared.

Dynamic memory allocations. In today's world utilization of memory resources is needed for efficient programming. We may come across situations where we may have to deal with data, which is dynamic in nature. The number of data items may change during the executions of a program. The number of customers change during the process at any time. When the list grows we need to allocate more memory space to accommodate additional data items. Such situations can be handled more easily by using dynamic allocation. Dynamic data items at run times, thus optimizing file usage of memory space.

The process of allocating memory at run time is known as dynamic memory allocation. Although "C" does not inherently have this facility there are four library routines which allow this function.

Many languages permit a programmer to specify an array size at run time. Such languages have the ability to calculate and assign during executions, the memory space required by the variables in the program. But "C" inherently does not have this facility but supports with memory management functions, which can be used to allocate and free memory during the program execution. The following functions are used in "C" for purpose of memory management.

Function	Task
malloc	Allocates memory requests size of bytes and returns a pointer of the allocated space
calloc	Allocates space for an group of elements initializes them to zero and returns a pointer to the memory
free	De allocates previously allocated space
realloc	Modifies the size of previously allocated memory.

Memory allocations process. According to the conceptual view the memory is partitioned in four different parts: in first part *program instructions* are stored, second *global variables,* third is used for *function calls* return address, arguments and local variables which are stored in stacks and last is called *heap* and is used for dynamic allocation during the execution of the program.

Allocating a block of memory. A block of memory may be allocated using the function *malloc*. The *malloc* function allocates a block of memory of specified size and returns a pointer of type void.

```
ptr= (type*) malloc (size);
```

`ptr` is a pointer of type the malloc returns a pointer (of type) to an area of memory with size. The general form of calloc is:

Example:

```
x= (int*) malloc (2* sizeof (int) );
```

On successful execution of this statement a memory equivalent to two times the area of int bytes is reserved and the address of the first byte of memory allocated is assigned to the pointer *x* of type int.

Allocating multiple blocks of memory. Calloc is function that is normally used to allocate multiple blocks of storage each of the same size and then sets all bytes to zero. The general form of calloc is:

```
ptr= (type*) calloc (n, elem-size);
```

The above statement allocates contiguous space for *n* blocks each size of elements size bytes. All bytes are initialized to zero and a pointer to the first byte of the allocated region is returned. If there is not enough space a null pointer is returned.

Freeing the used space. Compile time storage of a variable is allocated and released by the system in accordance with its storage class. With the dynamic runtime allocation, it is responsibility of programmer to free the allocated space when it is not re- quired. The release of storage space becomes important when the storage is limited. When we no longer need the data we stored in a block of memory and we do not intend to use that block for storing any other information, we may release that block of memory for future use, using the free function.

```
free (ptr);
```

`ptr` is a pointer that has been created by using malloc or calloc.

Data structure. *In computer science, a data structure is a particular way of storing and organizing data in a computer so that it can be used efficiently.*

Different kinds of data structures are suited to different kinds of applications, and some are highly specialized to specific tasks. For example, B-trees are particularly well- suited for implementation of databases, while compiler implementations usually use hash tables to look up identifiers.

Data structures are used in almost every program or software system. Specific data structures are essential ingredients of many efficient algorithms, and make possible the management of huge amounts of data, such as large databases and Internet indexing services. Some formal design methods and programming languages emphasize data structures, rather than algorithms, as the key organizing factor in software design.

Arrays are used to store a large set of data and manipulate them but the disadvantage is that all the elements stored in an array are of the same data type. If we need to use a collection of different data type items it is not possible using an array. When we require using a collection of different data items of different data types we can use a structure. *Structure is a method of packing data of different types.* A structure is a convenient method of handling a group of related data items of different data types.

```
structure definition:
general format:
struct tag_name
{
data type member1;
data type member2;
...
...
} Example: struct books
{
char title [30]; char author [25]; int page;
float price;
};
```

The keyword struct declares a structure to hold the details of four fields namely title, author, pages, and price. These are members of the structures. Each member may belong to different or same data type. The tag name can be used to define objects that have the tag names structure. The structure we just declared is not a variable by itself but a template for the structure.

We can declare structure variables using the tag name any where in the program. For example the statement,

```
struct lib_books book1,book2,book3;
```

declares book1,book2,book3 as variables of type struct lib_books each declaration has four elements of the structure lib_books. The complete structure declaration might look like this

```
struct lib_books
```

```
{
char title [20];
 char author [15];
int pages;
float price;
};
struct lib_books book1, book2, book3;
```

Structures do not occupy any memory until it is associated with the structure variable such as book1; the template is terminated with a semicolon. While the entire declaration is considered as a statement, each member is declared independent for its name and type in a separate statement inside the template. The tag name such as lib_books can be used to declare structure variables of its data type later in the program.

We can also combine both template declaration and variables declaration in one statement, the declaration

```
struct lib_books
{
char title [20]; char author [15]; int pages;
float price;
} book1, book2,book3;
is valid. The use of tag name is optional for example struct
{
 . . .
 . . .
 . . .
}
```

book1, book2, book3 declares book1,book2,book3 as structure variables representing three books but does not include a tag name for use in the declaration.

A structure is usually defineds before the main along with macro definitions. In such cases the structure assumes global status and all the functions can access the structure.

Union. Unions like structure contain members whose individual data types may differ from one another. However the members that compose a union all share the same storage area within the computer's memory where as each member within a structure is assigned its own unique storage area. Thus unions are used to observe memory. They are useful for application, involving multiple members. Values need not be assigned to all the members at any one time. Like structures, union can be declared using the key-word union as follows:

```
union item
{
int m; float p; char c;
}
code;
PROGRAMS OF STANDARD METHODS IN "C" LANGUAGE
```

14.3 Bisection Method (Section 2.7)

Flow-chart

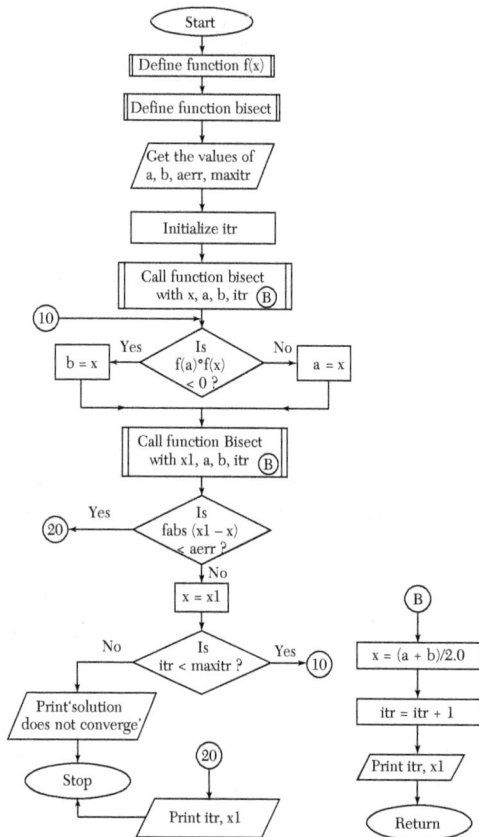

> **NOTES:** *a, b* are the limits in which the root lies
> *aerr* is the allowed error
> *itr* is a counter which keeps track of the number of iterations performed
> *maxitr* is the maximum number of iterations to be performed
> *x* is the value of root at the *n*th iteration
> *x1* is the value of root at $(n + 1)$th iteration.
> *Function Bisect:*
> *Purpose:* Performs and prints the result of one iteration
> *Variables:* *x* is the result of the current iteration.

Program

```
/* Bisection Method */
#include <stdio.h>
#include <math.h>
float f(float x)
{
 return (x*x*x - 4*x - 9);
}
void bisect(float *x,float a,float b,int *itr)
{
 *x = (a + b)/2;
 ++(*itr);
 printf("Iteration no. %3d X = %7.5f\n",*itr,*x);
}
main()
{
 int itr = 0, maxitr;
 float x, a, b, aerr, x1;
 printf("Enter the values of a,b,"
   "allowed error, maximum iterations\n");
 scanf("%f %f %f %d",&a,&b,&aerr,&maxitr);
 bisect(&x,a,b,&itr);
 do
 {
   if (f(a)*f(x) < 0)
       b = x;
   else
       a = x;
   bisect (&x1,a,b,&itr);
   if (fabs(x1-x) < aerr)
   {
       printf("After %d iterations, root <169>
         "= %6.4f\n",itr,x1);
```

```
        return 0;
    }
x = x1;
} while (itr < maxitr);
printf("Solution does not converge,"
  "iterations not sufficient");
  return 1;
}
```

Computer Solution of Example 2.15 (a)

```
Enter the values of a, b, allowed error, maximum iterations
3 2.0001 20
Iteration No. 1 X = 2.50000
Iteration No. 2 X = 2.75000
Iteration No. 3 X = 2.62500
Iteration No. 4 X = 2.68750
Iteration No. 5 X = 2.71875
Iteration No. 6 X = 2.70313
Iteration No. 7 X = 2.71094
Iteration No. 8 X = 2.70703
Iteration No. 9 X = 2.70508
Iteration No. 10 X = 2.70605
Iteration No. 11 X = 2.70654
Iteration No. 12 X = 2.70630
Iteration No. 13 X = 2.70642
Iteration No. 14 X = 2.70648
After 14 iterations, root = 2.7065
```

14.4 Regula-Falsi Method (Section 2.8)

Flow-chart

> **NOTES:** $f(x) = 0$ is the equation whose root is to be found
> $x0, x1$ are units in which root lies
> *aerr* is allowed error
> *maxitr* is maximum number of iterations to be performed
> *itr* is a counter which keeps track of the number of iterations performed
> $x2$ is value of root at nth iteration
> $x3$ is value of root at $(n + 1)$th iteration
> *Function Regula:*
> *Purpose:* Performs and prints the results of one iteration.
> *Variables:* x is value of root at nth iteration
> $fx0, fx1$ are values of $f(x)$ at $x0$ and $x1$, respectively.

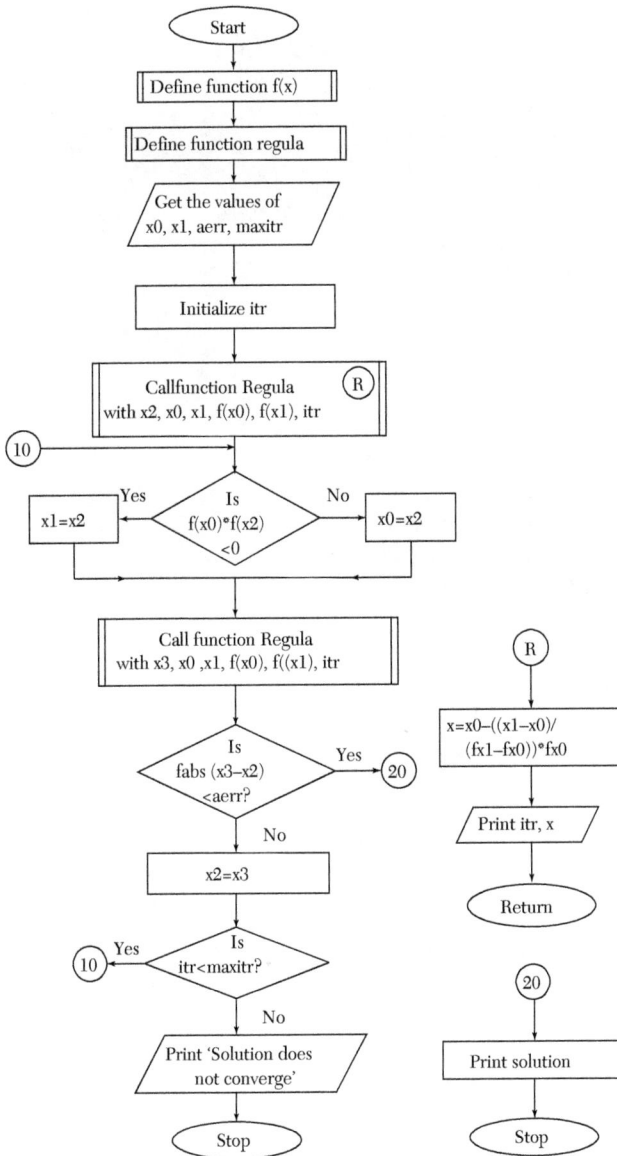

Program

```
/* Regula Falsi Method */
#include <stdio.h>
#include <math.h>
float f(float x)
```

```
{
 return cos(x)-x*exp(x);
}
void regula (float *x, float x0, float x1,
       float fx0, float fx1, int *itr)
 {
  *x = x0-((x1-x0)/(fx1-fx0))*fx0;
  ++(*itr);
  printf("Iteration no. %3d X = %7.5f\n",
     *itr,*x);
 }
 main()
 {
  int itr=0, maxitr;
  float x0,x1,x2,x3,aerr;
  printf("Enter the values for x0,x1,"
     "allowed error,maximum iterations\n•);
  scanf("%f %f %f %d",&x0,&x1,&aerr,&maxitr);
  regula(&x2,x0,x1,f(x0),f(x1),&itr);
  do
  {
    if (f(x0)*f(x2) < 0)
       x1 = x2;
    else
       x0 = x2;
    regula(&x3,x0,x1,f(x0),f(x1),&itr);
    if (fabs(x3-x2) < aerr)
    {
       printf("After %d iterations,"
          "root = %6.4f\n", itr,x3);
       return 0;
    }
    x2=x3;
 } while(itr < maxitr);
 printf("Solution does not converge,"
   "iterations not sufficient\n");
 return 1;
}
```

Computer Solution of Example 2.20

```
Enter the values for x0, x1, allowed error, maximum iterations
0 1.0001  20
```

```
Iteration No. 1 X = 0.31467
Iteration No. 2 X = 0.44673
Iteration No. 3 X = 0.49402
Iteration No. 4 X = 0.50995
Iteration No. 5 X = 0.51520
Iteration No. 6 X = 0.51692
Iteration No. 7 X = 0.51748
Iteration No. 8 X = 0.51767
Iteration No. 9 X = 0.51773
After 9 iterations, root = 0.5177
```

14.5 Newton Raphson Method (Section 2.11)

Flow-chart

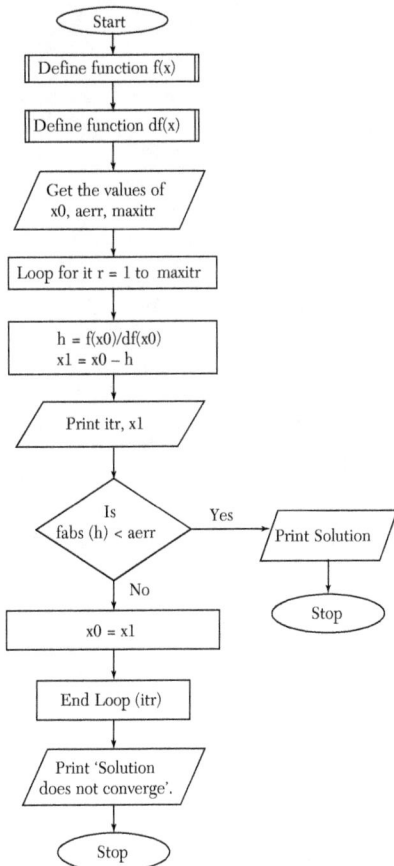

> **NOTES:** $F(x) = 0$ is the equation whose root is to be found
> $df(x)$ is the derivatives of $f(x)$ w.r.t. x
> $x0$ is value of root of nth iteration
> $x1$ is value of root of $(n + 1)$th iteration
> *aerr* is allowed error
> *maxitr* is maximum number of iterations to be performed
> *itr* is a counter which keeps track of the number of iterations performed.

Program

```c
/* Newton Raphson Method */
#include <stdio.h>
#include <math.h>
float f(float x)
{
  return x*log10(x)-1.2;
}
float df(float x)
{
  return log10(x) + 0.43429;
}
main()
{
  int itr,maxitr;
  float h,x0,x1,aerr;
  printf("Enter x0,allowed error,"
    "maximum iterations\n");
  scanf("%f %f %d",&x0,&aerr,&maxitr);
  for (itr=1;itr<=maxitr;itr++)
  {
    h = f(x0)/df(x0);
    x1 = x0-h;
    printf("Iteration no. %3d,"
       "x = %9.6f\n",itr,x1);
    if (fabs(h) < aerr)
    {
      printf("After %3d iterations,"
        "root = %8.6f\n", itr,x1);
      return 0;
       }
      x0 = x1;
  }
  printf("Iterations not sufficient,"
```

```
        "solution does not converge\n");
    return 1;
}
```

Computer Solution of Example 2.32

```
Enter x0, allowed error, maximum iterations
2.000001  10
Iteration No.        1 X = 2.813170
Iteration No.        2 X = 2.741109
Iteration No.        3 X = 2.740646
Iteration No.        4 X = 2.740646
After 4 iterations, root = 2.740646
```

14.6   Muller's Method (Section 2.13)

Flow-chart

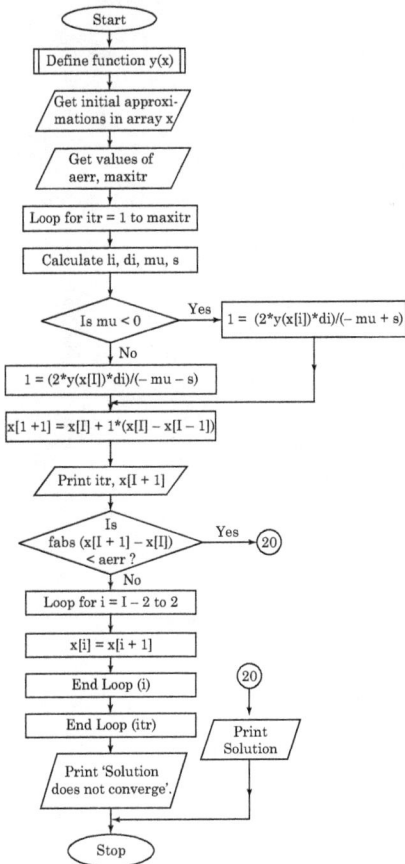

NOTES: $y(x) = 0$ is the equation whose root is to be found
x is an array which holds the three approximations to the root and the new improved value
I is defined as 2 in the program. This has been done because in C, array subscripts always start from zero and cannot be negative. Use of I facilitates more readable expressions. For *e.g.*, $x[0]$ can be written as $x[I - 2]$ which looks more close to x_{i-2} it actually represents.
li is λ_i
di is δ_i
mu is μ_i
s is $\sqrt{[\mu_i^2 - 4yi\delta_i\lambda_i(y_{i-2}\lambda i - y_{i-1}\delta i + y_i)]}$
l is λ

Program

```
/* Muller's Method */
#include <stdio.h>
#include <math.h>
#define I 2
float y(float x)
{
 return cos(x)-x*exp(x);
}
main()
{
 int i,itr,maxitr;
 float x[4],li,di,mu,s,l,aerr;
 printf("Enter the initial <169>
     "approximations\n");
 for (i = I-2;i<3;i++)
   scanf("%f",&x[i]);
 printf("Enter allowed error,"
    "maximum iterations\n");
 scanf("%f %d",&aerr,&maxitr);
 for(itr = 1;itr <= maxitr;itr++)
 {
   li = (x[I]-x[I-1])/(x[I-1]-x[I-2]);
   di = (x[I]-x[I-2])/(x[I-1]-x[I-2]);
   mu = y(x[I-2])*li*li
      - y(x[I-1])*di*di
      + y(x[I])*(di+li);
   s = sqrt((mu*mu - 4*y(x[I])*di*li
```

```
        *(y(x[I-2])*li-y(x[I-1])
            *di + y(x[I])))));
   if (mu < 0)
     l = (2*y(x[I])*di)/(-mu+s);
   else
     l = (2*y(x[I])*di)/(-mu-s);
     x[I+1] = x[I]+l*(x[I] - x[I-1]);
     printf("Iteration no. % 3d,"
         "x = %7.5f\n",itr,x[I+1]);
     if (fabs(x[I+1]-x[I]) < aerr)
     {
       printf("After %3d iterations,"
          "the solution is %6.4f\n",
          itr,x[I+1]);
       return 0;
     }
     for (i=I-2;i<3;i++)
         x[i] = x[i+1];
 }
 printf("Iterations not sufficient,"
   "solution does not converge\n");
 return 1;
}
```

Computer Solution of Example 2.34

```
Enter the initial approximations
-1 0 1
Enter allowed error, maximum iterations
.0001 10
Iteration No. 1 X = 0.44152
Iteration No. 2 X = 0.51255
Iteration No. 3 X = 0.51769
Iteration No. 4 X = 0.51776
After 4 iterations, the solution is 0.5178
```

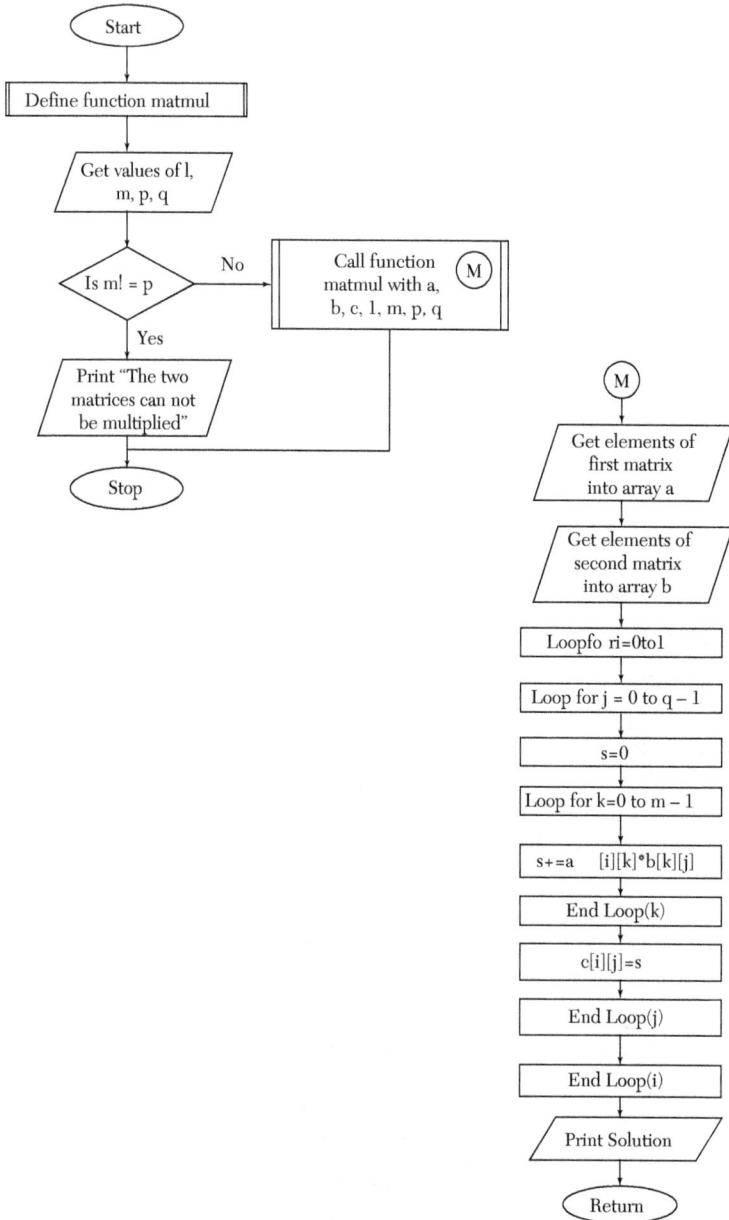

14.7   Multiplication of Matrices [Section 3.2 (3)4]

Flow-chart

```
                    ┌─────────┐
                    │  Start  │
                    └─────────┘
                         │
        ┌────────────────────────────────┐
        │    Define function matmul       │
        └────────────────────────────────┘
                         │
                  ╱─────────────╲
                 ╱ Get values of l,╲
                 ╲   m, p, q      ╱
                  ╲─────────────╱
                         │
                                      No      ┌──────────────────┐
                    ◇─────────────◇ ────────→ │  Call function   │  (M)
                    ◇  Is m! = p  ◇           │  matmul with a,  │
                    ◇─────────────◇           │  b, c, l, m, p, q│
                         │                    └──────────────────┘
                        Yes
                  ╱─────────────╲
                 ╱ Print "The two ╲
                 ╲ matrices can not╱
                 ╲ be multiplied"  ╱
                  ╲─────────────╱
                         │
                    ┌─────────┐
                    │  Stop   │
                    └─────────┘
```

```
                              (M)
                               │
                      ╱─────────────────╲
                     ╱  Get elements of  ╲
                     ╲   first matrix     ╱
                     ╲   into array a     ╱
                      ╲─────────────────╱
                               │
                      ╱─────────────────╲
                     ╱  Get elements of  ╲
                     ╲  second matrix     ╱
                     ╲  into array b      ╱
                      ╲─────────────────╱
                               │
                   ┌───────────────────────┐
                   │  Loopfo ri=0to1       │
                   └───────────────────────┘
                               │
                   ┌───────────────────────┐
                   │  Loop for j = 0 to q – 1│
                   └───────────────────────┘
                               │
                   ┌───────────────────────┐
                   │         s=0           │
                   └───────────────────────┘
                               │
                   ┌───────────────────────┐
                   │  Loop for k=0 to m – 1 │
                   └───────────────────────┘
                               │
                   ┌───────────────────────┐
                   │  s+=a  [i][k]*b[k][j]  │
                   └───────────────────────┘
                               │
                   ┌───────────────────────┐
                   │      End Loop(k)       │
                   └───────────────────────┘
                               │
                   ┌───────────────────────┐
                   │       c[i][j]=s        │
                   └───────────────────────┘
                               │
                   ┌───────────────────────┐
                   │      End Loop(j)       │
                   └───────────────────────┘
                               │
                   ┌───────────────────────┐
                   │      End Loop(i)       │
                   └───────────────────────┘
                               │
                      ╱─────────────────╲
                     ╱  Print Solution    ╱
                      ╲─────────────────╱
                               │
                          ┌─────────┐
                          │ Return  │
                          └─────────┘
```

> **NOTES:** *MAX* is largest number of rows or columns any matrix can have. (If MAX. = 20, 11 × 20 and 20 × 13 matrices can be multiplied but 1 × 21, 22 × 1 matrices cannot be multiplied until MAX > = 22
> *A, B* are arrays which contain the matrices to be multiplied
> *C* is array which contains the result of multiplication
> *L, M* are respectively the rows, columns of first matrix
> *P, Q* are respectively the rows, columns of second matrix
> *Function getelems*
> *Purpose:* To input a *m* × *n* matrix
> *Function Matmul.*
> *Purpose:* It performs the multiplication of matrices after taking them from the user and prints the result.
> *Variables:* i, j, k are loop control variables.

Program

```
/* Multiplication of matrices */
#include <stdio.h>
#define MAX 20
typedef float matrix[MAX][MAX];
void getelems(matrix x,int m,int n)
{
  int i,j;
  for(i=0;i<m;i++)
    for(j=0;j<n;j++)
      scanf("%f",&x[i][j]);
}
void printsol(matrix x,int m,int n)
{
 int i,j;
 for (i=0;i<m;i++)
 {
  for (j=0;j<n;j++)
    printf("%5.1f",x[i][j]);
  printf("\n");
 }
}
void matmul(matrix a,matrix b,matrix c,
      int l, int m,int p, int q)
{
 float s;
 int i,j,k;
```

```
   printf("Enter the elements of the"
     "first matrix\n");
  getelems(a,l,m);
  printf("Enter the elements of the"
     "second matrix\n");
 getelems(b,p,q);
 for (i=0;i<l;i++)
   for (j=0;j<q;j++)
   {
     s = 0;
     for (k=0;k<m;k++)
       s += a[i][k]*b[k][j];
     c[i][j] = s;
   }
 printf("The solution is \n");
 printsol(c,l,q);
}
main()
{
   matrix a,b,c;
   int l,m,p,q;
   printf("Enter the row, coloumn of the"
     "first matrix\n");
   scanf("%d %d",&l,&m);
   printf("Enter the row, coloumn of the"
     "second matrix\n");
   scanf("%d %d",&p,&q);
   if (m!=p)
     printf("The two matrices cannot"
        "be multiplied\n");
   else
     matmul(a,b,c,l,m,p,q);
}
```

Computer Solution of Example 3.7

```
Enter the row, column of the first matrix
3 3
Enter the row, column of the second matrix
3 2
Enter the elements of the first matrix
0 1 2
1 2 3
2 3 4
```

```
Enter the elements of the second matrix
    1    -2
   -1     0
    2    -1
The solution is
3.0   -2.0
5.0   -5.0
7.0   -8.0
```

14.8   Gauss Elimination Method [Section 3.4(3)]

Flow-chart

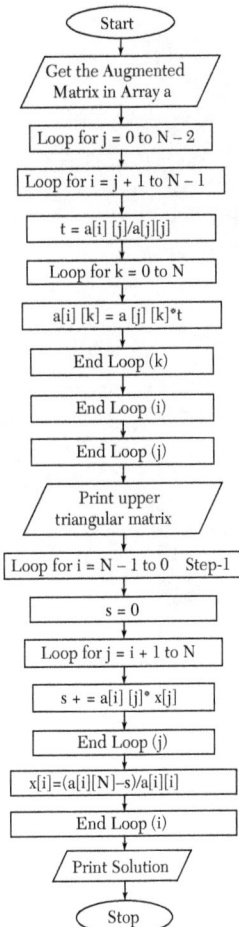

> **NOTES:** N is the number of unknowns
> a is an array which holds the Augmented Matrix
> x is an array which will contain values of unknowns
> i, j, k are loop control variables.

Program

```
/* Gauss elimination method */
#include <stdio.h>
#define N 4 main()
{
 float a[N][N+1],x[N],t,s;
 int i,j,k;
 printf("Enter the elements of the"
   "augmented matrix rowwise\n");
 for (i=0;i<N;i++)
  for (j=0;j<N+1;j++)
    scanf("%f",&a[i][j]);
 for (j=0;j<N-1;j++)
  for (i=j+1;i<N;i++)
  {
   t = a[i][j]/a[j][j];
   for (k=0;k<N+1;k++)
    a[i][k]=a[j][k]*t;
  }
 /* now printing the
 upper triangular matrix */
 printf("The upper triangular matrix"
        "is:-\n");
 for (i=0;i<N;i++)
{
for (j=0;j<N+1;j++)
  printf("%8.4f",a[i][j]);
printf("\n");
}
/* now performing back substitution */
for (i=N-1;i>=0;i- -)
{
 s = 0;
 for (j=i+1;j<N;j++)
   s += a[i][j]*x[j];
 x[i] = (a[i][N]-s)/a[i][i];
}
 /* now printing the results */
```

```
   printf("The solution is:- \n");
   for (i=0;i<N;i++)
     printf("x[%3d] = %7.4f\n",i+1,x[i]);
}
```

Computer Solution of Example 3.19

```
Enter the elements of augmented matrix rowwise
    10     -7      3      5      6
    -6      8     -1     -4      5
     3      1      4     11      2
     5     -9     -2      4      7
 10.000   -7.0000    3.0000    5.0000    6.0000
 0.0000    3.8000    0.8000   -1.0000    8.6000
-0.0000   -0.0000    2.4474   10.3158   -6.8158
 0.0000   -0.0000   -0.0000    9.9247    9.9247
The solution is:-
X[ 1] = 5.0000
X[ 2] = 4.0000
X[ 3] = -7.0000
X[ 4] = 1.0000
```

14.9   Gauss-Jordan Method [Section 3.4(4)]

Flow-chart

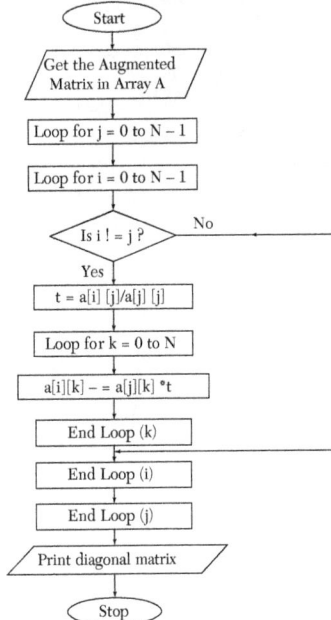

> Notes: *a* is an array which holds the Augmented Matrix
> *N* is the numbrt of unknowns. *e.g.* if it is a 3 × 3 system of equations,
> N = 3, and if 5 × 5 system take N = 5.
> *i, j, k* are loop variables.

Program

```
/* Gauss jordan method */
#include <stdio.h>
#define N 3 main()
{
 float a[N][N+1],t;
 int i,j,k;
 printf("Enter the elements of the "
        "augmented matrix rowwise\n");
 for (i=0;i<N;i++)
   for (j=0;j<N+1;j++)
    scanf("%f",&a[i][j]);
/* now calculating the values of x1,x2,....,xN */
 for (j=0;j<N;j++)
   for (i=0;i<N;i++)
     if (i!=j)
     {
       t = a[i][j]/a[j][j];
       for (k=0;k<N+1;k++)
          a[i][k] -= a[j][k]*t;
     }
/* now printing the diagonal matrix */
 printf("The diagonal matrix is:-\n");
 for (i=0;i<N;i++)
 {
   for (j=0;j<N+1;j++)
     printf("%9.4f",a[i][j]);
   printf("\n");
 }
/* now printing the results */
 printf("The solution is:- \n");
 for (i=0;i<N;i++)
    printf("x[%3d] = %7.4f\n",
         i+1,a[i][N]/a[i][i]);
}
```

Computer Solution of Example 3.22

```
Enter elements of augmented matrix rowwise
   10    -7    3    5    6
   -6     8   -1   -4    5
    3     1    4   11    2
    5    -9   -2    4    7
The diagonal matrix is:-
   10.0000      -0.0000    -0.0000    -0.0000    50.0000
    0.0000       3.8000    -0.0000     0.0000    15.2000
   -0.0000       0.0000     2.4474     0.0000   -17.1316
    0.0000      -0.0000     0.0000     9.9247     9.9247
The solution is:-
X[ 1] =    5.0000
X[ 2] =    4.0000
X[ 3] =   -7.0000
X[ 4] =    1.0000
```

14.10   Factorization Method [Section 3.4(5)]

Flow-chart

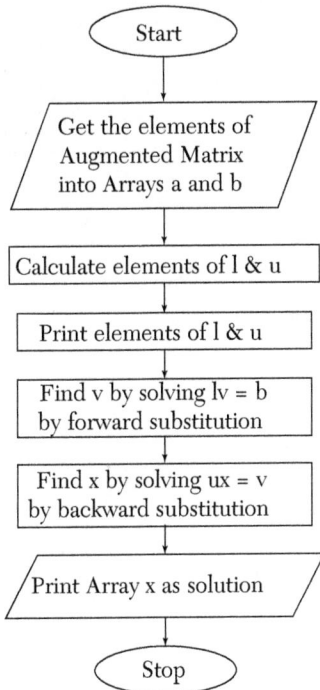

NOTES: $u_{ij} = a_{ij} - \sum_{k=1}^{i-1} u_{kj} l_{ik},$

$$l_{ij} = \left(a_{ij} - \sum_{k=1}^{i-1} u_{kj} l_{ik} \right) \Big/ u_{jj}$$

N is the number of unknowns
l is the lower triangular matrix u is the upper triangular matrix a is the coefficient matrix
b is the constant matrix (Column matrix)
v is a matrix such that $lv = b$
x will contain the values of unknowns
i, j, m are loop control variables
Function urow (I)
Purpose: Calculates elements of ith row of u
Variables: m is the number of unknowns
j, k are loop control variables
Function Lcol (J)
Purpose: Calculates elements of jth column of l
Variables: m is the number of unknowns
i, k are loop control variables.
Function Printmat
Purpose: To print an $N \times N$ matrix.

Program

```
/* Crout triangularization method */
#include <stdio.h>
#define N 4
typedef float matrix[N][N];
matrix l,u,a;
float b[N],x[N],v[N];
void urow(int i)
{
        float s;
    int j,k;
    for (j=i;j<N;j++)
    {
            s = 0;
            for (k=0;k<N-1;k++)
```

```
                    s += u[k][j]*l[i][k];
              u[i][j] = a[i][j]-s;
      }
}
void lcol(int j)
{
  float s;
  int i,k;
  for (i=j+1;i<N;i++)
  {
    s = 0;
    for (k=0;k<=j-1;k++)
      s += u[k][j]*l[i][k];
      l[i][j] = (a[i][j]-s)/u[j][j];
  }
}
void printmat(matrix x)
{
 int i,j;
 for (i=0;i<N;i++)
 {
   for (j=0;j<N;j++)
     printf("%8.4f",x[i][j]);
   printf("\n");
 }
}
 main()
 {
   int i,j,m;
   float s;
   printf("Enter the elements of augmented"
     "matrix rowwise\n");
   for (i=0;i<N;i++)
   {
     for (j=0;j<N;j++)
       scanf("%f",&a[i][j]);
     scanf("%f",&b[i]);
   }
   /* now calculating the elements of
   l and u */
   for (i=0;i<N;i++)
     l[i][i] = 1.0;
```

```
for (m=0;m<N;m++)
{
  urow(m);
   if (m < N-1) lcol(m);
}
 /* now printing the elements of l and u */
 printf("\t\tU\n"); printmat(u);
 printf("\t\tL\n"); printmat(l);
 /* now solving LV=B
 by forward substitution */
 for (i=0;i<N;i++)
 {
    s = 0;
    for (j=0;j<=i-1;j++)
      s += l[i][j]*v[j];
    v[i] = b[i]-s;
 }
 /* now solving UX=V
 by backward substitution */
 for (i=N-1;i>=0;i- -)
 {
    s = 0;
    for (j=i+1;j<N;j++)
      s += u[i][j]*x[j];
    x[i] = (v[i]-s)/u[i][i];
 }
 /* printing the results */
 printf("The solution is:-\n");
 for (i=0;i<N;i++)
   printf("x[%3d] = %6.4f\n",i+1,x[i]);
}
```

Computer Solution of Example 3.23

```
Enter the elements of augmented matrix rowwise
3          2     7     4
2          3     1     5
3          4     1     7
                 U
3.0000          2.0000      7.0000
0.0000          1.6667     -3.6667
0.0000          0.0000     -1.6000
                 L
1.0000          0.0000      0.0000
0.6667          1.0000      0.0000
1.0000          1.2000      1.0000
```

```
The solution is:
x[1] = 0.8750
x[2] = 1.1250
x[3] = -.1250
```

Computer Solution of Example 3.24

```
Enter the elements of augmented matrix rowwise
   10      -7       3      5      6
   -6       8      -1     -4      5
    3       1       4     11      2
    5      -9      -2      4      7
                    U
   10.0000        -7.0000        3.0000    5.0000
    0.0000         3.8000        0.8000   -1.0000
    0.0000         0.0000        2.4474   10.3158
    0.0000         0.0000        0.0000    9.9247
                    L
    1.0000         0.0000        0.0000    0.0000
 -  0.6000         1.0000        0.0000    0.0000
    0.3000         0.8158        1.0000    0.0000
    0.5000        -1.4474       -0.9570    1.0000
The solution is:-
x[ 1] =    5.0000
x[ 2] =    4.0000
x[ 3] = -7.0000
x[ 4] =    1.0000
```

14.11   Gauss-Seidal Iteration Method [Section 3.5(2)]

Flow-chart

> Notes: N is the number of unknowns
> a is an array which holds the augmented matrix
> x is an array which will hold the values of unknowns
> *aerr* is allowed error
> *maxitr* is the maximum number of iterations to be performed
> *itr* is the counter which keeps track of number of iterations performed
> *err* is error in value of x_i
> *maxerr* is maximum error in any value of x_i after an iteration.

Program

```
/* Gauss Seidal method */
#include <stdio.h>
#include <math.h>
#define N 4 main()
```

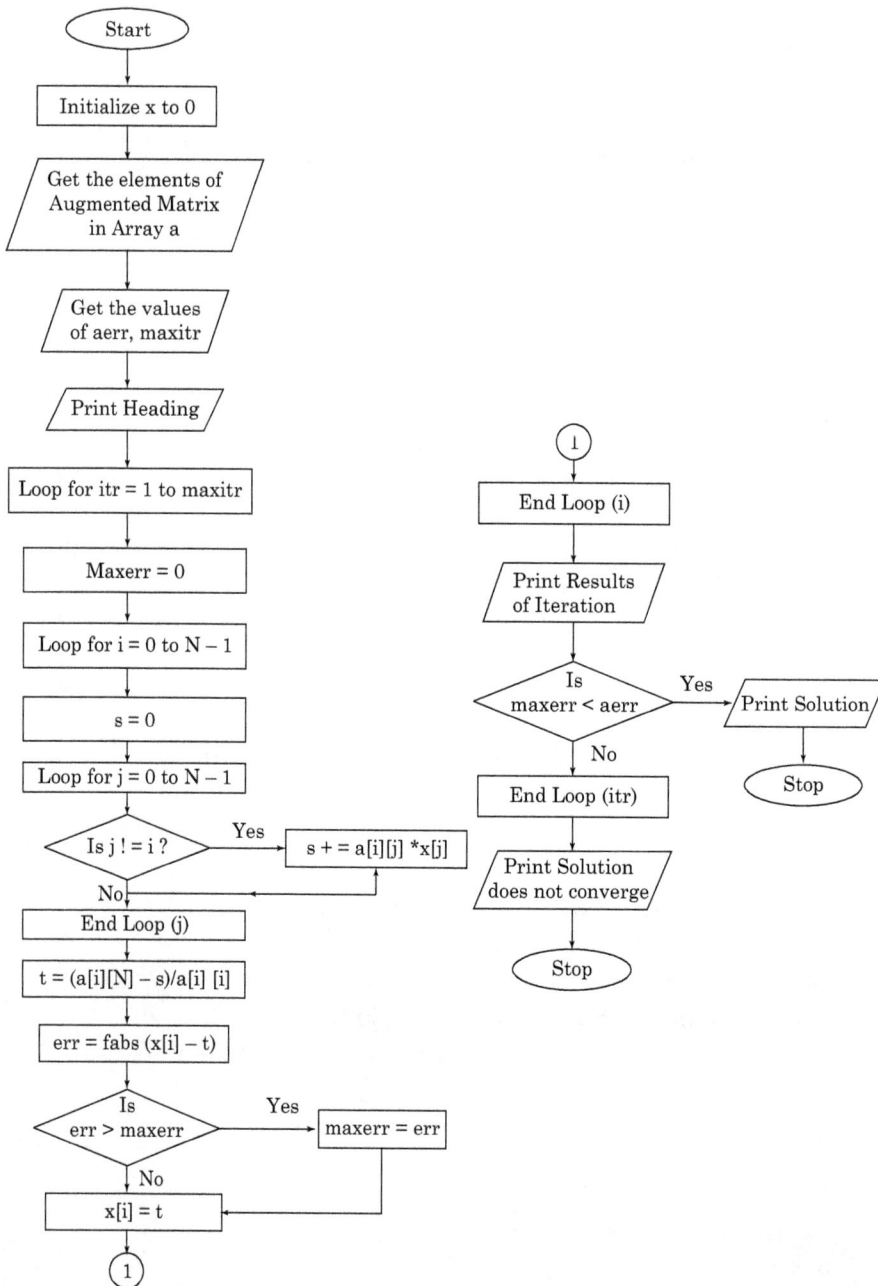

```
                    ┌─────────┐
                   (   Start   )
                    └─────────┘
                         │
              ┌──────────────────┐
              │ Initialize x to 0 │
              └──────────────────┘
                         │
            ╱──────────────────────╱
           ╱ Get the elements of  ╱
          ╱  Augmented Matrix    ╱
         ╱   in Array a         ╱
        ╱──────────────────────╱
                         │
            ╱──────────────────╱
           ╱ Get the values   ╱
          ╱  of aerr, maxitr ╱
         ╱──────────────────╱
                         │
           ╱───────────────╱
          ╱ Print Heading ╱
         ╱───────────────╱
                         │
        ┌──────────────────────────┐
        │ Loop for itr = 1 to maxitr│
        └──────────────────────────┘
                         │
            ┌──────────────────┐
            │    Maxerr = 0    │
            └──────────────────┘
                         │
        ┌──────────────────────┐
        │ Loop for i = 0 to N − 1│
        └──────────────────────┘
                         │
            ┌──────────────────┐
            │       s = 0      │
            └──────────────────┘
                         │
        ┌──────────────────────┐
        │ Loop for j = 0 to N − 1│
        └──────────────────────┘
```

Loop for itr = 1 to maxitr

Maxerr = 0

Loop for i = 0 to N − 1

s = 0

Loop for j = 0 to N − 1

Is j ! = i ? —— Yes ——→ s += a[i][j] *x[j]

No

End Loop (j)

t = (a[i][N] − s)/a[i] [i]

err = fabs (x[i] − t)

Is err > maxerr —— Yes ——→ maxerr = err

No

x[i] = t

(1)

(1)

End Loop (i)

Print Results of Iteration

Is maxerr < aerr —— Yes ——→ Print Solution

No

End Loop (itr)

Print Solution does not converge

Stop

Stop

```c
{
 float a[N][N+1],x[N],aerr,maxerr, t,s,err;
 int i,j,itr,maxitr;
 /* first initializing the array x */
 for (i=0;i<N;i++) x[i]=0;
 printf("Enter the elements of the"
       "augmented matrix rowwise\n");
 for (i=0;i<N;i++)
  for (j=0;j<N+1;j++)
    scanf("%f",&a[i][j]);
 printf("Enter the allowed error,"
    "maximum iterations\n");
 scanf("%f %d",&aerr,&maxitr);
 printf("Iteration x[1] x[2]" "x[3]\n");
 for (itr=1;itr<=maxitr;itr++)
 {
   maxerr = 0;
   for (i=0;i<N;i++)
   {
     s = 0;
     for (j=0;j<N;j++)
       if (j!=i) s += a[i][j]*x[j];
     t = (a[i][N]-s)/a[i][i];
     err = fabs(x[i]-t);
     if (err > maxerr) maxerr = err;
     x[i] = t;
   }
   printf("%5d",itr);
   for (i=0;i<N;i++)
      printf("%9.4f",x[i]);
   printf("\n");
   if (maxerr<aerr)
   {
     printf("Converges in %3d"
        "iterations\n",itr);
     for (i=0;i<N;i++)
       printf("x[%3d] = %7.4f\n",
           i+1,x[i]);
     return 0;
   }
 }
 printf("Solution does not converge,"
```

```
   "iterations not sufficient\n");
 return 1;
}
```

Computer Solution of Example 3.28

```
Enter the elements of augmented matrix rowwise
20  1  -2 17
3   20 -1 -18
2   - 3 20 25
Enter the allowed error, maximum iterations
.0001 10
```

Iteration	X(1)	X(2)	X(3)
1	0.8500	- 1.0275	1.0109
2	1.0025	- 0.9998	0.9998
3	1.0000	- 1.0000	1.0000
4	1.0000	- 1.0000	1.0000

```
Converges in 4 iterations
X[1] = 1.0000
X[2] = -1.0000
X[3] = 1.0000
```

Computer Solution of Example 3.30

```
Enter the elements of the augmented matrix rowwise
    10    -2   - 1   -1    3
  - 2    10   - 1   -1   15
  - 1    -1    10   -2   27
  - 1    -1    -2   10   -9
Enter the allowed error, maximum iterations
.0001 15
```

Iteration	x[1]	x[2]	x[3]	x[4]
1	0.3000	1.5600	2.8860	-0.1368
2	0.8869	1.9523	2.9566	-0.0248
3	0.9836	1.9899	2.9924	-0.0042
4	0.9968	1.9982	2.9987	-0.0008
5	0.9994	1.9997	2.9998	-0.0001
6	0.9999	1.9999	3.0000	-0.0000
7	1.0000	2.0000	3.0000	-0.0000

```
Converges in 7 iterations
x[ 1] = 1.0000
x[ 2] = 2.0000
x[ 3] = 3.0000
x[ 4] = 0.0000
```

14.12 Power Method (Section 4.11)

Flow-chart

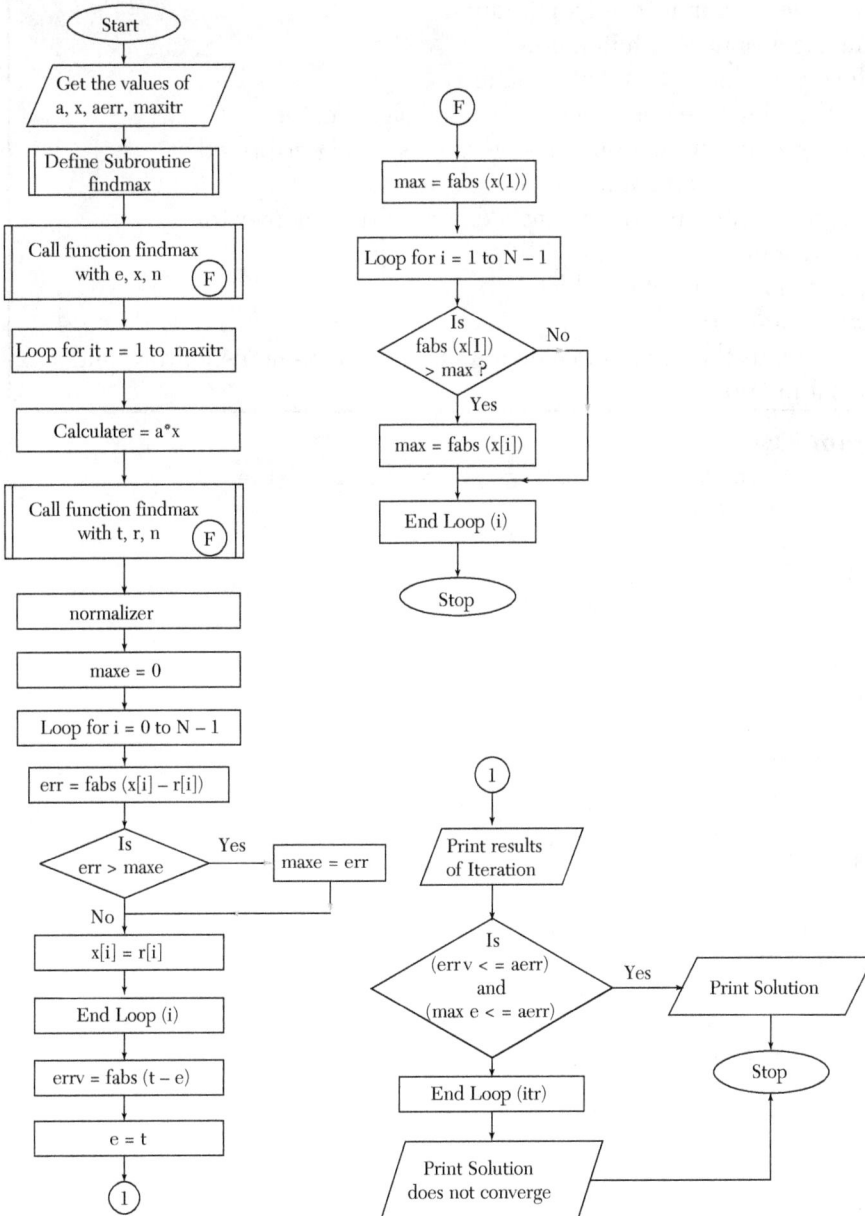

Notes: N is number of rows (or columns) in square matrix
a is the square matrix
x is the eigenvector at nth iteration
r is the eigenvector at $(n + 1)$th iteration
e is the eigenvalue at nth iteration
t is the eigenvalue at $(n + 1)$th iteration
aerr is the allowed error in eigenvalue and eigenvector
maxitr is the maximum number of iterations to be performed
err is error in an element of the eigenvector
maxe is the maximum error in any element of the eigenvector
errv is error in the eigenvalue
itr, i, k are loop control variables.
Function findmax:
Purpose: Finds the maximum element in array x(a N-element array) and
returns it in *Max*.

Program

```
/* Power method for finding largest eigenvalue */
#include <stdio.h>
#include <math.h>
typedef float array[N];
void findmax(float *max,array x)
{
 int i;
 *max = fabs(x[0]);
  for (i=1;i<N;i++)
    if (fabs(x[i]) > *max)
      *max = fabs(x[i]);
}
main()
{
  float a[N][N],x[N],r[N],maxe,
    err,errv,aerr,e,s,t;
  int i,j,k,itr,maxitr;
  printf("Enter the matrix rowwise\n");
  for (i=0;i<N;i++)
    for (j=0;j<N;j++)
      scanf("%f",&a[i][j]);
  printf("Enter the initial approximation"
    "to the eigen vector\n");
  for (i=0;i<N;i++)
```

```
      scanf("%f",&x[i]);
  printf("Enter the allowed error,"
    "maximum iterations\n");
  scanf("%f %d",&aerr,&maxitr);
  printf("Itr no. Eigenvalue"
    "EigenVector\n");
  /* now finding the largest eigenvalue in
  the initial approx. to eigen vector */
  findmax(&e,x);
  /* now starting the iterations */
  for (itr=1;itr<=maxitr;itr++)
  {
    /* loop to multiply the matrices
    a and x */
    for (i=0;i<N;i++)
    {
      s = 0;
      for (k=0;k<N;k++)
        s += a[i][k]*x[k];
      r[i]=s;
    }
    findmax(&t,r);
    for (i=0;i < N;i++) r[i] /= t;
    maxe = 0;
    for (i=0;i<N;i++)
    {
      err = fabs(x[i]-r[i]);
       if (err > maxe) maxe = err;
       x[i] = r[i];
    }
    errv = fabs(t-e);
    e = t;
    printf("%4d %12.4f",itr,e);
    for (i=0;i<N;i++)
      printf("%9.3f",x[i]);
    printf("\n");
    if ((errv <= aerr) && (maxe <= aerr))
    {
      printf("Converges in %d"
        "iterations\n",itr);
      printf("Largest eigen value"
        "= %6.2f\n",e);
```

```
    printf("Eigenvector:-\n");
    for (i=0;i<N;i++)
      printf("x[%3d] = %6.2f\n",
      i+1,x[i]);
    printf("\n"); return 0;
  }
}
 printf("Solution does not converge,"
   "iterations not sufficient\n");
 return 1;
}
```

Computer Solution of Example 4.11

```
Enter the matrix rowwise
 2  -1    0
-1    2   -1
 0  -1    2
Enter the initial approximation to the eigenvector
1 0 0
Enter the allowed error, maximum iterations
.01 10
    Itr No.       Eigen Value          Eigen Vector
          1         2.0000       1.000   - 0.500    0.000
          2         2.5000       1.000   - 0.800    0.200
          3         2.8000       1.000    -1.000    0.429
          4         3.4286       0.875    -1.000    0.542
          5         3.4167       0.805    -1.000    0.610
          6         3.4146       0.764    -1.000    0.650
          7         3.4143       0.741    -1.000    0.674
          8         3.4142       0.727    -1.000    0.688
          9         3.4142       0.719    -1.000    0.696
Converges in 9 iterations
Largest eigenvalue = 3.41
Eigenvector:-
X[1] =  0.72
X[2] = -1.00
X[3] =  0.70
```

14.13   Method of Least Squares (Section 5.5)

Flow-chart

```
                    ┌─────────────┐
                   (    Start     )
                    └─────────────┘
                           │
                           ▼
              ┌───────────────────────┐
              │ Initialize all elements│
              │   of augm to zero      │
              └───────────────────────┘
                           │
                           ▼
                ┌─────────────────────┐
               /   Get the value of n  /
                └─────────────────────┘
                           │
                           ▼
              ┌───────────────────────┐
              │  Read in the date points│
              │ and increment appropriate│
              │   elements of augm      │
              └───────────────────────┘
                           │
                           ▼
              ┌───────────────────────┐
              │ Assign values to non-  │
              │ unique elements of augm│
              └───────────────────────┘
                           │
                           ▼
                ┌─────────────────────┐
               /     Print augm        /
                └─────────────────────┘
                           │
                           ▼
              ┌───────────────────────┐
              │  Solve for a, b, c by  │
              │  Gauss Jordan Method   │
              └───────────────────────┘
                           │
                           ▼
                ┌─────────────────────┐
               /     Print a, b,       /
               /   c as solution       /
                └─────────────────────┘
                           │
                           ▼
                    ┌─────────────┐
                   (    Stop      )
                    └─────────────┘
```

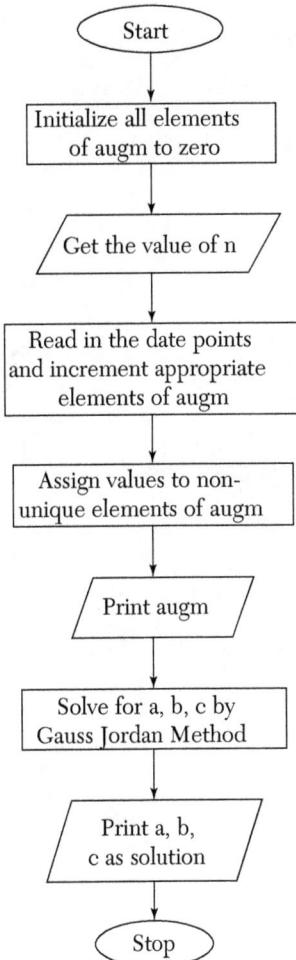

> Notes: *augm* is the augmented Matrix.
>
> *n* is the number of data points.

Program

```c
/* Parabolic fit by least squares */
#include <stdio.h>
main()
```

```
{
 float augm[3][4]={{0,0,0,0},{0,0,0,0},{0,0,0,0}};
 float t,a,b,c,x,y,xsq;
 int i,j,k,n;
 puts("Enter the no. of pairs of"
      "observed values:");
 scanf("%d",&n);
 augm[0][0] = n;
 for (i=0;i<n;i++)
 {
   printf("pair no. %d\n",i+1);
   scanf("%f %f",&x,&y);
   xsq = x*x;
   augm[0][1] += x;
   augm[0][2] += xsq;
   augm[1][2] += x*xsq;
   augm[2][2] += xsq*xsq;
   augm[0][3] += y;
   augm[1][3] += x*y;
   augm[2][3] += xsq*y;
 }
 augm[1][1] = augm[0][2];
 augm[2][1] = augm[1][2];
 augm[1][0] = augm[0][1];
 augm[2][0] = augm[1][1];
 puts("The augmented matrix is:-");
 for (i=0;i<3;i++)
 {
    for (j=0;j<4;j++)
       printf("%9.4f",augm[i][j]);
    printf("\n");
 }
 /* Now solving for a,b,c
 by Gauss Jordan Method */
 for (j=0;j<3;j++)
   for (i=0;i<3;i++)
     if (i!=j)
     {
        t = augm[i][j]/augm[j][j];
        for (k=0;k<4;k++)
           augm[i][k]
             -= augm[j][k]*t;
```

```
        }
    a = augm[0][3]/augm[0][0];
    b = augm[1][3]/augm[1][1];
    c = augm[2][3]/augm[2][2];
    printf("a = %8.4f b = %8.4f "
        "c = %8.4f\n",a,b,c);
}
```

Computer Solution of Example 5.7

```
Enter the no. of pairs of observed values:
7
Pair No. 1
1 1.1
Pair No. 2
1.5 1.3
Pair No. 3
2 1.6
Pair No. 4
2.5 2
Pair No. 5
3.0 2.7
Pair No. 6
3.5 3.4
Pair No. 7
4.0 4.1
The augmented matrix is:-
        7.0000      17.5000      50.7500      16.2000
       17.5000      50.7500     161.8750      47.6500
       50.7500     161.8750     548.1875     154.4750
     a = 1.0357    b = -0.1929   c = 0.2429
```

14.14   Method of Group Averages (Section 5.9)

Flow-chart

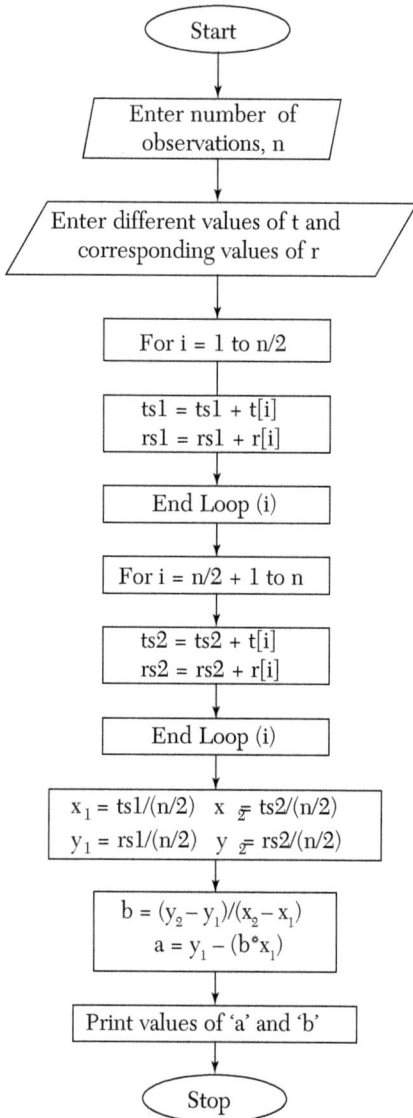

Program

```
#include<conio.h>
#include<stdio.h>
void main()
```

```
{
int t[10],n,i,ts1=0,ts2=0;
float a,b,rs1=0,rs2=0,r[10],x1,y1,x2,y2;
clrscr();
printf("enter the no. of observations\n");
scanf("%d",&n);
printf("enter the different values of t\n");
for (i=1;i<=n;i++)
{
scanf("%d",&t[i]);
}
printf("\n enter the corresponding values of r\n");
for (i=1;i<=n;i++)
{
scanf("%f",&r[i]);
}
for (i=1;i<=(n/2);i++)
{ ts1+=t[i]; rs1+=r[i];
}
for (i=((n/2)+1);i<=n;i++)
{ ts2+=t[i]; rs2+=r[i];
}
x1=ts1/(n/2);
y1=rs1/(n/2);
x2=ts2/(n/2);
y2=rs2/(n/2);
b=(y2-y1)/(x2-x1);
a=y1-(b*x1);
printf("the value of a&b comes out to be\n");
printf("a=%6.3f\nb=%6.3f",a,b);
getch();
}
```

Computer Solution of Example 5.16

```
Enter the no. of observations
8
enter the different values of t
40 50 60 70 80 90 100 110
enter the corresponding values of r
1069.1 1063.6 1058.2 1052.7 1049.3 1041.8 1036.3 1030.8
the values of a&b come out to be
a=1090.256
b=-0.534
```

14.15   Method of Moments (Section 5.11)

Flow-chart

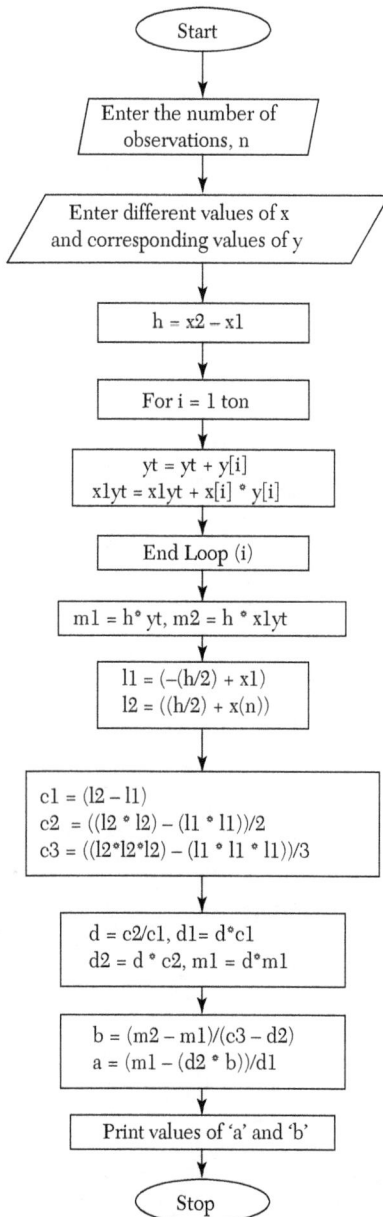

Start

Enter the number of observations, n

Enter different values of x and corresponding values of y

h = x2 – x1

For i = 1 ton

yt = yt + y[i]
x1yt = x1yt + x[i] ° y[i]

End Loop (i)

m1 = h° yt, m2 = h ° x1yt

l1 = (–(h/2) + x1)
l2 = ((h/2) + x(n))

c1 = (l2 – l1)
c2 = ((l2 ° l2) – (l1 ° l1))/2
c3 = ((l2°l2°l2) – (l1 ° l1 ° l1))/3

d = c2/c1, d1= d°c1
d2 = d ° c2, m1 = d°m1

b = (m2 – m1)/(c3 – d2)
a = (m1 – (d2 ° b))/d1

Print values of 'a' and 'b'

Stop

Program

```
#include <stdio.h>
#include <conio.h>
void main()
{
int x[10],y[10],i,n,yt=0,x1yt=0;
float a,b,11,12,c1,c2,c3,d,d1,d2,m1,m2,h;
clrscr();
printf("enter the no. of observations\n");
scanf("%d",&n);
printf("enter the different values of x");
for (i=1;i<n;i++)
{
scanf("%d",&x[i]);
}
printf("\nenter the corresponding values of y\n");
for (i=1;i<n;i++)
{
scanf("%d",&y[i]);
}
h=x[2]-x[1];
for(i=1;i<=n;i++)
{ yt+=y[i]; x1yt+=x[i]*y[i];
} m1=h*yt; m2=h*x1yt;
11=(-(h/2)+x[1]);
12=((h/2)+x[n]);
c1=(12-11);
c2=((12*12)-(11*11))/2;
c3=((12*12*12)-(11*11*11))/3;
printf("The observed equations are\n");
printf("%5.2fa+%5.2fb=%5.2f\n%5.2fa+5.2fb=%5.2f",
   c1,c2,m1,c2,c3,m2);
d=c2/c1;
d1=d*c1;
d2=d*c2;
m1=d*m1;
b=(m2-m1)/(c3-d2);
a=(m1-(d2*b))/d1;
printf("\nOn solving these equations
   we get a=%5.2f&b=%5.2f\n",a,b);
printf("hence the required equation is y=%5.2f+%5.2fx",a,b);
```

```
getch ();
}
```

Computer Solution of Example 5.20

```
Enter the no. of observations
4
enter the different values of x
1 2 3 4
enter the corresponding values of y
16 19 23 26
the observed equations are
4.00a+10.00b=84.00
10.00a+30.33b=227.00
on solving these equations we get a = 13.03&b=3.19
hence the required equation is y=13.03+3.19x
```

14.16   Newton's Forward Interpolation Formula (Section 7.2)

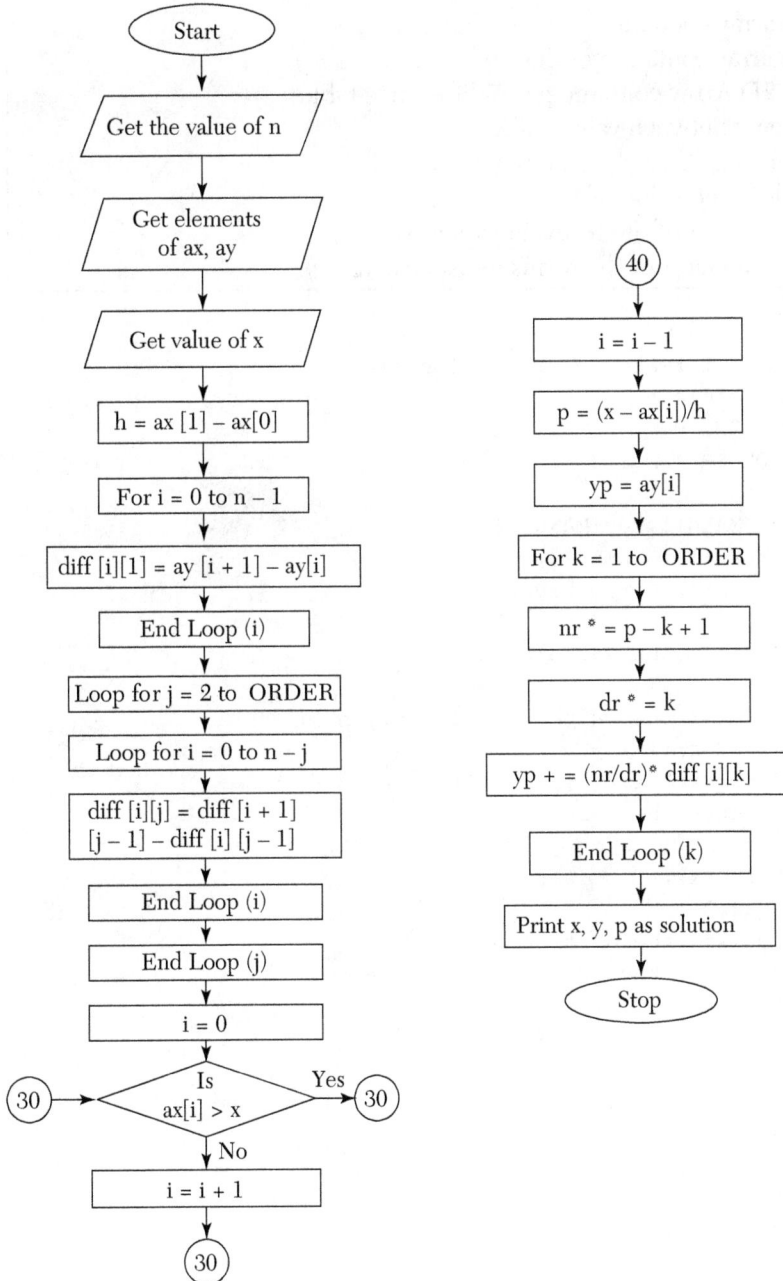

Start

Get the value of n

Get elements of ax, ay

Get value of x

h = ax [1] – ax[0]

For i = 0 to n – 1

diff [i][1] = ay [i + 1] – ay[i]

End Loop (i)

Loop for j = 2 to ORDER

Loop for i = 0 to n – j

diff [i][j] = diff [i + 1] [j – 1] – diff [i] [j – 1]

End Loop (i)

End Loop (j)

i = 0

30

Is ax[i] > x   Yes   30

No

i = i + 1

30

40

i = i – 1

p = (x – ax[i])/h

yp = ay[i]

For k = 1 to ORDER

nr ° = p – k + 1

dr ° = k

yp + = (nr/dr)° diff [i][k]

End Loop (k)

Print x, y, p as solution

Stop

> **NOTES:** *MAXN* is the maximum value of *N*
> *ORDER* is the maximum order in the difference table
> *ax* is an array containing values of *x* ($x_0, x_1, ..., x_n$)
> *ay* is an array containing values of $y(y_0, y_1, ..., y_n)$
> *diff* is a 2D Array containing the difference table
> *h* is spacing between values of X
> *x* is value of x at which value of y is wanted
> *yp* is calculated value of Y
> *nr* is numerator of the terms in expansion of y_P
> *dr* is denominator of the terms in expansion of y_P

Program

```
/* Newton's forward interpolation */
#include <stdio.h>
#define MAXN 100
#define ORDER 4 main()
{
 float ax[MAXN+1],ay[MAXN+1],
   diff[MAXN+1][ORDER+1],
    nr=1.0,dr=1.0,x,p,h,yp;
 int n,i,j,k;
 printf("Enter the value of n\n");
 scanf ("%d",&n);
 printf("Enter the values in form x,y\n");
 for (i=0;i<=n;i++)
   scanf("%f %f",&ax[i],&ay[i]);
 printf("Enter the values of x"
   "for which value of y is wanted \n");
 scanf("%f",&x);
 h=ax[1]-ax[0];
 /* now making the diff. table */
 /* calculating the 1st order differences */
 for (i=0;i<=n-1;i++)
   diff[i][1] = ay[i+1]-ay[i];
 /* calculating the second &
 higher order differences.*/
 for (j=2;j<=ORDER;j++)
     for (i=0;i<=n-j;i++)
     diff [i][j] = diff[i+1][j-1]
         -diff[i][j-1];
 /* now finding x0 */
```

```
i-0;
while (!(ax[i] > x)) i++;
/* now ax[i] is x0 & ay[i] is y0 */
i--;
p = (x-ax[i])/h; yp=ay[i];
/* Now carrying out interpolation */
for (k=1;k<=ORDER;k++)
{
    nr *= p-k+1; dr *=k;
    yp += (nr/dr)*diff[i][k];
}
printf ("when x = %6.1f, y = %6.2f\n"
     ,x,yp);
}
```

Computer Solution of Example 7.1

```
Enter the value of n
6
Enter the values in form x, y
100     10.63
150     13.03
200     15.04
250     16.81
300     18.42
350     19.90
400     21.27
Enter the values of x for which value of y is wanted
218
When x = 218.0, y = 15.70
```

14.17   Lagrange's Interpolation Formula (Section 7.12)

Flow-chart

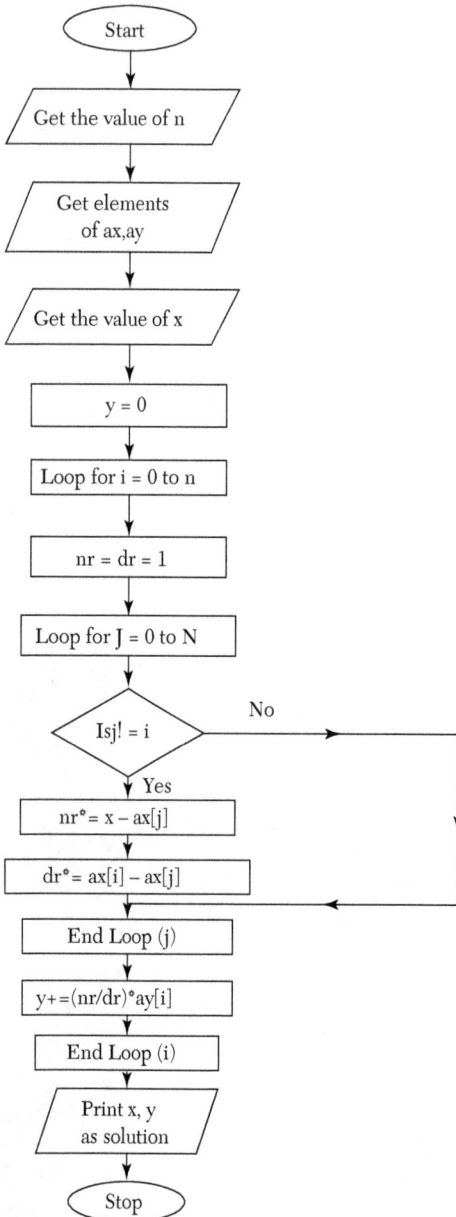

> **NOTES:** *MAX* is the maximum value of n
> ax is an array containing values of $x(x_0, x_1,....., xn)$
> ay is an array containing values of $y(y_0, y_1,......, yn)$
> x is the value of x at which value of y is wanted
> y is the calculated value of y
> nr is numerator of the terms in expansion of y
> dr is denominator of the terms in expansion of y.

Program

```
/*Lagrange's Interpolation*/
#include <stdio.h>
#define MAX 100
main()
{
 float ax [MAX+1],ay[MAX+1],nr,dr,x,y=0;
 int i,j,n;
 printf ("Enter the value of n\n");
 scanf("%d",&n);
 printf ("Enter the set of values\n");
 for (i=0;i<=n;i++)
   scanf ("%f%f",&ax[i],&ay[i]);
 puts("Enter the value of x for which"
      "value of y is wanted");
 scanf("%f",&x);
 for (i=0;i<=n;i++)
 {
   nr=dr=1;
   for(j=0;j<=n;j++)
     if (j!=i)
     {
       nr *= x-ax[j];
       dr *= ax[i]-ax[j];
     }
     y += (nr/dr)*ay[i];
 }
 printf ("When x=%4.1f y=%7.1f\n",x,y);
}
```

Computer Solution of Example 7.17

```
Enter the value of n
4
```

```
Enter the set of values
  5      150
  7      392
 11     1452
 13     2366
 17     5202
Enter the value of x for which value of y is wanted
9
When x = 9.0 y = 810.0
```

14.18   Newton's Divided Difference Formula (Section 7.14)

Flow-chart

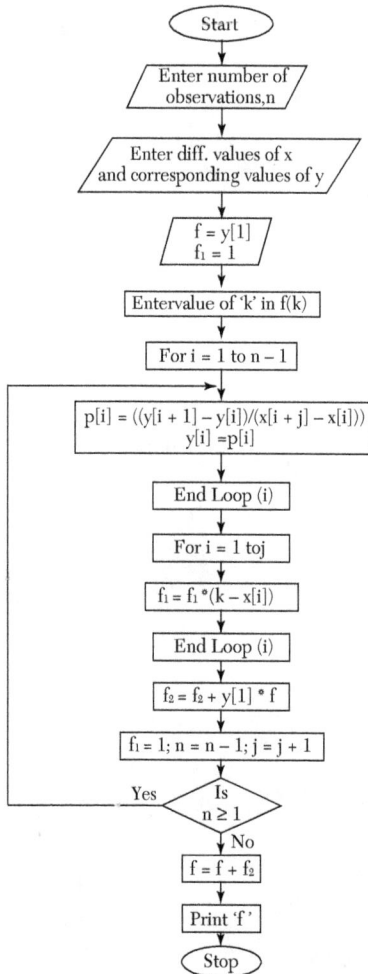

Program

```
#include<stdio.h>
#include<conio.h>
void main()
{
int x[10], y[10], p[10];
int k,f,n,i,j=1,f1=1,f2=0;
clrscr();
printf("enter the no. of observations\n");
scanf("%d",&n);
printf("enter the different values of x\n");
for (i=1;i=<n;i++)
{
scanf(''%d'',&x[i]);
}
printf("enter the corresponding values of y\n");
for(i=1;i<=n;i++)
{
scanf("%d",&y[i]);
}
f=y[1];
printf("enter the value of 'k' in f(k) you want to evaluate\n");
scanf("%d",&k);
do
{
for(i=1;i<=n-1;i++)
{
p[i]=((y[i+1]-y[i])/(x[i+j]-x[i]));
y[i]=p[i];
}
f1=1;
for(i=1;i<=j;i++)
{
f1*=(k-x[i]);
}
f2+=(y[1]*f1);
n--;
j++;
} while(n!=1);
f+=f2;
printf(''f(%d)=%d",k,f);
getch();
}
```

Computer Solution of Example 7.23

```
Enter the no. of observations
5
enter the different values of x
5 7 11 13 17
enter the corresponding values of y
150 392 1452 2366 5202
enter the value of 'k' in f(k) you want to evaluate
9
f(9) = 810
```

14.19 Derivatives Using Forward Difference Formulae [Section 8.2 (1)]

Flow-chart

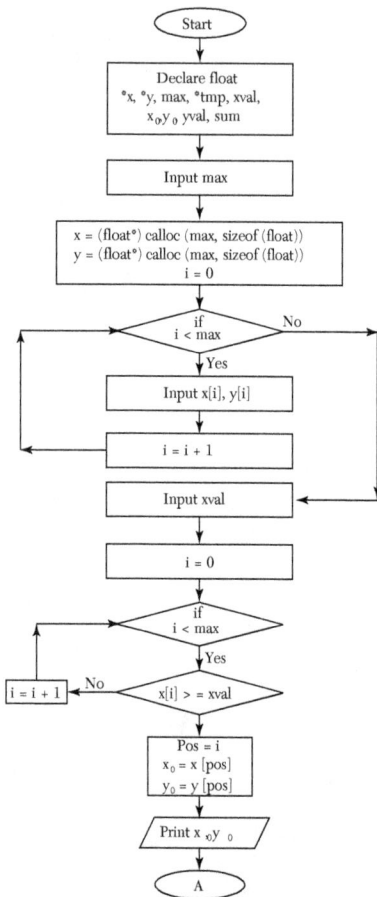

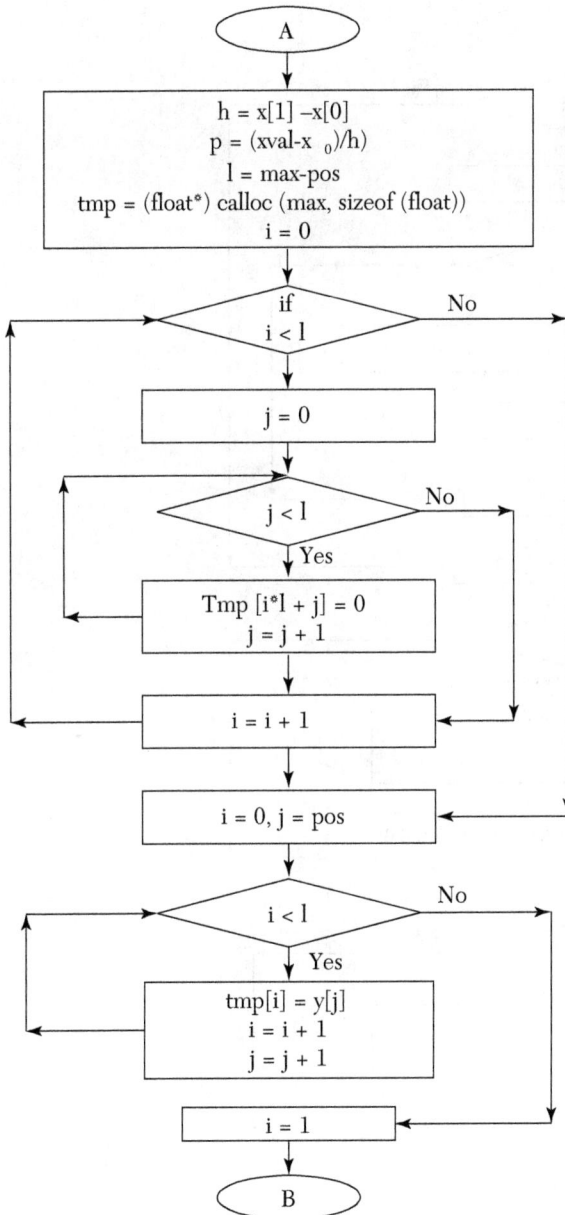

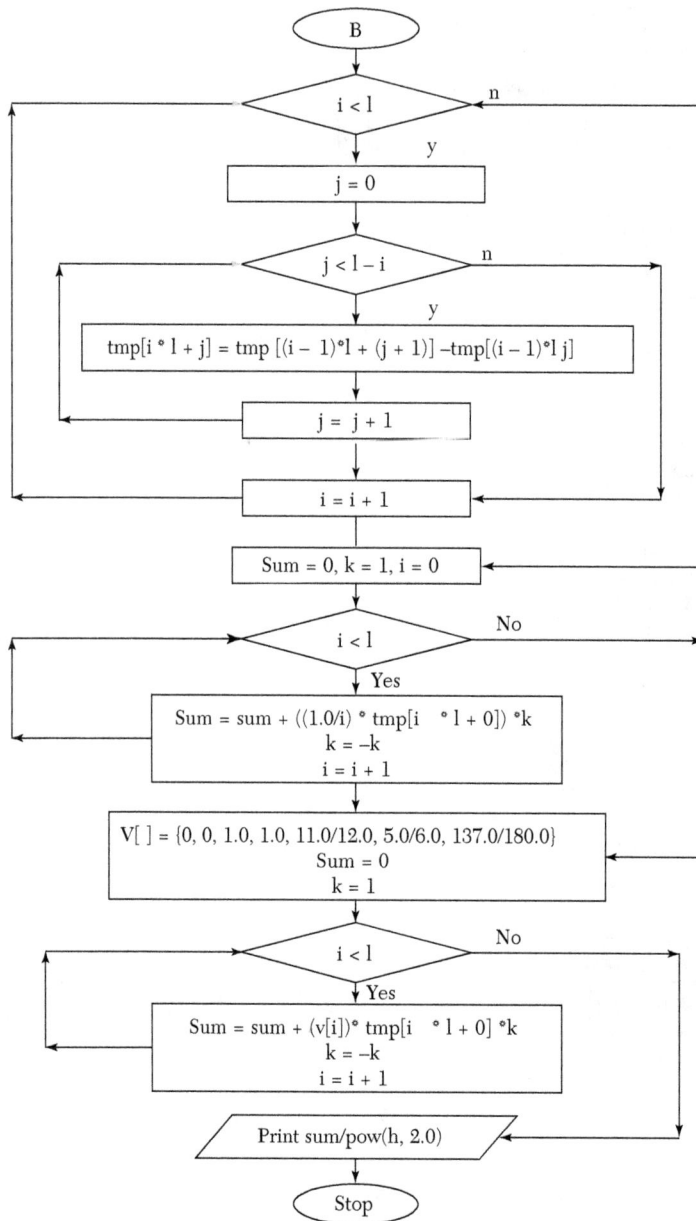

```
                    ┌─────────┐
                    │    B    │
                    └─────────┘
                         │
                  ╱──────────────╲      n
             ────<      i < l      >────────
            │     ╲──────────────╱          │
            │            │ y                 │
            │     ┌──────────────┐           │
            │     │    j = 0      │           │
            │     └──────────────┘           │
            │            │                    │
            │     ╱──────────────╲    n        │
          ──┼────<    j < l – i    >───────    │
         │  │     ╲──────────────╱        │   │
         │  │            │ y                │   │
         │  │  ┌────────────────────────────┐ │ │
         │  │  │ tmp[i ° l + j] = tmp [(i – 1)°l + (j + 1)] –tmp[(i – 1)°l j] │
         │  │  └────────────────────────────┘ │ │
         │  │            │                    │   │
         │  │     ┌──────────────┐           │   │
         │  └─────│   j = j + 1   │           │   │
         │        └──────────────┘           │   │
         │               │ ◄─────────────────┘   │
         │        ┌──────────────┐               │
         └────────│   i = i + 1   │◄──────────────┘
                  └──────────────┘
```

$$tmp[i \circ l + j] = tmp\,[(i - 1)°l + (j + 1)] - tmp[(i - 1)°l\,j]$$

Sum = 0, k = 1, i = 0

i < l   No

Yes

Sum = sum + ((1.0/i) ° tmp[i   ° l + 0]) °k
k = –k
i = i + 1

V[ ] = {0, 0, 1.0, 1.0, 11.0/12.0, 5.0/6.0, 137.0/180.0}
Sum = 0
k = 1

i < l   No

Yes

Sum = sum + (v[i])° tmp[i   ° l + 0] °k
k = –k
i = i + 1

Print sum/pow(h, 2.0)

Stop

Program

```c
/* Derivatives using forward difference */
#include<stdio.h>
#include<math.h>
#include<conio.h>
void main( )
{
    float *x=NULL, *y=NULL;
    float *tmp=NULL, *tmp1=NULL;
    float xval,h,p,x0,y0,yval,sum;
    int pos,i,k,max;
    int v[]={0,0,1.0,1.0,11.0/12.0,5.0/6.0,137.0/180.0};
    printf ("Enter the no of comparisons");
    scanf("%d",&max);
    x=(float*) malloc(max);
    x=(float*) malloc(max);
    tmp=(float*) malloc(max);
    printf("Enter the values in cv table for x and y");
    for (i=0;i<max;i++)
    {
    printf("\n value for %d x",i);
    scanf("%f",&x[i]);
    }
    for(i=0;i<max;i++)
    {
    printf("\n value for %d y",i);
    scanf("%f",&y[i]);
    }
    printf("Enter the value of x");
    scanf("%f",&xval);
    for(i=0;i<max;i++)
    {
        if(x[i]>=xval)
        {
            pos=i;
            break;
        }
    }
    x0=x[pos];
    y0=y[pos];
    printf("\n x0 is %f y0 is %f at %d",x0,y0,pos);
```

```
h=x[1]-x[0];
p=(xval-x0)/h);
if(pos<(max))
{
    int fact=1,i,l, j;
    // calculating no of elemets in array
    l=max-pos;
    tmp=(float*)malloc(l*l);
    printf("\n");
    for(i=0;i<l;i++)
    {
        for(j=0; j<=l; j++)
        {
            tmp[i*l+j]=0;
        }
        printf("\n");
    }
    printf("\n size of new array %d\n",l);
    // copying values of y in array
    for(i=0, j=pos;i<l;i++, j++)
    {
        tmp[i] = y[j];
    }
    printf("\n");
    for(i=1;i<l;i++)
    {
        for(j=0; j<l-i; j++)
        {
            tmp[i*l+j]=tmp[(i-1)*l+(j+1)] -tmp[(i-1)*l+(j)];
        }
    }
    printf("\nvalues are \n");
    for(i=0;i<l;i++)
    {
        for(j=0; j<l; j++)
        {
            printf("%.3f\t|",tmp[j*l+i]);
        }
        printf("\n");
    }
    // appling newtons forward differnation using first
    derivates sum=0;
```

```
        k=1;
        for(i=1;  i<l;  i++)
        {
            sum=sum+((1.0/i)*tmp[i*l+0])*k;
            k=-k;
        }
        printf("\n\n first (dy/dx): %f ",sum/h);
        sum=0;
        fact=1;
        k=1;
         for(i=2;i<l;i++)
        {
            sum=sum+(v[i]*tmp[i*l+0]*k;
            k= -k;
        }
        printf("\n\n second (dy/dx): %f ",sum/pow(h,2.0));
    }
}
```

Computer Solution of Example 8.1

```
value for 0x1.0
value for 0y7.989
value for 1x1.1
value for 1y8.403
value for 2x1.2
value for 2y8.781
value for 3x1.3
value for 3y9.129
value for 4x1.4
value for 4y9.451
value for 5x1.5
value for 5y9.750
value for 6x1.6
value for 6y10.031
Enter the value of x1.1
x0 is 1.1y0 is 8.403 at 1
size of new aray 6
values are
1.1 8.403  |0.378 |-0.03 |0.004 |-0.001 |0.003      |
1.2 8.781  |0.348 |-0.026 |0.003 |0.002 |0   |
1.3 9.129  |0.322 |-0.023 |0.005 |0    |0      |
1.4 9.451  |0.299 |-0.018 |0    |0      |0      |
```

```
1.5 9.75    |0.281 |0      |0      |0      |0      |
1.6 10.031 |0      |0      |0      |0      |0      |
first (dy/dx): 3.952
second (dy/dx): -3.74
```

14.20   Trapezoidal Rule (Section 8.5—I)

Flow-chart

```
        ( Start )
           │
    ┌───────────────────┐
    │ Define function y(x) │
    └───────────────────┘
           │
   ╱───────────────────╱
  ╱ Get values of x0, xn, n ╱
 ╱───────────────────╱
           │
    ┌───────────────────┐
    │   h = (xn – x0)/n   │
    └───────────────────┘
           │
    ┌───────────────────┐
    │  s = y(x0) + y(xn)  │
    └───────────────────┘
           │
    ┌───────────────────┐
    │ Loop for i = 1 to n – 1 │
    └───────────────────┘
           │
    ┌───────────────────┐
    │  s + = 2°y(x0 + i°h)  │
    └───────────────────┘
           │
    ┌───────────────────┐
    │    End Loop (i)     │
    └───────────────────┘
           │
   ╱───────────────────────╱
  ╱ Print(h/2)° s as solution ╱
 ╱───────────────────────╱
           │
        ( Stop )
```

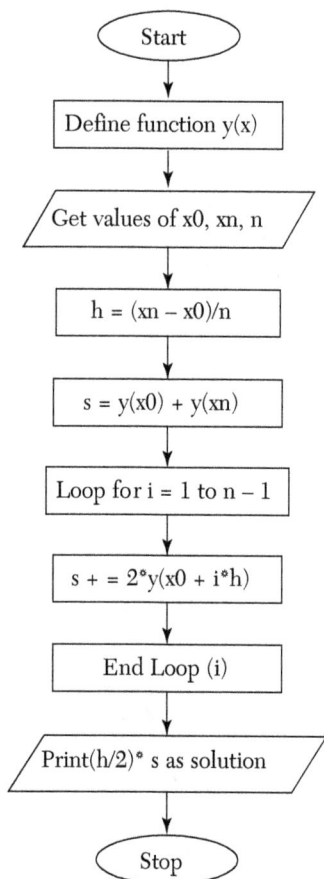

> **NOTES:** $y(x)$ is the function to be integrated
>
> x0 is x_0
>
> xN is x_n.

Program

```
/* Trapezoidal rule.*/
#include <stdio.h>
float y(float x)
{
 return 1/(1+x*x);
}
main()
{
 float x0,xn,h,s;
 int i,n;
 puts("Enter x0,xn,no. of subintervals");
 scanf ("%f %f %d",&x0,&xn,&n);
 h = (xn-x0)/n;
 s = y(x0)+y(xn);
 for (i=1;i<=n-1;i++)
  s += 2*y(x0+i*h);
 printf ("Value of integral is % 6.4f\n",
 (h/2)*s);
}
```

Computer Solution of Example 8.10 (I)

```
Enter x0, xn, no. of subintervals
0 6 6
Value of integral is 1.4108
```

14.21   Simpson's Rule (Section 8.5—II)

Flow-chart

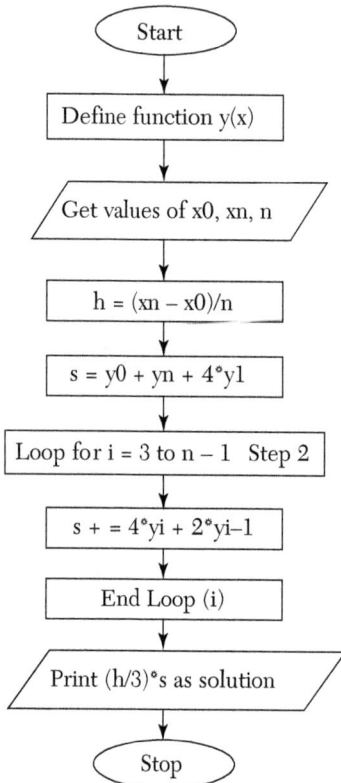

NOTE: $y(x)$ is the function to be integrated so that $y_i = y(x_i) = y(x_0 + i*h)$.

Program

```
/* Simpson's rule */
#include <stdio.h>
float y(float x)
{
 return 1/(1+x*x);
}
main()
{
```

```
float x0,xn,h,s;
int i,n;
puts("Enter x0,xn. no. of subintervals");
scanf("%f %f %d",&x0,&xn,&n);
h = (xn-x0)/n;
s = y(x0)+y(xn)+4*y(x0+h);
for (i=3;i<=n-1;i+=2)
  s += 4*y(x0+i*h)+2*y(x0+(i-1)*h);
printf("Value of integral is %6.4f\n", (h/3)*s);
}
```

Computer Solution of Example 8.10 (ii)

```
Enter x0, xn, no. of subintervals
0 6 6
Value of integral is 1.3662
```

14.22   Euler's Method (Section 10.4)

Flow-chart

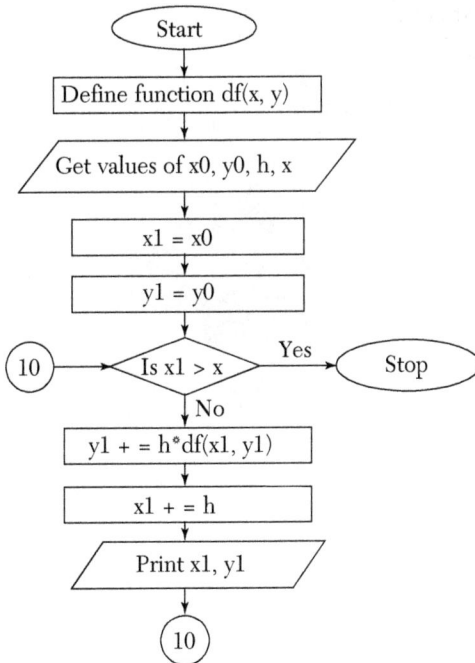

NOTES: $df(x, y)$ is dy/dx

$x0$ is x_{n+0} i.e., x_n

$x1$ is x_{n+1}

$y0$ is y_{n+0} i.e., y_n

$y1$ is y_{n+1}

Program

```
/*Euler's Method*/
#include <stdio.h>
float df(float x,float y)
{
 return x+y;
}
main()
{
 float x0,y0,h,x,x1,y1;
 puts("Enter the values of x0,y0,h,x");
 scanf("%f %f %f %f",&x0,&y0,&h,&x);
 x1=x0;y1=y0;
 while(1)
 {
   if(x1>x) return;
   y1 += h*df(x1,y1);
   x1 += h;
   printf("When x = %3.1f "
       "y = %4.2f\n",x1,y1);
 }
}
```

Computer Solution of Example 10.8

```
Enter the values of x0, y0, h, x
0 1.1 1
When x = 0.1 y = 1.10
When x = 0.2 y = 1.22
When x = 0.3 y = 1.36
When x = 0.4 y = 1.53
When x = 0.5 y = 1.72
When x = 0.6 y = 1.94
When x = 0.7 y = 2.20
When x = 0.8 y = 2.49
```

```
When x = 0.9 y = 2.82
When x = 1.0 y = 3.19
```

14.23 Modified Euler's Method (Section 10.5)

Flow-chart

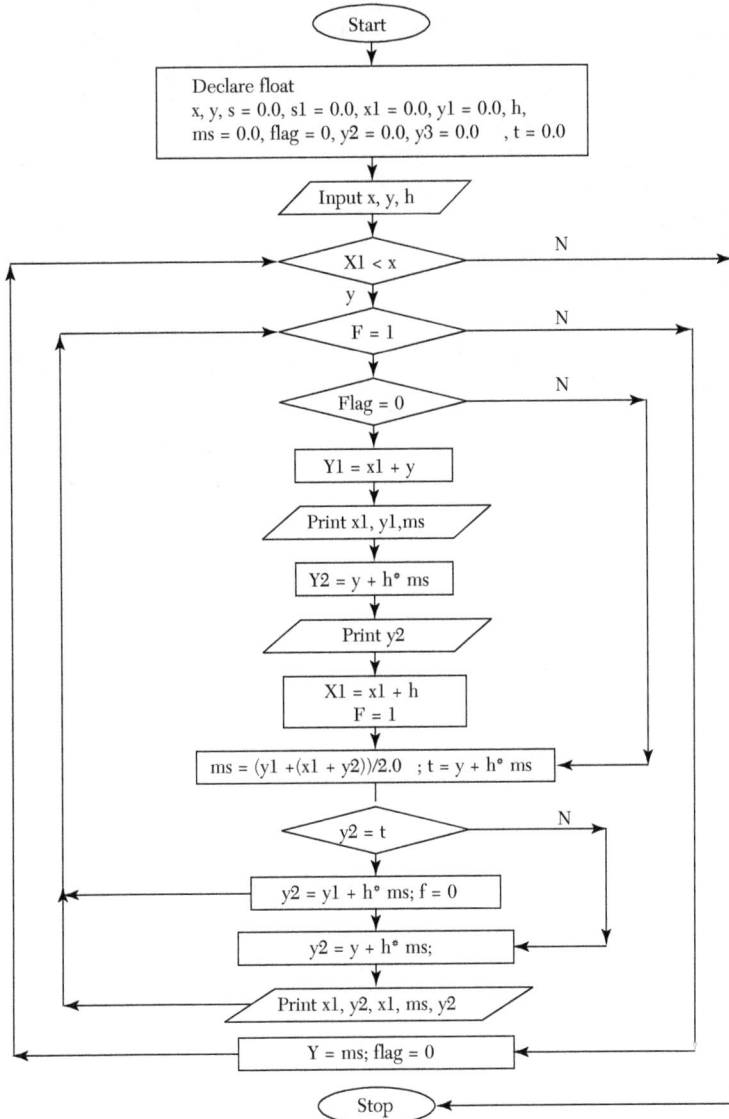

Program

```
/* Modified Euler's Method */
#include<stdio.h>
#include<math.h>
#include<conio.h>
void main( )
{
    float x,y,x1=0.0,y1=0.0,h,ms=0.0,flag=0,y2=0.0,t=0.0;
    int i,j;
    clrscr( );
    printf("\n Enter the value of x");
    scanf("%f",&x);
    printf("Enter the value of y");
    scanf("%f",&y);
    printf("enter the height");
    scanf("%f",&h);
    i=7;
 printf("x");gotoxy(10,i);printf("x+y=y1");gotoxy(28,i);
 printf ("mean slope");gotoxy(45,i);
 printf("old y+.1(mean slope)=new y");
    while(x1<x)
    {
    i++;

        do
        {
        i++;

            if(flag==0)
            {

                y1=x1+y;
gotoxy(2,i);printf("%.1f",x1);gotoxy(10,i);printf("%.5f",y1);g
otoxy(28,i);printf("%.5f",ms);
                m5=y1;
                y2=y+h*ms;
                gotoxy(45,i);printf("%.5f",y2);
                x1=x1+h;
                flag=1;

            }
```

```
            else
            {
                ms=(y1+(x1+y2))/2.0;
                t=y+h*ms;
                if(y2==t)
                {
                    y2=y+h*ms;
                    break;
                }
            }
gotoxy(2,i);printf("%.1f",x1);gotoxy(10,i);printf("%.1f+%.5f",
x1,y2);y2=y+h*ms;
gotoxy(28,i);printf("%.5f",ms);gotoxy(45,i);printf("%.5f",y2);
            }
        }while(1);
        y=y2;
        printf("\n\n");
        flag=0;
    }
}
```

Computer Solution of Example 10.10

```
enter the value of x.3
enter the value of y1
enter the height.1
```

x	x + y = y1	mean slope	old y +.1 (mean slope) = new y
0	1	0	1.1
0.1	0.1 + 1.1	1.1	1.11
0.1	0.1 + 1.11	1.105	1.1105
0.1	0.1 + 1.1105	1.10525	1.110525
0.1	0.1 + 1.110525	1.105263	1.110526
0.1	0.1 + 1.110526	1.105263	1.110526
0.1	1.2105261.105263	1.231579	
0.2	0.2 + 1.231579	1.321053	1.242632
0.2	0.2 + 1.242632	1.326579	1.243184
0.2	0.2 + 1.243184	1.326855	1.243212
0.2	0.2 + 1.243212	1.326869	1.243213
0.2	1.443213	1.32687	1.387535
0.3	0.3 + 1.387535	1.565374	1.399751
0.3	0.3 + 1.399751	1.571482	1.400362
0.3	0.3 + 1.400362	1.571787	1.400392
0.3	0.3 + 1.400392	1.571803	1.400394

14.24   Runge-Kutta Method (Section 10.7)

Flow-chart

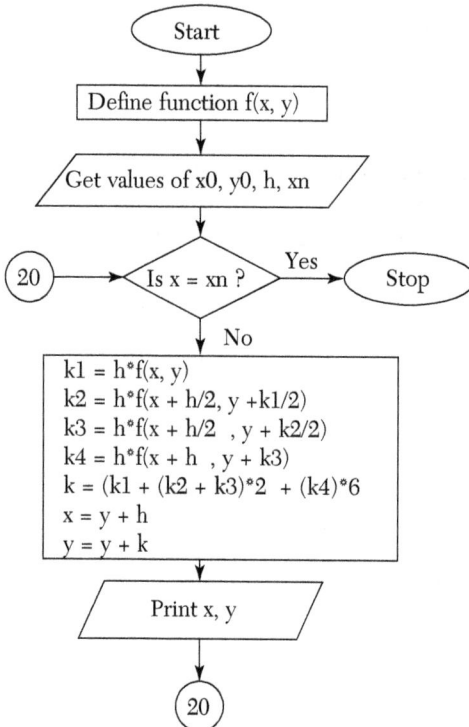

NOTES: *x0* is starting value of x, *i.e.*, x_0

xn is the value of x for which y is to be determined

Program

```
/* Runge Kutta Method */
#include <stdio.h>
float f(float x,float y)
{
 return x+y*y;
}
main()
{
 float x0,y0,h,xn,x,y,k1,k2,k3,k4,k;
 printf("Enter the values of x0,y0," "h,xn\n");
```

```
scanf ("%f %f %f %f",&x0,&y0,&h,&xn);
x = x0; y = y0;
while (1)
{
 if (x == xn) break;
 k1 = h*f(x,y);
 k2 = h*f(x+h/2,y+k1/2);
 k3 = h*f(x+h/2,y+k2/2);
 k4 = h*f(x+h,y+k3);
 k = (k1+(k2+k3)*2+k4)/6;
 x += h; y += k;
 printf("When x = %8.4f"
 "y = %8.4f\n",x,y);
 }
}
```

Computer Solution of Example 10.15

```
Enter the values of x0, y0, h, xn
0.0 1.0 0.2 0.2
When x = 0.1000 y = 1.1165
When x = 0.2000 y = 1.2736
```

14.25 Milne's Method (Section 10.9)

Flow-chart

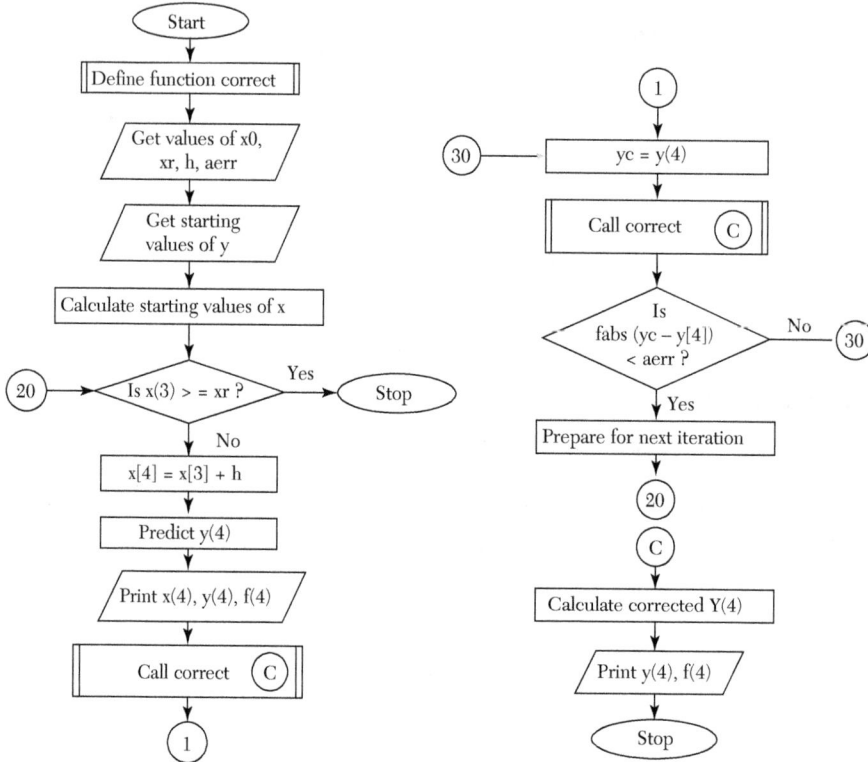

NOTES: x is an array such that $x[i]$ represents x_{n+i} for *e.g.* $x[0]$ represent xn

y is an array such that $y[i]$ represents y_{n+i}
xr is the last value of x at which value of y is required
h is spacing in values of x
$aerr$ is the allowed error in value of y
yc is the latest corrected value for y
f is the function which returns value of y'
corect is a subroutine that calculates the corrected value of y and prints it.

Program

```
/*Milne predictor corrector*/
#include <stdio.h
#include <math.h float x[5],y[5],h; float f(int i)
{
```

```c
  return x[i]-y[i]*y[i];
}
void corect()
{
 y[4] = y[2]+(h/3)*(f(2)+4*f(3)+f(4));
 printf("%23s %8.4f %8.4f \n", "",y[4],f(4));
}
main()
{
 float xr,aerr,yc;
 int i;
 puts("Enter the values of x0,xr,h,"
      "allowed error");
 scanf("%f %f %f %f",
       &x[0],&xr,&h,&aerr);
 puts("Enter the value of y[i], i=0,3");
 for (i=0;i<=3;i++) scanf("%f",&y[i]);
 for (i=1;i<=3;i++) x[i] = x[0]+i*h;
 puts(" x Predicted"
      " Corrected");
 puts(" y f" "y f");
 while (1)
 {
   if(x[3] = xr) return;
   x[4] = x[3]+h;
   y[4] = y[0]+
    (4*h/3)*(2*(f(1)+f(3))-f(2));
   printf("%6.2f %8.4f %8.4f\n",
          x[4],y[4],f(4));
   corect();
   while (1)
   {
     yc = y[4];
     corect();
     if(fabs(yc-y[4]) <= aerr) break;
   }
   for (i=0;i<=3;i++)
   {
      x[i] = x[i+1];
      y[i] = y[i+1];
   }
 }
}
```

Computer Solution of Example 10.19

```
Enter the values of x0, xr, h, allowed error
0 1.2.0001
Enter values of y[i]; i = 0, 3
0.02.0795.1762
x    Predicted     Corrected
y   f   y   f
0.80   0.3049      0.7070
       0.3046      0.7072
       0.3046      0.7072
1.00   0.4554      0.7926
       0.4556      0.7925
       0.4556      0.7925
```

14.26   Adams-Bashforth Method (Section 10.10)

Flow-chart

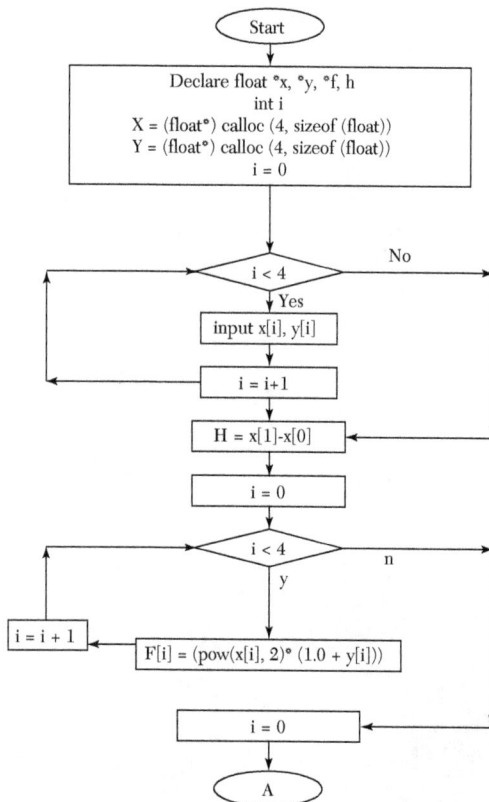

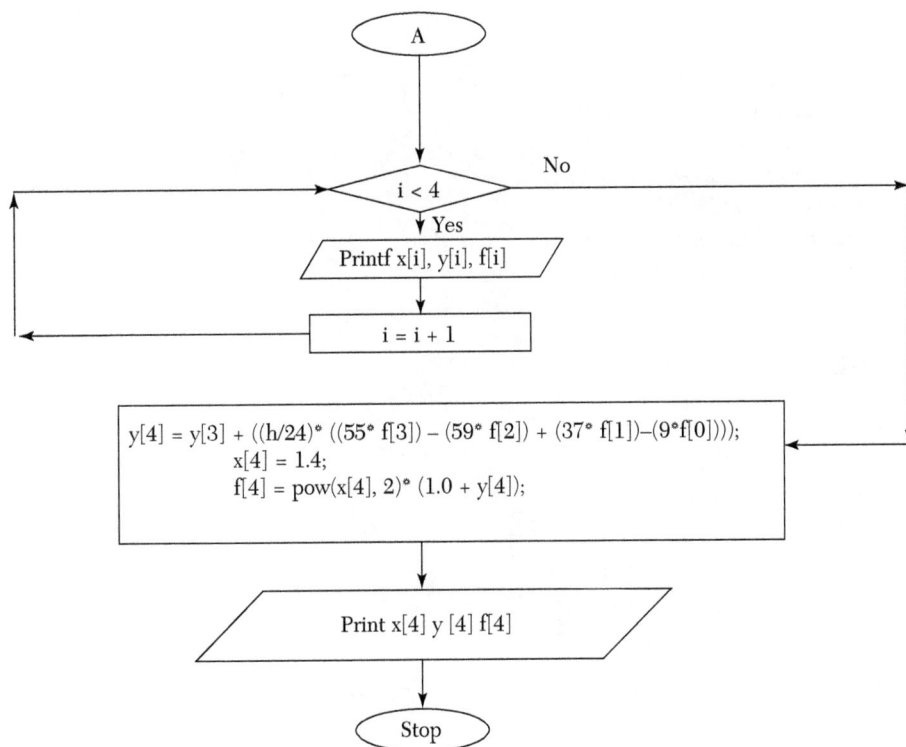

```
        ( A )
          |
          v
    No <--- i < 4 --->
          | Yes
          v
   / Printf x[i], y[i], f[i] /
          |
          v
      | i = i + 1 |
          |
          v
y[4] = y[3] + ((h/24)° ((55° f[3]) – (59° f[2]) + (37° f[1])–(9°f[0])));
       x[4] = 1.4;
       f[4] = pow(x[4], 2)° (1.0 + y[4]);
          |
          v
   / Print x[4] y [4] f[4] /
          |
          v
       ( Stop )
```

Program

```c
/*Adams-Bashforth Method*/
#include<stdio.h>
#include<malloc.h>
#include<math.h>
#include<conio.h>
void main( )
{
    float *x, *y, *f, *f1;
    float h;
    int i,size,row;
    clrscr( );
    printf("enter the size");
    scanf("%d",&size);
    x=(float*)malloc(size + 1);
    y=(float*)malloc(size+1);
    f1=(float*)malloc(size+1);
    f=(float*)malloc(size + 1);
```

```
  for (i=0;i<size;i++)
   {
     printf("enter the value for x[%d]",i);
     scanf("%f",&x[i]);
    }
    for(i=0;i<size;i++)
{
  printf("enter the value for y[%d]",i);
  scanf("%f",&y[i]);
}
h=x[1]-x[0];
// calculating values (f)
for(i=0;i<size;i++)
{
  float tx,ty,tf;
  fflush(stdin);
  tx=x[i];
  ty=y[i];
  tf=(pow(tx,2)*(1.0+ty));
  f[i]=tf;
}
printf("\nvalues for (x) (y) and (f) are\n");
row = 16;
for(i=0;i<=3;i++)
{
gotoxy(2,row);printf("x=");  gotoxy(6,row);printf("%.1f",x[i]);
gotoxy(13,row);  printf("y%d",i-3);gotoxy(16,row);printf("=");
gotoxy(18,row);printf("%f",y[i]);   gotoxy(28,row);printf("f%d"
,i-3);
gotoxy(32,row);printf("=");gotoxy(35,row);printf("%f",f[i]);
row++;
}
//using predicator
y[size]=y[size-1]+((h/24)*((55*f[size-1])-59*f[size-
2])+37*f[size-3])
-(9*f[size-4]))); x[size] = 1.4;
  f[size]=pow(x[size],2)*(1.0+y[size]);
  gotoxy(2,row);printf("x=");
gotoxy(6,row);printf("%.1f",x[size]);gotoxy(13,row);printf("y1
");gotoxy(16,row);printf("=");   gotoxy(18,row);printf("%f",y[s
ize]);gotoxy(28,row);printf("f1");gotoxy(32,row);printf("=");
gotoxy(35,row);printf("%f",f[size]);
}
```

Computer Solution of Example 10.23

```
enter the size 4
enter the value for x[0]1.0
enter the value for y[0]1.000
enter the value for x[1]1.1
enter the value for y[1]1.233
enter the value for x[2]1.2
enter the value for y[2]1.548
enter the value for x[3]1.3
enter the value for y[3]1.979

values for(x) (y) and (f) are

x = 1   y-3 = 1          f-3 = 2
x = 1.1 y-2 = 1.233      f-2 = 2.70193
x = 1.2 y-1 = 1.548      f-1 = 3.66912
x = 1.3 y0 = 1.979       f0 = 5.03451
x = 1.4 y1 = 2.572297f1 = 7.001702
```

14.27   Solution of Laplace Equation (Section 11.5)

Flow-chart

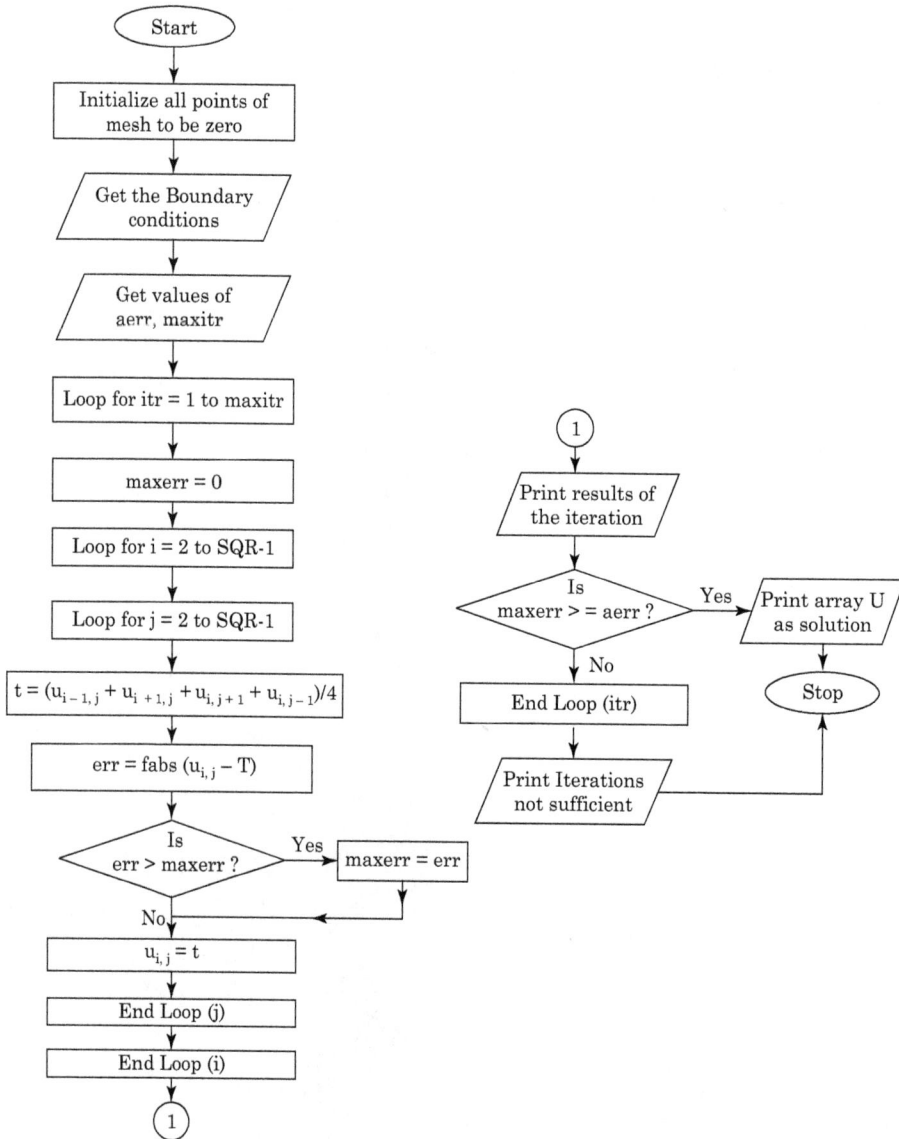

> **NOTES:** *SQR* is the size of the square mesh
> *u* is a 2D Array representing the square mesh
> *aerr* is the allowed error
> *maxitr* is the maximum allowed iterations
> *itr* is a counter which keeps track of the number of iterations performed
> *maxerr* is the maximum error in the mesh in an iteration
> *err* is error in a particular point of the mesh
> *f* is the execution time format
> *getrow* is a subroutine that inputs the ith row of the mesh
> *getcol* is a subroutine that inputs jth column of the mesh.

Program

```
/* Laplace's Equation */
#include <stdio.h>
#include <math.h>
#define SQR 4
typedef float array[SQR+1][SQR+1];
void getrow(int i,array u)
{
 int j;
 printf("Enter the values of u[%d,j],"
   "j=1,%d\n",i,SQR);
 for (j=1;j<=SQR;j++)
   scanf("%f",&u[i][j]);
 }
 void getcol(int j,array u)
 {
  int i;
  printf("Enter the values of u[i,%d],"
        "i=2,%d\n",j,SQR-1);
  for (i=2;i<=SQR-1;i++)
    scanf ("%f",&u[i][j]);
 }
 void printarr(array u,int width,int precision)
 {
  int i,j;
  for (i=1;i<=SQR;i++)
  {
    for (j=1;j<=SQR;j++)
      printf("%7.2f%7.2f%7.2f",width,precision, u[i][j]);
```

```c
      printf("\n");
  }
}
main ()
{
  array u;
 float maxerr,aerr,err,t;
 int i,j,itr,maxitr;
 for (i=1;i<=SQR;i++)
   for(j=1;j<=SQR;j++)
     u[i][j]=0;
 puts ("Enter the boundary conditions");
 getrow(1,u); getrow(SQR,u);
 getcol(1,u); getcol(SQR,u);
 puts ("Enter allowed error,"
        "maximum iterations");
 scanf ("%f %f",&aerr,&maxitr);
   for (itr=1;itr<=maxitr;itr++)
   {
     maxerr=0;
     for (i=2;i<=SQR-1;i++)
       for(j=2;j<=SQR-1;j++)
       {
          t=(u[i-1][j]+u[i+1][j]+
            u[i][j+1]+u[i][j-1])/4;
          err=fabs(u[i][j]-t);
          if (err > maxerr)
            maxerr = err;
          u[i][j]=t;
       }
     printf("Iteration no. %d \n",itr);
     printarr(u,9,2);
     if (maxerr <= aerr)
     {
       printf ("After %d iterations \n"
          "The solution:-\n",itr);
       printarr(u,8,1);
       return 0;
     }
   }
  puts ("Iterations not sufficient.");
  return 1;
```

}

Computer Solution of Example 11.3 (A)

```
Enter the boundary conditions
Enter the values of u[1, j], j = 1, 4
1000 1000 1000 1000
Enter the values of u[4, j], j = 1, 4
1000 500 0 0
Enter the values of u[i, 1], i = 2, 3
2000 2000
Enter the values of u[i, 4], i = 2, 3
500 0
Enter allowed error, maximum iterations
.1 10
Iteration No.1
1000.00      1000.00      1000.00      1000.00
2000.00       750.00       562.50       500.00
2000.00       812.50       343.75         0.00
1000.00       500           0.00          0.00
Iteration No. 2
1000.00      1000.00      1000.00      1000.00
2000.00      1093.75       734.38       500.00
2000.00       984.38       429.69         0.00
1000.00       500.00         0.00         0.00
Iteration No.3
1000.00      1000.00      1000.00      1000.00
2000.00      1179.69       777.34       500.00
2000.00      1027.34       451.17         0.00
1000.00       500.00         0.00         0.00
Iteration No. 4
1000.00      1000.00      1000.00      1000.00
2000.00      1201.17       788.09       500.00
2000.00      1038.09       456.54         0.00
1000.00       500.00         0.00         0.00
Iteration No. 5
1000.00      1000.00      1000.00      1000.00
2000.00      1206.54       790.77       500.00
2000.00      1040.77       457.88         0.00
1000.00       500.00         0.00         0.00
Iteration No. 6
1000.00      1000.00      1000.00      1000.00
2000.00      1207.89       791.44       500.00
```

```
2000.00        1041.44      458.22       0.00
1000.00        500.00       0.00         0.00
Iteration No. 7
1000.00        1000.00      1000.00      1000.00
2000.00        1208.22      791.61       500.00
2000.00        1041.61      458.30       0.00
1000.00        500.00       0.00         0.00
Iteration No. 8
1000.00        1000.00      1000.00      1000.00
2000.00        1208.31      791.65       500.00
2000.00        1041.65      458.33       0.00
1000.00        500.00       0.00         0.00
After 8 iterations
The solution:-
1000.0         1000.0       1000.0       1000.0
2000.0         1208.3       791.7        500.0
2000.0         1041.6       458.3        0.0
1000.0         500.0        0.0          0.0
The solution:-
1000.0         1000.0       1000.0       1000.0
2000.0         1208.3       791.7        500.0
2000.0         1041.6       458.3        0.0
1000.0         500.0        0.0          0.0
```

14.28   Solution of Heat Equation (Section 11.9)

Flow-chart

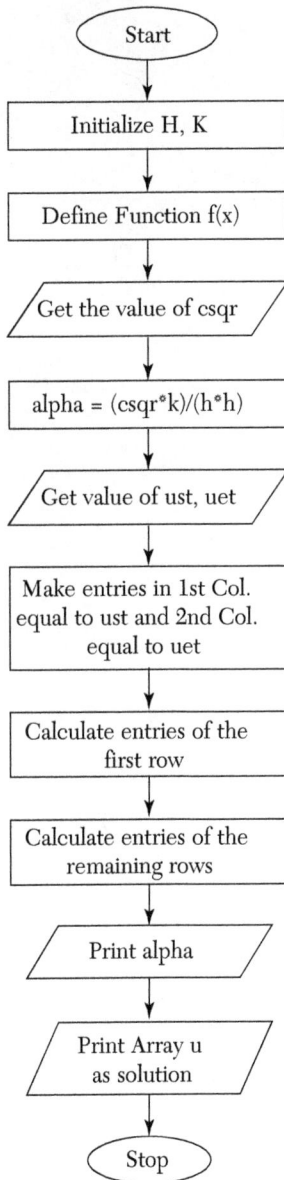

Start

Initialize H, K

Define Function f(x)

Get the value of csqr

alpha = (csqr°k)/(h°h)

Get value of ust, uet

Make entries in 1st Col. equal to ust and 2nd Col. equal to uet

Calculate entries of the first row

Calculate entries of the remaining rows

Print alpha

Print Array u as solution

Stop

NOTES: *XEND* is the ending value of x
TEND is the ending value of t
h is the spacing in values of x
k is the spacing in values of y
f(x) is value of u(x, 0)
csqr is value of C^2
alpha is α
ust is the value in the first column
uet is the value in the last column.

Program

```c
/*Solution of parabolic equations by Bendre
Schmidt method*/
#include <stdio.h>
#define XEND 8
#define TEND 5 float f(int x)
{
 return 4*x-(x*x)/2.0;
}
main()
{
 float u[XEND+1][TEND+1],h=1.0,k=0.125,
     csqr,alpha,ust,uet;
   int i,j;
   puts("Enter the square of 'c'");
   scanf("%f",&csqr);
   alpha = (csqr*k)/(h*h);
   puts ("Enter the value of u[0,t]");
   scanf ("%f",&ust);
   printf ("Enter the value of u[%d,t]\n",
     XEND); scanf("%f",&uet);
   for (j=0;j<=TEND;j++)
     u[0][j]=u[XEND][j]=ust;
   for (i=1;i<=XEND-1;i++)
     u[i][0]=f(i);
   for (j=0;j<=TEND-1;j++)
     for (i=1;i<=XEND-1;i++)
       u[i][j+1]=
         alpha*u[i-1][j]
         +(1-2*alpha)*u[i][j]
         +alpha*u[i+1][j];
```

```
printf("The value of alpha is %4.2f\n", alpha);
puts("The values of u[i,j] are:-");
for (j=0;j<TEND;j++)
{
  for (i=0;i<XEND;i++)
    printf("%7.4f",u[i][j]);
  printf("\n");
}
}
```

Computer Solution of Example 11.11

```
Enter the square of "c"
4
Enter value of u(0, t)
0
Enter value of u(8, t)
0
The value of alpha is 0.50
The values of u(i, j) are:-
0.0000 3.5000 6.0000 7.5000 8.0000 7.5000 6.0000 3.5000 0.0000
0.0000 3.0000 5.5000 7.0000 7.5000 7.0000 5.5000 3.0000 0.0000
0.0000 2.7500 5.0000 6.5000 7.0000 6.5000 5.0000 2.7500 0.0000
0.0000 2.5000 4.6250 6.0000 6.5000 6.0000 4.6250 2.5000 0.0000
0.0000 2.3124 4.2500 5.5625 6.0000 5.5625 4.2500 2.3125 0.0000
0.0000 2.1250 3.9375 5.1250 5.5625 5.1250 3.9375 2.1250 0.0000
```

14.29   Solution of Wave Equation (Section 11.12)

Flow-chart

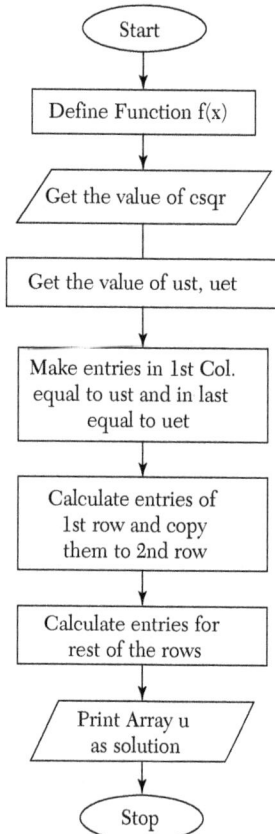

Notes: *XEND* is the ending value of *x*

TEND is the ending value of *t* *f(x)* is value of $u(x, 0)$

csqr is value of C^2

ust is the value in the first column

uet is the value in the last column

Program

```
/* Solution of Hyperbolic equation */
#include <stdio.h>
#define XEND 5
#define TEND 5 float f(int x)
{
  return x*x*(5-x);
```

```
}
 main()
{
 float u[XEND+1][TEND+1],csqr,ust,uet;
 int i,j;
 puts("Enter the square of 'c'");
 scanf("%d",&csqr);
 printf("Enter the value of u[0][t]\n");
 scanf("%f",&ust);
 printf("Enter the value of u[%d][t]\n",
   XEND); scanf("%f",&uet);
 for (j=0;j<=TEND;j++)
 {
   u[0][j] = ust; u[XEND][j] = uet;
 }
 for (i=1;i<=XEND-1;i++)
   u[i][1] = u[i][0] = f(i);
 for (j=1;j<=TEND-1;j++)
   for (i=1;i<=XEND-1;i++)
     u[i][j+1] = u[i-1][j]+u[i+1][j]
       -u[i][j-1];
 puts("The values of u[i][j] are:-");
 for (j=0;j<=TEND;j++)
 {
   for (i=0;i<=XEND;i++)
     printf("%6.1f",u[i][j]);
   printf("\n");
 }
}
```

Computer Solution of Example 11.14

```
Enter the square of "c"
16
Enter value of u(0, t)
0
Enter value of u(5, t)
0
The values of u(i, j) are:-
   0.0     4.0    12.0    18.0    16.0   0.0
   0.0     6.0    11.0    14.0     9.0   0.0
   0.0     7.0     8.0     2.0    -2.0   0.0
   0.0     2.0    -2.0    -8.0    -7.0   0.0
   0.0    -9.0   -14.0   -11.0    -6.0   0.0
   0.0   -16.0   -18.0   -12.0    -4.0   0.0
```

14.30   Linear Programming—Simplex Method (Section 12.8)

Flow-chart

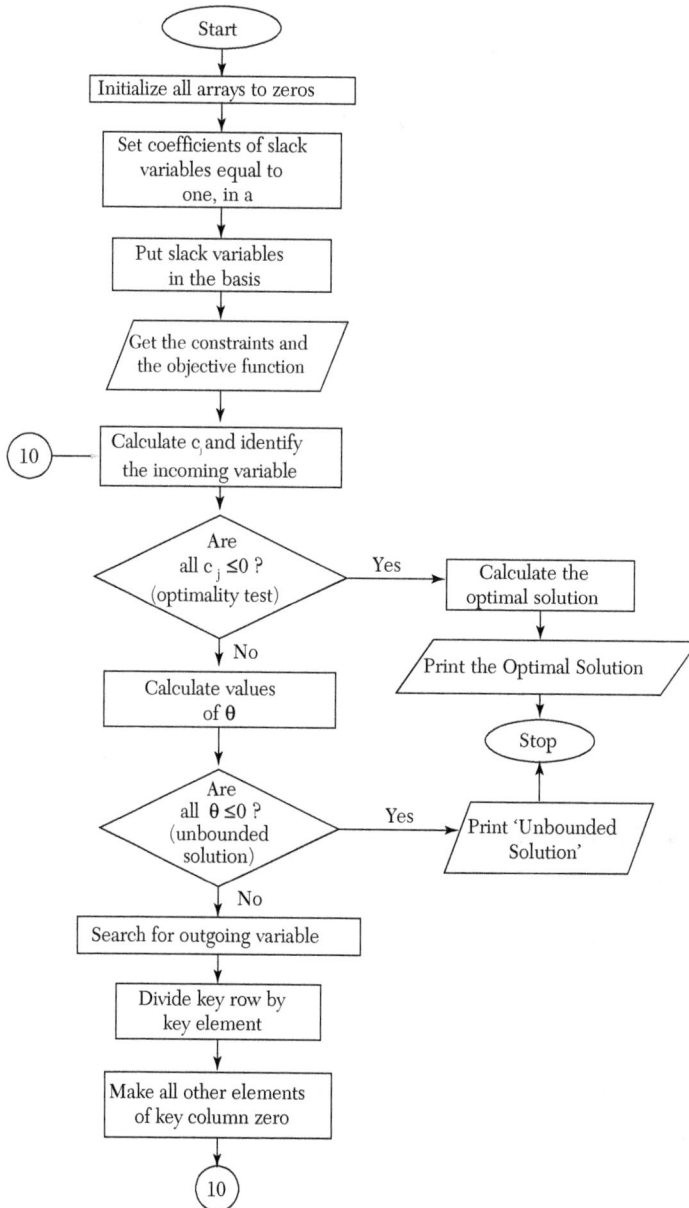

> **NOTES:** *ND* is number of decision variables
> *NS* is number of slack variables
> *a* is the array containing Body Matrix, Unit Matrix and b_i's
> *c* is an array containing values of c_j's
> *cb* is an array containing values of c_B's
> *th* is an array containing values of β's
> *bas* is basis. For xi's basis contains i, for si's basis contains $i + ND$
> *ki* is the key row
> *kj* is the key column.

Program

```
/* Linear programming by simplex method */
#include <stdio.h>
#define ND 2
#define NS 2
#define N (ND+NS)
#define N1 (NS*(N+1))
void init(float x[],int n)
{
 int i=0;
 for (;i<n;i++) x[i] = 0;
}
main()
{
 int i,j,k,kj,ki,bas[NS];
 float a[NS][N+1],c[N],cb[NS],th[NS],
   x[ND],cj,z,t,b,min,max;
 /* Initializing the arrays to zero */
 init(c,N); init(cb,NS);
 init(th,NS); init(x,ND);
 for (i=0;i<NS;i++) init(a[i],N+1);
 /* Now set coefficients for slack
  Variables equal to one */
  for (i=0;i<NS;i++) a[i][i+ND] = 1.0;
 /* Now put the slack variables in the basis */
 for (i=0;i<NS;i++) bas[i] = ND+i;
 /* Now get the constraints
 and the objective function */
 puts("Enter the constraints");
 for (i=0;i<NS;i++)
```

```
{
  for (j=0;j<ND;j++)
    scanf("%f",&a[i][j]);
  scanf("%f",&a[i][N]);
}
puts("Enter the objective function");
for (j=0;j<ND;j++)
  scanf("%f",&c[j]);
/* Now calculate cj and identify the incoming variable */
while (1)
{
 max = 0; kj = 0;
 for (j=0;j<N;j++)
 {
    z = 0;
    for (i=0;i<NS;i++)
       z += cb[i]*a[i][j];
    cj = c[j]-z;
    if(cj > max)
      {max = cj; kj = j;}
 }
 /* Apply the optimality test */
 if(max <= 0) break;
 /* Now calculate thetas */
 max = 0;
 for (i=0;i<NS;i++)
   if(a[i][kj] != 0)
   {
      th[i] = a[i][N]/a[i][kj];
      if(th[i] > max) max=th[i];
   }
 /* Now check for unbounded soln. */
 if(max <= 0)
 {
   puts("Unbounded solution");
  return 2;
 }
 /* Now search for the outgoing variable */
 min = max; ki = 0;
 for (i=0;i<NS;i++)
   if ((th[i] < min)&&(th[i] != 0))
   {
```

```
        min = th[i]; ki = i;
     }
  /*Now a[ki][kj] is the key element*/
  t = a[ki][kj];
    /*Divide the key row by key element*/
    for (j=0;j<N+1;j++) a[ki][j] /= t;
    /* Make all other elements of key coloumn zero */
    for (i=0;i<NS;i++)
      if(i != ki)
      {
         b = a[i][kj];
         for (k=0;k<N+1;k++)
            a[i][k]-=a[ki][k]*b;
      }
    cb[ki] = c[kj];
    bas[ki] = kj;
  }
  /* Now calculating the optimum value */
  for (i=0;i<NS;i++)
  if ((bas[i] >= 0) && (bas[i]<ND))
    x[bas[i]] = a[i][N];
  z = 0;
  for (i=0;i<ND;i++)
    z += c[i]*x[i];
  for (i=0;i<ND;i++)
    printf("x[%3d] = %7.2f\n",i+1,x[i]);
  printf("Optimal value = %7.2f\n",z);
}
```

Computer Solution of Example 12.4

```
Enter the constraints
4    2    80
2    5    180
Enter the objective function
3    4
x[ 1] = 2.50
x[ 2] = 35.00
Optimal value = 147.50
```

Computer Solution of Example 12.16

```
Enter the constraints
2    3    2    440
```

```
4    0    3    470
2    5    0    430
Enter the objective function
4    3    6
x[ 1] = 0.00
x[ 2] = 42.22
x[ 3] = 156.67
Optimal value = 1066.67
```

Exercises 14.1

1. Write a C program which prints all odd positive integers less than 100, omitting those integers divisible by 7.

2. Write a C program to convert a binary number to its equivalent decimal number.

3. Write a program to calculate N ! and use this to evaluate

$$N_{C_K} = \frac{N!}{K!(N-K)!}$$

4. Determine the number of integers n, $1 \le n \le 2000$, that are not divisible by 2, 3 or 5 but are divisible by 7.

5. Write a program to evaluate the roots of the equation $ax^2 + bx + c = 0$.

6. Write a computer program in "C" for finding out a real root of the equation $f(x) = 0$ by bisection method.

7. Write a C program to find a real root of $x^3 - 4x - 9 = 0$ using the method of false position.

8. Write an algorithm for the Newton-Raphson method to solve the equation $f(x) = 0$. Apply the same to solve the $\cos x - xe^x = 0$ near $x = 0.5$ correct to three decimal places.

9. Write a C program to solve the following equations by the Gauss-Seidal method: $83x + 11y - 4z = 95$; $7x + 52y + 13z = 104$; $3x + 8y + 29z = 71$.

10. With the help of a flow chart, write a C program to solve:
$7.5x + 3.8y + 2.9z = 15$; $3.2x + 6.8y + 7.4z = 37$; $1.3x + 2.1y + 3.2z = 7$, using the triangularization method.

11. Write a complete C program to (*i*) add two matrices (*ii*) multiply two matrices.

12. Given the data:

x:	5	10	15	20	25	30
y:	17	25	30	33	36	38

Write a C program to fit a quadratic relation using the least squares criterion.

13. Write a program in C to estimate $f(0.6)$ by the Lagrange interpolation for the following values:

x:	0.4	0.5	0.7	0.8
f(x):	-0.916	-0.693	-0.357	-0.223

14. Write a C program to evaluate $\int_{2}^{10} x(x^2 + 2)dx$ using the Simpson's rule.

15. Write a C program for evaluation of $\int_{2}^{10} f(x)dx$ by the Simpson's 3/8th rule.

16. Write a program in C for the second order Runge-Kutta method.

17. Develop a "C" program for solving differential equations using the Runge-Kutta fourth order formulae.

18. Write a C program to find $y(0.8)$ for the differential equation

$dy/dx = \dfrac{1}{2}(x + y)$, given the following table, using Milne's Predictor-Corrector method:

x:	0	0.2	0.4	0.6
y:	2	2.636	3.595	4.968

19. Write a computer program in C to maximize

$z = 6x_1 + 4x_2$

subject to $\quad 2x_1 + 3x_2 \le 100,\ 4x_1 + 2x_2 \le 120,\ x_1, x_2 \ge 0$, where x_1, x_2 are the number of items to be produced.

20. Develop a computer program in C for Example 12.17 and solve it.

CHAPTER 15

NUMERICAL METHODS USING C++ LANGUAGE

Chapter Objectives

- Introduction
- An overview of C++ features
- Programs of standard methods in C++ language

15.1   Introduction

C++ is a general purpose programming language, originally designed by Bjarne Stroustrup.

C++, a powerful language, is used for many purposes like writing operating systems, business, and scientific applications. It is not tied to any particular hardware and programs written in C++ are portable across any system. The programs written in C++ are efficient and fast. It is one of the most popular computer languages today.

An overview of C++ features is given below for ready reference. It is followed by Programs of Standard Numerical methods in C++ alongwith input/output of numerous examples solved in the Chapters 1 to 12.

15.2   An Overview of C++ Features

C++ constants are numbers, which do not change during execution of a program. These may be of three types:

Type	Example
Integer	27, 10897 etc.
Floating point (Real)	2.723, – 0.123 etc.
String	"Enter the value"

NOTE   *The string constants are enclosed in double quotes (").*

C++ variables can contain different C++ constants during the execution of the program. These are declared in a C++ program by first specifying the type *int* for an integer and *float* for the floating point and then the variable names separated by commas. The general format is

　　　　type　　　　　　*list of variables*

For *e.g.,* to declare integer variables the statement is

$$\text{int } a, b, c;$$

and to declare a floating point variables it is

$$\text{float } a, b, c;$$

Variables can be initialized at the same time as they are declared. For example,

$$\text{float } a = 1.5;$$

declares *a* as a float variable having a value 1.5.

Rules for naming C++ variables:

(*i*) A variable name may contain only alphabets, digits, and the underscore (_).

(*ii*) It must begin with a alphabet or an underscore.

(*iii*) It can be as long as you wish, but on some C++ systems only the first thirty-one characters are considered.

NOTE   *Lower-case and upper case alphabets are treated as different in C. For e.g., Num and num are two entirely different variable names. As a matter of convention, lower- case alphabets are used.*

Arrays. An array is an aggregate of variables of the same type. These variables are called *elements* of the array. The following statements declare arrays in C:

int $b[10]$;

float $c[2][2]$;

The first statement creates a one dimensional array named b having ten elements, each element being referred by an appropriate subscript in rectangle brackets, *i.e.*, $b[0], b[1],......, b[9]$.

The second statement creates a 2-dimensional array named c having four elements $c[0][0], c[0][1], c[1][0], c[1][1]$.

Rules for the naming of arrays are same as those for variable names.

NOTE   *Subscripts always start from zero in C++.*

User defined types. Apart from the built in types *int* and *float* C++ allows users to define an identifier that can represent an existing data type.

The syntax is

     typedef type identifier

For *e.g.*,

       typedef int number;

       typedef float matrix [2][2];

The first statement defines *number* to mean the same as *int*. The second defines *matrix* to be mean the same as 2×2 array of float.

The above two statements enable declarations of the form number a, b, c;

which declares three integers a, b, and c, and

       matrix x;

which declares a two-dimensional array x having four elements $x[0][0], x[0][1], x[1][0]$ and $x[1][1]$.

Initialization of arrays at the time of declaration

The syntax is

       type array-name [size] = {list of values}

for *e.g.*,   int $a[2] = \{2, 1\}$;

initializes $a[0]$ to 2 and $a[1]$ to 1.

int $a[2][2] = \{\{0, 1\}, \{3, 5\}\};$

initializes $a[0][0]$ to 0, $a[0][1]$ to 1, $a[1][0]$ to 3 and $a[1][1]$ to 5.

Arithmetic operators. These are as follows:

Symbol	Use
+	Addition
−	Subtraction
⚬	Multiplication
/	Division

while using the operators, the following order of precedence is adopted

(i) *, /   (ii) +, −

In this case, the order of operators is that different circular brackets are used.

There is no exponentiation operator in C, but there are various *library functions* avail- able for the same.

For *e.g.*, to calculate the square root *sqrt* function is used.

Further details on functions are presented later.

Mathematical expressions consist of a sequence of arithmetic operators and variable names, For *e.g.*,

(i) $a + b$ is written as $a + b$.

(ii) $\dfrac{a}{b} + c$ is written as $a/b + c$

(iii) $\dfrac{a}{b+c}$ is written as $a/(b + c)$

(iv) $\sqrt{b^2 - 4ac}$ is written as **sqrt** $(b*b - 4*a*c)$.

(v) $\alpha(\beta + \gamma)$ is written as alpha * (beta + gamma).

(vi) a^b is written as **exp** $(b*\textbf{ln}\,(a))$

NOTE   *The multiplication operator * has to be written explicitly. In C++, its presence is never assumed.*

exp *and* **ln** *are* **library functions**.

Arithmetic statements are of the form

$$var = exp;$$

where *var* is an integer or a floating point variable.

exp is a mathematical expression written in C++ format.

The = sign has a special meaning. It tells C++ to calculate the value of *exp.* and *assign* it to **var**.

For *e.g.*, $n = i * i;$

calculates the value of $i * i$ and assigns the result to the variable *n*. If $i = 10$, then *n* gets the value 100.

A C++ statement is always terminated by a semi colon (;).

C++ also permits statements of the type $k = n = i * i;$

This is equivalent to the following statements $n = i * i; k = n;$

NOTE *To test the equality of two expressions C++ uses "= =".*

Shorthand assignment operators

Apart from the assignment operator =, C++ can also support certain short hand assignment operators (+ +, − −, + =, − =, * =, / =). Their use is illustrated by the following examples.

Statement using the assignment operator	Statement using the shorthand assignment operator
$a = a + 1$	$a + +$ or $+ + a$
$a = a - 1$	$a - -$ or $- - a$
$a = a + 4$	$a + = 4$
$a = a - 4$	$a - = 4$
$a = a * 4$	$a * = 4$
$a = a / 4$	$a / = 4$

NOTE *The use of shorthand assignment operators not only results in concise programs but more efficient programs also.*

Comments in C++ are of two types. One is *single line* and other is *multi line*. Single line comments start with "//"

Multiline comments start with "/*" and end with "*/".

NOTE *Multiline comment can be used for one or more lines. This is the style that has been followed in this book.*

e.g.,

Single line comment   //Euler's Method

Multi line comment   /*Euler's Method*/

Input statement is **cin.**

> *Syntax* **cin** >> variable1 >> variable2......;

An example of the cin statement:

Assuming the declarations

> int *c;*
>
> float *a*[3][3];

the statement

> cin >> *c* >> *a* [1][2];

takes input from the user and stores it in the corresponding variables.

Output statement is **cout**

Syntax **cout** << argument1 << manipulator1 << argument2 <<......;

The arguments can be C++ constants or variables.

Manipulators

These are used for formatting the output. The manipulators used in this book are explained below:

> *setw (w)*

where *w* specifies the minimum width (*i.e.*, number of digits) for output. The width set with *setw* only applies to the next argument printed. So *setw* must be used prior to each argument where a specific width is desired.

e.g.,

the statement

> cout << setw (3) << 1

will output

ƀƀ1

where ƀ is a blank space.

Setprecision (p)

where p is the precision with which numbers are output.

For fixed format, the precision is the number of digits in the fractional part, while for scientific format, precision is the total number of digits (both before and after the decimal point). The default precision in C++ is 6.

The precision set with *setprecision* remains in effect, until the next *setprecision*.

fixed

By default scientific format is used in C++, *i.e.*, the precision set with *setprecision* applies to the entire number. In this book, we will be using the fixed format. This is done by using the *fixed* manipulator.

e.g.,

(*i*) for the following statement

cout << 97.0/7.0;

the output will be according to C++ default precision, *i.e.*, 6

13.8571 (total number of digits is 6)

(*ii*) for the following statements

cout << setprecision (5);

cout < 97.0/7.0;

the output will be

13.857 (total number of digits is 5)

(*iii*) Now to get 5 digits in the fractional part, consider the following statements cout << fixed;

cout << setprecision (5);

cout < 97.0/7.0;

the output will be

13.85714

endl—This causes the output to start from the next line.

NOTE *To use the manipulators include the statement # include < iomanip.h > in the beginning of your program.*

Relational operators available in C++ are:

Mathematical symbol	C++ symbol
>	>
≥	> =
<	<
≤	< =
=	= =
≠	! =

Logical expressions are mathematical expressions connected by relational op- erators. Their value is either true or false.

Examples of logical expressions: Assuming $i = 2, j = 3$

$i < j$ is true

$i = = j$ is false

$(i * j) > (i + j)$ is true.

In C++ the result of a logical expression is an integer. 0 is taken as false, any non-zero integer is taken as true.

Logical operators are used to test more than one conditions, *i.e.*, to combine more than one logical expressions.

Logical operator	C symbol
AND	&&
OR	||
NOT	!

The following tables illustrates their use.

AND

Logical expression 1	Logical expression 2	Result
True	True	True
True	False	False
False	True	False
False	False	False

OR

Logical expression 1	Logical expression 2	Result
True	True	True
True	False	True
False	True	True
False	False	False

NOT

Logical expression	Result
True	False
False	True

Decision making statement—If

Syntax

if (Lexp)

{Tstatements}

else

{Fstatements}

where *Lexp* is a logical expression.

Tstatments are C++ statements executed when value of Lexp is true.

Fstatments are C++ statements executed when value of Lexp is false.

The *else* part is optional.

Loops

(*i*) *While Loop*

Syntax

(*a*) **while** *(Lexp)*

{statements}

(*b*) **do**

{statements}

while *(Lexp)*

Both these forms of the while loop cause execution of *statements* while value of *Lexp* is true. The difference between the two forms is that in the latter, the statements are executed at least once irrespective of the value of

Lexp.

(ii) For loop

Syntax

for *(initialization statement; Lexp; increment statement)*

{statements}

The loop is best explained by the following flow chart:

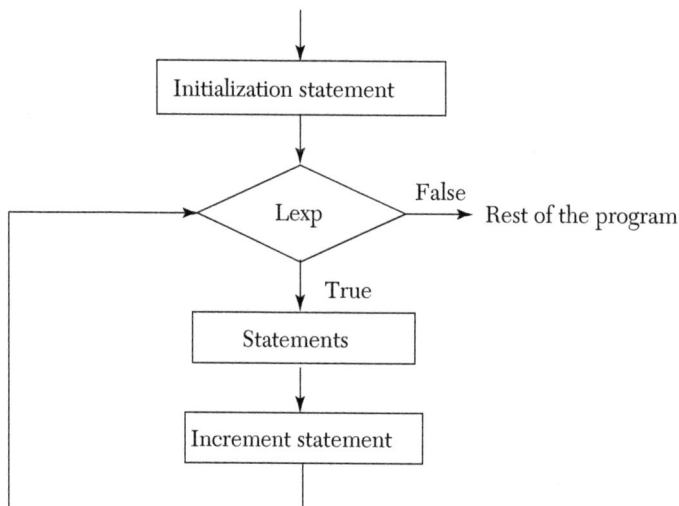

Break statement. When a break statement is encountered inside a loop, that loop is exited, *irrespective* of the value of *Lexp*, and the program continues with the statement immediately following the loop.

Functions. These are the basic block of a C++ program. *Functions* contain *State- ments* that specify what is to be done.

Every program *has to* contain a function named *main*. The program begins executing at the first statement of *main*. Apart from *main* the C++ functions are classified into

— *Library functions*

— *User defined functions*

These functions are called from *main* to accomplish various tasks.

(i) *Library functions* are already available and we just have to use them, *e.g.*, cout, cin, sqrt, cos, sin, fabs (used to get the absolute value of a floating point variable), etc.

(*ii*) *User defined functions* have to be written by the user in the program.

Syntax

> *return-type function-name (Argument-list)*
>
> {
>
> *statements*
>
> }

Program for understanding the various terms and concepts related with functions:

1. / * Sample program * /

2. # include <isostream.*h*>

3. float add (float *a*, float *x*);

4. void half (float *x*)

5. {

6. *x* / = 2;

7. return;

8. }

9. int main ( )

10. {

11. float *a* = 2, *b* = 2, *c*;

12. *c* = add (*a*, *b*);

13. cout << *a* << *b* << *c* << endl;

14. half (&*a*); half (&*b*);

15. cout << *a* << *b* << *c* << endl;

16. return 0;

17. }

18. float add (float *a*, float *x*)

19. {

20. float sum;

21. sum = $a + x$;

22. $a = 20, x = 20$; /*changing the formal arguments*/

23. return sum;

24. }

[Line numbers have been added for reference purpose and are not part of the program.]

(a) Declaration and Definition

Line number 3 is the *declaration* of the function named *add*. It indicates that there is a function *add* which takes two arguments, *a* and *b* both of type float and returns a float, *i.e.*, the *Argument list* is *float a, float b and return-type* is *float* (which can be *void* if function doesn't return anything—see line 4. It also indicates that the function is defined later in the program).

Lines 18–24 are the *definition* of the function *add i.e.*, they define how the function will make use of the *arguments* it received and return the required sum.

Lines 4–8 constitute both the *declaration* and *definition* of the function *half*.

(b) Calling a function

Line no. 12 calls the function *add* with *a* and *b* as arguments and stores the value returned by it into the variable *c*.

(c) Actual and Formal arguments

The variables *a* and *x* in the *declaration* of function *add* are called the *formal arguments* (line 3).

The variables *a* and *b* in the call to the function (line 12) are the *actual arguments*.

(d) Call by value/Call by reference

In C++ language the values of actual arguments are *always copied* to the formal arguments when a function is called. This way of passing arguments is called *call by value*. In this any change made to the formal arguments in the function does *not* affect the value of actual arguments.

But in other languages, notably FORTRAN any change made to the formal arguments *is re- flected* in the actual arguments. This is called *call by reference*, as the formal argument is treated just another name for the actual argument. Both of them *refer* to the same location in the computer's memory.

(*e*) **Simulating a call by reference in C++**

The following thumb rule can be followed to make the formal argument *refer* to the actual argument (and not just receive a copy of it)

"Precede the actual argument with an ampersand **&** (line 14) and precede the formal argument with an asterisk * (line 4)."

The actual working of this involves the concepts of *pointers*, which are another data type in

C++. The reader can refer to any standard C++ book for a complete understanding.

(*f*) **Return statement** in a function passes the control back to the calling function along with the calculated value (line 23)

Syntax return expression;

The *expression* can be omitted, in this case the return statement causes the function to just terminate then and there and pass control back to the calling function (line 7).

In case no return statement is present in a function, an implicit return takes place on encoun- tering the right curly brace }.

Preprocessor directives. The lines in a C++ program that begin with a hash (#) sign are called *preprocessor directives*. The two most commonly used are # define and # include.

NOTE   *There is no semi colon (;) after the directive.*

(*i*) **# define**

syntax

 # define *name replacement*

It instructs the computer to replace all occurences of *name* with the *replacement* even before the program is processed, *i.e.*, checked for syntax.

e.g., consider the following statements

define N 2 int *a*[N];

Before the program is processed by the compiler, the second line, *i.e.*, int *a*[N]; is changed to int *a*[2]; and the first line is removed.

∴ The resulting statement that is processed is int *a*[2].

(ii) **# include**

syntax

 # include *< header-file-name >*

This instructs the computer to insert the contents of the mentioned header file at the place where the directive appeared.

NOTE *The header file contains declarations of various functions and many preprocessor directives.*

Object-oriented Programming (OOP) is a programming paradigm that uses "objects and classes"—data structures consisting of datafields and methods together with their interactions to design applications and computer programs. Programming techniques may include features such as, *data abstraction, encapsulation, modularity, polymorphism, and inheritance*. It was not commonly used in mainstream software application development until the early 1990s. Many modern programming languages now support OOP.

An **object** is actually a identifiable identity with some characteristics and behavior, all relating to a particular real-world concept such as a bank account holder or player in a computer game. Other pieces of software can access the object only by calling its functions and procedures that have been allowed to be called by outsiders.

For example, the player's functions might include one to reveal the player's configuration on the field. The account holder's functions include one to reveal the current balance or to withdraw out or to deposit a sum.

Procedural vs. OPP Programming. In procedural programming the emphasis is on the programming where each statement tells the computer to do something. The focus is on processing the algorithm needed to perform the desire computation.

Also procedural programming is not according to the real world, *i.e.*, it is not close to the real world, "The world in which we live the activities we perform."

Limitation of procedural programming

(*i*) Emphasis on algorithm rather than data

(*ii*) No reusability of code

(*iii*) No overloading

(*iv*) No real world model.

OOPs Programming: Object-oriented programming has roots that can be traced to the 1960s. As hardware and software became increasingly complex, quality was often compromised. Researchers studied ways to maintain software quality and developed object- oriented programming in part to address common problems by strongly discrete, reusable block of programming logic. It focuses on data rather than processes, with programs composed of self-sufficient modules (objects) each containing all the information needed to manipulate its own data structure. This is in contrast to the existing modular programming which had been dominant for many years that focused on the *function* of a module, rather than specifically the data, but equally provided for code reuse, and self-sufficient reusable units of programming logic, enabling collaboration through the use of linked modules (subroutines). This more conventional approach, which still persists, tends to consider data and behavior separately.

OPP Terminology and Features: The OOP approach (based on certain concepts)

helps to overcome the drawbacks of procedural programming.

1. **Data abstraction**—refers to the act of representing essential features without including the background detail of explanations. For example, a class Car would be made up of an Engine, Gearbox, Steering objects, and many more components. To build the Car class, one does not need to know how the different components works internally, but only how to interface with them.

2. **Encapsulation**—refers to wrapping up of data and function (that operate on the data) into a single unit called *class*.

3. **Modularity**—is the property of a system that has been decomposed into a set of cohesive and loosely coupled modules.

4. **Inheritance**—is the capability of one class of thing to inherit capabilities of properties from another class. Members are often specified as

public, protected or **private,** determining whether they are available to all classes, sub-classes or only the defining class.

5. Polymorphism—is the ability for a message or data to be processsed in more than one form. Polymorphism is a property by which the same message can be sent to objects of several different classes.

Data Types

Fundamental data types: When programming, we store the variables in our computer's memory, but the computer has to know what kind of data we want to store in them, since it is not going to occupy the same amount of memory to store a simple number than to store a single letter or a large number, and they are not going to be interpreted the same way.

The memory in our computers is organized in bytes. A byte is the minimum amount of memory that we can manage in C++. A byte can store a relatively small amount of data: one single character or a small integer (generally an integer between 0 and 255). In addition, the computer can manipulate more complex data types that come from grouping several bytes, such as long numbers or non-integer numbers.

Next you have a summary of the basic fundamental data types in C++, as well as the range of values that can be represented with each one:

Name	Description	Size*	Range*
char	Character or small integer	1 byte	signed: – 128 to 127 unsigned: 0 to 255
short int (short)	Short Integer	2 bytes	signed: – 32768 to 32767 unsigned: 0 to 65535
int	Integer	4 bytes	signed: – 2147483648 to 2147483647 unsigned: 0 to 4294967265
long int (long)	Long integer	4 bytes	signed: 2147483648 to 2147483647 unsigned: 0 to 4294967295
bool	Boolean value. It can take one of two values: true or false	1 byte	true or false
float	Floating point number	4 bytes	+/– 3.4e + /– 38 (~ 7 digits)

Name	Description	Size*	Range*
`double`	Double precision floating point number	8 bytes	+/– 1.7e + /– 308 (~ 15 digits)
`long dou-ble`	Long double precision floating point number	8 bytes	+/– 1.7e + /– 308 (~ 15 digits)
`wchar_t`	Wide character	2 or 4 bytes	1 wide character

Integer Type: Integers are whole numbers with a machine dependent range of values. A good programming language as to support the programmer by giving a control on a range of numbers and storage space. C++ has three classes of integer storage namely short int, int and long int. All of these data types have signed and unsigned forms. A short int requires half the space than normal integer values. Unsigned numbers are always positive and consume all the bits for the magnitude of the number. The long and unsigned integers are used to declare a longer range of values.

Floating Point Types: Floating point number represents a real number with six digits precision. Floating point numbers are denoted by the keywords float. When the accuracy of the floating point number is insufficient, we can use the double to define the number. The double is same as float but with longer precision. To extend the precision further we can use long double which consumes 80 bits of memory space.

Void Type: Using void data type, we can specify the type of a function. It is a good practice to avoid functions that does not return any values to the calling function.

Character Type: A single character can be defined as a defined character type of data. Characters are usually stored in 8 bits of internal storage. The qualifier signed or unsigned can be explicitly applied to char. While unsigned characters have values between 0 and 255, signed characters have values from – 128 to 127.

User defined type declaration: In C++ language a user can define an identifier that represents an existing data type. The user defined datatype identifier can later be used to declare variables. The general syntax is

```
typedef type identifier;
```

here type represents existing data type and "identifier" refers to the "raw" name given to the data type.

Example

```
typedef int salary;
typedef float average;
```

Here salary symbolizes int and average symbolizes float. They can be later used to declare variables as follows:

```
Salary dept1, dept2;
Average section1, section2;
```

Therefore dept1 and dept2 are indirectly declared as integer datatype and section 1 and section 2 are indirectly float data type.

Declaration of Storage Class: Variables in C++ have not only the data type but also storage classes that provides information about their location and visibility. The storage class divides the portion of the program within which the variables are recognized.

auto: It is local variable known only to the function in which it is declared. Auto is the default storage class.

static: Local variable which exists and retains its value even after the control is transferred to the called function.

extern: Global variable known to all functions in the file

register: Social variables which are stored in the register.

Defining Symbolic Constants: A symbolic constant value can be defined as a preprocessor statement and used in the program as any other constant value. The general form of a symbolic constant is

```
# define symbolic_name value of constant
```

Valid examples of constant definitions are:

```
# define marks 100
# define total 50
# define pi 3.14159
```

These values may appear anywhere in the program, but must come before it is referenced in the program.

It is a standard practice to place them at the beginning of the program.

Declaring Variable as Constant

The values of some variable may be required to remain constant throughout the program. We can do this by using the qualifier const at the time of initialization.

Example:

```
Const int class_size = 40;
```
The const data type qualifier tells the compiler that the value of the int variable class_size may not be modified in the program.

Derived Data Types: The C++ programming language allows programmers to separate program-specific datatypes through the use of **classes**. Instances of these datatypes are known as objects and can contain member variables, constants, member functions, and overloaded operators defined by the programmer. Syntactically, classes are extensions of the C struct, which cannot contain functions or overloaded operators.

Differences between struct in C and classes in C++: In C++, a structure is a class defined with the `struct` keyword. Its members are by *default public*. A class defined with the `class` keyword has by *default private* members.

C++ classes have their own members. These members include variables (including other structures and classes), functions (specific identifiers or overloaded operators) known as method, construtors and destructors. Members are declared to be either publicly or privately accessible using the `public:` and `private:` access specifiers respectively. Any member encountered after a specifier will have the associated access until another specifier is encountered. There is also inheritance between classes which can make use of the `protected:` specifier.

```
#include<iostream.h>
#include<stdio.h>
class play
{
    int playcode;
    char playtitle [25];
    float duration;
    int noofscenes;
public:
play()
{
    duration=45;
    noofscenes=5;
}
void newplay()
{
```

```
        cout<<"\n enter the play code";
        cin>>playcode;
        cout<<"\n enter the play title";
        gets(playtitle);
    }
    void moreinfo(float a,int b)
    {
        duration=a;
        noofscenes=b;
    }
    void showplay()
    {
        cout<<playcode<<playtitle<<duration<<noofscenes;
    }
};
```

PROGRAMS OF STANDARD METHODS IN C++ LANGUAGE

15.3   Bisection Method (Section 2.7)

Flow-chart

> *Refer to Section 14.3, page 674*

Program

```
/* Bisection Method */
#include <iostream.h>
#include <iomanip.h>
#include <math.h>
float f(float x)
{
 return (x*x*x - 4*x - 9);
}
void bisect(float *x,float a,float b,int *itr)
{
 *x = (a + b)/2;
 ++(*itr);
 cout << "Iteration no." <<setw(3) << *itr
  << "X = " << setw(7) << setprecision(5)
  << *x << endl;
}
int main()
{
```

```
int itr = 0, maxitr;
float x, a, b, aerr, x1;
cout << "Enter the values of a,b,"
   << "allowed error, maximum iterations" << endl;
cin >> a >> b >> aerr >> maxitr;
cout << fixed;
bisect(&x,a,b,&itr);
do
{
  if (f(a)*f(x) < 0)
      b = x;
  else
      a = x;
  bisect (&x1,a,b,&itr);
  if (fabs(x1-x) < aerr)
  {
cout << "After" << itr << "iterations, root"
   << "=" << setw(6) << setprecision(4)
   << x1 << endl;
   return 0;
   }
   x = x1;
} while (itr < maxitr);
cout << "Solution does not converge,"
   << "iterations not sufficient" << endl;
 return 1;
}
```

> **NOTES:** *a*, *b* are the limits in which the root lies
> *aerr* is the allowed error
> *itr* is a counter which keeps track of the number of iterations performed
> *maxitr* is the maximum number of iterations to be performed
> *x* is the value of root at *n*th iteration
> *x1* is the value of root at $(n + 1)$th iteration.
> *Function Bisect:*
> *Purpose:* Performs and prints the result of one iteration
> *Variables: x* is the result of the current iteration.

Computer Solution of Example 2.15 (a)

```
Enter the values of a, b, allowed error, maximum iterations
3 2.0001 20
```

```
Iteration No. 1 X = 2.50000
Iteration No. 2 X = 2.75000
Iteration No. 3 X = 2.62500
Iteration No. 4 X = 2.68750
Iteration No. 5 X = 2.71875
Iteration No. 6 X = 2.70313
Iteration No. 7 X = 2.71094
Iteration No. 8 X = 2.70703
Iteration No. 9 X = 2.70508
Iteration No.10 X = 2.70605
Iteration No.11 X = 2.70654
Iteration No.12 X = 2.70630
Iteration No.13 X = 2.70642
Iteration No.14 X = 2.70648
After 14 iterations, root = 2.7065
```

15.4 Regula-Falsi Method (Section 2.8)

Flow-chart

> *Refer to Section 14.4, page 676*

Program

```cpp
/* Regula Falsi Method */
#include <isostream.h>
#include <iomanip.h>
#include <math.h>
 float f(float x)
 {
  return cos(x)-x*exp(x);
 }
 void regula (float *x, float x0, float x1,
     float fx0, float fx1, int *itr)
 {
  *x = x0-((x1-x0)/(fx1-fx0))*fx0;
  ++(*itr);
cout << "Iteration no." << setw(3) << *itr
     << "X = " << setw(7) << setprecision(5)
     << *X << endl;
 }
 int main()
 {
```

```cpp
int itr=0, maxitr;
float x0,x1,x2,x3,aerr;
cout << "Enter the values for x0,x1,"
     << "allowed error,maximum iterations" << endl;
cin >> x0 >> x1 >> aerr >> maxitr;
regula(&x2,x0,x1,f(x0),f(x1),&itr);
cout << fixed;
do
{
  if (f(x0)*f(x2) < 0)
     x1 = x2;
  else

  x0 = x2;
  regula(&x3,x0,x1,f(x0),f(x1),&itr);
  if (fabs(x3-x2) < aerr)
  {
     cout << "After" << itr << "iterations,"
          << "root = " << setw(6) << setprecision(4)
          << x3 << endl;
     return 0;
  }
  x2=x3;
} while(itr < maxitr);
cout << "Solution does not converge,"
     << "iterations not sufficient" << endl;
return 1;
}
```

Notes: $f(x) = 0$ is the equation whose root is to be found
x0, x1 are units in which root lies
aerr is allowed error
maxitr is maximum number of iterations to be performed
itr is a counter which keeps track of the number of iterations performed
x2 is value of root at *n*th iteration
x3 is value of root at $(n + 1)$th iteration
Function Regula:
Purpose: Performs and prints the results of one iteration.
Variables: x is value of root at *n*th iteration
fx0, fx1 are values of $f(x)$ at *x0* and *x*, 1 respectively.

Computer Solution of Example 2.20

```
Enter the values for x0, x1, allowed error, maximum iterations
0 1.0001 20
Iteration No. 1 X = 0.31467
Iteration No. 2 X = 0.44673
Iteration No. 3 X = 0.49402
Iteration No. 4 X = 0.50995
Iteration No. 5 X = 0.51520
Iteration No. 6 X = 0.51692
Iteration No. 7 X = 0.51748
Iteration No. 8 X = 0.51767
Iteration No. 9 X = 0.51773
```
   After 9 iterations, root = 0.5177

15.5   Newton Raphson Method (Section 2.11)

Flow-chart

   Refer to Section 14.5, page 679

Program

```
/* Newton Raphson Method */
#include <iostream.h>
#include <iomanip.h>
#include <math.h>
float f(float x)
{
 return x*log10(x)-1.2;
}
 float df(float x)
{
 return log10(x) + 0.43429;
}
int main()
{
 int itr,maxitr;
 float h,x0,x1,aerr;
 cout << "Enter x0,allowed error,"
      << "maximum iterations" << endl;
 cin >> x0 >> aerr >> maxitr;
 cout << fixed;
```

```
for (itr=1;itr<=maxitr;itr++)
{
  h = f(x0)/df(x0);
  x1 = x0-h;
  cout << "Iteration no." << setw(3) << itr
       << "X = " << setw(9) << setprecision(6)
       << x1 << endl;
  if (fabs(h) < aerr)
  {
    cout << "After" << setw(3) << itr
         << "iterations, root = "
         << setw(8) << setprecision(6) << x1;
    return 0;
}
  x0 = x1;
}
cout << "Iterations not sufficient,"
     << "solution does not converge" << endl;
return 1;
}
```

Notes: $F(x) = 0$ is the equation whose root is to be found
$df(x)$ is the derivatives of $f(x)$ w.r.t. x $x0$ is value of root of nth iteration
$x1$ is value of root of $(n + 1)$th iteration
$aerr$ is allowed error
$maxitr$ is maximum no. of iterations to be performed
itr is a counter which keeps track of the number of iterations performed.

Computer Solution of Example 2.32

```
Enter x0, allowed error, maximum iterations
2.000001 10
Iteration No. 1 X = 2.813170
Iteration No. 2 X = 2.741109
Iteration No. 3 X = 2.740646
Iteration No. 4 X = 2.740646
After 4 iterations, root = 2.740646
```

15.6   Muller's Method (Section 2.13)

Flow-chart

> *Refer to Section 14.6, page 681*

Program

```
/* Muller's Method */
#include <iostream.h>
#include <iomanip.h>
#include <math.h>
#define I 2
float y(float x)
{
 return cos(x)-x*exp(x);
}
int main()
{
 int i,itr,maxitr;
 float x[4],li,di,mu,s,l,aerr;
 cout << "Enter the initial"
      "approximations" << endl;
 for (i = I-2;i<3;i++)
   cin >> x[i];
 cout << "Enter allowed error,"
      "maximum iterations" << endl;
 cin >> aerr >> maxitr;
 cout << fixed;
 for(itr = 1;itr <= maxitr;itr++)
 {
   li = (x[I]-x[I-1])/(x[I-1]-x[I-2]);
   di = (x[I]-x[I-2])/(x[I-1]-x[I-2]);
   mu = y(x[I-2])*li*li
        - y(x[I-1])*di*di
          + y(x[I])*(di+li);
   s = sqrt((mu*mu - 4*y(x[I])*di*li
        *(y(x[I-2])*li-y(x[I-1])
           *di + y(x[I])))));
   if (mu < 0)
    l = (2*y(x[I])*di)/(-mu+s);
   else
    l = (2*y(x[I])*di)/(-mu-s);
   x[I+1] = x[I]+l*(x[I] - x[I-1]);
   cout << "Iteration no. " << setw(3) << itr
        << "X = " << setw(7) << setprecision(5)
        << x[I+1] << endl;
   if (fabs(x[I+1]-x[I]) < aerr)
```

```
    {
      cout << "After" << setw(3) << itr
           << "iterations, the solution is"
           << setw(6) << setprecision(4)
           << x[I+1] << endl;
      return 0;
    }
      for (i=I-2;i<3;i++)
        x[i] = x[i+1];
    }
cout << "Iterations not sufficient,"
     << "solution does not converge" << endl;
  return 1;
    }
```

> **NOTES:** $y(x) = 0$ is the equation whose root is to be found
> x is an array which holds the three approximations to the root and the new improved value
> I is defined as 2 in the program. This has been done because in C, array subscripts always start from zero and cannot be negative. Use of I facilitates more readable expressions. For *e.g.*, $x[0]$ can be written as $x[I-2]$ which looks more close to x_{i-2} it actually represents.
> *li* is λ_i
> *di* is δ_i
> *mu* is μ_i
> *s* is $\sqrt{[\mu_i^2 - 4y_i\,\delta_i\,\lambda_i(y_{i-2}\,\lambda_i - \lambda_{i-1}\delta_i + y_i)]}$
> *l* is λ

Computer Solution of Example 2.34

```
Enter the initial approximations
-1 0 1
Enter allowed error, maximum iterations
.0001 10
Iteration No. 1 X = 0.44152
Iteration No. 2 X = 0.51255
Iteration No. 3 X = 0.51769
Iteration No. 4 X = 0.51776
After 4 iterations, the solution is 0.5178
```

15.7 Multiplication of Matrices [Section 3.2 (3)4]

Flow-chart

> *Refer to Section 14.7, page 684*

Program

```
/* Multiplication of matrices */
#include <iostream.h>
#include < iomanip.h>
#include <math.h>
#define MAX 20
typedef float matrix[MAX][MAX];
void getelems(matrix x,int m,int n)
{
 int i,j;
 for(i=0;i<m;i++)
   for(j=0;i<m;i++)
     cin >> x[i][j];
}
void printsol(matrix x,int m,int n)
{
 int i,j;
 for (i=0;i<m;i++)
{
    for (j=0;j<n;j++)
      cout << setw(5) << setprecision(1)
           << x[i][j];
    cout << endl;
}
}
void matmul(matrix a,matrix b,matrix c,
        int l, int m,int p, int q)
{
 float s;
 int i,j,k;
 cout << "Enter the elements of the"
      << "first matrix" << endl;
 getelems(a,l,m);
 cout << "Enter the elements of the"
      << "second matrix" << endl;
 getelems(b,p,q);
 for (i=0;i<l;i++)
   for (j=0;j<q;j++)
```

```
    {
       s = 0;
       for (k=0;k<m;k++)
         s += a[i][k]*b[k][j];
       c[i][j] = s;
     }
 cout << "The solution is" << endl;
 printsol(c,l,q);
}
int main()
{
 matrix a,b,c;
 int l,m,p,q;
 cout << "Enter the row, column of the"
      << "first matrix" << endl;
 cin >> l >> m;
 cout << "Enter the row, column of the"
               "second matrix" << endl;
 cin >> p >> q;
 cout << fixed;
 if (m!=p)
 {
    cout << "The two matrices cannot"
         << "be multiplied" << endl;
    return 1;
 }
 else
 {
    matmul(a,b,c,l,m,p,q);
    return 0;
 }
}
```

> **Notes:** *MAX* is largest number of rows or columns any matrix can have.
> (If MAX. = 20, 11 × 20, and 20 × 13 matrices can be multiplied but 1 ×
> 21, 22 × 1 matrices cannot be multiplied till MAX > = 22
> A, B are arrays which contain the matrices to be multiplied
> C is array which contains the result of multiplication
> L, M are respectively the rows, columns of first matrix
> P, Q are respectively the rows, columns of second matrix
> *Function getelems*
> *Purpose:* To input a *m* × *n* matrix

Function Matmul.

Purpose: It performs the multiplication of matrices after taking them from the user and prints the result.

Variables: i, j, k are loop control variables.

Computer Solution of Example 3.7

```
Enter the row, column of the first matrix
3 3
Enter the row, column of the second matrix
3 2
Enter the elements of the first matrix
0 1 2
1 2 3
2 3 4
Enter the elements of the second matrix
 1 -2
-1  0
 2 -1
The solution is
3.0 -2.0
5.0 -5.0
7.0 -8.0
```

15.8   Gauss Elimination Method [Section 3.4 (3)]

Flow-chart

   Refer to Section 14.8, page 687

Program

```cpp
/* Gauss elimination method */
#include <iostream.h>
#include <iomanip.h>
#include <math.h>
#define N 4
int main()
{
 float a[N][N+1],x[N],t,s;
 int i,j,k;
 cout << "Enter the elements of the"
         "augmented matrix rowwise" << endl;
 cout << fixed;
```

```
for (i=0;i<N;i++)
  for (j=0;j<N+1;j++)
    cin >> a[i][j]);
for (j=0;j<N-1;j++)
  for (i=j+1;i<N;i++)
  {
    t  = a[i][j]/a[j][j];
    for (k=0;k<N+1;k++)
      a[i][k] -= a[j][k]*t;
  }
/* now printing the
upper triangular matrix */
cout << "The upper triangular matrix"
                     "is:-" << endl;
for (i=0;i<N;i++)
{
  for (j=0;j<N+1;j++)
    cout << setw(8) << setprecision(4) << a[i][j];
  cout << endl;
}
/* now performing back substitution */
for (i=N-1;i>=0;i--)
{
  s = 0;
  for (j=i+1;j<N;j++)
    s += a[i][j]*x[j];
  x[i] = (a[i][N]-s)/a[i][i];
}
/* now printing the results */
cout << "The solution is:- " << endl;
for (i=0;i<N;i++)
   cout << "x[" << setw(3) << i+1 << "] = "
   << setw(7) << setprecision(4) << x[i] << endl;
return 0;
}
```

> **NOTES:** N is the number of unknowns
> a is an array which holds the Augmented Matrix
> x is an array which will contain values of unknowns
> i, j, k are loop control variables.

Computer Solution of Example 3.19

```
Enter the elements of augmented matrix rowwise
10    -7     3     5     6
-6     8    -1    -4     5
 3     1     4    11     2
 5    -9    -2     4     7
The upper triangular matrix is:-
10.000      -7.0000      3.0000      5.0000      6.0000
 0.0000      3.8000      0.8000     -1.0000      8.6000
-0.0000     -0.0000      2.4474     10.3158     -6.8158
 0.0000     -0.0000     -0.0000      9.9247      9.9247
The solution is:-
X[ 1] = 5.0000
X[ 2] = 4.0000
X[ 3] = -7.0000
X[ 4] = 1.0000
```

15.9   Gauss-Jordan Method [Section 3.4 (4)]

Flow-chart

   Refer to Section 14.9, page 689

Program

```
/* Gauss jordan method */
#include <iostream.h>
#include <iomanip.h>
#define N 4
int main()
{
 float a[N][N+1],t;
 int i,j,k;
 cout << "Enter the elements of the"
      << "augmented matrix rowwise" << endl;
 for (i=0;i<N;i++)
    for (j=0;j<N+1;j++)
      cin >> a[i][j];
 /* now calculating the values
      of x1,x2,....,xN */
 cout << fixed;
     for (j=0;j<N;j++)
        for (i=0;i<N;i++)
        if (i!=j)
```

```
          {
            t = a[i][j]/a[j][j];
            for (k=0;k<N+1;k++)
                a[i][k] -= a[j][k]*t;
          }
/* now printing the diagonal matrix */
cout << "The diagonal matrix is:-" << endl;
for (i=0;i<N;i++)
{
   for (j=0;j<N+1;j++)
     cout << setw(9) << setprecision(4) << a[i][j];
   cout << endl;
}
/* now printing the results */
cout << "The solution is:- " << endl;
for (i=0;i<N;i++)
   cout << "x[" << setw(3) << i+1 << "] ="
        << setw(7) << setprecision(4)
        << a[i][N]/a[i][i] << endl;
return 0;
}
```

NOTES: a is an array which holds the augmented matrix
N is the number of unknowns. *e.g.*, if it is a 3×3 system of equations,
$N = 3$, and if 5×5 system take $N = 5$.
i, j, k are loop variables.

Computer Solution of Example 3.22

```
Enter elements of augmented matrix rowwise
10    -7     3      5      6
-6     8    -1     -4      5
 3     1     4     11      2
 5    -9    -2      4      7
The diagonal matrix is:-
10.0000     -0.0000     -0.0000  -0.0000     50.0000
 0.0000      3.8000     -0.0000   0.0000     15.2000
-0.0000      0.0000      2.4474   0.0000    -17.1316
 0.0000     -0.0000      0.0000   9.9247      9.9247
The solution is:-
X[  1] = 5.0000
X[  2] = 4.0000
X[  3] = -7.0000
X[  4] = 1.0000
```

15.10   Factorization Method [Section 3.4 (5)]

Flow-chart

>   *Refer to Section 14.10, page 691*

Program

```
/* Factorization method */
#include <iostream.h>
#include <iomanip.h>
#define N 3
typedef float matrix[N][N];
matrix l,u,a;
float b[N],x[N],v[N];
void urow(int i)
{
 float s;
 int j,k;
 for (j=i;j<N;j++)
 {
   s = 0;
   for (k=0;k<N-1;k++)
     s += u[k][j]*l[i][k];
   u[i][j] = a[i][j]-s;
 }
}
void lcol(int j)
{
 float s;
 int i,k;
 for (i=j+1;i<N;i++)
 {
   s = 0;
   for (k=0;k<=j-1;k++)
     s += u[k][j]*l[i][k];
   l[i][j] = (a[i][j]-s)/u[j][j];
 }
}
void printmat(matrix x)
{
 int i,j;
 for (i=0;i<N;i++)
 {
```

```
    for (j=0;j<N;j++)
      cout << setw(8) << setprecision(4) << x[i][j];
    cout << endl;
 }
}
int main()
{
 int i,j,m;
 float s;
 cout << "Enter the elements of augmented"
      << " matrix rowwise" << endl;
 for (i=0;i<N;i++)
 {
   for (j=0;j<N;j++)
     cin >> a[i][j];
   cin >> b[i];
 }
cout << fixed;
 /* now calculating the elements of l and u */
 for (i=0;i<N;i++)
   l[i][i] = 1.0;
 for (m=0;m<N;m++)
 {
   urow(m);
   if (m < N-1) lcol(m);
 }
 /* now printing the elements of l and u */
 cout << setw(14) << "U" << endl; printmat(u);
 cout << setw(14) << "L" << endl; printmat(l);
 /* now solving LV=B
    by forward substitution */
 for (i=0;i<N;i++)
 {
   s = 0;
   for (j=0;j<=i-1;j++)
     s += l[i][j]*v[j];
   v[i] = b[i]-s;
 }
/* now solving UX=V
   by backward substitution */
for (i=N-1;i>=0;i--)
{
```

```
  s = 0;
  for (j=i+1;j<N;j++)
    s += u[i][j]*x[j];
  x[i] = (v[i]-s)/u[i][i];
}
/* printing the results */
cout << "The solution is:-" << endl;
for (i=0;i<N;i++)
   cout << "x[" << setw(3) << i+1 << "] = "
        << setw(6) << setprecision(4)
        << x[i] << endl;
return 0;
}
```

Notes:
N is the no. of unknowns
l is the lower triangular matrix
u is the upper triangular matrix
a is the coefficient matrix
b is the constant matrix (Column matrix)
v is a matrix such that $lv = b$
x will contain the values of unknowns
i, j, m are loop control variables
Function urow (I)
Purpose: Calculates elements of ith row of u
Variables: m is the no. of unknowns
j, k are loop control variables
Function Lcol (J)
Purpose: Calculates elements of jth column of l
Variables: m is the no. of unknowns
i, k are loop control variables.
Function Printmat
Purpose: To print an $N \times N$ matrix.

Computer Solution of Example 3.23

```
Enter the elements of augmented matrix rowwise
3   2   7   4
2   3   1   5
3   4   1   7
```

```
              U
  3.0000    2.0000    7.0000
  0.0000    1.6667   -3.6667
  0.0000    0.0000   -1.6000


              L
  1.0000    0.0000    0.0000
  0.6667    1.0000    0.0000
  1.0000    1.2000    1.0000
```

The solution is:
x[1] = 0.8750
x[2] = 1.1250
x[3] = - 0.1250

Computer Solution of Example 3.24

Enter the elements of augmented matrix rowwise
```
10      -7       3      5      6
-6       8      -1     -4      5
 3       1       4     11      2
 5      -9      -2      4      7
                     U
 10.0000     -7.0000      3.0000       5.0000
  0.0000      3.8000      0.8000      -1.0000
  0.0000      0.0000      2.4474      10.3158
  0.0000      0.0000      0.0000       9.9247
                     L
  1.0000      0.0000      0.0000       0.0000
 -0.6000      1.0000      0.0000       0.0000
  0.3000      0.8158      1.0000       0.0000
  0.5000     -1.4474     -0.9570       1.0000
```
The solution is:-
x[1] = 5.0000
x[2] = 4.0000
x[3] = -7.0000
x[4] = 1.0000

15.11 Gauss-Seidal Iteration Method [Section 3.5 (2)]

Flow-chart

   Refer to Section 14.11, page 695

Program

```
/* Gauss Seidal method */
#include <iostream.h>
#include <iomanip.h>
#include <math.h>
#define N 3 int main()
{
 float a[N][N+1],x[N],aerr,maxerr, t,s,err;
 int i,j,itr,maxitr;
 /* first initializing the array x */
 for (i=0;i<N;i++) x[i]=0;
 cout << "Enter the elements of the"
      << "augmented matrix rowwise" << endl;
 for (i=0;i<N;i++)
    for (j=0;j<N+1;j++)
      cin >> a[i][j];
  cout << "Enter the allowed error,"
       << "maximum iterations" << endl;
   cin >> aerr >> maxitr;
  cout << fixed;
  cout << "Iteration" << setw(6) << "x[1]"
       << setw(11) << "x[2]"
       << setw(11) << "x[3]" << endl;
  for (itr=1;itr<=maxitr;itr++)
  {
    maxerr = 0;
    for (i=0;i<N;i++)
    {
      s = 0;
      for (j=0;j<N;j++)
        if (j!=i) s += a[i][j]*x[j];
      t = (a[i][N]-s)/a[i][i];
      err = fabs(x[i]-t);
      if (err >> maxerr) maxerr = err;
      x[i] = t;
    }
    cout << setw(5) << itr;
    for (i=0;i<N;i++)
      cout << setw(11) << setprecision(4) << x[i];
    cout << endl;
    if (maxerr << aerr)
```

```
    {
      cout << "Converges in" << setw(3) << itr
           << "iterations" << endl;
      for (i=0;i<N;i++)
        cout << "x[" << setw(3) << i+1 << "] = "
             << setw(7) << setprecision(4) << x[i]
             << endl;
      return 0;
    }
}
cout << "Solution does not converge,"
    << "iterations not sufficient" << endl;
 return 1;
}
```

> **NOTES:** *N* is the number of unknowns
> *a* is an array which holds the augmented matrix
> *x* is an array which will hold the values of unknowns
> *aerr* is allowed error
> *maxitr* is the maximum no. of iterations to be performed
> *itr* is the counter which keeps track of no. of iterations performed
> *err* is error in value of xi
> *maxerr* is maximum error in any value of xi after an iteration.

Computer Solution of Example 3.28

```
Enter the elements of augmented matrix rowwise
20      1     -2      17
 3     20     -1     -18
 2    - 3     20      25
Enter the allowed error, maximum iterations
.0001 10
      Iteration      X(1)        X(2)         X(3)
              1      0.8500    - 1.0275      1.0109
              2      1.0025    - 0.9998      0.9998
              3      1.0000    - 1.0000      1.0000
              4      1.0000    - 1.0000      1.0000
Converges in 4 iterations
X[1] =   1.0000
X[2] = -1.0000
X[3] =   1.0000
```

Computer Solution of Example 3.30

```
Enter the elements of the augmented matrix rowwise
10    -2    -1    -1      3
-2    10    -1    -1     15
-1    -1    10    -2     27
-1    -1    -2    10     -9
Enter the allowed error, maximum iterations
.0001 15
Iteration      x[1]        x[2]        x[3]        x[4]
    1         0.3000      1.5600      2.8860     -0.1368
    2         0.8869      1.9523      2.9566     -0.0248
    3         0.9836      1.9899      2.9924     -0.0042
    4         0.9968      1.9982      2.9987     -0.0008
    5         0.9994      1.9997      2.9998     -0.0001
    6         0.9999      1.9999      3.0000     -0.0000
    7         1.0000      2.0000      3.0000     -0.0000
Converges in 7 iterations
x[ 1] = 1.0000
x[ 2] = 2.0000
x[ 3] = 3.0000
x[ 4] = 0.0000
```

15.12   Power Method (Section 4.11)

Flow-chart

Refer to Section *4.12, page 699*

Program

```c
/* Power method for finding largest eigen value */
#include <iostream.h>
#include <iomanip.h>
#include <math.h>
#define N 3
typedef float array[N];
void findmax(float *max,array x)
{
 int i;
 *max = fabs(x[0]);
 for (i=1;i<N;i++)
    if (fabs(x[i]) > *max)
       *max = fabs(x[i]);
```

```cpp
}
int main()
{
 float a[N][N],x[N],r[N],maxe,
    err,errv,aerr,e,s,t;
 int i,j,k,itr,maxitr;
 cout << "Enter the matrix rowwise" << endl;
 for (i=0;i<N;i++)
    for (j=0;j<N;j++)
      cin << a[i][j];
 cout << "Enter the initial approximation"
      << "to the eigen vector" << endl;
 for (i=0;i<N;i++)
   cin >> x[i];
 cout << "Enter the allowed error,"
      << "maximum iterations" << endl;
 cin >> aerr >> maxitr;
 cout << fixed;
 cout << "Itr no." << setw(11) << "Eigenvalue"
      << setw(19) << "EigenVector" << endl;
 /*now finding the largest eigenvalue in
   the initial approx. to eigen vector */
findmax(&e,x);
/* now starting the iterations */
for (itr=1;itr<=maxitr;itr++)
{
 /* loop to multiply the matrices a and x */
 for (i=0;i<N;i++)
 {
  s = 0;
   for (k=0;k<N;k++)
     s += a[i][k]*x[k];
   r[i]=s;
}
findmax(&t,r);
for (i=0;i<N;i++) r[i] /= t;
maxe = 0;
for (i=0;i<N;i++)
{
  err = fabs(x[i]-r[i]);
  if (err > maxe) maxe = err;
  x[i] = r[i];
```

```
  }
errv = fabs(t-e);
e = t;
cout << setw(4) << itr
      << setw(12) << setprecision(4)
      << e;
for (i=0;i<N;i++)
  cout << setw(9) << setprecision(3)
        << x[i];
 cout << endl;
 if ((errv <= aerr) && (maxe <= aerr))
{
  cout << "Converges in" << itr
        << "iterations" << endl;
  cout << "Largest eigen value ="
        << setw(6) << setprecision(2)
        << e << endl;
  cout << "Eigen Vector:-" << endl;
  for (i=0;i<N;i++)
    cout << "x[" << setw(3) << i+1 << "] = "
          << setw(6) << setprecision(2)
          << x[i] << endl;
    cout << endl;
    return;
 }
}
    cout << "Solution does not converge,"
          << "iterations not sufficient" << endl;
    return 1;
}
```

NOTES: N is number of rows (or columns) in square matrix
a is the square matrix
x is the eigenvector at nth iteration
r is the eigenvector at (n + 1)th iteration
e is the eigenvalue at nth iteration
 t is the eigenvalue at (n + 1)th iteration
aerr is the allowed error in eigenvalue and eigenvector
maxitr is the maximum number of iterations to be performed
err is error in an element of the eigenvector
maxe is the maximum error in any element of the eigenvector

errv is error in the eigenvalue

itr, i, k are loop control variables.

Function findmax:

Purpose: Finds the maximum element in array x(a N-element array) and returns it in *Max*.

Computer Solution of Example 4.11

```
Enter the matrix rowwise
 2    -1     0
-1     2    -1
 0    -1     2
Enter the initial approximation to the eigen vector
1 0 0
Enter the allowed error, maximum iterations
.01 10
   Itr No.   Eigen Value                    Eigen Vector
      1         2.0000           1.000      -0.500     0.000
      2         2.5000           1.000      -0.800     0.200
      3         2.8000           1.000      -1.000     0.429
      4         3.4286           0.875      -1.000     0.542
      5         3.4167           0.805      -1.000     0.610
      6         3.4146           0.764      -1.000     0.650
      7         3.4143           0.741      -1.000     0.674
      8         3.4142           0.727      -1.000     0.688
      9         3.4142           0.719      -1.000     0.696
Converges in 9 iterations
Largest eigen value = 3.41
Eigen Vector:-
X[1] = 0.72
X[2] = -1.00
X[3] = 0.70
```

15.13   Method of Least Squares (Section 5.5)

Flow-chart

> *Refer to Section 4.13, page 703*

Program

```
/* Parabolic fit by least squares */
#include <iostream.h>
#include <iomanip.h>
int main()
```

```
{
 float augm[3][4]={{0,0,0,0},{0,0,0,0},{0,0,0,0}};
 float t,a,b,c,x,y,xsq;
 int i,j,k,n;
 cout << "Enter the no. of pairs of"
      <<"observed values:" << endl;
 cin >> n;
 cout << fixed;
 augm [0] [0] = n;
 for (i=0;i<n;i++)
 {
   cout << "Pair no. " << i+1 << endl;
   cin >> x >> y;
   xsq = x*x;
   augm[0][1] += x;
   augm[0][2] += xsq;
   augm[1][2] += x*xsq;
   augm[2][2] += xsq*xsq;
   augm[0][3] += y;
   augm[1][3] += x*y;
   augm[2][3] += xsq*y;
 }
 augm[1][1] = augm[0][2];
 augm[2][1] = augm[1][2];
 augm[1][0] = augm[0][1];
 augm[2][0] = augm[1][1];
 cout << "The augmented matrix is:-" << endl;
 for (i=0;i<3;i++)
 {
    for (j=0;j<4;j++)
       cout << setw(9) << setprecision (4) << augm[i][j];
     cout << endl;
 }
 /* Now solving for a,b,c
    by Gauss Jordan Method */
 for (j=0;j<3;j++)
   for (i=0;i<3;i++)
      if (i!=j)
      {
         t = augm[i][j]/augm[j][j];
         for (k=0;k<4;k++)
            augm[i][k] -= augm[j][k]*t;
```

```
        }
a = augm[0][3]/augm[0][0];
b = augm[1][3]/augm[1][1];
c =  augm[2][3]/augm[2][2];
cout << setprecision(4)
        << "a = " << setw(8) << a
        << "b = " << setw(8) << b
        << "c = " << setw(8) << c
        << endl;
 return 0;
}
```

> **Notes:** *augm* is the augmented matrix.
> *n* is the number of data points.

Computer Solution of Example 5.7

```
Enter the no. of pairs of observed values:
7
Pair No. 1
1 1.1
Pair No. 2
1.5 1.3
Pair No. 3
2 1.6
Pair No. 4
2.5 2
Pair No. 5
3.0 2.7
Pair No. 6
3.5 3.4
Pair No. 7
4.0 4.1
The augmented matrix is:-
        7.0000          17.5000        50.7500        16.2000
        17.5000         50.7500       161.8750        47.6500
        50.7500        161.8750       548.1875       154.4750
a = 1.0357          b = -0.1929   c = 0.2429
```

15.14   Method of Group Averages (Section 5.9)

Flow-chart

   Refer to Section 14.14, page 706

Program

```
#include<iostream.h>
#include<conio.h>
#include<iomanp.h>
void main()
{
int t[10],n,i,ts1=0,ts2=0;
float a,b,rs1=0,rs2=0,r[10],x1,y1,x2,y2;
clrscr();
cout<<"enter the no. of observations"<<endl;
cin>>n;
cout<<"enter the different values of t"<<endl;
 for(i=1;i<=n;i++)
{
cin>>t[i];
}
cout<<"\nenter the corresponding values of r"<<endl;
for(i=1;i<=n;i++)
{
cin>>r[i];
}
for(i=1;i<=(n/2);i++)
{
ts1+=t[i];
rs1+=r[i];
}
for(i=((n/2)+1);i<=n;i++)
{
ts2+=t[i];
rs2+=r[i];
}
x1=ts1/(n/2);
y1=rs1/(n/2);
x2=ts2/(n/2);
y2=rs2/(n/2);
b=(y2-y1)/(x2-x1);
a=y1-(b*x1);
cout<<"the value of a&b comes out to be
"<<endl<<"a="<<setw(5)<<setprecision(3)<<a<<"\n"<<"b="<<
setw(5)<<setprecision(3)<<b;
getch();
}
```

Computer Solution of Example 5.16

```
Enter the no. of observations
8
enter the different values of t
40 50 60 70 80 90 100 110
enter the corresponding values of r
1069.1 1063.6 1058.2 1052.7 1049.3 1041.8 1036.3 1030.8
the value of a&b comes out to be
a=1090.256
b=-0.534
```

15.15 Method of Moments (Section 5.11)

Flow-chart

   Refer to Section 14.15, page 708

Program

```cpp
#include<iostream.h>
#include<conio.h>
#include<iomanip.h>
void main()
{
int x[10],y[10],i,n,yt=0,x1yt=0;
float a,b,l1,l2,c1,c2,c3,d,d1,d2,m1,m2,h;
clrscr();
cout<<"enter the no. of observations"<<endl;
cin>>n;
cout<<"enter the different values of x"<<endl;
for(i=1;i<=n;i++)
{
cin>>x[i];
}
cout<<"\nenter the corresponding values of y"<<endl;
for(i=1;i<=n;i++)
{
cin>>y[i];
}
h=x[2]-x[1];
for(i=1;i<=n;i++)
{
yt+=y[i];
```

```
x1yt+=x[i]*y[i];
}
m1=h*yt;
m2=h*x1yt;
l1=(-(h/2)+x[1]);
l2=((h/2)+x[n]);
c1=(l2-l1);
c2=((l2*l2)-(l1*l1)/2);
c3=((l2*l2*l2)-(l1*l1*l1*)/3);
cout<<"The observed equations
are"<<endl<<c1<<"a+"<<c2<<"b"<<endl<<"&"<<c2<<"a+"<<c3<<"b";
d=c2/c1;
d1=d*c1;
d2=d*c2;
m1=d*m1;
b=(m2-m1)/(c3-d2);
a=(m1-(d2*b))/d1;
cout<<"\nOn solving these equations we get
 a="<<setw(5)<<setprecision(2)<<a<<"
&b="<<setw(5)<<setprecision(2)<<b<<endl;
cout<<"hence the required equation is
 y="<<setw(5)<<setprecision(2)<<a<<"+"<<
setw(5)<<setprecision(2)<<b<<"x";
getch();
}
```

Solution of Example 5.20

```
Enter the no. of observations
4
enter the different values of x
1 2 3 4
enter the corresponding values of y
16 19 23 26
the observed equations are
4.00a+10.00b=84.00
&10.00a+30.33b=227.00
on solving these equations we get a=13.03&b=3.19
hence the required equation is y=13.03+3.19x
```

15.16   Newton's Forward Interpolation Formula (Section 7.2)

Flow-chart

   Refer to Section 4.16, page 711

Program

```
/* Newton's forward interpolation */
#include <iostream.h>
#include <iomanip.h>
#define MAXN 100
#define ORDER 4 int main()
{
 float ax[MAXN+1],ay[MAXN+1],
   diff[MAXN+1][ORDER+1],
   nr=1.0,dr=1.0,x,p,h,yp;
 int n,i,j,k;
 cout << "Enter the value of n" << endl;
 cin >> n;
 cout << "Enter the values in form x,y" << endl;
 for (i=0;i<=n;i++)
  cin >> ax[i] >> ay[i];
 cout << "Enter the values of x"
      << "for which value of y is wanted" << endl;
 cin >> x;
 cout << fixed;
 h=ax[1]-ax[0];
 /* now making the diff. table */
 /* calculating the 1st order differences */
for (i=0;i<=n-1;i++)
  diff[i][1] = ay[i+1]-ay[i];
/* calculating the second & higher order differences.*/
  for (j=2;j<=ORDER;j++)
    for (i=0;i<=n-j;i++)
      diff [i][j] = diff[i+1][j-1]
                    -diff[i][j-1];
 /* now finding x0 */
 i=0;
 while (!(ax[i] > x)) i++;
 /* now ax[i] is x0 & ay[i] is y0 */
 i--;
 p = (x-ax[i])/h; yp=ay[i];
 /* Now carrying out interpolation */
```

```
for (k=1;k<=ORDER;k++)
{
 nr *= p-k+1; dr *=k;
 yp += (nr/dr)*diff[i][k];
}
cout << "When x = "
     << setw(6) << setprecision(1)
     << x
     << " y = "
     << setw(6) << setprecision(2)
     << yp << endl;
return 0;
}
```

NOTES: *MAXN* is the maximum value of *N*
ORDER is the maximum order in the difference table
ax is an array containing values of *x* (*x*0, *x*1,......, *x*n)
ay is an array containing values of *y*(*y*0, *y*1,......, *y*n)
diff is a 2D Array containing the difference table
h is spacing between values of X
x is value of x at which value of y is wanted
yp is calculated value of Y
nr is numerator of the terms in expansion of *yP*
dr is denominator of the terms in expansion of *yP*.

Computer Solution of Example 7.1

```
Enter the value of n
6
Enter the values in form x, y
100   10.63
150   13.03
200   15.04
250   16.81
300   18.42
350   19.90
400   21.27
Enter the values of x for which value of y is wanted
218
When x = 218.0, y = 15.70
```

15.17 Lagrange's Interpolation Formula (Section 7.12)

Flow-chart

Refer to Section 4.17, page 714

Program

```
/*Lagrange's Interpolation*/
#include <iostream.h>
#include <<iomanip.h>
#define MAX 100 int main()
{
 float ax [MAX+1],ay[MAX+1],nr,dr,x,y=0;
 int i,j,n;
 cout << "Enter the value of n" << endl;
 cin >> n;
 cout << "Enter the set of values" << endl;
 for (i=0;i<=n;i++)
   cin >> ax[i] >> ay[i];
 cout << "Enter the value of x for which"
      << "value of y is wanted" << endl;
 cin >> x;
 cout << fixed;
 for (i=0;i<=n;i++)
 {
    nr=dr=1;
    for(j=0;j<=n;j++)
      if (j!=i)
      {
         nr *= x-ax[j];
         dr *= ax[i]-ax[j];
      }
      y += (nr/dr)*ay[i];
 }
 cout << "When x="
      << setw(4) << setprecision(1)
      << x << "y="
      << setw(7) << setprecision(1)
      << y << endl;
 return 0;
}
```

> **NOTES:** *MAX* is the maximum value of *n*
> *ax* is an array containing values of $x(x_0, x_1,....., x_n)$
> *ay* is an array containing values of $y(y_0, y_1,......, y_n)$
> *x* is the value of *x* at which value of *y* is wanted
> *y* is the calculated value of *y*
> *nr* is numerator of the terms in expansion of *y*
> *dr* is denominator of the terms in expansion of *y*.

Computer Solution of Example 7.17

```
Enter the value of n
4
Enter the set of values
5    150
7    392
11   1452
13   2366
17   5202
Enter the value of x for which value of y is wanted
9
When x = 9.0 y = 810.0
```

15.18   Newton's Divided Difference Formula (Section 7.14)

Flow-chart

   Refer to Section 14.18, page 716

Program

```
#include<iostream.h>
#include<conio.h>
void main()
{
int x[10],y[10],p[10];
int k,f,n,i,j=1,f1=1,f2=0;
clrscr();
cout<<"enter the no. of observations\n";
cin>>n;
cout<<"enter the different values of x\n";
for(i=1;i<=n;i++)
{
cin>>x[i];
}
cout<<"enter the corresponding values of y\n";
```

```
for(i=1;i<=n;i++)
{
cin>>y[i];
}
f=y[1];
 cout<<"enter the value of 'k' in f(k) you want to evaluate\n";
cin>>k;
do
{
for(i=1;i<=n-1;i++)
{
p[i]=((y(i+1)-y[i])/(x[i+j]-x[i]));
y[i]=p[i];
}
for(i=1;i<=j;i++)
{
f1*=(k-x[i]);
} f2+=(y[1]*f1);  f1=1;
n--;
j++;
} while(n!=1);  f+=f2;  cout<<"f("<<k<<")="<<f;  getch();
}
```

Computer Solution of Example 7.23

```
Enter the no. of observations
5
Enter the different values of x
5 7 11 13 17
enter the corresponding values of y
150 392 1452 2366 5202
enter the value of 'k' in f(k) you want to evaluate
9 f(9)=810
```

15.19 Derivatives Using Forward Difference Formulae (Section 8.2)

Flow-chart

> *Refer to Section 14.19, page 718*

Program

```
/* derivatives using forward difference*/
#include<iostream.h>
#include<math.h>
```

```
#include<iomanip.h>
#include<conio.h>
void main( )
{
    float *x=NULL, *y=NULL,max;
    float *tmp = NULL;
    float xval,h,p,x0,y0,yval,sum;
    int pos,i;
    clrscr( );
    cout<<"enter the no of comparisons";
    cin>>max;
    x=new float[max];
    y=new float[max];
    cout<<"enter the values in cv table for x and y";
    for (i-0;i<max;i++)
    {
        cout<<"\n value for "<<i<<"x";
        cin>>x[i];
        cout<<"\n value for "<<i<<"y";
        cin>>y[i];
    }
    cout<<"enter the value of x";
    cin>>xval;
    for(i=0;i<max;i++)
    {
        if(x[i]>=xval)
        {
            pos=i;
            break;
        }
    }
    x0=x[pos];
    y0=y[pos];
    cout<<"\nx0 is "<<x0<<"y0 is "<<y0<<"at"<<pos;
    h=x[1]-x[0];
    p=(xval-x0)/h;
    if(pos<(max))
    {
    int i,l,j;
    // calculating no of elements in array
    l=max-pos;
    tmp= new float [l* l];
```

```
cout<<"\n";
for(i=0;i<l;i++)
{
    for(j=0;j<=l;j++)
    {
        tmp[i*l+j]=0;
    }
    cout<<"\n";
}
cout<<"\n size of new array" <<l<<"\n";
//copying values of y in array
for(i=0, j=pos;i<l;i++, j++)
{
    tmp[i]=y[j];
}
cout<<"\n";
for(i=1;i<l;i++)
{
    for(j=0; j<l-i; j++)
    {
        tmp[i*l+j]=tmp[(i-1)*l+(j+1)]-tmp[(i-1)*l+(j)];
    }
}
cout<<"\nvalues are \n";
for(i=0;i<l;i++)
{     cout<<x[i+pos]<<"\t";
      for(j=0; j<l; j++)
      {
            cout<<setprecision(3)<<tmp[j*l+i]<<"\t|";
      }
      cout<<"\n;
}
//appling newtons forward diffenation using first derivates
sum=0;
int k=1;
for(i=1;i<l;i++)
{
    sum=sum+((1.0/i)*tmp[i*tmp[i*l+0]*k);
    k=-k;
}
cout<<"\n\n first (dy/dx): "<<sum/h;
int v[]={0,0,1.0, 1.0, 11.0/12.0,5.0/6.0,137.0/180.0};
```

```
    sum=0;
    k=1;
    for(i=2;i<l;i++)
     {
        sum=sum+(v[i]*tmp[i*l+0]*k);
        k=-k;
     }
     cout<<"\n\n second (dy/dx): "<<sum/pow(h,2.0);
}
```

Computer Solution of Example 8.1

```
value for 0x1.0
value for 0y7.989
value for 1x1.1
value for 1y8.403
value for 2x1.2
value for 2y8.781
value for 3x1.3
value for 3y9.129
value for 4x1.4
value for 4y9.451
value for 5x1.5
value for 5y9.750
value for 6x1.6
value for 6y10.031
Enter the value of x1.1
x0 is 1.1y0 is 8.403at1

size of new array 6

values are
1.1 8.403  |0.378 |-0.03 |0.004 |-0.001        |0.003 |
1.2 8.781  |0.348 |-0.026       |0.003 |0.002 |0       |
1.3 9.129  |0.322 |-0.023       |0.005 |0      |0       |
1.4 9.451  |0.299 |-0.018       |0      |0      |0       |
1.5 9.75   |0.281 |0       |0      |0      |0      |
1.6 10.031 |0       |0       |0      |0      |0      |
First (dy/dx): 3.952

second (dy/dx): -3.74
```

15.20   Trapezoidal Rule (Section 8.5—I)

Flow-chart

   Refer to Section 14.20, page 724

Program

```
/* Trapezoidal rule.*/
#include <iostream.h>
#include <iomanip.h>
float y(float x)
{
 return 1/(1+x*x);
}
int main()
{
 float x0,xn,h,s;
 int i,n;
 cout << "Enter x0,xn,no. of subintervals" << endl;
 cin >> x0 >> xn >> n;
 cout << fixed;
 h = (xn-x0)/n;
 s = y(x0)+y(xn);
 for (i=1;i<=n-1;i++)
   s += 2*y(x0+i*h);
 cout << "Value of integral is"
      << setw(6) << setprecision(4)
      << (h/2)*s << endl;
 return 0;
}
```

> **NOTES:** $y(x)$ is the function to be integrated
> $x0$ is $x0$
>
> xN is xn.

Computer Solution of Example 8.10 (i)

```
Enter x0, xn, no. of subintervals
0 6 6
Value of integral is 1.4108
```

15.21   Simpson's Rule (Section 8.5—II)

Flow-chart

> *Refer to Section 14.21, page 726*

Program

```
/* Simpson's rule */
#include <iostream.h>
#include <iomanip.h>
float y(float x)
{
 return 1/(1+x*x);
}
int main()
{
 float x0,xn,h,s;
 int i,n;
 cout << "Enter x0,xn, no.of subintervals"
      << endl;
cin >> x0 >> xn >> n;
 cout << fixed;
 h = (xn-x0)/n;
 s = y(x0)+y(xn)+4*y(x0+h);
 for (i=3;i<=n-1;i+=2)
   s += 4*y(x0+i*h)+2*y(x0+(i-1)*h);
 cout << "Value of integral is"
      << setw(6) << setprecision(4)
      << (h/3)*s << endl;
 return 0;
}
```

> **NOTE:** $y(x)$ is the function to be integrated so that yi = y(xi) = y(x0 + i*h)

Computer Solution of Example 8.10 (ii)

```
Enter x0, xn, no. of subintervals
0 6 6
Value of integral is 1.3662
```

15.22   Euler's Method (Section 10.4)

Flow-chart

   (Refer to Section 14.22, page 727

Program

```
/*Euler's Method*/
#include <iostream.h>
#include <iomanip.h>
float df(float x,float y)
{
return x+y;
}
int main()
{
 float x0,y0,h,x,x1,y1;
 cout << "Enter the values of x0,y0,h,x" << endl;
 cin >> x0 >> y0 >> h >> x;
 cout << fixed;
 x1=x0;y1=y0;
 while(1)
 {
    if(x1>x) return 0;
    y1 += h*df(x1,y1);
    x1 += h;
    cout << "When x = "
         << setw(3) << setprecision(1)
         << x1 << " y = "
         << setw(4) << setprecision(2)
         << y1 << endl;
 }
}
```

> **NOTES:** *df(x, y)* is dy/dx
> *x0* is *xn*+0 *i.e., xn*
> *x1* is *xn*+1
> *y0* is *yn*+0 *i.e., yn*
> *y1* is *yn*+1

Computer Solution of Example 10.8

```
Enter the values of x0, y0, h, x
0 1.1 1
```

```
When x = 0.1 y = 1.10
When x = 0.2 y = 1.22
When x = 0.3 y = 1.36
When x = 0.4 y = 1.53
When x = 0.5 y = 1.72
When x = 0.6 y = 1.94
When x = 0.7 y = 2.20
When x = 0.8 y = 2.49
When x = 0.9 y = 2.82
When x = 1.0 y = 3.19
```

15.23 Modified Euler's Method (Section 10.5)

Flow-chart

> *Refer to Section 14.23, page 729*

Program

```
/* Modified Euler's Method*/
#include<iostream.h>
#include<math.h>
#include<iomanip.h>
#include<conio.h>
void main( )
{
    clrscr ( );
    int i,j;
    float x,y,x1=0.0,y1=0.0,h,ms=0.0,flag=0,y2=0.0,t=0.0;
    cout<<"\nenter the value of x";
    cin>>x;
    cout<<"enter the value of y";
    cin>>y;
    cout<<"enter the height";
    cin>>h;
    i=7;
    j=2;
    gotoxy(2,i);
    cout<<"x";gotoxy(10,i);cout<<"x+y=y1";gotoxy(28,i);
      cout<<"mean slope";gotoxy(45,i);cout<<"old y+.1(mean
      slope)new y"; while(x1<x)
    {
    i++;
```

```
            do
            {
             i++;
                       if(flag==0)
                       {
                            y1=x1+y;
                            gotoxy(2,i);cout<<x1;gotoxy(10,i);
                               cout<<y1;gotoxy(28,i);cout<<ms;
                            ms=y1;
                            y2=y+h*ms;
                            gotoxy(45,i);cout<<y2;
                            x1=x1+h;
                            flag=1;
                       }
                       else
                       {
                            ms=(y1+(x1+y2))/2.0;
                            t=y+h*ms;
                            if(y2==t)
                       {
                            y2=y+h*ms;
                            break;
                       }
gotoxy(2,i);cout<<x1;gotoxy(10,i);cout<<x1<<"+"<<x2;y2=y+h*ms;
gotoxy(28,i);cout<<ms;gotoxy(45,i);cout<<y2;
                       }
         }while(1);
         y=y2;
         cout<<"\n";
         flag=0;
    }
    getch( );
}
```

Computer Solution of Example 10.10

```
enter the value of x.3
enter the value of y1
enter the height.1
```

x	x+y=y1	mean slope	old y+.1(mean slope)new y
0	1	0	1.1
0.1	0.1 + 1.1	1.1	1.11
0.1	0.1 + 1.11	1.105	1.1105
0.1	0.1 + 1.1105	1.10525	1.110525
0.1	0.1 + 1.110525	1.105263	1.110526
0.1	0.1 + 1.110526	1.105263	1.110526
0.1	1.210526	1.105263	1.231579
0.2	0.2 + 1.231579	1.321053	1.242632
0.2	0.2 + 1.242632	1.326579	1.243184
0.2	0.2 + 1.243184	1.326855	1.243212
0.2	0.2 + 1.243212	1.326869	1.243213
0.2	1.443213	1.32687	1.387535
0.3	0.3 + 1.387535	1.565374	1.399751
0.3	0.3 + 1.399751	1.571482	1.400362
0.3	0.3 + 1.400362	1.571787	1.400392
0.3	0.3 + 1.400392	1.571803	1.400394

15.24 Runge-Kutta Method (Section 10.7)

Flow-chart

   Refer to Section 14.24, page 732

Program

```
/* Runge Kutta Method */
#include <iostream.h>
#include <iomanip.h>
float f(float x,float y)
{
 return x+y*y;
}
int main()
{
 float x0,y0,h,xn,x,y,k1,k2,k3,k4,k;
 cout << "Enter the values of x0,y0,"
      << "h,xn" << endl;
 cin >> x0 >> y0 >> h >> xn;
 x = x0; y = y0;
 cout << fixed;
```

```
while (1)
{
    if (x == xn) break;
    k1 = h*f(x,y);
    k2 = h*f(x+h/2,y+k1/2);
    k3 = h*f(x+h/2,y+k2/2);
    k4 = h*f(x+h,y+k3);
    k = (k1+(k2+k3)*2+k4)/6;
    x += h; y += k;
    cout << "When x = " << setprecision(4)"
        << setw(8) << x
        << " y = " << setw(8) << y << endl;
}
return 0;
}
```

Notes: *x0* is starting value of *x i.e.*, *x0*
xn is the value of *x* for which *y* is to be determined.

Computer Solution of Example 10.15

```
Enter the values of x0, y0, h, xn
0.0 1.0 0.2 0.2
When x = 0.1000 y = 1.1165
When x = 0.2000 y = 1.2736
```

15.25   Milne's Method (Section 10.9)

Flow-chart

  Refer to Section 14.25, page 734

Program

```
/*Milne predictor corrector*/
#include <iostream.h>
#include <iomanip.h>
#include <math.h>
float x[5],y[5],h;
float f(int i)
{
 return x[i]-y[i]*y[i];
}
void corect()
{
```

```
   y[4] = y[2]+(h/3)*(f(2)+4*f(3)+f(4));
   cout << setw(23) << ""
        << setprecision(4)
        << setw(8) << y[4]
        << setw(8) << f(4) << endl;
        << "",y[4],f(4)
}
int main()
{
 float xr,aerr,yc;
 int i;
 cout << "Enter the values of x0,xr,h,"
      << "allowed error" << endl;
 cin >> x[0] >> xr >> h >> aerr;
 cout << "Enter the value of y[i], i=0, 3" << endl;
 for (i=0;i<=3;i++) cin >> y[i];
 cout << fixed;
 for (i=1;i<=3;i++) x[i] = x[0]+i*h;
 cout << setw(5) << "x" << setw(15) << "Predicted"
      << setw(17) << "Corrected" << endl;
 cout << setw(11) << "y" << setw(10) << "f"
      << setw(7) << "y" << setw(10) << "f" << endl;
  while (1)
  {
    if(x[3] = xr) return 0;
    x[4] = x[3]+h;
    y[4] = y[0]+
              (4*h/3)*(2*(f(1)+f(3))-f(2));

    cout << setw(6) << setprecision(2) << x[4]
         << setprecision(4)
         << setw(8) << y[4]
         << setw(8) << f(4) << endl;
    corect(1);
    while (1)
    {
      yc = y[4];
      corect();
      if(fabs(yc-y[4]) <= aerr) break;
    }
    for (i=0;i<=3;i++)
    {
```

```
      x[i] = x[i+1];
      y[i] = y[i+1];
    }
  }
}
```

> **NOTE:** *x* is an array such that *x*[*i*] represents *xn*+*i* for *e.g.*, *x*[0] represent *xn*
> *y* is an array such that *y*[*i*] represents *yn*+*i*
> *xr* is the last value of *x* at which value of *y* is required
> *h* is spacing in values of *x*
> *aerr* is the allowed error in value of *y* *yc* is the latest corrected value for *y*
> *f* is the function which returns value of *y′*
> *corect* is a subroutine that calculates the corrected value of *y* and prints it.

Computer Solution of Example 10.19

```
Enter the values of x0, xr, h, allowed error
0 1.2.0001
Enter values of y[i]; i = 0, 3
0.02.0795.1762
```

	Predicted			Corrected	
X	y	f		y	f
0.80	0.3049	0.7070			
				0.3046	0.7072
				0.3046	0.7072
1.00	0.4554	0.7926			
				0.4556	0.7925
				0.4556	0.7925

15.26   Adams-Bashforth Method

Flow-chart

   Refer to Section 14.26, page 736

```
/*Adams-Bashforth Method*/
#include<iostream.h>
#include<stdio.h>
#include<malloc.h>
#include<math.h>
#include<conio.h>
void main( )
{
```

```
float *x, *y, *f;
float h;
int i,size,row;
clrscr( );
cout <<"enter the size";
cin>>size;
x=new float[size+1];
y=new float[size+1];
f=new float[size+1];
for (i=0;i<size;i++)
{
    cout<<"enter the value for x["<<i<<"]";
    cin>>x[i];
    cout<<"enter the value for y["<<i<<"]";
    cin>>y[i];
}
    h=x[1]-x[0];
// calculating values [f]
for(i=0;i<4;i++)
{
    f[i]=pow(x[i],2)*(1.0+y[i]);
}
cout<<"\nvalues for (x) (y) and (f) are\n";
row=16;
for(i=0;i<4;i++)
{
gotoxy(2,row);cout<<"x=";gotoxy(6,row);cout<<x[i];gotoxy(13,ro
w);cout<<"y"<<i-3;gotoxy(16,row);cout<<"=";gotoxy(18,row);cout
<<y[i];gotoxy(28,row);cout<<"f"<<i-3;gotoxy(32,row);cout<<"=";
gotoxy(35,row);cout<<f[i];
    row++;
}
//using predicator
    y[size]=y[size-1]+((h/24)*((55*f[size-1])-(59*f[size-
2])+(37*f[size-
3])-(9*f[size-4]))));
    x[size]=1.4;
    f[size]=pow(x[size],2)*(1.0+y[size]);
    gotoxy(2,row);cout<<"x=";
gotoxy(6,row);cout<<x[size];gotoxy(13,row);cout<<"y1";
gotoxy(16,row);cout<<"=";  gotoxy(18,row);cout<<y[size
];gotoxy(28,row);cout<<"f1";gotoxy(32,row);cout<<"=";
```

```
gotoxy(35,row);cout<<f[size];
}
```

Computer Solution of Example 10.23

```
enter the size 4
enter the value for x[0]1.0
enter the value for y[0]1.000
enter the value for x[1]1.1
enter the value for y[1]1.233
enter the value for x[2]1.2
enter the value for y[2]1.548
enter the value for x[3]1.3
enter the value for y[3]1.979


values for (x) (y) and (f) are


x = 1        y-3 = 1          f-3 = 2
x = 1.1      y-2 = 1.233      f-2 = 2.70193
x = 1.2      y-1 = 1.548      f-1 = 3.66912
x = 1.3      y0 = 1.979       f0 = 5.03451
x = 1.4      y1 = 2.572297    f1 = 7.001702
```

15.27   Solution of Laplace's Equation (Section 11.5)

Flow-chart

Refer to Section 14.27, page 740

```
/* Laplace's Equation */
#include <iostream.h>
#include <iomanip.h>
#include <math.h>
#define SQR 4
typedef float array[SQR+1][SQR+1];
void getrow(int i,array u)
{
 int j;
 cout << "Enter the values of u["
     << i << ",j], j=1, " << SQR << endl;
 for (j=1;j<=SQR;j++)
    cin >> u[i][j];
}
void getcol(int j,array u)
```

```
{
 int i;
 cout << "Enter the values of u[i," << j
      << "], i=2," << SQR-1 << endl;
 for (i=2;i<=SQR-1;i++)
    cin >> u[i][j];
}
void printarr(array u,int width,int precision)
{
 int i,j;
 for (i=1;i<=SQR;i++)
 {
   for (j=1;j<=SQR;j++)
     cout << setw(width) << setprecision(precision)
          << u[i][j];
     cout << endl;
 }
}
int main ()
{
 array u;
 float maxerr,aerr,err,t;
 int i,j,itr,maxitr;
 for (i=1;i<=SQR;i++)
  for(j=1;j<=SQR;j++)
    u[i][j]=0;
 cout << "Enter the boundary conditions" << endl;
 getrow(1,u); getrow(SQR,u);
 getcol(1,u); getcol(SQR,u);
 cout << "Enter allowed error,"
      << "maximum iterations" << endl;
 cin >> aerr >> maxitr;
 cout << fixed;
 for (itr=1;itr<=maxitr;itr++)
 {
    maxerr=0;
    for (i=2;i<=SQR-1;i++)
       for(j=2;j<=SQR-1;j++)
       {
         t=(u[i-1][j]+u[i+1][j]+
           u[i][j+1]+u[i][j-1])/4;
         err=fabs(u[i][j]-t);
```

```
      if (err > maxerr)
         maxerr = err;
      u[i][j]=t;
    }
    cout << "Iteration no. " << itr << endl;
    printarr(u,9,2);
    if (maxerr <= aerr)
  {
    cout << "After " << itr << " iterations"
         << endl
         << "The solution:-" << endl;
    printarr(u,8,1);
    return 0;
  }
 }
 cout << "Iterations not sufficient." << endl;
 return 1;
}
```

NOTES: *SQR* is the size of the square mesh
u is a 2D Array representing the square mesh
aerr is the allowed error
maxitr is the maximum allowed iterations
itr is a counter which keeps track of number of iterations performed
maxerr is the maximum error in the mesh in an iteration
err is error in a particular point of the mesh
f is the execution time format
getrow is a subroutine that inputs the ith row of the mesh
getcol is a subroutine that inputs jth column of the mesh.

Computer Solution of Exmaple 11.3 (a)

```
Enter the boundary conditions
Enter the value of u[1, j], j = 1, 4
1000    1000    1000    1000
Enter the values of u[4, j], j = 1, 4
1000    500    0  0
Enter the values of u[i, 1], i = 2, 3
2000   2000
Enter the values of u[i, 4], i = 2, 3
500   0
Enter allowed error, maximum iterations
.1    10
```

```
Iteration No. 1
1000.00          1000.00          1000.00          1000.00
2000.00           750.00           562.50           500.00
2000.00           812.50           343.75             0.00
1000.00           500.00             0.00             0.00
Iteration No. 2
1000.00          1000.00          1000.00          1000.00
2000.00          1093.75           734.38           500.00
2000.00           984.38           429.69             0.00
1000.00           500.00             0.00             0.00
Iteration No. 3
1000.00          1000.00          1000.00          1000.00
2000.00          1179.69           777.34           500.00
2000.00          1027.34           451.17             0.00
1000.00           500.00             0.00             0.00
Iteration No. 4
1000.00          1000.00          1000.00          1000.00
2000.00          1201.17           788.09           500.00
2000.00          1038.09           456.54             0.00
1000.00           500.00             0.00             0.00
Iteration No. 5
1000.00          1000.00          1000.00          1000.00
2000.00          1206.54           790.77           500.00
2000.00          1040.77           457.88             0.00
1000.00           500.00             0.00             0.00
Iteration No. 6
1000.00          1000.00          1000.00          1000.00
2000.00          1207.89           791.44           500.00
2000.00          1041.44           458.22             0.00
1000.00           500.00             0.00             0.00
Iteration No. 7
1000.00          1000.00          1000.00          1000.00
2000.00          1208.22           791.61           500.00
2000.00          1041.61           458.30             0.00
1000.00           500.00             0.00             0.00
Iteration No. 8
1000.00          1000.00          1000.00          1000.00
2000.00          1208.31           791.65           500.00
2000.00          1041.65           458.33             0.00
1000.00           500.00             0.00             0.00
After 8 iterations
The solution:-
1000.0           1000.0           1000.0           1000.0
2000.0           1208.3            791.7            500.0
```

2000.0	1041.6	458.3	0.0
1000.0	500.0	0.0	0.0

15.28   Solution of Heat Equation (Section 11.9)

Flow-chart

> *Refer to Section 14.28, page 745*

Program

```
/*Solution of parabolic equations by
Bendre Schmidt method*/
#include <iostream.h>
#include <iomanip.h>
#define XEND 8
#define TEND 5 float f(int x)
{
return 4*x-(x*x)/2.0;
}
int main()
{
 float u[XEND+1][TEND+1],h=1.0,k=0.125,
    csqr,alpha,ust,uet;
 int i,j;
 cout << "Enter the square of 'c'" << endl;
 cin >> csqr;
 alpha = (csqr*k)/(h*h);
 cout << "Enter the value of u[0,t]" << endl;
 cin >> ust;
 cout <<"Enter the value of u[" << XEND
    << ",t]" << endl;
 cin >> uet;
 cout << fixed;
 for (j=0;j<=TEND;j++)
   u[0][j]=u[XEND][j]=ust;
 for (i=1;i<=XEND-1;i++)
   u[i][0]=f(i);
 for (j=0;j<=TEND-1;j++)
   for (i=1;i<=XEND-1;i++)
     u[i][j+1]=
     alpha*u[i-1][j]
     +(1-2*alpha)*u[i][j]
     +alpha*u[i+1][j];
 cout << "The value of alpha is"
```

```
        << setw(4) << setprecision(2)
        << alpha << endl;
    cout << "The values of u[i,j] are:-"
        << endl;
    for (j=0;j<TEND;j++)
    {
        for (i=0;i<=XEND;i++)
            cout << setw(7) << setprecision(4)
                << u[i][j];
        cout << endl;
    }
    return 0;
}
```

> **Notes:** *XEND* is the ending value of x
> *TEND* is the ending value of t
> *h* is the spacing in values of x
> *k* is the spacing in values of y
> *f(x)* is value of u(x, 0)
> *csqr* is value of C²
> *alpha* is α
> *ust* is the value in the first column
> *uet* is the value in the last column.

Computer Solution of Example 11.11

```
Enter the square of "c"
4
Enter value of u(0, t)
0
Enter value of u(8, t)
0
The value of alpha is 0.50
The values of u(i, j) are:-
 0.0000  3.5000  6.0000  7.5000  8.0000  7.5000  6.0000  3.5000  0.0000
 0.0000  3.0000  5.5000  7.0000  7.5000  7.0000  5.5000  3.0000  0.0000
 0.0000  2.7500  5.0000  6.5000  7.0000  6.5000  5.0000  2.7500  0.0000
 0.0000  2.5000  4.6250  6.0000  6.5000  6.0000  4.6250  2.5000  0.0000
 0.0000  2.3124  4.2500  5.5625  6.0000  5.5625  4.2500  2.3125  0.0000
 0.0000  2.1250  3.9375  5.1250  5.5625  5.1250  3.9375  2.1250  0.0000
```

15.29   Solution of Wave Equation (Section 11.12)

Flow-chart

    Refer to Section 14.29, page 748

Program

```
/* Solution of Hyperbolic equation */
#include <iostream.h>
#include <iomanip.h>
#define XEND 5
#define TEND 5 float f(int x)
{
 return x*x*(5-x);
}
int main()
{
 float u[XEND+1][TEND+1],csqr,ust,uet;
 int i,j;
 cout << "Enter the square of 'c'" << endl;
 cin >> csqr;
 cout << "Enter the value of u(0, t)" << endl;
 cin >> ust;
 cout << "Enter the value of u("
      << XEND << ", t)" << endl;
 cin >> uet;
 cout << fixed;
 for (j=0;j<=TEND;j++)
 {
   u[0][j] = ust; u[XEND][j] = uet;
 }
 for (i=1;i<=XEND-1;i++)
    u[i][1] = u[i][0] = f(i);
 for (j=1;j<=TEND-1;j++)
    for (i=1;i<=XEND-1;i++)
      u[i][j+1] = u[i-1][j]+u[i+1][j]
                 -u[i][j-1];
 cout << "The values of u(i, j) are:-" << endl;
 for (j=0;j<=TEND;j++)
 {
  for (i=0;i<=XEND;i++)
    cout << setw(6) << setprecision(1)
         << u[i][j];
```

```
      cout << endl,
  }
  return 0;
}
```

> **NOTES:** *XEND* is the ending value of *x*
> *TEND* is the ending value of *t f(x)* is value of *u(x, 0)*
> *csqr* is value of C^2
> *ust* is the value in the first column
> *uet* is the value in the last column.

Computer Solution of Example 11.14

```
Enter the square of "c"
16
Enter value of u(0, t)
0
Enter value of u(5, t)
0
The values of u(i, j) are:-
```

0.0	4.0	12.0	18.0	16.0	0.0
0.0	4.0	12.0	18.0	16.0	0.0
0.0	8.0	10.0	10.0	2.0	0.0
0.0	6.0	6.0	-6.0	-6.0	0.0
0.0	-2.0	-10.0	-10.0	-8.0	0.0
0.0	-16.0	-18.0	-12.0	-4.0	0.0

15.30   Linear Programming—Simplex Method (Section 12.8)

Flow-chart

  Refer to Section 14.30, page 750

Program

```
/* Linear programming by simplex method */
#include <iostream.h>
#include <iomanip.h>
#define ND 2
#define NS 2
#define N (ND+NS)
#define N1 (NS*(N+1))
void init(float x[],int n)
{
  int i=0;
```

```cpp
 for (;i<n;i++) x[i] = 0;
}
int main()
{
 int i,j,k,kj,ki,bas[NS];
 float a[NS][N+1],c[N],cb[NS],th[NS],
  x[ND],cj,z,t,b,min,max;
 /* Initializing the arrays to zero */
 init(c,N); init(cb,NS);
 init(th,NS); init(x,ND);
 for (i=0;i<NS;i++) init(a[i],N+1);
 /* Now set coefficients for slack
    variables equal to one */
 for (i=0;i<NS;i++) a[i][i+ND] = 1.0;
 /* Now put the slack variables in the basis */
 for (i=0;i<NS;i++) bas[i] = ND+i;
 /*Now get the constraints
   and the objective function */
 cout << "Enter the constraints" << endl;
 for (i=0;i<NS;i++)
 {
   for (j=0;j<ND;j++)
   cin >> a[i][j];
   cin >> a[i][N];
 }
 cout << "Enter the objective function"
      << endl;
 for (j=0;j<ND;j++)
   cin >> c[j];
 cout << fixed;
 /*Now calculate cj and identify the incoming variable */
 while (1)
 {
   max = 0; kj = 0;
   for (j=0;j<N;j++)
   {
     z = 0;
     for (i=0;i<NS;i++)
        z += cb[i]*a[i][j];
     cj = c[j]-z;
     if(cj > max)
         {max = cj; kj = j;}
```

```
      }
      /* Apply the optimality test */
      if(max <= 0) break;
      /* Now calculate thetas */
      max = 0;
      for (i=0;i<NS;i++)
         if(a[i][kj]!= 0)
         {
            th[i] = a[i][N]/a[i][kj];
            if(th[i] > max) max=th[i];
         }
   /* Now check for unbounded soln. */
   if(max <= 0)
   {
      cout << "Unbounded solution";
      return 1;
   }
   /*Now search for the outgoing variable */
   min = max; ki = 0;
   for (i=0;i<NS;i++)
      if ((th[i] < min)&&(th[i]!= 0))
      {
         min = th[i]; ki = i;
      }
   /*Now a[ki][kj] is the key element*/
   t = a[ki][kj];
   /*Divide the key row by key element*/
   for (j=0;j<N+1;j++) a[ki][j] /= t;
   /*Make all other elements of key column zero */
   for (i=0;i<NS;i++)
    if(i!= ki)
    {
      b = a[i][kj];
      for (k=0;k<N+1;k++)
         a[i][k]-=a[ki][k]*b;
    }
    cb[ki] = c[kj];
    bas[ki] = kj;
}
/* Now calculating the optimum value */
for (i=0;i<NS;i++)
   if ((bas[i] >= 0) && (bas[i]<ND))
```

```
    x[bas[i]] = a[i][N];
z = 0;
for (i=0;i<ND;i++)
   z += c[i]*x[i];
for (i=0;i<ND;i++)
   cout << "x[" << setw(3) << i+1 << "] = "
        << setw(7) << setprecision(2)
        << x[i] << endl;
   cout << "Optimal value = "
        << setw(7) << setprecision(2)
        << z << endl;
 return 0;
}
```

> **NOTES:** *ND* is no. of decision variables.
> *NS* is no. of slack variables.
> *a* is the array containing Body Matrix, Unit Matrix and b_i's
> *c* is an array containing values of c_j's
> *cb* is an array containing values of c_B's
> *th* is an array containing values of θ's
> *bas* is basis. For x_i's basis contains *i*, for s_i's basis contains $i + ND$
> *ki* is the key row.
> *kj* is the key column.

Computer Solution of Example 12.4

```
Enter the constraints
4   2   80
2   5   180
Enter the objective function
3   4
x[ 1] = 2.50
x[ 2] = 35.00
Optimal value = 147.50
```

Computer Solution of Example 12.16

```
Enter the constraints
2   3   2   440
4   0   3   470
2   5   0   430
Enter the objective function
4   3   6
x[ 1] = 0.00
```

```
x[ 2] = 42.22
x[ 3] = 156.67
Optimal value = 1066.67
```

Exercises 15.1

1. Write a C++ program which prints all odd positive integers less than 100, omitting those integers divisible by 7.

2. Write a C++ program to convert a binary number to its equivalent decimal number.

3. Write a C++ program to calculate N! and use this to evaluate

$$N_{C_K} = \frac{N!}{K!(N-K)!}$$

4. Determine the number of integers n, $1 \leq n \leq 2000$, that are not divisible by 2, 3 or 5 but are divisible by 7.

5. Write a C++ program to evaluate the roots of the equation $ax^2 + bx + c = 0$.

6. Write a computer program in "C++" for finding a real root of the equation $f(x) = 0$ by the bisection method.

7. Write a C++ program to find a real root of $x^3 - 4x - 9 = 0$ using the method of false position.

8. Write an algorithm for the Newton-Raphson method to solve the equation $f(x) = 0$. Apply the same to solve $\cos x - xe^x = 0$ near $x = 0.5$ correct to three decimal places.

9. Write a C++ program to solve the following equations by the Gauss-Seidal method:
$83x + 11y - 4z = 95$; $7x + 52y + 13z = 104$; $3x + 8y + 29z = 71$.

10. With the help of a flow chart, write a C++ program to solve:
$7.5x + 3.8y + 2.9z = 15$; $3.2x + 6.8y + 7.4z = 37$; $1.3x + 2.1y + 3.2z = 7$, using the factorization method.

11. Write a complete C++ program to (i) Add two matrices (ii) Multiply two matrices.

12. Given the data:

x:	5	10	15	20	25	30
y:	17	25	30	33	36	38

Write a C++ program to fit a quadratic relation using least square criterion.

13. Write a program in C++ to estimate $f(0.6)$ by the Lagrange interpolation for the following values:

x	:	0.4	0.5	0.7	0.8
f(x):	-0.916	-0.693	-0.357	-0.223	

14. Write a C++ program to evaluate $\int_{2}^{10} x(x^2 + 2)dx$ using Simpson's rule.

15. Write "C++" program for evaluation of $\int_{0}^{4} f(x)$ Simpson's 3/8th rule.

16. Write a program in "C++" for second order Runge-Kutta method.

17. Develop a "C++" program for solving differential equations using the Runge-Kutta fourth order formulae.

18. Write a C++ program to find $y(0.8)$ for the differential equation

$dy/dx = \dfrac{1}{2}(x+y)$, given the following table using Milne's Predictor-

Corrector method:

x:	0	0.2	0.4	0.6
y:	2	2.636	3.595	4.968

19. Write a computer program in C++ to maximize

$z = 6x_1 + 4x_2$

subject to $2x1 + 3x_2 \le 100$, $4x_1 + 2x_2 \le 120$, $x_1, x_2 \ge 0$, where x_1, x_2 are the number of items to be produced.

20. Develop a computer program in C++ for Example 12.17 and hence solve it.

16

NUMERICAL METHODS USING MATLAB

Chapter Objectives

- Introduction
- An overview of MATLAB features
- 3 to 30 Programs of standard methods in MATLAB

16.1   Introduction

MATLAB is a numerical computing, fourth generation programming language built around an interactive programming environment. There is no need to compile, link, and execute after each correction, thus MATLAB programs can be developed in much shorter time than equivalent C or C++ programs. It has many built-in functions that make the learning of numerical methods much easier and interesting. Developed by Math Works, MATLAB allows matrix manipulation, plotting of functions and data, implementation of algorithms, creation of user interfaces, and interfacing with programs of other languages. Although it is numeric only, an optional toolbox uses the MuPAD symbolic engine, allowing access to other algebraic capabilities. An additional Package, Simulink, adds graphical multidomain simulation and model based design for dynamic and embedded systems. MATLAB (meaning "matrix laboratory") was created in the late 1970s, by Cleve Moler, then chairman of the computer science department at the University of New Mexico.

16.2   An Overview of MATLAB Features

Variables. Variables are defined with the assignment operator, "=". MATLAB is a weakly dynamically typed programming language. It is a weakly typed language because types are implicitly converted. It is a dynamically typed language because variables can be assigned without declaring their type, except if they are to be treated as symbolic objects and that their type can change. Values can come from constants, from computation involving values of other variables, or from the output of a function. For example:

```
>>x=17 x=
17
>>x='hat'
x=
hat
>>x=[3*4, pi/2]
x=
12.0000 1.5708
>>y=3*sin(x)
y=
-1.6097 3.0000
```

Variable names, which must start with a letter, are *case sensitive*. Hence sun and Sun represent two different variables. Variables that are defined within a MATLAB function are local in their scope. They are not available to other parts of the program and do not remain in memory after exiting the function (this applies to most programming languages). However, varables can be shared between a function and the calling program if they are declared global.

Vectors/Matrices. MATLAB is a "Matrix Laboratory" and as such it offers many ways to create vectors, matrices, and multidimensional arrays in a convenient way. In the MATLAB vernacular, a vector refers to a one-dimensional ($1 \times N$ or $N \times 1$) matrix, commonly referred to as an array in other programming languages. A matrix generally refers to a two dimensional *i.e.* $m \times n$ array where m and n are greater than or equal to one. Arrays with more than two dimensions are referred to as multidimensional arrays.

MATLAB provides a simple way to define simple arrays using the syntax: *init: increment: terminator.* For example:

```
>>a=[1:2:10]
a=
 1 3  5 7  9
```

defines a variable named "*a*" (or assigns a new value to an existing variable "*a*") which is an array consisting of elements 1, 3, 5, 7, and 9. That is, the array starts at l (init value), increments uniformly by 2 (increment value) until it reaches its final value (terminator value). The increment value can actually be left out of this syntax (along with one of the colons), to use a default value of one. For example:

```
>>a=[1:5]
a=
 1 2 3 4 5
```

Indexing is one based, which is usually the convention in mathematics, but not in some programming languages.

Matrices can be defined by separating elements of a row by blank space or comma and terminating a row by a semicolon. The list of elements should be surrounded by square brackets: [ ]. Parenthesis: ( ) are used to access elements and subarrays. They are also used to denote a function argument list.

```
>>A=[1 2 3 4; 2 3 4 5; 5 6 7 8; 3 5 2 6]
A =
    1    2    3    4
    2    3    4    5
    5    6    7    8
    3    5    2    6
>>A(3, 3)
  ans=

7
```

Sets of indices can be specified by expressions such as "2:4", which evaluates to (2, 3, 4). For example, a submatrix taken from rows 2 to 4 and columns 3 to 4 can be written as:

```
>>A(2:4, 3:4)

ans=

4 5

7 8

2 6
```

A square identity matrix of size n can be generated using the function *eye* and matrices of any size containing all zeros or ones can be generated by using functions *zeros* and *ones* respectively. For example:

```
>>A=eye(3)  A=
1 0   0

0 1   0

0 0   1

>>B=zeros(2, 3) B=
0 0   0

0 0   0

>>C=ones(3, 4)
C=
1 1   1   1

1 1   1   1

1 1   1   1
```

To know, the size of an already defined array, commands *length* and *size* are used. Most MATLAB functions can accept matrices and will apply themselves to each element. For example, mod ($2*J$, n) will multiply each element in J by 2 and then reduce each element modulo "n." MATLAB does include standard "for" and "while" loops, but using MATLAB's vectorized notation often produces code that is easier to read and faster to execute. This code excerpted from the function *magic.m,* creates a magic square M for odd values of n (MATLAB function *meshgrid* is used here to generate square matrices I and J containing $1: n$).

```
[J,I]=meshgrid (1:n);
A=mod(I+J-(n+3)/2,n);
B=mod(I+2*J-2,n);  M=n*A+B+1;
```

The apostrophe (prime) operator (') takes the complex conjugate transpose and has the same function as a transpose operator for real-valued matrices. For example:

```
>>A=[1 2 3; 3 4 5] A=
   1    2    3

   3    4    5
>>B=A'


B=

   1    3

   2    4

   3    5
```

The compatibility of dimensions must be observed while working with matrices, for example, while multiplication and extension of an existing matrix or defining another one based on it. For instance, if it is tried to annex a 4×1 matrix into the 3×1 matrix, MATLAB will reject it squarely, giving an error message.

Also, a dot (.) must be put in front of the operator for termwise (element-by-element) operations. For example:

```
>>A=[2 5 3; 3 4 6];

>>B=[3 4 8; 3 5 6];

>>C=A.*B
C=
    6    20    24

    9    20    36
```

Semicolon. Unlike many other languages, where the semicolon is used to terminate commands, in MATLAB the semicolon serves the purpose of suppressing the output of the line that it concludes.

Arithmetic Operators. All usual arithmetic operators such as (+) Addition, (−) Subtraction, (*) Multiplication and (^) Exponentiation are supported by MATLAB. Their matrix operation is as illustrated below.

```
>>A=[2 4 6; 1 2 5]
A=
    2    4    6
```

```
>>B=[1  7  8;  3  2  7]  B=
     1     7     8

     3     2     7

>>C=A+B
C=
     3    11    14

     4     4    12
>>D=A*B'


D=

    78    56

    55    42
```

There are two division operators in MATLAB: / Right division and\Left division. The right division x/y results in x divided by y, where x and y are scalars whereas the left division is equivalent to y/x. In the case where A and B are matrices, A/B returns the solution of X * A = B and A\B yields the solution of A * X = B.

Logical Operators. The various logical operators in MATLAB are (&) AND, (|) OR and (~) NOT. The related examples are shown below.

```
>>A=[2  3  8;3  2  5];

>>B=[2  4  6;  2  4  7];

>>(A>B)|(B>6)

ans=

    0     0     1

    1     0     1
```

Relational Operators. The relational operators supported by MATLAB are < Less than, > Greater than, < = Less than or equal to, > = Greater than or equal to, = Equal to, ~ = Not equal to. These operators always act element-wise on matrices; hence they result in a matrix of logical type. These operators return 1 for true and 0 for false. For example,

```
>>A=[2 6 4; 3 7 5];

>>B=[8 4 3; 2 9 7];

>>A>B

ans=
1    0    0
```

Graphics. Function *plot* can be used to produce a graph from two vectors x and y. The code is as shown below produces the following figure of sine function.

```
>>x=0:pi/100:2*pi;

>>y=sin(x);

>>plot(x,y)
```

Three dimensional graphics can be produced using the functions *surf*, *plot 3*, or *mesh*.

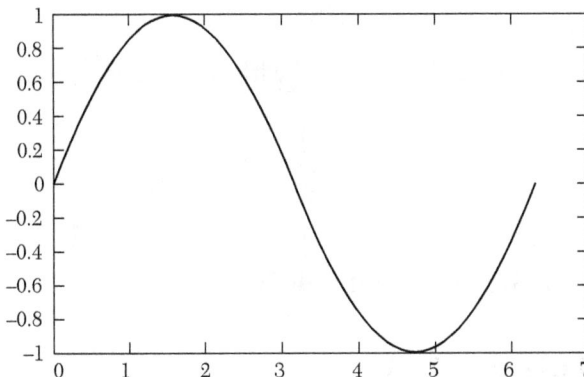

Structures. MATLAB supports structure data types. Since all variables in MATLAB are arrays, a more adequate name is "structure array," where each element of the array has the same field names. In addition, MATLAB supports dynamic field names (field look-ups by name, "field manipulations, etc.) Unfortunately, MATLAB JIT does not support MATLAB structure, therefore just a simple bundling of various variables in to a structure will come at a cost.

Function handles. MATLAB supports elements of lambda-calculus by introducing function handles, a references to functions, which are implemented either in files or anonymous/nested functions.

Classes. MATLAB supports classes, however the syntax and calling conventions are _ significantly different than in other languages, because MATLAB does not have reference data types. For example, a call to a method

```
object.method ();
```

cannot normally alter any variables of *object* variable. To create an impression that the method alters the state of variable, MATLAB toolboxes use the *evalin ( )* command, which has its own restrictions.

Object-Oriented Programming. MATLAB's support for object-oriented programming includes classes, inheritance, virtual dispatch, packages, pass-by-value semantics, and pass-by-reference semantics.

```
classdef byee
        methods
                function output1 (this)
                        disp ('byee')
                end
        end
end
```

When put in a file named m, this can be executed with the following commands:

```
>>x=byee;
>>x.output1;
Byee
```

Programs of Standard Methods In Matlab

16.3   Bisection Method (Section 2.7)

Flow-chart

   Refer to Section 14.3, page 674

Program

```
function[]=Bisection_Method()
clc
```

```
itr=0;
a=input('Enter the value of a:');
b=input('Enter the value of b:');
aerr= input('Enter the allowed error:');
maxitr=input {'Enter the maximum Iterations:');
  [x itr]=bisect(a,b,itr);
while(itr<maxitr)
  if(f(a)*f(x)<0)
        b=x;
    else
        a=x;
    end
[xl itr]=bisect(a,b,itr);
if(abs(xl-x)<aerr)
        fprintf{'After %d iteration ,root = %f \n',itr,xl)
        return
end
x=xl;
end
fprintf('Iterations not sufficient,solution does not
        converge \n);
function[x itr_r]=bisect(a,b,itr)
        if nargin <2, b=2; end
        x=(a+b)/2;
        itr_r=itr+1;
        fprintf('Iteration no. %d X = %f \n',itr,x)
end

        function[y]=f(x)
                y=(x*x*x-4*x-9);
        end
end
```

> **NOTES:** a, b are the limits in which the root lies
> aerr is the allowed error
> itr is the counter which keeps track of the number of iterations performed
> maxitr is the maximum number of iterations to be performed
> x is the value of root at nth iteration.
> xl is the value of the root at $(n + 1)$th iteration
> Function Bisect:
> Purpose: Performs and prints the result of one iteration
> Variables: x is the result of the current iteration.

Computer Solution of Example 2.15 (a)

```
Enter the value of a:3
Enter the value of b:2
Enter the allowed error:0.0001
Enter the maximum Iterations:20
Iteration no. 0 X=2.500000
Iteration no. 1 X=2.750000
Iteration no. 2 X=2.625000
Iteration no. 3 X=2.687500
Iteration no. 4 X=2.718750
Iteration no. 5 X=2.703125
Iteration no. 6 X=2.710938
Iteration no. 7 X=2.707031
Iteration no. 8 X=2.705078
Iteration no. 9 X=2.706055
Iteration no. 10 X=2.706543
Iteration no. 11 X=2.706299
Iteration no. 12 X=2.706421
Iteration no. 13 X=2.706482
After 14 iteration, root=2.706482
```

16.4   Regula-Falsi Method (Section 2.8)

Flow-chart

   Refer to Section 14.4, page 676

Program

```
function[]=Regula_Falsi()
clc
clear all itr=0;
x0=input('Enter the value of x0:');
xl=input('Enter the value of xl:');
aerr=input('Enter the allowed error:');
maxitr=input ('enter the maximum no. of iterations:');
  [x2 itr]=regula(x0,xl,f(x0),f(xl),itr);
while(itr<maxitr)
 if(f(x0)*f(x2)<0)
      xl=x2;
else
      x0=x2;
   end
```

```
    [x3 itr]=regula(x0,xl,f(x0),f(xl),itr);
    if(abs(x3-x2)<aerr)
        fprintf('After %d iteration, roots %f\n',itr,x3)
        return
    end
x2=x3;
end
fprintf('Iterations not sufficient, solution does not
        converge \n');
function [x itr_r]=regula(x0,x1,fx0,fx1,itr)
        x=x0-((x1-x0)/(fx1-fx0)) *fx0;
        itr_r=itr+1;
            fprintf ('iteration no. %d X=%f/n',itr,x)
end
    function [y]=f (x)
            y=(cos (x)-x*exp(x));
        end
```

> **NOTES:** $f(x) = 0$ is the equation whose root is to be found
> $x0$, x_1 are units in which root lies
> *aerr* is allowed error
> *maxitr* is maximum number of iterations to be performed
> *itr* is the counter which keeps track of the number of iterations performed
> x_2 is value of root at nth iteration
> x_3 is the value of root at $(n + 1)$ th iteration
> *Function Regula:*
> Purpose: Performs and prints the result of one iteration
> Variables: x is value of root at nth iteration
> fx_0, fx_1 are value of $f(x)$ at x_0 and x_1, respectively.

Computer Solution of Example 2.20

```
Enter the value of x0:0
Enter the value of xl:1
Enter the allowed error:0.0001
enter the maximum no. of iterations:20
iteration no. 0 X=0.314665
iteration no. 1 X=0.446728
iteration no. 2 X=0.494015
iteration no. 3 X=0.509946
iteration no. 4 X=0.515201
iteration no. 5 X=0.516922
```

```
iteration no. 6 X=0.517485
iteration no. 7 X=0.517668
iteration no. 8 X=0.517728
After 9 iteration, roots 0.517728
```

16.5   Newton Raphson Method (Section 2.11)

Flow-chart

> *Refer to Section 14.5, page 679*

Program

```
function[]=Newton_Raphson ()
clc
clear all
f=inline('x*log10(x)-1.2');
  df=inline('log10(x)+.43429');
x0=input('Enter the value of x0:');
aerr=input('Enter the allowed error:');
maxitr=input ('enter the maximum no. of iterations:');
for itr=1:1:maxitr
h=f(x0)/df (x0);
x1=x0-h;
  fprintf('iteration no. %d X=%f \n',itr,x1)
if(abs(h)<aerr)
    fprintf('After %d iteration, roots %f\n',itr,x1)
  return
  end
x0=x1;
end
fprintf('Iterations not sufficient,solution does not
        converge \n')
end
```

NOTES: $F(x) = 0$ is the equation whose root is to be found
$df(x)$ is the derivative of $f(x)$ w.r.t. x *aerr* is allowed error
maxitr is maximum number of iterations to be performed
itr is the counter which keeps track of the number of iterations performed
$x0$ is value of root at nth iteration
$x1$ is the value of root at $(n + 1)$th iteration

Computer Solution of Example 2.32

```
Enter the value of x0:2
Enter the allowed error:0.000001
```

```
enter the maximum no. of iterations:10
iteration no. 1 X=2.813170
iteration no. 2 X=2.741109
iteration no. 3 X=2.740646
iteration no. 4 X=2.740646
After 4 iteration, roots 2.740646
```

16.6   Muller's Method (Section 2.13)

Flow-chart

Refer to Section 14.6, page 681

Program

```
function[]=Mullers_method ()
clc
clear all
I = 3;
y=inline{'cos(x)-x*exp(x) ' );
disp('Enter the initial approximations');
for i=I-2:1:3
 x(i)=input('');
end
aerr= input{'Enter the allowed error:');
maxitr=input ('enter the maximum no. of iterations:');
  for itr=1:1:maxitr
    li=(x(I)-x(1-1) )/(x (1-1)-x{I-2));
  di=(x(I)-x(I-2))/{x(1-1)-x{I-2));
  mu=y(x(I-2))*li*li-y(x(1-1))*di*di+y(x(I))*(li+di);
s=sqrt((mu*mu-4*y{x(I))*di*li*(y(x(1-2))*
       li-y(x(1-1))*di+y(x(I)))));
if(mu<0)
     l=(2*y(x(I)>*di)/(-mu+s);
else
     l=(2*y(x(I)}*di)/(-mu-s);
end
     x(I+1)=x(I)+l*(x(I)-x(I-l));
  fprintf{'iteration no. %d X = %f \n',itr,x(1+1)}
if(abs(x(1+1)-x(I))<aerr)
   fprintf('After %d iterations, the solution is %f\n',itr,x(1+1))
    return
  end
```

```
for i=I-2:1:3
   x(i)=x(i+1);
  end
end
fprintf{'Iterations not sufficient,solution does not
        converge \n');
```

> **NOTES:** $y(x) = 0$ is the equation whose root is to be found
> x is na array which holds the three approximations to the root and the new improved value
> I is defined as 3 in the program. This has been done because in MATLAB, array subscripts always start from 1 and cannot be negative. Use of 1 facilitates more readable expressions. For e.g., $x[0]$ can be written as $x[I-3]$ which looks more like X_{i-3}, which it actually represents.
> li is λ_i
> di is δ_i
> mu is μ_i
> s is $\sqrt{[\mu_i^2 - 4yi\,\delta_i\,\lambda_i\,(y_{i-2}\,\lambda_i - y_{i-1}\,\delta_i + y_i]}$
> l is λ

Computer Solution of Example 2.34

```
Enter the initial approximations
-1
0
1
Enter the allowed error:0.0001
enter the maximum no. of iterations:10
iteration no. 1 X=0.441517
iteration no. 2 X=0.512546
iteration no. 3 X=0.517693
iteration no. 4 X=0.517757
After 4 iterations, the solution is 0.517757
```

16.7 Multiplication of Matrices [Section 3.2 (3)4]

Flow-chart

> *Refer to Section 14.7, page 684*

Program

```
clc
   Ml=input('Enter the element of first matrix');
```

```
   M2=input{'Enter the element of second matrix');
   mul=M1*M2-M2;
disp('Result after multiplication is:')
disp(mul)
```

Computer Solution of Example 3.7

```
Enter the element of first matrix [0 1 2; 1 2 3; 2 3 4]
Enter the element of second matrix [1 -2; -1 0; 2 -1]
Result after multiplication is:
3    -2
5    -5
7    -8
```

16.8   Gauss Elimination Method [Section 3.4 (3)]

Flow-chart

Refer to Section 14.8, page 687

Program

```
function[]=gauss_elimination_method()
clc
 N=4;
 a=input('Enter the element of matrix:-')
 for j=1:N-1
 for i=j+1:N
      t=a(i,j)/a (j,j);
      for k=1:N+1
      a(i,k)=a(i,k)-a(j,k)*t;
      end
 end
 end
for i=1:N
    for j=1:N+1
        fprintf('%8.4f',a(i,j));
    end
 fprintf('\n');
 end
   for i=N:-1:1
       s=0;
       for j=i+1:N
           s=s+a(i,j+1).*x(j);
   end
```

```
      x(i)=(a(i,N)-s)/a (i,i);
  end
end
```

> **Notes:** N is the number of unknowns
> a is an array which holds the augmented matrix
> x is an array which will contain values of unknowns
> i, j, k are loop control variables.

Computer Solution of Example 3.19

```
Enter the element of matrix:-[10 -7 3 5 6; -6 8 -1 -4 5; 3 1 4
11 2; 5 -9 -2 4 7]
10.0000 -7.0000    3.0000       5.0000      6.0000
0.0000   3.8000    0.8000      -1.0000      8.6000
0.0000   0.0000    2.4474      10.3158     -6.8158
0.0000   0.0000    0.0000       9.9247      9.9247
x=
-3.3947 -0.6842 7.0000 1.0000
```

16.9   Gauss-Jordan Method [Section 3.4 (4)]

Flow-chart

Refer to Section 14.9, page 689

Program

```
function[]=gauss_jordan_method()
N=4;
a=input('Enter the element of matrix:-\n');
for j=1:N
  for i=1:N
        if(i~=j)
                t=a(i,j)/a(j,j);
                for k=1:N+1
                        a(i,k)=a(i,k>-a(j,k).*t;
                end
        end
    end
end
fprintf('\nThe diagonal matrix is:-\n')
disp(a)
fprintf('\nThe solution is:-\n')
for i=1:N
```

```
        fprintf('x[%d]=%f\n',i,a(i,N + 1)./a(i,i));
end
```

> **Notes:** *a* is an array which holds the Augmented Matrix
> *N* is the number of unknowns *e.g.* if it is a 3 × 3 system of equations,
> *N* = 3 and if 5 × 5 system take N = 5
> *i, j, k* are loop variables.

Computer Solution of Example 3.22

```
Enter the element of matrix:-
[10 -7 3 5 6; -6 8 -1 -4 5; 3 1 4 1 1 2; 5 -9 -2 4 7]
The diagonal matrix is:
    10.0000      0          0          0      50.00.00
        0      3.8000       0          0      15.2000
        0        0       2.4474        0     -17.1316
        0        0          0       9.9247     9.9247
The solution is:-
x[1]=5.000000
x[2]=4.000000
x[3]=-7.000000
x[4]=1.000000
```

16.10 Factorization Method [Section 3.4 (5)]

Flow-chart

   Refer to Section 14.10, page 691

Program

```
function[]=Factorization_method()
clc
clear all
N=3;
u=zeros(N,N);
v=zeros(N,1);
x=ones(N,1);
a=input('Enter the element of matrix');
b=a(:,N);
l=zeros(N);
for m=1:N
    urow(m)
    lcol(m)
end
```

```
disp(u);
disp(1);
for i=1:N
    s=0;
    for j=1:i-1
        s=s+[l(i,j)*v(j)];
     end
       v(i)=b(i)-s;
end
for i=N:-1:1
    s=0;
  for j=i+1:N
        s=s+[u(i,j)*x(j)];
  end
  x(i)=(v(i)-s)/u(i,i);
  end
disp(x)
function[]=urow(i)
   for j=i:N
   s=0;
   for k=1:N
      s=s+[u(k,j).*l(i,k)];
      u(i, j)=a (i, j) -s;
end
end
end
function[]=lcol(j)
for i=j:N
    s=0;
     for k=1:j
     s=s+[u(k,j).*l(i,k)];
     if i==j
          1(i,j)=1;
     else
     1(i,j)=(a(i,j)-s)/u(j,j);
  end
  end
end
end
end
```

> **NOTES:** $u_{ij} = a_{ij} \sum_{k=1}^{i-1} (u_{kj} l_{ik})$
>
> $l_{ij} = \left[a_{ij} - \sum_{k=1}^{j-1} (u_{kj} l_{ik}) \right] l j_{ij}$
>
> N is the number of unknowns
> l is the lower triangular matrix
> u is the upper triangular matrix
> a is the coefficient matrix
> b is the constant matrix (column matrix)
> v is a matrix such that lv = b
> x will contain the values of unknowns
> i, j, m are loop control variables

Computer Solution of Example 3.23

```
Enter the element of matrix [3 2 7 4; 2 3 1 5; 3 4 1 7]
   3.0000        2.0000         7.0000
        0        1.6667        -3.6667
        0             0        -1.6000
   1.0000             0              0
   0.6667        1.0000              0
   1.0000        1.2000         1.0000
   0.8750
   1.1250
  -0.1250
```

Computer Solution of Example 3.24

```
Enter the element of matrix [10 -7 3 5 6; -6 8 -1 -4 5; 3 1 4
11 2; 5 -9 -2 4 7]
   10.0000       -7.0000        3.0000        5.0000
        0        3.8000        0.8000       -1.0000
        0             0        2.4474       10.3158
        0             0             0        9.9247
   1.0000             0             0             0
  -0.6000        1.0000             0             0
   0.3000        0.8158        1.0000             0
   0.5000       -1.4474       -0.9570        1.0000
   5.000
   4.000
  -7.000
   1.000
```

16.11   Gauss Siedel Iteration Method [Section 3.5 (2)]

Flow-chart

Refer to Section 14.11, page 695

Program

```
function[]=Gauss_Seidal_Method()
clear all clc
a=input('Enter the element of matrix:\n');
aerr=input ('Enter the allowed error:');
maxitr=input ('enter the maximum no. of iterations:');
N=4;
x=zeros(1,N);
fprintf('iterations  x[1]  x[2]  x[3]  x[4]\n'
for itr=1:maxitr
      maxerr=0;
      for i=1:N
            s=0;
            for j=1:N
                if (j~=i)
                      s=s+a(i,j)*x(j);
                end
        end
            t=(a(i,N+1)-s)/a(i,i);
            err=abs(x(i)-t);
            if(err>maxerr)
                    maxerr=err;
            end
            x(i)=t;
      end
      fprintf ('%d',itr)
        for i=1:N
        fprintf('%f',x(i))
        end
        fprintf ('\n')
        if(maxerr<aerr)
              fprintf('Converges in %d, iteration \n',itr)
                  for i=1:.N
                        fprintf('x(%d)=%2.4f \n',i,x(i))
                  end
              return
        end
```

```
    end
  fprintf('Solution does not converge, iteration not sufficient
\n')
  return
end
```

> **NOTES:** N is the number of unknowns
> a is an array which holds the augmented matrix
> x is an array which hold the values of unknowns
> aerr is allowed error
> maxitr is the maximum no. of iterations to be performed
> itr is the counter which keeps track of number of iterations performed
> err is the error in value of xi
> maxerr is maximum error in any value of xi after an iteration

Computer Solution of Example 3.28

```
Enter the element of matrix:
[20 1 -2 17; 3 20 -1 -18; 2 -3 20 25]
Enter the allowed error:0.0001
enter the maximum no. of iterations:10
```

iterations	x[1]	x[2]	x[3]	x[4]
1	0.850000	-1.027500	1.010875	
2	1.002463	-0.999826	0.999780	
3	0.999969	-1.000006	1.000002	
4	1.000001	-1.000000	1.000000	

```
Converges in 4, iteration
x(1)=1.0000
x(2)=-1.0000
x(3)=1.0000
```

Computer Solution of Example 3.30

```
Enter the element of matrix:
[10 -2 -1 -1 3;-2 0 -1 -1 15;-1 -1 10 -2 27;- 1 -1 -2 10 -9]
Enter the allowed error:0.0001
enter the maximum no. of iterations:15
```

iterations	x[1]	x[2]	x[3]	x[4]
1	0.300000	1.560000	2.886000	-0.136800
2	0.886920	1.952304	2.956562	-0.024765
3	0.983641	1.989908	2.992402	-0.004165
4	0.996805	1.998185	2.998666	-0.000768
5	0.999427	1.999675	2.999757	-0.000138
6	0.999897	1.999941	2.999956	-0.000025

16.2 An Overview of MATLAB Features

Variables. Variables are defined with the assignment operator, "=". MATLAB is a weakly dynamically typed programming language. It is a weakly typed language because types are implicitly converted. It is a dynamically typed language because variables can be assigned without declaring their type, except if they are to be treated as symbolic objects and that their type can change. Values can come from constants, from computation involving values of other variables, or from the output of a function. For example:

```
>>x=17 x=
17
>>x='hat'
x=
hat
>>x=[3*4, pi/2]
x=
12.0000 1.5708
>>y=3*sin(x)
y=
-1.6097 3.0000
```

Variable names, which must start with a letter, are *case sensitive.* Hence sun and Sun represent two different variables. Variables that are defined within a MATLAB function are local in their scope. They are not available to other parts of the program and do not remain in memory after exiting the function (this applies to most programming languages). However, varables can be shared between a function and the calling program if they are declared global.

Vectors/Matrices. MATLAB is a "Matrix Laboratory" and as such it offers many ways to create vectors, matrices, and multidimensional arrays in a convenient way. In the MATLAB vernacular, a vector refers to a one-dimensional ($1 \times N$ or $N \times 1$) matrix, commonly referred to as an array in other programming languages. A matrix generally refers to a two dimensional *i.e.* $m \times n$ array where m and n are greater than or equal to one. Arrays with more than two dimensions are referred to as multidimensional arrays.

MATLAB provides a simple way to define simple arrays using the syntax: *init: increment: terminator.* For example:

```
>>a=[1:2:10]
a=
 1 3 5 7 9
```

```
          end
    fprintf('\n*)
          if((errv<=aerr)&&(maxe<=aerr))
          fprintf{'Converges in %d iterations \n',itr);
          fprintf('Largest eigenvalue=%1.2f \n',e);
          fprintf('Eigen Vector:-\n');
          fprintf('%1.2f \n',x);
          return end
          end
end
```

NOTES: N is the number of rows (or columns) in square matrix
a is the square matrix
x is the eigenvector at nth iteration
r is the eigenvector at $(n + 1)$th iteration
e is the eigenvalue at nth iteration
t is the eigenvalue at $(n + 1)$th iteration
aerr is allowed error in eigenvalue and eigenvector
maxitr is the maximum number of iterations to be performed
errv is the error in eigenvalue
itr, i are loop control variables

Computer Solution of Example 4.11

```
Enter the element of matrix:
[2-1 0;-1 2-1;0-1 2]
Enter the initial approximation to the eigenvector:
[10 0]
Enter the allowed error:0.01
enter the maximum no. of iterations:10
  itr No.        Eigenvalue                    Eigenvector
      1       2.000000    1.000000      -0.500000    0.000000
      2       2.500000    1.000000      -0.800000    0.200000
      3       2.800000    1.000000      -1.000000    0.428571
      4       3.428571    0.875000      -1.000000    0.541667
      5       3.416667    0.804878      -1.000000    0.609756
      6       3.414634    0.764286      -1.000000    0.650000
      7       3.414286    0.740586      -1.000000    0.673640
      8       3.414226    0.726716      -1.000000    0.687500
      9       3.414216    0.718593      -1.000000    0.695621
Converges in 9 iterations
Largest eigenvalue=3.41
Eigenvector:-
```

```
0.72
-1.00
0.70
```

16.13   Method of Least Squares (Section 5.5)

Flow-chart

   Refer to Section 14.13, page 703

Program

```
function[]=Least_square_Method()
clear all
clc
augm=zeros(3,4);
n=input('Enter the number of pair of observation value:-\n');
augm(1,1)=n;
for i=1:n
      fprintf('Pair no. %d \n',i)
       x=input(' ');
       xsq=x(1)*x(1);
       augm(1,2)=augm(1,2)+x(1);
          augm(1,3)=augm(1,3)+xsq;
             augm(2,3)=augm(2,3)+x(1)*xsq;
                augm(3,3)=augm(3,3)+xsq*xsq;
                   augm(1,4)=augm(1,4)+x(2);
          augm(2,4)=augm(2,4)+x(1)*x(2);
augm(3,4)=augm(3,4)+xsq*x(2);
  end
     augm(2,2)=augm(1,3);
augm(3,2)=augm(2,3);
augm(2,1)=augm(1,2);
augm(3,1)=augm(2,2);
disp('The augmentd matrix is:-')
disp(augm)
for j=1:3
    for i=1:3
        if(i~=j)
             t=augm(i,j)/augm(j,j);
             for k=1:4
                 augm(i,k)=augm(i,k)-augm(j,k)*t;
             end
```

```
        end
    end
end
        a=augm(1,4)/augm(1,1);
        b=augm(2,4)/augm(2,2);
        c=augm(3,4)/augm(3,3);
        fprintf('a=%f b=%f c=%f \n',a,b,c)
```

> **Notes:** augm is the augmented matrix
> *n* is the number of data points

Computer Solution of Example 5.7

```
Enter the number of pair of observation value:-
7
Pair no.1
[1 1.1]
Pair no.2
[1.5 1.3]
Pair no.3
[2 1.6]
Pair no.4
[2.5 2]
Pair no.5
[3 2.7]
Pair no.6
[3.5 3.4]
Pair no.7
[4 4.1]
The augmentd matrix is:-
7.0000        17.5000       50.7500       16.2000
17.5000       50.7500       161.8750      47.6500
50.7500       161.8750      548.1875      154.4750
a=1.035714 b=-0.192857 c=0.242857
```

16.14   Method of Group Averages (Section 5.9)

Flow-chart

   Refer to Section 14.14 , page 706

Program

```
function []=Method_Of_Averages()
clc
```

```
format compact
format short g
n=input('Enter the No of Observations:-');
 t=input('Enter the different values of t:-')
disp(The Values of t are:')
disp(t);
r=input('Enter the Corresponding values of r')
disp('The Values of r are:')
disp(r);
  ts1=0;rs1=0;ts2=0;rs2=0;
  for i=1:(n/2)
    ts1=ts1+t(i);
    rs1 = rs1+r(i);
end
  for i=(n/2)+1:n
    ts2=ts2+t(i);
    rs2=rs2+r(i);
end
  x1=ts1/(n/2);
y1=rs1/(n/2);
x2=ts2/(n/2);
y2=rs2/(n/2);
b=(y2-y1)/(x2-x1);
a=y1-(b*x1);
disp('The values of a&b comes out to be:')
a
b
end
```

Computer Solution of Example 5.16

```
Enter the No of Observations:-8
Enter the different values of t:-[40 50 60 70 80 90 100 110]
t=
40 50 60 70 80 90 100 110
The Values of t are:
40 50 60 70 80 90 100 110
Enterthe Corresponding values of r[1069.1 1063.6 1058.2 1052.7
1049.3 1041.8 1036.3 1030.8]
r=
Columns 1 through 7
1069.1 1063.6 1058.2 1052.7 1049.3 1041.8 1036.3
Column 8
```

```
1030.8
The Values of r are:
Columns 1 through 7
1069.1 1063.6 1058.2 1052.7 1049.3 1041.8 1036.3
Column 8
1030.8
The values of a&b comes out to be:
a=1090.3
b=-0.53375
```

16.15   Method of Moments (Section 5.11)

Flow-chart

   Refer to Section 14.15, page 708

Program

```
function [ ]=Method_Of_Moments()
clc
format compact
n=input('Enter the No of Observations:-');
x=input('Enter the different values of x:-');
y=input('Enter the Corresponding values of y')
h=x(2)-x(1);
xlyt=0;yt=0;
for i=1:n
  yt=yt+y(i);
  xlyt=xlyt+x(i).*y(i);
end
ml=h.*yt;
m2=h.*xlyt;
11=(-(h/2)+x(1));
12=((h/2)+x(n));
cl=12-11;
c2=((12.*12)-{11.*11))/2;
c3=((12.*12.*12)-(11.*11.*11))/3;
fprintf('The Observed Equations are:\n')
fprintf('%5.2fa+%5.2fb=%5.2f\n *%5.2fa+%5.2fb=%5.2f\
        n',cl,c2,ml,c2,c3,m2)
d=c2/cl;
dl=d*cl;
d2=d*c2;
```

```
ml=d*ml;
b=(m2-ml)/(c3-d2);
a=(ml-(d2*b))/dl;
fprintf('\non solving these equations we get a=%5.2f & b=%5.2f\
n',a,b)
fprintf('hence the required equation is: y = %5.2f +%5.2fx\
n',a,b)
end
```

Computer Solution of Example 5.20

```
Enter the No of Observations:-4
Enter the different values of x:-[1 2 3 4]
Enter the Corresponding values of y [16 19 23 26]
 y=16  19 23    26
The Observed Equations are:
4.00a+10.00b=84.00
*10.00a+30.33b=227.00
On solving these equations we get a= 13.03 & b= 3.19
hence the required equation is: y = 13.03 + 3.19x
```

16.16   Newton's Forward Interpolation Formula (Section 7.2)

Flow-chart

   Refer to Section 14.16, page 714

Program

```
function []=Newtons_Forward_Interpolation_Formula()
clc
format compact
MAXN=100;
ORDER=4;
nr=1;,dr=1;
n=input('Enter the value of n:-');
  ax=input('Enter the values in form of x:-');
  ay=input('Enter the values in form of y:-');
  disp ([ax'     ayl])
x=input('Enter the value of x for which value of y is wanted:-')
h=ax(2)-ax(1);
for i=1:n
    diff(i,2)=ay(i)-ay(i);
end
for j=3:ORDER+1
```

```
    for i=1:n-j
    diff(i, j)-diff(i+1,j-1)-diff(i,j-1);
       end
end
i=1;
while(~(ax(i)>x))
i=i+1;
end
i=i-1;
p=(x-ax(i))/h;
yp=ay(i);
for k=2:ORDER+1
    nr=nr*(p-k+1);
    dr=dr-*'k;
    yp=yp+((nr/dr)*diff(i,k));
end
fprintf('%4.%f \n',yp)
end
```

Notes: MAXN is the maximum value of N
ORDER is the maximum order in the difference table
ax is an array containing values of *x* $(x_0, x_1,...,xn)$
ay is an array containing values of *y* $(y_0, y_1,...,yn)$
diff is a 2D array containing the difference table
h is spacing between values of *X*
x is value of *x* at which value of *y* is wanted
yp is calculated value of *Y*
nr is numerator of the terms in expansion of y_p
dr is denominator of the terms in expansion of y_p

Computer Solution of Example 7

```
Enter the value of n:-6
Enter the values in form of x:-[100 150 200 250 300 350 400]
Enter the values in form of y:-[ 10.63 13.03 15.04 16.81 18.42
19.90 21.27]
       100     10.63
       150     13.03
       200     15.04
       250     16.81
       300     18.42
       350     19.9
       400     21.27
```

```
Enter the value of x for which value of y is wanted:-218
x=218 15.0
```

16.17 Lagrange's Interpolation Formula (Section 7.12)

Flow-chart

> *Refer to Section 14.17, page 714*

Program

```
function[] = Lagranges_Interpolation_Formula()
clc
MAX=100;
n=input('Enter value of n:-');
ax=input{'Enter values of x:-');
ay=input{'Enter values of y:-');
x=input{'Enter value of which y is wanted:-');
y=0;
for i=1:n+1
    dr=1;
    nr=1;
    for j=1:n+1
        if(j~=i)
            nr=nr*(x-ax(j));
            dr=dr*(ax(i)-ax(j));
        end
    end
    y=y+((nr/dr)*ay(i));
end
fprintf('When x=%4.1f ,y=%4.1f \n',x,y);
end
```

> **Notes:** MAX is the maximum value of n
> ax is an array containing values of x $(x_0, x_1,....xn)$
> ay is an array containing values of y $(y_0, y_1,...,y_n)$
> x is value of x at which value of y is wanted
> y is calculated value of y
> nr is numerator of the terms in expansion of y
> dr is denominator of the terms in expansion of y

Computer Solution of Example 7.17

```
Enter value of n:-4
Enter values of x:-[5 7 1113 17]
Enter values of y:-[150 392 1452 2366 5202]
Enter value of which y is wanted:-9
When x=9.0,y=810.0
```

16.18   Newton's Divided Difference Formula (Section 7.14)

Flow-chart

   Refer to Section 14.18, page 716

Program

```
function [] = Newtons_ Divided- Difference Formula()
clc
n=input('Enter value of observation n:-'};
x=input('Enter values of x:-');
y=input('Enter values of y:-');
k=input('Enter value of which y is wanted:-');
j=i;
f=y(1);
f2=0;
while(n~=l)
    for i=l:n-l
        p(i)=(y(i+l)-y(i))/(x(i+j)-x(i));
        y(i)=p(i);
    end
    fl=l;
    for i=l:j
      fl=fl*(k-x (i));
    end
    f2-f2+(y(l)*fl);
    n=n-l;
    (j=j+1);
end
  f=f+f2;
  fprintf('f(%d)=%d\n',k,f)
end
```

Computer Solution of Example 7.23

```
Enter value of observation n:-5
```

```
Enter values ofx:-[5 7 11 13 17]
Enter values of y:-[150 392 1452 2366 5202]
Enter value of which y is wanted:-9
f(9)=810
```

16.19 Derivatives Using Forward Difference Formula [Section 8.2]

Flow-chart

Refer to Section 14.19, page 718

Program

```
function[]=Forward_Difference_Formula()
clc
v=[0 0 1 1 11/12 5/6 137/180];
max=8;
x=[1 1.1 1.2 1.3 1.4   1.5 1.6];
y=[7.989 8.403 8.781   9.129 9.451   9.750 10.031];
xval=1.1;
disp(['The values of x are: ', num2str(x)]);
disp(['The values of y are: ', num2str(y)]);
disp(['The value of x for evaluation is: ', num2str{xval)]);
for i=0<max
if(x(i+1)>=xval)
    pos=i+1;
    break;
end
end
x0=x(pos);
y0=y(pos);
fprintf(' \n x0 is %f y0 is %f at %d' , x0, y0, pos)
h=x(2)-x{1);
p=((xval-x0)/h);
if(pos<max)
    fact=1;
    l=max-pos;
    fprintf('\n' );
    for i=0<l
        for j=0:i
            tmp((i+1)*l+(j+1))=0;
        end
```

```
    fprintf('\n')
    end
    fprintf('\n size of new array %d \n',l);
    i=0;
j=pos;
    while(i<l)
    tmp(i+1)=y(j);
    i=i+1;
    j=j+1;
    end
    fprintf('\n');
    for i=1<l
        for j=0<l
          tmp((i+1)*l+(j+1))=tmp((i)*l+(j+2))-tmp((i)*l+(j+1));
          end
    end
    fprintf('\n values are \n');
    for i=0<l
        for j=0<l
            fprintf{'%f\t|',tmp((j+1)*l+(i+1)));
        end
            fprintf('\n')
    end
    sum=0;
    k=1;
    for i=1<l
        sum=sum+((1.0/(i+1))*tmp((i+1)*l+0))*k;
        k=-k;
    end
    fprintf('\n\n first (dy/dx):%f',sum/h);
    sum=0;
    fact=1;
    k=1;
    for i=2<l
        sum=sum+(v((i+1))*tmp((i+l)*l+0)*k);
        k=-k;
    end
    fprintf('\n\n second (dy/dx): %f',sum/(h^2))
end
```

```
end
```

Computer Solution of Example 8.1

The values of x are:	1	1.1	1.2	1.3	1.4	1.5	1.6
The values of y are:	7.989	8.403	8.781	9.129	9.451	9.75	10.031

The value of x for evaluation is: 1.1
x_0 *is 1.100000 y0 is 8.403000 at 1*
size of new array 6
first (dy/dx): 3.952600
second (dy/dx): –3.741200

16.20   Trapezoidal Rule (Section 8.5-1)

Flow-chart

> *Refer to Section 14.20, page 724*

Program

```
function[]=Trapezoidal_Rule()
format compact
clc
y=inline('1/(1+x.*x)');
x0=input('Enter x0:');
xn=input('Enter xn:');
n=input('Enter no of subintervals:');
h=(xn-x0)/n;
s=y(x0)+y(xn);
for i=1:n-1
    s=s+2.*y(x0+i.*h);
end
fprintf('The value of integral is: %f\n',h/2.*s)
end
```

NOTES: $y(x)$ is the function to be integrated
x_0 is x_0
x_n is x_n

Computer Solution of Example 8.10 (i)

```
Enter x0:0
Enter xn:6
Enter no of subintervals:6
The value of integral is:1.410799
```

16.21 Simpson's Rule (Section 8.5-II)

Flow-chart

Refer to Section 14.21, page 736

Program

```
function[]=Simpsons_Rule()
clc
y=inline('1/(1+x.*x)');
x0=input('Enter x0:');
xn=input('Enter xn:');
n=input('Enter no of subintervals:');
h=(xn-x0)/n;
s=y(x0)+y(xn)+4.*y(x0+h);
for i=3:2:n-1
    s=s+4.*y(x0+i.*h)+2.*y(x0+(i-1).*h);
end
fprintf('The value of integral is: %f\n',h/3.*s)
```

NOTE: $y(x)$ is the function to be integrated so that $yi = y(y_i) = y(x_0 + i*h)$

Computer Solution of Example 8.10 (ii)

```
Enter x0:0
Enter xn:6
Enter no of subintervals:6
The value of integral is: 1.366173
```

16.22 Euler's Method (Section 10.4)

Flow-chart

Refer to Section 14.22, page 727

Program

```
function[]=Eulers_Method()
format compact
format
clc df=inline{'x+y');
x0=input(' Enter value of x0:-');
y0=input('Enter value of y0:-');
h=input('Enter value of h:-');
x=input('Enter value of x:-'); xl=x0;
```

```
yl=y0;
while(x>xl)
    yl=yl+h.*df(xl,yl);
    xl=xl+h;
    fprintf('When x=%2.2f y=%2.2f\n',xl,yl)
end
end
```

> **NOTE:** $df(x, y)$ is dy/dx
> x_0 is x_{n+0}, *i.e.*, x_n
> x_1 is x_{n+1}
> y_0 is y_{n+0}, *i.e.*, y_n
> y_1 is $yn+1$

Computer Solution of Example 10.8

```
Enter value of x0:-0
Enter value of y0:-1
Enter value of h:-0.1
Enter value of x:-1
When x=0.10 y=1.10
When x=0.20 y=1.22
When x=0.30 y=1.36
When x=0.40 y=1.53
When x=0.50 y=1.72
When x=0.60 y=1.94
When x=0.70 y=2.20
When x=0.80 y=2.49
When x=0.90 y=2.82
When x=1.00 y=3.19
When x=1.10 y=3.61
```

16.23   Modified Euler's Method (Section 10.5)

Flow-chart

   Refer to Section 14.23, page 729

Program

```
function[]=Eulers_Method()
format compact
format short g
clc
df=inline('x+y');
```

```
x0=input('Enter x0:');
y0=input{'Enter y0:');
h=input{'Enter h:');
x=input('Enter x:');
x1=x0;
y1=y0;
while (1)
    if(x<x1)
        return;
    end
    y1=y1+h.*df(x1,y1);
    x1=x1+h;
    fprintf ('When x=%3.1f    y=%a4.2f\n',x1,y1)
end
end
```

Computer Solution of Example 10.10

```
Enter x0:0
Enter y0:1
Enter h:0.1
Enter x:0.3
Whenx=0.1 y=1.10
Whenx=0.2 y=1.22
Whenx=0.3 y=1.36
```

16.24   Runge-Kutta Method (Section 10.7)

Flow-chart

> *Refer to Section 14.24, page 732*

Program

```
function[]=Runga_Kutta_Method()
clc
format compact
format short g
f=inline('x+y*y');
x0=input('Enter the value x0:');
y0=input('Enter the value y0:');
h=input('Enter the value h:');
xn=input('Enter the value xn:'); x=x0;
y=y0;
```

```
while (1)
    if(x==xn)
        break
    end
    kl=h*f(x,y);
    k2=h*f(x+h/2,y+kl/2);
    k3=h*f(x+h/2,y+k2/2);
    k4=h*f(x+h,y+k3);
    k=(kl+(k2+k3)*2+k4)/6;
    x=x+h;
    y=y+k;
    fprintf('When x=%f y=%f \n',x,y)
  end
end
```

NOTES: x_0 is starting value of x, *i.e.*, x_0
x_n is the value of x for which y is to be determined

Computer Solution of Example 10.15

```
Enter the value x0:0
Enter the value y0:1
Enter the value h:0.2
Enter the value xn:0.2
When x=0.200000 y=1.273536
```

16.25   Milne's Method (Section 10.9)

Flow-chart

   Refer to Section 14.25, page 734

Program

```
function,[]=Milne_Method()
clc
format compact
format short g
global x y;
global h
x=[0 0 0 0 0 ];
y=[0 0 0 0 0];
x(1)=input('Enter the value x0:');
xr=input('Enter the last value of x:');
h=input{'Enter the spacing value:');
```

```
aerr=input('Enter the allowed error:');
y=input('Enter the value of y(i),i=0,3:-');
for i=1:3
    x(i+1)=x(1)+i*h;
    x(2:3,:)=x(2:3,:)+x(1,1)*6
end
disp('x Predicted Corrected');
disp('y f y f');
while (1)
  if(x(4)==xr)
      return
  end
  x(5)=x(4)+h;  y(5)=y(1)+(4*h/3)*(2*(f(2)+f(4))-f(3));
  fprintf('%f      %f      %f      \n', x(5), y(5), f(5) );
  correct();
  while(1)
      yc=y(5);
      corect();
      if(abs(yc-y(5)<=aerr))
          break;
      end
  end
  for i=1:4
      x(i)=x(i+1);
      y(i)=y(i+1);
  end
end
  function [z]=f(i)
        z=x(i)-y(i)*y(i);
  end
function[]=correct()
  y(5)=y(3)+(h/3)*(f(3)+4*f(4)+f(5));
  fprintf('%f %f\n', y(5),f(5))
  end
end
```

NOTES: x is an array such that $x[i]$ represents x_{n+i} for e.g., $x[0]$ represent x_n

y is an array such that $y[i]$ represents y_{n+i}
xr is the last value of x at which value of y is required
h is spacing in values of x
aerr is the allowed error in value of y yc is the latest corrected value for y
f is the function which returns value of y'
correct calculates the corrected value of y and prints it

Computer Solution of Example 10.19

```
Enter the value x0:0
Enter the last value of x:1
Enter the spacing value:0.2
Enter the allowed error:0.0001
Enter the value of y(i),i=0,3:-[0 0.2 0.0795 0.1762]
```

x	Predicted		Corrected	
	y	f	y	f
0.800000	0.283794	0.719461	0.305430	0.706712
			0.304580	0.707231
			0.304615	0.707210
1.000000	0.635420	0.596241		
			0.442469	0.804221
			0.456334	0.791759

16.26 Adams-Bashforth Method (Section 10.10)

Flow-chart

  Refer to Section 14.26, page 736

Program

```
function[]=Adams_Bashforth_Method()
clc
format compact
format short g
x=input{'Enter Values of x\n');
y=input('Enter Values of y\n');
sz=size(x);
sz=sz(2);
h=x(2)-x(1);
for i=1:sz
```

```
    tx=x(i);
    ty=y(i);
    tf=(tx^2*(1.0+ty));
    f(i)=tf;
end
for i=1:sz
x(i);
y(i);
f(i);
end
x(sz+1)=1.4;
y(sz+1)=y(sz)+(h/24)*((55*f(sz))-(59*f(sz-1))+(37*f(sz-2))-
(9*f<sz-3));
f(sz+1)=(x(sz+1)^2)*(1.0+y(sz+1));
for i=1:sz+1
fprintf('x=%4.If   y%d=%4.3f   f%d=%4.5f\n',x(i),   i-sz,y(i),i-
sz,f(i))
end
end
```

Computer Solution of Example 10.23

```
Enter Values of x
[1 1.1 1.2 1.3]
Enter Values of y
[1 1.233 1.548 1.979]
x=1.0 y-3=1.000 f-3=2.00000
x=1.1 y-2=1.233 f-2=2.70193
x=1.2 y-1=1.548 f-1=3.66912
x=1.3 y0=1.979 f0=5.03451
x=1.4 y1=2.572 f1=7.00170
```

16.27   Solution of Laplace's Equation (Section 11.5)

Flow-chart

  Refer to Section 14.27, page 740

Program

```
function[]=Laplace_Equation()
global SQR u
clc
SQR=4;
```

```
u=zeros(SQR);
disp('Enter the boundry conditions')
getrow(1,u);
getrow(SQR,u);
getcol(1,u);
getcol(SQR,u);
aerr=.1;
maxitr=10;
for itr=1:maxitr
    maxerr=0;
    for i=2:SQR-1
        for j=2:SQR-1
            t=(u(i-1,j)+u(i+1,j)+u(i,j+1)+u(i,j-1))/4;
            err=abs(u(i,j)-t);
            if(err>maxerr)
                maxerr=err;
            end
            u(i,j)=t;
        end
    end
fprintf('iteration No.%d \n',itr);
disp(u)
if(maxerr<=aerr)
   fprintf('After %d iterations    \n the  solution  is
\n',itr)
disp(u)
return
end
end
    function[]=getrow(i,u)
       global u
       fprintf('Enter the values of u[%d,j],j-1,%d \n',i,SQR);
       for j=1:SQR
           u(i, j)=input ('');
       end
    end
    function []=getcol(j,u)
       global u
       fprintf{'Enter the values of u[i,%d],i=2,%d \n',j,SQR-1);
          for i=2:SQR-1
              u(i,j)=input{'');
          end
```

```
      end
end
```

> **NOTES:** SQR is the size of the square mesh
> u is a 2D array representing the square mesh
> aerr is allowed error
> maxitr is the maximum number of iterations to be performed
> itr is the counter which keeps track of number of iterations performed
> err is the error in a particular point of the mesh
> maxerr is maximum error in the mesh in an iteration
> f is the execution time format
> getrow inputs the ith row of the mesh
> getcol inputs the jth column of the mesh

Computer Solution of Example 11.3 (a)

```
Enter the boundry conditions

Enter the values of u[1,j],j=1,4
The elements of the matrix 1000
The elements of the matrix 1000
The elements of the matrix 1000
The elements of the matrix 1000

Enter the values of u[4,j],j=1,4
The elements of the matrix 1000
The elements of the matrix 500
The elements of the matrix 0
The elements of the matrix 0

Enter the values of u[i,1],i=2,3
The elements of the matrix 2000
The elements of the matrix 2000

Enter the values of u[i,4], i=2,3
The elements of the matrix 500
The elements of the matrix 0
iteration No. 1
1.0e+003 *
        1.0000      1.0000      1.0000      1.0000
        2.0000      0.7500      0.5625      0.5000
        2.0000      0.8125      0.3438           0
        1.0000      0.5000           0           0
```

```
iteration No.2
1.0e+003*
    1.0000      1.00       00      1.0000    1.0000
    2.0000      1.09       38      0.7344    0.5000
    2.0000      0.98       44      0.4297    0
    1.0000      0.50       00      0         0
iteration No.3
1.0e+003*
    1.0000      1.00       00      1.0000    1.0000
    2.0000      1.17       97      0.7773    0.5000
    2.0000      1.02       73      0.4512    0
    1.0000      0.50       00      0         0
iteration No.4
1.0e+003 *
    1.0000      1.00       00      1.0000    1.0000
    2.0000      1.20       12      0.7881    0.5000
    2.0000      1.03       81      0.4565    0
    1.0000      0.50       00      0         0
iteration No.5
1.0e+003 *
    1.0000      1.0000     1.0000     1.0000
    2.0000      1.2065     0.7908     0.5000
    2.0000      1.0408     0.4579     0
    1.0000      0.5000     0          0
iteration No.6
1.0e+003 *
    1.0000      1.0000     1.0000     1.0000
    2.0000      1.2079     0.7914     0.5000
    2.0000      1.0414     0.4582     0
    1.0000      0.5000     0          0
iteration No.7
1.0e+003 *
    1.0000      1.0000     1.0000     1.0000
    2.0000      1.2082     0.7916     0.5000
    2.0000      1.0416     0.4583     0
    1.0000      0.5000     0          0
iteration No.8
1.0e+003 *
    1.0000      1.0000     1.0000     1.0000
    2.0000      1.2083     0.7917     0.5000
    2.0000      1.0417     0.4583     0
    1.0000      0.5000     0          0
```

```
After 8 iterations the solution is
1.0e+003*
    1.0000    1.0000    1.0000    1.0000
    2.0000    1.2083    0.7917    0.5000
    2.0000    1.0417    0.4583       0
    1.0000    0.5000       0         0
```

16.28   Solution of Heat Equation (Section 11.9)

Flow-chart

   Refer to Section 14.28, page 745

Program

```
function[]=Heat_Equation()
clc
format compact
format short g
XEND=8;
TEND=5;
u=zeros(XEND+1,TEND+1);
h=1.0;k=0.125;
f=inline('4.*x-(x.*x)/2. 0);
csqr=input('Enter the square' of c:\n');
alpha=(csqr.*k)/(h.*h);
ust=input('Enter the value of u[0,t]:');
fprintf('Enter the value of u[%d,t]\n',XEND);
uet=input('');
for j=1:TEND+1
    u(XEND,j)=ust;
    u(1,j)=u(XEND,j);
end
for i=1:XEND-1
u(i+1,1)=f(i);
end
for j=1:TEND
    for i=2:XEND
    u(i,j+1)=alpha*u(i-1,j) + (1-2*alpha)*u(i,j)+alpha*u(i+1,j);
    end
end
fprintf('The value of alpha is %4.2f\n1,alpha)
disp('The value of u(i,j) are:-')
```

```
disp(u');
end
```

> **NOTES:** XEND is the ending value of x
> TEND is the ending value of t
> h is the spacing in values of x
> k is the spacing in values of y
> $f(x)$ is value of $u(x, 0)$
> csqr is value of C^2
> alpha is a
> ust is the value in first column
> uet is the value in the last column

Computer Solution of Example 11.11

```
Enter the square of c:4
Enter the value of u[0,t]:0
Enter the value of u[8,t] 0
The value of alpha is 0.50
The value of u(i,j) are:-
Columns 1 through 7
```

0	3.5	6	7.5	8	7.5	6
0	3	5.5	7	7.5	7	5.5
0	2.75	5	6.5	7	6.5	5
0	2.5	4.625	6	6.5	6	4.625
0	2.3125	4.25	5.5625	6	5.5625	4.25
0	2.125	3.9375	5.125	5.5625	5.125	3.9375

```
Columns 8 through 9
```

3.5	0
3	0
2.75	0
2.5	0
2.3125	0
2.125	0

16.29   Solution of Wave Equation (Section 11.12)

Flow-chart

   Refer to Section 14.29, page 748

Program

```
function[]=Wave_Equation()
clc
```

```
XEND=5;
TEND=5;
f=inline('x*x*(5-x)1);
csqr=input('Enter the square of c\n');
ust=input('Enter the value of u[0][t]\n');
fprintf('Enter the value of u[%d][t]\n',XEND)
uet=input('');
for j=1:TEND+1
    u(1,j)=ust;
    u(XEND+1,j)=uet;
end
for i=1:XEND-1
    u(i+1,1)=f(i);
    u(i+1,2)=f(i);
end
for j=2:TEND
    for i=2:XEND
      u(i, j+1)=u(i-1,j)+u(i+1,j)-u(i,j-1);
    end
end
u'
end
```

NOTES: XEND is the ending value of x
TEND is the ending value of t
$f(x)$ is value of $u(x, 0)$
csqr is value of C^2
ust is the value in first column
uet is the value in the last column

Computer Solution of Example 11.14

```
Enter the square of c 16
Enter the value of u[0][t] 0
Enter the value of u[5j[t] 0
ans =
      0    4    12    18    16    0
      0    4    12    18    16    0
      0    8    10    10     2    0
      0    6     6    -6    -6    0
      0   -2   -10   -10    -8    0
      0  -16   -18   -12    -4    0
```

16.30   Linear Programming-Simplex Method (Section 12.8)

Flow-chart

Refer to Section 14.30, page 750

Program

```
function [ ]=Linear_Programming_Simplex_Method()
clc
ND=2;
NS=2;
N=ND+NS;
N1=NS*(N+1);
c=zeros(1,N);
cb=zeros(1,NS);
th=zeros(1,NS);
x=zeros(1,ND);
a=zeros(NS,N+1);
bas=zeros{1,NS);
for i=1:NS
      a(i,i+ND)=1.0;
end
for i=1:NS
    bas(i)=i+ND;
end
disp('Enter the constraints')
for i=1:NS
    for j=1:ND
        a(i, j)=input ('');
    end
    a(i,N+1)=input ('');
end
disp('Enter the objective function')
for i=1:ND
c(i)=input(1');
end
while(1)
  max=0;
  kj = 0;
```

```
    for j=1:N
        z=0;
        for i=1:NS
            z=z+cb (i) *a(i, j);
        end
            cj=c(j)-z;
            if(cj>max)
                max=cj;
                kj=j;
            end
end
        if(max<=0)
            break;
        end
        max=0;
        for i=1:NS
            if(a(i,kj)~=0)
                th(i)=a(i,N)/a{i,kj);
                if(th(i)>max)
                    max=th (i);
                end
            end
        end
        if(max<=0)
            disp('Unbounded solution');
            return;
        end
        min=max;
        ki=l;
        for i=1:NS
          if((th(i)<min)& & <th(i)~=0))
                min=th(i);
                ki=i;
          end
        end
        t=a(ki, kj );
        for j=1:N+1
          a(ki,j)=a(ki,j)/t;
        end
            for i=1:NS
                if(i~=ki)
                    b=a(i,kj);
```

```
                    for k=1:N+1
                      a(i,k)=a (i,k)-a(ki,k)*b;
                    end
               end
           end
           cb(ki)=c(kj);
           bas(ki)=kj;
    end
    for i=1:NS
         if((bas(i)>=0)&&(bas(i)<ND))
             x(bas(i))=a(i,N);
         end
    end
           z=0;
           for i=1:ND
               z=z+c(i)*x(i);
           end
           for i=1:ND
                fprintf('X(%d)=%7.2f \n',i+1,x(i))
           end
           fprintf('Optimal value =%7.2f \n',z)
    end
```

NOTES: ND is the number of decision variables
NS is the number of slack variables
a is the array containing body matrix, unit matrix and bi's
c is an array containing values of c_j's
cb is an array containing values of C_B'S
th is an array containing values of e's
bas is the basis. For x_i's basis contains i, for s_i's basis contains i+ND
ki is the key row
kj is the key column

Computer Solution of Example 12.4

```
Enter the constraints
4
2
80
2
5
180
Enter the objective function
```

```
3
4
X(1)= 2.50
X(2)= 35.00
Optimal value =  147.50
```

Computer Solution of Example 12.16

```
Enter the constraints
2
3
2
440
4
0
3
430
Enter the objective function
4
3
6
X(1)= 0.00
X(2)= 42.22
X(3)= 156.67
Optimal value= 1066.67
```

Exercises 16.1

1. Let $x = [1\,2\,3\,4]$.

    (*a*) Add five to each element

    (*b*) Add three to just the even-index elements

    (*c*) Compute the square root and square of each element

2. Create the vector $x = $ randperm (50) and then evaluate the following function using only logical indexing:

$$y(x) = 2 \text{ if } x < 8$$

$$= x - 9 \text{ if } 9 < = x < 35$$

3. Create a vector x with the elements,

$$x_n = (-1)^{n+1} (2n - 1)$$

4. Given the arrays $x = [1\ 2\ 3]$, $y = [2\ 4\ 5]$ and A = $[3\ 8\ 6; 5\ 4\ 3]$, find
(a) $x + y$ (b) $[x; y']$ (c) $[x; y]$ (d) A $- 3$

5. Write a MATLAB code to plot the function
$$y = x^3 - x^2 + 6x \sin (5x) - 9x$$

6. Write a MATLAB program to evaluate the roots of the equation $ax^2 + bx + c = 0$.

7. Write a program in MATLAB for finding a real root of the equation $f(x) = 0$ by the bisection method.

8. Write a MATLAB program to find a real root of $x3 - 4x - 9 = 0$ using the method of false position.

9. Write an algorithm for the Newton-Raphson method to solve the equation $f(x) = 0$. Apply the same to solve $\cos x - xe^x = 0$ near $x = 0.5$ correct to three decimal places.

10. Write a MATLAB program to solve the following equations by the Gauss-Seidel method: $83x + 11y - 4z = 95$; $7x + 12y + 13z = 104$; $3x + 8y + 29z = 71$.

11. With the help of a flow chart, write a MATLAB program to solve: $7.5x + 3.8y + 2.9z = 15$; $3.2x + 6.8y + 7.4z = 37$; $1.3x + 2.1y + 3.2z = 7$, using the factorization method.

12. Given the data:

x	5	10	15	20	25	30
y	17	25	30	33	36	38

Write a MATLAB program to fit a quadratic relation using least square criterion.

13. Write a program in MATLAB to estimate $f(0.6)$ by the Lagrange interpolation for the following values:

x	0.4	0.5	0.7	0.8
$f(x)$	−0.916	−0.693	−0.357	−0.223

14. Write a MATLAB program to evaluate $\int_2^{10} x(x^2 + x)dx$ using Simpson's rule.

15. Write a MATLAB program to evaluate $\int_a^4 f(x)$ using the Simpson's 3/8 rule.

16. Write a MATLAB for the second order Runge-Kutta method.

17. Write a MATLAB for solving differential equations using the Runge-Kutta fourth order formulae.

18. Write a MATLAB program to find $y(0.8)$ for the differential equation $dy/dx = \frac{1}{2}(x + y)$ Given the following table using Milne's Predictor-Corrector method:

x	0	0.2	0.4	0.6
y	2	2.636	3.595	4.968

19. Write a MATLAB program to maximize $z = 6x_1 + 4x_2$ subject to $2x_1 + 3x_2 \leq 100$, $4x_1 + 2x_2 \leq 120$, $x_1, x_2 \geq 0$ where $x1$, x_2 are number of items to be produced.

20. Write a MATLAB program to solve Example 12.17.

USEFUL INFORMATION

I Basic Information and Errors

1. Useful Data

$e = 2.7183$	$1/e = 0.3679$	$\log_e 2 = 0.6931$	$\log_e 3 = 1.0986$
$\pi = 3.1416$	$1/\pi = 0.3183$	$\log_e 10 = 2.3026$	$\log_{10} e = 0.4343$
$\sqrt{2} = 1.4142$	$\sqrt{3} = 1.732$	1 rad. = 57° 17′ 45″	1° = 0.0174 rad.

2. Conversion Factors

1 ft. = 30.48 cm = 0.3048 m	1 m = 100 cm = 3.2804 ft.
$1\ \text{ft}^2 = 0.0929\ \text{m}^2$	$1\ \text{acre} = 4840\ \text{yd}^2 = 4046.77\ \text{m}^2$
$1\ \text{ft}^3 = 0.0283\ \text{m}^3$	$1\ \text{m}^3 = 35.32\ \text{ft}^3$
1 m/sec = 3.2804 ft/sec.	1 mile/h = 1.609 km/h.

3. Some Notations

$\in$	belongs to	$\cup$	union
$\notin$	doesnot belong to	$\cap$	intersection
$\Rightarrow$	implies	$\ni$	such that
$\Leftrightarrow$	implies and implied by		

Factorial n, i.e., $n! = n(n-1)(n-2)\cdots 3.2.1$.

Double factorials: $(2n)!! = 2n(2n-2)(2n-4)\cdots 6.4.2.$
$(2n-1)!! = (2n-1)(2n-3)(2n-5)\cdots 5.3.1.$

Stirling's approximation. When n is large $n! \sim \sqrt{2\pi n}\cdot n^n e^{-n}$.

4. If X is the true value of a quantity and X' is its approximate value, then

    (*i*)   Absolute error = $|X - X'|$

    (*ii*)   Relative error = $\left| \dfrac{X - X'}{X} \right|$

    (*iii*)   Percentage error = $100 \left| \dfrac{X - X'}{X} \right|$

5. If Δy is the error in the function $y = f(x_1, x_2, \cdots, x_n)$ corresponding to the *errors* $\Delta x_1, \Delta x_2 \cdots, \Delta x_n$, *then*

$$\delta y = \frac{\partial y}{\partial x_1} \delta x_1 + \frac{\partial y}{\partial x_2} \delta x_2 + \ldots + \frac{\partial y}{\partial x_n} \delta x_n.$$

6. Relative error of a product of n numbers
            = Algebraic sum of their relative errors approximately

II   Solution of Algebraic and Trancendental Equations

1. Intermediate value property: If $f(x)$ is continuous in the interval $[a, b]$ and $f(a), f(b)$ have different signs, then the equation $f(x) = 0$ has at least one root between $x = a$ and $x = b$.

2. Descartes rule of signs: The equation $f(x) = 0$ cannot have more positive roots than the change of signs in $f(x)$ and cannot have more negative roots than the change of signs in $f(-x)$.

3. If $\alpha_1, \alpha_2, \alpha_3, \cdots$ be the roots of the equation $a_0 x^n + a_1 x^{n-1} + a_2 x^{n-2} + a_3 x^{n-3} + \cdots = 0$, then

$$\sum \alpha_1 = -\frac{a_1}{a_0}; \ \sum \alpha_1 \alpha_2 = \frac{a_2}{a_0}; \ \sum \alpha_1 \alpha_2 \alpha_3 = -\frac{a_3}{a_0}; \ \text{etc.}$$

4. Bisection method: Iteration formula *is* $x_3 = \dfrac{1}{2}\left(x_1 + x_2\right)$

This process is continued till the difference between two consecutive values is negligible.

5. Method of false-position or Regula falsi method: Iteration formula is

$$x_2 = x_0 - \frac{x_1 - x_0}{f(x_1) - f(x_0)} f(x_0)$$

This process is repeated till the difference between two consecutive values is negligible.

6. Secant method:

Iteration formula is $x_2 = x_1 - \dfrac{x_1 - x_0}{f(x_1) - f(x_0)} f(x_1)$

NOTE **Obs.** *If secant method once converges, its rate of convergence is 1.6 which is faster than that of method of false position.*

7. Iteration method: Writing $f(x) = 0$ as $x = \phi(x)$ and taking x_0 as the initial root of the given equation, the approximations to the root are $x_i = \phi(x_i)$ such that $\phi'(x) < 1$.

8. Newton-Raphson method *algorithm is*

$$x_{n+1} = x_n - \frac{f(x_n)}{f'(x_n)} \ (n = 0, 1, 2, ...)$$

NOTE **Obs.** *Condition for its convergence is $|f(x) f''(x)| < |f'(x)|2$. Newton's method has a second order of convergence. If this method once converges, it converges faster than the Regula-falsi method and is preferred.*

9. Iterative formula to find 1/N is $x_{n+1} = x_n (2 - Nx_n)$

10. Iterative formula to find $\sqrt{N}$ is $x_{n+1} = \dfrac{1}{2}(x_n + N / x_n)$

11. Method of Least squares: (*i*) *Curve of best fit $y = a + bx$*
Normal equations: $\Sigma y = na + b\Sigma x$, $\Sigma xy = a\Sigma x + b\Sigma x^2$
To find *a*, *b*, solve these equations.
(*ii*) *Curve of best fit $y = a + bx + cx^2$*
Normal equations: $\Sigma y = na + b\Sigma x + c\Sigma x^2$
$\Sigma xy = a\Sigma x + b\Sigma x^2 + c\Sigma x^3$, $\Sigma x^2 y = a\Sigma x^2 + b\Sigma x^3 + c\Sigma x^4$.
To find *a*, *b*, *c*, solve these equations.

III   Solution of Simultaneous Algebraic Equations

1. Numerical solution of linear simultaneous equations are
(*i*) *Direct methods*   (*ii*) *Indirect (or Iterative) methods*
Method of Determinants, Matrix Inversion method, Gauss-elimination method, Gauss-Jordan method and Factorization method are *direct*

methods; Gauss-Jacobi method, Gauss-Seidal method, and Relaxation methods are *indirect methods*.

2. Method of determinants–Cramer's rule. For the equations
$a_1 x + b_1 y + c_1 z = d_1, a_2 x + b_2 y + c_2 z = d_2, a_3 x + b_3 y + c_3 z = d_3,$

$$x = \frac{1}{\Delta} \begin{bmatrix} d_1 & b_1 & c_1 \\ d_2 & b_2 & c_2 \\ d_3 & b_3 & c_3 \end{bmatrix}, y = \frac{1}{\Delta} \begin{bmatrix} a_1 & d_1 & c_1 \\ a_2 & d_2 & c_2 \\ a_3 & d_3 & c_3 \end{bmatrix}, z = \frac{1}{\Delta} \begin{bmatrix} a_1 & b_1 & d_1 \\ a_2 & b_2 & d_2 \\ a_3 & b_3 & d_3 \end{bmatrix}$$

where $\Delta = \begin{bmatrix} a_1 & b_1 & c_1 \\ a_2 & b_2 & c_2 \\ a_3 & b_3 & c_3 \end{bmatrix}$

3. Matrix Inversion method. For the equations:
$a_1 x + b_1 y + c_1 z = d_1, a_2 x + b_2 y + c_2 z = d_2, a_3 x + b_3 y + c_3 z = d_3,$

if $\qquad A = \begin{bmatrix} a_1 & b_1 & c_1 \\ a_2 & b_2 & c_2 \\ a_3 & b_3 & c_3 \end{bmatrix}, X = \begin{bmatrix} x \\ y \\ z \end{bmatrix}$ and $D = \begin{bmatrix} d_1 \\ d_2 \\ d_3 \end{bmatrix}$

then $\qquad X = \begin{bmatrix} x \\ y \\ z \end{bmatrix} = \frac{1}{|A|} \begin{bmatrix} A_1 & A_2 & A_3 \\ B_1 & B_2 & B_3 \\ C_1 & C_2 & C_3 \end{bmatrix} \times \begin{bmatrix} d_1 \\ d_2 \\ d_3 \end{bmatrix}$

where A_1, B_1, etc. are the cofactors of a_1, b_1 etc. in the determinant $|A|$.

4. Gauss-elimination method. In the Gauss elimination method, the coefficient matrix is transformed to **upper triangular matrix.**

5. Gauss-Jordan method. In Gauss-Jordan method, the coefficient matrix is transformed to **diagonal matrix.**

6. Gauss-Jordan method of finding the inverse of a matrix A. The matrices A and I are written side by side and the same row transformations are performed on both. As soon as A is reduced to *I*, the other matrix represents A^{-1}.

7. The convergence in Gauss-Seidal method is thrice as fast as in Jacobi's method.

8. The condition for Gauss-Jacobi's method to converge is that the coefficient matrix should be diagonally dominant.

IV Finite Differences and Interpolation

1. **Forward differences:** $\Delta y_r = y_{r+1} - y_r$.
   Backward differences: $\nabla y_r = y_r - y_{r-1}$
   Central differences: $\Delta y_{x-1/2} = y_x - y_{x-1}$

2. **Relations between operators:**

   (i) $\Delta = E - 1$ (ii) $\nabla = 1 - E^{-1}$

   (iii) $\Delta = E^{1/2} - E^{-1/2}$ (iv) $\mu = \dfrac{1}{2}\left(E^{1/2} + E^{-1/2}\right)$

   (v) $\Delta = E\nabla = \nabla E = \Delta E^{1/2}$ (vi) $E = e^{hD}$

3. **Factorial notation.** The product $x(x-1)(x-2) \cdots (x-r+1)$ is denoted by $[x]^r$ and is called a factorial.
   Factorial polynomial is defined as $[x]^n = x(x-h)(x-2h)\cdots[x-(n-1)h]$. The result of differencing $[x]^n$ is analogous that of differentiating x^r.
   Important Result
   $$\Delta[x]^n = n[x]^{n-1}$$
   $$\Delta[ax + b]^n = na\,[ax + b]^{n-1}$$

4. **Reciprocal Factorial notation.** *The function* $\{(x + h)(x + 2h)...(x + nh)\}^{-1}$ *is denoted by* $[x]^{-n}$ *and is called a reciprocal factorial function.*
   Important Result
   $$\Delta[x]^{-n} = -n[x]^{-(n+1)}$$
   $$\Delta[ax + b]^{-n} = -na\,[ax + b]^{-(n+1)}$$

5. **Inverse Operator of** Δ. *If* $\Delta y_x = v_x$, *then* $y_x = \Delta^{-1} u_x$, Δ^{-1} or $1/\Delta$ is called the inverse operator of Δ and is analogous to $1/D$ or integration in calculus.
   Important Result
   $$\Delta^{-1}[x]^n = [x]^{n+1}/(n + 1)$$
   $$\Delta^{-1}[x]^{-n} = [x]^{-n+1}/(-n + 1)$$

V Interpolation

1. **Newton's forward interpolation formula:**
   $$y_p = y_0 + p\Delta y_0 + \frac{p(p-1)}{2!}\Delta^2 y_0 + \frac{p(p-1)(p-2)}{3!}\Delta^3 y_0 + ...$$
   where $p = (x - x_0)/h$.

2. Newton's backward interpolation formula:

$$y_p = y_n + p\nabla y_n + \frac{p(p+1)}{2!}\nabla^2 y_n + \frac{p(p+1)(p+2)}{3!}\nabla^3 y_n + \cdots$$

where $p = (x - x_n)/h$

3. Gauss forward interpolation formula:

$$y_p = y_0 + p\Delta y_0 + \frac{p(p-1)}{2!}\Delta^2 y_{-1} + \frac{p(p+1)(p-1)}{3!}\Delta^3 y_{-1}$$

$$+ \frac{(p+1)(p-1)(p-2)}{4!}\Delta^4 y_{-2} + \cdots$$

4. Gauss's backward interpolation formula:

$$y_p = y_0 + p\Delta y_{-1} + \frac{(p+1)p}{2!}\Delta^2 y_{-1} + \frac{(p+1)(p-1)}{3!}\Delta^3 y_{-2}$$

$$+ \frac{(p+2)(p+1)(p-1)}{4!}\Delta^4 y_{-2} + \cdots$$

5. Stirling's formula:

$$y_p = y_0 + p\left(\frac{\Delta y_0 + \Delta y_{-1}}{2}\right) + \frac{p^2}{2!}\Delta^2 y_{-1} + \frac{p(p^2-1)}{3!}\left(\frac{\Delta^3 y_{-1} + \Delta^3 y_{-2}}{2}\right)$$

$$+ \frac{p^2(p^2-1)}{4!}\Delta^4 y_{-2} + \cdots$$

6. Bessel's formula:

$$y_p = y_0 + p\Delta y_0 + \frac{p(p-1)}{2!}\frac{\Delta^2 y_{-1} + \Delta^2 y_0}{2} + \frac{(p-\frac{1}{2})p(p-1)}{3!}\Delta^3 y_{-1}$$

$$+ \frac{(p+1)p(p-1)(p-2)}{4!}\frac{\Delta^4 y_{-2} + \Delta^4 y_{-1}}{2} + \cdots$$

7. Laplace-Everett's formula:

$$y_p = qy_0 + \frac{q(q^2-1^2)}{3!}\Delta^2 y_{-1} + \frac{q(q^2-1^2)(q^2-2^2)}{5!}\Delta^4 y_{-2} + \cdots$$

$$+ py_1 + \frac{p(p^2-1^2)}{3!}\Delta^2 y_0 + \frac{p(p^2-1^2)(p^2-2^2)}{5!}\Delta^4 y_{-1} + \cdots [q = 1-p]$$

8. Lagrange's interpolation formula:

$$y = \frac{(x-x_1)(x-x_2)\ldots(x-x_n)}{(x_0-x_1)(x_0-x_2)\ldots(x_0-x_n)}y_0 + \frac{(x-x_0)(x-x_2)\ldots(x-x_n)}{(x_1-x_0)(x_1-x_2)\ldots(x_1-x_n)}y_1$$

$$+\ldots+ \frac{(x-x_0)(x-x_1)\ldots(x-x_{n-1})}{(x_n-x_0)(x_n-x_1)\ldots(x_n-x_{n-1})}y_n.$$

9. Lagrange's inverse interpolation formula:

$$x = \frac{(y-y_1)(y-y_2)\ldots(y-y_n)}{(y_0-y_1)(y_0-y_2)\ldots(y_0-y_n)}x_0 + \frac{(y-y_0)(y-y_2)\ldots(y-y_n)}{(y_1-y_0)(y_1-y_2)\ldots(y_1-y_n)}x_1$$

$$+\ldots+ \frac{(y-y_0)(y-y_1)\ldots(y-y_{n-1})}{(y_n-y_0)(y_n-y_1)\ldots(y_n-y_{n-1})}x_n.$$

10. Newton's divided difference formula:

$$y = f(x) = y_0 + (x-x_0)\,[x_0, x_1] + (x-x_0)\,(x-x_1)\,[x_0, x_1, x_2]$$
$$+ (x-x_0)\,(x-x_1)\,(x-x_2)\,[x_0, x_1, x_2, x_3] +\cdots$$

11. Hermite interpolation formula:

$$P(x) = [1-2(x-x_0)\,L'_0\,(x_0)]\,[L_0\,(x)]^2\,y(x_0) + (x-x_0)\,[L_0\,(x)]^2\,y'(x_0)$$
$$+ [1-2(x-x_1)\,L'_1\,(x_1)]\,[L_1\,(x)]^2\,y(x_1) + (x-x_1)\,[L_1(x)]^2\,y'(x_1)$$
$$+ [1-2(x-x_2)L'_2\,(x_2)]\,[L_2(x)]^2\,y(x_2) + (x-x_2)\,[L_2(x)]^2\,y'(x_2) +\ldots\ldots$$

12. Cubic Spline interpolation formula:

$$f(x) = \frac{1}{6h}\left[\left(x_{i+1}-x\right)^3 M_i + \left(x-x_i\right)^3 M_{i+1}\right]$$

$$+\frac{1}{h}\left[\left(x_{i+1}-x\right)\left(y_i - \frac{h^2}{6}M_i\right) + \left(x-x_i\right)\left(y_{i+1} - \frac{h^2}{6}M_{i+1}\right)\right]$$

where $\quad M_{i-1} + 4M_i + M_{i+1} = \dfrac{6}{h^2}\left(y_{i-1} - 2y_i + y_{i+1}\right), i = 1,2,3,\ldots,(n-1)$

and $\quad M_0 = 0,\ M_n = 0.$

VI   Numerical Differentiation

1. Forward difference formulae:

$$\left(\frac{dy}{dx}\right)_{x_0} = \frac{1}{h}\left[\Delta y_0 - \frac{1}{2}\Delta^2 y_0 + \frac{1}{3}\Delta^3 y_0 - \frac{1}{4}\Delta^4 y_0 +\ldots\right]$$

$$\left(\frac{d^2 y}{dx^2}\right)_{x_0} = \frac{1}{h^2}\left[\Delta^2 y_0 - \Delta^3 y_0 + \frac{11}{12}\Delta^4 y_0 -\ldots\right] \text{ and so on.}$$

2. Backward difference formulae:

$$\left(\frac{dy}{dx}\right)_{x_n} = \frac{1}{h}\left[\nabla y_n + \frac{1}{2}\nabla^2 y_n + \frac{1}{3}\nabla^3 y_n + \frac{1}{4}\nabla^4 y_n + \ldots\right]$$

$$\left(\frac{d^2y}{dx^2}\right)_{x_n} = \frac{1}{h^2}\left[\nabla^2 y_n + \nabla^3 y_n + \frac{11}{12}\Delta^4 y_n + \ldots\right] \text{ and so on.}$$

3. Central difference formulae:

(*i*) *Stirling's formula* gives

$$\left(\frac{dy}{dx}\right)_{x_0} = \frac{1}{h}\left[\frac{\Delta y_0 + \Delta y_{-1}}{2} - \frac{1}{6}\frac{\Delta^3 y_{-1} + \Delta^3 y_{-2}}{2} + \frac{1}{30}\frac{\Delta^5 y_{-2} + \Delta^5 y_{-3}}{2} + \ldots\right]$$

$$\left(\frac{d^2y}{dx^2}\right)_{x_0} = \frac{1}{h^2}\left[\Delta^2 y_{-1} - \frac{1}{12}\Delta^4 y_{-2} + \frac{1}{90}\Delta^6 y_{-3} - \ldots\right]$$

(*ii*) *Bessel's formula* gives

$$\left(\frac{dy}{dx}\right)_{x_0} = \frac{1}{h}\left[\Delta y_0 - \frac{1}{2}\left(\frac{\Delta^2 y_{-1} + \Delta^2 y_0}{2}\right) + \frac{1}{12}\Delta^3 y_{-1}\right.$$

$$\left. + \frac{1}{12}\left(\frac{\Delta^2 y_{-2} + \Delta^4 y_{-1}}{2}\right) - \frac{1}{12}\Delta^5 y_2 + \ldots\right]$$

$$\left(\frac{d^2y}{dx^2}\right)_{x_0} = \frac{1}{h^2}\left[\left(\frac{\Delta^2 y_{-1} + \Delta^2 y_0}{2}\right) - \frac{1}{2}\Delta^3 y_{-1} - \frac{1}{12}\left(\frac{\Delta^4 y_{-2} + \Delta^4 y_{-1}}{2}\right)\right.$$

$$\left. + \frac{1}{24}\Delta^5 y_{-2} + \ldots\right]$$

VII   Numerical Integration

1. Trapezoidal rule:

$$\int_{x_0}^{x_0+nh} f(x)dx = \frac{h}{2}\left[(y_0 + y_n) + 2(y_1 + y_2 + \ldots + y_{n-1})\right]$$

2. Simpson's 1/3rd rule:

$$\int_{x_0}^{x_0+nh} f(x)dx = \frac{h}{3}\left[(y_0 + y_n) + 4(y_1 + y_3 + \cdots + y_{n-1})\right.$$

$$\left. + 2(y_2 + y_4 + \cdots + y_{n-2})\right]$$

(Number of sub-intervals should be taken as even)

3. Simpson's 3/8th rule:

$$\int_{x_0}^{x_0+nh} f(x)dx = \frac{3h}{8}\Big[(y_0 + y_n) + 3(y_1 + y_2 + y_4 + y_5 + \cdots + y_{n-1})$$
$$+ 2(y_3 + y_6 + \cdots + y_{n-3})\Big]$$

(Number of sub-intervals should be taken as a multiple of 3)

4. Boole's rule:

$$\int_{x_0}^{x_0+nh} f(x)dx = \frac{2h}{45}[7y_0 + 32y_1 + 12y_2 + 32y_3 + 14y_4$$
$$+ 32y_5 + 12y_6 + 32y_7 + 14y_8 + \cdots]$$

(Number of sub-intervals should be taken as multiple of 4)

5. Weddle's rule:

$$\int_{x_0}^{x_0+nh} f(x)dx = \frac{3h}{10}\Big[y_0 + 5y_1 + y_2 + 6y_3 + y_4 + 5y_5 + 2y_6 + 5y_7 + \ldots\Big]$$

(Number of sub-intervals should be taken as a multiple of 6)

6. Errors:

Rule	No. of intervals (multiples of)	Error	Order of error
Trapezoidal	Any	$-\dfrac{h^2}{12}y''$	h^2
Simpson's 1/3	2	$-\dfrac{h^4}{180}y^{iv}$	h^4
Simpson's 3/8	3	$-\dfrac{3h^5}{80}y^{iv}$	h^5
Weddle's	6	$-\dfrac{h^7}{140}y_0^{vi}$	h^7

7. Romberg's method:

$$I(h,h/2) = \frac{1}{3}\Big[4I(h/2) - I(h)\Big]$$

The computation is continued till two successive values are equal.

8. Gaussian integration:

(*i*) Two point formula: $\int_{-1}^{1} f(x)dx = f\left(\dfrac{-1}{\sqrt{3}}\right) + f\left(\dfrac{1}{\sqrt{3}}\right)$

(*ii*) Three point formula: $\int_{-1}^{1} f(x)dx = \dfrac{8}{9}f(0) + \dfrac{5}{9}\left[f\left(-\sqrt{\dfrac{3}{5}}\right) + f\left(\sqrt{\dfrac{3}{5}}\right)\right]$

(*iii*) To apply Gaussian integration, the limits of integration a, b are

changed to -1, 1 by the transformation $x = \dfrac{1}{2}(b-a)u + \dfrac{1}{2}(b+a)$.

9. Double integration:
(*i*) *Trapezoidal rule:*

$$I = \dfrac{hk}{4}\Big[\big(f_\infty + f_{0m}\big) + 2\big(f_{01} + f_{02} + \cdots + f_{0,m-1}\big)$$

$$+ \big(f_{n0} + f_{nm}\big) + 2\big(f_{n1} + f_{n2} + \ldots + f_{n,m-1}\big)\Big]$$

$$+ 2\sum_{i=1}^{n-1}\Big\{\big[\big(f_{i0} + f_{im}\big) + 2\big(f_{i1} + f_{i2} + \cdots + f_{i,m-1}\big)\big]\Big\}\Big\}$$

where $f_{ij} = f(x_i, y_j)$

(*ii*) *Simpson's rule:*

$$\int_{y_{j-1}}^{y_{j+1}} \int_{x_{i-1}}^{x_{i+1}} f(x,y)dxdy = \dfrac{hk}{9}\Big[\big(f_{i-1,j-1} + 4f_{i-1,j} + f_{i-1,j+1}\big)$$

$$+ 4\big(f_{i,j-1} + 4f_{i,j} + 4f_{i,j+1}\big) + \big(f_{i+1,j-1} + 4f_{i+1,j} + f_{i+1,j+1}\big)\Big]$$

Adding all such intervals, we get *I*.

VIII   Number Solution of Ordinary Differential Equations

1. Picard's method: $y_1 = y_0 + \displaystyle\int_{x_0}^{x} f(x,y_0)dx$

$$y_2 = y_0 + \int_{x_0}^{x} f(x,y_1)dx \ \text{etc.}$$

2. Taylor's method:

$$y = y_0 + (x-x_0)(y')_0 + \dfrac{(x-x_0)^2}{2!}(y'')_0 + \dfrac{(x-x_0)^2}{2!}(y''')_0 + \ldots$$

3. Euler's method: $y_2 = y_1 + hf(x_0 + h, y_1)$
Repeat this process till y_2 is stationary. Then calculate y_3 and so on.

4. Modified Euler's method: $y_2 = y_1 + \dfrac{h}{2}\left[f\left(x_0 + h, y_1\right) + f\left(x_0 + 2h, y_2\right)\right]$

Repeat this step, until y_2 becomes stationary. Then calculate y_3 and so on.

5. Runge Kutta method: $y_1 = y_0 + k$, where $k = \dfrac{1}{6}\left(k_1 + 2k_2 + 2k_3 + k_4\right)$

such that $\quad k_1 = hf(x_0, y_0), k_2 = hf(x_0 + h/2, y_0 + k_{1/2})$

$$k_3 = hf(x_0, h/2, y_0 + k_2/2), k_4 = hf(x_0 + h, y_0 + k_3)$$

6. Milne's method:

(*i*) *Predictor formula:* $y_4 = y_0 + \dfrac{4h}{3}\left(2f_1 - f_2 + 2f_3\right)$

(*ii*) *Corrector formula:* $y_4 = y_2 + -\left(f_2 + \ f_3 + f_4\right)$

7. Adams-Bashforth method:

(*i*) *Predictor formula:* $y_1 = y_0 + \dfrac{h}{24}\left(55f_0 - 59f_{-1} + 37f_{-2} - 9f_{-3}\right)$

(*ii*) *Corrector formula:* $y_1 = y_0 + \dfrac{h}{24}\left(9f_1 + 19f_0 - 5f_{-1} + f_{-2}\right)$

(*Four prior values are required to find the next values by Milne's or Adams-Bashforth method*)

8. Central-difference approximations:

$$y_1' = \frac{1}{2h}\left(y_{i+1} - y_{i-1}\right)$$

$$y_1'' = \frac{1}{h^2}\left(y_{i+1} - 2y_i + y_{i-1}\right)$$

$$y_1''' = \frac{1}{2h^3}\left(y_{i+2} - 2y_{i+1} + 2y_{i-1} - y_{i-2}\right)$$

$$y_1^{iv} = \frac{1}{h^4}\left(y_{i+2} - 4y_{i+1} + 6y_i - 4y_{i-1} + y_{i-2}\right)$$

IX   Number Solution of Partial Differential Equations

1. Classification of second order equation:

$$A(x,y)\frac{\partial^2 u}{\partial x^2} + B(x,y)\frac{\partial^2 u}{\partial x \partial y} + C(x,y)\frac{\partial^2 u}{\partial y^2} + F\left(x, y, u, \frac{\partial u}{\partial x}, \frac{\partial u}{\partial y}\right) = 0$$

is said to be
 (*i*) *elliptic* if $B^2 - 4AC < 0$
 (*ii*) *parabolic* if $B^2 - 4AC = 0$
 (*iii*) *hyperbolic* if $B^2 - 4AC > 0$.

2. Laplace equation:

(*i*) Standard five point formula:

$$u_{i,j} = \frac{1}{4}\left[u_{i-1,j} + u_{i+1,j} + u_{i,j+1} + u_{i,j-1}\right]$$

(*ii*) Diagonal five point formula:

$$u_{i,j} = \frac{1}{4}\left[u_{i-1,j+1} + u_{i+1,j-1} + u_{i+1,j+1} + u_{i-1,j-1}\right]$$

(Four conditions are required to solve Laplace equation.)

3. Poisson's equation: $\dfrac{\partial^2 u}{\partial x^2} + \dfrac{\partial^2 u}{\partial y^2} = f(x,y)$

Standard five point formula:
$$u_{i-1,j} + u_{i+1,j} + u_{i,j+1} + u_{i,j-1} - 4u_{i,j} = h^2 f(ih, jh)$$

4. One-dimensional Heat equation:

$$\frac{\partial u}{\partial t} = c^2 \frac{\partial^2 u}{\partial x^2}$$

(*i*) Schmidt formula: $u_{i,j+1} = \alpha u_{i-1,j} + (1 - 2\alpha)\, u_{i,j} + \alpha u_{i+1,j}$,
$$\text{where } \alpha = kc^2/h^2$$

(*ii*) Bendre-Schmidt relation: $u_{i,j+1} = \frac{1}{2}\left(u_{i-1,j} + u_{i+1,j}\right)$

[when $\alpha = 1/2$ (*i*) reduces to (*ii*)]

(*iii*) *Crank-Nicolson formula:*
$$\alpha(u_{i+1,j+1} + u_{i-1,j+1}) - 2(\alpha + 1)\, u_{i,j+1} = 2(\alpha - 1)\, u_{i,j} - \alpha(u_{i+1,j} + u_{i-1,j})$$

5. Wave equation: $\dfrac{\partial^2 u}{\partial t^2} = c^2 \dfrac{\partial^2 u}{\partial x^2}$

(*i*) *Explicit formula for solution is*
$$u_{i,j+1} = 2(1 - \alpha^2 c^2)\, u_{i,j} + \alpha^2 c^2\,(u_{i-1,j} + u_{i+1,j}) - u_{i,j-1} \text{ where } \alpha = k/h$$
(*ii*) If α is so choosen that coefficient of $u_{i,j}$ is zero, then
$$\alpha(= k/h) = 1/c \text{ i.e. } k = h/c, \text{ then (\textit{i}) takes the simplified form}$$
$$u_{i,j+1} = u_{i-1,j} + u_{i+1,j} - u_{i,j-1}$$

which provides as *explicit scheme* for the solution of the wave equation.

ANSWERS TO EXERCISES

Exercises 1.1

1. 3.264, 35.45, 4986000, 0.7004, 0.0003222, 18.26.

2. 0.7546; $- 0.0002 \times 105$; 0.00265.　　　**3.** 0.0003, 0.001.

4. 0.077.　　　**5.** $- 0.0004$　　　**6.** 4.44%.

7. 0.01%; 10%　　　**8.** 1.65.　　　**9.** 1.149×10^{-4}; 4.836×10^{-4}.

10. 600.0002.　　　**11.** 165.55.　　　**12.** 0.17312; 0.0003178.

13. 0.5.　　　**14.** 0.0005.

Exercises 1.2

1. 0.0025.　　　**2.** 76.　　　**3.** 75.6; 0.7.　　**5.** 0.23; 0.14.

6. 0.7721.　　　**7.** 0.423　　　**8.** 10.

9. $\log_e (1.2) = 0.1823215$; $n = 9$.

10. $e^h + \cos(h) = 2 + h + \dfrac{h^3}{3!} + O(h^4)$; $e^h \cos(h) = 1 + h - \dfrac{h^3}{3} + O(h^4)$.

11. $\sin(t) + \cos(t) = 1 + t - \dfrac{t^2}{2} - \dfrac{t^3}{6} + \dfrac{t^4}{24} + \dfrac{t^5}{120} + O(t^6)$

$\sin(t)\cos(t) = t - \dfrac{2}{3}t^3 + \dfrac{2}{15}t^5 + O(t^6)$.

Exercises 1.3

1. (b)　　　**2.** 0.000005.　　　**3.** (b).　　　**4.** 0.00049.

5. 0.007.　　　**6.** 43.38; 0.63264; 0.2538.　**7.** 0.004.　　　**8.** 0.0058

9. (c).　　　**10.** 0.0015.　　　**11.** 0.0496.　　　**12.** 0.33.

13. 0.0005.　　　**14.** $< 1 \ (p \times 10^{n-1})$.

Exercises 2.1

1. $x^4 - 6x^3 + 3x^2 + 42x - 70 = 0$.
2. $(i) - 2, 1 \pm 3i$. (ii) $2 \pm \sqrt{3}, 3, -5$.
4. $a = 2, b = 1$.   **5.** $1, \frac{1}{2}, \pm\sqrt{5}$.   **6.** $-7, 2, 6$.   **7.** $1, 4, 7$.
9. $4/3$.

Exercises 2.2

1. $x^3 + 6x^2 - 36x + 27 = 0$.   **2.** $x^7 + 3x^5 + x^3 + x^2 + 7x - 1 = 0$.
3. $10x^4 + 9x^3 + 8x^2 - 7x + 1 = 0$.   **4.** $2/9, 2/3, -2/3$.

5. (i) $2, \dfrac{1}{2}, -3, -\dfrac{1}{3}$; (ii) $2, 2, \dfrac{1}{2}, \dfrac{1}{2}$.   **6.** $x^4 + 13x^3 + 60x^2 + 116x + 80 = 0$.

7. $\dfrac{1}{2}\left(5 \pm \sqrt{21}\right); \dfrac{1}{2}\left(-3 \pm \sqrt{5}\right)$ **8.** $y^2 - 30y^2 + 225y - 68 = 0$.

9. $3x^3 - 11x^2 + 9x - 2 = 0$.
10. Quotient = $15x^5 - x^4 + 14x^3 + 12x^2 - 7x - 21$, Remainder = $-9x + 29$.

Exercises 2.3

1. 1.32.   **2.** 0.45.   **3.** 0.71 rad.   **4.** 1.81 rad.

Exercises 2.4

1. (i) 1.321 (ii) 1.46 (iii) 2.875 (iv) 1.855.
2. (i) 0.0625. (ii) 0.567. (iii) 0.367.
3. (i) 2.128. (ii) 2.7065. (iii) 1.4036.
4. (i) 0.853. (ii) 0.6071. $(iii) - 0.134$ (iv) 2.798.
   (v) 3.789. (vi) 0.3604.
5. (i) 1.861.   **6.** (i) 0.99976. (ii) 0.99931.
7. $(i) - 2.0625$ (ii) 0.567 (iii) 3.496.
8. (i) 0.6071. (ii) 2.9428. (iii) 1.4973. (iv) 4.4346.
   (v) 0.2591. (vi) 2.8625.
9. (i) and (ii) 5.4772. **10.** (i) & (ii) 1.524.   **11.** 0.477.

Exercises 2.5

1. (i) 1.532. (ii) 2.095. (iii) 1.834. (iv) 1.226.
2. (i) 1.856. (ii) 2.198.   **3.** $- 16.56$.
4. $(i) - 1.9338$. (ii) 2.798. (iii) 4.545. (iv) 0.052.
   (v) 0.518. (vi) 0.695.

5. Root in interval $(-0.8, 0.5) = 0.77009$, Root in interval $(0, 1) = 0.76839$.

6. 6.889.   **7.** 0.5886.   **8.** 0.033 sec.

9. $x_{n+1} = \frac{1}{2}(x_n + N/x_n), x_{n+1} = \frac{1}{3}(2x_n + N/x_n^2); (a)$ 3.162. (b) 2.5713

10. $x_{n+1} = \frac{1}{4}\left(3x_n + \frac{N}{x_n^3}\right)$, 2.3784

11. (i) 0.0555. (ii) 0.2582. (iii) 0.4347.   **12.** 0.51776.

Exercises 2.6

1. 2.26   **2.** 1.839   **3.** 1.3688   **4.** 3.14.

Exercises 2.7

1. (i) 1.532 (ii) 0.684 (iii) 1.168.   **2.** 1.674.

3. 2.231.   **4.** $- 1.328$.   **5.** 2.924.

Exercises 2.8

1. 2.   **2.** $m = 2$, 3.973.   **3.** $- 0.573 \pm 0.89i$

4. $x^2 + 2.9026x - 4.9176$.   **5.** $- 0.759, - 1.42, - 3.411 \pm 2.903i$.

6. $(x^2 - 2x + 2)(x^2 - 6x + 25)$.   **7.** 5, 2.001, 0.9995.

8. 3, 2, 1.   **9.** 7.018, $- 2.974$, 0.958.   **10.** 2, 1, 1.

11. 6.3, 2.3, 0.4.

Exercises 2.9

1. (a).   **2.** $x_{n+1} = \frac{1}{3}\left(2x_n + N/x_n^2\right)$.

3. $x_{n+1} = \frac{1}{2}\left(2x_n + x_{n-1}\right)$.   **4.** $x_2 = x_0 - \frac{x_1 - x_0}{f(x_1) - f(x_0)} f(x_0)$.

5. (c).   **6.** Chord AB.   **7.** $x_{n+1} = \frac{1}{2}(x_n + N/x_n)$.

8. 1.79.

9. Initial approximation is chosen sufficiently close to the root.   **10.** (c)

11. Newton-Raphson method.   **12.** 0.657.

13. (c).   **14.** $|\phi'(x)| < 1$.

15. If we start with a smaller interval for the root.

16. True   **17.** 2.1.   **18.** $(2, 3)$

19. $x_n (2 - Nx_n)$ **20.** If we start with a smaller interval for the root.

Exercises 3.1

5. $0, -\dfrac{1}{2}$. **6.** 5.

Exercises 3.2

1. $x = 2, y = 4, z = 1, w = 3$.

4. $\begin{bmatrix} 0 & 3 & 0.5 \\ 3 & 1 & 3 \\ 0.5 & 3 & 9 \end{bmatrix} + \begin{bmatrix} 0 & 2 & -3.5 \\ -2 & 0 & -2 \\ 3.5 & 2 & 0 \end{bmatrix}$

4. $\begin{bmatrix} -3 & 1 & 7 \\ -1 & -1 & -5 \\ 5 & 1 & -13 \end{bmatrix}; \begin{bmatrix} -3/4 & 1/4 & 7/4 \\ -1/4 & -1/4 & 5/4 \\ 5/4 & 1/4 & -13/4 \end{bmatrix}$ **5.** $\begin{bmatrix} 7 & -3 & -3 \\ -1 & 1 & 0 \\ -1 & 0 & 1 \end{bmatrix}$

6. $\begin{bmatrix} 1 & 0 & 0 \\ 7/5 & 1 & 0 \\ 3/5 & 41/19 & 1 \end{bmatrix}; \begin{bmatrix} 5 & -2 & 1 \\ 0 & 19/5 & -32/5 \\ 0 & 0 & 327/19 \end{bmatrix}$ **7.** (i) 2; (ii) 3.

8. (i) Inconsistent (ii) Inconsistent (iii) Consistent, $x = -1, y = 1, z = 2$.

9. (i) $\lambda = 3, \mu \neq 10$; (ii) $\lambda \neq 3$; (iii) $\lambda = 3, \mu = 10$.

10. $\lambda = 1, -9$. For $\lambda = 1$, sol. is $x = k, y = -k, z = 2k$.

For $\lambda = -9$, sol. is $x = 3k, y = 9k, z = -2k$.

Exercises 3.3

1. $x = 2, y = -1, z = 1/2$. **2.** $x = 1.2, y = 2.2, z = 3.2$. **3.** $x = y = z = e^2$.

4. $u = 1, v = 1/2, w = 1/3$. **5.** $x = 19/50, y = -29/50, z = -51/50, t = 0$.

6. $x = 2, y = 1, z = 0$. **7.** $x = 1, y = -5, z = 5$.

8. $x = y = z = 2$. **9.** $i_1 = 1.5, i_3 = 2.5$.

10. $i_1 = 1.5; i_3 = 2.5$. **11.** $x = 1, y = 3, z = 5$.

12. $x = -12.75, y = 14.375, z = 8.75$ **13.** $x = 1, y = 2, z = 3$.

14. $x_1 = 2, x_2 = -1, x_3 = 3$. **15.** $x_1 = 1, x_2 = 2, x_3 = -1, x_4 = -2$.

16. $x = 1, y = 3, z = 5$. **17.** $x = 8.7, y = 5.7, z = -1.3$.

18. $x = 1, y = 3, z = 5$. **19.** $x = 1, y = 2, z = 3$.

20. $x_1 = 2, x_2 = 1/5, x_3 = 0, x_4 = 4/5$. **21.** $x = 35/18, y = 29/18, z = 5/18$.

22. $x = y = z = 1$. **23.** $x = 1, y = 3, z = 5$.

24. $x_1 = -1, x_2 = 0, x_3 = 1, x_4 = 2$.

Exercises 3.4

1. $x = 2.556, y = 1.722, z = -1.055.$ **2.** $x = 1, y = 2, z = 3, u = 4.$

3. $x = 2.426, y = 3.573, z = 1.926.$ **4.** $x = 1, y = 1, z = 1.$

5. $x = 0.998, y = 1.723, z = 2.024.$ **6.** $x = y = z = 1.$

7. $x_1 = 1.058, x_2 = 1.367, x_3 = 1.962.$ **8.** $x_1 = 3, x_2 = -2.5, x_3 = 7.$

9. $x = 1.35, y = 2.103, z = 2.845.$ **10.** $x = y = z = 1.$

11. $x = 52.5, y = 44.5, z = 59.7.$ **12.** $x = 1.93, y = 2.57, z = 2.43.$

Exercises 3.5

1. Ill-conditioned.   **2.** $x = 2, y = -1, z = 1.$

Exercises 3.6

1. $x = 2, y = 1; x = -1.683, y = 2.164.$ **2.** $x = -1.853, y = -1.927.$

3. $x = 3, y = 4.$ **4.** $x = 0.7974, y = 0.4006.$ **5.** $x = 3.162, y = 6.45.$

Exercises 3.7

1. (b) **2.** Section 3.5

3. Diagonal

4. The absolute value of the largest coefficient is greater than the sum of the absolute values of all the remaining coefficients.

5. diagonal matrix.   **6.** (b) **7.** Upper triangular matrix.

8. False. In fact, the rate of convergence of Gauss-Seidal method is twice as fast as that of Gauss-Jacobi method.

9. Section 3.4(3).   **10.** (a) **11.** $x = 1, y = 1.$

Exercises 4.1

1. $\begin{bmatrix} 8 & -1 & -3 \\ -5 & 1 & 2 \\ 10 & -1 & -4 \end{bmatrix}$ **2.** $\begin{bmatrix} 1/2 & -1/2 & 1/2 \\ -1/2 & 3 & -1 \\ 5/2 & -3/2 & 1/2 \end{bmatrix}$ **3.** $\dfrac{1}{21}\begin{bmatrix} 1 & 10 & -7 \\ 1 & -11 & 14 \\ -3 & 12 & 0 \end{bmatrix}$

4. $\dfrac{1}{3}\begin{bmatrix} 2 & -2 & 1 \\ -2 & 5 & -4 \\ 1 & -4 & 5 \end{bmatrix}$ **5.** $\begin{bmatrix} -0.5 & 0.2 & -1.6 \\ 0 & 0.2 & 0.4 \\ 0.5 & 0 & 1 \end{bmatrix}$ **6.** $\begin{bmatrix} 1/8 & -1/8 & 3/8 \\ -1/8 & 1/8 & 5/8 \\ 3/8 & 5/8 & 23/8 \end{bmatrix}$

7. $\dfrac{1}{2078}\begin{bmatrix} 206 & -17 & -24 \\ -16 & 102 & 22 \\ 46 & 34 & 196 \end{bmatrix}$ **8.** $\begin{bmatrix} 1.4 & 0.2 & -0.4 \\ -1.5 & 0 & 0.5 \\ 1.1 & -0.2 & -0.1 \end{bmatrix}$

9. $\dfrac{1}{9}\begin{bmatrix} -15 & -14 & 12 & 13 \\ 3 & 7 & -6 & -2 \\ 9 & -3 & 0 & -3 \\ -6 & -1 & 3 & 1 \end{bmatrix}$

10. $\begin{bmatrix} 0.091 & 0.182 \\ 0.273 & -0.454 \end{bmatrix}$

11. $\begin{bmatrix} 0.429 & 2.429 & -1.429 \\ 0.143 & 0.143 & -0.143 \\ -0.857 & -3.857 & 2.857 \end{bmatrix}$

Exercises 4.2

1. (a) 1, 6; $(1, -1), (4, 1)$ $(b) -1, 6; (1, 1), (2, -5)$.

2. (a) 1, 2, 3; $(1, 0, -1), (0, 1, 0), (1, 0, 1)$

(b) 5, $-3, -3$; $(1, 2, -1), (2, -1, 0), (3, 0, 1)$

(c) 8, 2, 2; $(2, -1, 1), (1, 0, -2), (1, 2, 0)$.

3. (i) $\begin{bmatrix} 1 & 1/2 & -2/3 \\ 0 & -1/2 & 0 \\ 0 & 0 & 1/3 \end{bmatrix}$ (ii) $\dfrac{1}{3}\begin{bmatrix} 2 & -1 & -1 \\ 0 & 3 & 0 \\ -1 & -1 & 2 \end{bmatrix}$; (iii) $\dfrac{1}{4}\begin{bmatrix} 3 & 1 & -1 \\ 1 & 3 & 1 \\ -1 & 1 & 3 \end{bmatrix}$.

4. $0 \le \lambda \le 57$ **5.** (a) 5.38, $\begin{bmatrix} 0.46 \\ 1 \end{bmatrix}$; (b) 4.618, $\begin{bmatrix} 1 \\ 0.618 \end{bmatrix}$

6. (a) 11.66, $[0.025, 0.422, 1]'$; (b) 7; $[2.099/7, 0.467/7, 1]'$

(c) 25.182, $[1, 0.045, 0.068]'$.

Exercises 4.3

1. (a) 4, -2, 6; $\left[1/\sqrt{2}, 0, -1/\sqrt{2}\right]'$, $[0, 1, 0]'$, $\left[1/\sqrt{2}, 0, 1/\sqrt{2}\right]'$,

(b) 0.3856, -1.3126, 5.9269; $[0.5654, -0.2949, -0.8243]'$,

$[0.5457, -0.736, 0.4006]'$, $[-0.6185, -0.6763, -0.4001]'$.

2. $\begin{bmatrix} 1 & -1.12 & 0 \\ -1.12 & 1.4 & -0.55 \\ 0 & -0.55 & 1.6 \end{bmatrix}$ **3.** 0, 3, 3. **4.** $\begin{bmatrix} 1 & 2\sqrt{2} & 0 \\ 2\sqrt{2} & 3 & 0 \\ 0 & 0 & -1 \end{bmatrix}$; 5$\begin{bmatrix} 1 \\ 1 \\ 1 \end{bmatrix}$ **5.** 3, 1, 3.

Exercises 4.4

1. Zero or unity.   **2.** $\begin{bmatrix} 2 & 3 \\ 3 & 5 \end{bmatrix}$.   **3.** $\begin{bmatrix} 1 \\ 1 \end{bmatrix}$ **4.** 1, 1, 1/5.

5. $5, -2$.   **6.** A^{-1} **7.** 2.   **8.** largest.

9. $1 \pm \sqrt{2}$.   **10.** eigenvalue.   **11.** 5.38.

12. either zero or unity. **13.** -4

14. $\begin{bmatrix} 0.8 & -0.4 & -0.2 \\ -0.2 & 0.6 & -0.2 \\ -0.2 & -0.4 & 0.8 \end{bmatrix}$

15. the largest   **16.** $\dfrac{1}{\lambda} X = A^{-1} X$.

Exercises 5.1

1. $a = 2.28, b = 6.19, p = 30.46$.

2. (i) $Y = a + bX$, where $X = x, Y = y/x$.   (ii) $Y = a + bX$ where $X = 1/xy$, $Y = x$

3. $a = 1120, b = 55.1$.   **4.** $a = 0.2, b = 0.0044$.   **5.** $n = 1.3, c = 200$.

6. $a = 0.5012, n = 0.5$.   **7.** $a = 4.1, b = 0.43$.   **8.** $a = 0.05, b = -0.02$.

Exercises 5.2

1. $y = 13.6x$.   **2.** $v = 1.758 - 0.053\theta$.

3. $Y = 0.004 P + 0.048$.   **4.** $R = 70.052 + 0.29\,t$.   **5.** $y = 48.9 + 0.5067x$.

6. $y = 1.243 - 0.004x + 0.22x^2$.   **7.** $y = -0.703 - 0.858x + 0.992\,x^2$.

8. $y = -0.98x^2 + 3.55x - 27x^2$.   **9.** $V = 2.593 - 0.326\,T + 0.023T^2$.

10. $R = 3.48 - 0.002\,V + 0.003V^2$.

Exercises 5.3

1. $a = 6.32, b = 0.0095$.   **2.** $a = 1.52, b = 0.49$.

3. $a = 3, b = 2$.   **4.** $y = 7.187 - 5.16/x;\ 4.894$.

5. $a = 0.5012, b = 1.9977$.   **6.** $y = 2.978x^{0.5143},\ 5.8769$

7. $k = 7.17, m = 1.95$.   **8.** $a = 9.484, b = 0.315$

9. $y = 0.1839\,e^{0.0221x}$.   **10.** $a = 32.15, b = 1.43, N = 387$.

11. $a = 146.3, k = -0.412$.   **12.** $f(t) = 0.678e^{-3t} + 0.312e^{-2t}$

13. $x = 0.999, y = 2.004$.   **14.** $x = 1.17, y = -0.75, z = 2.08$.

Exercises 5.4

1. $a = 11.1, b = 0.71$.   **2.** $a = 2.1, b = 0.19, p = 30.6$.

3. $y = 46.05 + 6.1\,x$.   **4.** $a = 6.73, b = 0.0092$.

5. $c = 2.6, n = 2.5$.   **6.** $a = 0.0028, b = 0.01, c = 4.18$.

7. $a = 15.8, b = 2.1, c = -0.5$.   **8.** $a = 1.459, b = 0.062$

9. $a = 23.4, b = 97.7, k = -0.45$. **10.** $a = 10, b = 3.1, n = -0.1$.

Exercises 5.5

1. $y = 0.12 + 0.47x$.   **2.** $y = 1.184 + 0.523\,x$

3. $y = 1.53 + 0.063x + 0.074x^2$.   **4.** $y = 0.485 + 0.397x + 0.124x^2$.

Exercises 5.6

1. zero.   **2.** $y = aX + c$, where $X = x^b$.   **3.** (ii)

4. $\Sigma y = nA + B\Sigma x$, $\Sigma xy = A\Sigma x + B\Sigma x^2$ where $y = \log_{10} y$, $A = \log_{10} a$, $B = \log_{10} b$.

5. Section 5.9.   **6.** Section 5.6.   **7.** $Y = aX + b$ where $X = x^2/\log_{10} x$, $Y = y/\log_{10} x$.

8. $a = 0.0167, b = 1.05$.

9. The moments of the observed values of y are respectively equal to the moments of the calculated values of y.

10. $a = 1.7, b = 1.26$.

11. $y = a + bx$ where $x = \log_{10} p, y = \log_{10} v,\ a = \dfrac{1}{\gamma}\log k, a = -\dfrac{1}{\gamma}$

12. $Y = a + bX$, where $X = 1/x$, $Y = 1/y$.

13. (C) **14.** (C) **15.** (B).   **16.** (A).

Exercises 6.1

2. 0.4.   **3.** $- 7459$.   **5.** 241.   **6.** 239.

7. 4.68, 2.68; 55.8, 99.88.

9. (i) $1 - 2\sin (x + 1/2) \sin 1/2, (ii)$ $\tan^{-1}\left(\dfrac{1}{2n^2}\right)$, (iii) $e^{3x} \{e^3 \log (1 + 1/x) + (e^3 - 1)$

$\log 2x\}$ (iv) $2^x (1 - x)(1 + x)$ (v) $192/[x(x + 4) (x + 8) (x + 12) (x + 16)]$
$(vi) - 2/[(x + 2)(x + 3)(x + 4)]$.

10. (i) $(e^2 - 1)^n\, e^{2x+3}$; (ii) $(- 1)^n\, n!/[x(x + 1)(x + 2)\cdots (x + n)]$

(iii) $\left(2\sin\dfrac{\alpha}{2}\right)^n \sin\left(ax + b + \dfrac{na + n\pi}{2}\right)$ (iv) $(e^{nh} - 1)^n\, e^{2x + 3}$

(v) $(- 1)^n\, n!\, \{x(x + 1)(x + 2)... (x + n)\}^{-1}$.

14. (i) 576; (ii) $24 \times 210 \times 10\,!$.

Exercises 6.2

1. $[x]^3 + [x]^2 - 1$.   **2.** $y = 3[x]^4 + 14[x]^3 + 15[x]^2 + 7[x] + 1, \Delta^4 y = 72$.

3. $4x^3 - 12x^2 + 8x + 1; 12x(x - 1)$.

4. (i) $\dfrac{1}{2}[x]^4 + 3[x]^3 + 4[x] + c.$ (ii) $\dfrac{1}{60}\left(12x^5 - 105x^4 + 170x^3 + 15x^2 + 148\right)$

6. $80640(2x + 9)(2x + 11)\ldots(2x + 19).$

7. $192\{(4x + 1)(4x + 5)\ldots(4x + 17)\}^{-1}; \dfrac{1}{8}\{(4x+1)(4x+5)\}^{-1}$

8. 2.0086.   **9.** 15.   **10.** $x^3 + 2x^2 + 3x.$

Exercises 6.3

2. (i) $2(\cos h - 1)\sin x;$ (ii) 8.

  (iii) $6h^2(x + h)^{-2};$ (iv) $2(\cos h - 1)[\sin(x + h) + 1].$

10. 31.   **11.** $f(1.5) = 0.222, f(5) = 22.022$

12. $y(4) = 74, y(6) = 261.$ **13.** $y(2004) = 306, y(2006) = 390.$

14. $-99.$ **15.** $y_4 = 1$ approx.

Exercises 6.4

1. $n(3n^2 + 6n + 1).$ **2.** $\dfrac{n(n+1)(n+2)(n+3)}{n}.$ **3.** $\dfrac{1}{4}\left\{1 - \dfrac{2}{(n+1)(n+2)}\right\}$

4. $\dfrac{1}{2}\left\{\dfrac{1}{4.5} - \dfrac{1}{(x+4)(x+5)}\right\}$ **5.** $\dfrac{n(n+1)(2n+1)}{6}.$

6. (i) $e^x(1 + 7x + 6x^2 + x^3);$ (ii) $e^x\left(1 + 3x + \frac{3}{2}x^2 + \frac{1}{6}x^3\right).$

7. (i) $(3 + 6x - x^2)(1 - x)^{-3};$ (ii) $(1 + x)(1 - x)^{-3}.$ **8.** $\dfrac{n}{30}(6n^4 + 15n^3 + 10n^2 - 1).$

9. $2[(x - 2)^n - (x - 3)^n].$

Exercises 6.5

1. $(a).$ **2.** $(b).$ **3.** 18.   **4.** $E = e^{hD}.$

5. zero.   **6.** $3x(x - 1)$ **7.** $[x]^3 + [x]^2 - 1.$ **8.** $6h^2(x + h).$

9. $1 - E^{-1}.$ **10.** a polynomial of the 6th degree.   **11.** zero.

12. $4x^3 - 3x^2 - 5x.$ **13.** 1, 3, 7.   **14.** -90

15. $\tan^{-1}\left(\dfrac{h}{1 + hx + x^2}\right).$ **16.** 2.

17. $\Delta = E - 1.$ **18.** Constant.   **19.** $\nabla^r y_{k+r}$ **20.** $e^{-h}\Delta^2 e^x.$

21. $\nabla^2 = (1 - E^{-1})^2.$ **22.** $(e - 1)^n e^x.$ **23.** $\Delta = E\nabla.$ **24.** 5.

25. $\nabla - \nabla^2.$ **26.** $(x - 2)^3.$ **27.** 16.5.   **28.** $(d).$ **29.** $(c).$

30. $(c).$ **31.** $(d).$ **32.** True   **33.** False.

34. False.   **35.** True.   **36.** True.   **37.** True.

Exercises 7.1

1. 5.54.   **2.** 6.36; 11.02

3. $\theta = 0.01\,x^2 + 1.01x + 130.1$; $\theta(x = 43) = 192.02$.   **4.** 0.788.

5. 4.43.   **6.** 8666.   **7.** 0.9623, 0.2903.

8. $(i)\ I_c = 0.5878$ $(ii)\ f_b = 0.363$.   **9.** 352.   **10.** 24.

11. 14706 approx.   **12.** 1.625.   **13.** 33.   **14.** 0.1955.

15. $y = \dfrac{2}{3}x^4 - 8x^3 + \dfrac{100}{3}x^2 - 56x + 31$.   **16.** 2530.   **17.** 0.1; 100.

18. 369 metric tons.   **19.** $u_2 = 42$, $u_4 = 49$.   **20.** 10, 22.   **21.** 755.

Exercises 7.2

1. 32.95.   **2.** $f(x) = \dfrac{2}{3}x^4 - 8x^3 + \dfrac{100}{3}x^2 - 56x + 31$.   **3.** 19.4.

4. 3.2219   **5.** 54000.   **6.** 0.70711.   **7.** 395.

8. 3.0375   **9.** 0.934.   **10.** 9.   **11.** 32.945.

12. 3.347.   **13.** 14.368   **14.** 3250.875.

15. 2.5283 by all formulae.

Exercises 7.3

1. 14.63   **2.** 7.03.   **3.** 2.8168.   **4.** 0.89.

5. 100.   **6.** $648 + 30x - x^2$.   **7.** $x^3 - 3x^2 + 5x - 6$.

8. $x^5 - 9x^4 + 18x^3 - x^2 + 9x - 18$.   **9.** 3.

10. $\dfrac{0.5}{x-1} - \dfrac{0.5}{x+1} + \dfrac{1}{x-2}$ **11.** $\dfrac{1}{5}\dfrac{1}{x-1} + \dfrac{3}{35}\dfrac{1}{x+1} - \dfrac{13}{10}\dfrac{1}{x-4} + \dfrac{71}{70}\dfrac{1}{x-6}$.

Exercises 7.4

1. 1   **2.** 133.19.   **3.** 100.   **4.** 1.48 mV.

5. $f(x) = \dfrac{1}{24}(x^3 - 25x + 24)$.   **6.** $f(x) = \dfrac{1}{20}x^3 - \dfrac{7}{6}x^2 + \dfrac{557}{60}x - 25$.

7. $f(x) = x^4 - 3x^3 + 5x^2 - 6$.   **8.** 147   **9.** $y = \dfrac{1}{6}(x^3 - x^2 + 4x - 6)$.

10. 31.

Exercises 7.5

1. $P(x) = x^4 + x^2 + 1.$ **2.** $P(x) = \dfrac{1}{16}(6-x)x^2.$ **3.** 1.1631.

Exercises 7.6

1. $y(x) = \begin{bmatrix} 3x^3 - 9x^2 + 11x - 11, & 1 \le x \le 2 \\ -3x^3 + 27x^2 - 61x + 37, & 2 \le x \le 3 \end{bmatrix}, y(1.5) = -4.625;\ y'(2) = 11.$

3. $y(x) = \frac{1}{6}\left[-142.9x^3 + 1058.4x^2 - 2475.2x + 1950\right]$

   (i) y(2.5) = -24.03, (ii) y′(3) = 2.817.

Exercises 7.7

1. 11.5   **2.** 6.304   **3.** 37.23.   **4.** 2.3.

5. 0.2679   **6.** 1.3714.

Exercises 7.8

1. Section 7.3.   **2.** (b)

3.

x	$f(x)$	I.D.D.	II.D.D.
5	7		
		2.9	
15	36		0.87
		17.7	
22	160		

4. Intermediate value of the variable.   **5.** Section 7.8.

6. $\dfrac{\left[x_1,x_2,x_3,x_4\right] - \left[x_0,x_1,x_2,x_3\right]}{x_4 - x_0}$ **7.** $\dfrac{-1}{4}$ and $\dfrac{1}{4}$.

8. Section 7.14

9. $f(x) = \dfrac{\left(x-x_1\right)\left(x-x_2\right)}{\left(x_0-x_1\right)\left(x_0-x_2\right)} + \dfrac{\left(x-x_0\right)\left(x-x_2\right)}{\left(x_1-x_0\right)\left(x_1-x_2\right)} + \dfrac{\left(x-x_0\right)\left(x-x_1\right)}{\left(x_2-x_0\right)\left(x_2-x_1\right)}$

10. $\dfrac{13}{5}.$

11. Lagrange's interpolating polynomial $P(x)$ agrees with $y(x)$ at the points x_0, $x_1,\dots, x_n$ whereas Hermite's interpolating polynomial $P(x)$ and $y(x)$ as well as $P'(x)$ and $y'(x)$ agree at the said $(n+1)$ points.

12. 1.857

13. *Extrapolation* is the process of estimating the value of a function outside the given range of values. **14.** $1/(abc)$. **15.** (a).

16. $x^3 - 7x^2 + 18x - 12$. **17.** (b). **18.** (c).

Exercises 8.1

1. -27.9, 117.67. **2.** 4.75, 9. **3.** 0.63, 6.6.

4. (a) 0.493, -1.165 **(b)** 0.4473, -0.1583; **(c)** 0.4662, -0.2043.

5. 2.8326. **6.** 1.4913. **7.** -0.06; 0.5.

8. (a) 0.3907; **(b)** 0.9848; **(c)** 0.342.

9. 7.956

10. (i) -52.4, (ii) -0.0191. **11.** 44.92.

12. 3. **13.** 3.82 rad./sec., 6.75 rad./sec.^{2}. **14.** 0.5403.

15. 0.2561. **17.** 0.1086. **18.** $y'(4) = 2.883$.

19. 135. **20.** $y_{max}(1) = 0.25$, $y_{min}(0) = 0$.

21. 0.692, 0.6137. **22.** Max $f(10.04) = 1340.03$.

Exercises 8.2

1. 0.26. **2.** (i) 0.695 (ii) 0.693

(iii) 0.693. **3.** (i) 0.7854, (ii) 0.7854,

(iii) 0.78535. (iv) 0.7854.

4. 1.61. **5.** 53.87, 53.6. **6.** 70.16.

7. 0.635. **8.** (i) 2.0009 (ii) 1.1873.

9. (i) 1.1249 (ii) 0.9744 (iii) 0.0911 (iv) 14.51086.

10. (a) $1.8276551, .0001924$; **(b)** $1.8278472, .0000003$;

(c) $1.8278470, .0000005$; **(d)** $1.8278474, .0000001$.

11. 1.3028. **12.** 403.67. **13.** 7.78. **14.** 710 sq. ft.

15. 3.032. **16.** 3.032. **17.** 408.8 cub. cm.

18. 1.063 sec; 1.064 sec. **19.** 552 m.; 3 m./sec.^{2}. **20.** 30.87 m/sec.

21. 29 min. nearly.

Exercises 8.3

1. $n = 8$ **2.** 0.3927-; $\pi = 3.1416$ **3.** 1.8278

4. 1.000003. **6.** (a) 0.01138; **(b)** 0.00083.

7. $\displaystyle\int_0^h y\,dx = \frac{h}{2}(y_0 + y_1) - \frac{h^2}{2}(y_0' + y_1')$. **8.** (i) 4.685 (ii) 1.00002

9. (i) 1.6027 (ii) 0.2376.   **10.** (i) 200.4014 (ii) 0.2666.   **11.** 0.4999.

Exercises 8.4

1. 0.876.

2. (i) 3.076   (ii) 0.31913.

3. 25.375.

4. 4.134.

5. (i) 0.49.   (ii) 0.3844.

Exercises 8.5

1. (c)

2. $\dfrac{1}{h}\left[\Delta f(a)-\dfrac{1}{2}\Delta^2 f(a)+\dfrac{1}{3}\Delta^3 f(a)-\ldots\right]$

3. h should be small.   **4.** 0.775.   **5.** $2\dfrac{2}{3}$ **6.** Section 8.4 (III).

7. (b) Section 8.4 (I).   **8.** larger number of sub-intervals. **9.** 0.7854.

10. Section 8.4 (II).

11. $I=\dfrac{hk}{2}\left[\left(f_{00}+f_{02}\right)+2f_{01}+\left(f_{10}+f_{12}+2f_{11}\right)+\left(f_{20}+f_{22}+2f_{21}\right)\right].$

12. $\displaystyle\int_{-1}^{1}f(x)dx=f\left(-1/\sqrt{3}\right)+f\left(1/\sqrt{3}\right).$ **13.**

14. a multiple of 6.   **15.** (b). **16.** $I=\dfrac{I_1 h_2^2-I_2 h_1^2}{h_2^2-h_1^2}.$

17. (c).   **18.** $\dfrac{-h^2}{12}\left(y_n'-y_0'\right)+\dfrac{h^4}{720}\left(y_n'''-y_0'''\right)+\ldots$

19. $\dfrac{8}{9}f(0)+\dfrac{5}{9}\left[f\left(-\sqrt{\dfrac{3}{5}}\right)+f\left(\sqrt{\dfrac{3}{5}}\right)\right].$ **20.** second and fourth.

21. 0.783 ($b-a$)   **22.**   where $nh=b-a$.

23. 0.69.   **24.** 1.36.

25. If the entire curve is itself a parabola.   **26.** False.

Exercises 9.1

1. $y_{x+3}-2y_{x+2}+2y_{x+1}=0.$ **2.** $\Delta y_n=(-1)^{n+1}/(n+1).$ **3.** $u_{n+1}-2u_n=0$

4. (i) $(x+2)y_{x+2}-2(x+1)y_{x+1}+xy_x=0;$

   (ii) $(x^2+x)y_{x+2}-(2x^2+4x)y_{x+1}+(x^2+3x+2)y_x=0.$

5. (i) $y_{n+2}-8y_{n+1}+15y_n=0;$ (ii) $y_{n+2}-6y_{n+1}+4y_n=0.$

6. (i) $(x-1)y_{x+2}-(3x-2)y_{x+1}+2xy_x=0;$

   (ii) $y_{x+2}-4y_x=0;$ (iii) $y_{x+3}-6y_{x+2}+11y_{x+1}-6y_x=0.$

Exercises 9.2

1. $u_n = (1-n)^{2n}$.

2. $y_n = c_1 \cos \dfrac{2n\pi}{3} + c_2 \sin \dfrac{2n\pi}{3}$.

3. $u_n = c_1 \cos n\pi/2 + c_2 \sin n\pi/2$.

4. $y_n = c_1 . 2^n + c_2 . 3^n$.

5. $y_n = (2)^{n-1} + (-2)^{n-1}$.

6. $u_k = c_1(-1)^k + (c_2 + c_3 k)2^k$.

7. $f(x) = (c_1 + c_2 x)(-1)^x + c_3 . 2^x$.

8. $u_n = 2n + (-2)^n$.

9. $y_n = 6 + (n-3)2^n$.

10. $u_n = 2^{n/2}\{c_1 \cos n\pi/4 + c_2 \sin n\pi/4\}$.

11. $y_m = 2^m\left\{c_1 \cos\dfrac{m\pi}{4} + c_2 \sin\dfrac{m\pi}{4} + c_3 \cos\dfrac{3m\pi}{4} + c_4 \sin\dfrac{3m\pi}{4}\right\}$

15. $y_n = c_1(-1)^n + c_2(10)^n$.

Exercises 9.3

1. $y_n = c_1(-1)^n + c_2(6)^n - 2^n/12$.

2. $y_n = \left(\dfrac{n}{15} - \dfrac{1}{25}\right)(-3)^n + \dfrac{2^n}{25}$

3. $y_p = c_1 + c_2 p + c_3 p^2 + \dfrac{1}{6}p(p-1)(p-2)$.

4. $y = c_1 + c_2 \cdot 3^x + \dfrac{1}{2}x \cdot 3^{x+1}$.

5. $u_x = c_1 . 2^x + c_2 . 5^x - 6 . 4^x$.

6. $y_x = (c_1 + c_2 x)2^x + 3x(x-1)2^{x-3} + 5 . 4^{x-1}$.

7. $u_n = c_1 + c_2(-1)^n + \dfrac{1}{2}\dfrac{\cos\left(\dfrac{n}{2}-1\right) - \cos\dfrac{n}{2}}{1 - \cos 1}$.

8. $y_p = c_1 \cos\dfrac{p}{2} + c_2 \sin\dfrac{p}{2} + \dfrac{p\cos\left(p-\frac{1}{2}\right)}{2\sin\frac{1}{2}}$.

9. $y_n = 2^n\left(\dfrac{2}{\sqrt{3}}\sin\dfrac{n\pi}{3} - 2\cos\dfrac{n\pi}{3}\right) + 2$.

10. $y_x = c_1 + 2^x + c_2(-2)^x - \dfrac{1}{27}\left(9x^2 + 12x + 11\right)$.

11. $y_n = c_1(-1)^n + c_2 \cos\dfrac{n\pi}{3} + c_3 \sin\dfrac{n\pi}{3} + \dfrac{1}{2}n(n-3)$.

12. $y_n = \left(c_1 + c_2 n\right)(3)^n + c_3(-1)^n + \dfrac{1}{3}(2)^n - \dfrac{3n}{4}$.

13. $y_n = \left(c_1 + c_2 n\right)2^{-n} + \dfrac{2^n}{9} + n(n-1)\left(\dfrac{1}{2}\right)^{n-1}$.

14. $y_n = c_1(-2)^n + c_2(-3)^n + \dfrac{n}{12} - \dfrac{7}{144}$.

15. $u_x = \left(c_1 + c_2 x\right)(-3)^x + \dfrac{2^x}{25}(5x - 2) + \dfrac{2}{4}x^{x-2} + \dfrac{7}{16}.$

16. $y_n = c_1(-2)^n + 2^n\left(c_2 \cos n\pi/3 + c_3 \sin n\pi/3\right) + \dfrac{3}{16}(2)^n + 2^{n-4}(2n + 3).$

17. $u_n = \left\{c_1 + c_2 n + \dfrac{1}{48}n(n-1)^2(n-2)\right\}2^n.$

18. $y_k = c_1 \cdot 2^k + c_2 \cdot 3^k + \dfrac{4^k}{2}(k^2 - 13k + 61).$

19. $y_n = 2^n \left\{(c_1 + n)\cos\dfrac{n\pi}{3} + c_2 \sin\dfrac{n\pi}{3}\right\}.$

Exercises 9.4

1. $y_{x+1} = ay_x,$ Sol. is $y_x = ca^x.$ **2.** $y_x = 2e^{2x-1}.$

3. $y_x = c\left(\dfrac{1}{2}\right)^x$ or $y_x = c(-1)^x.$ **4.** $y_n = e^{c(2)^{-n}}$

5. $y_x = \dfrac{c_1 + c_2 / 2^{x+1}}{c_1 + c_2 / 2^x} \cdot .$ **6.** $y_x = \dfrac{c_1 + c_2(x+1)2^{x+1}}{c_1 + c_2(x)2^x} - 5.$

Exercises 9.5

1. $y_x = a + b(-1)^x + x,\ z_x = a + b(-1)^{x+1} - (x + 1).$

2. $y_x = (a + bx)(-1)^x - \dfrac{1}{9}\cdot 2^{x+2},\ z_x = \dfrac{2^x}{9} - (-1)^x\left[a + b\left(x - \tfrac{1}{2}\right)\right].$

3. $u_n = 2 \cdot 4^n - 2 - \tfrac{1}{2}n(n-1),\ v_n = 4^n + 2 + \tfrac{1}{2}n(n+1).$

4. $u_x = -2a + b(-2)^x - c + \tfrac{1}{2}x(3 - x),\ v_x = a + c + b(-2)^x,$

$w_x = a + b(-2)^x + \tfrac{1}{2}x(x\_1).$

Exercises 9.6

1. $y_{i+1} - 2y_i + y_{i-1} = -\dfrac{l_m}{P}y_i.$ Solve it for $y_i.$

Exercises 9.7

1. $y_{n+2} - 5y_{n+1} + 6y_n = 0.$ **2.** $u_n = c_1 + c_2 n + c_3 n^2.$ **3.** $u_n = c_1 + c_2(-2)^n + c_3(3)^n.$

4. $y_n = 1 + 2^n.$ **5.** $y_n = c(2)^n - (n + 1).$

6. $(x^2 + x)y_{x+2} - (2x^2 + 4x)y_{x+1} + (x^2 + 3x + 2)y_x = 0.$ **7.** $y_n = c(2)^k + 1.$

8. $y_n = (2)^{n-1} + (-2)^{n-1}.$ **9.** Third.

10. $(x + 2)y_{n+2} - 2(n + 1)\, y_{n+1} + ny_n = 0$

11. Second. **12.** $(C_1 + C_2 n)\, 2^n$. **13.** $\dfrac{1}{2}x(x-1)(3)^{x-2}$.

14. $y_{n+2} - 6y_{n+1} + 9y_n = 0$ **15.** True.

Exercises 10.1

1. $y = 1 - \dfrac{x^2}{2} + \dfrac{x^4}{8} - \dfrac{x^6}{48}$. **2.** 0.0214. **3.** $y = \dfrac{1}{3}x^3 - \dfrac{1}{81}x^9 + \ldots$

4. (a) and (b) 0.9138.

5. $y(1.1) = 0.1103$, $y(0.2) = 0.2428$, Exact values $y(1.1) = 0.1103$, $y(1.2) = 0.2428$.

6. 2.02061. **7.** 1.1053425. **8.** 1.1164; 1.2725.

9. 1.00035. **10.** 1.005.

Exercises 10.2

1. 1.1831808. **2.** 0.4748. **3.** 1.1448.

4. $y(0.1) = 0.095$, $y(0.2) = 0.181$, $y(0.3) = 0.259$. **5.** 2.2352

6. $y(0.2) = 1.2046$, $y(0.4) = 1.4644$ **7.** 1.0928. **8.** 5.051.

Exercises 10.3

1. 1.7278. **2.** 2.5005. **3.** 1.0207, 1.038.

4. 2.5005 **5.** $y(0.2) = 2.44$, $y(0.4) = 2.99$, $y(0.6) = 3.68$.

6. $y(0.1) = 0.9052$, $y(0.2) = 0.8213$. **7.** $y(0.1) = 2.9917$, $y(0.2) = 2.9627$.

8. 1.1678. **9.** 1.1749. **10.** $y(0.5) = 3.219$, $y(1) = 3$.

11. 0.3487. **12.** 1.0911, 1.1677, 1.2352, 1.2902, 1.338

Exercises 10.4

1. 3.795. **2.** 1.2797 **3.** 1.5 approx.

4. $y(1.4) = 3.0794$, **5.** 1.837 **6.** $y(0.4) = 2.162$

7. 0.441.

Exercises 10.5

1. 0.2416. **2.** 1.0408 **3.** 0.6897.

4. $y(4.4) = 1.019$. **5.** 2.5751. **6.** $y(1.4) = 0.949$.

Exercises 10.6

1. $y_3 = 1 + \dfrac{x}{2} + \dfrac{3}{40}x^5 + \dfrac{1}{40}x^6 + \dfrac{1}{192}x^9.$

$z_3 = \dfrac{1}{2} + \dfrac{3}{8}x^4 + \dfrac{1}{10}x^5 + \dfrac{3}{34}x^8 + \dfrac{7}{340}x^9 + \dfrac{1}{256}x^{12}.$

2. $x(0.4) = 0.5024, y(0.4) = 0.6012.$

3. $y(0.1) = 0.105, y(0.2) = 0.22, z(0.1) = 0.999, z(0.2) = 0.997.$

4. $y(0.1) = 2.084, z(0.1) = 0.587.$

5. $y_2 = 1 + \dfrac{1}{2}x + \dfrac{3}{40}x^5.$ **6.** $0.5075.$ **7.** $1.1404.$

8. $y(0.2) = 0.9802, y'(0.2) = -0.196$ **9.** $-0.5159.$

10. $\theta(0.2) = 0.8367, (d\theta/dt)_{0.2} = 3.6545.$

11. $v(0.02) = 0.9965\, v_0, (dv/dt)_{0.02} = -0.3292\, v_0.$

Exercises 10.7

5. $(a)\ 0 < h < 0.2;$ $(b)\ 0 < h < 0.278.$

Exercises 10.8

1. $0.14031.$ **2.** $y(.25) = y(.75) = 2.4, y(.5) = 3.2.$

3. $y(1.25) = 1.3513, y(1.5) = 1.635, y(1.75) = 1.8505.$

4. $y(.25) = -0.3473, y(.5) = -0.9508, y(.75) = -1.7257.$

5. $n = 2, y(0.5) = 0.1389,$ true value $= 0.1505; n = 4, y(0.5) = 0.147.$

6. $y(.25) = 0.062, y(.5) = 0.25, y(.75) = 0.562.$ **7.** $y(1) = 1.0171, y(2) = 1.094.$

8. $y(1) = 7.4615.$ **9.** $0.189, 0.642, 1.217, 1.740.$

10. Taking $m_0 = 0.8, m_1 = 0.9,$ we get $m_2 = 0.9998$ and $y(1) = 1.174.$

11. Taking $m_0 = -1.8, m_1 = -1.9,$ we get $m_2 = -2$ and $y(0.5) = 0.4441.$

Exercises 10.9

1. $(b).$ **2.** $y_{i+1} - 2y_i + y_{i-1} + h^2 y_i = 5h^2.$

3. $1 + x + x^2 + x^3/6.$ **4.** Section 10.4.   **5.** Section 10.3.

6. That it can be applied to those equations only in which succesive integrations can be performed easily.

7. $y_4 = y_2 + \dfrac{h}{3}(f_2 + 4f_3 + f_4).$

8. Modified Euler's method. **9.** $y_1 = y_0 + \dfrac{h}{24}\left(55f_0 - 59f_{-1} + 37f_{-2} - 9f_{-3}\right)$.

10. It does not require prior calculations of higher derivatives, as the Taylor's method does.

11. four.   **12.** $y = 1 + \dfrac{x^2}{2} + \dfrac{x^4}{8}$ **13.** Runge's method

14. $y_4 = y_0 + \dfrac{4h}{3}\left(2f_1 - f_2 + 2f_3\right)$

15. $y_1 = y_0 + \dfrac{h}{24}\left(9f_1 + 19f_0 - 5f_{-1} + f_{-2}\right)$ **16.** 1.1818

17. $dy/dx = z, dz/dx + y(1 + yz) = 0$ **18.** Section 10.7(iv).

19. starting values.   **20.** Picard's and Runge-Kutta methods.

21. It agrees with Taylor's series solution upto the terms in $h4$.

22. 1.2.   **23.** Section 10.17 (3) and (4)

24. (a).   **25.** Section 10.1 (3).   **26.** (c).

27. $y_{i+1} + 2\,(h^2 - 1)\,y_i + y_i - 1 = 0$.   **28.** ($b$).

29. Milne's method & Adam-Bashforth method.   **30.** (b).

31. True.   **32.** False.   **33.** False.

Exercises 11.1

1. Parabolic   **2.** Hyperbolic.

3. (i) Parabolic.   (ii) Elliptic.   (iii) Elliptic.

4. Outside the ellipse $(x/0.5)^2 + (y/0.25)^2 = 1$.

Exercises 11.2

1. $u_1 = 7.9, u_2 = 13.7, u_3 = 17.9, u_4 = 6.6, u_5 = 11.9, u_6 = 16.3, u_7 = 6.6, u_8 = 11.2,$
$u_9 = 14.3$.

2. $u_1 = 2.38, u_2 = 5.6, u_3 = 9.87, u_4 = 2.89, u_5 = 6.14, u_6 = 9.89, u_7 = 3.02, u_8 = 6.17,$
$u_9 = 9.51$.

3. $u_1 = 26.66, u_2 = 33.33, u_3 = 43.33, u_4 = 46.66.$ **4.** $u_1 = 0.99, u_2 = 1.49, u_3 = 0.49$.

5. $u_1 = 1.999, u_2 = 2.999, u_3 = 3.999, u_4 = 2.999$.

6. (a) $u_1 = 0.126, u_2 = 0.126, u_3 = 0.376, u_4 = 0.376$.

  (b) $u_1 = 0.126, u_2 = 0.126, u_3 = 0.376, u_4 = 0.376$.

7. $u_1 = 1.57, u_2 = 3.71, u_3 = 6.57, u_4 = 2.06, u_5 = 4.69, u_6 = 8.06, u_7 = 2, u_8 = 4.92,$
$u_9 = 9$.

8. $u_1 = -3, u_2 = -2, u_3 = -2.$ **9.** $u_1 = u_4 = -4.5, u_2 = -6.25, u_3 = -2.75$.

Exercises 11.3

1. $u_1 = 1.9$, $u_2 = 4.9$, $u_3 = 9.1$; $u_4 = 2.1$, $u_5 = 4.7$, $u_6 = 8.4$; $u_7 = 1.6$, $u_8 = 3.9$, $u_9 = 6.7$.

2. $u_1 = 275$, $u_2 = 350$, $u_3 = 275$; $u_4 = 350$, $u_5 = 450$, $u_6 = 350$; $u_7 = 275$, $u_8 = 350$, $u_9 = 275$.

3. $u_1 = 1$, $u_2 = 1.3$, $u_3 = 0.7$, $u_4 = 1$.

Exercises 11.4

1.

i \ j	0	1	2	3	4
0	0	3	4	3	0
1	0	2	3	2	0
2	0	1.5	2	1.5	0
3	0	1	1.5	1	0
4	0	0.75	1	0.75	0
5	0	0.5	0.75	0.5	0

2.

j \ i	0	1	2	3	4	5	6	7	8	9	10
0	0	0.09	0.16	0.21	0.24	0.25	0.24	0.21	0.16	0.09	0
1	0	0.08	0.15	0.20	0.23	0.24	0.23	0.20	0.15	0.08	0
2	0	0.075	0.14	0.19	0.22	0.23	0.22	0.19	0.14	0.075	0
3	0	0.07	0.13	0.18	0.21	0.22	0.21	0.18	0.13	0.07	0

3.

i \ j	0	1	2	3	4	5
0	0	24	84	144	144	0
1	0	42	84	114	144	0
2	0	42	78	78	72	0
3	0	39	60	67.5	57	0
4	0	30	53.25	49.5	39	0
5	0	26.6	39.75	43.5	33.75	0
6	0	19.88	35.06	32.25	24.75	0

4.

j \ i	0	1	2	3	4
0	0	5	8	7	0
1	0	3.14	4.57	3.14	0
2	0	1.75	2.45	1.75	0

5.

j \ i	0	1	2	3	4
0	0	0.5	1	0.5	0
1	0	0.5	0.5	0.5	0
2	0	0.25	0.5	0.25	0
3	0	0.25	0.25	0.25	0

6. $u_{1,1,1} = u_{2,1,1} = u_{1,2,1} = u_{2,2,1} = 9/16$, $u_{1,1,2} = u_{2,1,2} = u_{1,2,2} = u_{2,2,2} = 27/64$.

Exercises 11.5

1.

$t = 0.3, x =$	0.1	0.2	0.3	0.4	0.5
Num. sol. $u =$	0.02	0.04	0.06	0.075	0.08
Exact sol. $u =$	0.02	0.04	0.06	0.075	0.08

2.

j \ i	0	1	2	3	4	5
0	0	20	15	10	5	0
1	0	7.5	15	10	5	0
2	0	− 5	2.5	10	5	0
3	0	− 5	− 10	− 2.5	5	0
4	0	− 5	− 10	− 15	− 7.5	0
5	0	− 5	− 10	− 15	− 20	0

3.

t \ x	0	0.1	0.2	0.3	0.4	0.5
0.1	0	0.037	0.07	0.096	0.113	0.119
0.2	0	0.031	0.059	0.082	0.096	0.101
0.3	0	0.023	0.043	0.059	0.07	0.074
0.4	0	0.012	0.023	0.031	0.037	0.039
0.5	0	0	0	0	0	0

4.

i / j	0	1	2	3	4
0	0	0	0	0	0
1	0	0	0	0	70.7
2	0	0	0	70.7	100
3	0	0	70.7	100	70.7
4	0	70.7	100	70.7	0

Exercises 11.6

1. (b).   **2.** False.   **3.** a hyperbolic equation.

4. Poisson's equation.   **5.** $u_{i,j+1} = u_{i-1,j} + u_{i+1,j} - u_{i,j-1}$.

6. $(u_{i-1,j} - 2u_{i,j} + u_{i+1,j})/h^2$ **7.** Section 11.9 (1).

8. $\frac{1}{4}\left(u_{i-1,j+1} + u_{i+1,j-1} + u_{i+1,j+1} + u_{i-1,j-1}\right)$

9. hyperbolic   **10.** Bendre-Schmidt.

11. $u_{i,j+1} = 2\,(1 - 4\alpha^2)\,u_{i,j} + 4\alpha^2(u_{i-1,j} + u_{i+1,j} - u_{i,j-1})$.

12. $u_{i,j+1} = \frac{1}{2}\left(u_{i-1,j} + u_{i+1,j}\right)$ **13.** Section 11.5 (2).

14. Section 11.9 (2).   **15.** $\lambda < \frac{1}{2}$ **16.** one dimensional heat

17. Schmidt method and Crank-Nicolson method   **18.** Elliptic.   **19.** $O(h)^2$.

20. Section 11.6 (2)

21. $u_{i,j+1} = u_{i+1,j} - u_{i-1,j} - u_{1,j-1}$.   **22.** $\dfrac{\partial^2 u}{\partial x^2} + \dfrac{\partial^2 u}{\partial y^2} = f(x,y)$.

23. 100.   **24.** $\dfrac{u_{i-1,j} - 2u_{i,j} + u_{i+1,j}}{h^2} + \dfrac{u_{1,j-1} - 2u_{i,j} + u_{i,j+1}}{h^2} = 0$.

25. $y < 0$.   **26.** $k = \dfrac{1}{4}$

27. The solution value at any point on the $(j + 1)$th level is dependent on the solution values at the neighboring points on the same level and on three values on the jth level.

Exercises 12.1

1. Max. $Z = 1.2x_1 + 1.4x_2$; subject to $40x_1 + 25x_2 \le 1000$,
$35x_1 + 28x_2 \le 980$, $25x_1 + 35x_2 \le 875$ and $x_1, x_2 \ge 0$.

2. Max. $Z = 3x_1 + 2x_2 + 4x_3$; subject to $4x_1 + 3x_2 + 5x_3 \leq 2000$,
$2x_1 + 2x_2 + 4x_3 \leq 2500$, $100 \leq x_1 \leq 150$, $200 \leq x_2$ and $50 \leq x_3$.

3. Max. $Z = 3x_1 + 2x_2 + x_3$; subject to $3x_1 + 4x_2 + 3x_3 \leq 42$,
$5x1 + 3x_3 \leq 45$, $3x_1 + 6x_2 + 2x_3 \leq 41$ and $x_1, x_2, x_3 \geq 0$.

4. Max $Z = 400\, x + 300\, y$; subject to $x + y \leq 200$, $x \geq 20$, $y \geq 4x$, $x \geq 0$, $y \geq 0$.

5. Min. $Z = x_1 + x_2$; subject to $2x_1 + x_2 \geq 12$, $5x_1 + 8x_2 \geq 74$,
$x_1 + 6x_2 \geq 24$ and $x_1, x_2, x_3 \geq 0$.

6. Min. $Z = 41x_1 + 35x_2 + 96x_3$; subject to $2x_1 + 3x_2 + 7x_3 \geq 1250$, $x_1 + x_2 \geq 250$,
$5x_1 + 3x_2 \geq 900$, $6x_1 + 25x_2 + x_3 \geq 232.5$ and $x_1, x_2, x_3 \geq 0$.

7. Min. $Z = 100x_1 + 250x_2 + 160x_3$; subject to $0.94x_1 + x_2 + 1.04x_3 \leq 0.98$,
$10x_1 + 15x_2 + 17x_3 \geq 14$, $470x_1 + 500x_2 + 520x_3 \geq 495$, $x_1 + x_2 + x_3 = 1$ and $x_1, x_2, x_3 \geq 0$.

8. Max. $Z = 10,000x_1 + 9,000x_2 + 2,800x_3$; subject to $5,000x_1 + 4,500x_2 + 4,250x_3 \leq 100,000$, $x_1 \leq 6$, $x_2 \geq 2$, $x_3 \geq 2$ and $x_1, x_2, x_3 \geq 0$.

Exercises 12.2

1. $x_1 = 20/19$, $x_2 = 45/19$; max. $Z = 5/19$.   **2.** $x_1 = 8/15$, $x_2 = 12/5$; max. $Z = 24.8$.

3. $x_1 = 6$, $x_2 = 12$; min. $Z = 240$.   **4.** $x_1 = 500$, $x_2 = 600$; max. $Z = 1200$.

5. $x_1 = 50$, $x_2 = 0$; max. $Z = 200$.

6. 450 units of product B only; max. profit = US$1800.

7. $X = 2$, $Y = 4.5$; max. profit = US$37.

8. $A = 1.18$ units, $B = 0.53$ units; max. profit = US$14.50 approx.

9. $M = 200$ metric tons, $N = 300$ metric tons; max. profit = US$19,000.

10. 2000/11 units of product A and 1000/11 units of B; max. profit = US$10,000.

11. $x_1 = 4$, $x_2 = 0$; max. $Z = 8$.

12. 20 to 35 runs in magazine A and 10 to 16 runs in magazine B.

13. Unbounded solution.   **14.** Infinite number of optimal solutions; max. $Z = 8$.

15. $x_1 = 2$, $x_2 = 4$; min. $Z = 64$.

16. Production cost will be min. if G and J run for 12 and 4 days respectively.

Exercises 12.3

1. Max. $Z = 3x_1 + 5x_2 + 8x_3$; subject to $2x_1 - 5x_2 + s_1 = 6$,
$3x_1 + 2x_2 + x_3 - s_2 = 5$, $3x_1 + 4x_3 + s_3 = 3$; $x_1, x_2, x_3, s_1, s_2, s_3 \geq 0$.

2. Min. $Z = 3x_1 + 2x_2 + 5x_3$, subject to $-5x_1 + 2x_2 + s_1 = 5$,

$2x_1 + 3x_2 + 4x_3 - s_2 = 7$, $2x_1 + 5x_3 + s_3 = 3$, $x_1, x_2, x_3, s_1, s_2, s_3 \geq 0$.

3. Max. $Z = 3x_1 - 2x_2 + 4x_4 - 4x_5$; subject to $x_1 + 2x_2 + x_4 - x_5 + s_1 = 8$,

$2x_1 - x_2 + x_4 - x_5 - s_2 = 2$, $-4x_1 + 2x_2 + 3x_4 - 3x_5 = 6$; $x_1, x_2, x_4, x_5, s_1, s_2 \geq 0$.

4. (*i*) $x_1 = 2$, $x_3 = 1$ (Basic); $x_3 = 0$ (Non-basic) (*ii*) $x_1 = 5$, $x_3 = -1$ (Basic); $x_2 = 0$

(Non-basic) (*iii*) $x_2 = 5/3$, $x_3 = 2/3$ (Basic); $x_1 = 0$ (Non-basic)

All the three basic solutions are non-degenerate.

6. $x_1 = x_3 = x_4 = 0$ and $x_2 = 1/2$.

7. Basic solutions are (*i*) $x_1 = 2$, $x_2 = 1$ (Basic) and $x_3 = 0$; (*ii*) $x_1 = x_3 = 1$ (Basic) and

$x_2 = 0$; (*iii*) $x_2 = -1$, $x_3 = 2$ (Basic) & $x_1 = 0$.

(*a*) First two solutions are non-degenerate basic feasible solutions.

(*b*) First solution is optimal & $Z_{max} = 5$.

Exercises 12.4

1. $x_1 = 2$, $x_2 = 4$; max. $Z = 14$.   **2.** $x_1 = 0$, $x_2 = 20$; max. $Z = 200$.

3. $x_1 = 7/3$, $x_2 = 4/3$; max. $Z = 16$.   **4.** $x_1 = 5$, $x_2 = x_3 = 0$; max. $Z = 50$.

5. $x_1 = 0 = x_2$, $x_3 = 1$; max. $Z = 3$.   **6.** $x_1 = 0$, $x_2 = 6$, $x_3 = 4$; max. $Z = 6$.

7. $x_1 = 89/41$, $x_2 = 50/41$, $x_3 = 62/41$; max. $Z = US\$765/41$.

8. $x_1 = 4$, $x_2 = 5$, $x_3 = 0$; min. $Z = -11$.

9. $x_1 = 280/13$, $x_2 = 0$, $x_3 = 20/13$, $x_4 = 180/13$; max. $Z = 2280/13$.

10. $x_1 = 0$, $x_2 = 400$ units; max. profit = US\$1200.

11. $x_1 = 125$, $x_2 = 250$ units; max. profit = US\$2250.

12. $x_1 = 400$ gms, $x_2 = 0$; min. cost = US\$2.

13. $x_1 = 0$, $x_2 = x_3 = 50$; max. profit = US\$700.

14. $x_1 = 0.5$, $x_2 = x3 = 0.04$ units; min. cost = US\$5.80.

15. Acrages for corn, wheat, soybeans are 250, 625, zero respectively to achieve a max. profit of US\$32,500.

Exercises 12.5

1. $x_1 = 0$, $x_2 = 2$, $x_3 = 0$; max. $Z = 4$.   **2.** $x_1 = 3$, $x_2 = 2$, $x_3 = 0$; max. $Z = 8$.

3. $x_1 = -6/5$, $x_2 = -6/5$; max. $Z = -48/5$.

4. $x_1 = 0$, $x_2 = 10/3$, $x_3 = 4/3$; min. $Z = 34/3$.

5. $x_1 = x_2 = x_3 = 5/2$, $x_4 = 0$; max. Z = 15.

6. $x_1 = 21/13$, $x_2 = 10/13$; max. Z = 31/13.

  7. Infeasible.   **8.** $x_1 = 23/3$, $x_2 = 5$, $x_3 = 0$; max. Z = 85/3.

9. $x_1 = 55/7$, $x_2 = 30/7$, $x_3 = 0$; max. Z = 155/7. **10.** $x_1 = 2$, $x_2 = 0$; max. Z = 18.

11. Degenerate solution: $x_1 = 0$ (non-basic); $x_2 = 1$, $x_3 = 0$ (basic); max. Z = 3.

12. $x_1 = x_3 = 0$, $x_2 = 4$; max. Z = 24.

Exercises 12.6

1. Min. $W = 26y_1 + 7y_2$; subject to $6y_1 + 4y_2 \geq 10$,

  $5y_1 + 2y_2 \geq 13$, $3y_1 + 5y_2 \geq 19$; $y_1, y_2, y_3 \geq 0$.

2. Max. $W = 11y_1 + 7y_2 + y_3 + 5y_4$; subject to $3y_1 + 2y_2 - y_3 + 3y_2 \leq 2$,

  $4y_1 + 3y_2 + 2y_3 + 2y_4 \leq 4$, $y_1 - 2y_2 + 3y_3 + 2y_4 \leq 3$; $y_1, y_2, y_3, y_4 \geq 0$.

3. Min. $W = 6y_1 + 9y_2 + 10y_3$; subject to $4y_1 + 2y_2 + y_3 \geq 3$, $-5y_1 + 3y_2 + y_3 \geq 1$,

  $9y_1 - 4y_2 + 5y_3 \leq -4$, $y_1 - 5y_2 - 7y_3 \geq 1$, $-2y_1 + y_2 + 11\ y_3 \geq 9$; $y_1, y_2, y_3 \geq 0$.

4. Min. $W = -3y_1 + y_2 + 4y_3$; subject to $y_1 + 3y_2 - 2y_3 \leq -3$,

  $y_1 + y_3 \geq 16$, $y_1 - 2y_2 + y_3 \leq -7$; $y_1, y_2 \geq 0$, y_3 unrestricted in sign.

5. Min. $W = -6y_1 + 3y_2 + 4y_3$; subject to $-y_1 + 3y_2 - 4y_3 \geq 3$,

  $-y_1 - 2y_2 + 3y_3 \geq 1$, $-y_1 + 3y_2 - 6y_3 \geq 2$; $y_1 \geq 0$, y_2, y_3 unrestricted in sign.

6. Max. $W = 2y_1 + 3y_2 - 5y_3$; subject to $2y_1 + 3y_2 - y_3 \leq 2$,

  $3y_1 + y_2 - 4y_3 \leq 3$, $5y_1 + 7y_2 - 6y_3 = 4$; $y_1, y_2 \geq 0$, y_3 unrestricted.

Exercises 12.7

1. $x_1 = x_2 = 0$, $x_3 = 5/2$; min. Z = 2.5.   **2.** $x_1 = 4$, $x_2 = 2$; max. Z = 10.

3. $x_1 = 7$, $x_2 = 0$; max. Z = 21.   **4.** $x_1 = 0$, $x_2 = 100$, $x_3 = 230$; max. Z = 1350.

5. $y_1 = 2/3$, $y_2 = y_3 = 0$, min. $W = -4/3$.

Exercises 12.8

1. $x_1 = 0$, $x = 1$; max. Z = -1.   **2.** $x_1 = 3/5$, $x_2 = 6/5$; min. Z = 12/5.

3. $x_1 = 6$, $x_2 = 2$, $x_3 = 0$; min. Z = 10.

4. $x_1 = 0$, $x_2 = 30/11$, $x_3 = 16/11$, $x_4 = 0$; min. Z = 258/11.

5. $x_1 = 65/23$, $x_2 = 0$, $x_3 = 20/23$, min. Z = 215/23.

Exercises 12.9

1. $x_{11} = 200$, $x_{12} = 50$, $x_{22} = 175$, $x_{24} = 125$, $x_{33} = 275$, $x_{34} = 125$;
  min. cost = US$ 12075.

2. $x_{13} = 14, x_{21} = 6, x_{22} = 5, x_{23} = 1, x_{32} = 5$; min cost = US\$143.

3. $x_{11} = 50, x_{12} = 100, x_{21} = 150, x_{33} = 150, x_{42} = 100, x_{43} = 50$;
max. tonnage =US\$ 3300.

4. $x_{11} = 20, x_{13} = 10, x_{22} = 20, x_{23} = 20, x_{24} = 10, x_{32} = 20$; min. cost = US\$180.

5. $x_{11} = 140, x_{13} = 60, x_{21} = 40, x_{22} = 120, x_{33} = 90$; min. cost = US\$5920.

6. $x_{11} = 5, x_{14} = 2, x_{22} = 2, x_{23} = 7, x_{32} = 6, x_{34} = 12$; min. cost = US\$743.

7. $x_{11} = 150, x_{13} = 20, x_{22} = 160, x_{24} = 40, x_{33} = 90, x_{34} = 90$; max. profit = US\$4920.

8. $x_{13} = 2, x_{22} = 1, x_{23} = 2, x_{31} = 4, x_{33} = 1$; min. cost = US\$33.

9. $x_{13} = 60, x_{21} = 50, x_{23} = 20, x_{32} = 80$; min. cost = US\$750.

10. $x_{15} = 800, x_{21} = 400, x_{24} = 100, x_{32} = 400, x_{33} = 200, x_{34} = 300, x_{43} = 300$;
min. cost = 9200.

Exercises 12.10

1. $x_{11} = x_{22} = x_{33} = 1$; min. cost = US\$18.

2. $A \rightarrow 2, B \rightarrow 3, C \rightarrow 4, D \rightarrow 1$; min. $Z = 38$.

3. $I \rightarrow B, II \rightarrow A, III \rightarrow D, IV \rightarrow C$; min. cost = US\$49.

4. $A \rightarrow$ Dyn. Prog., $B \rightarrow$ Queuing Th., $C \rightarrow$ Reg. Analysis, $D \rightarrow$ L.P.;
min. time = 28 hrs.

5. $A \rightarrow 5, B \rightarrow 1, C \rightarrow 4, D \rightarrow 3, E \rightarrow 2$; min. cost = US\$9.

6. $A \rightarrow X, B \rightarrow W, C \rightarrow V, D \rightarrow Y, E \rightarrow Z$, min. time = 45 hrs.

7. $1 \rightarrow IV, 2 \rightarrow II, 3 \rightarrow VI, 4 \rightarrow I, 5 \rightarrow III, 6 \rightarrow V$; min. profit = US\$270.

8. $1 \rightarrow A, 2 \rightarrow B, 3 \rightarrow C$, or $1 \rightarrow A, 2 \rightarrow C, 3 \rightarrow B$; min. cost = US\$41.

9. $A \rightarrow 1, B \rightarrow 4, C \rightarrow 6, D \rightarrow 3, E \rightarrow 2$ or $A \rightarrow 3, B \rightarrow 4, C \rightarrow 1, D \rightarrow 6, E \rightarrow 2$;
min. cost = US\$52.

Exercises 12.11

1. Section 12.5 Def. 2. **2.** it provides an optimality test.   **3.** Section 12.11

4. Section 12.17 (1).   **5.** Section 12.14.   **6.** Section 12.6 (1).

7. Min. $W = 7y_1 + 5y_2$, subject to $2y_1 + 3y_2 \le 4, 3y_1 - 2y_2 \le 9, 2y_1 + 4y_2 \le 2$,
$y_1 \ge 0, y_2$ is unrestricted in sign.

8. Section 12.12 (2).   **9.** Section 12.14.

10. Minimize $Z = (2x_{11} + 3x_{12} + 11x_{13} + 4x_{14}) + (5x_{21} + 6x_{22} + 8x_{23} + 7x_{24})$,
subject to $x_{11} + x_{12} + x_{13} + x_{14} = a_1 (= 15), x_{21} + x_{22} + x_{23} + x_{24} = a_2 (= 20)$,
$x_{11} + x_{21} = b_1 (= -10), x_{12} + x_{22} = b_2 (= 5); x_{13} + x_{23} = b_3 (= 12)$,
$x_{14} + x_{24} = b_4 (= 8)$ and $x_{ij} \ge 0$. $[\because \Sigma a_i = \Sigma b_j = 35$

11. (i) $x_1 = 3, x_2 = 5, x_3 = 0$; (ii) $x_1 = 0.5, x_2 = 0, x_3 = 2.5$.

12. Section 12.5 (Def. 4)

13. Section 12.13.   **14.** balanced.   **15.** Section 12.9.

16. Section 12.7 (3).   **17.** optimal.

18. Minimize $y = 5y_4 - 3y_3$, subject to $y_4 + y_3 = 5$, $2y_4 - 5y_3 \geq 6$, $y_3 \geq 0$ and y_4 unrestricted.

19. 5.   **20.** Max. Z = 5/19.   **21.** Section 12.7.

22. Section 12.16.   **23.** Section 12.7 (2-ii)

24. Min. $W = 2y_1 + 4y_2 + 3y_3$, subject to $-y_1 + y_2 + y_3 \geq 2$, $2y_1 + y_2 \geq 1$, $y_1, y_2 \geq 0$.

25. North-West corner rule and Vogel's approximation method.

26. Slack or surplus variables.

Exercises 13.1

1. (a) 21 (b) 0.84375 (c) 0.6640625 (d) 1.4140625.

2. (a) 10110.101 (b) – 1010.001.   **4.**   1000101; 101111.

5. (a) 0.009 (b) 0.106 × 10–4.   **6.**   0.001; 0.0000111.

Exercises 14.1

7. 2.7065   **8.** 0.5177

10. $x = 1.052, y = 1.369, z = 1.962$.   **13.** 2591.87.

Exercises 15.1

7. 2.7065   **8.** 0.5177

10. $x = 1.052, y = 1.369, z = 1.962$.   **13.** 2591.87.

Exercises 16.1

8. 2.7065   **9.** 0.5177

11. $x = 1.052, y = 1.369, z = 1.962$.   **13.** 2591.87.

BIBLIOGRAPHY

1. 1. Anita, H.M. (1991). *Numerical Methods for Scientists and Engineers,* Tata McGraw- Hill Publishing Company, New Delhi.

2. Balagurusamy, E. (1999). *Numerical Methods*, Tata McGraw-Hill, New Delhi.

3. Chapra, S.C. and Canale, R.P. (1989). *Numerical Methods for Engineers*, McGraw- Hill Book Company.

4. Conte, S.D. and Carl de Boor (1981). *Elementary Numerical Analysis*, McGraw-Hill Book Company.

5. Gerald, C.F. and Wheatly, P.O. (1994). *Applied Numerical Analysis*, Addison-Wesley Publishing Company.

6. Hamming, R.W. (1973). *Numerical Methods for Scientists and Engineers,* McGraw- Hill, New York.

7. Jain, M.K., Iyengar S.R.K. and Jain R.K. (2003). *Numerical Methods for Scientific and Engineering Computation,* New Age International Publishers, India.

8. Maron, M.J. and Robert J.L. (1991). *Numerical Analysis, a Practical Approach*, Walsworth, Belmont, CA.

9. Mathews, J.H. (1994). *Numerical Methods for Mathematics, Science and Engineering*, Prentice-Hall of India.

10. Pearson, C.E. (1986). *Numerical Methods for Engineering and Science*, Van Nostrand Reinhold, New York.

11. Salvadori, M.G. and Baron, M.L. (1961). *Numerical Methods in Engineering*, Prentice Hall, Englewood Cliffs, NJ.

12. Scarborough, J.B. (1974): *Numerical Mathematical Analysis*, Oxford & IBH Publishing Co., New Delhi.

13. Scheid, F. (1990). *Numerical Analysis* (Schaum's Series), McGraw Hill Publishing Company, New York.

14. Schilling, R.J. and Harris S.L. (2007). *Applied Numerical Methods for Engineering using MATLAB and C*, Brooks/Cole.

15. Stanton, R.G. (1961). *Numerical Methods for Science and Engineering*, Prentice-Hall, Englewood Cliffs, NJ.

INDEX